ELECTRON SPIN RESONANCE

A Comprehensive Treatise
on Experimental Techniques

BOOKS WRITTEN AND EDITED BY THE AUTHOR

The author has written the following books:

Electron Spin Resonance, A Comprehensive Treatment of Experimental Techniques, C. P. Poole, Jr., Interscience, New York, 1967, 922 pages, (1st ed.); translated into Russian by MIP Press, Moscow, 1969.

Relaxation in Magnetic Resonance, C. P. Poole, Jr. and H. A. Farach, Academic Press, New York, 1971.

Theory of Magnetic Resonance, C. P. Poole, Jr. and H. A. Farach, Wiley-Interscience, New York, 1972; translated into Spanish by Teoria de la Resonancia Magnetica, Reverte, Spain, 1976.

Magnetic Resonance of Phase Transitions, Frank J. Owens, C. P. Poole, Jr., and H. A. Farach, Eds., Academic Press, New York, 1979.

The author has edited the English language editions of the following Russian books:

Electron Paramagnetic Resonance, S. A. Al'tshuler and B. M. Kozyrev, translated by Scripta Technica, Academic Press, New York, 1964.

Diamagnetism and the Chemical Bond, Ya. G. Dorfman, translated by Scripta Technica, Academic Press, New York, 1965.

Spectroscopic Analysis of Gaseous Mixtures, O. P. Bochkova and E. Ya. Schreyder, translated by Scripta Technica, Academic Press, New York, 1965.

The Theory of Nuclear Magnetic Resonance, I. V. Aleksandrov, translated by Scripta Technica, Edward Arnold, London, 1966.

The Chemical Bond in Semiconductors and Solids, N. N. Sirota, Ed., translated by Consultants Bureau, Plenum Press, New York, 1967.

Electron Spin Resonance
Second Edition
A Comprehensive Treatise on Experimental Techniques

CHARLES P. POOLE, Jr.

DEPARTMENT OF PHYSICS AND ASTRONOMY
UNIVERSITY OF SOUTH CAROLINA
COLUMBIA, SOUTH CAROLINA

A Wiley-Interscience Publication

JOHN WILEY & SONS

New York · Chichester · Brisbane · Toronto · Singapore

Library of Congress Cataloging in Publication Data:

Poole, Charles P.
    Electron spin resonance.

    "A Wiley-Interscience publication."
    Includes bibliographies and index.
    1.   Electron paramagnetic resonance.   I.   Title.

QC762.P6 1982   538'.364   82-6911
ISBN 0-471-04678-7   AACR2

Printed in the United States of America

10   9   8   7   6   5   4   3   2   1

221684

TO MY MOTHER AND FATHER

Preface

The first edition of this book appeared in 1967, about twenty years after the founding of the field of electron spin resonance (ESR). Its purpose was to present an overall summary and bibliography of experimental techniques. There were general background chapters on transmission lines, electromagnetic theory, guided waves, microwave network theory and electronic circuitry, followed by chapters on the various integral and ancillary parts of spectrometer systems. One chapter contained summary descriptions of various spectrometer systems, another treated sensitivity, and the last full-length chapter was on lineshapes. Now, fifteen years later, the time seems appropriate to attempt a more recent summary of experimental techniques that will bring the previous volume up to date with descriptions of new techniques that have been developed during recent years.

To gain some perspective on the use of the first edition a check was made of the Citation Index, which revealed that between 1965 and 1976 the work was cited 265 times, with 123 of the citations referring to individual chapters or pages thereof. The following table indicates the extent to which each chapter of the first edition was referred to in the literature.

Chapter Number

First Edition	Second Edition[a]	Russian Translation	Chapter Name in First Edition	Number of Citations	Comments on Second Edition
1	1	—	Introduction	1	Shortened somewhat
2	(2)	—	Transmission Lines	0	Mostly omitted
3	(2)	—	Electromagnetic Waves	1	Mostly omitted
4	2	1	Guided Electromagnetic Waves	0	Retained
5	—	—	Microwave Network Theory	0	Omitted
6	3	2	Microwave Generators	5	New section added
7	4	3	Waveguide Components	0	Shortened
8	5	4	Resonant Cavities	14	Retained
9	6	—	Magnetic Fields	3	Shortened
10	6	5	Magnetic Field Scanning and Modulation	10	Retained
11	7	6	Detectors	2	Shortened
12	7	—	Electronic Circuitry	1	Mostly omitted
13	(5)	—	Overall Spectrometer Systems	3	Mostly omitted

Chapter Number

First Edition	Second Edition[a]	Russian Translation	Chapter Name in First Edition	Number of Citations	Comments on Second Edition
14	11	13	Sensitivity	19	Retained
15	8	7	Vacuum Systems	4	Shortened somewhat
16	9	8	Variable Temperatures and Pressures	0	Updated
17	10	9	Irradiation	0	Shortened
18	13	11	Relaxation Times	16	Expanded
19	14	10	Double Resonance	3	Rewritten and expanded
20	12	12	The Characteristics of Spectral Line Shapes	41	Expanded
21	—	—	Summary of Spectrometer Operation	0	Omitted

[a] Numbers in parentheses denote short sections included in the indicated chapters of the second edition.

We see from these results that most of the citations are to the book itself without designating a part. Other citations are to particular pages or chapters. Certain chapters were frequently cited, and others were rarely or never referred to in the literature. The third column of the table lists the numbering of the chapters in the shortened 1970 Russian edition.

The data in this table were used as a guide in planning the second edition. Several of the earlier chapters were omitted in part or in full, and some of the latter ones were expanded. The second column of the table indicates where the material in the corresponding first-edition chapter was placed in the present edition. The comments in the last column indicate how the chapter was changed. The entry "updated" in this column indicates the substitution or addition of recently developed techniques in particular sections. Most of the chapters are, of course, updated by the addition of recent references and the discussion of recent work. Some chapters, however, have a much larger percentage of new work included in them since they cover fields where the technological improvements were the greatest.

In the preparation of this volume I have benefited greatly from the encouragement and criticism of my colleagues and students. I am especially thankful for all that I learned through my association with my co-worker Professor Horacio A. Farach. I wish to thank Dr. Marvin Abraham, Ray Du Varney, and James A. Norris for reading the last two chapters and giving me their comments. Professor Carl Clark carried out a helpful computer key-word reference search. The comments of Drs. Gareth R. Eaton and R. Talpe were appreciated.

I wish to thank many colleagues and the following corporations, journals, and publishing houses for granting their permission to reproduce a number of their figures: Academic Press, American Chemical Society, American Institute of Physics, Archives des Sciences, Bell Telephone Laboratories, Bruker Instruments, Cambridge University Press, Central Scientific Company, Centrale des Revues, Cornell University Press, Eastman Kodak, Elsevier Sequoia, English University Press, Faraday Society, Institute of Electronics and Electrical Engineers, Institute of Physics, Instruments Publishing Company, McGraw-Hill, National Bureau of Standards, National Research Council of Canada, New York Academy of Sciences, North Holland Publishing Co., Pergamon Press, Physical Society, Physical Society of Japan, Pitman Publishing, Radio Corporation of America, Reinhold Publishing Corporation, Royal Society of London, Societé Francaise de Physique, Springer-Verlag, Taylor and Francis, Ltd., Technology Press MIT, Van Nostrand, Varian Associates, VEB Deutscher Verlag der Wissenschaften, Verlag Birkhäuser, and *Z. Naturforschung*. Reprints sent to me by numerous colleagues were of inestimable value in the preparation of this volume.

I wish to thank my wife, Kathleen, for the various ways in which she assisted in the preparation of the manuscript. My former Chairman, Dr. O. F. Schuette, and present Chairman, Dr. F. T. Avignone, of the USC Physics Department kindly provided various departmental facilities that expedited the processing of the work. I wish to thank the members of the departmental secretarial staff, Mrs. Patricia Gonzalez, Mrs. Mildred Hedgepath, and Mrs. Lynn Waters, for their painstaking typing of the various revisions of the manuscript. Their patience with my cacography is certainly appreciated. Mr. Willie Zimmerman prepared the lists of references for the various chapters and ably handled much of the correspondence.

CHARLES P. POOLE, JR.

Columbia, South Carolina
August 1982

Preface to the First Edition

The field of electron spin resonance (ESR) was founded about twenty years ago. It has experienced a continuous growth since then, and is now expanding more rapidly than ever before. During the first ten years of its existence, the experimental techniques and the theoretical superstructure were developed to the point where extensive summaries of the field became necessary, and books on this subject made their appearance. Several of these books included excellent summaries of the experimental techniques in one or more introductory chapters. However, the main purpose of each volume was the presentation of the theory and experimental results. Those who wished to build entire spectrometers or elaborate auxiliary equipment found it necessary to consult the literature for the required information. In addition, expositions of the mechanics of recording spectra and obtaining data therefrom took a secondary place to the explication of data interpretation. During the past ten years the literature has continued to proliferate. The existing experimental techniques are being continuously refined, and occasionally new ones are introduced. As a result, the research worker who is interested in instrumentation has been forced to rely upon the journal literature.

This book was written in the belief that an overall summary and bibliography of ESR experimental techniques has been long overdue. It is an attempt to present a balanced treatment of both the theoretical and the practical aspects of the instrumentation. Although the presentation is in no sense historical, nevertheless some space has been reserved for the early apparatus which was built by the pioneers, and used to carry out the definitive experiments in ESR. Much of the discussion is devoted to the exposition of general principles which underlie ESR instrumentation, such as electromagnetic theory and the mechanism of magnetic field modulation. The *MIT Radiation Laboratory Series* (*RLS*) is extensively paraphrased. Recent developments such as enhancement techniques, helices, and acoustic spin resonance are covered. Thus this book is an attempt to present the general background material, the basic principles, and the main applications of ESR experimental techniques. It endeavors to show the reader how to design, build, and use an ESR spectrometer.

At the present time, scientists from many diverse disciplines such as physics, chemistry, biology, and medicine carry out ESR research. They approach the subject with a variety of backgrounds in mathematics, electronics, shop experience, etc. Therefore, to properly achieve the above objectives, it has been necessary to write for several disparate audiences. Some sections are of a theoretical nature, and will be of greatest interest to physicists. Other sections are of a more descriptive nature, and are written for those who are more interested in the qualitative results of the theory, or in the design details of a specific component. Numerous cross references and a minimum of presuppositions in each chapter should enable the reader to garner much information from the latter chapters without a prior reading of the earlier ones. The last paragraph of Sec. 1-A discusses the interrelationship between the subject matter in the various chapters. In terms of relative mathematical sophistication the chapters might be classified in difficult (Chaps. 2–5, 8, 10, 14, and 20), average (12, 13, 18, and 19), and easy (1, 6, 7, 9, 11, 15–17, and 21) groups.

This volume is intended mainly as a general reference book. However, the use of ESR spectrometers has become so widespread that many university departments will be interested in including some instrumentation instruction in their curriculum. A biennial special topics course in ESR instrumentation will greatly benefit those university departments which have active programs in the field. For this purpose one might wish to cover the first four introductory chapters and various sections in Chaps. 6–12. Extra time might be spent on the subject matter in Chap. 14 since this brings into focus much of the material covered earlier. Chapters 18–20 might be discussed next, and the course concluded with selected topics from Chaps. 13 and 15–17. The material in Chap. 5 constitutes good supplementary reading for the better students. The author taught such a course from this volume during the fall semester of 1965 at the University of South Carolina. Those professors who now teach a general introduction to magnetic resonance might consider the inclusion in the courses of material from Chaps. 1, 8, 10, 14, and 18–21, with an emphasis on Chaps. 14 and 20.

The principal references for each subject are arranged alphabetically by author at the end of the appropriate chapter. The literature is covered through most of 1965. *Physics Abstracts* was found to be an excellent source of references. Since this is an experimental techniques book, the following abbreviations were used for the principal instrumentation journals:

ETP   *Experimentelle Technik der Physik*

JSI   *Journal of Scientific Instruments*

PTE   *Pribory i Tekhnika Eksperimenta* [English trans.: *Instruments and Experimental Techniques*] This journal resembles the *Journal of the Chemical Society* in its inconvenient lack of volume numbers.

RSI   *Review of Scientific Instruments*

All the references are given in the standard form. The references to *Paramagnetic Resonance* (*1963*) refer to the *Proceedings of the 1962 Jerusalem Conference* edited by W. Low. When the page number is known in the English translation of a foreign language journal, such as *PTE*, it is often inserted in parentheses between original page and the year.

Parts of Chaps. 1, 18, and 20 originally appeared in an unpublished report, "The Dance of the Nucleons," written at the University of Maryland in 1957.

This book concerns electron spin resonance or ESR. This branch of science has several equivalent names. In the older literature, the field was often referred to as paramagnetic resonance, and some authors still use this name. A large number of contemporary scientists refer to the field as electron paramagnetic resonance or EPR. Several other names have appeared on occasion. The present author has adopted the name ESR more or less out of personal taste. *De gustibus non est disputandum.*

Many treatises on electromagnetic theory make use of the Meter-Kilometer-Second or MKS system of units, while most of the published ESR literature is written in the centimeter-gram-second or cgs system. We have adopted the MKS system for the introductory Chaps. 2–5 and also for much of Chaps. 8, 9, and 14 since these treat electromagnetic theory. We would have preferred to write the entire book using MKS units, but it seemed too much at variance with the generally accepted notation. Accordingly, cgs units appear throughout much of the remainder of the book. It is believed that very little confusion will result from this arrangement. Some of the principal units merely require powers of ten for their conversion (e.g., webers/meter^2 ↔ gauss, joule ↔ erg, meter ↔ centimeter, kilogram ↔ gram, and the Bohr magneton units joule meter^2/weber ↔ erg/gauss). Appendix A presents the principal equations in each system of units, and lists the main conversion factors. In other ways we have followed a uniform system of nomenclature throughout the book in conformity with the list of symbols found on p. xxi. Chapter 13 constitutes an exception because therein we present block diagrams of typical spectrometers in their original form. This serves to introduce the reader to the divergent notation that he will encounter throughout the literature. The unit of length, inch (1 in. = 2.54 cm) appears occasionally in the text.

In the preparation of this volume I have benefited greatly from the encouragement and criticism of my colleagues and students. Especial appreciation is due to O. F. Griffith, III for reading the entire manuscript and to D. A. Giardino and T. C. Sayetta for reading close to half of the text in manuscript form. Their critiques were very valuable. The following read particular chapters and offered helpful comments: W. E. Barr, R. L. Childers, C. W. Darden, III, R. D. Edge, W. R. Ferris, R. G. Fellers, F. H. Giles, Jr., T. J. Hardwick, J. S. Hyde, J. F. Itzel, W. K. Jackson, Jr., E. R. Jones, Jr., R. P. Kohin, S. Kumar, E. C. Lerner, E. E. Mercer, D. E. O'Reilly, G. S. Painter, O. F. Schuette, J. C. Schug, J. E. Sees, J. R. Singer, M. P. Stombler, E. F. Strother, J. Taylor, H. H. Tobin, and T. G. Weismann. Miss Elizabeth Obear kindly checked the accuracy of the long lists of references.

I wish to thank many colleagues and the following corporations, journals, and publishing houses for granting their permission to reproduce a number of their figures: Academic Press, American Chemical Society, American Institute of Physics, *Arch. Sci.*, Bell Telephone Laboratories, Cambridge University Press, Central Scientific Co., Cornell University Press, Eastman Kodak, Faraday Society, Institute of Electronics and Electrical Engineers, Institute of Physics, Instruments Publishing Co., McGraw-Hill, National Research Council of Canada, Pergamon Press, Physical Society, Physical Society of Japan, Radio Corp. of America, Reinhold Publishing Corp., Royal Society of London, Societé Française de Physique, Springer-Verlag, Taylor and Francis, Ltd., Technology Press MIT, Van Nostrand, Varian Assoc., VEB Deutscher Verlag der Wissenschaften, Verlag Birkhäuser, and *Z. Naturforschung.*

In addition, thanks are due to the Gulf Research and Development Company for the drafting on some of the figures. Reprints sent to me by numerous colleagues were of inestimable value in the preparation of this volume. Please continue to send me reprints of your current work.

I wish to express my appreciation to my wife, Kathleen, for her diligent typing of the first draft of the manuscript, and to my daughter Elizabeth whose birth during this period did not unduly delay the proceedings. I wish to thank the members of the secretarial staff here at the Physics Department of USC, Miss Patricia Acton, Mrs. Teresa Horton, Mrs. Jean Josey, and Mrs. Jean Padgett for their painstaking typing of the second draft and its revisions. Their patience with my cacography is certainly appreciated.

CHARLES P. POOLE, JR.

Columbia, South Carolina

Contents

Greek Symbols

α	Attenuation constant
α_ϵ	Attenuation constant due to dielectric losses
α_M	Mass absorption coefficient
α_R	Attenuation constant due to conductor losses
β	Bohr Magneton; cavity coupling coefficient; current sensitivity of a crystal detector; feedback ratio; phase constant
β_N	Nuclear magneton
Γ	Reflection coefficient
γ	Gyromagnetic ratio; propagation constant ($\alpha + j\beta$); polarization factor
Δ	Weiss constant; hyperfine structure anomaly
δ	Deviation ratio (in FM); skin depth
ϵ	Dielectric constant; extinction coefficient; population factor
ϵcd	Optical density
ϵ'	Real part of dielectric constant
ϵ''	Imaginary part of dielectric constant
ϵ_0	Dielectric constant of free space
η	Viscosity
$\hat{\theta}$	Unit vector in θ direction; spin echo pulse angle
θ_i	Temperature of ith spin system (used only in Sec. 13-A)
Λ	Shape constant of absorption curve
Λ'	Shape constant of absorption derivative
λ	Spin–orbit coupling constant; wavelength
λ_c	Cutoff wavelength
λ_g	Guide wavelength
μ	Amplification factor of vacuum tube; magnetic moment; muon; permeability
μ_c	Electric moment
μ_0	Permeability of free space
ν	Frequency; neutrino
Π_i	Formation of a product
π	3.14159
ρ	Charge density; electron density; mass density; resistivity
σ	Electrical conductivity
$\tau_{1/2}$	Time to scan between half power points
τ_0	Response time
τ_c	Correlation time

τ_D	Debye correlation time
τ_R	Rotational correlation time
$\hat{\varphi}$	Unit vector in φ direction
Φ	Scalar potential
ϕ	Magnetic flux
χ	Magnetic susceptibility
χ'	Real part of magnetic susceptibility
χ''	Imaginary part of magnetic susceptibility
χ_e	Electric susceptibility
χ_0	Static magnetic susceptibility
ψ	Wave function
Ω	Ohm
ω	Angular frequency $(2\pi f, 2\pi \nu)$
ω_0	Resonant frequency

Roman Symbols

A	Area; incident wave amplitude; acetonitrile
A, A_i	Hyperfine coupling constant
A_N	Hyperfine coupling constant of nitrogen
A_p	Hyperfine coupling constant of proton
A	Vector potential
$\mathcal{A}$	Attenuation
AM	Amplitude modulation
ATR	Anti-transmit–receive (tubes); attenuated total reflectance
af	Audio frequency
AFC	Automatic frequency control
At	Ampere turn
B	Reflected wave amplitude; susceptance; noise bandwidth
B	Magnetic induction or magnetic flux density
bcc	Body-centered cubic
BeV	Billion electron volt
C	Coulomb
C	Capacitance
c	Concentration; velocity of light *in vacuo*; curie
cgs	Centimeter–gram–second (system of units)
CRO	(Cathode ray) oscilloscope
CW	Continuous wave
D	Detectability; multiplicity factor; standard deviation; zero field splitting for axial symmetry
D	Electric displacement (electric flux density)
dc	Direct current; zero frequency (analogical use)
DME	Dimethoxyethane
DMF	Dimethylformamide
DMSO	Dimethylsulfoxide
dN	Noise power
DPPH	α,α'-Diphenyl-β-picryl hydrazyl
E	Zero field splitting for lower than axial symmetry
	Energy
E_i	Crystal field energy
E_m	Maximum value of electric field **E**
E	Electric field
ELDOR	Electron–electron double resonance
ENDOR	Electron nuclear double resonance

ESR	Electron spin resonance
e	Charge of electron
emf	Electromotive force
EPR	Electron paramagnetic resonance
eV	Electron volt
F	Force; noise figure
F_{amp}	Amplifier noise figure
F_K	Klystron noise figure
FM	Frequency modulation
f	Frequency
f_c	Cutoff frequency
f_{mod}	Modulation frequency
fcc	Face-centered cubic
G	Gauss
G	Conductance; gain
GeV	Gigaelectron volt (10^9 eV)
GHz	Gigahertz (10^3 MHz = 1 GHz)
g	Spectroscopic splitting factor, or g factor
H	Henry
H	Magnetic field strength
H_1	rf magnetic field
H_m	Maximum value of magnetic field $\mathbf{H}$
H_t	Magnetic field tangential to surface
$\mathcal{H}_N$	Nuclear spin energy (Hamiltonian)
$\mathcal{H}_Q$	Quadrupole energy (Hamiltonian)
$\mathcal{H}_{cf}$	Crystal field energy (Hamiltonian)
$\mathcal{H}_{elect}$	Electronic energy (Hamiltonian)
$\mathcal{H}_{hfs}$	Hyperfine structure energy (Hamiltonian)
$\mathcal{H}_{LS}$	Spin–orbit energy (Hamiltonian)
$\mathcal{H}_{SS}$	Spin–spin energy (Hamiltonian)
$\mathcal{H}_{Zee}$	Zeeman energy (Hamiltonian)
h	Planck's constant
$\hbar$	$h/2\pi$
hfs	Hyperfine structure
I	Electric current; nuclear spin; powder lineshape
Im	Imaginary part of
ir	Infrared
J	Exchange integral
J_m	mth-order Bessel function
$\mathbf{J}$	Electron current density; total angular momentum ($\mathbf{J} = \mathbf{L} + \mathbf{S}$)
JSI	*Journal of Scientific Instruments*
j	$(-1)^{1/2}$
K	A constant; voltage gain scans feedback
k	Boltzmann's constant; decay rate constant
k_0	Frequency factor

keV	Kiloelectron volt
kHz	Kilohertz
L	Conversion loss; inductance; insertion loss; load (subscript)
L	Orbital angular momentum
LET	Linear energy transfer
LO	Local oscillator
l	Length
lf	Low frequency
M	Figure of merit of a crystal detector; magnetization
M_I, m_i	Projection of nuclear spin I along magnetic field direction
MeV	Million electron volt
MHz	Megahertz
MKS	Meter-kilometer-second (system of units)
m	Meter
m	Mass (of electron); modulation index
mmf	Magnetomotive force
N	Noise power (also dN); number of spins
N_{hfs}	Number of hyperfine components
NBS	National Bureau of Standards
NMR	Nuclear magnetic resonance
Np/m	Nepers per meter
NQR	Nuclear quadrupole resonance
n	Index of refraction; transformer turns ratio
n_i	Number of hyperfine components; population of ith energy level
P	Polarization; power; transition probability
P_B	Bucking power
P_c	Power in cavity
P_w	Power in waveguide
P	Poynting vector
$\mathcal{P}$	Parity operator
PMR	Paramagnetic resonance
PTE	*Pribory i Tekhniki Exsperimenta*
ppm	Parts per million
Q	Electric charge; quality factor
Q_L	Loaded Q
Q_x	"Sample" Q
Q	Quadrupole moment (electric)
q	Electric charge
q_i	Excess charge density
R	Resistance; ELDOR reduction factor
R_c	Cavity resistance
R_{dc}	dc resistance (video resistance)
R_g	Generator resistance
R_s	Surface resistivity
$\mathcal{R}$	Reluctance

RBE	Relative biological effectiveness
rd	Rutherford
Re	Real part of
RSI	*Review of Scientific Instruments*
r	Normalized resistance R/Z_0; radius; reflected (subscript); ripple factor
$\hat{r}$	Unit vector in r direction
rf	radio frequency
rms	Root mean square
S	Area of surface; signal power; stabilization factor; slowing factor
$\mathbf{S}$	Spin angular momentum
SCE	Saturated calomel electrode
T	Temperature
T_1	Spin–lattice (longitudinal) relaxation time
T_2	Spin–spin (transverse) relaxation time
T_c	Transition temperature
T_d	Detector temperature
T_s	Sample temperature
T_x	Exchange relaxation time; T_x^{ij}
TE	Transverse electric
TEM	Transverse electromagnetic
TM	Transverse magnetic
TMAI	Tetramethylammoniumiodide
TNBAP	Tetra-*n*-butylammoniumperchlorate
TNPAP	Tetra-*n*-propylammoniumperchlorate
TR	Transmit–receive (microwave tube)
t	Noise temperature; time; transmitted (subscript)
U	Energy; energy density
U_E	Energy stored in electric fields
U_H	Energy stored in magnetic fields
uhf	Ultrahigh frequency
uv	Ultraviolet
V	Crystal field potential; potential due to electric (multipole) moment; voltage
V_c	Resonant cavity volume
V_e	Potential due to electric (multipole) moment
V_w	Volume of waveguide λg long
$\mathbf{V}$	Vector
VSWR	Voltage standing-wave ratio
v	Velocity
v_g	Group velocity
v_p	Phase velocity
W	Relaxation rate
Wb	Weber
WWV	NBS radio station

X	Reactance
x	Normalized reactance X/Z_0
$\hat{x}$	Unit vector in x direction
Y	Admittance; ESR amplitude or lineshape; signal-to-noise ratio
Y_l^m	Spherical harmonic
Y_0	Characteristic admittance
YIG	Yttrium iron garnet
$\hat{y}$	Unit vector in y direction
Z	Impedance

ELECTRON SPIN RESONANCE

A Comprehensive Treatise
on Experimental Techniques

CHAPTER 1

Introduction

A. Historical Background

In 1934 Cleeton and Williams constructed a primitive microwave spectrometer and detected the inversion of the ammonia molecule. Two years later Gorter (1936) attempted to detect nuclear magnetic resonance in solids by observing an increase in temperature, and he showed remarkable insight by attributing the negative result of his experiment to a long spin-lattice relaxation time ($T_1 > 10^{-2}$ sec). Both groups were severely hampered in their studies by the limitations of the available experimental equipment. During the 1930s the state of the art had not developed sufficiently to provide the instrumentation that is required for constructing magnetic resonance or microwave spectrometers. As a result these fields remained inactive until the middle of the next decade. The original Gorter type of resonance experiments have been successfully carried out in recent years (e.g., Guéron and Solomon, 1965; Schmidt and Solomon, 1966, Sujak et al., 1967).

During World War II a great deal of research was carried out in the development of radar. The technical problems that were solved by those engaged in this work included (1) the development of high-power microwave generators called magnetrons to produce the radar signal; (2) the design of highly directional antennas to transmit the signal and receive the echo; (3) the construction of sensitive (crystal) detectors to detect the echo; (4) the development of electronic methods for distinguishing the echo from the transmitted signal and for determining the distance of the target by the time delay of the echo after the transmitted pulse; (5) the perfection of narrow band amplifiers, lock-in detectors, and other noise-reducing circuits to increase the sensitivity of the radar system; and (6) the design of data display systems such as special oscilloscope arrangements. The wartime effort included fundamental studies of new fields of science such as microwave engineering and semiconductor devices in addition to the design of hardware.

At the close of the war, microwave and electronic technology had advanced to the point where electron spin resonance and microwave spectrometers could be constructed with the required sensitivity and resolution. Bleaney and Penrose (1946) and Good (1946) carried out more detailed microwave absorption studies on the ammonia molecule, while Zavoisky (1945) and Cummerow and Halliday (1946) detected electron spin resonance absorption in solids, and Griffiths (1946) observed ferromagnetic resonance. At the same time, Bloch

(1946); Bloch, Hansen, and Packard (1946); Purcell (1946); Purcell, Bloembergen, and Pound (1946); and Purcell, Torrey, and Pound (1946) founded the field of nuclear magnetic resonance. During the past 20 years these fields of research have grown tremendously, and their publications to date are numbered in the tens of thousands.

At the close of the war some of the principal scientists who directed and carried out the radar research at the Radiation Laboratory of the Massachusetts Institute of Technology collected and recorded the results of the concentrated radar research in 28 volumes called the *Radiation Laboratory Series* (*RLS*). These are listed at the end of Chap. 2. More than 30 years later these volumes are still used by workers in the fields of electron spin resonance and microwave spectroscopy. The basic problem in radar is the generation of very-high-power microwave pulses, and the detection of a very, very weak microwave echo signal. The fundamental problem in electron spin resonance, on the other hand, is the production of continuous (CW) intermediate power microwaves and the detection of a very small change in this power level. Despite this basic difference, many of the components that enter into the construction of a modern ESR spetrometer have their origin in the *Radiation Laboratory Series*.

During the past three decades a large number of spectrometers have been constructed at various universities and research laboratories and described in the literature, and several commercial models are on the market. Various types of auxiliary equipment have been designed for variable temperature investigations, high-pressure work, irradiation studies, sample rotations, and so on.

By the mid-1960s the field of ESR instrumentation had reached a state of maturity that warranted its summary in a single volume; this was done in the first edition of this book, which appeared in 1967. Since that time there have been a number of developments in the state of the art that warrant the publication of another survey of experimental techniques, and this volume will describe these developments.

In the preparation of the manuscript for the present edition some of the chapters of the first edition were considerably shortened (Chaps. 1 and 7) or eliminated (Chaps. 2, 3, 5, 13, and 21), combined together (Chaps. 4, 9, 10, 11, and 12), changed very little (Chaps. 8, 14, 15, and 17), considerably updated (Chaps. 6 and 16), expanded (Chap. 20), or completely rewritten (Chaps. 18 and 19). The ordering of the chapters is close to that in the first edition except that the sensitivity and lineshape chapters appear earlier in the text. In a number of cases reference is made to the first edition for material that has been eliminated. An attempt has been made to keep most of the chapters self-contained.

B. Spectroscopy

The general field of spectroscopy is subdivided into several regions depending on the energy involved in a typical quantum jump. A summary of the various branches of spectroscopy is presented in Table 1-1. Historically they developed as separate fields of research; each employed particular experimental techniques, and these instrumentation differences just happened to coincide with different physical phenomena such as the progressively increasing energies associated with rotational, vibrational, and electronic spectra.

Electron spin resonance (ESR) is frequently considered to be in the microwave branch of spectroscopy, and nuclear magnetic resonance (NMR) is usually classified in radiofrequency spectroscopy, but these are merely instrumental characterizations based, for example, on the last two columns of Table 1-1. In terms of the observed phenomena, ESR studies the interaction between electronic magnetic moments and magnetic fields. Occasionally, electron spin resonance studies are caried out with NMR instrumentation using magnetic fields of several gauss rather than several thousand gauss. The splitting of energy levels by a magnetic field is customarily referred to as the Zeeman effect, and so we may say that ESR is the study of direct transitions between electronic Zeeman levels, while NMR is the study of direct transitions between nuclear Zeeman levels. In concrete terms it may be said that ESR and NMR study the energy required to reorient electronic and nuclear magnetic moments, respectively, in a magnetic field.

Straight microwave spectroscopy uses apparatus similar to that employed in ESR, but in contrast to ESR, it measures molecular rotational transitions directly, and when it employs a magnetic field, it is usually for the purpose of producing only a small additional splitting of the rotational energy levels. In fact, in this branch of spectroscopy it is much more customary to produce Stark effect splittings by means of an applied electric field. In ESR, on the other hand, a strong magnetic field is an integral part of the experimental arrangement.

C. Magnetic Moments

The magnetic moment of an electron spin μ_S is given by

$$\mu_S = 2\beta\mathbf{S} \tag{1}$$

while the magnetic moment associated with orbital momentum μ_L is

$$\mu_L = \beta\mathbf{L} \tag{2}$$

where the Bohr magneton β defined by*

$$\beta = \frac{e\hbar}{2m} \tag{3}$$

is a convenient unit of magnetic moment, $\mathbf{S}$ is the spin angular momentum operator, and $\mathbf{L}$ is the orbital angular momentum operator. One may write Eqs.

---

*This is an MKS formula. In the cgs system, we have $\beta = e\hbar/2mc$.

Branch	Frequency, Hz	Wavelength	Typical Energy Unit Name	Value in Joules
Static	0–60		Joule	1
			Calorie	4.186
Low or audio frequency	10^3–10^5	3–300 km	kHz	6.62377×10^{-31}
Radio frequency	10^6–10^8	300–3 m	Joule	1
			cm^{-1}	1.98574×10^{-23}
Microwaves	10^9–10^{11}	30 cm to 3 mm	MHz	6.62377×10^{-28}
Infrared	10^{12} to 3×10^{14}	300–1 μm	cm^{-1}	1.98574×10^{-23}
			kcal/M	4.186×10^3
			Joule	1
Visible, ultraviolet	4×10^{14} to 3×10^{15}	0.8–0.1 μm	Erg	1×10^{-7}
			eV	1.60207×10^{-19}
			MHz	6.62377×10^{-28}
X rays	10^{16}–10^{19}	30–0.03 nm	eV	1.60207×10^{-19}
			keV	1.60207×10^{-16}
γ rays	10^{19}–10^{22}	3×10^{-9} to 3×10^{-12} cm	MeV	1.60207×10^{-13}
Low energy, nuclear	10^{19}–10^{23}	3×10^{-9} to 3×10^{-13} cm	MeV	1.60207×10^{-13}
High energy, nuclear	10^{23}–10^{26}	3×10^{-13} to 3×10^{-17} cm	BeV	1.60207×10^{-10}
			GeV	1.60207×10^{-7}
High-energy cosmic rays	$> 10^{25}$		BeV	1.60207×10^{-10}
			GeV	1.60207×10^{-7}

(1) and (2) in terms of the g factor:

$$\boldsymbol{\mu}_S = g\beta\mathbf{S} \tag{4}$$

$$\boldsymbol{\mu}_L = g\beta\mathbf{L} \tag{5}$$

where $g = 2$ and 1 for the spin and orbital motion, respectively. The g factor is the ratio of the magnetic moment to the angular momentum expressed in dimensionless units by means of the Bohr magneton. The ratio of the Bohr magneton to the unit of nuclear magnetic moments called the nuclear magne-

1-1
Branches of Spectroscopy

Phenomenon	Typical Radiation Generator	Typical Detector
	Battery	Ammeter Voltmeter
Dielectric absorption	Mechanical	Ammeter Voltmeter
NQR, NMR, dielectric absorption	Tuned circuit Crystal	Antenna
Molecular rotations, ESR	Klystron Magneton Solid State generator	Antenna Crystal Bolometer
Molecular vibrations	Heat source	Bolometer PbS cell
Electronic transitions	Incandescent lamp	Photocell
Electronic transitions	Discharge tube	Photocell
Inner shell electronic transitions	heavy element bombardment	Geiger counter Photomultiplier
Nuclear energy level transitions	Radioactive nuclei	Scintillation detector
Strange particle creation	Accelerator (e.g., synchrotron)	Bubble chamber Spark chamber
Extraterrestrial	Star, magnetic field in galaxy	Extensive shower detector

ton β_N is

$$\frac{\beta}{\beta_N} = \frac{e\hbar/2m}{e\hbar/2m_p} = \frac{9.2838 \times 10^{-24} \,\mathrm{J\,m^2/Wb}}{5.0508 \times 10^{-27} \,\mathrm{J\,m^2/Wb}} = \frac{9.2838 \times 10^{-21} \,\mathrm{erg/G}}{5.0508 \times 10^{-24} \,\mathrm{erg/G}}$$

$$= 1838 \tag{6}$$

which is the ratio of the rest mass m_p of the proton to the rest mass m of the electron. Thus ESR energies are generally about 2000 times as big as NMR energies.

In NMR one usually expresses the magnitude of a particular nuclear magnetic moment by its gyromagnetic ratio γ. The gyromagnetic ratio of an electron spin is related to its g factor by the expression

$$\gamma = \frac{g\beta}{\hbar} \tag{7}$$

$$= 8.809 \times 10^{10} g \text{ rad m}^2/\text{Wb} \tag{8a}$$

$$= 8.809 \times 10^6 g \text{ rad}/\text{G} \tag{8b}$$

where g is dimensionless, and γ has the units $(2\pi \times \text{Hz}/\text{magnetic field strength})$.

If an electron has both spin and orbital motion, then the total angular momentum $\mathbf{J}$ is obtained by the vector addition

$$\mathbf{J} = \mathbf{L} + \mathbf{S} \tag{9}$$

where $\mathbf{J}$ has the possible magnitudes $|L - S|, |L - S + 1|, \ldots, |L + S|$. For example, a single S electron has $S = \frac{1}{2}$, $L = 0$, and $J = \frac{1}{2}$ while a single D electron has $S = \frac{1}{2}$, $L = 2$, and $J = \frac{5}{2}$ or $\frac{3}{2}$.

As a result of Eqs. (1) and (2) the vector addition of the orbital and spin components to the magnetic moment gives a value

$$\boldsymbol{\mu} = g\beta\mathbf{J} \tag{10}$$

for the overall magnetic moment, where the Landé g factor has the form

$$g = \frac{3}{2} + \frac{S(S + 1) - L(L + 1)}{2J(J + 1)} \tag{11}$$

which is usually derived in texts on modern physics. Of course, for pure spin or orbital motion these expressions reduce to Eqs. (1) and (2), respectively.

In solids the electronic orbital motion interacts strongly with the crystalline electric fields and becomes decoupled from the spin, a process called "quenching." The more complete the quenching, the closer the g factor approaches the free-electron value. For example, $g = 2.0036$ in the free radical α, α'-diphenyl-β-picryl hydrazyl, which is very close to the free-electron value of 2.0023. It equals 1.98 in many chromium compounds, and it sometimes exceeds 6 for Co^{2+}. Thus the amount of quenching varies with the spin system.

The g factor is very often anisotropic and varies with the direction (x', y', z') in a single crystal. In the general case the g factor forms a symmetric tensor $\mathbf{g}'$ with six components g'_{ij}

$$\mathbf{g} = \begin{bmatrix} g'_{x'x'} & g'_{x'y'} & g'_{x'z'} \\ g'_{y'x'} & g'_{y'y'} & g'_{y'z'} \\ g'_{z'x'} & g'_{z'y'} & g'_{z'z'} \end{bmatrix} \tag{12}$$

where

$$g'_{ij} = g'_{ji} \tag{13}$$

It is always possible to find the principal axes (x, y, z) where the g tensor is diagonal

$$\mathbf{g} = \begin{bmatrix} g_{xx} & 0 & 0 \\ 0 & g_{yy} & 0 \\ 0 & 0 & g_{zz} \end{bmatrix} = \begin{bmatrix} g_x & 0 & 0 \\ 0 & g_y & 0 \\ 0 & 0 & g_z \end{bmatrix} \tag{14}$$

It is very common for the g tensor to have axial symmetry, in which case

$$g_{\parallel} = g_{zz} \tag{15}$$

$$g_{\perp} = g_{xx} = g_{yy} \tag{16}$$

where the z axis is taken as the symmetry axis. For an arbitrary orientation of a crystal in a magnetic field one obtains a resonance characterized by the g factor

$$g = \left(g_{xx}^2 \cos^2\theta_x + g_{yy}^2 \cos^2\theta_y + g_{zz}^2 \cos^2\theta_z \right)^{1/2} \tag{17}$$

where, for example, θ_x is the angle between the x axis and the magnetic field direction and $\cos\theta_x$ is called the direction cosine of x. The three direction cosines obey the relation

$$\cos^2\theta_x + \cos^2\theta_y + \cos^2\theta_z = 1 \tag{18}$$

so one of them may be easily eliminated from Eq. (16). In spherical coordinates Eq. (17) assumes the form

$$g = \left[g_{xx}^2 \sin^2\theta \cos^2\phi + g_{yy}^2 \sin^2\theta \sin^2\phi + g_{zz}^2 \cos^2\theta \right]^{1/2} \tag{19}$$

For axial symmetry Eq. (19) becomes

$$g = \left(g_{\perp}^2 \sin^2\theta + g_{\parallel}^2 \cos^2\theta \right)^{1/2} \tag{20}$$

where the subscript z is dropped since θ is understood to be the angle between the symmetry axis (along $g_{\parallel}$) and the magnetic field direction.

Electron spin resonance measurements from randomly oriented radicals produce ESR spectra whose lineshapes are powder patterns of Eqs. (19) or (20) (these are discussed in Chap. 12).

The sign of the g factor may be obtained with circularly polarized microwaves (see Sec. 5-O). In most ESR experiments only the magnitude of g is determined.

D. The Spin Hamiltonian

The interaction energy of a paramagnetic atom in a constant magnetic field H_0 is given by the spin Hamiltonian $\mathcal{H}$;

$$\mathcal{H} = \mathcal{H}_{elect} + \mathcal{H}_{cf} + \mathcal{H}_{LS} + \mathcal{H}_{SS} + \mathcal{H}_{Zee} + \mathcal{H}_{hfs} + \mathcal{H}_{Q} + \mathcal{H}_{N}$$

and the various terms have the following typical forms and magnitudes

$$\mathcal{H}_{elect} = \text{electronic energy} \approx 10^4 - 10^5 \text{ cm}^{-1} \text{ (optical region)} \tag{1}$$

$$\mathcal{H}_{cf} = \text{crystal field energy} \approx 10^3 - 10^4 \text{ cm}^{-1} \text{ (infrared or optical region)} \tag{2}$$

$$\mathcal{H}_{LS} = \text{spin–orbit interaction} = \lambda L \cdot S \approx 10^2 \text{ cm}^{-1} \tag{3}$$

$$\mathcal{H}_{SS} = \text{spin–spin interaction} = D\left[S_z^2 - \tfrac{1}{3}S(S+1)\right] = 0\text{–}1 \text{ cm}^{-1} \tag{4}$$

$$\mathcal{H}_{Zee} = \text{Zeeman energy} = \beta H \cdot (L + 2S) = \beta(g_x H_x S_x + g_y H_y S_y + g_z H_z S_z) = 0\text{–}1 \text{ cm}^{-1} \tag{5}$$

$$\mathcal{H}_{hfs} = \text{hyperfine structure} = (A_x S_x I_x + A_y S_y I_y + A_z S_z I_z) = 0\text{–}10^2 \text{ cm}^{-1} \tag{6}$$

$$\mathcal{H}_{Q} = \text{quadrupole energy} = \{3eQ/[4I(2I-1)]\} \times (\partial^2 V/\partial z^2)\left[I_z^2 - \tfrac{1}{3}I(I+1)\right] = 0\text{–}10^{-2} \text{ cm}^{-1} \tag{7}$$

$$\mathcal{H}_{N} = \text{nuclear spin energy} = \gamma \beta_N H \cdot I = 0\text{–}10^{-3} \text{ cm}^{-1} \tag{8}$$

Several of the symbols in these eight equations are defined as follows:

$\lambda = \text{spin–orbit coupling constant}$

$S_z, L_z = z$ component (along H) of the spin and orbital angular momenta, respectively

$D = \text{zero-field splitting constant}$

$\beta = \text{Bohr magneton}$

$g_z = zz$ component of g factor

$A_z = zz$ component of hyperfine coupling constant A

$I_z = z$ component of nuclear spin I

$e =$ electronic charge

$Q =$ nuclear electric quadrupole moment

$V =$ crystalline electric field potential

$\gamma =$ nuclear gyromagnetic ratio

$\beta_N =$ nuclear magneton

The energy $\mathcal{H}_{elect}$ is the electronic energy of the paramagnetic ion in the free state, and the energy $\mathcal{H}_{cf}$ is the interaction energy of the free ion's electronic structure with the crystalline electric field. This term helps to determine the g factor, as may be exemplified by Polder's values (Polder, 1942) for Cu^{2+} in a tetragonal crystalline electric field

$$g_\parallel = 2 - \frac{8\lambda}{(E_3 - E_1)} \tag{9}$$

$$g_\perp = 2 - \frac{2\lambda}{(E_4 - E_1)} \tag{10}$$

where λ is the spin–orbit coupling constant, and E_1, E_2, E_3, and E_4 are the four levels into which the crystal field splits the $3d^9$, ^2D Cu^{2+} ground electronic level. For Cu^{2+}, $\lambda = -852$ cm^{-1}, and the energy denominators are more than 10 times this value, so the g factors for Cu^{2+} vary from 2.15 to 2.4. Formulas similar in form to these relations may be obtained for other transition-metal ions in lattice sites of various symmetries.

The crystal field splitting between the ground orbital level E_1 and the next excited orbital level E_2 helps to determine the spin-lattice relaxation time T_1 (in seconds) produced by the direct and Raman processes, and for $S = \frac{1}{2}$, Kronig (1939) gave the relations

$$T_1 = \frac{10^4 (E_2 - E_1)^4}{\lambda^2 H^4 T} \qquad \text{direct process} \tag{11}$$

$$T_1 = \frac{10^4 (E_2 - E_1)^6}{\lambda^2 H^2 T^7} \qquad \text{Raman process} \tag{12}$$

where E_1, E_2, and λ are in cm^{-1}, H is in gauss, and T is the absolute

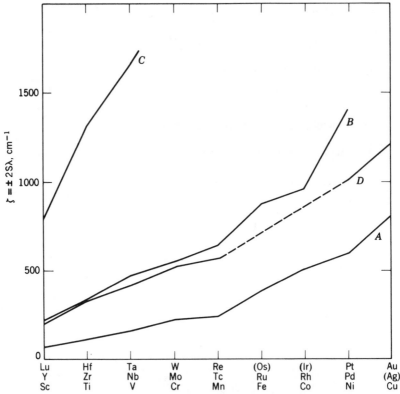

Fig. 1-1.   The spin–orbit coupling constant ζ for d electrons in neutral atoms. (A) First series; (B) second series; (C) third series; (D) third series scaled down by a factor of 4 (Griffith, 1961).

temperature. A direct process corresponds to a transition between two spin states with the emission or absorption of a phonon, while a Raman process entails the absorption of one phonon and the emission of another. (A phonon is a quantized lattice vibration.)

The spin–orbit interaction $\lambda L \cdot S$ further splits the optical energy levels, influences the g factor in the manner shown in Eqs. (9) and (10), and affects the spin-lattice relaxation time T_1 in accordance with Eqs. (11) and (12). The variation of the spin–orbit coupling constant λ is given in Fig. 1-1 for the neutral atoms in several transition series. The quantity ζ is related to λ by

$$\zeta = \pm 2S\lambda \qquad (13)$$

where S is the spin. The negative sign refers to a more than half-filled shell, and the positive sign refers to a half-filled or a less than half-filled shell. Some authors call ζ the spin–orbit coupling constant, while others refer to λ by the same name.

The spin–orbit interaction is sometimes the same order of magnitude as the Zeeman energy, and in this case it leads to a very complicated system of energy

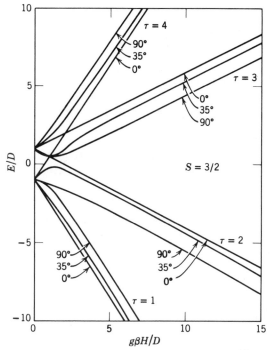

Fig. 1-2. Angular dependence of the energy levels (ordinate) of Cr^{+3} as a function of the normalized magnetic field strength (abscissa) (Davis and Strandberg, 1957).

levels that is strongly dependent on the orientation of the crystal in the magnetic field. Ruby (Cr^{3+}/Al_2O_3) is a satisfactory maser material because the spin–spin term ($D = 0.193$ cm^{-1}) is energetically comparable to the Zeeman energy, and the magnetic field strength and direction may be adjusted to give convenient maser operating conditions, as the energy-level scheme in Fig. 1-2 indicates. A strong exchange interaction or rapid reorientation effects, such as one finds in liquids, can average out the influence of the spin–spin interaction, and render the resonant line narrow and isotropic.

The Zeeman term given in Eq. (5) and its axially symmetric form were explained in the previous section. Electron spin resonance may be described as the measurement of the Zeeman energy $\mathcal{H}_{Zee}$, and in essence ESR does nothing more than study the manner in which the other Hamiltonian terms perturb or are perturbed by the Zeeman energy.

When the crystalline electric field has a symmetry lower than axial there is sometimes a lower symmetry spin–spin interaction of the form $E(S_x^2 - S_y^2)$, where $-1 \leqslant 3E/D \leqslant 0$ (Poole et al., 1974).

The interaction of nuclear spins with an unpaired electron produces hyperfine structure (hfs). When the electronic spin of a transition metal interacts with its own nuclear spin, the hfs is described by the Hamiltonian term given by Eq. (6),

$$\mathcal{H}_{\text{hfs}} = A I \cdot S \tag{14}$$

which for axial symmetry has the form

$$\mathcal{H}_{\text{hfs}} = A_\perp \left(S_x I_y + S_y I_y \right) + A_\parallel S_z I_z \tag{15}$$

In some systems, such as aromatic molecules, the unpaired electron circulates among several atoms, and the resulting hyperfine structure is the result of a

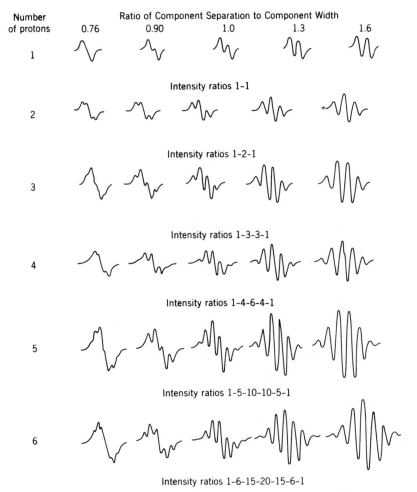

Fig. 1-3. Theoretical hyperfine structure curves for a Gaussian lineshape (Poole, 1958; Poole and Anderson, 1959).

Hamiltonian term of the form

$$\mathcal{H}_{hfs} = \sum A_i m_i \tag{16}$$

where the projection m_i of the ith nuclear spin on the magnetic field direction may take on the following $2I_i + 1$ values: $I_i, I_i - 1, I_i - 2, \ldots, 1 - I_i, -I$. The hyperfine coupling constant varies with the nuclear species, and it is a measure of the strength of the interaction between the nuclear and electronic spins.

When several $I = \frac{1}{2}$ nuclei are equally coupled (i.e., have same A_i), they produce an intensity ratio that follows the binomial coefficient distribution, as shown in Fig. 1-3. The figure indicates that spectral lines are well-resolved when their separation exceeds their width, and when this is not true they tend to merge together or coalesce into one line. This figure is useful for comparing with experimentally determined spectra (see also the more detailed spectra calculated by Lebedev, Chernikova, Tikhomirova, and Voevodskii, 1963). Jones et al. (1971) discussed lineshapes arising from unresolved metal hyperfine splittings of ion pairs. As shown in Tables 1-2 and 1-3, the intensity ratio of 1–3–3–1 is obtained with the methyl radical CH_3 ($I = \frac{1}{2}$), while the ratio of 1–2–3–2–1 is obtained with two equally coupled nitrogen nuclei ($I = 1$) such as the ones found in DPPH. A system containing three equally coupled protons (A_p) and two equally coupled nitrogens (A_N) possesses an hfs Hamiltonian with the form

$$\mathcal{H}_{hfs} = A_p(m_1 + m_2 + m_3) + A_N(m_4 + m_5) \tag{17}$$

where m_1, m_2, and m_3 assume the values $\pm \frac{1}{2}$ and m_4 and m_5 may equal 0 or ± 1. If the coupling constant $A_p \gg A_N$, the spectrum will consist of four widely

TABLE 1-2

Determination of Hyperfine Structure Intensity Ratios for Three Equally
Coupled $I = \frac{1}{2}$ Nuclei (e.g., Protons)

Spin Configuration			m_1	m_2	m_3	$M = m_1 + m_2 + m_3$	Intensity Ratio
↑	↑	↑	$\frac{1}{2}$	$\frac{1}{2}$	$\frac{1}{2}$	$\frac{3}{2}$	1
↑	↑	↓	$\frac{1}{2}$	$\frac{1}{2}$	$-\frac{1}{2}$		
↑	↓	↑	$\frac{1}{2}$	$-\frac{1}{2}$	$\frac{1}{2}$	$\frac{1}{2}$	3
↓	↑	↑	$-\frac{1}{2}$	$\frac{1}{2}$	$\frac{1}{2}$		
↑	↓	↓	$\frac{1}{2}$	$-\frac{1}{2}$	$-\frac{1}{2}$		
↓	↑	↓	$-\frac{1}{2}$	$\frac{1}{2}$	$-\frac{1}{2}$	$-\frac{1}{2}$	3
↓	↓	↑	$-\frac{1}{2}$	$-\frac{1}{2}$	$\frac{1}{2}$		
↓	↓	↓	$-\frac{1}{2}$	$-\frac{1}{2}$	$-\frac{1}{2}$	$-\frac{3}{2}$	1

TABLE 1-3

Determination of Hyperfine Structure Intensity Ratios for Two Equally
Coupled $I = 1$ Nuclei (e.g., Nitrogen) Such As the Ones Found in DPPH

Spin Configurations		m_1	m_2	$M = m_1 + m_2$	Intensity Ratio
↑	↑	1	1	2	1
↑	→	1	0	1	2
→	↑	0	1		
↑	↓	1	−1	0	3
→	→	0	0		
↓	↑	−1	1		
→	↓	0	−1	−1	2
↓	→	−1	0		
↓	↓	−1	−1	−2	1

separated groups of lines with the relative intensity ratio 1–3–3–1 each of which is split into a 1–2–3–2–1 quintet while when $A_p \ll A_N$, the main split is into a widely spaced 1–2–3–2–1 quintet with each of these components further split into a 1–3–3–1 quartet as shown in Fig. 1-4. One may easily compute that when $A_p = A_N$ in Eq. (17), the resulting spectrum has eight lines with the intensity ratio 1–5–12–18–12–5–1 as shown in Fig. 1-4c.

If there are n nuclei with $I = \frac{1}{2}$ contributing to the hfs, then there will be 2^n different hyperfine components if all the coupling constants differ and no degeneracy occurs. For example, five protons give the result from Fig. 1-3:

$$1 + 5 + 10 + 10 + 5 + 1 = 32 = 2^5 \tag{18}$$

If there are n nuclei with the nuclear spin I then there will be $(2I + 1)^n$ hyperfine components. For several nuclei with the individual values I_i and n_i the total number of hyperfine components N_{hfs} will be

$$N_{hfs} = \prod_i (2I_i + 1)^{n_i} \tag{19}$$

where the symbol $\prod_i$ denotes the formation of a product. For example, the system depicted in Fig. 1-4 has

$$I_p = \tfrac{1}{2} \quad n_p = 3$$
$$I_N = 1 \quad n_N = 2 \tag{20}$$

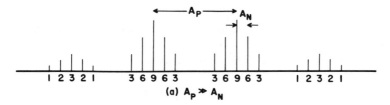

(a) $A_P \gg A_N$

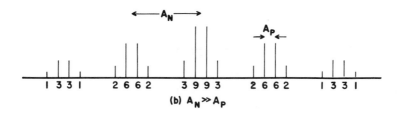

(b) $A_N \gg A_P$

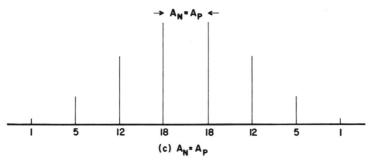

(c) $A_N = A_P$

Fig. 1-4. Hyperfine structure patterns for three equally coupled $I = \frac{1}{2}$ nuclei with coupling constant A_p and two equally coupled $I = 1$ nuclei with coupling constant A_N.

with the result that

$$N_{hfs} = \left[2\left(\tfrac{1}{2}\right) + 1\right]^3 (2 + 1)^2 = 72 \tag{21}$$

This may be checked by adding the intensities shown in Fig. 1-4c:

$$1 + 5 + 12 + 18 + 18 + 12 + 5 + 1 = 72 \tag{22}$$

When some nuclei are equivalent to others, the resulting degeneracy has the effect of decreasing the number and increasing the amplitude of the components in the hyperfine pattern without affecting the overall integrated intensity.

Usually, all the hyperfine components have the same linewidth and shape, but sometimes relaxation mechanisms cause deviations from this rule, as explained in Sec. 12-Q.

The quadrupolar interaction $\mathcal{H}_Q$ has the form

$$\left[I_z^2 - \tfrac{1}{3}I(I+1) \right] = \left[m_I^2 - \tfrac{1}{3}I(I+1) \right] \tag{23}$$

where $m_I = I_z$ is the z component of the nuclear spin I. The hyperfine structure Hamiltonian alone produces a symmetric pattern, and the quadru-

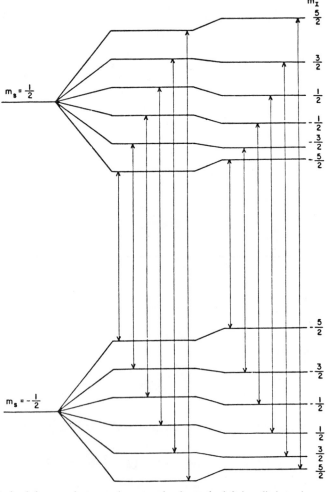

Fig. 1-5. Each of the two electron spin energy levels on the left is split into six equally spaced levels shown in the center by the hyperfine interaction for $I = \tfrac{5}{2}$. To first order the nuclear quadrupole interaction splits these levels unequally as indicated on the right, but the six allowed transitions denoted by the arrows are not changed in energy.

pole Hamiltonian produces energy level shifts that, in accordance with the selection rule $\Delta m_I = 0$, do not disturb the observed spectrum to first order. For example, when $I = \frac{5}{2}$,

$$
m_I^2 - \tfrac{1}{3}I(I + 1) =
\begin{aligned}
&= \tfrac{10}{3} \quad m_I = \pm\tfrac{5}{2} \\
&= -\tfrac{2}{3} \quad m_I = \pm\tfrac{3}{2} \\
&= -\tfrac{8}{3} \quad m_I = \pm\tfrac{1}{2}
\end{aligned}
\tag{24}
$$

and the levels for a given m_I are shifted equally, as shown in Fig. 1-5. If $\mathcal{H}_Q$ becomes appreciable relative to $\mathcal{H}_{Zee}$, then second-order quadrupole effects become important, and the observed hyperfine pattern becomes unsymmetrical. In nuclear quadrupole resonance (NQR) quadrupole splittings are measured directly.

The nuclear spin energy $\gamma \beta_N \mathcal{H} \cdot I$ is very small and can usually be neglected.

The optical spectra of solids containing transition metals arise from large crystal-field energy splittings perturbed by the spin–orbit coupling. Sometimes one can directly observe the spin–spin coupling energy $\mathcal{H}_{SS}$ or the nuclear quadrupole interaction $\mathcal{H}_Q$ in the absence of a magnetic field by employing the proper microwave frequency.

E. Information Gained from Electron Spin Resonance

The principal information gained from electron spin resonance spectra is an evaluation of the various terms in the spin Hamiltonian as discussed in the preceding section. Usually $\mathcal{H}_{Zee}$ and $\mathcal{H}_{hfs}$ are evaluated directly from ESR data, while the crystal field and spin–orbit energies are independently evaluated from optical spectra, and are then correlated with ESR data.

To be most informative an ESR spectrum of a particular paramagnetic system will be recorded at several temperatures, several frequencies, and several microwave powers. Sometimes one may employ the ESR spectrum to identify an unknown transition-metal ion or lattice defect, or it may distinguish between several valence states of the same ion. The ESR spectrum frequently identifies the lattice site and symmetry of the paramagnetic species, particularly if single crystal data are available. Considerable information can be obtained about the nuclei in the immediate neighborhood of the absorbing spin, and sometimes relaxation-time data detect long-range effects. Diffusion constants, correlation times, and the type of hydration can be determined from the ESR spectra of solutions. Chemical bonds in molecules and crystals sometimes may be characterized by ESR studies. The effective masses of atoms in semiconductors may be deduced. The concentrations of paramagnetic species may be determined. ESR studies furnish detailed information on ferromagnetic, antiferromagnetic, and ferrimagnetic materials.

The preceding is just a brief enumeration of some of the types of information that are furnished by electron spin resonance investigations. There are a number of books, proceedings of conferences, and review articles that devote considerable space to this topic.

F. Systems Studied by Electron Spin Resonance

A large number of systems have been studied by electron spin resonance, and this section will enumerate the principal ones. There is some overlap in these categories.

Biological Systems*

(1) hemoglobin (Fe); (2) nucleic acids; (3) enzymes; (4) chloroplasts when irradiated; (5) riboflavin (before and after uv irradiation).

Chemical Systems

(1) polymers; (2) catalysts; (3) rubber; (4) free radicals.

Conduction Electrons

(1) solutions of alkali metals in liquid ammonia; (2) alkali and alkaline earth metals (fine powders); (3) alloys (e.g., small amount of paramagnetic metal alloyed with another metal); (4) nonresonant absorption of microwaves by superconductors.

Free Radicals

(1) stable solid free radicals (a single exchange narrowed resonance); (2) stable free radicals in solution (hfs obtained); (3) free radicals produced by irradiation (usually at low temperature, sometimes single crystals); (4) condensed discharges (free radicals produced in gas and condensed on solid at low temperature); (5) biological systems; (6) biradicals; (7) electrochemical generation of ion radicals (polarography); (8) triplet states; (9) paramagnetic molecules (e.g., NO, NO_2, ClO_2).

Gases

Relatively few ESR studies have been made of gaseous systems, although some of the earliest spectrometers (compare first edition, Table 13-1) were designed for use with gases such as NO (Beringer and Castle, 1950; Jarke et al., 1976 NO_2 (Burch et al., 1974), and hydrogen atoms (Beringer and Heald, 1954). Selected recent studies of gases include halogens (Cook and Miller, 1973; Fisanick-Englot and Miller, 1976; McDowell and Tanaka, 1974; Miller, 1973; Miller and Freund, 1973; Tiedemann and Schindler, 1971; Ultee, 1971; and

*The presence of water produces large dielectric losses.

Zagarski et al., 1975), the radicals SO and SF (Takagi and Kojima, 1974), oxygen atoms (Tiedemann and Schindler, 1971), and oxygen molecules (Cook et al., 1973; Miller, 1971). Pittke et al., (1973) discussed gas analysis by ESR.

Irradiated Substances

(1) ionic crystals (e.g., alkali halides), (F centers and other centers); (2) solid organic compounds; (3) liquid organic compounds; (4) organic single crystals; (5) polymers; (6) semiconductors (e.g., Ge and Si); (7) photoconductors (e.g., dyes).

Naturally Occurring Substances

(1) minerals with transition elements [e.g., ruby (Cr/Al_2O_3), dolomite $Mn/(Ca, Mg)(CO_3)$]; (2) minerals with defects (e.g., quartz); (3) hemoglobin (Fe); (4) petroleum; (5) coal; (6) rubber.

Semiconductors

(1) cyclotron resonance (e.g., Ge, Si, InSb); (2) doped semiconductors (e.g., Si with As, Sb, P); (3) irradiated semiconductors; (4) graphite.

Transition Elements*

(1) single crystals (1% in diamagnetic crystal, anisotropic g factors, and hfs constants evaluated); (2) relaxation time studies (mostly liquid He temperature, low power); (3) chelates, sandwich compounds; (4) alloys.

Experimental data and theoretical discussions of these systems may be found in the books, proceedings of conferences, and review articles listed in the Selected Bibliography at the end of the first edition of this book.

G. Relationship of ESR with Magnetic Susceptibility and Other Techniques

An electron spin resonance measurement consists of the simultaneous determination of the microwave frequency ν and the magnetic field strength H. These data are used to calculate the g factor from the relation

$$g = \left(\frac{h}{\beta}\right)\left(\frac{\nu}{H}\right) \tag{1}$$

where the constant of proportionality is the ratio of Planck's constant to the Bohr magneton. A magnetic susceptibility measurement carried out with a paramagnetic sample entails the determination of the force exerted on a

---

*Most of the work has been with the first transition series (Ti to Cu).

sample in a magnetic field as a function of the absolute temperature T. The data are used to calculate the susceptibility χ_0 the Weiss constant Δ, and the g factor by means of the Curie–Weiss law

$$\chi_0 = \frac{N_g^2\beta^2 S(S+1)}{3k(T+\Delta)} \tag{2}$$

where S is the spin, k is Boltzmann's constant, and N is the number of spins in the sample. This is the spin-only formula used for a quenched orbital angular momentum. In the absence of quenching one replaces $\mathbf{S}$ by $\mathbf{J}$ where

$$\mathbf{J} = \mathbf{L} + \mathbf{S} \tag{3}$$

Thus both ESR and magnetic susceptibility methods provide the g factor. The ESR technique singles out each ion and electronic state and resolves the corresponding spectra, while the magnetic susceptibility technique measures an average of the susceptibilities of all the paramagnetic states in the sample. Electron spin resonance provides additional information concerning hyperfine interactions with nuclear spins. Both methods furnish information on zero-field splittings, with the ESR results being more specific. The Weiss constant Δ determined by magnetic susceptibility measurements may be employed to evaluate the exchange integral (exchange energy) J between two paramagnetic ions by the approximation

$$\Delta = \frac{\frac{2}{3}zS(S+1)J}{k} \tag{4}$$

One should not confuse this use of J with the total angular momentum. This same approximation may be employed with the Curie temperature T_C (or Néel temperature T_N), which separates paramagnetic and ferromagnetic or antiferromagnetic behavior

$$T_C = \frac{\frac{2}{3}zS(S+1)J}{k} \tag{5}$$

where z is the number of paramagnetic nearest-neighbor ions exchange-coupled to each paramagnetic ion. It must be emphasized that the last two equations are only rough approximations since usually $T_C \neq \Delta$, and often Δ exceeds T_C by a factor of 2 to 4.

Some authors have made ESR and magnetic susceptibility measurements on the same system such as chromia alumina (Poole and MacIver, 1967), V_2MoO_8 ($V_2O_5 \cdot MoO_3$) (Sperlich et al., 1974), a copper complex (Jones et al., 1981), and polyglycine (Itoh et al., 1976). Karthe and Veith (1971) determined χ_0 from ESR dispersion signals, and Seehra (1968) found that the extra broadening in samples with large magnetic losses is a monotomic function of χ_0.

A number of workers combine electron spin resonance with other instrumental techniques to obtain a better overall understanding of various systems under study (Stratouly, 1975). Examples of other instrumental methods that have been combined with ESR are chromatography (Kominami, 1977; Kominami et al., 1976), electrical conductivity (Houlihan and Mulay, 1974), dielectrics (Dietz et al., 1976), dielectric relaxation (Edgar and Welsh, 1975, 1979), electrochemistry (Allendoerfer et al., 1975), electron diffraction (Gerasimenko, 1977), electron microscopy (Grigorovskaya, 1977), ESCA (Gajardo, 1978; Ohdomari et al., 1975), the Hall effect (Bucker, 1975; Feichtinger et al., 1978), optical spectroscopy (see Sec. 14-H) (Abraham et al., 1972, 1973; Jain et al., 1970; Kishimoto and Morigaki, 1977; Murakami et al., 1977; Pontnau et al., 1975; Tohver et al., 1972; Venturini and Morosin, 1977; Weil, 1975; and Weltner, 1978), thermal studies (Beech and Thompson, 1971), and X-ray diffraction (Cordischi et al., 1978; Goeffroy et al., 1976; Poole and MacIver, 1967; and Zhurkov et al., 1971). Faudrowicz (1971) described a microwave method for investigating the quality of ruby boules. Zuckermann (1974) reported using methods analogous to those employed in crystallography for the study of anisotropic ESR spectra.

References

Abraham, M. M., Y. Chen, J. L. Kolopus, and H. T. Tohver, *Phys. Rev.* **5**, 4945 (1972).

Abraham, M. M., Y. Chen, J. T. Lewis, and F. A. Modine, *Phys. Rev. B* **7**, 2732 (1973).

Allendoerfer, R. D., G. A. Martinchek, and S. Bruckenstein, *Anal. Chem.* **47**, 890 (1975).

Beech, G., and F. Thompson, *Thermochim. Acta* **2** 195 (1971).

Beringer, R., and J. G. Castle, *Phys. Rev.* **78**, 581 (1950).

Beringer, R., and M. A. Heald, *Phys. Rev.* **95**, 1474 (1954).

Bleaney, B., and R. P. Penrose, *Nature* **157**, 339 (1946).

Bloch, F., *Phys. Rev.* **70**, 460 (1946).

Bloch, F., W. W. Hansen, and M. Packard, *Phys. Rev.* **69**, 127 (1946).

Bucker, W., *J. Non-Cryst. Solids* **18**, 11 (1975).

Burch, D. S., W. H. Tanittila, and M. Mizushima, *J. Chem. Phys.* **61**, 1607 (1974).

Cleeton, C. E., and N. H. Williams, *Phys. Rev.* **45**, 234 (1934).

Cook, T. J., B. R. Zegarski, W. H. Breckenridge, and T. A. Miller, *J. Chem. Phys.* **58**, 1548 (1973).

Cook, T. J., and T. A. Miller, *J. Chem. Phys.* **59**, 1342, 1352 (1973).

Cordischi, D., D. Gazzoli, and M. Valigi, *J. Solid State Chem.* **24**, 371 (1978).

Cummerow, R. L., and D. Halliday, *Phys. Rev.* **70**, 433 (1946).

Davis, C. F., and M. W. P., Strandberg, *Phys. Rev.* **105**, 447 (1957).

Dietz, W., H. Kilp, W. Noerpel, M. Strassmann, and M. Stockhausen, *Z. Naturforsch.* **31A**, 457 (1976).

Edgar, A., and H. K. Welsh, *J. Phys. C Solid State Phys.* **8**, 1336 (1975); **12**, 703 (1979).

Faudrowicz, A., A. Jelenski, A. Piotrowski, and J. Twarowski, *Electron Technol.* **4**, 31 (1971).

Feichtinger, H., J. Waltl, and A. Gschwandtner, *Solid State Comm.* **27**, 867 (1978).

Fisanick-Englot, G. J., and T. A. Miller, *J. Chem. Phys.* **64**, 786 (1976).

Gajardo, P., A. Mathieux, P. Grange, and B. Delmon, *C. R. Hebd, Seances Acad. Sci. Ser. C* **287**, 345 (1978).

Geoffroy, M., L. Ginet, and E. A. Lucken, *Mol. Phys.* **31**, 745 (1976).

Gerasimenko, N. N., A. V. Dvurechenskii, A. I. Mashin, and A. F. Khokhlov, *Fiz. Tekh. Poluprovodn.* **11**, 190 (1977).

Good, W. E., *Phys. Rev.* **69**, 539 (1946).

Gorter, C. J., *Physica* **3**, 995 (1936).

Griffith, J. S., *The Theory of Transition Metal Ions*, Cambridge Univ. Press, Cambridge, England, 1961, p. 113.

Griffiths, J. H. E., *Nature* **158**, 670 (1946).

Grigorovskaya, V. V., V. E. Basin, D. K. Khakimova, L. S. Lyubchenko, E. S. Cherepanova, and A. A. Berlin, *J. Polym. Sci. Polym. Phys. Ed.* **15**, 20765 (1977).

Guéron, M., and I. Solomon, *Phys. Rev. Lett.* **15**, 667 (1965).

Houlihan, J. F., and L. N. Mulay, *Phys. Stat. Sol. B* **65**, 513 (1974).

Ito, A. S., D. M. Esquivel, M. R. Kawamura, P. R. Crippa, and S. Iostani, *Radiat. Eff.* **29**, 75 (1976).

Jain, S. C., S. K. Agarwal, G. D. Sootha, and R. Chandler, *J. Phys. C* **3**, 1343 (1970).

Jarke, F. H., N. A. Ashford, and I. J. Solomon, *J. Chem. Phys.* **64**, 3097 (1976).

Jones, E. R., Jr., H. A. Farach, C. P. Poole, Jr., E. A. H. Griffith, and E. L. Amma, On the antiferromagnetic coupling of $CuCl_4$ ions (to be published).

Jones, M. T., M. Komarynsky, and R. D. Rataiczak, *J. Phys. Chem.* **75**, 2769 (1971).

Karthe, V. W., and G. Vieth, *ETP* **19**, 443 (1971).

Kishimoto, N., and K. Morigaki, *J. Phys. Soc. Japan* **42**, 137 (1977).

Kominami, S., *Radiat. Res.* **72**, 89 (1977).

Kominami, S., S. Rokushika, and H. Hatono, *Int. J. Radiat. Biol.* **30**, 525 (1976).

Kronig, R. del., *Physica* **6**, 33 (1939).

Lebedev, Ya. S., D. M. Chernikova, N. N. Tikhomirova, and V. V. Voevodskii, *Atlas of Electron Spin Resonance Spectra*, Consultants Bureau, New York, 1963.

Miller, T. A., and R. S. Freund, *J. Chem. Phys.* **58**, 2345 (1973).

Miller, T. A., *J. Chem. Phys.* **58**, 2358 (1973).

Miller, T. A., *J. Chem. Phys.* **54**, 330 (1971).

McDowell, C. A., and I. Tanaka, *Chem. Phys. Lett.* **26**, 463 (1974).

Murakami, K., S. Namba, N. Kishimoto, K. Masuda, and K. Gamo, *Sci. Pap. Inst. Phys. and Chem. Res.* **71**, 103 (1977).

Ohdomari, I., M. Ikeda, K. Ohno, and T. Itoh, *J. Vac. Soc. Japan* **18**, 300 (1975).

Pittke, E. C., G. Schoffa, and M. Birkle, *Messtechnik* **81**, 77 (1973).

Polder, D., *Physica* **9**, 709 (1942).

Pontnau, J., R. Adde, P. S. Allen, E. P. Andrew, and C. A. Bates, *Proc. 18th Congr. Magn. Res. and Related Phenom.*, Vol. 1, 1975, p. 133.

Poole, C. P., Jr., Thesis, Department of Physics, University of Maryland, 1958.

Poole, C. P., Jr., and R. S. Anderson, *J. Chem. Phys.* **31**, 346 (1959).

Poole, C. P., Jr., H. A. Farach, and W. K. Jackson, *J. Chem. Phys.* **61**, 2220 (1974).

Poole, C. P., Jr., and D. S. MacIver, *Adv. Catal.* **17**, 223 (1967).

Purcell, E. M., *Phys. Rev.* **69**, 681 (1946).

Purcell, E. M., N. Bloembergen, and R. V. Pound, *Phys. Rev.* **70**, 986, 988 (1946).

Purcell, E. M., H. C. Torrey, and R. V. Pound, *Phys. Rev.* **69**, 37 (1946).

Schmidt, J. and I. Solomon, *J. Appl. Phys.* **37**, 3719 (1966).

Seehra, M. S., *RSI* **39**, 1044 (1968).

Sperlich, G., P. H. Zimmermann, and G. Keller, *Z. Phys.* **270**, 267 (1974).

Stratouly, D. S., *Am. Lab.* **7**, 73 (1975).

Sujak, B., I. Bójko and W. Nosel, *Acta Phys. Polonica* **31**, 777 (1967).

Takagi, K., and T. Kojima, *Japan J. Appl. Phys.* **13**, 1195 (1974).

Tiedemann, P., and R. N. Schindler, *J. Chem. Phys.* **54**, 797 (1971).

Tiedemann, P., and R. N. Schindler, *Z. Naturforsch.* **26a**, 1090 (1971).

Tohver, H. T., B. Henderson, Y. Chen, and M. M. Abraham, *Phys. B* **5**, 3276 (1972).

Ultee, C. J., *J. Chem. Phys.* **54**, 5437 (1971).

Venturini, E. L., and B. Morosin, *Phys. Lett. A* **61**, 326 (1977).

Weltner, W., Jr., *Ber. Bunsenges. Phys. Chem.* **82**, 80 (1978).

Weil, J. A., *Radiat. Eff.* **26**, 261 (1975).

Zavoisky, E., *J. Phys. USSR* **9**, 211 (1945).

Zegarski, B. R., T. J. Cook, and T. A. Miller, *J. Chem. Phys.* **62**, 2952 (1975).

Zhurkov, S. N., V. A. Zakrevskii, V. E. Korsukov, and V. S. Kuksendo, *Fiz. Tver. Tela* **13**, 2004 (1971).

Zuckermann, B., *J. Chem. Phys.* **60**, 1189 (1974).

CHAPTER 2

Guided Electromagnetic Waves

A. Reflection of Plane Waves from Conductors

The reflection of a plane wave $E \exp[j(\omega t - \beta z)]$ vibrating in the x direction and normally incident on a perfect conductor is analogous to a transmission line terminated in a short circuit because the transverse **E** vector must be zero at the conductor. To satisfy this boundary condition the reflected electromagnetic wave sets up standing waves corresponding to $-E \exp[j(\omega t + \beta z)]$ so that

$$E_x = E\{\exp[j(\omega t - \beta z)] - \exp[j(\omega t + \beta z)]\} \tag{1}$$

$$= -2jE \sin \beta z e^{j\omega t} \tag{2}$$

as shown in Fig. 2-1. The **H** vector H_y vibrates at right angles to and in space quadrature with the **E** vector

$$H_y = \left(\frac{2E}{Z_0}\right) \cos \beta z e^{j\omega t} \tag{3}$$

where Z_0 is the characteristic impedance of the medium. The reflected wave carries away the same amount of energy that the incident wave brings to the conductor, so Poynting's vector at the surface averages to zero.

Another way to look at the problem is to consider the reflection coefficient Γ

$$\Gamma = \frac{(Z_L - Z_0)}{(Z_L + Z_0)} \tag{4}$$

where Z_L is the characteristic impedance of the conductor. One of Maxwell's equations is

$$\nabla \times \mathbf{H} = \mathbf{J} + \frac{\partial \mathbf{D}}{\partial t} \tag{5}$$

$$= (\sigma + j\omega\epsilon)\mathbf{E} \tag{6}$$

$$= j\omega\left[\epsilon - j\left(\frac{\sigma}{\omega}\right)\right]\mathbf{E} \tag{7}$$

24

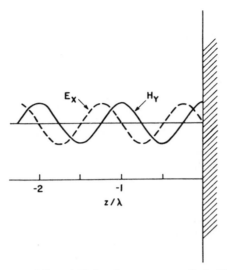

Fig. 2-1. Standing waves of E_x and H_y for plane wave normally incident on a conductor.

so the ratio of the conductivity to the angular frequency σ/ω constitutes an effective imaginary term in the dielectric constant. For the conductor as a termination and $\sigma/\omega \gg \epsilon$

$$Z_L = \left[\frac{\mu}{(\epsilon - j\sigma/\omega)} \right]^{1/2} = (1+j)\left(\frac{\omega\mu}{2\sigma}\right)^{1/2} = (1+j)R_s \qquad (8)$$

where R_s is the surface resistivity, which becomes very small when $\sigma/\omega \gg \epsilon$. Recall that for free space

$$Z_0 = \left(\frac{\mu_0}{\epsilon_0}\right)^{1/2} = 120\pi \ \Omega \qquad (9)$$

and μ_0 in a conductor is usually equal to the permeability of free space, so that

$$Z_L \ll Z_0 \qquad (10)$$

for the conductor. As a result

$$\Gamma \approx -1 \qquad (11)$$

For copper

$$R_s = \left(\frac{\omega\mu}{2\sigma}\right)^{1/2} = 2.61 \times 10^{-7}(f)^{1/2} \ \Omega \qquad (12)$$

where f is in cycles per second. Since for green light at a wavelength of 600 μm,

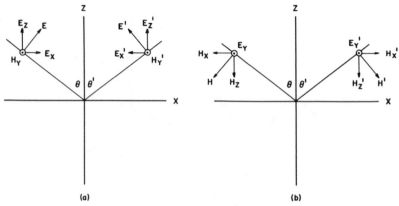

Fig. 2-2. The reflection of a plane wave at a conductor surface (a) with the **E** vector in the plane of incidence and **H** directed out of the paper, and (b) with the **H** vector in the plane of incidence and **E** directed out of the paper. The angle of incidence is θ and the angle of reflection is $\theta' = \theta$.

$f = 5 \times 10^{14}$ Hz, it follows that even at optical frequencies R_s for copper is less than 10 Ω, and Eqs. (10) and (11) are still valid approximations.

When a plane wave $E \exp[-jk(x \sin\theta + z \cos\theta)]$ is incident on a plane conductor at an arbitrary angle of incidence the analysis is somewhat more complicated. There are two cases to consider: (1) the electric vector **E** in the plane of incidence, and (2) the magnetic vector **H** in the plane of incidence as illustrated in Fig. 2-2. The mathematical treatment for these cases is found in most texts on electromagnetic theory such as Jackson's (1975), and only a qualitative discussion will be given here.

The first general characteristic of reflection at oblique incidence is that for both polarizations the angle of incidence θ equals the angle of reflection θ'. When the **E** vector is polarized in the plane of incidence it remains in that plane after reflection, as shown in Fig. 2-2a. The **H** vector remains pointing in the same direction so that both **E** and **H** are still perpendicular to the direction of propagation. The electromagnetic field above the conductor has the form of a standing wave with respect to the z direction, and the form of a traveling wave in the x direction. From the boundary condition, the component of **E** in the x direction E_x vanishes at the conductor surface and at points nd above the surface, where the distance nd is

$$nd = \frac{n\pi}{\omega(\mu\epsilon)^{1/2}\cos\theta} = \frac{n\lambda}{2\cos\theta} \tag{13}$$

When the **H** vector is polarized in the plane of incidence it remains in this plane after reflection, as shown in Fig. 2-2b. The **E** vector remains pointing out of the paper, and again traveling waves are set up in the x direction and standing waves in the z direction, with Eq. (13) satisfied for the vector **E**.

Whenever $\theta \neq 90°$ the phase velocities v_p in the x and z directions are

$$v_{px} = \frac{\omega}{\beta_x} = \frac{v}{\sin\theta} \qquad (14)$$

and

$$v_{pz} = \frac{\omega}{\beta_z} = \frac{v}{\cos\theta} \qquad (15)$$

where β_x and β_z are the real parts of the propagation constant in the x and z directions, respectively. It may seem paradoxical that this exceeds the velocity of light v

$$c = v = \frac{1}{(\mu\epsilon)^{1/2}} \qquad (16)$$

normal to the wavefront since it is generally believed that the velocity of light cannot be exceeded. However, a careful analysis will show that the electromagnetic energy flow $\mathbf{E} \times \mathbf{H}$ is normal to the wavefront and moves at the velocity v, so the phase velocity v_p is merely the velocity of points on the wavefront, and is not associated with the rate of propagation of electromagnetic energy.

The wave impedance Z_z associated with oblique incidence between two media is the ratio of the $\mathbf{E}$ to the $\mathbf{H}$ vectors tangential to the boundary. For the $\mathbf{E}$ vector polarized in the plane of incidence one has

$$Z_z = \frac{E_x}{H_y} = -\frac{E'_x}{H'_y} = \left(\frac{\mu}{\epsilon}\right)^{1/2}\cos\theta \qquad (17)$$

and for $\mathbf{H}$ in the plane of incidence

$$Z_z = -\frac{E_y}{H_x} = \frac{E'_y}{H'_x} = \left(\frac{\mu}{\epsilon}\right)^{1/2}\sec\theta \qquad (18)$$

where the primed components denote the waves after reflection. Thus one polarization increases the wave impedance and one decreases it from the intrinsic impedance of the medium. The wave impedance concept will be useful for the analysis of waveguides.

The electromagnetic wave configuration resulting from the plane wave

$$E\exp[-jk(x\sin\theta + z\cos\theta)] = E\exp[-j(\beta_x x + \beta_z z)] \qquad (19)$$

$$H = \frac{E}{(\mu/\epsilon)^{1/2}} \qquad (20)$$

with the propagation constants

$$\beta_x = k \sin \theta \tag{21}$$

$$\beta_z = k \cos \theta \tag{22}$$

$$k = \frac{2\pi}{\lambda} = \omega(\mu\epsilon)^{1/2} \tag{23}$$

incident on a conductor surface becomes the superposition of the incident and reflected waves. For the **E** field polarized in the plane of incidence it has the form

$$E_x = -2jE \cos \theta \sin \beta_z z \exp(-j\beta_x x) \tag{24}$$

$$E_z = - \left(\frac{\mu}{\epsilon}\right)^{1/2} \sin \theta \, H_y \tag{25}$$

$$H_y = \left[\frac{2E}{(\mu/\epsilon)^{1/2}}\right] \cos \beta_z z \exp(-j\beta_x x) \tag{26}$$

When the **H** vector is polarized in the plane of incidence the corresponding field configurations are

$$E_y = -2jE \sin \beta_z z \exp(-j\beta_x x) \tag{27}$$

$$H_x = - \left[\frac{2E}{(\mu/\epsilon)^{1/2}}\right] \cos \theta \cos \beta_z z \exp(-j\beta_x x) \tag{28}$$

$$H_z = \left[\frac{1}{(\mu/\epsilon)^{1/2}}\right] \sin \theta \, E_y \tag{29}$$

These wave configurations are useful for the analysis of waveguide propagation because the electromagnetic waves in waveguides may be considered as undergoing multiple reflections while they propagate.

When the incident wave is polarized in a direction other than the two discussed here it may be considered as a superposition of these two cases, and each component may be handled separately.

B. Electromagnetic Field of Guided Waves

Two of Maxwell's equations have the form:

$$\nabla \times \mathbf{E} = - \frac{\partial \mathbf{B}}{\partial t} \tag{1}$$

and

$$\nabla \times \mathbf{H} = \mathbf{J} + \frac{\partial \mathbf{D}}{\partial t} \tag{2}$$

and in the interior of the microwave transmission line $\mathbf{J} = 0$. An electromagnetic wave propagated in the z direction has the propagation factor $\exp(j\omega t - \gamma z)$, where the propagation constant γ is defined by

$$\gamma = \alpha + j\beta \tag{3}$$

All of the time and z dependence is contained in the propagation factor, which considerably simplifies the curl Eqs. (1) and (2). As a result they become

$$\frac{\partial E_z}{\partial y} + \gamma E_y = -j\omega\mu H_x \tag{4}$$

$$-\gamma E_x - \frac{\partial E_z}{\partial x} = -j\omega\mu H_y \tag{5}$$

$$\frac{\partial E_y}{\partial x} - \frac{\partial E_x}{\partial y} = -j\omega\mu H_z \tag{6}$$

$$\frac{\partial H_z}{\partial y} + \gamma H_y = j\omega\epsilon E_x \tag{7}$$

$$-\gamma H_x - \frac{\partial H_z}{\partial x} = j\omega\epsilon E_y \tag{8}$$

$$\frac{\partial H_y}{\partial x} - \frac{\partial H_x}{\partial y} = j\omega\epsilon E_z \tag{9}$$

There are three general types of wave configurations that may propagate along microwave transmission lines, namely, transverse electromagnetic (TEM) waves, transverse electric (TE) waves, and transverse magnetic (TM) waves. They have the following properties

$$H_z = E_z = 0 \qquad \text{TEM} \tag{10}$$

$$E_z = 0, \quad H_z \neq 0 \quad \text{TE} \tag{11}$$

$$H_z = 0, \quad E_z \neq 0 \quad \text{TM} \tag{12}$$

Some authors use the symbol H for TE waves and E for TM waves. A TEM mode is usually found in coaxial cables, while rectangular and cylindrical waveguides are unable to propagate TEM modes.

Maxwell's curl Eqs. (1) and (2) are considerably simplified for transverse electromagnetic waves, and in this case Eqs. (5), (7), (4), and (8), respectively, have the form

$$\gamma E_x = j\omega\mu H_y \tag{13}$$

$$j\omega\epsilon E_x = \gamma H_y \tag{14}$$

$$\gamma E_y = -j\omega\mu H_x \tag{15}$$

$$j\omega\epsilon E_y = -\gamma H_x \tag{16}$$

These four equations lead to the relation

$$\gamma^2 = \omega^2\mu\epsilon \tag{17}$$

with the velocity of propagation v and the characteristic impedance Z_0 given by

$$v = (\mu\epsilon)^{-1/2}$$

$$Z_0 = \left(\frac{\mu}{\epsilon}\right)^{1/2} \tag{18}$$

and as a result the six Maxwell curl relations (4) through (9) simplify to

$$E_x = -j\left(\frac{\mu}{\epsilon}\right)^{1/2} H_y \tag{19}$$

$$E_y = j\left(\frac{\mu}{\epsilon}\right)^{1/2} H_x \tag{20}$$

$$\frac{\partial E_x}{\partial y} = \frac{\partial E_y}{\partial x} \tag{21}$$

$$\frac{\partial H_x}{\partial y} = \frac{\partial H_y}{\partial x} \tag{22}$$

For a Cartesian coordinate solution to these equations, let

$$E_x + jE_y = E_0 e^{j\theta} = E_0(\cos\theta + j\sin\theta) \tag{23}$$

$$H_x + jH_y = H_0 e^{j\theta'} = H_0(\cos\theta' + j\sin\theta') \tag{24}$$

subject to the condition that

$$E_0 = Z_0 H_0 \tag{25}$$

and it follows from Eqs. (19) and (20) that

$$\cos \theta = -\sin \theta' \tag{26}$$

$$\sin \theta = \cos \theta' \tag{27}$$

$$\cot \theta = -\tan \theta' \tag{28}$$

with the solution

$$\theta = \theta' + \frac{\pi}{2} \tag{29}$$

Thus without loss of generality one may select **E** in the y direction, and the four Eqs. (19) through (21) reduce to

$$E = Z_0 H \tag{30}$$

where

$$E = E_y \exp(j\omega t - \gamma z) \tag{31}$$

$$H = H_x \exp(j\omega t - \gamma z) \tag{32}$$

which constitutes a plane wave. This wave configuration or mode will propagate between lossless parallel-plane conductors with the **E** vector perpendicular to the conductor surfaces. Of course, both vectors will be perpendicular to the direction of propagation, and the Poynting vector will move at the velocity of light $1/(\mu\epsilon)^{1/2}$, since it is a TEM mode. If the finite conductivity of the conducting planes is taken into account, a small electric field E_z will develop along the direction of propagation associated with the finite electric current that flows in the z direction on the guide walls. Aside from this weak axial electric field, the wave configuration remains nearly the same as in the lossless case.

A TEM mode in cylindrical coordinates is discussed in the next section on coaxial lines. It is not possible for TEM modes to propagate along hollow tubes such as rectangular, cylindrical, or elliptical waveguides. Both transverse electric and transverse magnetic waves are able to propagate between parallel planes, but they do not find much application in ESR, and so will not be discussed here. They are similar to the corresponding TE and TM waveguide modes, and may be derived therefrom by letting either the x or y waveguide dimension become much larger than the other dimension.

It is not possible for microwaves to propagate along a guiding system with both a longitudinal magnetic field component H_z and a longitudinal electric field component E_z in addition to the transverse **E** and **H** components. However it is possible to propagate wave configurations with either an H_z component (TE mode) or an E_z (TM mode) in addition to the transverse **E** and **H** components. The general boundary conditions on the three mode types (TEM, TE, and TM) are: (1) the magnetic field lines must form continuous closed loops that surround either a conduction current J or a displacement current $\partial D/\partial t$; (2) the electric field lines may form continuous closed loops that surround a changing magnetic field $\partial B/\partial t$; or (3) the electric field lines may end normally on a charge induced on the conductor surface. This last condition cannnot be satisfied by a TEM mode.

Several of the books listed at the end of this chapter discuss guided electromagnetic waves and microwaves.

C. Voltage Standing-Wave Ratio

If a transmission line is not matched there will be an incident wave with amplitudes E_1 and H_1 traveling to the right and a reflected wave with amplitudes E_2 and H_2 traveling to the left, and Eqs. (31) and (32) of the previous section become

$$E = E_1 \exp(j\omega t - \gamma z) + E_2 \exp(j\omega t + \gamma z) \tag{1}$$

$$H = H_1 \exp(j\omega t - \gamma z) - H_2 \exp(j\omega t + \gamma z) \tag{2}$$

The minus sign arises because the Poynting vector **E** × **H** reverses direction for the reflected wave. The ratio of E_2 to E_1 is the reflection coefficient Γ

$$\Gamma = \frac{E_2}{E_1} \tag{3}$$

Along a uniform transmission line the electric field goes through maxima and minima when one moves from points where E_1 and E_2 add, to positions where they subtract. Thus there are places of maximum electric field

$$E_{max} = |E_1| + |E_2| \tag{4}$$

and points of minimum electric field

$$E_{min} = |E_1| - |E_2| \tag{5}$$

The voltage standing-wave ratio VSWR is defined by

$$\text{VSWR} = \frac{E_{\max}}{E_{\min}} = \frac{(1 + |\Gamma|)}{(1 - |\Gamma|)} \tag{6}$$

which may be rearranged to

$$|\Gamma| = \frac{(\text{VSWR} - 1)}{(\text{VSWR} + 1)} \tag{7}$$

The magnetic field along the line reaches a maximum

$$H_{\max} = \frac{(|E_1| + |E_2|)}{Z_0} \tag{8}$$

at those points where the electric field reaches a minimum, and vice versa

$$H_{\min} = \frac{(|E_1| - |E_2|)}{Z_0} \tag{9}$$

and as a result the impedance Z passes through maxima and minima given by

$$Z_{\max} = (\text{VSWR})Z_0 \tag{10}$$

$$Z_{\min} = \frac{Z_0}{\text{VSWR}} \tag{11}$$

If a transmission line is "matched" or terminated by its characteristic impedance Z_0, then there is no reflected wave E_2, and the voltage standing-wave ratio is 1. Thus the amount by which the VSWR exceeds unity is a measure of

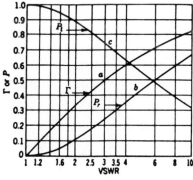

Fig. 2-3. Relation between VSWR and (a) reflection coefficient Γ, (b) relative reflected power P_r, and (c) relative transmitted power P_t (RLS-10, p. 17).

TABLE 2-1

Transmission Line Formulas[a] (Ramo et al., 1965, p. 444)

Shielded parallel wire line diagram: $p = s/d$, $q = s/D$

	Coaxial line	Parallel wire line	Shielded parallel wire line[b]	Parallel bar line — Formulae for $a \ll b$
Capacitance C, F/m	$\dfrac{2\pi\epsilon}{\ln(r_0/r_i)}$	$\dfrac{\pi\epsilon}{\cosh^{-1}(s/d)}$	—	$\epsilon b/a$
External inductance L, H/m	$\mu/2\pi \ln(r_0/r_i)$	$\mu/\pi\cosh^{-1}(s/d)$	—	$\mu(b/a)$
Conductance G, mhos/m	$\dfrac{2\pi\sigma}{\ln(r_0/r_i)} = \dfrac{2\pi\omega\epsilon_0''}{\ln(r_0/r_i)}$	$\dfrac{\pi\sigma}{\cosh^{-1}(s/d)} = \dfrac{\pi\omega\epsilon_0''}{\cosh^{-1}(s/d)}$	—	$\sigma b/a = \omega\epsilon_0\epsilon'a/b$
Resistance R, Ω/m	$\dfrac{R_s}{2\pi}\left(\dfrac{1}{r_0}+\dfrac{1}{r_i}\right)$	$\dfrac{2R_s}{\pi d}\left[\dfrac{s/d}{\sqrt{(s/d)^2-1}}\right]$	$\dfrac{2R_{s2}}{\pi d}\left[1+\dfrac{1+2p^2}{4p^4}\left(1-4q^2\right)\right]$ $+\dfrac{8R_{s3}}{\pi D}q^2\left[1+q^2-\dfrac{1+4p^2}{8p^4}\right]$	$\dfrac{2R_s}{b}$
Internal inductance L_i, H/m (for high frequency)	$\dfrac{R}{\omega}$			

$$\frac{R}{\omega} \longrightarrow$$

	Coaxial	Two-wire	Shielded pair[b]	Parallel plate
Characteristic impedance at high frequency Z_0, Ω	$\dfrac{\eta}{2\pi}\ln\left(\dfrac{r_0}{r_i}\right)$	$\dfrac{\eta}{\pi}\cosh^{-1}\left(\dfrac{s}{d}\right)$	$\dfrac{\eta_1}{\pi}\left\{\ln\left[2p\left(\dfrac{1-q^2}{1+q^2}\right)\right]-\dfrac{1+4p^2}{16p^4}\left(1-4q^2\right)\right\}$	$\dfrac{\pi a}{b}$
Z_0 for air dielectric	$60\ln\left(\dfrac{r_0}{r_i}\right)$	$120\cosh^{-1}(s/d)$ $\cong 120\ln(2s/d)$ if $s/d\gg1$	$120\left\{\ln\left[2p\left(\dfrac{1-q^2}{1+q^2}\right)\right]-\dfrac{1+4p^2}{16p^{4}}\left(1-4q^2\right)\right\}$	$120\pi(a/b)$

Attenuation due to conductor α_c $\qquad\qquad \dfrac{R}{2Z_0}$

Attenuation due to dielectric α_d $\qquad\qquad \dfrac{GZ_0}{2}=\dfrac{\sigma\eta}{2}=\dfrac{\pi\sqrt{\epsilon'\mu'}}{\lambda_0}\left(\dfrac{\epsilon''}{\epsilon'}\right)$

Total attenuation, dB/m $\qquad\qquad 8.686(\alpha_c+\alpha_d)$

Phase constant for low loss lines β $\qquad\qquad \omega\sqrt{\mu\epsilon}=2\pi/\lambda$

---

^{a}All units above are mks.

$\epsilon = \epsilon_0(\epsilon' - j\epsilon'') = $ dielectric constant, F/m
$\mu = \mu'\mu_0 = $ permeability, H/m $\qquad\Big\}$ for the dielectric
$\eta = (\mu/\epsilon)^{1/2}$ Ω

$\epsilon'' = $ loss factor of dielectric $= \sigma/\omega\epsilon_0$
$R_s = $ skin effect surface resistivity of conductor, Ω
$\lambda = $ wavelength in dielectric $= \lambda_0/\sqrt{\epsilon'\mu'}$

^{b}Formulas for shielded pair obtained from E. I. Green, F. A. Leibe, and H. E. Curtis, *Bell System Tech. Syst.* **15**, 248 (1936).

the amount of mismatch. A transmission line of infinite length is equivalent to one terminated by its characteristic impedance.

Some microwave engineers make use of the power standing-wave ratio PSWR, which is related to the voltage standing-wave ratio by

$$PSWR = (VSWR)^2 \tag{12}$$

The relationships between the reflection coefficient Γ, the relative reflected power P_r, and the relative transmitted or absorbed power P_t are shown in Fig. 2-3 (see RLS-8, p. 64) where

$$P_r = |\Gamma|^2 \tag{13}$$

$$P_t = 1 - |\Gamma|^2 \tag{14}$$

and of course

$$P_t + P_r = \text{incident power} \tag{15}$$

D. Coaxial Lines

The usual mode found in coaxial lines is transverse electromagnetic (TEM_{01}). The electric field E_r consists of radial lines of force extending from the inner conductor to the outer conductor, and the magnetic field H_φ consists of concentric circles that surround the inner conductor. The electric field E_r begins and ends on charges induced on the inner and outer conductors, and the magnetic field circles H_φ surround the conduction current J_z in the center conductor (J_z also flows in the outer conductor) to satisfy boundary conditions (1) and (3) mentioned at the end of Sec. 2-B. Since this is a TEM mode it travels at light velocity v

$$v = \frac{1}{(\mu\epsilon)^{1/2}} \tag{1}$$

and the guide wavelength λ_g equals the free space wavelength λ. Because the guide wavelength is independent of the waveguide dimensions the cutoff wavelength λ_c is infinite, and there is no lower limit on the frequency that will propagate in this mode. The capacitance per unit length, inductance per unit length, and conductance per unit length are related to the natural logarithm of the ratio of the inner conductor radius r_i to the outer conductor radius r_0 as shown in Table 2-1.

Of more practical importance is the characteristic impedance Z_0, which is

$$Z_0 = \left(\frac{1}{2\pi}\right)\left(\frac{\mu}{\epsilon}\right)^{1/2}\ln\left(\frac{r_0}{r_i}\right) \; \Omega \tag{2}$$

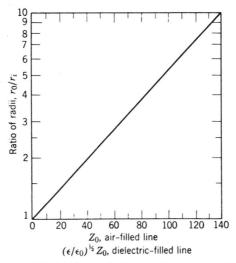

Fig. 2-4. Characteristic impedance Z_0 of coaxial line in ohms (RLS-9, p. 183).

and for air or vacuum as the dielectric between the two conductors this simplifies to

$$Z_0 = 60 \ln\left(\frac{r_0}{r_i}\right) \ \Omega \tag{3}$$

as shown graphically in Fig. 2-4. The resistance per unit length R is

$$R = \left(\frac{R_s}{2\pi}\right)\left(\frac{1}{r_0} + \frac{1}{r_i}\right) \tag{4}$$

where R_s is the surface resistivity. The attenuation due to the conductor losses α_R is

$$\alpha_R = \frac{R}{2Z_0} \ \text{Np/m} \tag{5a}$$

$$= \frac{4.343R}{Z_0} \ \text{dB/m} \tag{5b}$$

Typical characteristic impedances for commercially available coaxial lines lie in the range 30 to 100 Ω. For a typical coaxial cable $r_i = 5 \times 10^{-4}$ m and $Z_0 = 50 \ \Omega$, giving

$$R \approx 320R_s \tag{6}$$

and

$$\alpha_R \approx 28R_s \text{ dB/m} \tag{7}$$

For copper

$$R_s = 2.61 \times 10^{-4}f^{1/2} \ \Omega \tag{8}$$

so that

$$\alpha_R \approx 7 \times 10^{-3}f^{1/2} \text{ dB/m} \tag{9}$$

where the frequency f is in megahertz. For a typical X-band frequency $f = 10^4$ MHz

$$\alpha_R \approx 0.7 \text{ dB/m} \tag{10}$$

and a coaxial cable 4 m long will attenuate the microwave power by about 50%. The calculation for α_R is mainly illustrative because the actual values of α_R for coaxial cables exceed calculated values owing to the effect of dielectric losses, and increased conduction losses arising from braiding of the outer conductor [compare Eq. (7-B-8) of the first edition]. In practical cases, tabulated values from engineering handbooks should be employed in determining attenuation constants. Typical coaxial cables attenuate X-band microwaves 10 times as effectively as waveguides.

For comparative purposes Table 2-1 gives the electrical characteristics of a coaxial cable, of a parallel wire transmission line with and without a shield, and of a parallel bar transmission line. Parallel wire transmission lines tend to have higher characteristic impedances than coaxial lines, and they are balanced while the coaxial line is not. As an example of the application of each: (1) a shielded parallel wire line is often used to connect the output of a push–pull field modulation amplifier to the magnetic field modulation coils since one needs only to reverse the cable connector to reverse the phase of the modulation, and the entire system is balanced, while (2) a coaxial line often carries the signal from the output of the crystal to the preamplifier, since both the crystal and the preamplifier are electrically unbalanced.

In conclusion, it should be mentioned that higher-order waves of TE and TM type can propagate along coaxial lines, and the formulas in Table 2-1 do not apply to such modes (see RLS-8, p. 41; RLS-9, Chap. 2). For most coaxial line applications the wavelength exceeds the average diameter (or circumference), and so these higher-order modes are beyond cutoff and cannot propagate. In these applications they only become important in the neighborhood of discontinuities. The wave configurations of several higher-order coaxial modes are shown in Fig. 2-5.

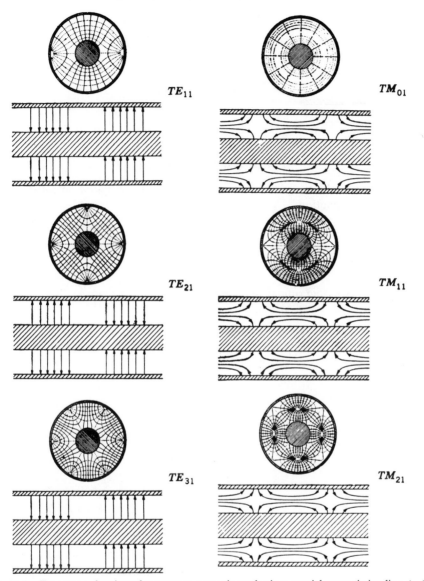

Fig. 2-5.   Transverse electric and transverse magnetic modes in a coaxial transmission line. (—) Electric field E; (...) magnetic field H (RLS-10, Sec. 2.4).

E. Rectangular Waveguides

It was mentioned above that hollow waveguides cannot support TEM modes, so the rectangular waveguide shown in Fig. 2-6 will transmit only TE and TM modes. The generalized TE mode wave configuration may be obtained by setting $E_z = 0$ in Eqs. (2-B-4) to (2-B-9), and the generalized TM mode results when $H_z = 0$ in these equations. Each mode type will be discussed in turn.

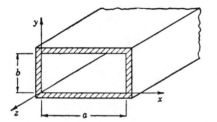

Fig. 2-6. Rectangular waveguide coordinates (RLS-9, p. 38).

Transverse Electric Modes

The magnetic field component H_z of the TE wave configuration satisfies the Laplace wave equation

$$\nabla^2 H_z = \mu\epsilon \left(\frac{\partial^2 H_z}{\partial t^2} \right) \tag{1}$$

and for the harmonic time dependence $e^{j\omega t}$

$$\nabla^2 H_z = -k^2 H_z \tag{2}$$

where

$$k^2 = \omega^2 \mu\epsilon \tag{3}$$

The entire z dependence of the H_z vector is contained in the propagation factor $e^{-\gamma z}$, so

$$\frac{\partial^2 H_z}{\partial z^2} = \gamma^2 H_z \tag{4}$$

and

$$\frac{\partial^2 H_z}{\partial x^2} + \frac{\partial^2 H_z}{\partial y^2} = -(k^2 + \gamma^2) H_z \tag{5}$$

It is now convenient to rewrite Eqs. (2-B-4) and (2-B-5) with $E_z = 0$

$$E_y = -j\left(\frac{k}{\gamma}\right)\left(\frac{\mu}{\epsilon}\right)^{1/2} H_x \tag{6}$$

$$E_x = j\left(\frac{k}{\gamma}\right)\left(\frac{\mu}{\epsilon}\right)^{1/2} H_y \tag{7}$$

Equations (2-B-7) and (2-B-8) may now be solved for H_x and H_y

$$H_x = \left(\frac{\gamma}{k_c^2}\right)\left(\frac{\partial H_z}{\partial x}\right) \tag{8}$$

$$H_y = \left(\frac{\gamma}{k_c^2}\right)\left(\frac{\partial H_z}{\partial y}\right) \tag{9}$$

where

$$\gamma = \alpha + j\beta = \left(k_c^2 - k^2\right)^{1/2} \tag{10}$$

If we neglect the conductor attenuation for the present (i.e., $\alpha \ll \beta$), then

$$k_c = \frac{2\pi}{\lambda_c} = \left(k^2 + \gamma^2\right)^{1/2} = 2\pi f_c(\mu\epsilon)^{1/2} \tag{11}$$

and

$$\gamma = j\beta = \frac{2\pi j}{\lambda_g} = \left(k_c^2 - k^2\right)^{1/2} = j\omega(\mu\epsilon)^{1/2}\left[1 - \left(\frac{f_c}{f}\right)^2\right]^{1/2} \tag{10'}$$

for a perfect dielectric with zero loss tangent.

The boundary condition that the normal derivative of H_z

$$\frac{\partial H_z}{\partial n} = 0 \tag{12}$$

must vanish at the walls limits the possible functional dependence of the wave vector H_z. The solution of Eqs. (6)–(9) with this boundary condition is

$$H_x = \left(\frac{\gamma k_x}{k_c^2}\right) H_0 \sin k_x x \cos k_y y \tag{13}$$

$$H_y = \left(\frac{\gamma k_y}{k_c^2}\right) H_0 \cos k_x x \sin k_y y \tag{14}$$

$$H_z = H_0 \cos k_x x \cos k_y y \tag{15}$$

$$E_x = Z_{TE} H_y \tag{16}$$

$$E_y = -Z_{TE} H_x \tag{17}$$

TE_{10}

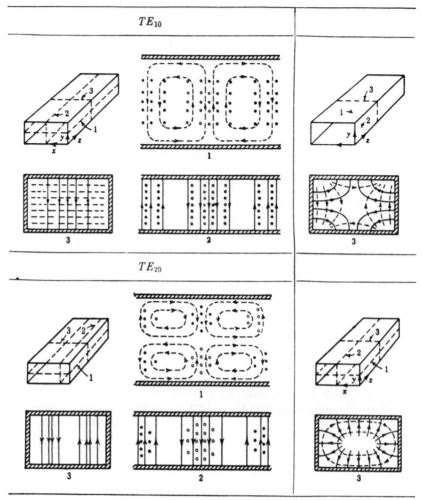

TE_{20}

Fig. 2-7. Summary of wave types for rectangular guides. The symbol (●) denotes a vector directed up out of the paper and (○) one into the paper. Dotted lines refer to **H** and solid lines to **E** (Ramo et al., 1965).

where the characteristic impedance Z_{TE} for a TE mode is

$$Z_{TE} = \left(\frac{\mu}{\epsilon}\right)^{1/2}\left[1 - \left(\frac{f_c}{f}\right)^2\right]^{-1/2} \tag{18}$$

Several mode configurations are sketched in Fig. 2-7. The cutoff frequency f_c is defined by Eq. (11), and it is the lower limit of the frequencies that will propagate for a given mode. The boundary conditions on k_x and k_y are

$$k_x = \frac{m\pi}{a}; \quad k_y = \frac{n\pi}{b} \tag{19}$$

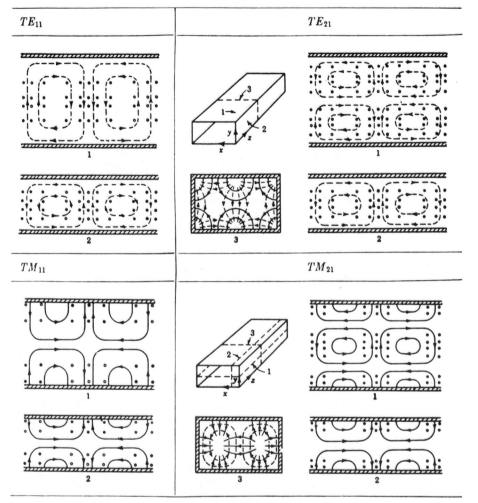

Fig. 2-7. (*Continued*).

where *m* and *n* are integers. Therefore

$$k_c = \frac{2\pi}{\lambda_c} = \left[\left(\frac{m\pi}{a}\right)^2 + \left(\frac{n\pi}{b}\right)^2\right]^{1/2} = \left(k_x^2 + k_y^2\right)^{1/2} \tag{20}$$

where λ_c is the cutoff wavelength. One may write

$$\frac{1}{\lambda^2} = \frac{1}{\lambda_g^2} + \frac{1}{\lambda_c^2} \tag{21}$$

where λ is the free-space wavelength and $= [f(\mu\epsilon)^{1/2}]^{-1}$ in the dielectric, and

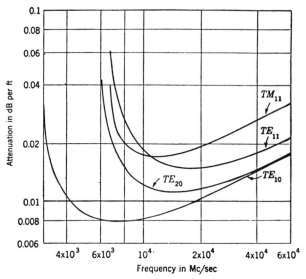

Fig. 2-8. Ohmic attenuation α_R in rectangular copper waveguide for several modes: $a = 2$ in. $= 5.08$ cm; $b = 1$ in. $= 2.54$ cm (1 ft. $= 304.8$ cm) (RLS-8, p. 48).

λ_g is the guide wavelength or wavelength in the direction of propagation z

$$\gamma = \alpha + \left(\frac{2\pi}{\lambda_g}\right) j \tag{22}$$

Thus for propagation to occur, it is necessary for the free-space wavelength to be less than the cutoff wavelength. When this is not the case then the guide wavelength λ_g becomes imaginary and contributes to the attenuation constant α in Eq. (22).

As a result of the ohmic losses in the conducting waveguide walls the microwave power is attenuated by the factor $\exp(-\alpha_R z)$. The attenuation constant α_R for the TE_{mn} mode is given by

$$\alpha_R = \frac{1}{2} \frac{\text{power dissipated in walls per meter}}{\text{power transmitted}} \tag{23}$$

$$= \frac{R_8}{2} \frac{\oint\left(|H_x|^2 + |H_y|^2 + |H_z|^2\right) dl}{\int \text{Re}(\mathbf{E} \times \mathbf{H}^*) \, dA} \tag{24}$$

$$= \frac{2R_8}{b(\mu/\epsilon)^{1/2}\left[1 - (f_c/f)^2\right]^{1/2}} \left[\left(1 + \frac{b}{a}\right)\left(\frac{f_c}{f}\right)^2 \right.$$

$$\left. + \left(1 - \left(\frac{f_c}{f}\right)^2\right)\left(\frac{b/a(b/am^2 + n^2)}{b^2m^2/a^2 + n^2}\right)\right] \text{Np/m} \tag{25}$$

which has the simpler form for the TE_{m0} mode (Ramo et al., 1965, p. 422)

$$\alpha_R = \frac{R_8}{b(\mu/\epsilon)^{1/2}\left[1 - (f_c/f)^2\right]^{1/2}}\left[1 + \frac{2b}{a}\left(\frac{f_c}{f}\right)^2\right] \text{ Np/m} \qquad (26)$$

The frequency dependence of the attenuation constant α_R for several modes is shown in Fig. 2-8. One may convert Eqs. (23)–(26) to the units dB/m by multiplying the right-hand side by 8.686. The transverse electric modes TE_{mn} are labeled by their m, n values, and the mode configurations of four of them are shown in Fig. 2-7.

The TE₁₀ Mode

The dominant mode is the one with the lowest cutoff frequency, and from Eq. (20) it is apparent that for rectangular waveguides the TE_{10} mode is dominant. The field configurations are obtained from Eqs. (13)–(19) by letting $m = 1$ and $n = 0$, and renormalizing as follows

$$H_x = +H_0 \sin\left(\frac{\pi x}{a}\right) \qquad (27)$$

$$H_z = H_0\left(\frac{\lambda_g}{2a}\right)\cos\left(\frac{\pi x}{a}\right) \qquad (28)$$

$$E_y = -Z_{TE}H_0 \sin\left(\frac{\pi x}{a}\right) \qquad (29)$$

$$H_y = E_x = E_z = 0 \qquad (30)$$

These equations are normalized differently in Sec. 5-H. The cutoff wavelength λ_c, characteristic impedance Z_{TE}, group velocity v_g, and phase velocity v_p of this mode are given by

$$\lambda_c = 2a \qquad (31)$$

$$Z_0 = \frac{(\mu/\epsilon)^{1/2}}{\left[1 - (\lambda/2a)^2\right]^{1/2}} \qquad (32)$$

$$v_g = \frac{1}{(\mu\epsilon)^{1/2}}\left[1 - \left(\frac{\lambda}{2a}\right)^2\right]^{1/2} \qquad (33)$$

$$v_p = \frac{1}{(\mu\epsilon)^{1/2}}\left[1 - \left(\frac{\lambda}{2a}\right)^2\right]^{-1/2} \qquad (34)$$

The attenuation α_R due to the ohmic losses in the guide walls was given by

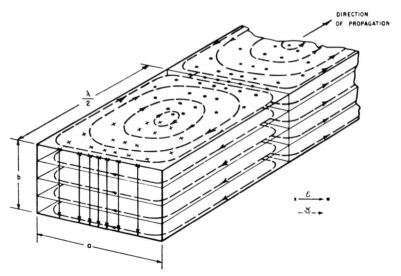

Fig. 2-9. Mode configuration for rectangular TE_{10} mode. The symbol (●) denotes a vector directed up from the waveguide, (×) refers to one aimed down into the waveguide, and (λ) corresponds to the guide wavelength (Reintjes and Coate, 1952, p. 573).

Eq. (26), and that due to dielectric losses α_ϵ is

$$\alpha_\epsilon = \frac{\omega(\mu\epsilon)^{1/2}(\epsilon''/\epsilon')}{2\left[1 - (\lambda/2a)^2\right]^{1/2}} \ \text{Np/m} \tag{35}$$

$$= \frac{4.343\omega(\mu\epsilon)^{1/2}(\epsilon''/\epsilon')}{\left[1 - (\lambda/2a)^2\right]^{1/2}} \ \text{dB/m} \tag{36}$$

Three-dimensional sketches of several TE electromagnetic mode configurations are shown in Figs. 2-9 and 2-10, and the current distribution in the waveguide walls is shown in Fig. 2-11 for the TE_{10} mode. The top and bottom faces of the waveguide have the charge density ϵE_y since the electric field E_y V/m begins on the charge ϵE_y and ends on the charge $-\epsilon E_y$ C/m. The electric current density J A/m equals the tangential magnetic field component H_t at the walls. The magnetic field loops surround the displacement current $\partial D/\partial t$ A/m, which completes the circuit from the top to the bottom of the waveguide. The guide wavelength and characteristic impedance of this mode are independent of the guide dimension b, but as b increases at a constant power level both the attenuation constant and the electric field strength decrease. For small b the electric field $\mathbf{E}$ can become so strong at the very high microwave power levels used in radar installations that arcing across the waveguide is likely to occur, but in ESR spectrometers this is almost never a problem. For most ESR applications one uses standard commercially available rectangular waveguides

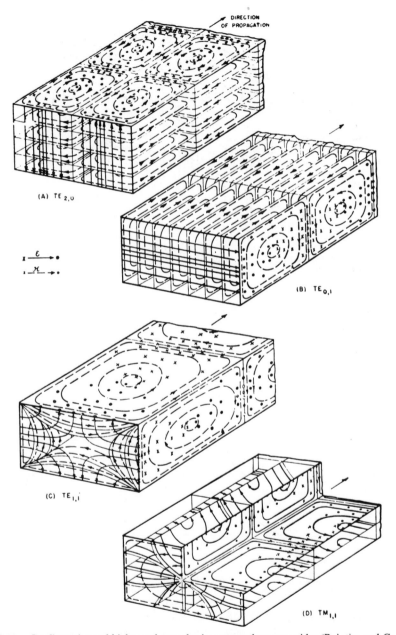

Fig. 2-10. Configurations of higher-order modes in rectangular waveguides (Reintjes and Coate, 1952, p. 589); notation as in Fig. 2-9.

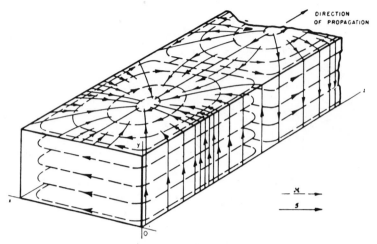

Fig. 2-11. Current distribution (—) in waveguide walls for the rectangular TE$_{10}$ mode (Reintjes and Coate, 1952, p. 576).

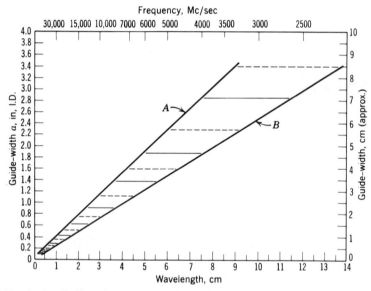

Fig. 2-12. Preferred dimensions of rectangular waveguides suitable for various wavelength ranges: (A) limit dictated by higher-order modes; (B) limit dictated by proximity to cutoff (Southworth, 1950, p. 82).

48

operating in the TE_{10} mode, and the recommended frequency ranges of these waveguides are shown in Fig. 2-12.

Transverse Magnetic Modes

The electric vector E_z of a TM wave configuration satisfies Laplace's equation

$$\nabla^2 E_z = \mu\epsilon\left(\frac{\partial^2 E_z}{\partial t^2}\right) \tag{37}$$

The solution of this equation is obtained in a manner analogous to the solution of Eq. (1) with the aid of the boundary condition that the electric field vector parallel to the boundary vanishes. The mathematical manipulations will be left as an exercise for the reader, and only the conclusions will be presented. The electromagnetic wave configurations for the TM modes are

$$E_x = -\left(\frac{\gamma k_x}{k_c^2}\right) E_0 \cos k_x x \sin k_y y \tag{38}$$

$$E_y = -\left(\frac{\gamma k_y}{k_c^2}\right) E_0 \sin k_x x \cos k_y y \tag{39}$$

$$E_z = E_0 \sin k_x x \sin k_y y \tag{40}$$

$$H_x = -\frac{E_y}{Z_{TM}} \tag{41}$$

$$H_y = \frac{E_x}{Z_{TM}} \tag{42}$$

$$H_y = 0 \tag{43}$$

The TM_{11} mode is sketched in Fig. 2-10.

The characteristic impedance Z_{TM} of a TM mode is given by

$$Z_{TM} = \left(\frac{\mu}{\epsilon}\right)^{1/2}\left[1 - \left(\frac{f_c}{f}\right)^2\right]^{1/2} \tag{44}$$

and the attenuation constant α_R arising from ohmic losses in the waveguide walls is (Ramo et al., 1965, p. 423)

$$\alpha_R = \frac{2R_s}{b(\mu/\epsilon)^{1/2}\left[1 - (f_c/f)^2\right]^{1/2}}\left[\frac{m^2(b/a)^3 + n^2}{m^2(b/a)^2 + n^2}\right] \text{ Np/m} \tag{45}$$

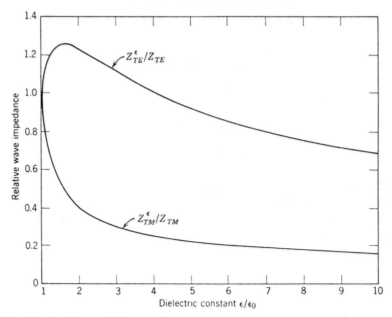

Fig. 2-13.   Relative characteristic (wave) impedance as a function of the dielectric constant ϵ/ϵ_0 for TE modes (Z_{TE}^ϵ/Z_{TE}) and TM modes (Z_{TM}^ϵ/Z_{TM}) (RLS-8, p. 371).

The frequency dependence of Eq. (45) for the TM_{11} mode is shown in Fig. 2-8. The remaining expressions (3), (10), (11), and (19)–(23) are valid for both TE and TM modes. It should be noted that by comparing Eq. (18) to Eq. (44) the characteristic impedance of the transverse electric wave exceeds that in the unbounded dielectric $(\mu/\epsilon)^{1/2}$, while that of a transverse magnetic wave is less than $(\mu/\epsilon)^{1/2}$. There exist TE_{mn} waves with either m or n equal to zero (but not both), while for TM modes both m and n must be positive integers.

In a dielectric-filled waveguide the cutoff wavelength λ_c and k_c are still defined in terms of the waveguide dimensions by Eq. (20) so that the cutoff frequency f_c is inversely proportional to the dielectric constant ϵ in accordance with Eq. (11)

$$f_c \lambda_c (\mu\epsilon)^{1/2} = 1 \tag{46}$$

As a result the characteristic impedances Z_{TE}^ϵ and Z_{TM}^ϵ for TE and TM modes, respectively, in a dielectric-filled waveguide are

$$Z_{TE}^\epsilon = \left(\frac{\epsilon_0}{\epsilon}\right)^{1/2}\left[1 - \left(\frac{\epsilon_0}{\epsilon}\right)\left(\frac{\lambda}{\lambda_c}\right)^2\right]^{-1/2} \tag{47}$$

$$Z_{TM}^\epsilon = \left(\frac{\epsilon_0}{\epsilon}\right)^{1/2}\left[1 - \left(\frac{\epsilon_0}{\epsilon}\right)\left(\frac{\lambda}{\lambda_c}\right)^2\right]^{1/2} \tag{48}$$

where $\mu = \mu_0$ in the dielectric. Figure 2-13 is a graph of the relative characteristic impedances Z_{TE}^{ϵ}/Z_{TE} and Z_{TM}^{ϵ}/Z_{TM} versus the relative dielectric constant ϵ/ϵ_0 for $(\lambda/\lambda_c)^2 = 0.8$. Further details are given in RLS-8 (p. 369ff.).

Sometimes space limitations such as those that exist in Dewars require the use of a waveguide with as small a cross section as possible. The cross section may be decreased by filling the waveguide with a dielectric material such as Teflon and scaling the dimensions in the ratio $(\epsilon/\epsilon_0)^{1/2}$, where ϵ_0 is the dielectric constant of air (or vacuum), and ϵ is the dielectric constant of Teflon.

The characteristics of standard rectangular waveguides are given in Table 4-3.

F. Cylindrical Waveguides

The electromagnetic wave configurations for the allowed TE (or TM) modes in cylindrical waveguides are obtained by solving the Laplace equation for H_z (or E_z) in cylindrical coordinates

$$\nabla^2 H_z = \frac{1}{r}\frac{\partial}{\partial r}\left(r\frac{\partial H_z}{\partial r}\right) + \frac{\partial^2 H_z}{\partial z^2} + \frac{1}{r^2}\frac{\partial^2 H_z}{\partial \varphi^2} = \mu\epsilon\frac{\partial^2 H_z}{\partial t^2} \tag{1}$$

The same equation holds for the other field components. As in the rectangular waveguide case one assumes the time and z dependence $\exp(j\omega t - \gamma z)$, where z is the direction of propagation. This gives

$$\frac{1}{r}\frac{\partial}{\partial r}\left(r\frac{\partial H_z}{\partial r}\right) + \frac{1}{r^2}\frac{\partial^2 H_z}{\partial \varphi^2} + k_c^2 H_z = 0 \tag{2}$$

where

$$k^2 = \left(\frac{2\pi}{\lambda}\right)^2 = \omega^2\mu\epsilon \tag{3}$$

$$k_c = \frac{2\pi}{\lambda_c} = (k^2 + \gamma^2)^{1/2} = 2\pi f_c(\mu\epsilon)^{1/2} \tag{4}$$

The φ dependence for TE modes is of the form

$$A \sin m\varphi + B \cos m\varphi \tag{5}$$

221684

and so for the z component of the magnetic field one obtains the expression

$$\frac{1}{r}\frac{\partial}{\partial r}\left(r\frac{\partial H_z}{\partial r}\right) + \left(k_c^2 - \frac{m^2}{r^2}\right)H_z = 0 \tag{6}$$

which is Bessel's equation. For TM modes the corresponding equation holds for E_z

$$\frac{1}{r}\frac{\partial}{\partial r}\left(r\frac{\partial E_z}{\partial r}\right) + \left(k_c^2 - \frac{m^2}{r^2}\right)E_z = 0 \tag{7}$$

The solution of Bessel's equation is the mth-order Bessel function $J_m(k_c r)$, and the electromagnetic field distributions are in terms of the Bessel function $J_m(k_c r)$ and its derivative $J_m'(k_c r)$

$$\frac{dJ_m(k_c r)}{d(k_c r)} = J_m'(k_c r) \tag{8}$$

Several miscellaneous relations that are important for cylindrical waveguide modes above cutoff ($f > f_c$) are

$$\beta_z = \left(k_c^2 - k^2\right)^{1/2} = j\omega(\mu\epsilon)^{1/2}\left[1 - \left(\frac{f_c}{f}\right)^2\right]^{1/2} \tag{9}$$

$$\gamma = \alpha + j\beta \tag{10}$$

$$\frac{1}{\lambda^2} = \frac{1}{\lambda_g^2} + \frac{1}{\lambda_c^2} \tag{11}$$

$$\lambda f = \frac{1}{(\mu\epsilon)^{1/2}} = v \tag{12}$$

$$\lambda_g = \frac{2\pi}{\beta_z} \tag{13}$$

$$\lambda_c = \frac{2\pi}{k_c} = \frac{1}{\left[f_c(\mu\epsilon)^{1/2}\right]} \tag{14}$$

where $(\mu\epsilon)^{1/2}$ is assumed to be real.

For transverse electric modes the electromagnetic wave configurations are

$$H_r = -\frac{\gamma}{k_c} H_0 J'_m(k_c r) \begin{cases} \cos m\varphi \\ \sin m\varphi \end{cases} \tag{15}$$

$$H_\varphi = \frac{m\gamma}{k_c^2 r} H_0 J_m(k_c r) \begin{cases} \sin m\varphi \\ -\cos m\varphi \end{cases} \tag{16}$$

$$H_z = H_0 J_m(k_c r) \begin{cases} \cos m\varphi \\ \sin m\varphi \end{cases} \tag{17}$$

$$E_r = Z_{\mathrm{TE}} H_\varphi \tag{18}$$

$$E_\varphi = -Z_{\mathrm{TE}} H_r \tag{19}$$

$$E_z = 0 \tag{20}$$

where the factor $\exp(j\omega t - \gamma z)$ is understood in these expressions. Several TE mode configurations are sketched in Figs. 2-14 and 2-15. For the TE modes the characteristic impedance is

$$Z_{\mathrm{TE}} = \frac{(\mu/\epsilon)^{1/2}}{\left[1 - (f_c/f)^2\right]^{1/2}} \tag{21}$$

The boundary condition on the radial component of the magnetic field is $H_r = 0$ at the conductor surface $r = a$, or

$$J'_m(k_c a) = 0 \tag{22}$$

There are an infinite number of solutions of this equation, and a typical solution is the nth-order root $(k_c a)'_{mn}$. These roots are listed in Table 5-1 of Chap. 5 because their principal application in ESR is in the design of resonant cavities. From Eq. (14) it follows that

$$\lambda_c = \frac{2\pi a}{(k_c a)'_{mn}} = \frac{1}{\left[f_c(\mu\epsilon)^{1/2}\right]} \tag{23}$$

and the attenuation constant α_R arising from ohmic losses in the guide walls is, in nepers per meter (RLS-10, p. 70)

$$\alpha_R = \frac{R_s}{a(\mu/\epsilon)^{1/2}\left[1 - (f_c/f)^2\right]^{1/2}} \left[\left(\frac{f_c}{f}\right)^2 + \frac{m^2}{(k_c a)'^2_{mn} - m^2}\right] \tag{24}$$

Wave Type	TM_{01}	TM_{02}
Field distributions in cross-sectional plane, at plane of maximum transverse fields		
Field distributions along guide	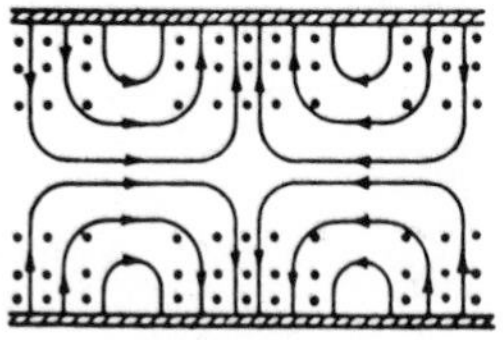	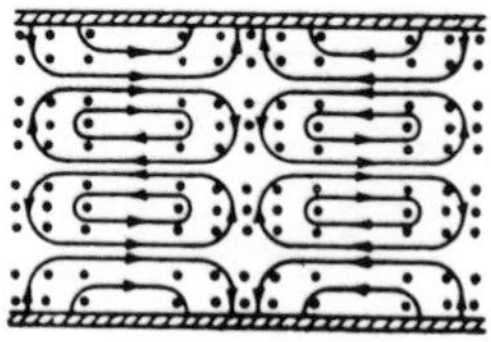
Field components present	$E_s,\ E_r,\ H_\phi$	$E_s,\ E_r,\ H_\phi$
p_{nl} or p'_{nl}	2.405	5.52
$(k_c)_{nl}$	$\dfrac{2.405}{a}$	$\dfrac{5.52}{a}$
$(\lambda_c)_{nl}$	$2.61a$	$1.14a$
$(f_c)_{nl}$	$\dfrac{0.383}{a\sqrt{\mu\,\epsilon}}$	$\dfrac{0.877}{a\sqrt{\mu\,\epsilon}}$
Attenuation due to imperfect conductors	$\dfrac{R_s}{a\eta}\ \dfrac{1}{\sqrt{1-(f_c/f)^2}}$	$\dfrac{R_s}{a\eta}\ \dfrac{1}{\sqrt{1-(f_c/f)^2}}$

Fig. 2-14. Summary of wave types for cylindrical-waveguides. The notation follows Fig. 2-7 and $\eta = \sqrt{\mu/\epsilon}$ (Ramo et al., 1965, Table 8.04).

For transverse magnetic modes the electromagnetic wave configurations are

$$E_r = -\frac{\gamma}{k_c} E_0 J'_m(k_c r) \begin{cases} \cos m\varphi \\ \sin m\varphi \end{cases} \tag{25}$$

$$E_\varphi = -\frac{m\gamma}{k_c^2 r} E_0 J_m(k_c r) \begin{cases} \sin m\varphi \\ -\cos m\varphi \end{cases} \tag{26}$$

$$E_z = E_0 J_m(k_c r) \begin{cases} \cos m\varphi \\ \sin m\varphi \end{cases} \tag{27}$$

$$H_r = -\frac{E_\varphi}{Z_0} \tag{28}$$

TM_{11}	TE_{01}	TE_{11}
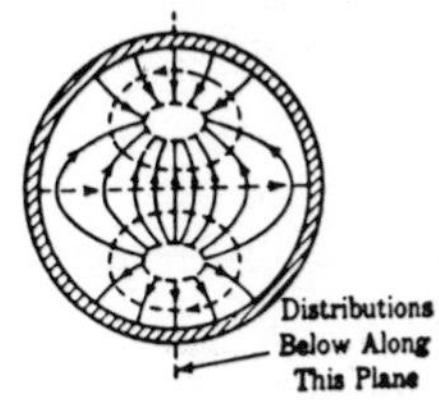 Distributions Below Along This Plane	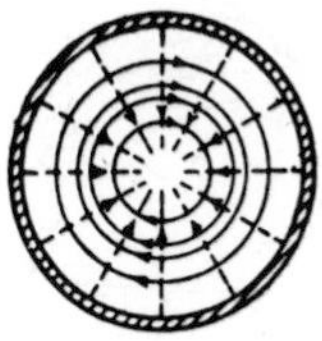	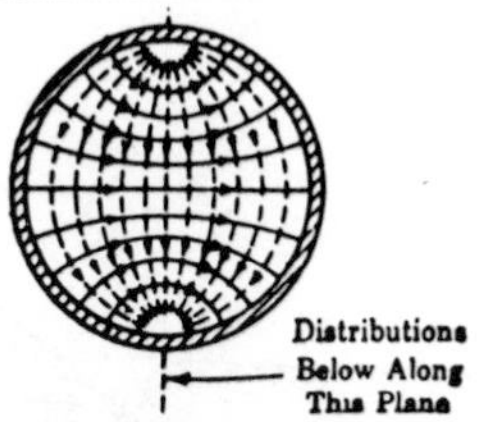 Distributions Below Along This Plane
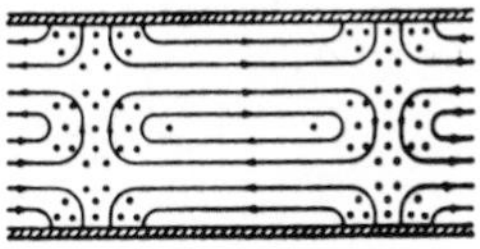	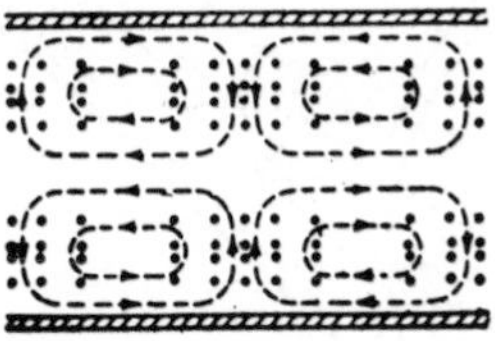	
$E_z, E_r, E_\phi, H_r, H_\phi$	H_z, H_r, E_ϕ	$H_z, H_r, H_\phi, E_r, E_\phi$
3.83	3.83	1.84
$\dfrac{3.83}{a}$	$\dfrac{3.83}{a}$	$\dfrac{1.84}{a}$
$1.64a$	$1.64a$	$3.41a$
$\dfrac{0.609}{a\sqrt{\mu\,\epsilon}}$	$\dfrac{0.609}{a\sqrt{\mu\,\epsilon}}$	$\dfrac{0.293}{a\sqrt{\mu\,\epsilon}}$
$\dfrac{R_s}{a\eta}\dfrac{1}{\sqrt{1-(f_c/f)^2}}$	$\dfrac{R_s}{a\eta}\dfrac{(f_c/f)^2}{\sqrt{1-(f_c/f)^2}}$	$\dfrac{R_s}{a\eta}\dfrac{1}{\sqrt{1-(f_c/f)^2}}\left[\left(\dfrac{f_c}{f}\right)^2+0.420\right]$

Fig. 2-14. (Continued).

$$H_\varphi = \frac{E_r}{Z_0} \qquad (29)$$

$$H_z = 0 \qquad (30)$$

where again the term $\exp(j\omega t - \gamma z)$ is understood, and several modes are sketched in Figs. 2-14 and 2-15. In addition

$$Z_{TM} = \left(\frac{\mu}{\epsilon}\right)^{1/2}\left[1-\left(\frac{f_c}{f}\right)^2\right]^{1/2} \qquad (31)$$

The boundary condition requires the transverse electric field to be zero at the

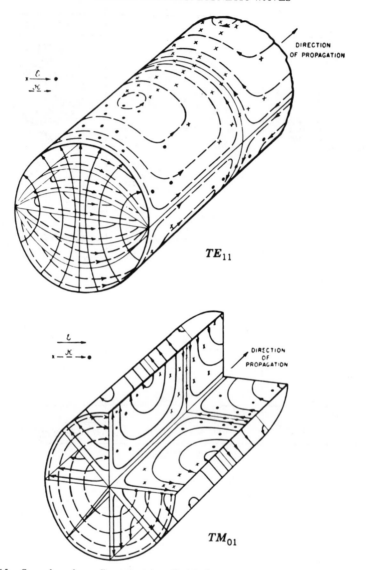

Fig. 2-15. Several mode configurations in cylindrical waveguides (Reintjes and Coate, 1952, pp. 608, 609, 610).

boundary, so that

$$J_m(k_c a) = 0 \tag{32}$$

with the roots $(k_c a)_{mn} = k_c a$ listed in Table 5-1. Therefore

$$\lambda_c = \frac{2\pi a}{(k_c a)_{mn}} = \frac{1}{\left[f_c (\mu \epsilon)^{1/2} \right]} \tag{33}$$

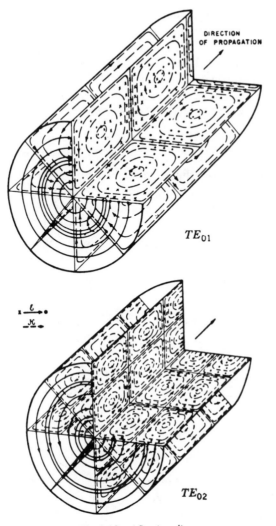

Fig. 2-15. (*Continued*).

and

$$\alpha_R = \frac{R_s}{a(\mu/\epsilon)^{1/2}} \frac{1}{\left[1 - (f_c/f)^2\right]^{1/2}} \text{ Np/m} \tag{34a}$$

$$= \frac{8.686 R_s}{a(\mu/\epsilon)^{1/2}} \frac{1}{\left[1 - (f_c/f)^2\right]^{1/2}} \text{ dB/m} \tag{34b}$$

As in the case of rectangular waveguides, frequencies that exceed the cutoff frequency f_c are propagated, while others are exponentially attenuated (compare Sec. 2-H). The TE_{11} mode is the dominant mode because it has the lowest cutoff frequency. It is the circular analog of the rectangular waveguide TE_{10} mode, as may be seen by comparing Figs. 2-7, 2-9, 2-14, and 2-15. When a centered circular iris is placed in a rectangular TE_{10} mode waveguide one may consider it as propagating the circular TE_{11} mode, and if the iris diameter is less than $0.293\lambda_c$ cm, as in a resonant cavity iris, then the iris waveguide will be operating below cutoff.

G. Miscellaneous Channels for Microwaves

The preceding sections contained a detailed treatment of waveguides with rectangular and cylindrical cross sections. These are the principal waveguides to be encountered in both ESR and radar applications. Nevertheless, it is useful for ESR spectroscopists to be acquainted with other types of guiding systems.

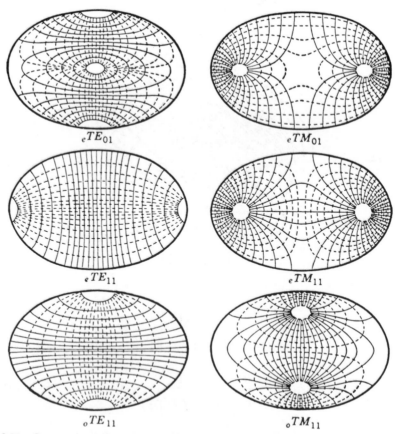

$_eTE_{01}$ $_eTM_{01}$

$_eTE_{11}$ $_eTM_{11}$

$_oTE_{11}$ $_oTM_{11}$

Fig. 2-16.   Cross-sectional view of field distribution of modes in elliptical waveguide (RLS-10, p. 84).

Elliptic Waveguides

Several wave configurations in waveguides of elliptical cross section are shown graphically in Fig. 2-16. They are expressed mathematically by means of Mathieu functions (Stratton, 1941; RLS-10, p. 80), which employ elliptic coordinates (see Margenau and Murphy, 1956, Sec. 5.9). In the limit of small ellipticity the elliptic modes degenerate into the cylindrical modes discussed in the previous section.

Radial Transmission Lines

These consist of two plane concentric circular parallel conductors separated by a dielectric. The generator is placed in the center, and the microwave energy is propagated radially. The sectorial horn antenna discussed in Sec. 4-M is a wedge-shaped waveguide, which may be considered as an intermediate case between a rectangular and a radial transmission line.

Conical Waveguides

Microwaves may be propagated along a conical waveguide with and without the outer conductor. When the conical solid angle is 4π, the waveguide becomes all space, and spherical waves propagate radially in all directions (see RLS-10, p. 98).

Helical Waveguides

A wire wound in the form of a helix can propagate microwaves with a phase velocity less than the velocity of light. They are used for antennas, and are employed in traveling-wave tubes as slow wave structures (see, Sec. 3-B).

Inclined Plane Waveguides

It was mentioned earlier that microwaves can propagate between parallel plane conductors. Propagation can also occur when the planes are not parallel.

Guidance by a Single Cylindrical Conductor

Stratton (1941, p. 524) discusses the propagation of a TEM wave along a wire, and mentions that TE and TM modes are rapidly attenuated.

Dielectric Rod Waveguide

The dielectric rod waveguide is known in optics as the light pipe. The dielectric constant of the rod is assumed to exceed that of the dielectric outside, and while the microwaves (or light) pass down the rod they undergo multiple reflections at an angle that exceeds the critical angle θ_c. As a result, total internal reflection occurs, and none of the energy is lost from the rod. There is a critical frequency f_c below which energy is lost to the outside medium. For

$f > f_c$ there are no losses to the outside, and for a perfect dielectric ($\epsilon'' = 0$) the wave propagates unattenuated. In practice the dielectric will have a finite loss tangent that results in attenuation. For the TM_{0n} mode in a dielectric rod characterized by the parameters μ_r, ϵ_r, and radius a, one has

$$f_c = \frac{X_{0n}}{2\pi a(\mu_r \epsilon_r - \mu_e \epsilon_e)^{1/2}}$$ (1)

where μ_e and ϵ_e characterize the external medium. From Eq. (2-F-33) expression (1) reduces to that of a cylindrical waveguide bounded by a conductor when $\mu_r \epsilon_r \gg \mu_e \epsilon_e$.

When $f \gg f_c$ the electromagnetic fields attenuate very rapidly in the external medium and are only appreciable near the interface. This is the case with a light pipe. When f is only slightly above f_c, the microwave energy storage extends far into the external medium, but for perfect dielectrics inside and outside, none is lost from the rod to this medium.

A dielectric slab with a dielectric constant exceeding that of the external medium can conduct electromagnetic radiation in a manner analogous to the dielectric rod. The rod and slab waveguides are discussed by Ramo et al., (1965, p. 448) and by Subrahmaniam (1962). Brackett et al. (1957) have made use of dielectric waveguide cells in their microwave spectrometer.

Stripline

Microwaves may be guided along metallic foil mounted on dielectrics in the manner of printed circuits. Johansson et al. (1974) describe a stripline resonator.

H. Waveguides beyond Cutoff

When the microwave frequency f is less than the cutoff frequency f_c then the phase constant β becomes imaginary and is given by

$$\beta = -j\left(\frac{2\pi}{\lambda_c}\right)\left[1 - \left(\frac{f}{f_c}\right)^2\right]^{1/2}$$ (1)

Under these conditions the propagation constant γ becomes real and is denoted by

$$\gamma = \alpha + j\beta$$ (2)

$$= \alpha + \alpha_c$$ (3)

where $\alpha_c = j\beta$. Since both α and α_c are positive definite when $f < f_c$ it follows that the electromagnetic energy is exponentially attenuated. When $f \ll f_c$ one obtains the simplified expression

$$\alpha_c = 2\pi f_c (\mu\epsilon)^{1/2} = \frac{2\pi}{\lambda_c} \tag{4}$$

A typical application of these formulas is the estimation of the insertion loss of a thick centered circular iris of radius a and thickness l across a rectangular waveguide that propagates the TE_{10} rectangular mode. This iris will sustain the circular TE_{11} mode for which

$$\alpha_c = \frac{(k_c a)'_{11}}{a} \tag{5}$$

from Eq. (2-F-23), where $(k_c a)'_{11}$ is the Bessel function root and $f \ll f_c$ in the iris. Therefore the iris exponentially attenuates the microwaves in accordance with the relation $\exp(-\alpha_c l)$.

I. Imperfect Dielectrics

Maxwell's fourth equation

$$\nabla \times \mathbf{H} = \mathbf{J} + \frac{\partial \mathbf{D}}{\partial t} \tag{1}$$

for a temporal dependence $e^{j\omega t}$ has the form

$$\nabla \times \mathbf{H} = j\omega \left(\epsilon - \frac{j\sigma}{\omega} \right) \mathbf{E} \tag{2}$$

where use was made of Ohm's law $J = \sigma E$. It is more customary to consider the relative dielectric constant ϵ/ϵ_0 as made up of a real part ϵ' and an imaginary part ϵ'' defined by

$$\frac{\epsilon}{\epsilon_0} = \epsilon' - j\epsilon'' \tag{3}$$

Physically, ϵ''/ϵ' is the dimensionless ratio of the conduction current J to the displacement current $\partial D/\partial t$ in the dielectric, and it is sometimes referred to as the loss tangent or $\tan \delta$

$$\tan \delta = \frac{\epsilon''}{\epsilon'} = \frac{36\pi\sigma}{\omega\epsilon'} \times 10^9 \tag{4}$$

where σ is in mhos per meter. Values of $\tan \delta$ and ϵ'/ϵ_0 are listed in Table 2-2 for a number of typical dielectric materials.

TABLE 2-2

Dielectric Constant ϵ'/ϵ_0 and Loss Tangent $\tan \delta$ for Several Solids at Room Temperature[a]

Material	10^8 cps		10^{10} cps	
	ϵ'/ϵ_0	$\tan \delta$	ϵ'/ϵ_0	$\tan \delta$
Alumina, Al_2O_3	$\left\{ \begin{array}{c} 9.34 \\ 11.54 \end{array} \right\}$	—	—	—
Bakelite (BM-16981, not preformed or pretreated)	4.7	1×10^{-2}	4.5	1.2×10^{-2}
Lucite HM-119 (polyacrylate)	2.58	6.7×10^{-3}	2.57	4.9×10^{-3}
Plexiglass	—	—	2.59	6.7×10^{-3}
Polyamide resin (Nylon 66)	3.16	2.1×10^{-2}	—	—
Polyisobutylene	2.23	3×10^{-4}	—	—
Polytetrafluoroethylene (Teflon)	2.1	$<2 \times 10^{-4}$	2.08	3.7×10^{-4}
Pyrex glass (Corning 7740)	4.52	4.5×10^{-3}	4.52	8.5×10^{-3}
Rubber; natural (vulcanized)	2.42	1.8×10^{-2}	—	—
Silica, fused	3.78	3×10^{-5}	3.78	1.7×10^{-4}
Styrofoam (polystyrene)	—	—	1.03	1.5×10^{-4}

[a]Data obtained from *American Institute of Physics Handbook*, 1979, Sec. 5.

When an electromagnetic wave propagates in the z direction through an unbounded dielectric medium the z dependence may be of the form

$$E = E_0 \exp(j\omega t - j\beta z - \alpha z)$$ (5)

where for simplicity only the positively traveling wave will be considered. In the absence of losses ($\epsilon'' = 0$ and $\alpha = 0$) the wave is undiminished in amplitude as it proceeds in the z direction with the velocity **v**

$$v = \frac{1}{(\mu\epsilon)^{1/2}}$$ (6)

the propagation constant β

$$\beta = \omega(\mu\epsilon)^{1/2} = \frac{2\pi}{\lambda}$$ (7)

and the characteristic impedance Z_0

$$Z_0 = \left(\frac{\mu}{\epsilon} \right)^{1/2}$$ (8)

When losses are present ϵ is complex and

$$v = \text{Re}(\mu\epsilon)^{-1/2}$$ (9)

$$\alpha = -\omega \, \text{Im}(\mu\epsilon)^{1/2}$$ (10)

and

$$\beta = \omega \, \mathrm{Re}(\mu\epsilon)^{1/2} \tag{11}$$

In addition, the impedance Z has both a real part and an imaginary part. More explicitly

$$(\mu\epsilon)^{1/2} = (\mu\epsilon_0)^{1/2}\left(\epsilon' - \left[\frac{j\sigma}{\omega\epsilon_0}\right]\right)^{1/2} \tag{12}$$

For a good conductor $\sigma \gg \omega\epsilon'\epsilon_0$, so

$$\gamma = \omega(\mu\epsilon)^{1/2} = \frac{(1-j)}{\delta} \tag{13}$$

and

$$\alpha = \beta = \frac{1}{\delta} \tag{14}$$

where δ is the skin depth, which will be explained in the next section. Further discussion of Eq. (14) will be postponed until then. For a low loss dielectric

$$\frac{\sigma}{\omega\epsilon_0} \ll \epsilon' \tag{15}$$

and the square root in Eq. (12) may be approximated by a binomial expansion

$$(\mu\epsilon)^{1/2} = (\mu\epsilon_0\epsilon')^{1/2}\left[1 - \frac{j\sigma}{2\omega_0\epsilon_0\epsilon'} + \frac{1}{8}\left(\frac{\sigma}{\omega_0\epsilon_0\epsilon'}\right)^2\right] \tag{16}$$

As a result

$$\alpha = \frac{\pi\sigma}{\lambda\omega\epsilon_0\epsilon'} = \left(\frac{\pi}{\lambda}\right)\tan\delta \tag{17}$$

$$\beta = \left(\frac{2\pi}{\lambda}\right)\left[1 + \frac{1}{8}\left(\frac{\sigma}{\omega\epsilon_0\epsilon'}\right)^2\right] \tag{18}$$

and

$$Z_0 = \left(\frac{\mu}{\epsilon_0\epsilon'}\right)^{1/2}\left[1 - \frac{3}{8}\left(\frac{\sigma}{\omega\epsilon_0\epsilon'}\right)^2 + j\left(\frac{\sigma}{2\omega\epsilon_0\epsilon'}\right)\right] \tag{19}$$

One should bear in mind that

$$\epsilon = \epsilon_0(\epsilon' - j\epsilon'')$$

$$\epsilon'' = \frac{\sigma}{\omega\epsilon_0} \tag{20}$$

J. Skin Effect

In a good conductor the conductivity σ is very high, and there are no free charges present. If any free charges were generated in a good conductor they would move to the surface with a time constant ϵ/σ, which is extremely small. The conduction current J obeys Ohm's law $J = \sigma E$, and the displacement current $\partial D/\partial t$ is negligible compared with the conduction current J when $\omega\epsilon \ll \sigma$.

If we assume that the displacement current may be neglected, then the following equations hold within the conductor.

$$\nabla^2 \mathbf{H} = \sigma\mu\left(\frac{\partial \mathbf{H}}{\partial t}\right) = j\omega\sigma\mu\mathbf{H} \tag{1}$$

$$\nabla^2 \mathbf{E} = \sigma\mu\left(\frac{\partial \mathbf{E}}{\partial t}\right) = j\omega\sigma\mu\mathbf{E} \tag{2}$$

$$\nabla^2 \mathbf{J} = \sigma\mu\left(\frac{\partial \mathbf{J}}{\partial t}\right) = j\omega\sigma\mu\mathbf{J} \tag{3}$$

where a sinusoidal time variation is assumed. In the case of a plane conductor with no spatial variation in the current flow on the surface Eq. (3) may be written

$$\frac{\partial^2 \mathbf{J}}{\partial z^2} = j\omega\sigma\mu\mathbf{J} \tag{4}$$

where the z direction is perpendicular to the conductor surface. This equation has the solution

$$J = J_0 \exp\left[-(1+j)\left(\frac{z}{\delta}\right)\right] \tag{5}$$

where J_0 is the current density on the surface where $z = 0$, and the skin depth δ

$$\delta = \left(\frac{2}{\omega\mu\sigma}\right)^{1/2} \text{ m} \tag{6}$$

is the depth at which the current decays to $1/e = 0.369$ of its value at the

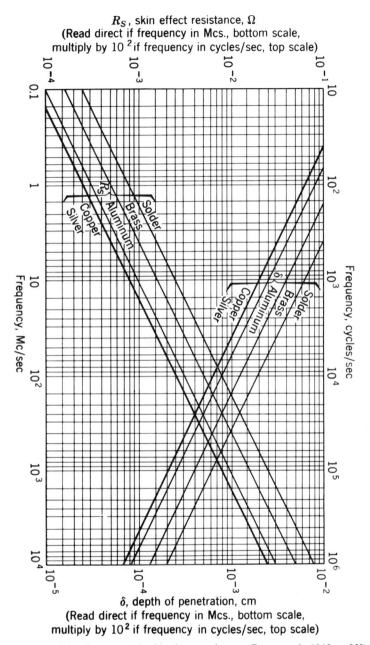

R_S, skin effect resistance, Ω
(Read direct if frequency in Mcs., bottom scale,
multiply by 10^2 if frequency in cycles/sec, top scale)

δ, depth of penetration, cm
(Read direct if frequency in Mcs., bottom scale,
multiply by 10^2 if frequency in cycles/sec, top scale)

Fig. 2-17. Skin effect quantities for plane conductors (Ramo et al., 1965, p. 252).

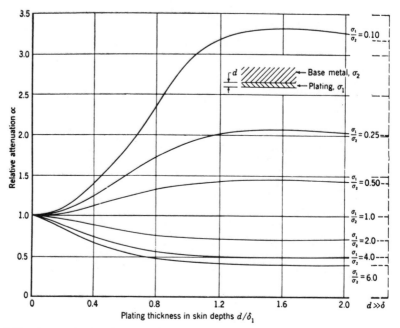

Fig. 2-18. Attenuation in plated conductors. The relative attenuation α is the ratio of the attenuation of the combination to the attenuation of the base metal (RLS-9, p. 128).

surface. The imaginary part of the exponential (5) indicates that there is also a phase change with depth. Numerical values of the skin depth for several important conductors may be obtained from Fig. 2-17. Included on this figure are values of the skin effect resistance R_S.

$$R_S = \frac{1}{\sigma\delta} \tag{7}$$

which is the "surface" resistance of the conductor. From this figure we see that at the typical modulation frequency of 100 kc, the skin depth for copper is 0.2 mm, at the typical NMR frequency of 10 Mc, $\delta = 0.02$ mm for copper, and at the typical ESR frequency of 10^{10} cps the corresponding value is 6.6×10^{-4} mm. Thus microwaves only require conductors with a very thin metallic layer of high conductivity, and the electromagnetic energy is essentially confined to the surface. Figure 2-18 shows the attenuation resulting from a thin layer of one metal on the surface of a second (see RLS-9, p. 128). This is important because microwave cavities are frequently plated in this manner. The skin depth of 6.5 mm at 100 cps exceeds the 1.3-mm wall thickness of RG-52/U X-band waveguide. Thus, low-frequency audio modulation coils may be effectively located outside a cavity constructed from such a waveguide.

The derivation of Eq. (4) assumes that the skin depth is much smaller than any curvature on the conductor surface. It should be emphasized that the current density J does actually penetrate below the skin depth, but its magnitude decreases exponentially with depth so that almost none is found below several skin depths.

References

Included in this list are several books, that are not referred to in the text, and the entire MIT *Radiation Laboratory Series* (*RLS*) which was published by the McGraw-Hill Book Co., Inc., New York and reprinted by Boston Technical Publishers, Inc., Boston, Mass., 1964.

Adv. Microwaves, L. Young and H. Sobol, Eds., Academic Press, New York, continuing series starting in 1966.

American Institute of Physics Handbook, 2nd ed., McGraw-Hill, New York, 1972.

Atwater, H. A., *Introduction to Microwave Theory*, McGraw-Hill, New York, 1962.

Baranski, S., and P. Czerski, *Biological Effects of Microwaves*, Academic Press, New York, 1976.

Brackett, E. B., P. H. Kasai, and R. J. Myers, *RSI* **28**, 699 (1957).

Ginzton, E. L., *Microwave Measurements*, McGraw-Hill, New York, 1957.

Groll, H., *Mikrowellen Messtechnik*, Friedr. Vieweg & Sohn, Braunschweig, 1969.

Gupta, K. C., *Microwave Integrated Circuits*, Amarjit Singh, Wiley, New York, 1974.

Jackson, J. D., *Classical Electrodynamics*, Wiley, New York, 1975.

Johansson, B., S. Haraldson, L. Pettersson, and O. Beckman, *RSI* **45**, 1445 (1974).

Laverghetta, T., *Microwave Measurements and Techniques*, Artech. House, 1976.

Margenau, H., and G. M. Murphy, *The Mathematics of Physics and Chemistry*, 2nd. ed., Van Nostrand, Princeton, N.J., 1964.

Nergaard, L. S., and M. Glicksman, Eds., *Microwave Solid State Engineering*, Princeton, N.J., 1964.

Ramo, S., J. R. Whinnery, and T. Van Duzer, *Fields and Waves in Communication Electronics*, Wiley, New York, 1965.

Reintjes, J. F., and G. T. Coate, *Principles of Radar*, 3rd ed., The Technology Press, MIT, and McGraw-Hill, New York, 1952.

RLS-1, L. N. Ridenour, Ed., *Radar System Engineering*, 1947.

RLS-2, J. S. Hall, Ed., *Radar Aids to Navigation*, 1947.

RLS-3, A. Roberts, Ed., *Radar Beacons*, 1947.

RLS-4, J. A. Pierce, A. A. McKenzie, and R. H. Woodward, Eds., *Long Range Navigation* (*Loran*), 1948.

RLS-5, G. N. Glasoe and J. V. Lebaċqz, Eds., *Pulse Generators*, 1948.

RLS-6, G. B. Collins, Ed., *Microwave Magnetrons*, 1948.

RLS-7, D. R. Hamilton, J. K. Knipp, and J. B. H. Kuper, Eds., *Klystrons and Microwave Triodes*, 1948.

RLS-8, C. G. Montgomery, R. H. Dicke, and E. M. Purcell, Eds., *Principles of Microwave Circuits*, 1948.

RLS-9, G. L. Ragan, Ed., *Microwave Transmission Circuits*, 1948.

RLS-10, N. Marcuvitz, Ed., *Waveguide Handbook*, 1951.

RLS-11, C. G. Montgomery, Ed., *Technique of Microwave Measurements*, 1947.

RLS-12, S. Silver, Ed., *Microwave Antenna Theory and Design*, 1949.

RLS-13, D. E. Kerr, Ed., *Propagation of Short Radio Waves*, 1951.

RLS-14, L. D. Smullin and C. G. Montgomery, Eds., *Microwave Duplexers*, 1948.

RLS-15, H. C. Torrey and C. A. Whitmer, Eds., *Crystal Rectifiers*, 1948.

RLS-16, R. V. Pound, Ed., *Microwave Mixers*, 1948.

RLS-17, J. F. Blackburn, Ed., *Components Handbook*, 1949.

RLS-18, G. E. Valley and H. Wallman, Eds., *Vacuum Tube Amplifiers*, 1948.

RLS-19, B. Chance, V. W. Hughes, E. F. MacNichol, D. Sayre, and F. C. Williams, *Waveforms*, 1949.

RLS-20, B. Chance, R. I. Hulsizer, E. F. MacNichol, and F. C. Williams, *Electronic Time Measurements*, 1949.

RLS-21, I. A. Greenwood, J. V. Holdan, and D. MacRae, *Electronic Instruments*, 1948.

RLS-22, T. Soller, M. A. Starr, and G. E. Valley, *Cathode Ray Tube Displays*, 1948.

RLS-23, S. N. Van Voorhis, Ed., *Microwave Receivers*, 1948.

RLS-24, J. L. Lawson and G. E. Uhlenbeck, Eds., *Threshold Signals*, 1950.

RLS-25, H. M. James, N. B. Nichols, and R. S. Phillips, Eds., *Theory of Servomechanisms*, 1947.

RLS-26, W. M. Cady, M. B. Karelitz, and L. A. Turner, Eds., *Radar Scanners and Radomes*, 1948.

RLS-27, A. Svoboda and H. M. James, Eds., *Computing Mechanisms and Linkages*, 1948.

RLS-28, K. Henney, *Index*, 1953.

Saad, T., Ed., *The Microwave Engineers' Handbook*, Vols. 1 and 2, Archtech. Press, 1971.

Stanforth, J. A., *Microwave Transmission*, English Univ. Press, 1972.

Southworth, G. C., *Principles and Applications of Waveguide Transmission*, Van Nostrand, Princeton, N.J., 1950, p. 689.

Stratton, J. A., *Electromagnetic Theory*, McGraw-Hill, New York, 1941.

Subrahamaniam, V., *J. Indian Inst. Sci.* **44**, 148 (1962).

Sucher, M., and J. Fox, Eds., *Handbook of Microwave Measurements*, 3rd ed., Interscience, New York, 1963.

Thomas, H. E., *Handbook of Microwave Techniques and Equipment*, Halsted Press, 1972.

CHAPTER 3

Microwave Generators

A. Magnetrons

The principal application of magnetrons is to systems that require the generation of very high microwave power outputs, particularly if the power is pulsed rather than continuous. Some magnetrons furnish in excess of 1 MW of peak output power during the pulse. In a typical case the pulse will be 0.1 or 1 μsec long, and the duty cycle will be 10^{-3}. The latter means that the magnetron is only producing microwave power 0.1% of the time.

In a magnetron, the electrons that are emitted by the cathode travel along cycloids in a magnetic field until they either return to the cathode or reach the anode. An example of such a cycloidlike electron path is shown in Fig. 3-1. In this figure the anode is in the center, and eight resonators are arranged radially around it. The electrons moving in the magnetic field may be made to travel in suitable orbits and interact with a resonant cavity so as to supply energy to the cavity, thereby sustaining oscillations.

B. Traveling-Wave Tubes and Backward-Wave Oscillators

The traveling-wave tube is a special type of microwave tube that has a much broader bandwidth than a klystron or magnetron. An example of a traveling-wave tube is shown in Fig. 3-2. The cathode emits electrons, which are accelerated and focused by the electron gun. These electrons pass down the helix to the collector at 1/13 of the velocity of light when a 15,000-V accelerating potential is employed. The actual length of the wire in the helix is 13 times as long as the distance along the axis of the helix. As a result the electromagnetic wave that travels along the wire at the velocity of light actually progresses along the helical axis at 1/13 of the velocity of light while it interacts with the electron beam. The helix may thus be referred to as a slow wave structure, and it operates over a very wide frequency band because it is nonresonant. The electromagnetic wave interacts with the electrons in such a way that it bunches them and extracts energy from them.

When the traveling-wave tube is employed as an amplifier, the electromagnetic wave enters by the input waveguide in Fig. 3-2. Subsequently, it travels along the helix while interacting with the electronic beam, extracting energy from it, and gradually growing in amplitude. When it reaches the output waveguide, the wave will be perhaps 20 or 30 dB stronger, which means that

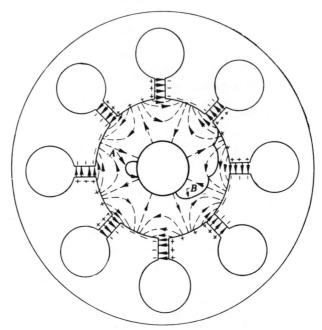

Fig. 3-1. Paths followed by electrons in oscillating magnetron (RLS-6, p. 26).

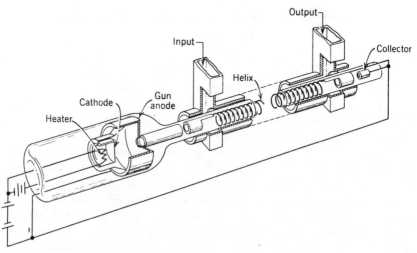

Fig. 3-2. Schematic of traveling wave amplifier (Pierce, 1950, p. 7).

the output power from the traveling wave tube may be 100 or 1000 times the input power.

Aside from their high power gain and wide bandwidth, traveling-wave tubes also have desirable noise figures, a wide range of design frequencies [200 to 50,000 MHz according to Pierce (1950)], and simplicity of construction. When a traveling-wave tube is converted to an oscillator, it is referred to as a backward-wave oscillator. It may be tuned in frequency by adjusting the beam voltage.

Erikson (1966) constructed a zero-field spectrometer that employed a resonant cavity, a traveling-wave amplifier and a slide-trombone line stretcher as a regenerative microwave oscillator. The frequency was scanned by a clock motor, which slowly moved the upper wall of the cavity. The line stretcher was servocontrolled to maintain the impedance matching of the system during the frequency scan.

Some of the spectrometers described in various chapters of this monograph incorporate traveling wave tubes, generally as amplifiers. The previous edition mentions several references that treat traveling-wave tubes (Dreizler et al., 1960; Gittins, 1965; Hok, 1956; Kondrat'ev, 1963; Mouthaan, 1965; Ober, 1965 and their use in spectrometers (Bogle et al., 1961; Mock, 1960; Uchida and Buyle-Bodin, 1962; Webb, 1962; and Wilmshurst et al., 1962).

C. Klystrons

Introduction

Klystrons (RLS-7) are employed as generators of microwave power in the frequency range from 500 to 35,000 MHz. Some have been made to produce continuous wave, or CW, power up to 25 kW, and others to supply pulsed power up to 10^4 W, but in electron spin resonance spectrometers the klystron usually provides less than 1 W of CW output. Since the klystron finds more widespread use in ESR than any other generator of microwaves, it will be discussed in more detail than the other microwave sources.

Reflex Klystrons

A schematic diagram of a reflex klystron is shown in Fig. 3-3. The cathode supplies electrons by thermionic emission just as in a conventional vacuum tube, and these electrons are accelerated toward the rf gap (anode) by the beam or resonator voltage V_B. In the rf gap the electron beam is subjected to an rf electric field that alternately slows down and speeds up the electrons in the beam, thereby velocity-modulating them. After traversing the rf gap the velocity-modulated electron beam enters the drift space in which the fast electrons move away from the slower ones behind them and catch up with the slower ones in front of them. The net effect is the formation of groups or bunches of electrons. Since the reflector is negative relative to the anode or rf

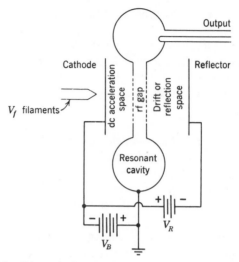

Fig. 3-3.   Schematic diagram of a reflex klystron (RLS-11, p. 25).

gap, the bunched electron beam is turned around and returned to the gap. When the radio frequency, beam voltage, and reflector voltage are properly adjusted, the bunched electrons will return to the gap with the proper phase, and as a result they will be "debunched" in the process of giving up energy to the resonant cavity. This cavity is coupled to a transmission line in such a way that the energy extracted from the electron beam is made to propagate down the transmission line. If the reflector voltage is not properly chosen, then the electrons will reenter the rf gap with incorrect phases, and no energy will be extracted from them. A necessary condition for a reflex klystron to oscillate is that the transit time in the reflection drift space be in the vicinity of $(n + \frac{3}{4})$ rf periods, where n is an integer. This ensures that the electrons are properly bunched when they arrive back at the rf gap. In addition, the dc beam current I_B must exceed a minimum value called the "starting current" to ensure that the power extracted from the electron beam is greater than the circuit and load losses involved in maintaining the gap voltage. When these two conditions are satisfied, oscillations will occur.

For a given beam voltage V_B there are definite values of the reflector voltage V_R that correspond to stable oscillations, as illustrated in Fig. 3-4. The various modes that are labeled with their n values are characterized by different transit times (not by different frequencies as in resonant cavity modes), and usually they extend through zero to positive values of the reflector voltage. However, the figure only gives the regions of stable oscillation that correspond to zero reflector current. This is because the current collected by the reflector at positive and at slightly negative values of V_R may cause excessive heating and permanently damage the klystron. To obviate this possibility, it is advisable to install a diode that will short-circuit the reflector to the anode in the event that the negative voltage supply to the reflector fails to operate.

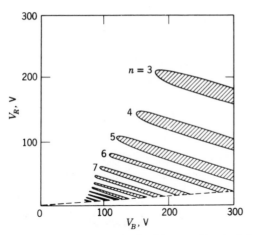

Fig. 3-4. Typical reflector-mode pattern of a reflex klystron. The shaded areas correspond to those combinations of beam voltage V_B and reflector voltage V_R at which oscillation occurs (RLS-11, p. 27).

In routine use, a klystron is operated at a predetermined beam voltage, and often this voltage is a value recommended by the manufacturer. For a particular value of beam voltage V_B and klystron cavity frequency f_0, the klystron frequency f and power output P will vary in the manner shown in Fig. 3-5. When the cavity frequency f_0 is altered, the mode patterns of Fig. 3-5 will shift to the right or left corresponding to variations of the transit time with frequency. A typical reflex klystron has a mechanical tuning range between 5 and 50%, which means that the klystron cavity frequency f_0 may be varied within this range. The electronic tuning range is the deviation from f_0 that may be obtained through variations in the reflector voltage only, as shown in Fig. 3-5. Typical electronic tuning ranges are from 0.2 to 0.8% of the frequency f_0. For example, the Varian X-13 klystron may be tuned mechanically from 8100 to 12,400 MHz and has an average electronic tuning range of 60 Mc at 8100 MHz and 45 MHz at 12,400 MHz. It is normal for both the electronic tuning range and the power output to be frequency dependent. Table 6-2 of the first edition gives the characteristics of several typical klystrons.

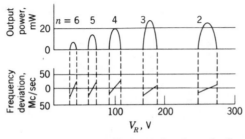

Fig. 3-5. Output power and frequency of oscillation as functions of reflector voltage in a reflex klystron (RLS-11, p. 27).

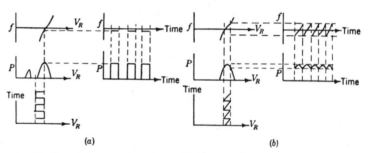

Fig. 3-6. Amplitude and frequency modulation characteristics of the reflex klystron: (*a*) for square-wave amplitude modulation; (*b*) for sawtooth frequency and amplitude modulation (RLS-11, p. 28).

Reflector Modulation

The effect of square-wave and sawtooth reflector voltage modulation on the klystron power output is shown in Fig. 3-6. The particular square wave illustrated has the effect of producing pulses of microwave power. If the amplitude of the sawtooth in Fig. 3-6*b* were increased to a value that extended from 20 to 300 V for the 2K25 klystron, then the entire mode pattern shown on the upper part of Fig. 3-5 would be displayed on the oscilloscope. This is a convenient technique to employ when tuning a klystron to an external resonant cavity. Often a low-amplitude sine-wave reflector voltage modulation is employed as an integral part of an automatic frequency control (AFC) system. Anomalous modulation effects occur at very high frequencies when the period of the modulation frequency $(1/f_m)$ becomes comparable to the decay time of the loaded klystron resonant cavity, or when f_m itself becomes comparable to the cavity bandwidth. Sushkov and Meos (1965) generated nanosecond pulses with a klystron.

Load

For fixed filament and beam voltages the power output of a klystron depends upon the load. Figure 3-7 shows the variation in the power output for three loads when the frequency is changed by varying the reflector voltage. The maximum power output is obtained with the optimum load that corresponds to terminating the transmission line with the complex conjugate of the generator

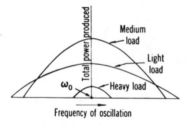

Fig. 3-7. Total rf power produced during electronic tuning for three types of loads (RLS-7, p. 316).

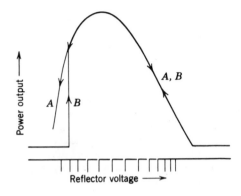

Fig. 3-8. The oscilloscope traces A, B show the hysteresis in the dependence of the klystron power on the reflector voltage—V_R. Equally spaced frequency markers are shown below (RLS-7, p. 391).

impedance. A greater range of frequencies may be obtained with a lighter load. There are some loads that prevent the klystron from oscillating. An isolater is generally connected immediately after the klystron to impedance match it to the waveguide and eliminate these load-dependency problems.

Hysteresis

When the output power and frequency at a given reflector voltage depend on the direction of approach to this reflector voltage, then the phenomenon of electronic tuning hysteresis occurs. An example of this is shown in Fig. 3-8, which presents the oscilloscope trace of a klystron mode produced by a sinusoidal sweep. The first half of the cycle A traces out the mode from right to left, and the retrace B fails to produce the extreme left side of the mode. Equally spaced frequency markers are shown. Electronic hysteresis may be caused by electrons making multiple transits through the rf gap, by transit angles varying with the rf voltage, and by an appreciable variation of the small-signal electronic transconductance across the width of a low mode.

D. Solid-State Generators

In recent years various solid-state devices have come into general use as microwave generators (Berteaud, 1976; Dascálu, 1974; De Loach, 1967; Eastman, 1972; Grivet, 1976; Hershberger, 1972; Shurmer, 1971). They have the advantage that they operate with low applied voltages, and hence they require simpler power supplies than klystrons. We describe several of these devices.

Varactors

A varactor diode is a semiconductor junction device with a barrier capacitance that varies nonlinearly with the applied voltage (Berteaud, 1976; Penfield and Rafuse, 1962; Shurmer, 1971; Uhlir, 1958). It contains a p–n junction formed from an upper layer of p-type material in contact with a lower n-type epitaxial

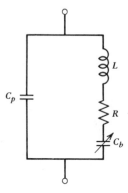

Fig. 3-9. Equivalent circuit showing the series resistance R representing junction losses, the series stray inductance L, the parallel capacitance C_p arising from the cartridge and stray effects, and the effective barrier capacitance C_b, whose value depends upon the applied voltage.

layer. A gold sphere rests on the p-type layer, and a heavily doped n^+-type substrate supports the junction from below. In the equivalent circuit illustrated in Fig. 3-9, C_b is the effective barrier capacitance that depends upon the applied voltage, the series resistance R accounts for all the losses in the junction, C_p represents the capacitance of the cartridge plus stray capacitance, and L results from stray inductance. Increases in applied voltage enlarge the area of activated junction material, and the result is an increase in the capacitance C_b.

Varactors are often employed as harmonic generators and as frequency multipliers. Varactors have also been used to provide the electronic tuning of Gunn diode oscillators, as will be mentioned below.

Tunnel Diode

The presence of a region of negative resistance over a limited range of voltage permits a tunnel diode to be employed as a microwave oscillator (Shurmer, 1971; Sterzer, 1967). We see from Fig. 3-10 that both germanium and gallium arsenide tunnel diodes exhibit such a negative resistance region at low forward biases. The equivalent circuit of such a diode contains a positive series resistance R and a negative resistance R_{neg} in parallel with the junction capacitance C and in series with an inductance L, as indicated in Fig. 3-11. The barrier layer of the diode must be quite thin ($\sim 10^{-8}$ m) to provide a large enough probability for the tunneling of electrons across the p-n junction.

Tunnel diodes are convenient to use as microwave oscillators because the conditions of stability are not difficult to maintain. To a first approximation it may be assumed that the negative resistance is linear for low rf voltages. Because of the relatively small range of negativity the maximum power output is considerably below 1 W.

There is a special form of tunnel diode called a backward diode in which the resistance is negative for a rather minimal current range, as indicated in Fig. 3-12. This diode is useful as a detector or mixer, but not as an oscillator. Figure 3-12 shows the characteristic curve of a typical point-contact diode detector for comparison.

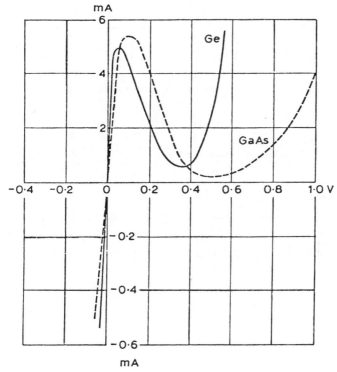

Fig. 3-10. Current–voltage characteristics of two tunnel diodes illustrating the region of negative resistance at small positive bias voltages (from Shurmer, 1971, p. 110).

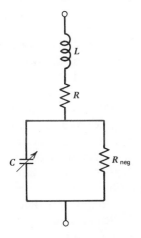

Fig. 3-11. Equivalent circuit of a tunnel diode showing the positive series resistance R, the series inductance L, the variable parallel junction capacitance C, and the parallel negative resistance R_{neg}.

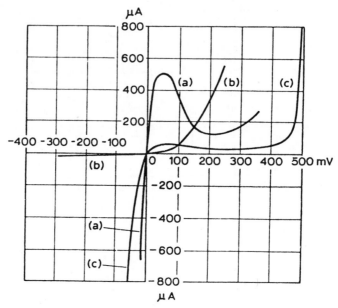

Fig. 3-12. Comparison of the current–voltage characteristics of (*a*) tunnel diode, (*b*) point contact diode, and (*c*) backward diode. Note the minimal region of negative resistance in the backward diode (from Shurmer, 1971, p. 124).

Gunn-Effect Oscillators

The Gunn effect corresponding to periodic high-frequency fluctuations in the current occurs in materials such as gallium arsenide and indium phosphide when the applied voltage is greater than a critical value (Berteaud, 1976; Bott and Fawcett, 1968; Eastman, 1972; Kroemer, 1972; Shurmer, 1971). These oscillations result from the formation of an accumulation layer of thermally excited electrons with an associated positively charged layer, and the movement of this dipole layer to the anode where it becomes extinguished. The velocity of the dipole domain is about 10^7 cm/sec at saturating applied voltages, as indicated in Fig. 3-13, and hence the transit time across a 10^{-3} cm thick specimen is 10^{-10} sec. In this mode one domain traverses the specimen per cycle, and the resulting X-band oscillations (10^{10} Hz) are maintained in a resonant cavity coupled to the diode. Transit time Gunn oscillators have efficiencies of several percent, which exceeds that of klystrons. X-band Gunn oscillators can have mechanical tuning ranges of 1 GHz and varactor diode assisted electronic tuning ranges of 50 MHz. A technique called injection phase locking may be employed for fixed-frequency high-stability operation. Gunn oscillators are less noisy than avalanche oscillators, which will be discussed next. Typical characteristics at X band are a CW power output of 0.5 W, power input at 6 V, a 20-cm^3 cavity, and a specimen mass of 28 g. Pulsed operation is also available.

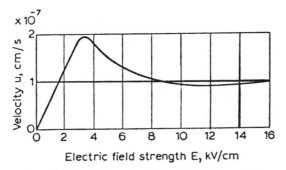

Fig. 3-13.   Dependence of drift velocity u of a Gunn effect GaAs oscillator on the applied electric field strength E. The device operates in the region of saturation beyond 8 kV/cm where the velocity is very little influenced by the field strength (from Shurmer, 1971, p. 168).

Gunn diodes can also oscillate in a limited space charge accumulation, or LSA, mode. In this mode the time is not sufficient for the accumulation of space charge, and so dipole domains do not form. The diode functions as a simple negative resistance device with oscillations that depend on an external cavity rather than on the specimen length for their frequency. As a result the power capabilities may be increased with the use of thicker specimens, and high power for both CW and pulsed operation may be achieved using the LSA mode. Operation up to 100 GHz is possible.

Zublin et al. (1975) developed a YIG tuned GaAs oscillator that operates from 26.5 to 4 GHz to replace a backward oscillator in a tunable receiver. Kohno (1976) discusses spin–lattice relaxation time measurements with Gunn oscillators.

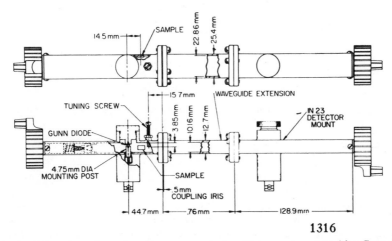

1316

Fig. 3-14.   Side and top views of a microwave marginal oscillator spectrometer with a Gunn diode source (Walsh and Rupp, 1970).

Walsh and Rupp (1970) designed a marginal oscillator ESR spectrometer that places a Gunn diode directly in the resonant cavity. The first version of their spectrometer is sketched in Fig. 3-14. The variable frequency cavity that contains both the sample and the Gunn oscillator is a standard waveguide detector mount. The biasing power comes from a 0–15-V regulated electronic supply or battery pack, and it delivers 150 mA. An improved, more compact microwave marginal oscillator spectrometer (Walsh and Rupp, 1971) illustrated in Fig. 3-15 employs the Gunn device as both the microwave source and the detector; that is, it is self-detecting. Similar waveguide structures were employed, but the piston and iris were replaced by fixed metal end plates, and the overall waveguide size was reduced from RG-52 (2.54 × 1.27 cm OD) to RG-91 (1.78 × 0.994 cm) waveguide. This raised the oscillation frequency to ~ 12.4 GHz. A cylindrical sample mounting stud permits rapid sample changes and facilitates angular rotation studies. The sensitivity of 3×10^{12} spins is comparable to that of conventional homodyne bridge spectrometers. More recently Rupp (1981) and Walsh and Rupp (1981) designed a self-detecting spectrometer that employs a BARITT diode (BARrier Injection Transit Time).

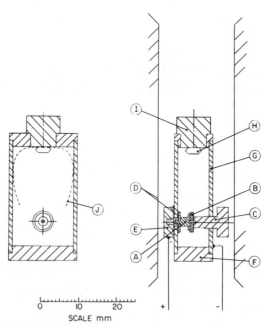

Fig. 3-15. Sketch of a self-detecting Gunn diode microwave marginal oscillator spectrometer. A—Gunn diode; B—copper coupling disk; C—copper mounting post; D—0.01-mm Mylar bypass disks; E—copper bias terminal threaded on copper diode-mounting bushing; F—optional end wall; G—RG-91 waveguide; H—paramagnetic sample; I—sample mounting and rotation stud; J—microwave magnetic field path leading to alternative sample locations on side walls (Walsh and Rupp, 1971).

Avalanche Diodes

If during a voltage cycle a sufficiently high potential is applied across a reverse biased p–n junction a voltage breakdown can occur that produces a relatively large supply of holes and electrons called an avalanche (Berteaud, 1976; Eastman, 1972; Evans, 1972; Read, 1958; Shurmer, 1971). An electric field impressed perpendicular to the avalanche producing electric field can cause the charge carriers to drift with an effective negative resistance and thereby produce microwave oscillations. In addition to this p–n junction type of avalanche diode, a combination of transit time effects and avalanche breakdown produce oscillations in what is called a Read diode or an impact ionization avalanche transit time (IMPATT) mode device. The current and voltage of an IMPATT diode are 180° out of phase corresponding to a negative resistance that produces the oscillations. A higher-efficiency type of operation may be obtained with what is referred to as the trapped plasma avalanche-triggered transit, or TRAPATT, mode. Space limitations do not permit a detailed discussion of the differences between these mechanisms.

Avalanche diodes can provide more than 1 W of power at X band, and they have been made to operate above 100 GHz. Wide tuning ranges are achievable, as indicated in Fig. 3-16 for a germanium IMPATT oscillator. Figure 3-17 illustrates a varactor-modulated IMPATT oscillator mount. Avalanche diodes tend to have high AM and FM noise characteristics. The former can be largely eliminated with the use of balanced mixers. Tuning ranges of 10% are achievable at X band. Hogg (1973a) discussed applications of IMPATT diodes as rf sources for ESR and published (1973b) the design of a low-cost X-band IMPATT diode marginal oscillator for ESR. Figure 3-18 shows the IMPATT diode holder, matching transformer and power supply. Care was taken to

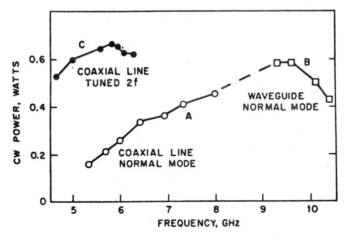

Fig. 3-16. Dependence of continuous wave output power of a germanium IMPATT oscillator on the frequency for oscillations in three modes (from Evans, 1972, p. 128).

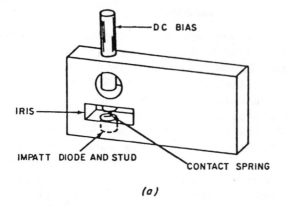

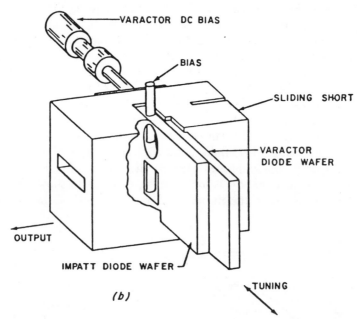

Fig. 3-17.   Sketch of a varactor-modulated IMPATT oscillator showing (*a*) the diode wafer and (*b*) the oscillator mount (from Evans, 1972, p. 174).

provide a heat sink, which keeps the diode junction considerably below 200°C. The diode must be protected against voltage transients, particularly when turning the equipment on and off. This device provided an ESR sensitivity of 2×10^{12} spins. Lee and Strandley (1969) reported an 86 GHz IMPATT diode oscillator.

PIN Diodes

A PIN diode consists of a high resistivity *p*- or *n*-type silicone layer (I) between heavily doped and hence low resistivity layers of the *p* and *n* types (Berteaud,

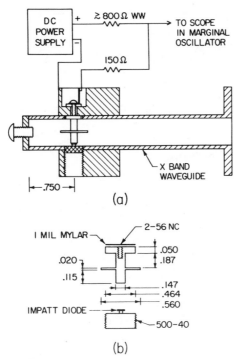

Fig. 3-18.   Sample holder and power supply for IMPATT diodes (Hogg, 1973a, 1973b).

1976; Shurmer, 1971). One pair of layers acts like a single $p–n$ junction with a low-resistivity back contact of the same conductivity type as the high-resistivity part of the junction.

The PIN junction behaves differently at low and high frequencies. At low frequencies it acts like an ordinary $p–n$ junction rectifier. At high frequencies the depletion region at the boundary cannot follow the rapid oscillations, and the impedance becomes that of the I region, which is high with zero or reverse bias, and low with forward bias.

The avalanche behavior of PIN diodes permits them to be operated as microwave oscillators. Their output and noise characteristics as well as their ease of fabrication are intermediate between those of IMPATT and $p–n$ avalanche diode oscillators.

Summary of Solid-State Oscillators

The various types of microwave generators tend to have an achievable power output that is inversely proportional to the square of the frequency. There is theoretical and experimental support for this. For example, Evans (1972) summarized De Loach's (1967) analysis of the maximum efficiency of an IMPATT oscillator, which showed that there is a maximum power that can be transferred to mobile charge carriers, and for a mode of operation limited by a

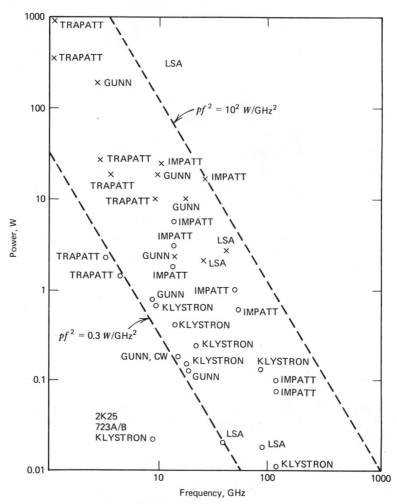

Fig. 3-19. Power outputs of solid state microwave oscillators with the highest Pf^2 ratings for pulsed ($\times$) and continuous wave ($\bullet$) operation. The points of several klystrons listed in Table 6-2 of the first edition are indicated for comparison. [Data from Evans (1972) and Shurmer (1971).]

minimum circuit impedance this maximum power is inversely proportional to the square of the frequency. In addition Evans showed that an f^{-2} maximum power limitation also occurs if thermal effects dominate. On the experimental side we plot in Fig. 3-19 the power output achieved with various types of solid-state oscillators (Evans, 1972; Shurmer, 1971) over three octaves of frequency, and we see that most data points lie between the products $Pf^2 = 0.3$ and $Pf^2 = 100$ W GHz^2, corresponding to a slope of -2 on the log P versus log f graph. The klystrons from Table 6-2 of the first edition with the highest Pf^2 ratings are plotted as points in Fig. 3-19, and they lie within the same range of Pf^2 values. We see from the figure that the army-navy surplus low-power 723 A/B or 2K25 klystrons that were employed in most of the X-band home built ESR spectrometers of the late 1940s and 1950s have a Pf^2 product that lies considerably below and to the left of the $Pf^2 = 0.3$ line. Modern ESR spectrometers always employ higher-power oscillators.

E. Klystron Power Supplies

Power Supply Requirements

There are three basic power supplies required for a klystron: (1) the filament voltage V_f, (2) the beam (resonator) voltage V_B, and (3) the reflector voltage V_R. The functions of these power supplies are shown diagrammatically in Fig. 3-3. The filament voltage heats the cathode and causes it to emit electrons thermionically, the beam voltage accelerates the electrons, and the reflector voltage reverses their direction. The reflector voltage is by far the most critical of the three voltages and requires the most stable source, while the filament supply is the least critical. Some klystrons also require an additional supply voltage for the focus grid.

The Filament Power Supply

The simplest way to obtain a constant filament voltage is to use a lead storage ("automobile") battery. This source of V_f is electrically more stable and ripple-free than many electronic sources of V_f now in use. The only liability is the necessity of periodically checking the level of the acid in the battery cells, and charging them regularly. One hazardous aspect of using a storage battery arises when the resonant cavity grid (anode) is grounded and the cathode is operated at the potential $-V_B$. In this case it is necessary to connect one terminal of the battery to the cathode in order to avoid damaging the klystron. This means that the storage battery may be riding at a lethal voltage, and it should be kept in an insulated box. A typical electronic filament power supply consists of a full-wave rectifier followed by a π filter. It is certainly not as ripple-free as a storage battery, but it does render satisfactory service. In general, increasing the filament voltage increases the klystron power output, but it also shortens the klystron's lifetime. It is best to employ the value of V_f that is recommended by the manufacturer. The fact that the output power increases with the filament voltage is merely mentioned in case one needs extra power and cannot obtain a more powerful klystron, since this is a convenient source of increased power. If the filament voltage is increased too much, one runs the risk of permanently damaging the klystron. Case and Larsen (1963) describe the circuit shown in Fig. 6-15 of the first edition which protects a klystron against overvoltage and undervoltage on its filament and also against failure of the klystron water cooling system.

The Beam Voltage Power Supply

It is desirable to have a beam or resonator voltage V_B with at most 1 or 2 mV of ripple. When the grid (anode) is grounded, V_B must be obtained from a negative power supply with the transistors or vacuum tubes operating "upside down," while when the cathode is grounded a standard or positive power supply may be employed. One simple type of beam supply for low-voltage klystrons consists of several large B batteries connected in series. Electronic power supplies are available commercially, and we shall not discuss any here. The first edition may be consulted for a description of several circuits. Figure 3-20 shows some typical waveforms of a regulated power supply. The klystron power output increases, and the klystron lifetime decreases with an increase in beam voltage. In this respect the beam voltage behaves like the filament voltage, and, again, too high a value can damage the tube.

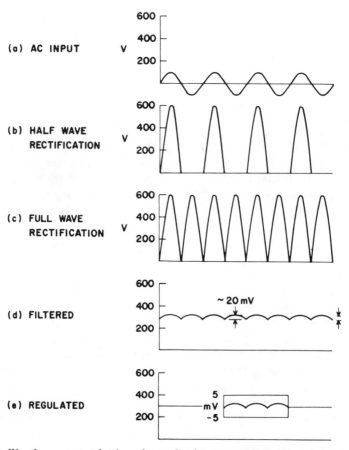

Fig. 3-20. Waveforms at several points of a regulated power supply. An expanded ordinate scale is presented on the insert of (e).

Reflector Voltage Power Supply

The reflector voltage is more critical than the beam voltage, and hence must be more stable in value and contain less ripple. Since the reflector does not draw any current, there are no power requirements for this supply. The first edition may be consulted for a description of several reflector supply circuits. A diode should be employed as a protective device that will conduct and short-circuit the reflector to the cathode if the reflector begins drawing current. Provision should be made for modulating the reflector. The reflector voltage can also be obtained from dry cells, and these will have a long lifetime since the reflector does not draw current.

F. Frequency Stabilization of Klystrons

The principal function that an electron spin resonance spectrometer must perform is to measure accurately the ratio of the frequency to the magnetic field strength corresponding to a resonance absorption. To attain high precision in such a measurement, it is necessary to stabilize both the frequency and the magnetic field to a sufficient extent so that the uncertainties in these two quantities are less than the desired uncertainty in the g-factor determination. Regulated power supplies alone do not produce sufficient stability, and this section will therefore discuss how to stabilize the frequency. Most of the general stabilization techniques to be described may be used with several types of rf oscillators, but for simplicity only the klystron will be discussed.

The object of a frequency stabilizer is to respond to all frequency changes by a mechanism that returns the frequency to its initial value. For example, the klystron frequency may be compared to the resonant frequency of a cavity, and when a change occurs in the klystron frequency a small error voltage is produced that has a positive polarity for a frequency decrease and a negative polarity for an increase (or vice versa). This error signal is amplified and applied to the klystron reflector voltage supply in such a way as to bring the frequency back to its initial value. It is generally unnecessary to stabilize a klystron to better than one part in 10^6. In the remainder of this section several practical ways of performing this stabilizing action will be described.

Most automatic frequency-control systems are of the reflector modulation type. The block diagram of such a stabilizer is presented in Fig. 3-21. The oscillator modulates the klystron by impressing a 13-kHz voltage on the reflector, and it also provides the reference signal on the grid of the mixer tube. The error signal is detected at the crystal detector, amplified, and impressed on the control grid of the mixer tube. This error signal interacts with the reference signal in such a way as to change the amount of plate current i_p through the tube. The voltage drop across the resistor R is $i_p R$, and being in series with the reflector supply it provides part of the total reflector voltage. The error signal changes this voltage drop in such a way as to counteract a frequency deviation and return the klystron to its proper frequency.

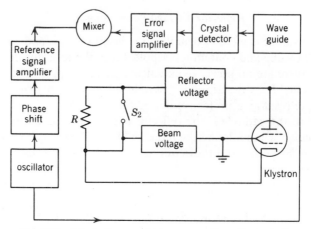

Fig. 3-21.   Block diagram of klystron stabilizer (Poole, 1958).

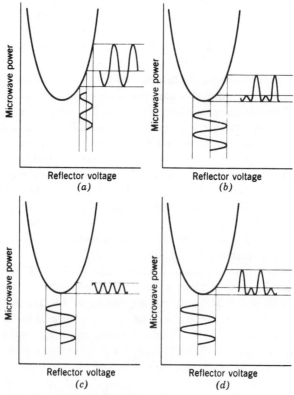

Fig. 3-22.   Error signal of klystron stabilizer in the neighborhood of resonance: (a) far off resonance ($f \gg f_0$); (b) close to resonance ($f > f_0$); (c) on resonance ($f = f_0$); (d) close to resonance ($f < f_0$) (Poole, 1958).

The mechanism whereby the error signal is produced is shown in Fig. 3-22, where the U-shaped part of the diagram is a spread-out picture of the bottom of the cavity pip shown in Fig. 3-23c. On resonance, the 13-kHz reflector modulation signal produces a 26-kHz output, and close to resonance, one obtains a 13-kHz output plus a 26-kHz signal with the 13-kHz component having the opposite phase on the two sides of resonance, as indicated by Fig. 3-23b and Fig. 3-23d. The greater the deviation from resonance, the larger the amplitude of the 13-kHz error signal. The phase of the reference signal can be adjusted by a phase reversal switch and phase shifting potentiometer in the reference signal amplifier. An RLC-tuned circuit is employed to reject the 26-kHz component of the error signal. As a result, the 13-kHz error signal mixes with the 13-kHz reference signal in the mixer, and a properly adjusted phase shift results in frequency stabilization. Three stabilizers that operate on this principle are described in the first edition (Jung, 1960; Poole, 1958; Waring, 1963). The Pound (1946, 1947) stabilizer (RLS-11) also described in that edition works on a different principle.

Andrist et al. (1976) designed the digital feedback system illustrated on Fig. 3-24, which stabilizes the ratio of the microwave frequency to the static field value. The HP5360A electronic computing counter with an associated HP5255A heterodyne converter samples the klystron (f_{mw}) and proton magnetometer (f_p) frequencies periodically for 100-msec intervals every 0.3 sec. The computing counter calculates the ratio f_{mw}/f_p, compares it to the desired preselected value, and produces a digital error signal to correct the magnetic field for deviations from this value. The digital error signal is then converted to analog form, amplified, and impressed on the magnet current stabilizer to return the magnetic field to its proper value. The system stabilizes the frequency ratio to within 1 part in 10^6.

Takao and Hayashi (1970) designed the highly sensitive low-field modulation (~ 80 Hz) X-band spectrometer illustrated in Fig. 3-25. It employs a parametron preamplifier and a cascaded homodyne receiver with a varactor reflection-type 30-MHz preswitch as a signal detector. The low noise detection system is phase sensitive and rejects frequency modulation noise 90° out of phase with the ESR signal. The AFC error signal detector consists of a phase-sensitive 21-MHz switching homodyne receiver that helps to provide the short time frequency stability of 1 part in 10^8.

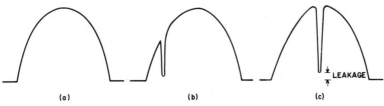

(a)   (b)   (c)

Fig. 3-23.   Klystron mode (a) far removed from cavity frequency, (b) with cavity frequency on side of mode, and (c) with cavity frequency centered on mode. The ordinate is klystron power, and the abscissa is reflector voltage.

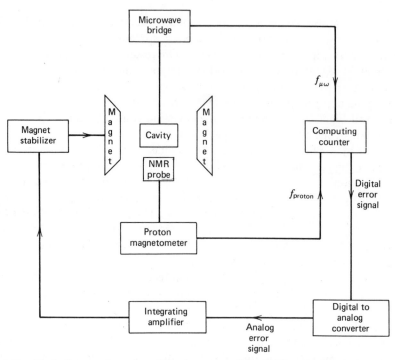

Fig. 3-24. Bloch diagram of a ratio stabilization system. [Adapted from Andrist et al. (1976).]

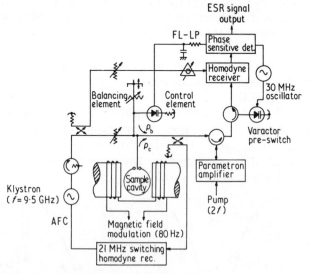

Fig. 3-25. Block diagram of an ESR spectrometer in which the signal detector is a parametron amplifier and a homodyne receiver with a 30-MHz varactor preswitch. The 21-MHz switching homodyne receiver provides the AFC and short-term stability. The automatic self-balancing element, the varactor control element, and the feedback path through the FL-LP low-pass filter provide the long-term stability (Takao and Hayashi, 1970).

The long-term stability of the system is provided by an automatic self-balancing circuit, which consists of (1) a varactor diode control element mounted in an X-band waveguide with a nonreflecting load and (2) a feedback loop from the phase-sensitive detector through the low-pass filter FL-LP to the control element. The low-pass filter has a cutoff frequency lower than the modulating frequency. When disturbances other than the ESR absorption change the output power then a signal is fed back to the balancing element to restore the bridge balance. This self-balancing circuit can also be used as a marker signal generator by impressing on the varactor diode a pulse train voltage in synchronism with the magnetic field modulation.

Many klystron stabilizers have difficulty operating at low microwave power levels. Mehlkopf et al. (1972) used a microwave transmission filter in the leakage branch of a homodyne spectrometer to achieve stabilization off the sample cavity for cavity power levels down to 10^{-10} W. The leakage branch provides bucking power to the crystal detector.

Ginodman et al. (1970) stabilized a klystron off a superconducting resonator. Marshall et al. (1973) achieved frequency stabilization to within a few parts in 10^8 for about an hour by phase-locking the klystron to a harmonic of a quartz-controlled 8-MHz oscillator.

Peter and Strandberg (1958) describe a phase stabilizer that can lock the klystron to a high harmonic of a quartz-controlled oscillator. Phase stabilization is capable of producing a more monochromatic output than other stabilizing schemes discussed above. Buckmaster and Dering (1966) described several electronic phase-lock oscillators.

Redhardt (1961) reduced the effect of klystron noise with the ESR cavity and a second nearly equivalent cavity both coupled to a microwave bridge. The system tolerates fluctuations in the source frequency up to 10^3–10^4 times as high as in the single cavity case before producing a noticeable change in the wave reflected from the ESR cavity. Mehlkopf and Smidt (1961) describe a similar system. Payne (1964) described an "electron oscillator" ESR spectrometer that achieves high sensitivity without a frequency stabilizer [see also Price and Anderson (1957) and Sloan et al. (1964)].

The original articles on klystron stabilizers (Pound, 1946, 1947; RLS-11, p. 58; Tuller et al., 1948) presented the fundamental principles of stabilization. The first edition mentions stabilizer designs reported by Berry and Benton (1965), Bouthinon and Coumes (1964), Bruin and van Ladesteyn (1959), Daniels and Farach (1961), Dayhoff (1951), Dušek (1961), Gambling and Wilmshurst (1961), Hervé et al. (1959), Kester (1965), Khaiken (1961), Narath and Gwinn (1962), Oliver (1962), Owston (1964), Peter and Strandberg (1958), Radford (1963), Sauzade (1961), Sirkis and Coleman (1954), Smith (1960), and Zimmerer (1959, 1962).

G. Frequency Determination

A cavity wavemeter is ordinarily employed for measuring a microwave frequency to about 1 part in 20,000, but Bussey and Estin (1960) have reported a precision of 1 or 2 parts in 10^6 [see also van den Bosch and Bruin (1953)]. If the microwave frequency is mixed with the appropriate harmonic of a crystal oscillator, the beat or intermediate frequency may be measured. This is capable of determining the microwave frequency to within one part per million. Frequency standards are discussed in the first edition of this work. Ramsey (1971, 1972) presented a history of automatic frequency standards.

References

Andrist, M., K. Loth, and Hs. H. Günthard, *J. Phys. E.* **9**, 947 (1976).

Berry, J. E., Jr., and A. Benton, *RSI* **36**, 958 (1965).

Berteaud, A. J., *Les Hyperfrequencies et Leurs Applications*, Presses Universitaires de France, 1976.

Bogle, G. S., H. F. Symmons, V. R. Burgess, and J. V. S. Sierins, *Proc. Phys. Soc.* **77**, 561 (1961).

Bott, I. B., and W. Fawcett, *Adv. Microwaves* **3**, 224 (1968).

Bouthinon, M., and A. Coumes, *J. Phys. Suppl.* **25**, 41A (1964).

Bruin, F., and D. van Ladesteyn, *Physica* **25**, 1 (1959).

Buckmaster, H. A., and J. C. Dering, *JSI* **43**, 554 (1966).

Bussey, H. E., and A. J. Estin, *RSI* **31**, 410 (1960).

Case, W. E., and N. T. Larsen, *RSI* **34**, 809 (1963).

Daniels, J. M., and H. A. Farach, *RSI* **32**, 1262 (1961).

Dascălu, D., *Transit-Time Effects in Unipolar Solid-State Devices*, Abacus Press, Tunbridge Wells, Kent, England, 1974.

Dayhoff, E. S., *RSI* **22**, 1025 (1951).

De Loach, B. C., *Adv. Microwaves* **2** (1967).

Dreizler, H., W. Maier, and H. D. Rudolph, *Arch. Sci.* **13**, 137 (1960).

Dušek, J., *Czech. J. Phys.* **B11**, 528 (1961).

Eastman, L. F., *Gallium Arsenide Microwave Bulk and Transit Time Devices*, Artech House, Dedham, Mass., 1972.

Erikson, L. E., *Phys. Rev.* **143**, 295 (1966).

Evans, W. J., Chap. 3 in *Topics in Solid State and Quantum Electronics*, W. D. Hershberger, Ed., Wiley, New York, 1972.

Gambling, W. A., and T. H. Wilmshurst, *JSI* **38**, 334 (1961).

Ginodman, V. V., P. S. Gladkov, Yu. V. Dedik, and Z. F. Kaplun, *PTE* **13**(4), 154 (1136), (1970).

Grivet, P., *Microwave Circuits and Amplifiers*, Volume 2 in *The Physics of Transmission Lines at High and Very High Frequencies*, Academic Press, New York, 1976.

Gittins, J. F., *Power Travelling Wave Tubes*, American Elsevier, New York, 1965.

Hershberger, W. D., Ed., *Topics in Solid State and Quantum Electronics*, Wiley, New York, 1972.

Hervé, J., J. Pescia, and M. Sauzade, *C. R. Acad. Sci.* **249**, 1486 (1959).

Hogg, R. D., *RSI* **44**, 582 (1973).

Hogg, R. D., *Am. J. Phys.* **41**, 224 (1973).

Hok, G., *Proc. IRE* **44**, 1061 (1956).

Jung, P., *JSI* **37**, 372 (1960).

Kester, T., *JSI* **42**, 442 (1965).

Khaikin, M. S., *PTE* **3**, 103 (1961).

Kohno, M., *Jeol News* **13A**, 10 (1976).

Kondrat'ev, B. V., *Ukr. Fiz. Zh.* **8**, 1203 (1963).

Kroemer, H., Chap. 2, in *Topics in Solid State and Quantum Electronics*, W. D. Hershberger, Ed., Wiley, New York, 1972.

Lee, T. P., and R. D. Strandley, *Bell. Syst. Tech. J.* **48**, 143 (1969).

Marshall, S. A., S. V. Nistor, Chao-Yuan Huang, and T. Marshall, *Phys. Status Solidi B* **59**, 55 (1973).

Mehlkopf, A. F., and J. Smidt, *RSI* **32**, 1421 (1961).

Mehlkopf, A. F., J. Smidt, and T. A. Tiggelman, *RSI* **43**, 693 (1972).

Mock, J. B., *RSI* **31**, 551 (1960).

Mouthaan, K., *Intern. J. Electron.* **18**, 301 (1965).

Narath, A., and W. D. Gwinn, *RSI* **33**, 79 (1962).

Ober, J., *Philips Res. Rep.* **20**, 357 (1965).

Olivier, M., *J. Phys. Radium* **23**, 144A (1962).

Owston, C. N., *JSI* **41**, 698 (1964).

Payne, J. B., *IEEE Trans. Microwave Theory Tech.* **MTT-12**, 48 (1964).

Penfield, P., and R. Rafuse, *Varactor Applications*, MIT Press, Cambridge, Mass., 1962.

Peter, M., and M. W. P. Strandberg, *Proc. IRE* **43**, 869 (1958).

Pierce, J. R., *Traveling Wave Tubes*, Van Nostrand, New York, 1950; *Proc. IRE* **50**, 978 (1962).

Poole, C. P., Jr., Ph.D. thesis, Department of Physics, University of Maryland, 1958.

Pound, R. V., *RSI* **17**, 490 (1946); *Proc. IRE* **35**, 1405 (1947); RLS-11, pp. 58–78.

Price, V. G., and C. T. Anderson, *IRE Natl. Conv. Record*, Part 3, **5**, 57 (1957).

Radford, H. E., *RSI* **34**, 304 (1963).

Ramsey, N. F., Programme of the 25th Annual Frequency Control Symposium. Atlantic City, N.J., Fort Monmouth, N.J., April 26–28 (1971): reported in *IEEE Trans. Instrum. Meas.* **IM-21**, 90 (1972).

Read, W. T., *Bell Syst. Tech. J.* **37**, 401 (1958).

Redhardt, A., *Z. Angew. Phys.* **13**, 108 (1961).

Rupp, L. W., Jr., *RSI*, in press.

RLS-6, see Chap. 2 for complete reference.

RLS-7, see Chap. 2 for complete reference.

RLS-11, see Chap. 2 for complete reference.

Sauzade, M., *Ann. Phys.* (*France*) **6**, 595 (1961).

Shurmer, H. V., *Microwave Semiconductor Devices*, Pitman, London, 1971.

Sirkis, M. D., and P. D. Coleman, *RSI* **25**, 401 (1954).

Sloan, E. L. III, A. Ganssen, and E. C. LaVier, *Appl. Phys. Lett.* **4**, 109 (1964).

Smith, M. I. A., *JSI* **37**, 398 (1960).

Sterzer, F., *Adv. Microwaves* **2** (1967).

Sushkov, A. D., and V. A. Meos, *Zh. Tekh. Fiz.* **35**, 723 (1965).

Takao, I., and T. Hayashi, *J. Phys. E Sci. Instrum.* **3**, 315 (1970).

Tuller, W. G., W. C. Galloway, and F. P. Zaffarano, *Proc. IRE* **36**, 794 (1948).

Uchida, T., and M. Buyle-Bodin, *Ampere Colloquium*, Eindhoven, 1962, p. 726.

Uhler, A., Jr., *Proc. Inst. Radio Eng.* **46**, 1099 (1958).

van den Bosch, J. C., and F. Bruin, *Physica* **19**, 705 (1953).

Walsh, W. M., Jr., and L. W. Rupp, Jr., *RSI* **41**, 1316 (1970); **42**, 468 (1971).

Walsh, W. M., Jr., and L. W. Rupp, Jr., *RSI*, in press.

Waring, R. K., Jr., *RSI* **34**, 1228 (1963).

Webb, R. H., *RSI* **33**, 732 (1962).

Wilmshurst, J. H., W. A. Gambling, and D. J. Ingram, *J. Electron. Control* **13**, 339 (1962).

Young, L., Ed., *Advances in Microwaves*, a continuing series, first appeared in 1967.

Zimmerer, R. W., *RSI* **30**, 1052 (1959); **33**, 858 (1962).

Zublin, K. E., W. T. Wilser, and W. R. Green, *Microwave J.* **18**, 33 (1975).

CHAPTER 4

Waveguide Components

A. Spectrometer Components

The block diagram of a typical unsophisticated electron spin resonance spectrometer is shown in Fig. 4-1. The waveguide components may be considered as bounded by the klystron, resonant cavity, and two crystals. They are connected to the electronic circuits by means of the cable from the klystron stabilizer-power supply, and through the two coaxial (or parallel wire) cables from the crystals. The magnet system is not connected to the waveguide, but modulation coils often are mounted on the cavity.

The purpose of the present chapter is to describe the operating principles and practical application of the various waveguide components that one customarily incorporates into ESR spectrometers. An attempt will be made to provide enough information to assist in the design of new equipment, and in the assembly of novel spectrometer arrangements. No mention will be made of those waveguide components such as TR (transmit–receive) or ATR (anti-transmit–receive) tubes that are widely used in radar (compare with RLS-14), but are seldom used in ESR spectrometers.

Much of the information contained in this chapter may be found in Volumes 8, 9, 10, 11, and 14 of the *MIT Radiation Laboratory Series* (RLS). A number of more recent developments, such as isolators and circulators, will also be described.

B. Attenuator

An attenuator is a circuit element that, when inserted between a generator and a load, reduces the amplitude and changes the phase of the radio frequency signal incident on the load. Usually the phase change is disregarded.

The microwave power is reduced from the value P_1 in the absence of the attenuator to the value P_2 when it is in place. This reduction in power is measured by a quantity called the attenuation $\mathcal{C}$, which is defined by

$$\mathcal{C} = 10 \log_{10}\left(\frac{P_1}{P_2}\right) \tag{1}$$

and is expressed in the units of decibels (dB). This expression assumes that both the generator and load ends of the transmission line are terminated in matched impedances. Under these conditions P_1 is the maximum available

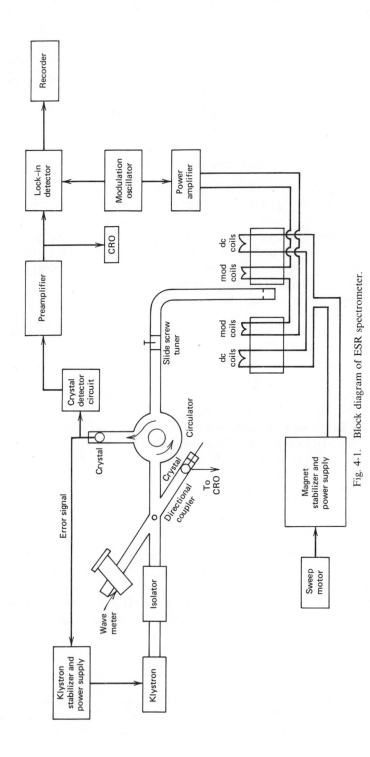

Fig. 4-1. Block diagram of ESR spectrometer.

power from the generator, and P_2 is the power delivered through the attenuator to a matched load.

The attenuation $\mathcal{C}$ should not be confused with the attenuation constant α. The latter is the dissipative part of the propagation constant γ

$$\gamma = \alpha + j\beta \tag{2}$$

which measures the amplitude $e^{-\gamma z}$ along the z direction in a transmission line.

A calibrated attenuator is one that has a dial and can be set for a predetermined attenuation constant $\mathcal{C}$. Most attenuators are designed to be broadband devices, and they are rated in terms of the maximum voltage standing-wave ratio (VSWR) that they produce when inserted in a matched line at any frequency within the predetermined frequency band. An attenuator frequently possesses a maximum power capacity, and it may be burned out if this input power is exceeded. The attenuator calibration curve (dB versus displacement) is ordinarily furnished by the manufacturer.

A simple waveguide attenuator is constructed from an IRC resistance card, which is a phenol fiber 0.08 cm thick on which is sprayed a mixture of graphite and a binder. The latter volatilizes at about 100°C, and leaves a carbon coating that absorbs microwave power. These resistance cards are discussed in Sec. 4-H. Uskon cloth and Polyiron may also be employed as attenuating materials.

A waveguide flap attenuator is shown in Fig. 4-2. It consists of a single strip of 200 Ω/square IRC resistance card cut in a circle to minimize the VSWR. The design in Fig. 4-2 provides 10–15 dB at X band. For greater rigidity, two

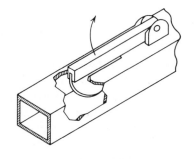

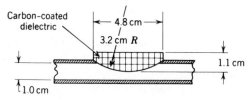

Fig. 4-2. Variable waveguide attenuator of the flap type.

200-Ω/square resistance cards may be glued back to back, and this results in a combined film resistivity of 100 Ω/square, and greater attenuation. The resistance card is inserted into the broad side of the waveguide so that the **E** vector will be parallel to the card. One liability of the flap attenuator is that the calibration curve of dB/insertion is very nonlinear.

C. Isolator

An isolator or gyrator is a two-terminal pair microwave ferrite device that makes use of the Faraday effect to permit the transmission of microwaves in one direction, and prevents their transmission in the opposite direction (Grivet, 1976; Helszain, 1975; Stanforth, 1972). Ferrites are magnetic materials with resistivities from 10^6 to 10^{13} times that of iron. As a result, no skin depth difficulties hinder the propagation of microwaves through them. The ferrite is placed in a circular waveguide, and the electron spins in the ferrite are aligned by a magnetic field directed along the microwave axis of propagation. As the microwaves propagate through the ferrite they interact with the spins, and the result of this interaction is that the plane of polarization of the microwaves is rotated. Another way of viewing this rotation is to consider the incident linearly polarized wave as resolved into a positive and a negative circularly polarized component. These two circular components travel at different velocities in the ferrite, and on emerging they unite to form a plane wave that is polarized in a different direction from the incident wave.

The most significant feature of this Faraday rotation is its antireciprocal character. Assume that a microwave plane wave traverses an isolator and has its plane of polarization rotated by the angle θ. If it is reflected and traverses the isolator in the opposite direction, its plane of polarization will be rotated by another angle θ for a total rotation of 2θ. If the isolator obeyed the reciprocity theorem the rotation of the returning wave would cancel that of the incident one, instead of adding to it.

Two examples of isolators are shown in Fig. 4-3. The input and output waveguides on the right are oriented at angles of 45° with respect to each other, and they are separated by a circular waveguide containing the ferrite. The longitudinal magnetic field is adjusted to the value that rotates the microwave plane of polarization by 45° so that in one direction the output wave is polarized properly relative to the waveguide, while in the other direction, it is 90° out of phase with the waveguide, and cannot propagate. The resistor cards help to match the ferrite by reducing reflections. A waveguide twist may be employed to rotate one of the waveguides back so it is parallel to the other, since it is convenient to be able to insert the isolator in a transmission line without having to rotate the source waveguide components relative to the remainder of the waveguide system. Another method is to orient the two rectangular waveguide terminals of the isolator at $\pm 22.5°$, respectively, relative to the waveguide transmission line into which it is inserted. In this case the H field from the source waveguide will be reduced to $H\cos(\pi/8)$, or $0.924H$,

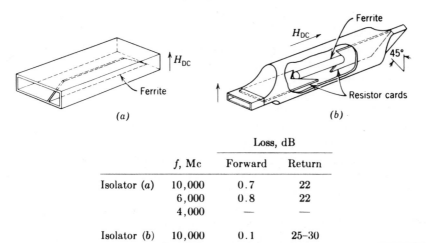

(a)   (b)

| | f, Mc | Loss, dB | |
		Forward	Return
Isolator (a)	10,000	0.7	22
	6,000	0.8	22
	4,000	—	—
Isolator (b)	10,000	0.1	25–30
	6,000	—	—
	4,000	0.2	25–30

Fig. 4-3. Two types of X-band isolators. (a) The rectangular type has a forward loss of 0.75 dB and a return loss of 22 dB, while (b) the cylindrical type has a forward loss of 0.15 dB and a return loss of about 28 dB (Rowen, 1953).

when it enters the isolator waveguide input, and after undergoing the Faraday rotation the H field will be farther reduced to $H(\cos \pi/8)^2$, or $0.854H$, when it leaves the isolator.

The amount of attenuation provided by the isolator in the backward direction is varied by changing the strength of the longitudinal magnetic field. The magnetic field may be produced by a coil and supplied by a storage battery in series with a potentiometer that varies the magnetic field strength. There is a very nonlinear relation between the strength of the magnetic field and the current in the coils. In addition, isolators exhibit a pronounced hysteresis because the amount of attenuation depends on the previous history of the applied current. A typical isolator hysteresis curve is shown in Fig. 4-4.

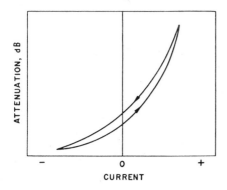

Fig. 4-4. Typical isolator hysteresis loop showing attenuation as a function of the direct current through the coil.

A nonvariable isolator is usually placed between the klystron and the rest of the waveguide components to prevent the waveguide network from acting back on and influencing the frequency stability of the klystron. The isolator is said to decouple or match the klystron to the transmission line. Such a nonvariable isolator usually employs a permanent magnet, and care must be exercised not to store it near ferromagnetic objects (e.g., on a steel shelf) where it may become demagnetized.

A variable isolator may be employed to adjust the amount of microwave power that is incident on the sample resonant cavity when very low power studies are being made. Such low power levels are often required for studies at liquid helium temperature. Crystal detectors do not function well when the overall microwave power drops below 1 or 2 mW because under these conditions sufficient leakage (crystal current) cannot be provided. Thus it is often desirable to use an isolator to lower the power incident on the cavity, but not the power incident on the crystal detector. This can be accomplished by placing the isolator between the microwave bridge (magic T) and the resonant cavity. In this position it will attenuate the incident microwave power, but will not attenuate the reflected signal. When using this experimental setup, the isolator current may be turned off while the klystron frequency (and reference cavity) are adjusted to the sample cavity frequency, since during this procedure it is desirable to display the cavity "pip" on the oscilloscope. Too much isolator attenuation will decouple the cavity from the rest of the waveguide system and prevent such oscilloscopic observation of the sample cavity.

Since the isolator is a nonreciprocal device it cannot be represented by an equivalent circuit consisting of resistors, inductors, and capacitors. The scattering matrix for an ideal isolator is

$$\begin{bmatrix} B_1 \\ B_2 \end{bmatrix} = \begin{bmatrix} 0 & S_{12} \\ 0 & 0 \end{bmatrix} \begin{bmatrix} A_1 \\ A_2 \end{bmatrix} = \begin{bmatrix} 0 & 1 \\ 0 & 0 \end{bmatrix} \begin{bmatrix} A_1 \\ A_2 \end{bmatrix} \tag{1}$$

In other words, when the microwave amplitudes A_1 and A_2 are incident on the two terminals, the respective transmitted amplitudes B_1 and B_2 are given by

$$B_1 = A_2 \tag{2}$$

$$B_2 = 0 \tag{3}$$

The impedance and admittance matrices are antisymmetric and have the form

$$Z = \begin{bmatrix} 0 & R \\ -R & 0 \end{bmatrix} \tag{4}$$

and

$$Y \begin{bmatrix} 0 & -1/R \\ 1/R & 0 \end{bmatrix} \tag{5}$$

which leads to the relations

$$V_1 = RI_2 \tag{6}$$

$$V_2 = -RI_1 \tag{7}$$

Chapter 7 of the book by Grivet (1976) may be consulted for more information on isolators.

D. Directional Coupler

A directional coupler is usually employed to monitor the power P_i in the incident or forward traveling wave of a transmission line by extracting a small amount of it. The total power P is the difference between the powers P_i and P_r in the forward and backward (reflected) traveling waves, respectively,

$$P = P_i - P_r \tag{1}$$

$$= P_i(1 - |\Gamma|^2) \tag{2}$$

where Γ is the complex voltage reflection coefficient of the terminating load.

A small probe or hole in the waveguide is sensitive to the total electric (or magnetic) field strength. If two holes are placed a quarter of a wavelength apart as shown in Fig. 4-5, then only the forward traveling wave in the main guide will be coupled to the coaxial output. The power coupled from the

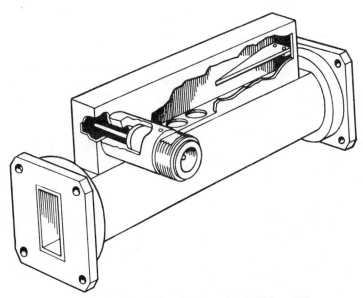

Fig. 4-5.   Two-hole directional coupler (RLS-8, p. 312).

backward traveling wave is dissipated in the matched load shown in the upper right of Fig. 4-5. The amount of power coupled from the main waveguide increases with increasing hole diameter. Directional couplers may also be made with one hole (the Bethe-hole coupler) and many holes. Sometimes other types of coupling techniques are used. The design specifications for a large number of directional couplers are given in Chap. 14 of RLS-11.

Directional couplers are usually characterized by their directivity D, which is a measure of the rejection of the backward traveling wave

$$D = 10 \log_{10}\left(\frac{P_f}{P_b}\right) \tag{3}$$

and the coupling C, which characterizes the amount of power extracted by the side or auxiliary arm

$$C = 10 \log_{10}\left(\frac{P_i}{P_f}\right) \tag{4}$$

The quantities in the arguments of the logarithms are defined in Fig. 4-6. The directivity of the two-hole coupler shown in Fig. 4-5 is

$$D = -20 \log_{10}\left(\frac{2\pi \Delta l}{\lambda_g}\right) \tag{5}$$

where Δl is the departure of the hole spacing from a quarter of a guide wavelength and

$$\Delta l \ll \lambda_g \tag{6}$$

The expression for the coupling C is considerably more complicated (RLS-11, p. 877). The first edition (p. 226) gives expressions for C and D in terms of the impedance matrix.

In electron spin resonance spectrometers, a directional coupler is often employed to couple a little power ($C \sim 20$ dB) from the main waveguide to an auxiliary arm equipped with a wavemeter and detector. To measure the frequency one may use a crystal detector in series with a microammeter. When

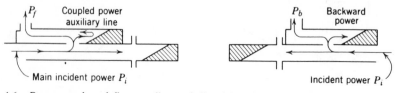

Fig. 4-6. Powers used to define coupling and directivity of a directional coupler. (RLS-11, p. 859).

the wavemeter is tuned to the proper frequency, it will absorb energy and decrease the detected current. The setting for the minimum current gives the microwave frequency. If power is to be measured, the auxiliary arm may be terminated by a bolometer or thermistor associated with a power bridge. When the directional coupler has a coupling of less than 20 dB the auxiliary arm may couple back into the main waveguide arm and produce interference and noise. In any event, it is best to minimize these effects and establish uniform experimental conditions by setting the detector-tuning plunger for zero detector current in the sidearm.

A directional coupler may be used as a substitute for a magic *T*. This has the principal benefit of rendering the bridge more compact in size. A circulator is to be preferred, however.

E. Magic *T*

A magic *T* is a directional coupler with equal power division, as discussed in RLS-8, Chaps. 9 and 12; RLS-10, Chap. 7; RLS-11, Chap. 9; and RLS-14, Chap. 8. Its low-frequency analogue is the hybrid coil of Fig. 4-7, and a sketch of the side outlet magic *T* is shown in Fig. 4-8. Arm 1 (sidearm) is opposite to arm 2 (sidearm), and arm 3 (H or shunt arm) is opposite to arm 4 (E or series arm), while arms 1 and 2 are adjacent to arms 3 and 4. The magic *T* is matched by terminating each arm with its characteristic impedance. It has the property that at match, a wave incident on any arm will split equally between the two adjacent arms, while no power will be reflected back or enter the opposite arm. Thus, at match the opposite arms of a magic *T* are decoupled from each other.

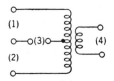

Fig. 4-7.   Circuit of a hybrid coil which is the low frequency analogue of a magic *T*. (RLS-8, p. 307).

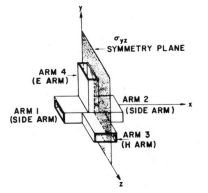

Fig. 4-8.   A magic *T*.

If A_1, A_2, A_3, and A_4 are the wave amplitudes incident on the four junctions, then at match the respective reflected wave amplitudes B_1, B_2, B_3, and B_4 are given by the scattering matrix

$$
\begin{bmatrix} B_1 \\ B_2 \\ B_3 \\ B_4 \end{bmatrix} = \frac{j}{2^{1/2}} \begin{bmatrix} 0 & 0 & 1 & 1 \\ 0 & 0 & 1 & -1 \\ 1 & 1 & 0 & 0 \\ 1 & -1 & 0 & 0 \end{bmatrix} \begin{bmatrix} A_1 \\ A_2 \\ A_3 \\ A_4 \end{bmatrix}
\tag{1}
$$

which is equivalent to the four equations:

$$
B_1 = \left(\frac{j}{2^{1/2}} \right)(A_3 + A_4)
$$

$$
B_2 = \left(\frac{j}{2^{1/2}} \right)(A_3 - A_4)
$$

$$
B_3 = \left(\frac{j}{2^{1/2}} \right)(A_1 + A_2)
$$

$$
B_4 = \left(\frac{j}{2^{1/2}} \right)(A_1 - A_2)
\tag{2}
$$

These four equations are a mathematical statement of the magic T property mentioned above.

The lack of coupling between arms 3 and 4 results from symmetry because the TE_{10} mode in arm 3 has even symmetry (i.e., is symmetric), and that in arm 4 has odd symmetry (i.e., is antisymmetric) with reference to the y, z symmetry plane σ_{yz}. To show this, consider the TE_{01} electromagnetic vectors in the two arms in question with x set equal to zero at the σ_{yz} symmetry plane. The electromagnetic fields transform in the manner shown in Table 4-1, where the arrow indicates the sign that results from the reflection. As explained in Sec. 7-F of the first edition, the solution in arm 3 is even, while that in arm 4 is odd. An incoming even wave cannot produce an outgoing odd wave and vice versa. The waveforms in arms 1 and 2 may be either symmetric or antisymmetric depending, respectively, on whether the symmetry plane occurs at the center of the **H** vector loops where H_x is maximum and H_z and E_y are zero, or between two such loops where H_z and E_y are maxima, and H_x is zero. More explicitly, for arms 1 and 2 one has the two solutions shown in Table 4-1. It should be noted that the even and odd solutions in arms 1 and 2 are $\lambda_g/4$, or 90°, out of phase with respect to each other. This phase difference is responsible for the two negative signs in Eq. (1).

The lack of coupling between arms 1 and 2 results from the use of matching devices to match the magic T into arms 3 and 4. This matching may be

TABLE 4-1

Symmetry Properties of the TE_{10} Mode Solutions of Maxwell's Equations in the Magic T Shown in Fig. 4-8 with Respect to the $\sigma_{y,\,z}$ Reflection Plane

Arm	Symmetric (Even) Solution	Antisymmetric (Odd) Solution
1 and 2	$H_x(x, z) \rightarrow H_x(-x, z)$ $E_y(x, z) \rightarrow E_y(-x, z)$ $H_z(x, z) \rightarrow -H_z(-x, z)$	$H_x(x, z) \rightarrow -H_x(-x, z)$ $E_y(x, z) \rightarrow -E_y(-x, z)$ $H_z(x, z) \rightarrow H_z(-x, z)$
3	$H_x(x, z) \rightarrow H_x(-x, z)$ $E_y(x, z) \rightarrow E_y(-x, z)$ $H_z(x, z) \rightarrow -H_z(-x, z)$	Forbidden
4	Forbidden	$E_x(y, z) \rightarrow E_x(y, z)$ $H_y(y, z) \rightarrow H_y(y, z)$ $H_z(y, z) \rightarrow H_z(y, z)$

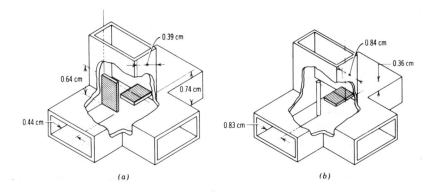

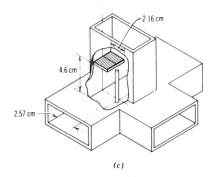

Fig. 4-9. Positions of irides and posts for matching waveguide Ts. Waveguides (a) 0.64×1.27 cm ($\frac{1}{4} \times \frac{1}{2}$ in.); (b) 1.27×2.54 cm ($\frac{1}{2} \times 1$ in.); (c) 3.81×7.62 cm ($1\frac{1}{2} \times 3$ in.) (RLS-11, pp. 526–528; RLS-14, pp. 361, 362).

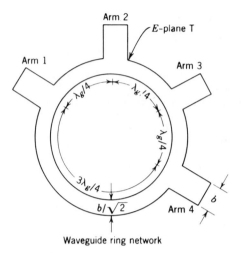

Waveguide ring network

Fig. 4-10.   Waveguide ring network (rat race) (RLS-11, p. 529; RLS-14, p. 359).

accomplished by means of an iris and post as shown in Fig. 4-9, or by other methods (e.g., RLS-14, p. 365). Miane (1973) discusses the use of an unmatched T.

An alternative type of magic T consists of the ring circuit, hybrid T or "rat race," and an example of one is shown in Fig. 4-10. This circuit may be analyzed by considering that the input from arm 1 splits, with half of the incident power going around the ring counterclockwise and the other half going around clockwise. At arm 2 the two paths correspond to $\lambda_g/4$ and $5\lambda_g/4$, and at arm 4 they both correspond to $3\lambda_g/4$, so in each case the signals add. However, at arm 3, the two paths correspond to $\lambda_g/2$ and λ_g, so the signals arrive out of phase, and cancel. A similar analysis may be made for inputs to the other arms. One concludes that the rat race resembles the side outlet magic T in that a signal incident on any arm splits equally between the two adjacent arms and does not couple at all to the opposite arm. In the waveguide case the ring is matched to the four-terminal waveguides by making the ring dimensions equal to $1/\sqrt{2}$ times the small dimensions of the waveguide, as indicated in Fig. 4-10 (compare Sec. 4-L). The rat race is frequency sensitive since the arms are a quarter of a wavelength apart at only one frequency.

A right-angle ring circuit magic T can also be analyzed in terms of the two possible paths from each input to output junction (see RLS-8, p. 309 and RLS-14, p. 367). Both the ring and right-angle ring circuits may be employed for coaxial magic Ts (compare Fig. 7-18 of the first edition). Section 2-D discusses the relationship between the radii and characteristic impedance of a coaxial line. A magic T with one waveguide and three coaxial terminations is shown in RLS-8 (p. 308) and RLS-11 (p. 529).

The magic T forms an integral part of many ESR spectrometers, and in this application one ordinarily places the resonant cavity and klystron in opposite

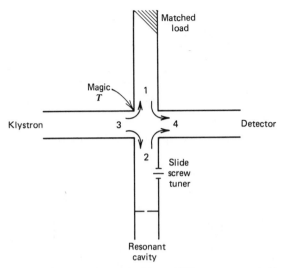

Fig. 4-11. The use of a magic T in an ESR spectrometer bridge.

arms, with the detector and a matched load terminating the remaining two arms, as suggested in Fig. 4-11. The function of the load is to match the magic T and dissipate half of the power from the klystron. The bridge is usually operated near match so that the incident klystron power splits equally between the cavity and load, and the reflected signal splits between the signal source and detector. Usually an isolator is placed in front of the source, and it acts as a matched load for the signal reflected into the klystron arm. Thus one of the inefficiencies of a magic T is the factor of 2 loss every time the microwaves pass through the T. Ordinarily the circulator is employed instead of a magic T since it does not entail such a loss. It is described in the next section. Magic T junctions are also employed in balanced mixers, the Pound frequency stabilizer (microwave discriminator), balanced duplexers, and other applications.

F. Circulator

A circulator is a nonreciprocal multiterminal pair network that transmits power from one terminal to the next in sequence (Helszain, 1975). In other words, a signal entering the ith arm of the circulator leaves by the $(i + 1)$th arm. In addition, a circulator appears matched looking into each arm.

Figure 4-12 shows two types of circulators: The rectangular circulator on the left has a device (π) that changes the phase of the microwaves by π, or 180°, when they traverse it in one direction. If the path length from terminals 1 and 2 to 3 and 4 is the same around both paths, and equal to an even number of guide wavelengths, then in the absence of the unidirectional element, a signal entering 1 will leave by 4 and one entering 3 will leave by 2. This device will be

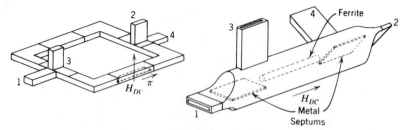

Fig. 4-12.   Two types of circulators (Rowen, 1953).

reciprocal and have the scattering matrix

$$
\begin{bmatrix} B_1 \\ B_2 \\ B_3 \\ B_4 \end{bmatrix} = \begin{bmatrix} 0 & 0 & 0 & 1 \\ 0 & 0 & 1 & 0 \\ 0 & 1 & 0 & 0 \\ 1 & 0 & 0 & 0 \end{bmatrix} \begin{bmatrix} A_1 \\ A_2 \\ A_3 \\ A_4 \end{bmatrix}
\tag{1}
$$

where A_i and B_i are the input and output waves, respectively, at the ith junction. When the unidirectional π phase shift is inserted, the signals entering terminals 2 and 4 will be unaffected, but these entering 1 and 3 will now leave by 2 and 4, respectively, which corresponds to

$$
\begin{bmatrix} B_1 \\ B_2 \\ B_3 \\ B_4 \end{bmatrix} = \begin{bmatrix} 0 & 0 & 0 & 1 \\ 1 & 0 & 0 & 0 \\ 0 & 1 & 0 & 0 \\ 0 & 0 & 1 & 0 \end{bmatrix} \begin{bmatrix} A_1 \\ A_2 \\ A_3 \\ A_4 \end{bmatrix}
\tag{2}
$$

As a result $B_{i+1} = A_i$, and the wave that enters the ith arm leaves by the $(i + 1)$th arm.

The circulator on the right-hand side of Fig. 4-12 may be analyzed by considering that the ferrite rotates the microwaves counterclockwise through $\pi/4$, or $45°$, relative to the direction looking along the magnetic field H_{DC}. The signal that enters terminal 1 cannot leave 3 as a result of symmetry. It traverses the ferrite and is rotated to the proper angle to leave by terminal 2. Similarly, the signal that enters by 3 is rotated into line with 4, where it leaves the isolator. One can similarly reason that the signals entering terminals 2 and 4 will leave by 3 and 1, respectively. Thus we see that the two circulators illustrated in Fig. 4-12 have the same scattering matrix (2).

In an ESR spectrometer one may make use of a three-terminal pair circulator by placing the klystron in arm 1, the resonant cavity in arm 2, and the detector in arm 3, as shown in Fig. 4-13. This arrangement allows the klystron power to go directly to the cavity, and the signal reflected at resonance to go

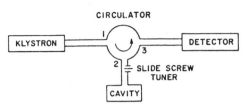

Fig. 4-13. The use of a circulator in an ESR spectrometer.

directly to the detector in accordance with the relation

$$
\begin{bmatrix} B_1 \\ B_2 \\ B_3 \end{bmatrix} = \begin{bmatrix} 0 & 0 & 1 \\ 1 & 0 & 0 \\ 0 & 1 & 0 \end{bmatrix} \begin{bmatrix} A_1 \\ A_2 \\ A_3 \end{bmatrix} \tag{3}
$$

In practice, A_1 will be large, A_2 will be small, and A_3 will be zero if the detector presents a matched load to the transmission line. The last condition is not ordinarily satisfied (compare following section and 11-C).

Circulators have small insertion losses, just as their isolator counterparts do. This will usually be specified by the manufacturer. The first edition (p. 237) contains references to the circulator literature.

G. Tuner

A tuner is a device that introduces reactance or susceptance into the transmission line. This has the effect of altering the impedance match and changing the standing-wave ratio. The most important tuning adjustment in an ESR spectrometer is provided by the slide screw tuner shown in Figs. 4-11 and 4-13. Ideally, before insertion of the tuner in Fig. 4-11, the entire microwave network consisting of the various magic T arms will be matched, and as a result, all the klystron power from arm 3 will split between arms 1 and 2 with none reaching the detector in arm 4. Since a crystal works best with a finite microwave power incident upon it, the slide screw tuner is introduced to mismatch the magic T slightly, and thereby allow a little klystron power to reach the detector. The present section describes the principle of tuning and discusses the details of several types of tuners (RLS-8, Chap. 5; see also RLS-9, Chap. 8; RLS-14, Chap. 2). If one employs the circulator shown in Fig. 4-13 the slide screw tuner may be put in the cavity arm (terminal 2) and adjusted for the proper crystal current.

Single Screw Tuner

Several types of tuners insert obstacles into the transmission line. These obstacles reflect the microwaves and thereby alter the impedance match and VSWR. When a screw is inserted into the center of the broad side of a

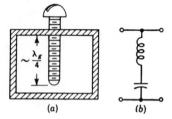

Fig. 4-14. (*a*) A single screw tuner and (*b*) its equivalent circuit (RLS-9, p. 705).

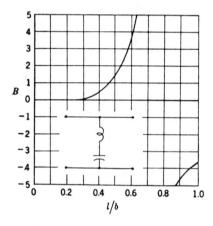

Fig. 4-15. The susceptance B of a tuning screw 0.13 cm in diameter as a function of the depth of insertion l in X-band waveguide of width $b = 1$ cm when $\lambda = 3.2$ cm (RLS-8, p. 169).

rectangular waveguide as shown in Fig. 4-14 it first acts like a shunt capacitance across the waveguide, with the susceptance $B = \omega C$ increasing with insertion in accordance with the curve of Fig. 4-15. When the length of the screw approximates a quarter of a guide wavelength, it becomes resonant, and the susceptance becomes infinite. Greater insertions produce an inductive effect $B = -1/\omega L$. Since rf electrical currents flow along the screw it is necessary for it to make good electrical contact with the waveguide. At X band a number 2 copper (or brass) screw is satisfactory.

Double and Triple Screw Tuner

The single screw tuner may be employed to match over a wide range of loads, but it suffers from the drawback that it does not cover much of the inductive range of susceptance. A pair of tuning screws separated by $\lambda_g/8$ or $5\lambda_g/8$, or three screws separated by $\lambda_g/4$ will allow tuning over a much wider range of susceptances (RLS-9, p. 507), and they are capable of achieving a VSWR < 2.

Slide Screw Tuner

A single screw that may be moved along the waveguide for at least a half of a guide wavelength will also match a transmission line over a wide range of susceptance. Such a device is illustrated in Fig. 4-16. To a first approximation the magnitude of the reflection from the screw depends on its insertion, and

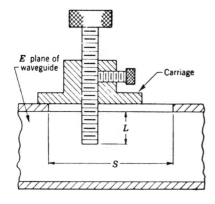

Fig. 4-16.   Slide screw tuner capable of moving a distance S along the waveguide (RLS-9, p. 486).

the phase of the reflection depends on its position. From Figs. 4-1 and 4-13 it may be seen that a slide screw tuner is customarily employed as an integral part of the ESR bridge. To use the tuner in this application, its longitudinal position is varied until one obtains a maximum signal (leakage) on the crystal, and its insertion is adjusted for the desired leakage. For small insertions this produces an undercoupled cavity match. It is possible to overcouple the cavity by adjusting the longitudinal position for a minimum in the leakage.

Tuning Plunger

Impedance matching is frequently accomplished by placing a tuning plunger at the end of a waveguide (RLS-8, p. 198). Two examples of such plungers are shown in Fig. 4-17. The plunger on the left of the figure has a folded-back coaxial linelike arrangement or choke joint that consists of two $\lambda_g/4$ line lengths, and it has the effect of reflecting a short circuit at the end of the plunger. The other plunger uses three quarter-wavelength sections (RLS-8, p. 198). A tuning plunger is frequently employed in conjunction with a detector mount in order to shift the maximum in the electric vector to the position of the detector.

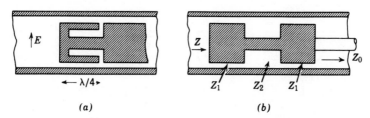

Fig. 4-17.   Plungers for use in rectangular waveguide; (a) folded back type; (b) dumbbell type (RLS-8, p. 198).

H. Matched Load

A load is a resistive impedance (admittance) that terminates a transmission line. Short-circuit terminations or plungers were discussed in the last section. A transmission line that is terminated in its characteristic impedance Z_0 is equivalent to a transmission line of infinite length, and is said to be matched. The physical significance of a matched transmission line is that it dissipates all the incident microwave power without producing any reflected wave. In other words, it produces a voltage standing-wave ratio (VSWR) of 1. In an ESR spectrometer a matched load is placed in one arm of the magic T, as shown in Fig. 4-11, and this helps to match the entire waveguide network. A resonant cavity with a properly matched iris is said to be critically coupled, and at its resonant frequency it becomes a matched load when placed at the end of a transmission line. Such a cavity is also shown in Fig. 4-11. Matched loads are discussed in the first few sections of RLS-11, Chap. 12.

Matched loads employed in resonance spectroscopy are ordinarily of the low power variety. The polyiron matched load shown in Fig. 4-18 produces a VSWR of 1.01 over the 3.13- to 3.53-cm band. Polyiron is capable of

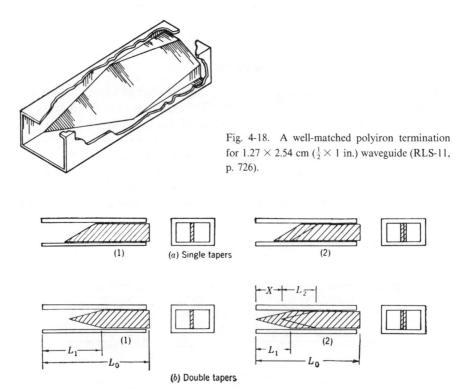

Fig. 4-18. A well-matched polyiron termination for 1.27×2.54 cm ($\frac{1}{2} \times 1$ in.) waveguide (RLS-11, p. 726).

(a) Single tapers

(1)   (2)

(b) Double tapers

(1)   (2)

Fig. 4-19. Various designs of IRC resistance loads for rectangular waveguide (a) single tapers; (b) double tapers (RLS-11, p. 729).

withstanding higher microwave powers than the other low-power loads to be discussed.

A fairly simple matched load may be constructed from an IRC resistance card, and several designs are shown in Fig. 4-19. The resistance card may be easily damaged by soldering operations and excessive power levels, but the latter limitation is not important at the ESR power levels currently in use ($P < 1$ W). It is fairly easy to construct a homemade matched load. Metalized glass may also be employed for the construction of matched loads. Another simple matched load is constructed from a wooden wedge two or three guide wavelengths long that is coated with graphite (i.e., Glyptal). Loads are ordinarily tapered to ensure a low VSWR. Coaxial matched loads are constructed on the same principles and from the same materials as waveguide loads.

I. Iris

An iris may be defined as a metallic partition extending partially across the waveguide in a plane perpendicular to the direction of propagation. Four commonly used irides are shown in Fig. 4-20. The two inductive irides have an inductance connected across the transmission line as their equivalent circuit, and the capacitive iris has a capacitance connected across the transmission line as its equivalent circuit. The resonant iris receives its name from the fact that its equivalent circuit is a combination of inductive and capacitive elements in parallel. The electrical characteristics of the symmetrical and asymmetrical shunt irides are shown graphically in RLS-9 (p. 212) and RLS-10 (p. 222).

The circular iris is frequently employed as a coupling hole from a waveguide to a resonant cavity; it is discussed in detail in Sec. 5-F. A circular iris across a waveguide may be employed as a narrow-band filter as shown in Fig. 4-21.

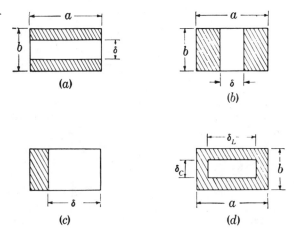

Fig. 4-20.   Windows for rectangular guides: (a) symmetrical capacitive; (b) symmetrical inductive; (c) asymmetrical inductive; (d) resonant (RLS-12, p. 230).

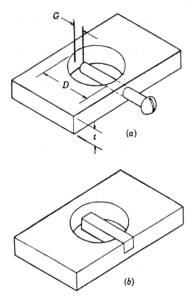

Fig. 4-21. Centered circular irides suitable for narrow band filters matched by (*a*) screw and (*b*) rectangular post (RLS-9, p. 690).

This iris resonates when the periphery of the opening is equal to approximately half a wavelength, and the frequency dependence of the voltage standing-wave ratio from such a filter is shown in RLS-9 (p. 690). The capacitive screw is only a trimmer for small frequency adjustments. When a thin resonant iris is employed to couple energy into a resonant cavity, a tuning screw such as the one shown in Fig. 4-21 may be used to match the cavity over a wide range of Q values. The equivalent circuit for a resonant iris used in this manner is a transformer, as shown in Figs. 5-24a, 5-24c, and 11-20.

J. Mode Transducer

A mode transducer is a device to transform a guided electromagnetic wave from one mode to another. This may take the form of a waveguide iris to transform from one rectangular waveguide mode to another, as Fig. 4-22

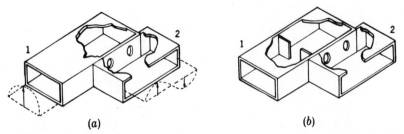

Fig. 4-22. Coupling from the TE_{10} mode in one waveguide to the TE_{20} mode in another by means of small holes: (*a*) shows **E** vector and (*b*) shows inductive iris (RLS-8, p. 337).

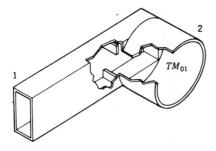

Fig. 4-23. Transducer for converting from the TE_{10} rectangular mode to the TM_{01} cylindrical mode (RLS-8, p. 339).

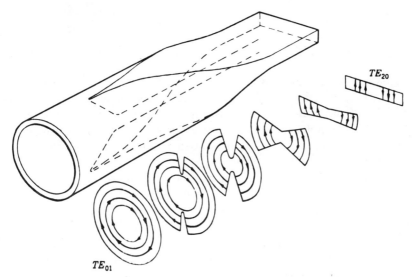

Fig. 4-24. Transducer for converting from the TE_{20} mode in rectangular guide to the TE_{01} mode in cylindrical guide (RLS-8, p. 340).

illustrates, or it may be a device to transform from a rectangular to a cylindrical waveguide mode, such as the ones illustrated in Figs. 4-23 and 4-24.

K. Slotted Section

In Sec. 4-G it was mentioned that slide screw tuners accomplish their task by moving a probe along the broad section of the waveguide to vary its coupling with the electric field vector. A slotted line, or slotted section, is a slide-screw tuner that is provided with a mechanism for measuring the strength of the microwave electric field at the probe. Figure 4-25 shows a sketch of a coaxial section and another of a waveguide slotted section. The probe insertion may be varied, and it is best to insert it as little as is feasible in order to avoid disturbing the electromagnetic field configuration. Ordinarily one employs an electric probe such as the one illustrated on the figure, but a magnetic probe consisting of a small loop antenna could also be used. The slot itself acts as a

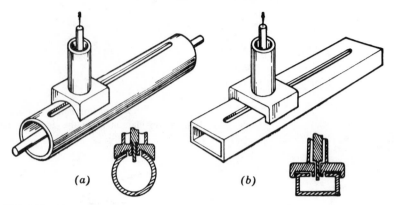

Fig. 4-25.   Slotted section with probe: (a) coaxial line; (b) waveguide (RLS-11, p. 479).

waveguide beyond cutoff, and so does not appreciably radiate energy to the exterior. When making use of the slotted section, one assumes that the slotted line itself may be treated as a lossless transmission line, and that the presence of the probe does not seriously modify the electromagnetic field configurations in the transmission line. The latter is often minimized through the use of a probe shield.

The probe of the slotted section extracts a small fraction of the power flowing in the transmission line, and it is connected to an external circuit containing a rectifier or detector. When the microwave source is unmodulated, the crystal detector may be followed by a dc meter or by a dc amplifier and a meter. A more sensitive system employs a superheterodyne receiver equipped with a crystal mixer, local oscillator, and second detector. When the microwave source is modulated, then a crystal or bolometer followed by a narrow band amplifier may be employed. When working into a low impedance load, a crystal detector produces about 1 μA of rectified current for 1 μW of incident rf power. At this and lower power levels the crystal is a square-law detector with the rectified current proportional to the microwave power.

By moving the probe along the slot, one may measure the power P_{max} and P_{min} at the maximum and minimum positions along the line, and deduce the VSWR from the formula

$$\text{VSWR} = \left(\frac{P_{max}}{P_{min}} \right)^{1/2} = \frac{|A| + |B|}{|A| - |B|} \tag{1}$$

where A and B are the amplitudes of the incident and reflected waves, respectively. The ratio P_{max}/P_{min} is sometimes referred to as the power standing-wave ratio. The magnitude of the reflection coefficient of the load $|\Gamma_L|$ may be deduced from the expression

$$|\Gamma_L| = \frac{|B|}{|A|} = \frac{\text{VSWR} - 1}{\text{VSWR} + 1} \tag{2}$$

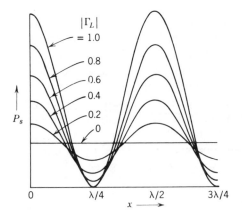

Fig. 4-26. Variation of probe power P_s with position of probe in slotted line. The parameter Γ_L is the load reflection coefficient, and the origin is taken at the point where Γ_L is a maximum (RLS-11, p. 475).

The complex reflection coefficient Γ_L of the load is defined by

$$\Gamma_L = |\Gamma_L| e^{j\theta} \tag{3}$$

and the phase angle θ may be determined by observing the distance $x_{\min}$ of the nearest minimum from the load. One obtains

$$\theta = 2\beta x_{\min} \pm \pi \tag{4}$$

where β is the phase constant (imaginary part of the propagation constant γ). Equation (2) may be reformulated in terms of the load impedance or admittance, $Z_0 = 1/Y_0$

$$\Gamma_L = \frac{Z_L - Z_0}{Z_L + Z_0} = \frac{Y_0 - Y_L}{Y_0 + Y_L} \tag{5}$$

Figure 4-26 gives a plot of the probe power P_S versus probe position x for several load reflection coefficients Γ_L. It is important to determine the phase of the reflection coefficient from the position of a minimum because the maxima are slightly shifted when the probe susceptance is not equal to zero, as is normally the case (see RLS-11, Chap. 8).

L. Quarter-Wavelength Transformer

A transmission line of characteristic impedance Z_1 may be matched to a transmission line of characteristic impedance Z_2 by interposing between them a quarter of a wavelength of a transmission line with the characteristic impedance Z_{12} given by (RLS-9, pp. 90 and 217)

$$Z_{12} = (Z_1 Z_2)^{1/2}. \tag{1}$$

Quarter-wavelength transformers of this type were employed to match the magic T illustrated in RLS-11 (p. 528) and RLS-14 (p. 365). The characteristic impedance Z_0 of a coaxial line given by Eq. (2-D-2) may be employed in designing coaxial quarter-wavelength transformers.

If two waveguides operating in the TE_{10} mode have the same wide dimension a but different small dimensions b_1 and b_2, respectively, they will each have the same characteristic impedance Z_0 [compare Eq. (2-E-32)] if they are filled with the same medium (e.g., air) since Z_0 is independent of b. Nevertheless, when they are joined together reflections will be set up at their interface, and a large VSWR will result. Their junction may be matched by an intervening quarter-wavelength section of length $\lambda_g/4$ and width b given by

$$b = (b_1 b_2)^{1/2} \qquad (2)$$

Thus for the TE_{10} mode, the narrow dimension of the waveguide acts like an effective characteristic impedance. This principle is applied to match the rat race illustrated in Fig. 4-10.

M. Antenna

Microwave antenna theory is discussed at great length in RLS-12. There are a large number of different types of microwave antennas, such as electric and magnetic dipoles, linear arrays, horns, parabolas, and especially shaped antennas. The electric and magnetic dipoles have electromagnetic field amplitudes that vary as r^{-3} (static field), r^{-2} (induction field), and the r^{-1} (radiation field) (RLS-12, p. 93). The radiation field dominates for distances $r \gg \lambda$ away from the dipole, and produces an inverse-square dependence of the radiated power on the distance. At distances less than a wavelength, the other nonradiative field configurations dominate. The existence of these near-field r^{-3} and r^{-2} terms is important in microwave setups in the laboratory, because radiation effects that result from faulty apparatus design or assembly will frequently manifest themselves in the near-field region. For example, if one failed to place the matched load on the magic-T of Fig. 4-11, and then walked past the waveguide opening, the effect on the recorder or oscilloscope display would result from disturbing the near-field electromagnetic configurations emanating from the waveguide. One might have to go to the other side of the room to reach the far-field (radiation-field) region.

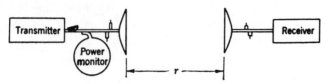

Fig. 4-27. Arrangement of antennas for transmitting microwaves over a short distance r (RLS-11, p. 907).

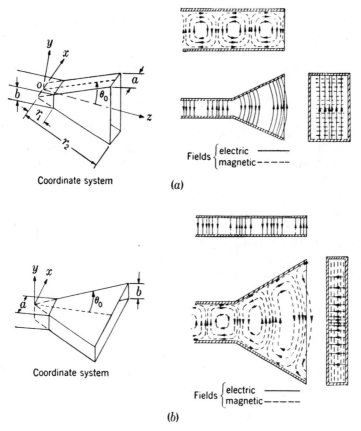

Coordinate system   (a)

Fields $\begin{cases} \text{electric} \text{ ———} \\ \text{magnetic} \text{ -----} \end{cases}$

Coordinate system

Fields $\begin{cases} \text{electric} \text{ ———} \\ \text{magnetic} \text{ -----} \end{cases}$

(b)

Fig 4-28.   Lowest-mode field configurations in sectoral horns: (a) E plane and (b) H plane (RLS-12, p. 351).

Sometimes it is desirable to do an ESR or microwave spectroscopy experiment by arranging two antennas as in Fig. 4-27 and placing the sample in the space between them. In such an experimental arrangement it is convenient to place the magnet and sample between two sectoral horns (RLS-12, Chap. 10; De Grasse et al., 1959; Bottreau and Marzat, 1964) such as those shown in Fig. 4-28. Paraffin or polystyrene lenses may be used to focus the microwaves.

N.   Phase Shifter

The guide wavelength λ_g in a waveguide depends upon the guide dimensions, and for a rectangular waveguide operating in the TE_{10} mode it is given explicitly by

$$\lambda_g = \frac{\lambda}{\left[1 - (\lambda/2a)^2\right]^{1/2}} \tag{1}$$

where λ is the free-space wavelength (for an air-filled waveguide), and a is the wide dimension of the waveguide. If some mechanism is employed to change this dimension, then the result is equivalent to changing the effective length of the waveguide, and the phase of the wave configuration in the waveguide system will be altered. Physically this means that the positions of the maxima and minima in the standing wave pattern will be shifted, and these positions may be measured by means of a slotted line, as discussed in Sec. 4-K.

A phase shifter or squeeze section may be constructed by cutting a slot in the two broad sides of the waveguide and employing a mechanical device to vary the broad waveguide dimension a. Unfortunately, the change in phase is not a linear function of the change in a (see RLS-11, Sec. 8-9).

A phase shifter is useful in an ESR low power bridge. In this application it may be placed between the circulator (or magic T) and sample cavity for adjusting the phase of the microwaves incident on the resonant cavity. This will have the effect of producing absorption, dispersion, or a combination of both. In other words, it permits one to observe either the real part χ' or the imaginary part χ'' of the magnetic susceptibility.

O. Flexible Coupling and Rotary Joint

It is axiomatic that coaxial cables are flexible and waveguides are rigid. Sometimes one requires a nonrigid waveguide connection, and for this purpose one may employ flexible waveguides (RLS-9, p. 287; RLS-11, p. 244). Flexible waveguides may be constructed from a spirally wrapped metal or from a series of choke and flange joints, called vertebrae, held in a rubber envelope. One ESR application for flexible waveguides is to connect a waveguide network to a resonant cavity suspended in a Dewar flask under close tolerances. This prevents one from tightening waveguide screws in a way that will strain or fracture the glassware.

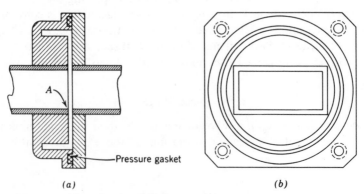

(a)   (b)

Fig. 4-29.   Choke flange waveguide coupling: (a) profile; (b) anterior (RLS-9, p. 194; RLS-11, p. 14).

TABLE 4-2

Standard Rectangular Waveguides and Couplings (RLS-11, p. 15)

RMA Designation	Waveguide Army–Navy Type No.	OD, in. (1 in. = 2.54 cm)	Wall, in.	Wavelength Band, cm	Wavelength for P_{max} and Loss, cm	P_{max}, MW	Loss for Copper, dB/m	Choke Coupling	Flange Coupling	Design Wavelength, cm	Bandwidth for VSWR > 1.05, %
WR 284	RG-48/U	3 × 1.5	0.080	7.1–13.0	10.0	10.5	0.020	UG-54/U	UG-53/U	10.7	±15
		ID2.75 × 0.375	0.049	7.0–12.6	10.0	2.77	0.058	-200/U	-214/U	9.0	±15
WR 187	RG-49/U	2 × 1	0.064	4.8–8.5	6.5	4.86	0.031	-148/U	-149/U		
WR 137	RG-50/U	1.5 × 0.75	0.064	3.6–6.3	5.0	2.29	0.063				
WR 112	RG-51/U	1.25 × 0.625	0.064	2.9–5.1	3.2	1.77	0.072	-52/U	-51/U	3.20	
WR 90	RG-52/U	1.0 × 0.5	0.050	2.3–4.1	3.2	0.99	0.117	-40/U	-39/U	3.20	±6
WR 42	RG-53/U	0.5 × 0.25	0.040	1.07–1.9	1.25	0.223	0.346	-117/U	-116/U	1.25	> ±2
WR 34		0.42 × 0.25	0.040	0.9–1.4							
WR 28	RG-96/U	0.36 × 0.22	0.040	0.75–1.1			0.56	-600/U	-599/U		
WR 22		0.304 × 0.192	0.040	0.6–0.9							
WR 19		0.268 × 0.174	0.040	0.5–0.75							
WR 15		0.228 × 0.154	0.040	0.4–0.6							
WR 12		0.202 × 0.141	0.040	0.33–0.5							
WR 10		0.180 × 0.130	0.040	0.27–0.4							

Rotary joints that permit two transmission lines to rotate relative to each other about their common direction of propagation are important in radar applications for rotating antennas. In ESR single-crystal studies it is useful to be able to rotate a resonant cavity about its axis. This is feasible when the sample is placed in the center of a cylindrical cavity operating in the TE_{011} mode and coupled with its symmetry axis along the waveguide propagation direction. To provide a proper impedance match, it is best to make the waveguide junction with the choke flange combination shown in Fig. 4-29. The cavity coupling may be a circular slot located at the junction, and concentric with the cavity symmetry axis.

P. Flange

In electron spin resonance studies, waveguides are usually coupled together by soldering the ends to flat flanges, and screwing the flanges together in order to obtain a good electrical contact. Satisfactory contact may be ensured by moving one flange back a few mils from the end of the waveguide and applying a greater pressure as shown in RLS-9 (p. 193), but this is normally not resorted to in ESR since contact couplings are ordinarily satisfactory. An ESR bridge cannot be matched sans good waveguide couplings. It is best to silver solder flanges, although soft soldering can also be satisfactory.

A choke flange such as the one shown in Fig. 4-29 is electrically equivalent to a series-branching transmission line one-quarter wavelength long that ends in a short circuit. Therefore a short circuit is reflected to the point A, and it is not necessary to have good electrical contact at the waveguide joint in order to achieve a low standing-wave ratio. The outer groove in the choke flange may be considered as a low-impedance coaxial line. A choke flange is usually threaded, and it is always mated with a flat or cover flange with the screw hole drilled through. Two choke flanges should never be joined together. Other choke flange designs are discussed in RLS-9, Chap. 4. Both choke and waveguide flanges are listed in Table 4-2.

References

Bottreau, A., and C. Marzat, *C. R. Acad. Sci.* **259**, 758 (1964).

DeGrasse, R. W., D. C. Hogg, E. A. Ohm, and H. E. D. Scovil, *J. Appl. Phys.* **30**, 2013 (1959).

Grivet, P., *Microwave Circuits and Amplifiers*, P. W. Hawkes, Trans., Academic Press, New York, 1976.

Helszain, J., *Nonreciprocal Microwave Junctions and Circulators*, Wiley-Interscience, New York, 1975.

Miane, J. L., *C. R. Acad. Sci.* **277B**, 17 (1973).

RLS-8. RLS-9. RLS-10. RLS-11. RLS-14.

Rowen, J. H., *Bell Syst. Tech. J.* **32**, 1333 (1953).

Stanforth, J. A., *Microwave Transmission*, English Univ. Press, 1972.

CHAPTER 5

Resonant Cavities

A resonant cavity is an integral part of almost all electron spin resonance spectrometers, and therefore, the present chapter will be devoted to an extensive study of the theory, design, and use of cavities. Resonant cavities have also been used occasionally in microwave spectrometers in place of the usual waveguide cell.

Some investigators have used helices instead of cavities. Wilmshurst et al. (1962) have compared the sensitivities obtained with spectrometers that employ cavities to those using helices, and their conclusions are discussed in Sec. 11-G. Theoretical discussions of cavity resonators have been given by Heer (1964), Sloan et al. (1964), and Lawson (1965). Some special purpose resonators are described in other parts of this book, such as those for variable temperatures and pressures (Chap. 9) and double resonance (Chap. 14). Most of the cavities to be described were designed for use with solid or perhaps liquid samples, although several have been constructed specifically for use with gases [e.g., Brown and Richardson (1973), de Groot et al. (1973), and Matenaar and Schindler (1973)].

A. Series RLC-Tuned Circuit

A resonant cavity is the microwave analogue of an rf-tuned circuit, and so the latter will be considered first. In Fig. 5-1 we see the series RLC-tuned circuit fed by the voltage V. The differential equation for the voltage V and current I in this circuit is given by

$$V = L\left(\frac{dI}{dt}\right) + RI + \frac{q}{C} \tag{1}$$

where I is the current, and q is the charge on the condenser. For a sinusoidally varying voltage $V = V_m c^{j\omega t}$ we have the steady-state solution

$$I = \left(\frac{V_m}{Z}\right)e^{j(\omega t - \theta)} \tag{2}$$

where $V_m = I_m |Z|$. The impedance Z has the magnitude

$$Z = \left[R^2 + \left(\omega L - \frac{1}{\omega C}\right)^2\right]^{1/2} \tag{3}$$

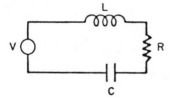

Fig. 5-1.   Series RLC circuit.

and the phase angle θ is

$$\theta = \tan^{-1}\left[\frac{(\omega L - 1/\omega C)}{R}\right] \tag{4}$$

At very low frequencies we may approximate

$$V \cong -\frac{jI}{\omega C} \tag{5}$$

and at very high frequencies

$$V \cong j\omega LI \tag{6}$$

In either case the current is very small and leads or lags the voltage by 90°. At the resonant frequency f_0

$$\omega_0 = 2\pi f_0 = \frac{1}{(LC)^{1/2}} \tag{7}$$

we have

$$V = RI \tag{8}$$

so that the current flow reaches a maximum and is in phase with the voltage. The quality factor Q is defined as

$$Q = \frac{\omega_0 L}{R} = \frac{1}{R\omega_0 C} \tag{9}$$

and this Q is very important in the theory of resonant cavities. Using the relation $V = ZI$ and the approximation $(\omega + \omega_0)/\omega \sim 2$, which is valid for a high Q, we can write

$$\frac{Z}{R} = 1 + 2jQ\frac{(\omega - \omega_0)}{\omega_0} \tag{10}$$

For ω close to ω_0 we let

$$Q = \frac{\omega_0}{\Delta\omega} \tag{11}$$

and

$$\frac{Z}{R} = 1 + J\left[\frac{(\omega - \omega_0)}{\frac{1}{2}\Delta\omega}\right] \tag{12}$$

Physically, $\frac{1}{2}\Delta\omega$ is the value of $\omega - \omega_0$ that makes the real part of the impedance equal to the imaginary part. At this value the current flow through the circuit of Fig. 5-1 falls to $1/2^{1/2}$ of its value at resonance, and the power dissipated (I^2R) is one half the value at resonance. Another important way of defining Q is

$$Q = \frac{2\pi(\text{energy stored})}{\text{energy dissipated per cycle}} = 2\pi\left[\frac{\frac{1}{2}LI_m^2}{RI_m^2/2f}\right] \tag{13}$$

When the energy source of a resonator is turned off the stored energy U decays exponentionally from the initial value U_0 according to the relation

$$U = U_0 e^{-\omega_0 t Q} = U_0 e^{-t\Delta\omega} \tag{14}$$

B. Microwave Resonant Cavity

At microwave frequencies it is not feasible to employ lumped circuit elements such as the RLC circuit just discussed, because the skin effect results in a very high effective resistance in ordinary copper wires, and the dimensions of the circuit elements become comparable to the wavelength, which causes them to lose energy by radiation. A microwave resonant cavity is a box fabricated from high-conductivity metal with dimensions comparable to the wavelength. At resonance, the cavity is capable of sustaining microwave oscillations, which form an interference pattern (standing wave configuration) from superposed microwaves mutliply reflected from the cavity walls. Each particular cavity size and shape can sustain oscillations in a number of different standing wave configurations called modes; these are discussed in the next few sections. The modes of rectangular and cylindrical resonant cavities may be derived from the waveguide modes discussed in Chap. 2.

When a waveguide is terminated in its characteristic impedance Z_0, the transverse electric field and transverse magnetic field vectors reach their maxima at the same longitudinal or z position. In other words they are in phase with respect to both space and time, as shown in Figs. 2-5, 2-7, 2-10, 2-14, and 2-15. This maximizes the time-averaged Poynting's vector **P** or vector

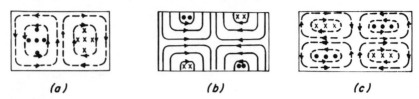

Fig. 5-2. Electromagnetic field patterns in several rectangular and cylindrical cavity resonators. (a) Cyl. TE_{112}; TE_{102}; (b) cyl. TM_{012}; (c) cyl. TE_{012}; rect. TE_{202}. (•) Direction up from the paper; (×) direction down into the paper; (--) H lines of force; (—) E lines of force.

rate of energy flow given by

$$\mathbf{P} = \tfrac{1}{2} E \times \mathbf{H} \tag{1}$$

In a resonant cavity, on the other hand, the electromagnetic field configurations are the result of standing waves with the transverse electric field maximum occurring $\lambda_g/4$ from the transverse magnetic field maximum, so they are in space quadrature. This makes Poynting's vector vanish, and consequently there is no net energy flow, but merely energy storage and dissipation. The wave configurations in Figs. 2-5, 2-7, 2-10, 2-14, and 2-15 may be applied to resonant cavities by displacing the electric field configuration $\lambda_g/4$ along the waveguide so that the right-hand rule is obeyed, as shown in Fig. 5-2 (e.g., by curling the fingers in the direction of circular H field lines of force and having the thumb point in the direction of the enclosed $\mathbf{E}$ vector). In resonant cavities the H field lines of force always form loops that enclose E field lines of force, and the E field lines of force either form loops that enclose H fields, or they terminate on induced surface charges.

Before discussing particular cases it will be helpful to say a few words about the general properties of electromagnetic waves confined in a metallic box. In a high-Q resonator the electric and magnetic fields are 90° out of time phase with each other. When the electric fields are maximum, the magnetic fields are zero, and vice versa. Hence the stored energy in the electric fields U_E

$$U_E = \frac{\epsilon}{2} \int | E_m |^2 \, d\tau \tag{2}$$

equals the stored energy in the magnetic fields U_H

$$U_H = \frac{\mu}{2} \int | H_m |^2 \, d\tau \tag{3}$$

when each is evaluated at the part of the cycle corresponding to the maximum value denoted by the subscript m. The losses in a cavity arise from the dissipation of heat by the surface current density J in the skin effect resistance

R_s, and this ohmic power loss P_L is given by

$$P_L = \frac{R_s}{2} \int |H_{tm}|^2 \, dS \tag{4}$$

where the maximum tangential field H_{tm} along the surface is integrated over all the cavity walls. H_t is numerically equal to and vectorially perpendicular to J, and both H_t and J are parallel to the surface. There may be additional energy dissipation arising from a lossy dielectric such as water, or a paramagnetic sample with a large loss tangent (see Sec. 2-I). In addition, there will be losses from radiation out of a cavity coupling hole. These topics will be elaborated upon later.

When the Q of a resonant cavity arises only from the ohmic losses in the walls it is called the unloaded Q, denoted by the symbol Q_u, and is given by

$$Q_u = \frac{\omega\mu}{R_s} \frac{\int |H_m|^2 \, d\tau}{\int |H_{tm}|^2 \, dS} \tag{5}$$

The overall or loaded Q (denoted by Q_L) may be computed by summing the reciprocals of the Q_ϵ, due to dielectric losses, the Q_r due to the cavity coupling hole, and Q_u, and hence it has the form

$$\frac{1}{Q_L} = \frac{1}{Q_u} + \frac{1}{Q_\epsilon} + \frac{1}{Q_r} \tag{6}$$

where the individual quality factors are defined by Eq. (5), and

$$Q_r = \frac{2\pi(\text{stored energy})}{\text{energy lost through coupling holes per cycle}} \tag{7}$$

$$Q_\epsilon = \frac{2\pi(\text{stored energy})}{\text{energy lost in dielectric per cycle}} = \frac{\mu\int |H_m|^2 \, d\tau}{\int \epsilon'' |E_m|^2 \, d\tau} \tag{8}$$

For the dielectric Q, the numerator is integrated over the regions of the cavity where the imaginary part of the dielectric constant is nonvanishing. At room temperature, Q values usually vary between a few thousand and 50,000, while superconducting cavities give Q's of many million (Maxwell, 1964, a review; Verkin et al. 1964; Viet 1964). For example, Wilson (1963) observed a Q of 2×10^8 in a lead-plated TE_{011} mode cylindrical cavity at 2856 Mc and 1.75° K.

Methods of measuring the Q of a microwave cavity resonator are given by RLS-11, Alaeva and Karasev (1961), and Teodoresku (1962). A microwave pulse experiences a time delay in traversing a transmission cavity (Mungall and Morris, 1960; see also Ruthberg, 1958).

The next three sections will be devoted to a detailed discussion of rectangular, cylindrical, and coaxial resonant cavities, respectively. Culshaw (1961), Boyd et al. (1961, 1962), Wagner and Birnbaum (1961), Lichtenstein et al. (1963), Ulrich et al. (1963), Krupnov and Skvortsov (1964), Strauch et al. (1964), and Lotsch (1965) discuss the use of a millimeter wave Fabry–Perot interferometer as a resonator. Many other shapes exist including a type without side walls (Szulkin 1960; Vainstein, 1963), a H-shaped resonator (Patrushev, 1956; Dunn et al., 1956), one with nonorthogonal boundaries (Ledinegg and Urgan, 1955), a millimeter wave disk resonator (Barchukov and Prokhorov, 1961), and the "echo box" type with dimensions much greater than the wavelength [see, e.g., Kahn et al. (1964), Meyer et al. (1960].

A number of recent articles treat the interaction between a plasma and a resonant cavity [see, e.g., Agdur and Enander (1962); Thomassen (1963)]. Zimmerer (1962) describes a 50- to 75-Gc wavemeter that uses a confocal resonator, and Reichert and Townsend (1965) discuss the line resonator (compare Sec. 5-P). Resonant cavities designed for use at high and low temperatures are discussed in Chap. 9, and those that are suitable for irradiation studies are described in Chap. 10. A dual cavity designed to hold both the sample under study and a standard sample is described in Secs. 5-M and 11-K. Vetter and Thompson (1962) designed a servocontrolled cavity to follow frequency changes in another cavity, and Erickson (1966) sweeps the ESR frequency by moving the cavity wall with a clock motor. Biomodal cavities are discussed in Sec. 5-K.

Lewis and Carver (1964) studied electron spin transmission through a 30-μm thick lithium sample using two rectangular TE_{101} transmission cavities in series. The reflected and transmitted ESR signals differed considerably, and this was attributed to spin diffusion.

In the magnetic resonance literature it is customary to denote the microwave or rf magnetic field at the sample by H_1, and we shall often adhere to this convention. Thus, we shall talk about the average value of $H_1^2 = \langle H_1^2 \rangle_c$ within the cavity, and the average value of $H^2 = \langle H^2 \rangle_w$ in the waveguide outside the cavity. This is a somewhat inconsistent notation, and the symbol H_1 would not be employed in this manner if usage did not sanction it.

C. Rectangular Resonant Cavities

Rectangular resonant cavities can support both TE_{mnp} modes and TM_{mnp} modes, where the subscripts m, n, and p are the number of half-wavelength variations in the standing-wave pattern in the x, y, and z directions, respectively, as defined in Fig. 5-3. These modes are derived from the TE_{mn} and TM_{mn} waveguide modes, respectively, by making the resonant cavity $p/2$ guide wavelengths long. The cavity dimensions are a, b, and d in the x, y and z directions, with the propagation constant k:

$$k = j\beta = \left(k_x^2 + k_y^2 + k_z^2\right)^{1/2} = \left(k_c^2 + k_z^2\right)^{1/2} \tag{1}$$

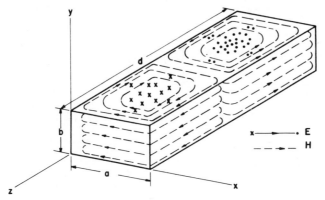

Fig. 5-3. Electromagnetic field configurations in a TE_{102} mode rectangular resonant cavity of dimensions a, b, and d.

which has the components

$$k_x = \frac{m\pi}{a}, k_y = \frac{n\pi}{b}, k_z = \frac{p\pi}{d} \qquad (2)$$

The resonant frequency $2\pi f_0 = \omega_0$ is given by

$$\omega_0(\mu\epsilon)^{1/2} = k = \frac{2\pi}{\lambda} = \pi\left(\frac{m^2}{a^2} + \frac{n^2}{b^2} + \frac{p^2}{d^2}\right)^{1/2} \qquad (3)$$

where λ is the free-space wavelength. The guide wavelength λ_g and cut-off wavelength λ_c are related to λ through Eq. (2-E-21):

$$\frac{1}{\lambda^2} = \frac{1}{\lambda_g^2} + \frac{1}{\lambda_c^2} \qquad (4)$$

where

$$\frac{1}{\lambda} = \frac{k}{2\pi} \qquad (5)$$

$$\frac{1}{\lambda_g} = \frac{k_z}{2\pi} \qquad (6)$$

$$\frac{1}{\lambda_c} = \frac{k_c}{2\pi} = \frac{\left(k_x^2 + k_y^2\right)^{1/2}}{2\pi} \qquad (7)$$

Keeping in mind the quantities defined in Eqs. (1)–(3) we may write down the wave configuration for the general TE_{mnp} mode with $p > 0$ and either m or

$n > 0$.

$$H_x = -H_0\left(\frac{k_x k_z}{k_x^2 + k_y^2}\right)\sin k_x x \cos k_y y \cos k_z z \tag{8}$$

$$H_y = -H_0\left(\frac{k_y k_z}{k_x^2 + k_y^2}\right)\cos k_x x \sin k_y y \cos k_z z \tag{9}$$

$$H_z = H_0 \cos k_x x \cos k_y y \sin k_z z \tag{10}$$

$$E_x = jH_0\left(\frac{\mu}{\epsilon}\right)^{1/2}\left(\frac{kk_y}{k_x^2 + k_y^2}\right)\cos k_x x \sin k_y y \sin k_z z \tag{11}$$

$$E_y = -jH_0\left(\frac{\mu}{\epsilon}\right)^{1/2}\left(\frac{kk_x}{k_x^2 + k_y^2}\right)\sin k_x x \cos k_y y \sin k_z z \tag{12}$$

$$E_z = 0 \tag{13}$$

The general TM_{mnp} mode with $m > 0$, $n > 0$ is

$$E_x = -H_0\left(\frac{\mu}{\epsilon}\right)^{1/2}\left(\frac{k_x k_z}{k_x^2 + k_y^2}\right)\cos k_x x \sin k_y y \sin k_z z \tag{14}$$

$$E_y = -H_0\left(\frac{\mu}{\epsilon}\right)^{1/2}\left(\frac{k_y k_z}{k_x^2 + k_y^2}\right)\sin k_x x \cos k_y y \sin k_z z \tag{15}$$

$$E_z = H_0\left(\frac{\mu}{\epsilon}\right)^{1/2}\sin k_x x \sin k_y y \cos k_z z \tag{16}$$

$$H_x = jH_0\left(\frac{kk_y}{k_x^2 + k_y^2}\right)\sin k_x x \cos k_y y \cos k_z z \tag{17}$$

$$H_y = -jH_0\left(\frac{kk_x}{k_x^2 + k_y^2}\right)\cos k_x x \sin k_y y \cos k_z z \tag{18}$$

$$H_z = 0 \tag{19}$$

TE_{mnp} and TM_{mnp} modes of the same order are degenerate because they have identical frequencies, and other cases of so-called accidental degeneracy may occur for certain ratios of the cavity dimensions. To increase the frequency in a resonant cavity of a given size, it is necessary to fit additional half-waves in one or more dimensions, which means that the order of the mode must increase. If the cavity dimensions are decreased and the mode type is unchanged, then the frequency increases.

The dimensionless quantity $Q_u \delta / \lambda$ for TE modes with m and $n > 0$ is (RSL-11, p. 296)

$$Q_u \frac{\delta}{\lambda} = \frac{(abd/4\pi)(k_x^2 + k_y^2)(k_x^2 + k_y^2 + k_z^2)^{3/2}}{ad\left[k_x^2 k_z^2 + (k_x^2 + k_y^2)^2\right] + bd\left[k_y^2 k_z^2 + (k_x^2 + k_y^2)^2\right] + abk_z^2(k_x^2 + k_y^2)}$$

(20)

while for $m = 0$, it is

$$Q_u \frac{\delta}{\lambda} = \frac{(abd/2\pi)(k_y^2 + k_z^2)^{3/2}}{k_y^2 d(b + 2a) + k_z^2 b(d + 2a)}$$

(21)

and for $n = 0$

$$Q_u \frac{\delta}{\lambda} = \frac{(abd/2\pi)(k_x^2 + k_z^2)^{3/2}}{k_x^2 d(a + 2b) + k_z^2 a(d + 2b)}$$

(22)

For TM modes and $p > 0$ we have

$$Q_u \frac{\delta}{\lambda} = \frac{(abd/4\pi)(k_x^2 + k_y^2)(k_x^2 + k_y^2 + k_z^2)^{3/2}}{k_x^2 b(a + d) + k_y^2 a(b + d)}$$

(23)

while when $p = 0$ the quality factor for TM modes is given by

$$Q_u \frac{\delta}{\lambda} = \frac{(abd/2\pi)(k_x^2 + k_y^2)^{3/2}}{k_x^2 b(a + 2d) + k_y^2 a(b + 2d)}$$

(24)

When using these formulas it should be recalled that the skin depth δ is defined in Sec. 2-J by

$$\delta = \left(\frac{2}{\omega \mu \sigma}\right)^{1/2}$$

(25)

Figure 2-17 shows the dependence of δ on ω for several materials.

The TE_{102} mode (sometimes called TE_{012}) is the dominant mode, and is also the most important one used in ESR spectrometers [see Heuer (1965)]. If we set

$m = 1$, $n = 0$, and $p = 2$ in eqs. (8)–(12) we obtain for the only nonvanishing field components

$$H_x = \frac{H_0}{\left[1 + (d/2a)^2\right]^{1/2}} \sin \frac{\pi x}{a} \cos \frac{2\pi z}{d} \tag{26}$$

$$H_z = \frac{-H_0}{\left[1 + (2a/d)^2\right]^{1/2}} \cos \frac{\pi x}{a} \sin \frac{2\pi z}{d} \tag{27}$$

$$E_y = j\left(\frac{\mu}{\epsilon}\right)^{1/2} H_0 \sin \frac{\pi x}{a} \sin \frac{2\pi z}{d} \tag{28}$$

where $k_y = 0$, and the quantity H_0 is slightly redefined to make the equations appear more symmetical. Essentially the same mode configurations may be obtained by letting $p = 0$ and then interchanging the roles of y and z in eqs. (14)–(18). The coefficients of these wave configurations [eqs. (26)–(28)] are related through the characteristic impedance Z_{TE} of the corresponding waveguide with width a:

$$Z_{TE} = \frac{(\mu/\epsilon)^{1/2}}{\left[1 - (f_c/f)^2\right]^{1/2}} = \left(\frac{\mu}{\epsilon}\right)^{1/2}\left[1 + \left(\frac{d}{2a}\right)^2\right]^{1/2}$$

$$= \frac{(\mu/\epsilon)^{1/2}\lambda_g}{\lambda} \tag{29}$$

The resonant frequency f is independent of the (narrow) b dimension of the cavity, although, of course, the unloaded Q is a function of all of the three cavity dimensions. The lack of dependence of b on f means that the resonant cavity may be made as narrow as one wishes to accommodate a small magnet gap. This is not possible with cylindrical cavities, so in general a cylindrical cavity requires a considerably wider gap. The b dimension may be increased to accommodate variable temperature apparatus within the cavity.

For a TE_{102} mode, the unloaded Q has the form

$$Q_u \frac{\delta}{\lambda} = \frac{4b\left[a^2 + (\frac{1}{2}d)^2\right]^{3/2}}{d^3(a + 2b) + 4a^3(d + 2b)} \tag{30}$$

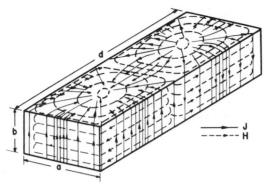

Fig. 5-4.   Current distribution J in a TE_{102} mode rectangular resonant cavity with dimensions a, b, and d.

For the TE_{011} mode, the unloaded Q is

$$Q_u \frac{\delta}{\lambda} = \frac{1}{2} \frac{b(a^2 + d^2)^{3/2}}{d^3(a + 2b) + a^3(d + 2b)} \tag{31}$$

and for a square cavity ($a = d$), this simplifies to

$$Q_u \frac{\delta}{\lambda} = \frac{1}{2^{1/2}} \frac{b}{a + 2b} \tag{32}$$

Figure 5-3 gives the configurations of the electric and magnetic fields in the TE_{102} mode, and Fig. 5-4 shows the current flow in the cavity walls. The electric field begins and ends on induced charges on the broad face of the cavity, and the magnetic field H_t tangential to the cavity walls induces currents in the walls perpendicular to the H_t direction, as shown in Fig. 5-4. The circuit for the electric current flow is completed by the displacement current that flows through the centers of the magnetic field loops, and induces charges on the wide surface (xz plane) near the centers of these loops. These charges reverse polarity every half-cycle, and the ohmic losses result from electrons rushing back and forth through the cavity walls every half-cycle to build them up.

It is desirable to design the sample cavity in such a way that (1) the rf magnetic field H_1 in the sample is perpendicular to the applied steady magnetic field H_0, (2) the sample is located at a point of maximum H_1, and (3) it is placed at a position of minimum E. The first requirement arises from the nature of the resonance condition for allowed transitions; the second requirement is because below saturation the amount of rf energy absorbed by the sample is proportional to H_1^2, and the greater the H_1, the greater the signal-to-noise ratio, while the third requirement minimizes the dielectric power loss,

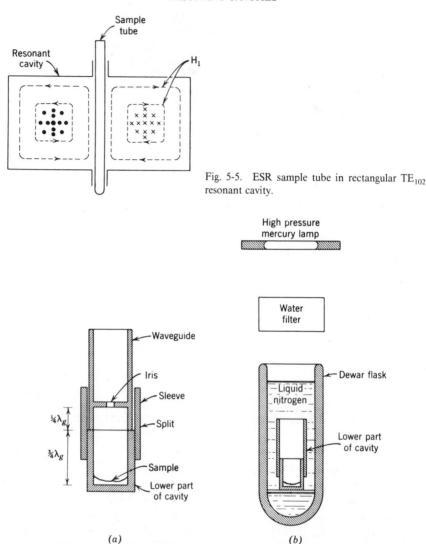

Fig. 5-5.   ESR sample tube in rectangular TE$_{102}$ resonant cavity.

Fig. 5-6.   Split rectangular cavity and irradiation technique. The right-hand illustration shows the cavity disassembled from the waveguide. (*a*) Resonant cavity assembled; (*b*) resonant cavity irradiated (Poole and Anderson, 1959).

which has a deleterious effect on the signal-to-noise ratio. These three requirements are met by the arrangement shown in Fig. 5-5, when the constant magnetic field is in the *y* direction of Fig. 5-3. Such a cavity may be provided with a mount for orienting single crystal samples and for variable temperature studies (see Chap. 9). Longer cavities are sometimes used with aqueous and other lossy samples (Estin, 1962; Wilmshurst, 1963; Stoodley, 1963).

A knowledge of mode configurations and current flow in resonant cavities is particularly useful for designing special purpose resonators without appreciably decreasing their Q. For example, it is possible to cut slots in the bottom of the rectangular TE_{10p} cavity parallel to the directions of current flow shown in Fig. 5-4, and these slots may be used for uv irradiation. In another arrangement a split cavity may be used as shown in Fig. 5-6 (Poole and Anderson, 1959), where it is not necessary to achieve good electrical contact along the split when the cavity is assembled.

D. Cylindrical Resonant Cavities

In analogy to rectangular resonant cavities, cylindrical cavities can support both TE_{mnp} and TM_{mnp} modes, where the subscripts m, n, and p refer to the number of half-cycle variations in the angular (ϕ), radial (r), and longitudinal (z) directions, respectively. These modes may be derived from the TE_{mn} and TM_{mn} cylindrical waveguide modes discussed in Chap. 2.

Let the resonant cavity radius be a, and let the length be d. In analogy to the rectangular cavity case, we have

$$k = j\beta = \omega_0(\mu\epsilon)^{1/2} = (k_c^2 + k_z^2)^{1/2} = \left[\frac{(k_c a)_{mn}^2}{a^2} + \left(\frac{p\pi}{d}\right)^2\right]^{1/2} \quad (1)$$

where $(k_c a)_{mn}$ is a Bessel function root, since cylindrical waveguide modes have Besseloid radial variations and $k_z = p\pi/d$. For transverse magnetic waves TM_{mnp}, the quantity $(k_c a)_{mn}$ is the nth root of the mth-order Bessel function $J_m(k_c r)$

$$J_m(k_c a) = 0 \quad (2)$$

For transverse electric modes the first derivative of the Bessel function $J_m(k_c r)$ must vanish at the surface

$$\frac{d}{d(k_c r)} J_m(k_c r)\big|_{r=a} = J_m'(k_c a) = 0 \quad (3)$$

and the nth root of this equation for the m-order Bessel function is denoted by $(k_c a)_{mn}'$. The most useful Bessel roots $(k_c a)_{mn}'$ and $(k_c a)_{mn}$ are listed in columns 2 and 4, respectively, of Table 5-1. For TE modes, $(k_c a)_{mn}'$ replaces $(k_c a)_{mn}$ in Eq. (1).

TABLE 5-1

The nth Roots of the mth-Order Bessel Functions $J'_m(k_c a)$ and $J_m(k_c a)$, and the Corresponding TE_{mnp} and TM_{mnp} Cylindrical Cavity Modes (RLS-11, p. 299)

TE_{mnp} Mode	nth Root of $J'_m(k_c a)$	TM_{mnp} Mode	nth Root of $J_m(k_c a)$
$11p$	1.841	$01p$	2.405
$21p$	3.054	$11p$	3.832
$01p$	3.832	$21p$	5.136
$31p$	4.201	$02p$	5.520
$41p$	5.318	$31p$	6.380
$12p$	5.332	$12p$	7.016
$51p$	6.415	$41p$	7.588
$22p$	6.706	$22p$	8.417
$02p$	7.016	$03p$	8.654
$61p$	7.501	$51p$	8.772
$32p$	8.016	$32p$	9.761
$13p$	8.536	$61p$	9.936
$71p$	8.578	$13p$	10.174
$42p$	9.283		
$81p$	9.648		
$23p$	9.970		
$03p$	10.174		

The mode configurations for the general cylindrical TE_{mnp} mode with $n > 0$ and $p > 0$ are

$$H_r = \frac{k_z H_0}{\left(k_c^2 + k_z^2\right)^{1/2}} J'_m(k_c r)\cos m\phi \cos k_z z \tag{4}$$

$$H_\phi = -\frac{mk_z H_0}{\left(k_c^2 + k_z^2\right)^{1/2}} \frac{J_m(k_c r)}{k_c r} \sin m\phi \cos k_z z \tag{5}$$

$$H_z = \frac{k_c H_0}{\left(k_c^2 + k_z^2\right)^{1/2}} J_m(k_c r)\cos m\phi \sin k_z z \tag{6}$$

$$E_r = -m\left(\frac{\mu}{\epsilon}\right)^{1/2} H_0 \left[\frac{J_m(k_c r)}{k_c r}\right] \sin m\phi \sin k_z z \tag{7}$$

$$E_\phi = -\left(\frac{\mu}{\epsilon}\right)^{1/2} H_0 J'_m(k_c r)\cos m\phi \sin k_z z \tag{8}$$

$$E_z = 0 \tag{9}$$

The mode configurations for the TE_{011} mode are given in Sec. 5-H. For the

TM_{mnp} modes with $n > 0$ we have

$$E_r = -\frac{(\mu/\epsilon)^{1/2} H_0 k_z}{(k_c^2 + k_z^2)^{1/2}} J_m'(k_c r) \cos m\phi \sin k_z z \tag{10}$$

$$E_\phi = \frac{(\mu/\epsilon)^{1/2} H_0 m k_z}{(k_c^2 + k_z^2)^{1/2}} \frac{J_m(k_c r)}{k_c r} \sin m\phi \sin k_z z \tag{11}$$

$$E_z = \frac{(\mu/\epsilon)^{1/2} H_0 k_c}{(k_c^2 + k_z^2)^{1/2}} J_m(k_c r) \cos m\phi \cos k_z z \tag{12}$$

$$H_r = -mH_0 \left[\frac{J_m(k_c r)}{k_c r} \right] \sin m\phi \cos k_z z \tag{13}$$

$$H_\phi = -H_0 J_m'(k_c r) \cos m\phi \cos k_z z \tag{14}$$

$$H_z = 0 \tag{15}$$

The dependence of the frequency f on the radius a and the length d of a cylindrical cavity in a vacuum (or air) is easily presented graphically by rearranging Eq. (1) to the form

$$(2af)^2 = \left(\frac{c(k_c a)_{mn}}{\pi} \right)^2 + \left(\frac{cp}{2} \right)^2 \left(\frac{2a}{d} \right)^2 \tag{16}$$

where $c = 1/(\mu_0 \epsilon_0)^{1/2}$ is the velocity of light *in vacuo*. For convenience one plots $(2af)^2$ against $(2a/d)^2$ as shown in Fig. 5-7. In designing a resonant cavity it is best to select values of a and d so that there are no extraneous modes that have resonant frequencies near the design point. In TE modes, $(k_c a)_{mn}'$ replaces $(k_c a)_{mn}$ in Eq. (16).

The Q factor for the TE_{mnp} cylindrical mode is (RLS-11, Sec. 5-5)

$$Q_u \frac{\delta}{\lambda} = \frac{(1/2\pi)[1 - m/(k_c a)_{mn}'^2][(k_c a)_{mn}'^2 + (p\pi a/d)^2]^{2/3}}{(k_c a)_{mn}'^2 + 2(2a/d)(p\pi a/d)^2 + (1 - 2a/d)[mp\pi a/d(k_c a)_{mn}']^2}. \tag{17}$$

and the corresponding formula for the TM_{mnp} mode is

$$Q_u \frac{\delta}{\lambda} = \frac{[(k_c a)_{mn}^2 + (p\pi a/d)^2]^{1/2}}{2\pi(1 + 2a/d)} \tag{18}$$

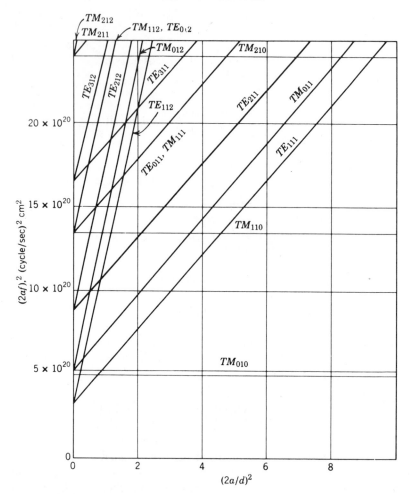

Fig. 5-7. Mode chart for right circular cylinder of radius a, length d, and resonant frequency f (RLS-11, p. 298).

These formulas are plotted in Figs. 5-8 to 5-10. One should note from Fig. 5-8 that a TE_{0np} mode has a maximum value of $Q\delta/\lambda$ when the cavity length d equals the diameter $2a$.

Gruber and Ritzmann (1974) describe a comparative method for determining the Q of a TE_{011} mode cylindrical resonator. Duchiweicz and Francik (1978) present methods for measuring the Q using a typical ESR spectrometer setup.

As Fig. 5-7 indicates, the dominant mode in a cylindrical cavity is the TE_{111} mode. Its configuration corresponds to half of the TE_{112} mode, which is sketched in Fig. 5-11. It is analogous to the TE_{101} rectangular cavity mode, and the two may be derived from one another by deforming the cylinder into a rectangular parallelopiped and vice versa. Since this mode is dominant, it may

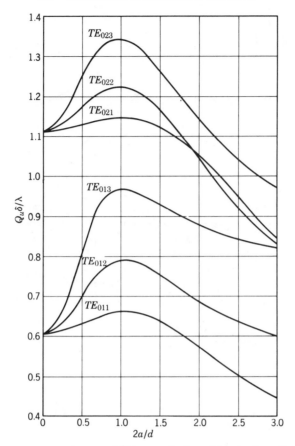

Fig. 5-8. $Q_u \delta / \lambda$ versus $2a/d$ for several TE_{0np} modes in a right circular cylinder (RLS-11, p. 300).

be used in situations where the magnet pole gap is too small to support higher-order modes. Unfortunately, the TE_{111} mode has the lowest Q factor of any transverse electric mode, as Fig. 5-9 indicates. If it is desired to employ a TE_{11p} cylindrical resonator for a sample cavity, the sketch of the TE_{112} mode in Fig. 5-11 may be used as a guide in positioning the sample. Rieckhoff and Weissbach (1962) use the TE_{112} mode for optical studies at liquid helium temperatures.

If one provides a tuning plunger on one end of a TE_{111} cavity, then it is necessary to build into the plunger a quarter-wave choke, which reflects a short circuit to the gap between the end of the plunger and the cavity walls. Such a plunger allows the rf current to flow at the gap, and this current flow is necessary for a high Q.

Most frequency meters (e.g., RLS-11, Rogers et al., 1950; Bussey and Estin, 1960) employ the cylindrical TE_{011} mode because it has a fairly high Q factor, and because there is no rf current flow between the cylinder walls and the end

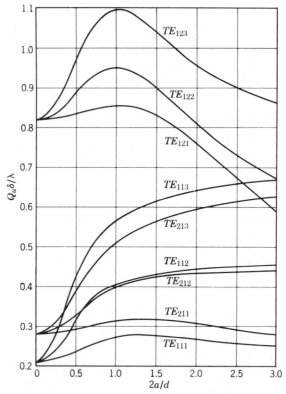

Fig. 5-9. $Q_u \delta/\lambda$ versus $2a/d$ for several TE_{mnp} modes in a right circular cylinder (RLS-11, p. 301).

plate. In fact, there is no electric current flow in either the radial (r) or longitudinal (z) direction, but only in the angular (ϕ) direction. This property enables one to use a pistonlike tuning plunger on the end plate, as shown in Fig. 5-12, so that the frequency may be varied by screwing the end plate in and out. The current flow property renders it unnecessary to achieve good electrical contact between the end plate and the cylinder. If one deliberately leaves the finite gap between the piston edges and the walls as shown in Fig. 5-12, then other modes will be suppressed because they require rf current to flow across this gap. In particular, this will suppress the TM_{111} mode that is degenerate with the TE_{011} mode (see Fig. 5-7). The cylindrical TE_{011} mode is particularly useful for a sample cavity since H_z is very strong along the cavity axis. More efficient mode suppression is obtained by means of a helical line of epoxy on the side wall (Estin, 1962). It is possible to have a very large hole in the end plate (e.g., see Fig. 5-13) without appreciably decreasing the Q. When employed as a sample cavity its plunger may be easily tuned for use at the same frequency with and without a Dewar insert or a quartz linear (Estin, 1962). This cavity may be used with a larger sample than the rectangular TE_{101} cavity,

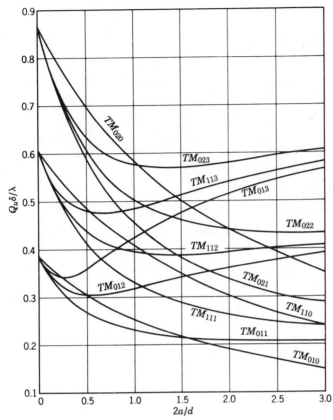

Fig. 5-10. $Q_u\delta/\lambda$ versus $2a/d$ for several TM_{mnp} modes in a right circular cylinder (RLS-11, p. 302).

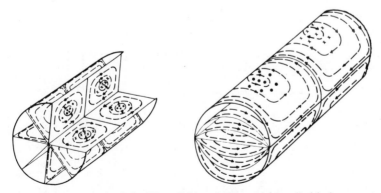

Fig. 5-11. Diagramatic sketch of the TE_{012} (left) and TE_{112} (right) cylindrical resonant cavity modes [adapted from Reintjes and Coate (1952), p. 608 and p. 610].

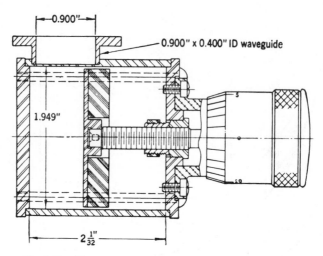

Fig. 5-12. Wavemeter for 10,000-MHz region (RLS-11, p. 323).

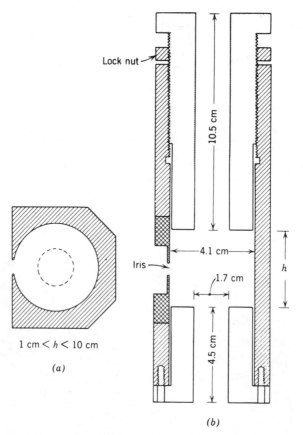

(a)

(b)

Fig. 5-13. Tunable TE_{011} mode cylindrical resonant cavity. This X-band copper cavity was provided with a tunable iris, and a scale plus a vernier for setting the frequency. Spacers may be used with the bottom end plate to properly center the iris. (a) Top view; (b) side view.

and in addition it has a considerably higher Q. Two disadvantages of this cavity are its large size (large magnet gap requirement) and the slight inconvenience of adapting it to 100-kc modulation.

E. Coaxial Resonant Cavities

Coaxial resonators are less important than waveguide resonators for ESR applications, and the reader is referred to other sources (e.g., RLS-11) for background information and design data. The TE_{mnp} and TM_{mnp} modes are labeled with the same convention that is employed in the cylindrical case. The resonant frequencies depend upon Bessel functions of both the first and the second kind. TEM modes also exist here. Reentrant cavities can be employed as coaxial types. Raoult and Fanguin (1960) describe a coaxial resonator that is excited by probes from a circular TE_{11} mode waveguide inside its center conductor, while Rajangam et al. (1965) frequency-modulated a reentrant coaxial cavity. Decorps and Fric (1964) designed a meter wave spectrometer that uses a coaxial cavity. Allendoerfer et al. (1975, 1981) designed a coaxial cavity for simultaneous ESR and electrochemical measurements (compare Sect. 5-P).

F. Coupling to Resonators

The resonant cavity may be coupled to the waveguide by means of a coupling hole or iris, so this section is prefaced by a few words about coupling holes or irides. Consider a waveguide with a generator at one end, a load at the other, and a thin copper sheet placed perpendicular to the waveguide axis blocking

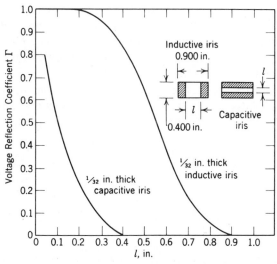

Fig. 5-14.   Reflection coefficient of inductive and capacitive irides in waveguide 1.02×2.29 cm at $\lambda = 3.2$ cm. (RLS-14, p. 52).

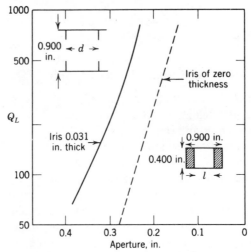

Fig. 5-15. Loaded Q as a function of iris width l for rectangular cavity in 1.02×2.29 cm guide (RLS-9, p. 655).

the passage of microwave power to the load. If a slot is cut in the center of the copper sheet parallel to the long waveguide dimension, a capacitive iris is formed, while if the slot is cut parallel to the short dimension, an inductive iris is formed. The voltage reflection coefficient Γ at the iris is defined as the ratio of the amplitude of the incident **E** vector to the amplitude of the reflected **E** vector. Thus Γ is a measure of the amount of power that fails to pass through the iris and enter a resonant cavity. Figure 5-14 shows the reflection coefficient Γ for inductive and capacitive irides as a function of the slot width. Figure 5-15 shows the effect of iris width on the cavity Q. When the hole thickness increases, the size of the coupling hole must be increased to maintain a constant coupling. The resonant cavity illustrated on Fig. 5-6 was used with a symmetical inductive iris 5.3 mm wide and 0.5 mm thick. Further details on inductive and capacitive irides are found in Sec. 4-1.

Of somewhat wider application is the centered circular iris shown in Fig. 5-16. The relationship between the iris diameter and thickness and its electrical properties (susceptance) are shown in Fig. 5-17 and RLS-14 (p. 53). These data

$$\omega_0 = 1/(LC)^{1/2}$$

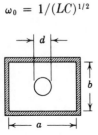

Fig. 5-16.   Centered circular iris.

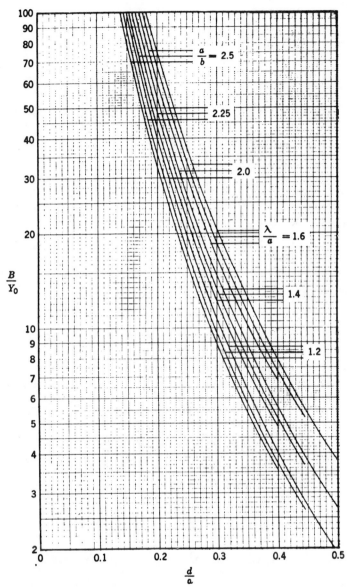

Fig. 5-17. The relative susceptible B/Y_0 of a centered circular iris of zero thickness, where d is the iris diameter, a and b are the waveguide dimensions defined in Fig. 5-16, and λ is the free-space wavelength (RLS-10, p. 240).

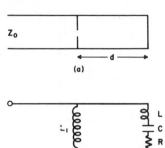

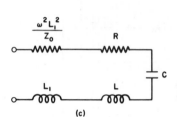

Fig. 5-18. (*a*) A waveguide reflection cavity and iris; (*b*) its equivalent circuit; (*c*) an alternative equivalent circuit (see RLS-8, p. 232; see also, Fig. 5-2).

are useful for designing coupling holes according to the procedure outlined in RLS-8, Sec. 7.10.

An equivalent circuit for the centered circular iris is shown in Fig. 5-18, where the reflection (reaction) resonant cavity near its resonant frequency ω_0 is represented by a series RLC circuit with the usual conditions

$$\omega_0 = (LC)^{-1/2} \tag{1}$$

and

$$Q = \frac{\omega_0 L}{R} = \frac{1}{R\omega_0 C} \tag{2}$$

and Z_0 is the characteristic impedance of the waveguide given by

$$Z_0 = \frac{1}{Y_0} = \left(\frac{\mu}{\epsilon}\right)^{1/2}\left[1 - \left(\frac{f_c}{f}\right)^2\right]^{-1/2} = \frac{120\pi}{\left[1 - (f_c/f)^2\right]^{1/2}} \tag{3}$$

for TE modes. Casini et al. (1976) used an elliptical iris.

The cavity loss resistance R may be evaluated by assuming that all the cavity ohmic losses are in the cavity side walls, so the input impedance of the cavity

Z_s in the absence of the iris would be (RLS-8, Sec. 7-10)

$$Z_s = Z_0 \tanh(\alpha + j\beta)d \tag{4}$$

$$= \frac{\alpha d + j \tan \beta d}{1 + j\alpha d \tan \beta d} \tag{5}$$

$$\approx Z_0(\alpha d + j \tan \beta d) \tag{6}$$

since $\beta d \approx n\pi$, where n is an integer and both $\alpha d \ll 1$ and $|\tan \beta d| \ll 1$. Near resonance we may write

$$\tan \beta d \approx \tan \pi \left(1 - \frac{\Delta\lambda_g}{\lambda_g}\right) \tag{7}$$

$$\approx -\pi \frac{\Delta\lambda_g}{\lambda_g} \tag{8}$$

$$\approx \pi \left(\frac{\lambda_g^2}{\lambda^2}\right) \frac{(\omega - \omega_0)}{\omega_0} \tag{9}$$

and Eq. (6) becomes

$$Z_s = Z_0 \left[\alpha d + j\left(\frac{\pi\lambda_g^2}{\lambda^2}\right) \frac{(\omega - \omega_0)}{\omega_0}\right] \tag{10}$$

which is the same form as a series RLC circuit input impedance from Eq. (5-A-10)

$$Z_s = R\left[1 + \frac{2jQ(\omega - \omega_0)}{\omega_0}\right] \tag{11}$$

Therefore

$$R = Z_0 \alpha d \tag{12}$$

the unloaded Q is given by

$$Q_u = \left(\frac{\pi}{2\alpha d}\right)\left(\frac{\lambda_g}{\lambda}\right)^2 \tag{13}$$

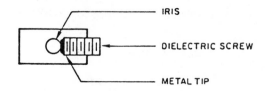

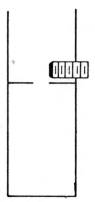

Fig. 5-19.   Method of varying cavity coupling to waveguide. (Chang et al., 1978).

and

$$L = \left(\frac{\pi}{\omega_0}\right)\left(\frac{Z_0}{2}\right)\left(\frac{\lambda_g}{\lambda}\right)^2 \tag{14}$$

From the equivalent circuit in Fig. 5-18 the radiation Q is

$$Q_r = \left(\frac{Z_0}{\omega_0 L_1}\right)\frac{(L + L_1)}{L_1} \tag{15}$$

$$= \left(\frac{Z_0^2}{\omega_0^2 L_1^2}\right)\left(\frac{\pi}{2}\right)\left(\frac{\lambda_g}{\lambda}\right)^2 \tag{16}$$

using the approximating $L_1 \ll L$. For the case of critical coupling when the resonant cavity is perfectly matched to the waveguide and the voltage standing-wave ratio is unity, one has

$$Q_r = Q_u \tag{17}$$

and

$$\alpha d \approx \left(\frac{\omega_0 L_1}{Z_0} \right)^2 = \left(\frac{Y_0}{B} \right)^2 \qquad (18)$$

where $Y_0 = 1/Z_0$ is the characteristic admittance, and $B = 1/\omega L_1$ is the susceptance.

A cavity $\lambda_g/2$ long constructed from X band ($ab = 2.25 \times 1$ cm) waveguide at $\lambda = 3.2$ cm has $\alpha d \approx 3 \times 10^{-4}$ Np (using α from Table 4-1, so the iris susceptance $B/Y_0 \approx 55$ is required for critical coupling. From Fig. 5-17 this corresponds to an iris diameter $d \sim 0.2a$ giving a hole diameter of 0.45 cm for a zero thickness iris. The apparent susceptance of this hole in a nonzero thickness iris will be greater as shown in RLS-14 (p. 53), so that a somewhat greater hole size is needed in practice. Since ohmic losses in the cavity ends were ignored, their inclusion tends to require an even larger hole, and a nominal diameter for a rectangular cavity is about 0.6 cm.

The coupling constant of the iris may be made variable by placing a capacitive tuning screw immediately outside the iris. The behavior of such a

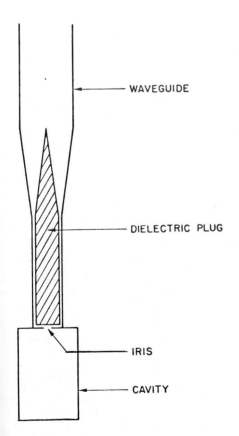

WAVEGUIDE

DIELECTRIC PLUG

IRIS

CAVITY

Fig. 5-20. Gordon dielectric plunger coupling scheme (Chang et al., 1978).

screw is discussed in Sec. 4-I. Here it will suffice to mention that for small insertions the screw acts like a capacitance in parallel with the equivalent inductance of an inductive or circular iris, and by varying its insertion a wide range of coupling constants may be obtained. Another way to vary the coupling constant is to gradually insert a piece of dielectric such as Teflon in front of the iris. The Teflon tuner may have a metal tip and be inserted perpendicular to the narrow wall of the waveguide, as shown in Fig. 5-19 (Chang et al. 1978), and this allows the iris of a rectangular TE_{101} cavity to be tuned conveniently while the cavity remains in the magnetic field. It does not provide as wide a range of tuning conditions as the screw arrangement.

The Gordon (1961) coupler (Chang et al. 1978) employs a segment of waveguide beyond cutoff filled with a moveable dielectric plunger, as illustrated in Fig. 5-20. The distance between the iris and the plunger determines

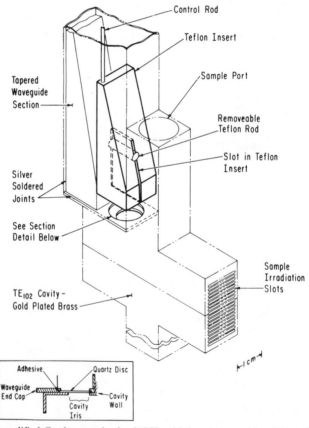

Fig. 5-21. A modified Gordon coupler for 9 GHz with low-noise and low-drift characteristics. It has a quartz disk seal to exclude liquid helium from the waveguide. The cavity is held in place by a spring-loaded Teflon stirrup. A quartz cavity replaces the metal one for ENDOR experiments (Isaacson, 1976).

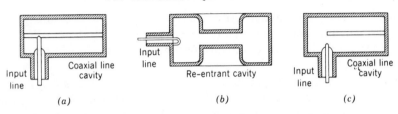

Fig. 5-22.   Methods of connecting a coaxial line to a cavity resonator. (*a*) Junction coupling; (*b*) loop coupling; (*c*) probe coubling (Reintjes and Coate, 1952, p. 668).

Fig. 5-23.   Loop-coupled rectangular cavity (RLS-14, p. 26).

the coupling. The modified Gordon coupler (Isaacson 1976) illustrated in Fig. 5-21 provides stable, low noise operation at liquid-helium temperatures. Isaacson et al. (1980) designed a coaxial coupler for an X-band ESR–ENDOR system operating between 1.4 and 300 K. Variable coupling to resonant cavities is discussed by Gould and Cunliffe (1956), Gordon (1961), Faulkner and Holman (1963), Ranon and Stamires (1970), and Whitehouse (1978) (see also Ager et al., 1963).

Goldberg (1978) notes that high sensitivity at successively lower power levels should be achieved by keeping the iris at a high sensitivity setting, close to critical coupling, and supplying the detector with bucking power of in-phase microwave radiation directly from the klystron. The alternative practice of opening the cavity iris at low power to maintain a constant crystal current has the effect of mismatching the cavity, lowering the power density at the sample, and decreasing the sensitivity.

Two single crystals of Al_2O_3 may be used to produce simultaneously a high pressure seal and an impedance match to a high Q cavity resonator (Lawson and Smith, 1959). The resonator supports pressures up to 10^4 bars (see Sec. 9-D).

To connect a coaxial line to a resonant cavity several types of coupling devices may be used, such as loops, probes, and irides. Some specific coupling arrangements are illustrated in Figs. 5-22 and 5-23.

Further information on irides will be found in Sec. 4-I.

G. Radiation Quality Factor

A resonant cavity without any connecting aperture to the outside world is characterized by its unloaded quality factor Q_u. To be useful it is necessary to connect a cavity to a waveguide by means of an iris or other coupling device, and this entails a lowering of the Q. The extent to which the Q is lowered is characterized by its radiation quality factor Q_r, and so Q_r will be discussed next.

Sometimes it is convenient to represent the circuit of a coupling hole by a transformer with turns ratio n, and the resulting equivalent circuits are shown in Figs. 5-24a and 5-24c for a reaction (reflection) and transmission cavity, respectively. For the reflection cavity circuit, the radiation Q is given by

$$Q_r = \frac{L\omega}{R_G n^2} \tag{1}$$

where R_G includes the characteristic impedance of the waveguide plus the generator impedance. The unloaded Q, including ohmic losses in the cavity walls and dielectric losses in the sample, is

$$Q_u = \frac{L\omega}{R_c} \tag{2}$$

The ratio of these Q's equals the coupling parameter β

$$\beta = \frac{Q_\omega}{Q_r} = \frac{R_G n^2}{R_c} \tag{3}$$

The radiation Q represents losses due to power that may be considered as leaving the cavity through the coupling hole to be dissipated in R_G, while the unloaded Q here represents losses due to the cavity alone. The overall or loaded Q_L is given by

$$\frac{1}{Q_L} = \frac{1}{Q_r} + \frac{1}{Q_u} \tag{4}$$

If the cavity is perfectly coupled or matched to the waveguide, then

$$\beta = \text{VSWR} = 1 \tag{5}$$

and at resonance the cavity effectively terminates the transmission line in its characteristic impedance Z_0. For the overcoupled cavity

$$\frac{Q_u}{Q_r} > 1 \tag{6}$$

and the voltage standing-wave ratio VSWR is given by

$$\beta = \text{VSWR} \tag{7}$$

since the transmission line is effectually terminated by a resistance greater than its characteristic impedance Z_0. The inverse is true in the undercoupled case, where

$$\frac{Q_u}{Q_r} < 1 \tag{8}$$

and

$$\frac{1}{\beta} = \text{VSWR} \tag{9}$$

The coupling parameter represents the efficacy with which energy stored in the cavity-coupling system is coupled to the external load, and dissipated there (RLS-11, p. 289).

The transmission cavity with the equivalent circuit shown in Fig. 5-24 (see also Roussy and Felden, 1965) may lose power through each coupling hole so the radiation Q has the form

$$Q_r = \frac{L\omega}{\left(R_G n_1^2 + R_L n_2^2\right)} \tag{10}$$

The input and output cavity coupling parameters denoted by β_1 and β_2, respectively, are defined by

$$\beta_1 = \frac{R_G n_1^2}{R_c} \tag{11}$$

$$\beta_2 = \frac{R_L n_2^2}{R_c} \tag{12}$$

Therefore the ratio of Q_u to Q_r is given by

$$\frac{Q_u}{Q_r} = \beta_1 + \beta_2 = \frac{\left(R_G n_1^2 + R_L n_2^2\right)}{R_c} \tag{13}$$

If both the generator and detector are matched to the waveguide of characteristic impedance Z_0 then

$$R_G = R_L = Z_0 \tag{14}$$

with the result that

$$\frac{Q_u}{Q_r} = \left(\frac{Z_0}{R_c}\right)\left(n_1^2 + n_2^2\right) \tag{15}$$

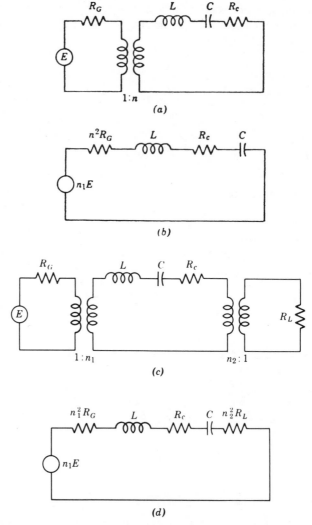

Fig. 5-24. (a) Equivalent circuit of a reflection cavity; (b) alternative form for the equivalent circuit of a reflection cavity. (c) Equivalent circuit of a transmission cavity; (d) alternative form for the equivalent circuit of a transmission cavity (RLS-11, p. 290).

and at the optimum operating conditions we have

$$R_G n_1^2 = R_c = R_L n_2^2 \tag{16}$$

When the resonant cavity terminates the waveguide transmission line, the iris is usually located at a point of maximum tangential H field in the waveguide. Maximum coupling to the resonant cavity occurs if the iris is also located at a point in the cavity where the H field is strong and oriented in the

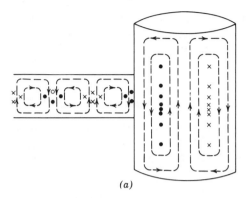

(a)

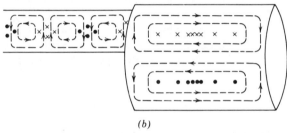

(b)

Fig. 5-25.   Two methods of coupling to the TE_{011} mode in a cylindrical resonant cavity: (a) lateral coupling; (b) end plate coupling.

same direction as the H field in the waveguide. In a picturesque way we may say that the **H** vector on the cavity side of the iris is a logical continuation of the **H** vector on the waveguide side. Two ways of coupling a TE_{011} cylindrical cavity resonator to the end of a rectangular waveguide operating in the TE_{10} mode are shown in Fig. 5-25. When the iris is inductive or capacitive, it may be considered as transmitting the rectangular TE_{10} mode, and when it is circular, it may be considered as transmitting the cylindrical TE_{11} mode. If the iris location is changed from the optimum locations shown in Fig. 5-25, the coupling constant decreases, and a larger opening must be used to preserve the same Q. If the resonant cavity in Fig. 5-25a is rotated by 90° about the waveguide axis, it will be properly oriented to excite the cylindrical TM_{111} mode, which is degenerate with the cylindrical TE_{011} mode.

The coupling parameters are of great practical importance because the ESR sensitivity depends upon whether or not the cavity is overcoupled or undercoupled, and also upon the extent of its deviation from perfect coupling or match (compare Sec. 11-A). Frequently wavemeters (frequency meters) are undercoupled to the sidewall of a waveguide so that when they are off resonance they produce very little disturbance in the waveguide VSWR and mode configurations. This absorption cavity coupling is seldom used for ESR sample cavity applications (see Sec. 11-G).

H. Filling Factors

The ESR signal at resonance is proportional to the amount of power absorbed by the sample, and this in turn is proportional to the average value of the microwave magnetic field $\langle H_1^2 \rangle_s$ at the sample where, in general,

$$\langle H_1^2 \rangle = \frac{\int H_1^2\, dV}{\int dV} = \frac{1}{V} \int H_1^2\, dV \tag{1}$$

More precisely, H_1 is the component of the rf magnetic field perpendicular to the static magnetic field. If the sample has an rf susceptibility χ'' and a volume V_s, then the amount of power absorbed is proportional to $\chi'' V_s \langle H_1^2 \rangle_s$. The resonance absorption will decrease the Q of the resonant cavity, and this may be expressed mathematically by adding the sample loss term Q_χ to Eq. (5-B-6) as follows:

$$\frac{1}{Q} = \frac{1}{Q_u} + \frac{1}{Q_\epsilon} + \frac{1}{Q_r} \qquad \text{off resonance} \tag{2}$$

$$\frac{1}{Q} = \frac{1}{Q_u} + \frac{1}{Q_\epsilon} + \frac{1}{Q_r} + \frac{1}{Q_\chi} \qquad \text{at resonance} \tag{3}$$

where Q_χ is the ratio of the power stored in the overall cavity volume V_c to the power dissipated in the sample

$$Q_\chi = \frac{\frac{1}{2}\mu_0 \int_{\text{cavity}} H_1^2\, dV}{\frac{1}{2}\mu_0 \int_{\text{sample}} \chi'' H_1^2\, dV} \tag{4}$$

$$= \frac{V_c \langle H_1^2 \rangle_c}{\chi'' V_s \langle H_1^2 \rangle_s} \tag{5}$$

Equation (5) may be expressed in terms of the filling factor η

$$Q_\chi = \frac{1}{(\chi'' \eta)} \tag{6}$$

where η is defined by (see also Talpe, 1971)

$$\eta = \frac{\displaystyle\int_{\text{sample}} H_1^2 \, dV}{\displaystyle\int_{\text{cavity}} H_1^2 \, dV} = \frac{V_s \langle H_1^2 \rangle_s}{V_c \langle H_1^2 \rangle_c} \tag{7}$$

In the absence of saturation effects the sensitivity of an ESR spectrometer is directly proportional to η as explained in Sec. 11-C.

It is usually easy to obtain the volume ratio V_s/V_c, but the ratio $\langle H_1^2 \rangle_s / \langle H_1^2 \rangle_c$ requires the integration of H_1^2 over both the sample and the cavity. This will be done for several cases that commonly arise in ESR experiments.

First consider a very small sample placed in the center of the sample tube shown in Fig. 5-5. The microwave field H_1 has a sinusoidal variation along the axis of the tube with a maximum in the center of the TE_{102} rectangular cavity, and zero at the top and bottom. The microwave magnetic field configurations in this TE_{102} mode resonator (see Heuer 1965) are given explicitly by Eqs. (5-C-26) to (5-C-28)

$$H_x = \frac{H_0}{\left[1 + (d/2a)^2\right]^{1/2}} \sin \frac{\pi x}{a} \cos \frac{2\pi z}{d} \tag{8}$$

$$H_z = -\frac{H_0}{\left[1 + (2a/d)^2\right]^{1/2}} \cos \frac{\pi x}{a} \sin \frac{2\pi z}{d} \tag{9}$$

$$E_y = \left(\frac{\mu}{\epsilon}\right)^{1/2} H_0 \sin \frac{\pi x}{a} \sin \frac{2\pi z}{d} \tag{10}$$

$$H_y = E_x = E_z = 0 \tag{11}$$

where $m = 1$, $n = 0$, and $p = 2$. The x direction is taken along the sample tube, and the z direction along the length of the cavity, in accordance with Fig. 5-3. The sample is located at $x = \frac{1}{2}a$, $y = \frac{1}{2}b$, and $z = \frac{1}{2}d$, and as a result at the sample position, H_z vanishes and $H_1 = H_x$, with the explicit value

$$H_1 = \frac{H_0}{\left[1 + (d/2a)^2\right]^{1/2}} \tag{12}$$

Since the sample is small (dimensions $\ll a$, b, and d), H_1 is essentially constant over its volume, and so

$$\langle H_1^2 \rangle_s = H_1^2 = \frac{H_0^2}{1 + (d/2a)^2} \tag{13}$$

To find $\langle H_1^2 \rangle_c$ it is necessary to integrate H_1^2

$$H_1^2 = H_x^2 + H_z^2 \tag{14}$$

over the cavity using the expression (1)

$$\langle H_1^2 \rangle_c = \frac{1}{abd} \left[\int_0^a \int_0^b \int_0^d H_x^2 \, dx \, dy \, dz + \int_0^a \int_0^b \int_0^d H_z^2 \, dx \, dy \, dz \right] \tag{15}$$

$$= \frac{1}{ad} \left[\int_0^a \int_0^d H_x^2 \, dx \, dz + \int_0^a \int_0^d H_z^2 \, dx \, dz \right] \tag{16}$$

since the electromagnetic fields are independent of y, and the cavity volume V_c equals abd. Each integration gives either $a/2$ or $d/2$, thus

$$\langle H_1^2 \rangle_c = \frac{1}{4} H_0^2 \left[\frac{1}{1 + (d/2a)^2} + \frac{1}{1 + (2a/d)^2} \right] \tag{17}$$

$$= \frac{1}{4} H_0^2 \tag{18}$$

independent of the ratio $2a/d$. The filling factor η then becomes

$$\eta = \frac{V_s}{V_c} \frac{4}{1 + (d/2a)^2} = 4 \frac{V_s}{V_c} \left(\frac{\lambda}{\lambda_g} \right)^2 \tag{19}$$

using Eq. (7). Usually $d \approx 2a$, and when d actually equals $2a$ then η becomes

$$\eta = 2 \frac{V_s}{V_c} \tag{20}$$

A comparison of Eqs. (13) and (18) indicates that, in a TE_{102} rectangular resonant cavity with $d = 2a$, the maximum squared rf field H_x at the sample is about twice the average value in the cavity.

When an ESR spectrometer is used to study samples of fairly high electrical conductivity (e.g., in cyclotron resonance), it is sometimes convenient to construct one wall of the cavity out of the sample itself. If the end of a rectangular TE_{102} cavity (the side opposite the iris) consists of a sample of effective thickness δ, then $z = d$, and Eqs. (8) and (9) become

$$H_x = \frac{H_0 \sin \pi x/a}{\left[1 + (d/2a)^2 \right]^{1/2}} \tag{21}$$

$$H_z = E_x = 0 \tag{22}$$

Consequently, $\langle H_1^2 \rangle_s$ is

$$\langle H_1^2 \rangle_s = \frac{H_0^2}{ab\delta\left[1 + (d/2a)^2\right]} \int_0^a \sin^2 \frac{\pi x}{a}\, dx \int_0^b dy \int_{d-\delta}^d dz$$

$$= \frac{1}{2} \frac{H_0^2}{1 + (d/2a)^2} \tag{23}$$

As a result

$$\eta = \frac{2V_s}{V_c} \frac{1}{1 + (d/2a)^2} \tag{24}$$

and if $2a = d$

$$\eta = \frac{V_s}{V_c} \tag{25}$$

One can easily show that for a "square mode" cavity ($2a = d$), this result is independent of z (Poole, 1958) since a uniform planar sample located across the waveguide in the xy plane has the filling factor

$$\eta = \frac{V_s}{V_c} \tag{26}$$

for all z. This case also corresponds closely to a flat quartz aqueous solution cell. Materials of high dielectric loss are efficiently studied with this arrangement. The Hedvig (1959) equation (5-J-17) modifies this result for high dielectric constants.

If the sample tube of radius r in Fig. 5-5 is completely filled, then one has $V_s = \pi r^2 a$, and

$$\langle H_1^2 \rangle_s = \frac{H_0^2}{\pi r^2 a} \left[\frac{1}{1 + (d/2a)^2} \int_0^a \sin^2 \frac{\pi x}{a}\, dx \int_{-r}^r \cos^2 \frac{2\pi z}{d}\, dz \int_{-r\sin\phi}^{r\sin\phi} dy \right.$$

$$\left. + \frac{1}{1 + (2a/d)^2} \int_0^a \cos^2 \frac{\pi x}{a}\, dx \int_{-r}^r \sin^2 \frac{2\pi z}{d}\, dz \int_{-r\sin\phi}^{r\sin\phi} dy \right] \tag{27}$$

where the relevant coordinate system is shown in Fig. 5-26. The integration over x gives $a/2$ immediately since the sample tube extends uniformly over the

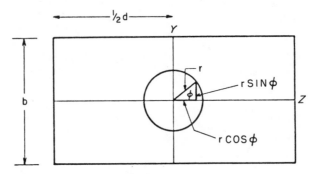

Fig. 5-26. Coordinate system for calculation of the filling factor η of a dielectric rod in a rectangular TE$_{102}$ mode resonant cavity.

entire cavity width from $x = 0$ to $x = a$, and one easily obtains

$$\langle H_1^2 \rangle_s = \frac{H_0^2}{\pi r} \left\{ \frac{1}{1 + (d/2a)^2} \int_{-r}^{r} \left[1 - \left(\frac{z}{r} \right)^2 \right]^{1/2} \cos^2 \frac{2\pi z}{d}\, dz \right.$$

$$\left. + \frac{1}{1 + (2a/d)^2} \int_{-r}^{r} \left[1 - \left(\frac{z}{r} \right)^2 \right]^{1/2} \sin^2 \frac{2\pi z}{d}\, dz \right\} \quad (28)$$

where $\sin \phi = [1 - (z/r)^2]^{1/2}$. If we assume $2a = d$ then this simplifies to

$$\langle H_1^2 \rangle_s = \frac{H_0^2}{\pi r} \int_{0}^{r} \left[1 - \left(\frac{z}{r} \right)^2 \right]^{1/2} dz \quad (29)$$

$$= \frac{1}{4} H_0^2 \quad (30)$$

and the filling factor is

$$\eta = \frac{V_s}{V_c} \quad (31)$$

Note that this result is independent of r (except for the dependence of V_s on r). It is identical to that obtained above in the parallel plane case for the same condition of $2a = d$. Equation (31) will of course be much more complicated when $2a \neq d$. If $2a \neq d$ then an approximate expression may be obtained by assuming $(2a/d)^{\pm 1} = 1 \pm \delta$, where $\delta \ll 1$ and $z/r \ll 1$ over the range of integration, and by expanding the sines and cosines in power series. This will be left as an *exercitatio lectori*.

All of the above filling factor calculations were for the TE_{102} mode rectangular resonant cavity. The second most important cavity resonator for use in ESR spectrometers is the cylindrical TE_{011} type, and its electromagnetic wave configurations are (see Sec. 5-D)

$$H_r = \frac{H_0 J_0'(k_c r)\cos(\pi z/d)}{\left\{1 + [(k_c a)_{01}' d/\pi a]^2\right\}^{1/2}} \tag{32}$$

$$H_z = \frac{H_0 J_0(k_c r)\sin(\pi z/d)}{\left\{1 + [\pi a/(k_c a)_{01}' d]^2\right\}^{1/2}} \tag{33}$$

$$E_\phi = -\left(\frac{\mu}{\epsilon}\right)^{1/2} H_0 J_0'(k_c r)\sin\left(\frac{\pi z}{d}\right) \tag{34}$$

$$H_\phi = E_r = E_z = 0 \tag{35}$$

where the root of $J_0'(k_c r)$ is given by $(k_c a)_{01}' = 3.832$. The electric field equations are not needed here, but are merely listed for completeness. The average value of H_1 over the volume of the cavity of length d and radius a is given by Eq. (1).

$$\langle H_1^2 \rangle_c = \frac{H_0^2}{\pi a^2 d}\left[\frac{1}{1 + [(k_c a)_{01}' d/\pi a]^2} \int_0^a J_0'^2(k_c r)r\,dr \int_0^d \cos^2\frac{\pi z}{d}\,dz \int_0^{2\pi} d\phi \right.$$

$$\left. + \frac{1}{1 + [\pi a/(k_c a)_{01}' d]^2} \int_0^a J_0^2(k_c r)r\,dr \int_0^d \sin^2\frac{\pi z}{d}\,dz \int_0^{2\pi} d\phi \right] \tag{36}$$

The second and third integrals of each pair give $d/2$ and 2π, respectively, so that

$$\langle H_1^2 \rangle_c = \frac{H_0^2}{a^2}\left[\frac{1}{1 + [(k_c a)_{01}' d/\pi a]^2} \int_0^a J_0'^2(k_c r)r\,dr \right.$$

$$\left. + \frac{1}{1 + [\pi a/(k_c a)_{01}' d]^2} \int_0^a J_0^2(k_c r)r\,dr \right] \tag{37}$$

These integrals may be evaluated from the useful general relations (e.g., see

Margenau and Murphy, 1956: Jahnke and Emde, 1933)

$$\int_0^r J_n^2(k_c r) r \, dr = \frac{r^2}{2} \left[J_n^2(k_c r) - J_{n-1}(k_c r) J_{n+1}(k_c r) \right] \tag{38}$$

$$J_n'(k_c r) = \left(\frac{n}{k_c r} \right) J_n(k_c r) - J_{n+1}(k_c r) \tag{39}$$

$$\left(\frac{2n}{k_c r} \right) J_n(k_c r) = J_{n-1}(k_c r) + J_{n+1}(k_c r) \tag{40}$$

which give the pertinent particular expressions

$$\int_0^r J_0^2(k_c r) r \, dr = \left(\frac{r^2}{2} \right) \left[J_0^2(k_c r) + J_1^2(k_c r) \right] \tag{41}$$

$$\int_0^r J_0'^2(k_c r) r \, dr = \left(\frac{r^2}{2} \right) \left[J_0^2(k_c r) - \left(\frac{2}{k_c r} \right) J_0(k_c r) J_1(k_c r) + J_1^2(k_c r) \right] \tag{42}$$

$$J_0'(k_c r) = -J_1(k_c r) = J_{-1}(k_c r) \tag{43}$$

$$J_2(k_c r) = \left(\frac{2}{k_c r} \right) J_1(k_c r) - J_0(k_c r) \tag{44}$$

The application of the boundary conditions for this mode yields

$$J_0'(k_c a) = J_{\pm 1}(k_c a) = 0 \tag{45}$$

with the Bessel function root $(k_c a)_{01}'$ obtained from Table 5-1:

$$(k_c a)_{01}' = 3.832 \tag{46}$$

$$J_0(k_c a) = 0.4028 \tag{47}$$

$$J_0(0) = 2 \lim_{r \to 0} \left[\frac{J_1(k_c r)}{k_c r} \right] = 1 \tag{48}$$

The maximum value of $J_0'(k_c r)$

$$\max J_0'(k_c r) = 0.5819 \tag{49}$$

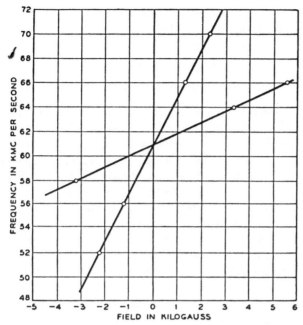

Fig. 5-48. Transitions in positive and negative magnetic fields for ruby subjected to circularly polarized microwaves (Mock, 1960).

by NMR techniques to determine the signs of gyromagnetic ratios. Dascola et al. (1965), Gharbage (1978, 1979), and Servant et al. (1978) discuss the Cotton Mouton-Voigt effect.

A number of authors have discussed methods of determining the relative signs of hyperfine coupling constants (Heller, 1965). For this purpose McConnell et al. (1960) and Heller and Cole (1962) used oriented free radicals; McConnell and Holm (1957), Forman et al. (1959), and Eaton et al. (1962) measured NMR shifts in organic complexes of transition elements; while Carrington and Longuet-Higgins (1962) and deBoer and Mackor (1962) deduced the relative signs from the variations in the hyperfine linewidths. Biehl et al. (1975) used triple resonance on free radicals in solution to determine relative signs of hyperfine coupling constants, and Nunome et al. (1975) evaluated these signs from nuclear spin-forbidden transitions. Adrian et al. (1973) compared observed and calculated [23]Na hyperfine couplings to deduce these signs, and Heller (1965) used zero-field resonance spectra (Cole et al., 1963) to evaluate them. Bernheim and Chien (1976) evaluated the absolute sign of the zero-field splitting of methylene. Schuch et al. (1974) determined the absolute sign of a triplet state zero-field parameter D by optically detected ESR. The fit of high-field experimental data to the $A = Q_\rho$ relationship of McConnell and Chesnut (1958) and the self-consistent field treatment of McLachlan (1960) may also be used for this purpose (see Sec. 1-G of the first edition). Gubanov and Chirkov (1973) determined the signs of the nitrogen hyperfine coupling constants in DPPH.

P. Summary and Applications

We shall say a few words by way of summary in order to place the various sections of this chapter in perspective. The first two sections presented some background material to give an insight into how resonators work. The three following sections give detailed discussions of rectangular, cylindrical, and coaxial cavities. These sections describe the electromagnetic field configurations, current distributions, quality factors, and so on, of the individual modes. This information is important when selecting a cavity type and modifying its design for a specific experimental application. For example, a knowledge of the field distribution is required to properly position the sample, to mount intramural NMR coils for double resonance experiments, and to incorporate a frequency changing mechanism into the cavity. A knowledge of the current distributions in the walls allows one to cut slots for uv irradiation and high frequency modulation purposes. The sensitivity is strongly dependent on the Q and the impedance match (coupling) between the resonator and the waveguide. Sections F and G provide an understanding of this relationship, and prepare the reader for the discussion of sensitivity that is found in Chap. 11.

Section 5-H presents a detailed treatment of filling factors. It shows one how to calculate the effectiveness of various sample shapes and orientations, and compares different types of cavities. This section also provides an understanding of the role that cavities play in the sensitivity of an ESR spectrometer. Section J explains how ideal cavities are perturbed by the insertion of various objects such as dielectrics and conductors. Since we frequently insert samples, Dewars, coils, and the like into cavities, it is important to acquire an understanding of the nature of the resultant disturbance. Section I explains how to measure the microwave magnetic field H_1 in the cavity.

Sections K, L, M, and N introduce the reader to specific types of cavities. Chapter 11 on sensitivity and Chap. 14 on double resonance contain several examples of bimodal resonators. Such cavities find extensive use in maser and laser applications. Section O discusses resonant cavities with circularly polarized mode patterns.

ESR spectroscopists often find it necessary to design resonators for special applications such as (1) wavemeters, (2) matched reference and sample cavities, (3) masers, (4) high- and low-temperature studies, (5) irradiation investigations, (6) double resonance, (7) high pressures, (8) refractometers, and (9) circular polarization modes. Diagrams, design data, and references for many such cavities will be found scattered throughout the book. Birnbaum (1950), Vetter and Thompson (1962), Roussy (1965), and Verweel (1965) discussed microwave refractometers.

In the first edition the present section (therein called Sec. 8-N) concluded with descriptions of several miscellaneous resonant cavities:

1. An ENDOR cavity by Kramer et al. (1965).
2. An ENDOR cavity by Holton and Blum (1962) [compare e.g., Brown and Sloop (1970), Davis and Mims (1978), Dinse et al. (1973), Kasatochkin et al. (1974), Schmalbein et al. (1972), Yagi et al. (1970), and Sec. 14-B].
3. A cavity for electric field studies by Mims (1964) [compare Sec. 14-G, Bhoraskar and Abe (1976), Kiel (1965), Legrand et al. (1977), Ludwig and Ham (1963), Teodorescu (1974)].
4. A high-temperature cavity by Walsh et al. (1965) [compare Sharpless (1959)].
5. Two low-temperature cavities by Llewellyn (1957) with novel modulation coils.
6. A line reservator for dynamic nuclear enhancement studies with H_1 fields up to 8 G, designed by Reichert and Townsend (1965) [compare Locher and Gorter (1962)].

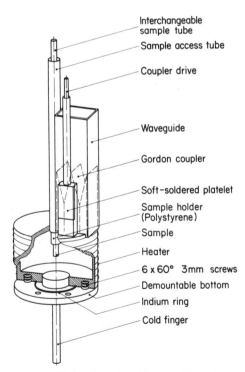

Interchangeable sample tube
Sample access tube
Coupler drive
Waveguide
Gordon coupler
Soft-soldered platelet
Sample holder (Polystyrene)
Sample
Heater
6 x 60° 3mm screws
Demountable bottom
Indium ring
Cold finger

Fig. 5-49. Low-temperature multipurpose cavity with central sample access tube and eccentric variable coupling arrangement. The Dewar head illustrated in Fig. 5-50 and the modular sample arrangements of Fig. 5-51 may be used with this cavity (Berlinger and Müller, 1977).

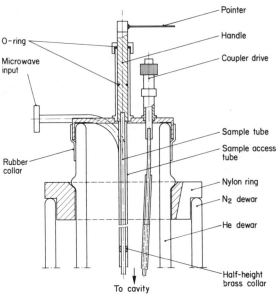

Fig. 5-50. Low-temperature Dewar arrangement used with the cavity sketched on Fig. 5-49 (Berlinger and Müller, 1977).

Instead of repeating the descriptions of these cavities we discuss several more recent resonators with various special features or useful applications. Then we conclude the section by mentioning several references to additional cavity designs.

Berlinger and Müller (1977) designed two multipurpose K-band (19 to 20 GHz) resonators for applications between 1.6 and 1300° K. The authors have used these cavities for a variety of experiments including electric field effects, *in situ* irradiation with light, ENDOR, uniaxial compression, and tension. These cavities have a modular design, are easy to construct, and permit rapid sample changing (Strandberg et al., 1956). They were used with a home-built single-klystron superheterdyne system working with a single sideband at 30 MHz (Brown et al., 1965). Two GaAs diodes mounted in push–pull provide the local oscillation power. The 19.5-GHz frequency is a good compromise between X and Q bands, allowing reasonable cavity and sample sizes.

The low-temperature cavity of Berlinger and Müller (1977) illustrated in Fig. 5-49 employs a Gordon (1961) coupler for impedance watching. A centrally located sample access hole permits easy sample interchange through the sample tube illustrated in Fig. 5-50 (Ager et al., 1963; Andrew and Kim, 1966; Aseltine and Kim, 1966; Strandberg et al., 1956). The various modular sample arrangements illustrated in Fig. 5-51 may be used with the cavity.

The high-temperature cavity and associated multistrip heating element designed by Berlinger and Müller (1977) are illustrated in Fig. 5-52. Most of the modular sample arrangements sketched in Fig. 5-51 may be used with the

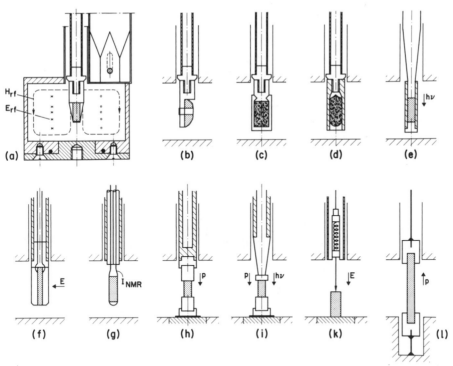

Fig. 5-51. Modular sample arrangements used in the low-temperature cavity of Fig. 5-49. The different mountings show (a)—sample with known plane; (b)—sample with known direction; (c) and (d)—powder samples; (e)—irradiation with light, $h\nu$; (f)—application of electric field $E \perp H_{rf}$; (g)—ENDOR experiment; (h)—uniaxial compression, p; (i)—uniaxial compression p under light excitation $h\nu$; (k)—electric field $E \parallel H_{rf}$; (l)—tension experiment (Berlinger and Müller, 1977).

high-temperature cavity when high-purity quartz is used for sample tubes instead of polystyrene.

Allendoerfer et al. (1975, 1981) designed the coaxial cavity sketched in Fig. 5-53, which improves the sensitivity obtained with lossy samples. It operates in the TE_{011} mode with the microwave magnetic field H_1 lines sketched in Fig. 5-54. This cavity is 2 to 3 times more sensitive than standard waveguide cavities when the sample temperature deviates from ambient by more than 25°C. The authors provide a detailed analysis of the cavity sensitivity for various experimental arrangements.

Johansson et al. (1974) designed a TEM mode stripline resonator. It consists of a center conducting resonant strip detailed in Fig. 5-55 separated from two parallel conducting ground planes by solid dielectric material, as illustrated in Fig. 5-56. Modulation coils attached to the sides operated at 100 kHz. It served as a half-wavelength resonator at S band (3.3 GHz) or as a three-half wavelength resonator at X band (9.8 MHz) with an unloaded Q of 1000. A

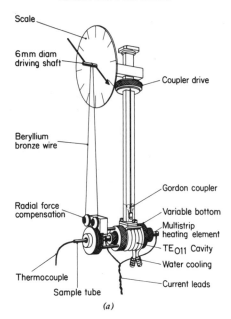

Scale

6mm diam
driving shaft

Coupler drive

Beryllium
bronze wire

Radial force
compensation

Gordon coupler

Variable bottom

Multistrip
heating element

TE$_{011}$ Cavity

Water cooling

Thermocouple

Current leads

Sample tube

(a)

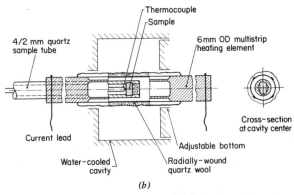

Thermocouple

Sample

4/2 mm quartz
sample tube

6mm OD multistrip
heating element

Cross-section
at cavity center

Current lead

Adjustable bottom

Water-cooled
cavity

Radially-wound
quartz wool

(b)

Fig. 5-52. (*a*) High-temperature multipurpose circular cavity assembly shown with a sample rotation drive on the left and a heater on the right. (*b*) Details of the heating element [Berlinger and Müller (1977)].

resonator with two crossed center strips fed with microwave power with the proper phase difference between strips provided circularly polarized microwaves in the center.

Brown (1974) built a 1-GHz spectrometer for aqueous samples that also employs a triplate stripline cavity. The spectrometer was designed to maximize the product ηQ_L for aqueous samples (Stoodley, 1963), which Brown claims was not satisfactorily done by maximizing ηQ at X band (Sogo et al., 1961) by operating close to cutoff (Rorke, 1966) or by using a five-wavelength-long cavity (Cook and Mallard, 1963). Stoodley showed that, taking into account

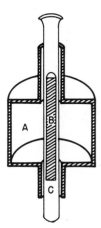

Fig. 5-53. Schematic diagram of a coaxial microwave cavity constructed from an X-band cylindrical cavity A with a cylindrical metallic conductor B located in a quartz tube C along its axis. This cavity is excited in the TE_{011} mode sketched in Fig. 5-54 (Allendoerfer and Carroll, 1981).

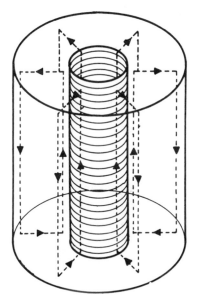

Fig. 5-54. The TE_{011} mode of the coaxial cavity illustrated in Fig. 5-53, where the dashed lines represent the magnetic field contours (Allendoerfer and Carroll, 1981).

the frequency variations of the dielectric constant of water (Collie et al., 1948), the optimum frequency is 1 GHz. This spectrometer had a minimum detectable spin density 10.8 times better than that of an optimally designed X-band spectrometer. Franconi and Bonari (1980) constructed an S band (2.14 GHz) spectrometer for biomedical applications.

Sorokin et al. (1960) reported a dual mode cavity for relaxation time measurements, and it is sketched in Fig. 5-57.

Wolfe and Jeffries (1971) designed the tunable cavity with a rotating crystal mount shown in Fig. 5-58. This is a cylindrical TE_{011} mode cavity arranged so that the sample crystal can be rotated about a horizontal axis and the magnetic field can be rotated in a horizontal plane. This permits the c axis of a crystal to

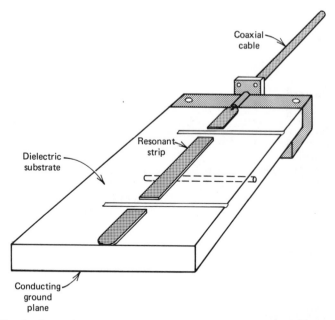

Fig. 5-55.   Details of central resonant conducting strip part of stripline resonator showing sample hole (a) and two holes (b) for quartz rods, that are used to adjust the coupling. Figure 5-56 shows how this center conducting strip is located between two ground planes (Johansson et al., 1974).

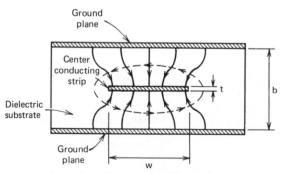

Fig. 5-56.   Cross section of stripline resonator showing central conducting strip of Fig. 5-55 located between ground planes. The sketch shows the electric field (—) and magnetic field (---) lines of the TEM modes (Johansson et al., 1974).

be precisely aligned along the field so that $g_{\parallel}$ can be determined accurately. The cavity also has the facility for optical pumping. Weil and Claridge (1970) described an electromagnetic goniometer that employs pickup coils to monitor the sample orientation relative to the cavity H_1 direction.

The remaining chapters of the book contain numerous descriptions of resonant cavities and their uses. We mention a few additional ones here. Evenson and Burch (1966) present an X-band cavity for studies of para-

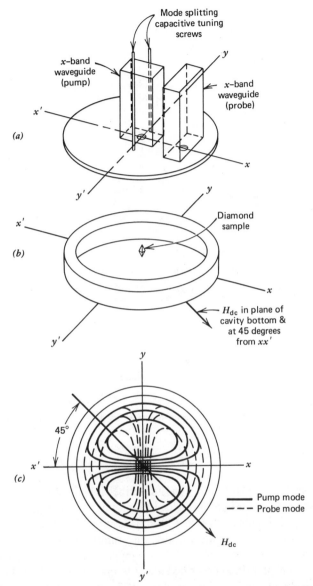

Fig. 5-57. Sketches of dual mode cavity used for relaxation time measurements (a) Top view of top plate showing waveguide coupling and tuning screws, (b) main body of cavity showing sample in position, and (c) magnetic field configuration of pump (—) and probe (---) modes (Sorokin et al., 1960).

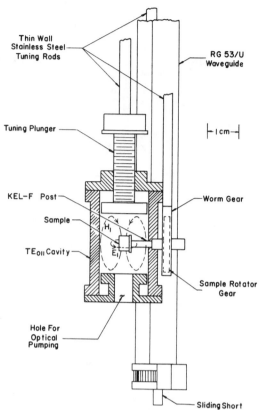

Fig. 5-58. Cylindrical TE_{011} mode resonant cavity used in a pulse spectrometer (Wolfe and Jeffries, 1971).

magnetic gases. Pol'skii (1963, 1964) describes a helically wound cavity with a very homogeneous modulating field. Lukin et al. (1974) reported a millimeter-range cavity. Bil'dyukevich (1964, 1965) designed an 8-mm low-temperature resonator with provision for rotating the sample in horizontal and vertical planes, and he also describes a low-temperature X-band resonator (1963, 1965). Casini et al. (1976) described a low-temperature cavity. The low-temperature cylindrical cavity of Ghidalia (1973) operated in the TE_{113} mode. Grosmaitre and Roy (1973) and Koningsberger et al. (1973) reported high-temperature cavity designs. Chamel et al. (1976) described a TE_{011} cylindrical resonator whose walls were made of a set of parallel plates that constitute transition lines beyond cutoff. The optical access of this cavity far exceeds that achieved by slotted walls. Shaw et al. (1965) present an X-band cavity with a lithium fluoride window to permit plasma production by ultraviolet irradiation down to 1100 Å. Livingston and Zeldes (1966) used a cavity with a grid on one face for uv irradiating liquids, and their detector was a Philco L4154 backward diode. Assmus and Silber (1975) described a cavity cryostat for investigating single crystals under uniaxial stress. Noon et al. (1965) used a waveguide cell for magnetoplasma studies. Redhardt (1965) discusses accurate measurements on reflection resonators. Cevc (1969) reported a plastic resonator suitable for high and low modulation frequencies. Dixon and Norman (1963), Hirshon and Fraenkel (1955), Piette et al. (1961), and Storey et al. (1962) describe aqueous solution cavities. Pristupa et al. (1975) discussed model cavities, Isaev-Ivanov and Fomichev (1976) described a balanced resonator, and Breslev et al. (1971) used balanced resonators to improve the sensitivity.

References

Adrian, F. J., E. L. Cochran, and V. A. Bowers, *J. Chem. Phys.* **59**, 56 (1973).

Agdur, B., and B. Enander, *J. Appl. Phys.* **33**, 575 (1962).

Ager, R., T. Cole, and J. Lambe, *RSI* **34** 308 (1963).

Alaeva, T. I., and V. Karasev, *PTE* **5** 183 (1002) (1961).

Allendoerfer, R. D., G. A. Martinchek, and S. Bruckenstein, *Anal. Chem.* **47**, 890 (1975).

Allendoerfer, R. D., and J. B. Carroll, *J. Magn. Res.*, in press.

Altschuler, H. M., *RSI* **34**, 1441 (1963).

Andresen, S. G., and J. de Prins, *Proc. IEEE* **53**, 511 (1965).

Andrew, R. A., and Y. W. Kim, *RSI* **34**, 308 (1963).

Artman, J. O., N. Bloembergen, and S. Shapiro, *Phys. Rev.* **109**, 1392 (1958).

Artman, J. O., and P. Tannenwald, *J. Appl. Phys.* **26**, 1124 (1955).

Aseltine, C. L., and Y. W. Kim, *RSI* **37**, 1270 (1966).

Auld, B. A., *J. Appl. Phys.* **34**, 1629 (1963).

Assmus, W., and D. Silber, *RSI* **46**, 299 (1975).

Bakker, M. J. A., and J. Smidt, *Appl. Sci. Res.* **B9**, 199 (1962).

Barchukov, A. I., and A. M. Prokhorov, *Arch. Sci. Spec.* **14**, 494 (1961).

Battaglia, A., M. Iannuzzi, and E. Polacco, *Ricerca Sci.* **11A**, (3) 119 (1963).

Bennett, R. G., P. C. Hoell, and P. R. Schwenker, *RSI* **29**, 659 (1958).

Berlinger, W., and K. A. Muller, *RSI* **48**, 1161 (1977).

Bergmann, S. M., *J. Appl. Phys.* **31**, 275 (1960).

Bernheim, R. A., and S. H. Chien, *J. Chem. Phys.* **65**, 2023 (1976).

Bersohn, M., *RSI* **34**, 107 (1963).

Bhoraskar, S., and R. Abe, *Jap. J. Appl. Phys.* **15**, 1471 (1976).

Bil'dyukevich, A. L., *PTE* **6**, 186 (1194) (1963); **2**, 185 (453) (1964); *Cryogenics* **5**, 205, 277 (1965).

Biehl, R., M. Plato, and K. Mobius, *J. Chem. Phys.* **63**, 3515 (1975).

Birnbaum, G., *RSI* **21**, 169 (1950).

Bloch, F., W. W. Hansen, and M. Packard, *Phys. Rev.* **70**, 474 (1946).

Bonori, M., C. Franconi, P. Galuppi, P. S. Allen, E. R. Andrew, and C. A. Bates, Proc. 18th Ampere Congr. *on* Magn. Reson. Relat. Phenom., **1**, 193 (1974).

Bonori, M., C. Franconi, P. Galuppi, and C. A. Tiberio, submitted for publication, 1981a.

Bonori, M., C. Franconi, P. Paluppi, C. A. Tiberio and W. Cecchetti, submitted for publication, 1981b.

Boudouris, G., *C.R. Acad. Sci.* **258**, 2499 (1964).

Bowers, K. D., and W. B. Mims, *Phys. Rev.* **115**, 285 (1959).

Breslev, S. E., E. N. Kazbekov, and V. N. Fomichev, *Zh. Tekh. Fiz.* **41**, 1237 (1971).

Boyd, G. D., and J. P. Gordon, *Bell Syst. Tech. J.* **40**, 489 (1961).

Boyd, G. D., and H. Kogelnik, *Bell Syst. Tech. J.* **41**, 1347 (1962).

Brodwin, M. E., and T. G. Burgess, *IEEE Trans. Instrum. Meas.* **IM-12**, 7 (1963).

Brodwin, M. E., and M. K. Parsons, *J. Appl. Phys.* **36**, 494 (1965).

Brown, G., *J. Phys. E. Sci. Instrum.* **7**, 635 (1974).

Brown, G., D. R. Mason, and J. S. Thorp, *JSI* **42**, 648 (1965).

Brown, I. M., and R. J. Richardson, *RSI* **44**, 77 (1973).

Brown, I. M., and D. J. Sloop, *RSI* **41**, 1774 (1970).

Brown, S. C., and D. J. Ross, *J. Appl. Phys.* **23**, 711 (1952).

Bussey, H. E., and A. J. Estin, *RSI* **31**, 410 (1960).

Callebaut, D. K., and M. E. Vanwormhoudt, *Physica* **26**, 255 (1960).

Carrington, A., and H. C. Longuet-Higgins, *Mol. Phys.* **5**, 447 (1962).

Casini, G., P. Graziani, R. Linari, and A. Tronconi, *Rev. Sci. Instrum.* **47**, 318 (1976).

Cevc, P., *RSI* **40**, 515 (1969).

Chang, T., D. Foster, and A. H. Kahn, *J. Res. NBS* **83**, 133 (1978).

Chang, Te-Tse, *Phys. Rev.* **136**, A1413 (1964).

Chang, W. S. C., J. Cromach, and A. E. Siegman, *J. Electron. Control* **6**, 508 (1959).

Chamel, R., R. Chicault, and Y. Merle d'Aubigne, *J. Phys. E. Sci. Instr.* **9**, 87 (1976).

Charru, A., *C. R. Acad. Sci.* **243**, 652 (1956).

Chester, P. F., P. E. Wagner, J. G. Castle, and G. Conn, *RSI* **30**, 1127 (1959).

Cole, T., T. Kuhida, and H. C. Heller, *J. Chem. Phys.* **38**, 2915 (1963).

Collie, C. H., J. B. Hasted, and D. M. Ritson, *Proc. Phys. Soc.* **60** (71), 145 (1948).

Conciauro, G., *RSI* **46**, 1253 (1975).

Conciauro, G., C. Franconi, P. Galuppi, M. Puglisi, and E. Randazzo, Abstracts of the 4th International Symposium on Magnetic Resonance, Rehovot Israel, 305 (1971).

Conciauro, G., M. Puglisi, C. Franconi, P. Galuppi, and E. Randazzo, *J. Magn. Res.* **9**, 363 (1973).

Conciauro, G., and E. Randazzo, *RSI* **44**, 1087 (1973).

Cook, P., and J. R. Mallard, *Nature* **198**, 145 (1963).

Cook, A. R., L. M. Matarrese, and J. S. Wells, *RSI* **35**, 114 (1964).

Copeland, E. S., *Rev. Sci. Instru.* **44**, 437 (1973).

Crain, C. M., and C. E. Williams, *RSI* **28**, 620 (1957).

Cross, L. G., *J. Appl. Phys.* **30**, 1459 (1959).

Culshaw, W., *IRE Trans. Microwave Theory Tech.* **MTT-9**(2), 135 (1961).

Dalal, D. P., S. S. Eaton and G. R. Eaton, *J. Magn. Reson.* **44**, 415 (1981).

Dascola, G., D. C. Giori, and V. Varacca, *Nuovo Cimento* **37**, 382 (1965).

Davis, J., and W. B. Mims, *RSI* **49**, 1095 (1978).

de Boer, E., and E. L. Mackor, *Mol. Phys.* **5**, 493 (1962).

Decorps, M., and C. Fric, *C. R. Acad. Sci.* **259**, 1394 (1964).

de Groot, M. S., C. A. deLange, and A. A. Monster, *J. Magn. Reson.* **10**, 51 (1973).

Diaz, J., S. Haraldson, U. Smith, and L. Pettersson, *RSI* **45**, 454 (1974).

Dinse, K. P., I. Möbius, and R. Biehl, *Z. Naturforsch.* **28A**, 1069 (1973).

Dixon, W. T., and R. O. C. Norman, *J. Chem. Soc.* 3119 (1963).

Duchiewicz, J., and A. Francik, *Pomiary Autom. Kontrola* (*Poland*) **24**, 166 (1978).

Dunn, P. D., C. S. Sable, and D. J. Thompson, *Atomic Energy Res. Establ.* Harwell, Rept., GP/R 1966 (1956).

Eaton, D. R., A. D. Josey, W. D. Phillips, and R. E. Benson, *J. Chem. Phys.* **37**, 347 (1962).

Eisinger, J., and G. Feher, *Phys. Rev.* **109**, 1172 (1958).

Erickson, L. E., *Phys. Rev.* **143**, 295 (1966).

Eshbach, J. R., and M. W. P. Strandberg, *RSI* **23**, 623 (1952).

Estin, A. J., *RSI* **33**, 369 (1962).

Evenson, K. M., and D. S. Burch, *RSI* **37**, 236 (1966).

Faulkner, E. A., and A. Holman, *JSI* **40**, 205 (1963).

Foerster, G. V., *Z. Natarforsch.* **15a**, 1079 (1960).

Forman, A., J. N. Murrell, and L. E. Orgel, *J. Chem. Phys.* **31**, 1129 (1959).

Forrer, J., A. Schweiger, N. Berchten, and Hs. H. Günthard, *J. Phys. E Sci. Instr.* **14**, 565 (1981).

Franconi, C., *Proc. Intern. Conf. Magn. Res.*, C. Franconi, Ed., Gordon and Breach, New York, 1970, p. 397.

Franconi, C., *JSI* **41**, 148 (1970a).

Franconi, C., and M. Bonori, Rendiconti della Lxxxl Reunione Annuale della Assoc. Electrotechica ed Electronica Italiana, Trieste, 1980, p. 1.61.

Freed, J. H., D. S. Leniart, and J. S. Hyde, *J. Chem. Phys.* **47**, 2262 (1967).

Galt, J. K., W. A. Yager, F. R. Merritt, B. B. Cetlin, and H. W. Dail, Jr., *Phys. Rev.* **100**, 718 (1955).

Ganssen, A., and J. C. Webster, *Proc. IEEE* **53**, 540 (1965).

Garstens, M. A., *J. Appl. Phys.* **30**, 976 (1959).

Gharbage, S., *C. R. Hebd. Seances Acad. Sci. Ser. B (France)* **286**, 275 (1978).

Gharbage, S., *C. R. Hebd. Seances Acad. Sci. Ser. B (France)* **288**, 17 (1979).

Gheorghiu, O. C., and E. I. Rodeanu, *Stud. Cercetari Fiz. (Romania)* **14**, 399 (1963).

Ghidalia, W., *J. Phys. E Sci. Instrum.* **6**, 1081 (1973).

Ginzton, E. L., *Microwave Measurements*, McGraw-Hill, New York, 1957, p. 445.

Giordano, M., M. Martinelli, L. Pardi, and S. Santucci, *RSI* **47**, 1402 (1976).

Goldberg, I. B., *J. Magn. Reson.* **32**, 233 (1978).

Goldberg, I. B., and H. R. Crowe, *Anal. Chem.* **49**, 1353 (1977).

Goldberg, I. B., H. R. Crowe, R. M. Housley, and E. H. Cirlin, *J. Geoph. Res.* **84**, 1025 (1979).

Gordon, J. P., *RSI* **32**, 658 (1961).

Gould, R. N., and A. Cunliffe, *Philos. Mag.* (8th Series) **1**, 1126 (1956).

Grosmaitre, C. S., and J. C. Roy, *RSI* **44**, 652 (1973).

Gruber, G., and F. Ritzmann, *ETP* **22**, 143 (1974).

Gubanov, V. A., and A. K. Chirkov, *Acta Phys. Pol.* **A43**, 361 (1973).

Hedvig, P., *Acta Phys. Hungaricae* **10**, 115 (1959).

Heer, C. V., *Phys. Rev.* **134**, A799 (1964).

Heller, H. C., *J. Chem. Phys.* **42**, 2611 (1965).

Heller, C., and T. Cole, *J. Chem. Phys.* **37**, 243 (1962).

Heuer, K., *Jenaer Jahrbuch (Germany)* **1965**, 169–179.

Hirshon, J. M., and G. K. Fraenkel, *Rev. Sci. Instr.* **26**, 34 (1955).

Holton, W. C., and Y. Blum, *Phys. Rev.* **125**, 89 (1962).

Holz, G., W. Köhnlein, A. Müller, and K. G. Zimmer, *Strahlentherapie* **120**, 161 (1963).

Hutchison, C. A., and B. Weinstock, *J. Chem. Phys.* **32**, 56 (1960).

Hyde, J. S. and H. W. Brown, *J. Chem. Phys.* **37**, 368 (1962).

Isaacson, R. A., *RSI* **47**, 973 (1976).

Isaacson, R. A., C. Lulich, S. B. Oseroff, and R. Calvo, *RSI* **51**, 1409 (1980).

Isaev-Ivanov, V. V., V. N. Fomichev, *PTE* **19**, 172 (1976).

Jahnke, E., and F. Emde, *Funktionentafln mit Formeln und Kurven*, Teubner, Leipzig, 1933, Vol. 2.

Janssen, L., and J. Witters, *J. Appl. Phys.* **41**, 2064 (1970).

Johansson, B., S. Haraldson, L. Pettersson, and Beckman, *RSI* **45**, 1445 (1974).

Kahn, W. K., L. Bergstein, H. Gamo, G. J. E. Goubau, J. T. Latourrette, and G. T. Di Francia, *Proc. Symp. Quasi Optics*, Brooklyn Polytechnic Press, 1964, p. 397.

Kastler, A., *C. R. Acad. Sci.* **238**, 669 (1954).

Kasatochkin, S. V., Yu, A. Timofeev, T. I. Alaeva, L. F. Vereshchagin, and E. N. Yakovlev, *PTE* **17**(6), 140 (1717) (1974).

Kemeny, G., *Phys. Rev.* **133**, A69 (1964).

Kiel, A., *Proc. Intern. Conf. Magnetism (London) (Inst. Phys. Phys. Soc.)* **1965**, 465.

Kikuchi, C., J. Lambe, G. Makhov, and R. W. Terhune, *J. Appl. Phys.* **30**, 1061 (1959).

Kohnlein, W., and A. Muller, *Free Radicals in Biological Systems*, M. S. Blois Jr., H. W. Brown, R. M. Lindblom, M. Weissbluth, and R. M. Lemmon, Eds., Academic Press, New York, 1961, p. 113.

Kooser, R. G., W. V. Volland, and J. H. Freed, *J. Chem. Phys.* **50**, 5243 (1969).

Koningsberger, D. C., G. J. Mulder, B. Pelupessy, *J. Phys. E* **6**, 306 (1973).

Kramer, K. D., and W. Müller-Warmuth, *Z. Angew. Phys.* **16**, 281 (1963).

Kramer, K. D., W. Müller-Warmuth, and J. Schindler, *J. Chem. Phys.* **43**, 31 (1965).

Krupnov, A. F., and V. A. Skvortsov, *Zh. Eksperim. i Teor. Fiz.* **47**, 1605 (1964).

Lambe, J., and R. Ager, *RSI* **30**, 599 (1959).

Lawson, A. W., and G. E. Smith, *RSI* **30**, 989 (1959).

Lawson, J. D., *Am. J. Phys.* **33**, 733 (1965).

Ledinegg, E., and P. Urgan, *Acta. Phys. Austriaca* **9**, 335 (1955).

Legrand, M., G. Dreyfus, and J. Lewiner, *J. Phys. Lett.* **38**, 439 (1977).

Lewis, R. B., and T. R. Carver, *Phys. Rev. Letters* **12**, 693 (1964).

Lichtenstein, M., J. J. Gallagher, and R. E. Cupp, *RSI* **34**, 843 (1963).

Livingston, R., and H. Zeldes, *J. Chem. Phys.* **44**, 1245 (1966).

Llewellyn, P. M., *JSI* **34**, 236 (1957).

Locher, P. R., and C. J. Gorter, *Physica* **28**, 797 (1962).

Lotsch, H. K. V., *Japan J. Appl. Phys.* **4**, 435 (1965).

Ludwig, G. W., and F. S. Ham, *Paramagnetic Resonance* **2**, 620 (1963).

Lukin, S. N., E. D., Nemchenko, and L. G. Oranskii, *PTE* **17**(4), 116 (1974).

MacDonald, J. R., *RSI* **26**, 433 (1955).

Maiman, T. H., *J. Appl. Phys.* **31**, 222 (1960).

Margenau, H., and G. M. Murphy, *The Mathematics of Physics and Chemistry*, Vol., 1, 2nd ed., Van Nostrand, Princeton, N.J., 1956.

Marr, G. V., and P. Swarup, *Can. J. Phys.* **38**, 495 (1960).

Martinelli, M., L. Pardi, C. Pinzino, and S. Santucci, *Phys. Rev.* **16B**, 164 (1977).

Matenaar, H., and R. N. Schindler, *Messtechnik* **81**(67), 104 (1973).

Maxwell, E., *Progr. Cryogenics* **4**, 123 (1964).

McConnell, H. M., C. Heller, T. Cole, and R. W. Fessenden, *J. Am. Chem. Soc.* **82**, 776 (1960).

McConnell, H. M., and C. H. Holm, *J. Chem. Phys.* **27**, 314 (1957).

McConnell, H. M., and D. B. Chesnut, *J. Chem. Phys.* **28**, 107 (1958).

McLachlan, A. D., *Mol. Phys.* **3**, 233 (1960).

Meyer, E., H. W. Helberg, and S. Vogel, *Z. Angew. Phys.* **12**, 337 (1960).

Meyer, J. W., *MIT Lincoln Lab. Rept. M35-46* (1955).

Mims, W. B., *Phys. Rev.* **133**, A835 (1964).

Misra, H., *RSI* **29**, 590 (1958).

Mock, J. B., *RSI* **31**, 551 (1960).

Mollenauer, L. F., and S. Pan, *Phys. Rev. B.* **6**, 772 (1972).

Moran, P. R., *Phys. Rev.* **135**, A247 (1964).

Mullett, L. B., *Atomic Energy Res. Establ. (Harwell) Rept. G/R 853* (1957).

Mungall, A. G., and D. Morris, *Can. J. Phys.* **38**, 1510 (1960).

Muromtsev, V., A. K. Piskunov, and N. V. Verein, *Radio Eng. Electron. Phys.* **7**, 1129 (1962).

Noon, J. H., E. H. Holt, and J. F. Reynolds, *RSI* **36**, 622 (1965).

Nunome, K., K. Toriyama, and M. Iwasaki, *J. Chem. Phys.* **62**, 2937 (1975).

Okada, F., *Mem. Defense Acad. (Japan)* **11**, 52 (1961).

Okaya, A., *Paramagnetic Resonance*, **2**, 687 (1963).

Okaya, A., and L. F. Barash, *Proc. IRE* **50**, 2081 (1962).

Patrushev, V. L., *Dokl. Akad. Nauk.* **107**, 409 (1956).

Paul, H., *Chem. Phys.* **15**, 115 (1976).

Pellegrini, C., *Alta Frequenza* **24**, 12 (1955).

Piette, L. H., I. Yamazaki, and H. S. Mason, in *Free Radicals in Biological Systems* Academic, New York, 1961, p. 195.

Pipkin, F. M., and J. W. Culvahouse, *Phys. Rev.* **106**, 1102 (1957); **109**, 319 (1958).

Pol'skii, Yu. E., *PTE* **3**, 181 (558) (1963); *Cryogenics* **4**, 141 (1964).

Poole, C. P., Jr., Thesis, University of Maryland, 1958.

Poole, C. P., Jr., and R. S. Anderson, *J. Chem. Phys.* **31**, 346 (1959).

Portis, A. M., *J. Phys. Chem. Solids* **8**, 326 (1959); *Phys. Rev.* **91**, 1071 (1953).

Portis, A. M., and D. T. Teaney, *J. Appl. Phys.* **29**, 1692 (1958); *Phys. Rev.* **116**, 838 (1959).

Pozzolo, V., and R. Zieh, *Atti Accad. Sci. Torino I (Italy)* **97**, 1056 (1962—1963).

Price, R., R. Reed, and C. A. Roberts, *IEEE Trans. Antennas Propagation* **AP-11**, 587 (1963).

Pristupa, A. I., *Zavod. Lab.* **42**, 1344 (1976).

Pristupa, A. I., D. I. Kadyrov, V. S. Sokolov, and V. M. Chibrikin, *Zavod. Lab.* **41**, 695 (1975).

Rajangam, K. B., F. Hai and K. R. MacKenzie, *RSI* **36**, 794 (1965).

Ramo, S., J. R. Whinnery, and J. van Duzer, *Fields and Waves in Communication Electronics*, John Wiley, New York, 1965.

Ranon, U., and D. N. Stamires, *RSI* **41**, 147 (1970).

Raoult, G., and R. Fanguin, *C. R. Acad. Sci.* **251**, 1169 (1960).

Raoult, R., J. Chandezon, M. T. Chenon, A. M. Duclauz, and M. Perrin, *Proc. 12th Colloque Ampere Bordeaux*, 1963, p. 167.

Rataiczak, R. D., and M. T. Jones, *J. Chem. Phys.* **56**, 3898 (1972).

Redhardt, A., *Z. Angew. Phys.* **19**, 310 (1965).

Reichert, J. F., and J. Townsend, *Phys. Rev.* **137**, A476 (1965).

Reintjes, J. F., and G. T. Coate, *Principles of Radar*, 2nd ed., McGraw-Hill, New York, 1952.

Rieckhoff, K. E., and R. Weissbach, *RSI* **33**, 1393 (1962).

RLS-8.

RLS-9.

RLS-10.

RLS-11.

RLS-14.

Robinson, L. C., *J. Appl. Phys.* **34**, 1495 (1963).

Rodbell, D. S., *J. Appl. Phys.* **30**, 1845 (1959).

Rogers, J. D., H. L. Cox, and P. G. Braunschweiger, *RSI* **21**, 1014 (1950).

Rorke, D., *JSI* **43**, 396 (1966).

Rosenbaum, F. J., *RSI* **35**, 1550 (1964).

Roussy, G., *J. Phys.* (*France*) **26**, 64A (1965).

Roussy, G., and M. Felden, *J. Phys.* (*France*) **26**, 11A (1965).

Rusiniak, L., *Phys. Stat. Sol. B* **78**, 265 (1976).

Ruthberg, S., *RSI* **29**, 999 (1958).

Saito, T., *J. Phys. Soc. Japan* **19**, 1232 (1964).

Scaglia, C., *Electron. Lett.* **1**(7), 200 (1965).

Schmalbein, D., A. Witte, R. Rodel, and G. Laukien, *RSI* **43**, 1164 (1972).

Schreurs, J. W. H., G. E. Blomgren, and G. K. Fraenkel, *J. Chem. Phys.* **32**, 1861 (1960).

Schuch, H., F. Seiff, R. Furrer, K. Mobius, and K. P. Dinse, *Z. Naturforsch. A* **29a**, 1543 (1974).

Schweiger, A., and Hs. H. Günthard, *Mol. Phys.* **42**, 283 (1981).

Seehra, M. S., *RSI* **39**, 1044 (1968).

Servant, Y., J. C. Bissey, J. Lissayou, and S. Gharbage, *C. R. Hebd. Seances Acad. Sci. Ser. B* **287**, 121 (1978).

Sharpless, W. M., *Bell System Tech. J.*, **38**, 259 (1959).

Shaw, T. M., G. H. Brooks, and R. C. Gunton, *RSI* **36**, 478 (1965).

Singer, L. S., W. H. Smith, and G. Wagoner, *RSI* **32**, 213 (1961).

Slater, J. C., *Rev. Mod. Phys.* **18**, 441 (1946).

Sloan III E. L., A. Ganssen, and E. C. LaVier, *Appl. Phys. Letters* **4** 109 (1964).

Snowden, D. P., *IRE Trans. Instr.* **I-11**, 156 (1962).

Sogo, P. B.., L. A. Carter, and M. Calvin, *Free Radicals in Biological Systems*, Academic Press, New York, (1961), p. 31.

Sorokin, P. P., G. J. Lasher, and I. L. Gelles, *Phys. Rev.* **118**, 939 (1960).

Spencer, E. G., R. C. LeCraw, and L. A. Ault, *J. Appl. Phys.* **28**, 130 (1957).

Steinert, L. A., *J. Appl. Phys.* **30**, 1109 (1959).

Stoodley, L. G., *Nature* **198**, 1077 (1963).

Stoodley, L. G., *J. Elect. Contr.* **14**, 431 (1963).

Storey, W. H., C. M. Monita, and D. G. Cadena, *Nature* **195**, 963 (1962).

Strandberg, M. W. P., M. Tinksham, I. H. Solt, and C. F. Davis, *RSI* **27**, 546 (1956).

Strauch, R. G., R. E. Cupp, M. Lichtenstein, and J. J. Gallagher, *Proc. Symposium on Quasi-Optics*, Brooklyn Polytechnic Press, Brooklyn, N. Y., 1964, p. 581.

Strnisa, F. V., and J. W. Corbett, *Cryst. Lattice Defects* **5**, 261 (1974).

Szulkin, P., *Bull. Acad. Polon. Sci. Ser. Sci. Tech.* **8**, 639 (1960).

Suematsu, H., and S. Tanuma, *J. Phys. Soc. Japan* **33**, 1619 (1972).

Suzuki, I., Y. Kaneko, and T. Yamamoto, *RSI* **49**, 1706 (1978).

Takach, Sh., and T. Tot, *Acta Tech. Hung.* **42**, 181 (1963).

Talpe, J., *Theory of Experiments in Paramagnetic Resonance*, Pergamon, New York, 1971.

Teaney, D. T., W. E. Blumberg, and A. M. Portis, *Phys. Rev.* **119**, 1851 (1960).

Teaney, D. T., M. P. Klein, and A. M. Portis, *RSI* **32**, 721 (1961).

Teodoresku, I., *Rev. Phys.* (*Rumania*) **7**, 45 (1962).

Teodorescu, M., *Rev. Roum. Phys.* **19**, 989 (1974).

Teodorescu, M., *Rev. Roum. Phys.* **19**, 989 (1974).

Tereshchenko, A. I., V. A. Korobkin, and N. M. Kovtun, *Zh. Tech. Fiz.* **31**, 1388 (1011) (1961).

Tereshchenko, A. I., and V. A. Korobkin, *Zh. Tech. Fiz.* **33**, 214 (154) (1963).

Thomassen, K. I., *J. Appl. Phys.* **34**, 1622 (1963).

Thompson, B. C., G. A. Persyn, and A. W. Nolle, *RSI* **34**, 943 (1963).

Thompson, M. C., Jr., F. E. Freethey, and D. M. Waters, *RSI* **29**, 865 (1958).

Tinkham, M., and M. W. P. Strandberg, *Proc. IRE* **43**, 734 (1955).

Troup, G. J., *Proc. Inst. Radio Engrs. Australia* **23**, 166 (1962).

Ulrich, R., K. F. Renk, and L. Genzel, *IEEE Trans.* **MTT-11, 5** 363 (1963).

Vainstein, L. A., *Zh. Eksperim. i Teor. Fiz.* **44**, 1050 (1963).

Verweel, J., *Philips Res. Rept.* **20**, 404 (1965).

Verkin, B. I., I. M. Dmitrenko, V. M. Dmitriev, G. E. Churilov, and F. F. Mende, *Zh. Tekh. Fiz.* **34**, 1709 (1320) (1964).

Vetter, M. J., and M. C. Thompson, Jr., *RSI* **33**, 656 (1962).

Viet, N. T., *C. R. Acad. Sci.* **4218** (1964).

Vigouroux, B., J. C. Gourdon, P. Lopez, and J. Pescia, *J. Phys. E Sci. Instr.* **6**, 557 (1973).

Villeneuve, A. T., *IRE Trans. Microwave Theory Tech.* **MTT-7**, 441 (1959).

Vonbun, F. O., *RSI* **31**, 900 (1960).

Wagner, W. G., and G. Birnbaum, *J. Appl. Phys.* **32**, 1185 (1961).

Walsh, W. M., Jr., J. Jeener, and N. Bloembergen, *Phys. Rev.* **139**, A1338 (1965).

Weil, J. A., and R. F. C. Claridge, *RSI* **41**, 140 (1970).

Whitehouse, J. E., *RSI* **49**, 61 (1978).

Wilmshurst, T. H., *Nature* **199**, 477 (1963).

Wilmshurst, T. H., W. A. Gambling, and D. J. E. Ingram, *J. Electron. Control* **13**, 339 (1962).

Wilson, P. B., *Nucl. Instr. Methods* **20**, 336 (1963).

Witte, A., *J. Phys. E Sci. Instrum.* **5**, 130 (1972).

Witte, A., G. Laukien, and P. Dullenkope, *Appl. Phys.* **2**, 63 (1973).

Wolfe, J. P., and C. D. Jeffries, *Phys. Rev. B* **4**, 731 (1971).

Yagi, H., M. Inque, T. Tatsukawa, and T. Yamamoto, *Japan J. Appl. Phys.* **9**, 1386 (1970).

Yee, H. Y., *IEEE Trans.* **MTT-13** 256 (1965).

Zimmerer, R. W., *RSI* **33**, 858 (1962).

Zverev, G. M., *PTE* **5**, 109 (930) (1961).

CHAPTER 6

Magnetic Field, Scanning and Modulation

A. Magnetic Field Requirements

In electron spin resonance a large constant magnetic field H is employed for producing resonance absorption in accordance with the resonance condition, which is usually written

$$h\nu = g\beta H_0 \tag{1}$$

in electron spin resonance studies and

$$\omega = \gamma H_0 \tag{2}$$

in nuclear magnetic resonance studies. In these equations g is the g factor, γ is the gyromagnetic ratio, β is the Bohr magneton, and

$$\hbar\gamma = g\beta \tag{3}$$

When one replaces Planck's constant $h = 2\pi\hbar$ and the Bohr magneton β by their numerical values, then one obtains the functional relationship between the resonant frequency ν and its associated magnetic field H_0

$$\nu = 1.400 \times 10^6 g H_0 \tag{4}$$

where ν is in cycles per second, H is in gauss, and g is a dimensionless factor, which is usually close to 2. Most ESR work is done in the X-band microwave region where the magnetic field $H_0 \sim 3400$ G is required for a typical frequency of 9.5 GHz and $g = 2$. Many ESR experiments are performed in the K-band region at 25 GHz (~ 9000 G) and at 35 GHz ($\sim 12,500$ G), and it is only in special applications that higher frequencies or fields are employed. Thus for almost all ESR applications a magnet that produces up to 15 kG is satisfactory. The resonant magnetic fields H_0 corresponding to $g = 2$ are listed in Table 6-1 for several ESR frequencies.

Sometimes one encounters a resonance whose width exceeds 1 kG, and when this is the case, it is preferable to use K band, since X band will give a more distorted lineshape, especially at the low field side. The necessity of observing broad resonance lines increases the required upper limit on the magnet since it

TABLE 6-1

Magnetic Field Requirements for Electron Spin Resonance and Nuclear Magnetic Resonance
(ESR Sample Lengths Are for TE_{102}, Rectangular Resonant Cavities)

Resonance Experiment	Frequency ν, MHz	Sample Length,[a] cm	Coil or Waveguide Width,[b] cm	Typical Width of Narrow Line, mG	Magnetic Field H_0, kG	Magnet Homogeneity Requirement	
						ΔH_0, mG	$\Delta H_0/H_0$, ppm
ESR–S band	3,000	7.2	3.8	100	1.1	10	10
ESR–X band	10,000	2.3	1.3	100	3.6	10	3
ESR–K band	24,000	1.1	0.63	100	8.6	10	1
ESR–Q band	70,000	0.3	0.35	100	25	10	0.4
ESR–1 mm band	300,000	0.08	0.3	100	110	10	0.1
NMR–^{1}H	10	1	1.5	1.0	2.4	0.1	0.04
NMR–^{13}C	10	1	1.5	1.0	9.3	0.1	0.01
NMR–^{39}K	10	1	1.5	1.0	50	0.1	0.002
NMR–^{1}H	100	1	1.5	1.0	24	0.1	0.004

[a] Large inner dimension in waveguide case (limits sample length).
[b] Small outer dimension in waveguide case (limits magnet gap).

is desirable to record the resonance line considerably beyond its inflection point $\left(\frac{1}{2}\Delta H_{pp} + H_0\right)$ on the high field side. A reasonable requirement is the ability to record a K-band resonance out to a field of 1.5 H_0, which corresponds to 13,000 G. This upper limit will allow one to record a K-band resonance with full width $\Delta H_{pp} = 5000$ G out to a field of $\frac{1}{2}\Delta H_{pp}$ beyond its inflection point (derivative half-width). Broader resonance lines are seldom encountered in ESR.

Several transition elements have g factors that deviate considerable from 2 (e.g., 0 to 9 for Fe^{2+} and 1.4 to 7 for Co^{2+}), and these deviations are frequently orientation-dependent. A 13,000-G magnet at $f = 24$ GHz will be unable to reach $g < 1.3$, while the same magnet at $f = 9.6$ GHz will have $g < 0.53$ beyond range. These g-factor limitations are of practical importance in only a few cases since the overwhelming percentage of the tabulated transition element spectra and all free radicals have $g > 1.3$. Thus both linewidth and g-factor considerations indicate that a magnet that reaches 13,000 G is adequate for almost all ESR experiments. In NMR studies such a magnet is capable of reaching most nonzero spin nuclei at 2 MHz (^{39}K is an exception), and in addition it will resonate protons up to 55 MHz. These NMR figures can be computed directly from a table of nuclear gyromagnetic ratios since NMR linewidths seldom exceed 50 G [e.g., see Poole et al. (1965)], and chemical shifts are rarely greater than one part per thousand.

To detect ESR signals it is not sufficient to have a magnet that will supply the required magnetic field strength; it is also necessary for this magnetic field to be very homogeneous over the sample volume. A rule of thumb is that the variations in the magnetic field strength over the sample ΔH_0 should be less than one tenth of the linewidth ΔH_{pp}, and in Table 6-1 values of $\Delta H_0 = \Delta H_{pp}/10$ are tabulated for a typical very narrow ESR line of width 0.1 G. Alkali metals in ammonia (Hutchison and Pastor, 1953) some free radical ions (Hausser, (1962) and radical ion salts (Eichele et al., 1981) have produced linewidths almost a factor 10 below this value, but at present these are exceptional cases, and so the value of $\Delta H_{pp} = 0.1$ G is used in Table 6-1. From this table it may be seen that a magnet for use at X band should have a magnetic field H that is constant to within 10 mG over the sample volume, and in column three we see that the sample is 0.9 in. or 2.3 cm long when it extends over the entire length of the broad dimension of the waveguide. The X-band entry in the last column of the table shows that the relative homogeneity should be

$$\frac{\Delta H_0}{H_0} = \frac{0.01}{3600} = 3 \times 10^{-6} = 3 \text{ ppm} \tag{5}$$

Since the widths of narrow resonance lines are frequently independent of the microwave frequency, in going from S band to the 1-mm band the absolute homogeneity requirement ΔH_0 remains constant at 10 mG, while the relative homogeneity requirement $\Delta H_0/H_0$ decreases from 10 to 0.1 ppm. This behavior is important in magnet stabilization systems since a magnet current stabilizer usually supplies a current I, which is stabilized to deviate by less than a certain $\Delta I/I$ value from the desired current. As a result, for a given stabilizer the absolute current fluctuation will increase with magnetic field strength. Sometimes the homogeneity of a magnet is improved by initially raising the field strength to a high value, and then decreasing it to the desired operating point. Goldberg and Crowe (1975) suggested an inexpensive method of improving magnet performance. Rupp et al. (1976) describe a miniature magnet for ESR. Ganapathy (1977), Sthanapati et al. (1977) and Venofrinovich et al. (1977) recently designed magnet power supplies.

It should be noted that the narrow NMR linewidths listed in Table 6-1 are typically two orders of magnitude below the ESR values, and this puts much more stringent requirements on the NMR magnet stabilization system. In addition, as mentioned above, ESR linewidths almost as narrow as 0.01 G have already been observed, and if still narrower ones are detected, it will become necessary to make use of NMR magnet systems with their increased stability.

The tens of kilogauss fields needed for spectrometers operating at millimeter wavelengths are obtained with superconducting magnets (Favrot et al., 1976; Krueger and Martinelli, 1972; Pavlovskii et al., 1979; Regel and Regel, 1972; Wagner and White, 1979; compare Sec. 9-G of the first edition). For example Galkin et al. (1977) designed a 2-mm-range spectrometer for chemical research.

B. The Magnetic Circuit

In an electrical circuit, Ohm's law relates the applied electromotive force (emf) or voltage V to the current I in terms of the resistance R:

$$\text{emf} = V = IR \tag{1}$$

The resistance of a rod of length l and uniform cross-sectional area A is

$$R = \frac{l}{\sigma A} = \rho \left(\frac{l}{A} \right) \tag{2}$$

where the conductivity σ is an intrinsic characteristic of a substance, and of course the reciprocal of the conductivity σ is the resistivity ρ. The electromotive force between two points a and b is the line integral of the electric field E between the points

$$\text{emf} = \int_a^b \mathbf{E} \cdot \mathbf{dl} \tag{3}$$

In a magnetic circuit the analogue of Ohm's law is

$$\text{mmf} = \mathcal{R} \phi \tag{4}$$

where the magnetic flux ϕ through the circuit is proportional to the applied magnetomotive force (mmf). The reluctance $\mathcal{R}$ of a rod of length l and cross-sectional area A is related to the permeability μ by the expression

$$\mathcal{R} = \frac{l}{\mu A} \tag{5}$$

$$= \frac{l \times 10^7}{4 \pi A (\mu / \mu_0)} \tag{6}$$

where μ_0 is the permeability of free space, and μ / μ_0 is relative permeability. Often the magnetic flux ϕ equals the magnetic induction B times the cross-sectional area A:

$$\phi = BA \tag{7}$$

If a magnetic circuit such as an iron ring is magnetized by n turns of wire carrying the current I, then the magnetomotive force in the circuit is nI:

$$\text{mmf} = nI \tag{8}$$

another definition of mmf is the analogue of Eq. (3):

$$\text{mmf} = \int_a^b \mathbf{H} \cdot \mathbf{dl} \tag{9}$$

The magnetic flux ϕ is constant in a series magnetic circuit, so for several successive cross sections $A_1, A_2, \ldots$ one has

$$\phi = A_1 B_1 = A_2 B_2, \ldots \tag{10}$$

and B is constant in a magnet with a constant cross-sectional area that employs cylindrical pole caps as shown in Figs. 6-1 and 6-2a. When tapered pole caps such as the ones shown in Fig. 6-2b are employed, the flux in the air gap is approximated by

$$\phi = \left(\frac{\pi D_2^2}{4} \right) B \tag{11}$$

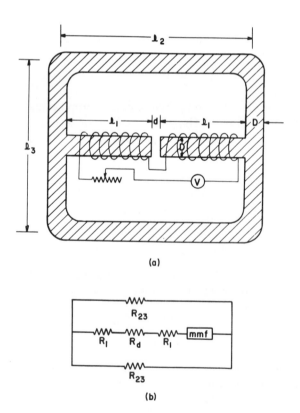

(a)

(b)

Fig. 6-1. Schematic diagram of (a) an electromagnet and (b) its equivalent circuit. The magnet consists of a yoke (ℓ_2, ℓ_3), center iron pieces (ℓ_1), and gap (d).

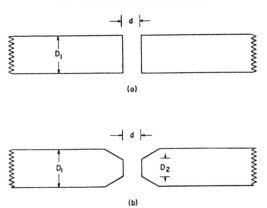

Fig. 6-2.   Magnet with (a) cylindrical pole caps and (b) tapered pole caps.

instead of the cylindrical value

$$\phi = \left(\frac{\pi D_1^2}{4}\right) B \qquad (12)$$

If ϕ remains constant in both cases, B for the tapered case will increase by the factor $(D_1/D_2)^2$ relative to the cylindrical case. Unfortunately ϕ does not remain constant, because the tapered air gap presents a greater reluctance than the cylindrical gap, and this limits ϕ in accordance with the relation

$$\text{mmf} = \phi\left[\mathcal{R}_{\text{gap}} + \mathcal{R}_{\text{yoke}}\right] \qquad (13)$$

for a constant applied magnetomotive force. When $\mathcal{R}_{\text{gap}} \ll \mathcal{R}_{\text{yoke}}$ the taper does not change ϕ, but in most practical magnet systems $\mathcal{R}_{\text{gap}}$ is comparable to $\mathcal{R}_{\text{yoke}}$, and the magnitude of the magnetic flux is strongly dependent on the amount of taper.

   The preceding paragraph defined the terms employed in a magnetic circuit analysis. These results are summarized in Table 6-2. This presupposes that all of the flux remains in the circuit just as Ohm's law is based on the fact that all of the electrical current remains in the conductor. Unfortunately a considerable amount of flux actually does leak out, so the concept of a magnetic circuit is only an approximation. This is particularly the case when a magnetic circuit has an air gap, since a great deal of flux leakage occurs at the gap, and the greater the gap the more the leakage. Nevertheless some insight can be gained into the way an electromagnet works by analyzing a typical electromagnet from the viewpoint of its magnetic circuit.

   Figure 6-1a shows an electromagnet with a double yoke of permeability μ, and a uniform cross section A in all of its parts. A voltage generator V actuates the coils, and a resistor permits an adjustment of the coil current. The

TABLE 6-2

Magnetic Circuit Quantities and Their Electrical Analogues

Magnetic Circuit			Electric Circuit		
Quantity	Unit, mks	Symbol	Quantity	Unit, mks	Symbol
Magnetomotive force	At	mmf	Electromotive force or voltage	V	emf, V
Magnetic flux	Wb	Φ	Current	A	I
Inverse permeability	M/H	$1/\mu$	Conductivity	$1/M\Omega$	σ
Reluctance	At/Wb	$\mathcal{R}$	Resistance	Ω	R

reluctance $\mathcal{R}_1$ of each center iron piece of length l_1 is

$$\mathcal{R}_1 = \frac{1}{\mu_0} \frac{l_1}{A(\mu/\mu_0)} \tag{14}$$

and the reluctance $\mathcal{R}_{23}$ of each outer half of the yoke of length $l_2 + l_3$ is

$$\mathcal{R}_{23} = \frac{1}{\mu_0} \frac{l_2 + l_3}{A(\mu/\mu_0)} \tag{15}$$

In the air gap $\mu = \mu_0$, so the reflectance $\mathcal{R}_d$ of the gap is

$$\mathcal{R}_d = \frac{1}{\mu_0} \frac{d}{A} \tag{16}$$

where the gap is assumed to have the same cross-sectional area A. This approximation is only valid for a gap-to-pole piece diameter ratio d/D much less than 1. The equivalent magnetic circuit is shown in Fig. 6-1b, where the mmf from the two activating coils is represented by one "mmf generator." The parallel arrangement of the two parts of the yoke $\mathcal{R}_{23}$ leads to the relation

$$\text{mmf} = \phi\left[2\mathcal{R}_1 + \mathcal{R}_d + \tfrac{1}{2}\mathcal{R}_{23}\right] \tag{17}$$

$$= \frac{B}{\mu_0}\left(\frac{2l_1 + \tfrac{1}{2}(l_2 + l_3)}{\mu/\mu_0} + d\right) \tag{18}$$

For a typical magnetic material

$$\frac{\mu}{\mu_0} \sim 10^3 \tag{19}$$

so the reluctance of the air gap is the dominant factor in limiting the available magnetic field strength in the air gap. For a typical electromagnet

$$d \ll l_1 \tag{20}$$

$$2l_1 \sim l_2 \sim l_3 \tag{21}$$

and using these relations in Eqs. (8) and (18), one obtains

$$B = \frac{\mu_0 n I}{l_2} \left[\frac{1}{d/l_2 + 2 \times 10^{-3}} \right] \tag{22}$$

Frequently $l_2 \sim 200$ cm, and under these conditions the gap contributes half of the reluctance when

$$d = 2l_2 \times 10^{-3} = 0.4 \text{ cm} \tag{23}$$

Usually d exceeds this value in ESR applications.

The preceding calculation assumed that the cross-sectional area A of the gap is the same as that of the section l_1. For this assumption to be tenable it is necessary for the gap radius $(A/\pi)^{1/2}$ to be considerably greater than the pole piece separation d.

$$d \ll \left(\frac{A}{\pi} \right)^{1/2} \tag{24}$$

One may estimate the gap diameter at 30 cm, which leads to the requirement

$$d \ll 15 \text{ cm} \tag{25}$$

In practice one usually finds d between 3 and 8 cm, or in other words

$$d \sim \frac{(A/\pi)^{1/2}}{3} \tag{26}$$

and as a result, a considerable amount of excess flux leaks out to the space surrounding the pole caps.

Since typical magnet gaps are close to 5 cm, one may conclude that: (1) from Eq. (23) the reluctance of the magnet gap is the dominant factor in determining the intensity of the magnetic field; (2) from Eq. (25) the assumption that the equivalent gap area approximates the pole piece area is not valid; (3) using tapered pole caps is less effective than the ratio $(D_1/D_2)^2$ obtained from Eqs. (11) and (12). Nevertheless, the magnetic circuit concept is helpful in providing a qualitative understanding of the way magnets function, and in describing qualitatively the dependence of the magnet gap on the magnetic field strength.

C. Magnetometers

The strength of a magnetic field is determined by means of a magnetometer or gaussmeter. In ESR work, NMR magnetometers are ordinarily employed, but before discussing them, a brief description of other magnetometer principles will be given.

The Hall Effect

When a metal of length l, width w, thickness t, and resistivity ρ, carrying a current I, is placed in a magnetic field B directed along t perpendicular to the current direction, a potential difference V_h appears across the width w at right angles to both I and B. The magnitude of this voltage is given by the equation

$$V_h = \frac{R_H I B}{t} \tag{1}$$

where R_H is the Hall coefficient in volt-centimeters/ampere-oersted, I is in amperes, B is in gauss, and t is in centimeters. In p-type or hole semiconductors, the Hall voltage is positive, and in n-type or electron semiconductors it is negative. Second-order effects cause the Hall voltage V_h to deviate slightly from the linear dependence on the magnetic field. Hall probes covering the range from 50 to 20,000 G are available commercially, and several manufacturers incorporate them into their magnet control systems. This eliminates hysteresis effects. In addition, it allows one to set the magnet to a particular field value and scan over a predetermined inverval in gauss.

Peaking Strips

This magnetometer employs a ferromagnetic alloy with an almost rectangular hysteresis loop.

Rotating and Vibrating Coils

A voltage is induced in a coil that moves (e.g., rotates or vibrates) in a magnetic field in such a manner that the total flux through the coil changes.

Forces on Conductors Carrying Currents

When a conductor carrying a current I is placed in a magnetic field B, it experiences a force proportional to $\mathbf{I} \times \mathbf{B}$.

Nuclear Magnetic Resonance Magnetometers

Nuclear magnetic resonance results from the interaction of a nuclear magnetic moment $g\beta_N I$ with a magnetic field H in the same way that electron spin resonance results from the interaction of an electronic magnetic moment $g\beta S$

with a magnetic field. The magnitudes of the interactions

$$h\nu = \hbar\gamma H = g\beta_N I \cdot \mathbf{H} \quad \text{NMR} \tag{2a}$$

$$h\nu = g\beta \mathbf{S} \cdot \mathbf{H} \quad\quad\quad \text{ESR} \tag{2b}$$

differ by three orders of magnitude since the nuclear magneton β_N incorporates the protonic mass m_p in place of the electronic mass m that enters the denominator of the expression for the Bohr magneton β [compare Eq. 1-C-3]. The nuclei ^{1}H, ^{2}H(D) and ^{7}Li used in magnetometers have nuclear spins $I = \frac{1}{2}$, 1 and $\frac{3}{2}$ respectively. The proton NMR signal corresponding to a typical $g = 2$ ESR X-band resonance (3300 G) is about 14 Mc. More precisely, in a 1000-G field the $g = 2.000$ ESR resonances occur at 2,799.4 Mc, proton (^{1}H) NMR comes at 4.2577 Mc, deuteron (D) NMR appears at 0.6536 Mc, and (^{7}Li) NMR is detected at 1.6547 Mc. Since the principles of NMR are no doubt familiar to most ESR specialists they are not reviewed here. For further details consult one of the books by Aleksandrov (1966), Ault and Dudek (1976), Emsley et al. (1965), Harris (1976), Jackman and Cotton (1975), Lynden-Bell and Harris (1975), or Poole and Farach (1972). Specific NMR magnetometers are described by Brandwein and Lipsicas (1970), de Martini and Lucchini (1964), Garwin and Patlach (1965), Gordienco and Antonenko (1963), Hartmann (1972), Idoine and Brandenberger (1971), Muha (1965), Niemela (1964), Pierce and Hicks (1965), and Wright and van der Beken (1973). Low-temperature NMR magnetometers for use with superconducting magnets have been designed that place only the rf coil in the cryostat (Higgins and Chang, 1968; Hill and Hwang, 1966; Maxfield and Merrill, 1965) while others refrigerate the whole oscillator (Alderman, 1970; Miyoshi and Cotts, 1968; Pierce and Hicks, 1965; Slavin, 1972; Yagi et al., 1973). Sari and Carver (1970) designed a proton magnetometer for room temperature access superconducting magnets.

Wampler et al. (1975) designed a tunable low-temperature NMR oscillator for measuring the magnetic fields of superconducting magnets. Figure 6-3 presents the circuit diagram of the NMR probe and the regulator amplifier of the detection system. The probe contains a Hartley marginal oscillator that employs two light-emitting GaAsP varactor diodes (LEDs) D1 and D2 and the coil LI as the tank circuit. The oscillator is driven by the Si n-channel dual-gate MOSFET 3NI transistor TI. The signal from the probe enters the regulator through a coaxial cable and is amplified 20 times by the broad-band μA 733 amplifier, decoupled by the emitter follower T3, demodulated by two 0A182 diodes, filtered in an RC filter with a time constant τ of 50 μsec and further amplified by transister T4. The output signal is fed to a lock-in detector for a further decrease in signal-to-noise ratio. The low gyromagnetic ratio nuclei ^{27}Al and ^{51}V that are employed in the probe permit the measurement of magnetic fields in superconducting magnets within an accuracy of 0.1G.

An oscilloscope trace of the nuclear resonance signal is presented in Fig. 6-4, and the appearance of the "wiggles" shown in Fig. 6-4a is a measure of the

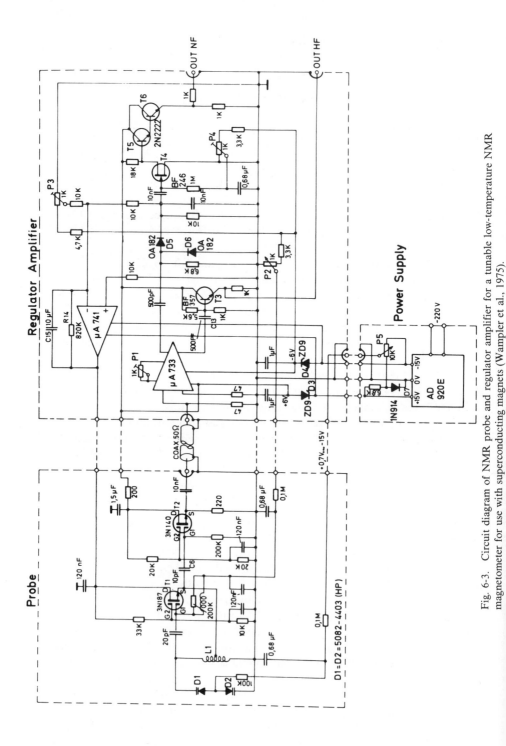

Fig. 6-3. Circuit diagram of NMR probe and regulator amplifier for a tunable low-temperature NMR magnetometer for use with superconducting magnets (Wampler et al., 1975).

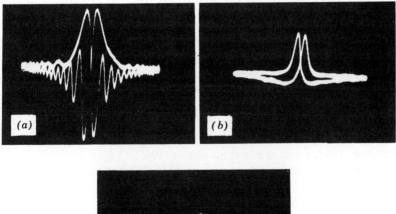

Fig. 6-4. Oscilloscopic presentation of an NMR magnetometer signal. (a) Trace on an oscilloscope of proton resonance in 0.5 cm^3 of mineral oil in 6400-G field, illustrating use as a field meter. Total sweep here is about 0.5 G. The "wiggles" are very prominent owing to the good homogeneity of the field over the sample. Sweep amplitude is about 0.5 G peak-to-peak. (b) Resonance in same magnet as (a), but in aqueous solution with added paramagnetic MnSO$_4$. Sweep amplitude is about 5 G, peak-to-peak. (c) Oscilloscope trace in field of 360 G with only about 0.5 G inhomogeneity over the sample. Sweep amplitude is about 10 G, peak-to-peak (Pound and Knight, 1950).

magnetic field homogeneity. The less homogeneous the magnetic field, the less prominent are the wiggles. Water samples for proton sources are normally doped with a paramagnetic salt such as 1N MnSO$_4$ to shorten the relaxation times. It is also convenient to saturate the solution with LiCl so that ^{7}Li NMR may be used in high magnetic fields. The ^{7}Li resonance is weaker because of the lower concentration of the ^{7}Li nuclei relative to the protons, and its presence in the solution does not effect the ^{1}H resonance. The magnetometer has been operated from 250 to 10,000 G (1 to 45 MHz) with protons, and if ^{7}Li is used, the range is extended to 25,000 G. The accuracy of the measurements is limited by the magnetic field homogeneity and the accuracy of the frequency measuring equipment. In practice, one may measure field values at points corresponding to 100-kHz or 1-MHz crystal marker frequencies. For less accurate work, the capacitor dial may be calibrated against crystal markers, a frequency meter such as the BC-221, or a commercial frequency counter (plus transfer oscillator).

NMR magnetometers are convenient for use with ESR spectrometers because they are usable over a very wide range of magnetic fields, and they can easily reach an accuracy of 1 part in 10^5. The frequency can be measured by standard crystal-controlled oscillators, and these can be calibrated against the National Bureau of Standards (Washington, D.C.) radio station WWV, which broadcasts standard frequencies at 5, 10, 15, and 20 MHz accurate to 2 parts in 10^8 (2.5-, 25-, 30-, and 35-MHz frequencies are also broadcast at much lower power by WWV). These frequencies are modulated at one Hz with a 5-msec pulse (with an accuracy of 1 μsec) at the start of each record. A proton magnetometer may be used to provide accurate field calibration over narrow ranges by utilizing sidebands generated by a frequency-modulation technique (Mito et al., 1964; Lancaster and Smallman, 1965).

An ESR magnetometer operating at typical NMR frequencies may be employed to measure weak magnetic fields. Bourdel et al. (1970) discussed optimization of an ESR magnetometer.

D. Magnet Scanning

The condition for energy absorption by the spin system is

$$h\nu \approx g\beta H \tag{1}$$

If the magnetic field H is kept constant and the frequency ν is varied, then the amount of energy absorbed will vary in accordance with the shape function $Y(\nu)$ where $Y(\nu)$ often has one of the two forms

$$Y(\nu) = y_m \exp\left[-0.693\left(\frac{\nu - \nu_0}{\frac{1}{2}\Delta\nu_{1/2}}\right)^2\right] \quad \text{Gaussian} \tag{2}$$

$$Y(\nu) = \frac{y_m}{1 + \left[(\nu - \nu_0)/\frac{1}{2}\Delta\nu_{1/2}\right]^2} \quad \text{Lorentzian} \tag{3}$$

In these expressions $\Delta\nu_{1/2}$ is the half-amplitude full linewidth in frequency units and y_m is the peak amplitude at $\nu = \nu_0$. If, on the other hand, the frequency is maintained constant and the magnetic field is varied, then one has the corresponding shape functions

$$Y(H) = y_m \exp\left[-0.693\left(\frac{H - H_0}{\frac{1}{2}\Delta H_{1/2}}\right)^2\right] \quad \text{Gaussian} \tag{4}$$

and

$$Y(H) = \frac{y_m}{1 + \left[(H - H_0)/\frac{1}{2}\Delta H_{1/2}\right]^2} \quad \text{Lorentzian} \tag{5}$$

where $\Delta H_{1/2}$ is the half-amplitude full linewidth in magnetic field units. These lineshapes are displayed graphically in Figs. 12-6 and 12-8. For most ESR resonances one can set

$$\Delta\nu_{1/2} = \left(\frac{g\beta}{h}\right)\Delta H_{1/2} = \left(\frac{\gamma}{2\pi}\right)\Delta H_{1/2} \tag{6}$$

where the gyromagnetic ratio γ and g factor are known from the resonance condition.

There are several reasons that it is not convenient to vary or scan the microwave frequency:

1. The microwave generator (klystron) power output is strongly dependent on the frequency, so for best results the instrumentation would become more complex by the addition of a power stabilizer. Individual klystrons of the same model number vary in their frequency-power characteristics.
2. The tuning of the microwave transmission line is frequency-sensitive, and it is not feasible to automatically tune the frequency-sensitive components while scanning since they have nonlinear responses. More specifically, some of the components that would have to be continuously retuned include the resonant cavity dimensions, the cavity iris, the slide screw tuner, the isolater (or attenuator), and the (crystal) detector tuning stubs and plunger.
3. If the generator is a klystron, the mechanical and electronic tuning would have to be synchronized to maintain the resonant cavity pip at the top of the klystron mode during the scan.
4. Typical klystrons can only be varied by 5 or 10% above and below their center frequency, and so very broad resonances can only be scanned in frequency over a fraction of their linewidth. A backward wave oscillator with its lack of resonant structures, and its much wider frequency range, would be preferable in this application.

When the magnetic field is scanned, on the other hand, all of the above difficulties are automatically eliminated: (1) The microwave generator output is independent of the magnetic field strength. (2) The microwave tuning adjustments are independent of the magnetic field setting. (3) No adjustments of the microwave generator are necessary. (4) It is possible to scan from zero to several times the resonant magnetic field strength. As a result of these considerations, the magnetic field is normally scanned in ESR studies. Erickson (1966), however, describes an ESR spectrometer in which the frequency is scanned by moving the upper wall of the resonant cavity. In straight microwave spectroscopy where resonant cavities generally are not employed, the frequency is usually scanned, and the resulting tuning problems often produce a drifting baseline.

A simple magnet scanner may consist of a variable resistance in series with the magnet coils, as shown in Fig. 6-5a. If the variable resistor or rheostat R is

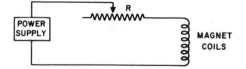

Fig. 6-5. Circuit for varying the current through the magnet coils in a linear manner by varying the resistance in series with them.

varied by a motor at a uniform rate the current through the magnet coils will also vary uniformly with time. It is preferable to scan the magnet linearly in field strength rather than linearly in current in order to avoid hysteresis effects. This can be done by making use of the voltage output of a Hall effect probe and varying the error voltage in the stabilization feedback loop of the magnet power supply. Electronic sweep units have been described by Blume and Williams (1964), Cook (1962), Cousins et al. (1963), Jung (1964), and Strandberg et al. (1956). Hikada (1973) designed a digital sweep unit. Štirand (1962) discusses the effect of the scanning rate on the recorded lineshape and position. For further details, see Sec. 11-F.

Dvornikov and Grebennik (1970) and Sura (1977) discussed automatic magnetic field calibration for ESR spectra recording, Ackerman et al. (1976) reported a numerical device for magnetic field calibration and Hatch et al. (1977) published an article on field sweep calibration. Andriessen (1976) described a magnetic field standard covering a wide field range. Ishchenko and Okulov (1975) designed an automatic system of markers for the field and frequency. Rupp et al. (1976) described a miniature magnet for ESR, and Goldberg and Crowe (1975) report an inexpensive method for improving magnet performance.

E. DC Detection

When the magnetic field is scanned through the region of resonance, the spin system in the resonant cavity absorbs a small amount of energy from the microwave (rf) magnetic field H_1, and also produces a slight change in the resonant frequency of the microwave cavity. These two factors produce a change in the amount of microwave power that is incident on the detector. For a strongly paramagnetic sample, the variation of the incident power will manifest itself by changing the detector output signal. For example, the rectified microwave power from a crystal detector may be observed by connecting a milliammeter in series with it, or the temperature change in a bolometer may be deduced from the resistance change measured by a voltmeter connected across it. These crude detection techniques can be employed with broad, intense lines (Shaffer et al. 1976), but for most samples the sensitivity is too low. Isaev-Ivanov et al. (1976) describe a method of recording spectra sans field modulation.

If the ESR sample is strong enough so that sensitivity is not a problem, then direct detection is capable of producing lineshapes that are free of modulation distortions. Both source and field modulation distort the lineshape when the modulation amplitude is not very small compared to the linewidth ΔH expressed in gauss, or when the modulation frequency is not much less than the linewidth $\Delta\nu$ expressed in frequency units [see Eq. (6-D-6)]. One instrumental shortcoming of the direct detection scheme is the fact that dc amplifiers tend to be more troublesome than ac amplifiers.

F. Source Modulation

A source-modulated spectrometer is one in which the microwave power is modulated and the ESR signal is detected and amplified at the modulation frequency. A large modulation amplitude may be obtained by switching an isolator on and off with a sawtooth voltage so that it alternately transmits first all of the microwave power, and then only a very small fraction of it. Square wave modulation of the reflector voltage shown in Fig. 3-6 can also turn the microwave power on and off (see Whitford, 1961). A combination of frequency and amplitude modulation will result if the klystron reflector has impressed on it a sinusoidal or sawtooth waveform whose amplitude is less than the width of the klystron mode, as Fig. 3-6 indicates. Coumes and Ligeon (1964) simultaneously modulated the beam and reflector voltages to achieve a broad bandwidth.

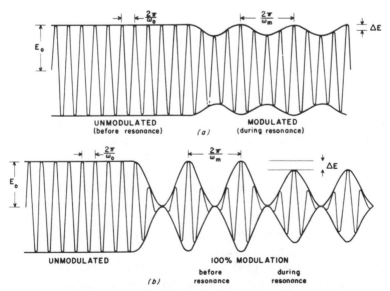

Fig. 6-6. Microwave signal before and during resonance absorption: (a) using magnetic field modulation with the condition that the modulation amplitude E_m equals the dc absorption ΔE, and (b) using 100% source power modulation. Time increases to the right, and the microwave ($2\pi/\omega_0$) and modulation ($2\pi/\omega_m$) periods are indicated.

Figure 6-6*b* shows the microwave waveform at the detector before source modulation, after source modulation, before resonance absorption, and after resonance absorption. At X band the microwave period $2\pi/\omega_0$ is about 10^{-10} sec, and a modulation frequency of 100 kHz corresponds to a period of $2\pi/\omega_m = 10^{-5}$ sec. The modulation shown on this figure corresponds to the output of a sine-wave modulated isolator. It should be noticed that the amplitude of the modulation envelope before resonance is E_0, and during resonance it becomes $E_0 - \Delta E$, where

$$\frac{\Delta E}{E_0} \ll 1 \tag{1}$$

A different tuning condition could cause E_0 to increase by ΔE instead of decreasing, as assumed in Fig. 6-6. The detector demodulates or removes the microwave carrier frequency, and only passes the signal at ω_m.

For optimum sensitivity, a source-modulated spectrometer requires the same precise power stability as the dc detection spectrometer described in the last section.

In order to obtain maximum sensitivity when using field modulation, it is necessary to employ modulation amplitudes comparable to the linewidth. In practice it is difficult to modulate at amplitudes exceeding 30 G, and so the sensitivity decreases for very broad lines. For example, if the linewidth exceeds 1,000 G, the sensitivity with a 30-G modulation amplitude decreases more than a factor of 30. Source modulation does not become insensitive when broad lines are measured, and so it is sometimes applied to such studies.

For accurate lineshape determinations, it is necessary to use a source modulation amplitude that is small relative to the linewidth, which necessitates a loss of sensitivity. This is particularly important when one suspects the presence of unresolved hyperfine structure.

The use of source modulation enables one to work at low temperatures without the inconvenience of inserting modulation coils into the Dewar. In the past, most low-temperature experimental setups have either incorporated such modulation coils within the Dewar or mounted them on the polecaps.

References to several source modulated spectrometers were given in Sec. 10-C of the first edition. Additional ones were described by Gourdon et al. (1970, 1973, 1973) and Pescia (1965). The spectrometer of Koepp (1969) employs modulated microwave field polarization. Gunthard (1974) discusses the response of chemical systems to modulated excitation. Vigouroux et al. (1976) employed source modulation and monitored the longitudinal conduction ESR (compare Sec. 12-I) to minimize the distortion of the lineshape. Source modulation is also useful for the study of slow molecular motions by saturation transfer methods (compare Sec. 13-G; Balasurbramanian et al., 1978). Rinehart and Legan (1964) describe a power leveler that regulates milliwatts of K-band power to within 0.5%.

Den Boef and Henning (1974) designed a strain-modulated electron spin resonance spectrometer that modulates the internal crystal field instead of the external magnetic field by means of an alternating mechanical stress impressed on the sample. Collins et al. (1971), den Boef and Henning (1974), Henning et al. (1974, 1975, 1976, 1978), Robinson et al. (1974, 1975, 1977, 1978), and Mitin (1975) investigated various materials with such a spectrometer (compare Sec. 14-J).

G. Magnetic Field Modulation

Most electron spin resonance spectrometers incorporate magnetic field modulation, and therefore it will be discussed more extensively than the dc detection and source modulation schemes. When the magnetic field is modulated at the angular frequency ω_m, an alternating field $\frac{1}{2}H_m \sin \omega_m t$ is superimposed on the constant magnetic field $(H_0 + H_\delta)$. This "constant" magnetic field is ordinarily swept linearly over the range ΔH_0 from $(H_0 - \frac{1}{2}\Delta H_0)$ to $(H_0 + \frac{1}{2}\Delta H_0)$ in a time t_0, where H_0 is the magnetic field strength at the center of the scan. At any time t during the scan, the instantaneous magnetic field strength H is given by

$$H = H_0 + H_\delta + H_{\text{mod}} \tag{1}$$

$$= H_0 + \Delta H_0 \left(\frac{t}{t_0} - \frac{1}{2} \right) + \frac{1}{2} H_m \sin \omega_m t \tag{2}$$

where

$$H_\delta = \Delta H_0 \left(\frac{t}{t_0} - \frac{1}{2} \right) \tag{3}$$

It is assumed that the scan is slow enough so that there are many cycles of the modulation frequency during the passage between the peak-to-peak (or half-amplitude) points of each resonant line, since under these conditions one may consider the magnetic field $(H_0 + H_\delta)$ as effectively constant. There are other conditions that must be satisfied for a true slow passage experiment, such as the necessity of scanning through the resonant line in a time that is long relative to the spin-lattice and spin–spin relaxation times, but these will not concern us here.

The mechanism whereby the magnetic field modulation is transformed to microwave power modulation is shown in Fig. 6-7. The field modulated sine wave $\sin \omega_m t$ is converted by the nonlinear lineshape to a complex signal $F(H)$ that is a superposition of the fundamental modulation frequency ω_m and a

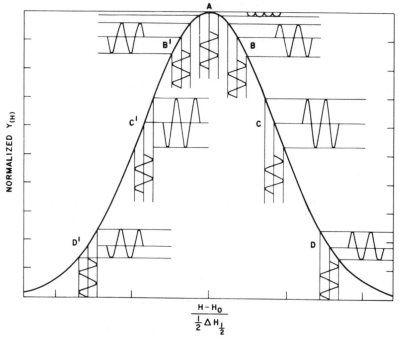

Fig. 6-7. The ESR signal produced at various points on the resonant line in a magnetic-field-modulated spectrometer. The vertical magnetic field modulation interacts with the bell-shaped absorption curve [χ' or $Y_{(H)}$] to produce the horizontal ESR signal.

large number of harmonics of ω_m:

$$F(H) = \sum_{n=0}^{\infty} \left[a_n(H)\cos n\omega_m t + b_n(H)\sin n\omega_n t \right] \qquad (4)$$

The signal C at the inflection point of Fig. 6-7 is composed primarily of the fundamental $\sin \omega_m t$, while that at the center of the line A has the second harmonic $\cos 2\omega_m t$ as its principal component. Between these two points at B, the ESR signal has both the fundamental, the second harmonic, and higher harmonic terms in its Fourier series. Below the inflection point at D, the waveform resembles that at B, but is inverted and shifted by 180°. Figure 6-6a shows the actual waveform at the detector, with the microwave carrier ω_0 modulated by the modulation fundamental ω_m at the inflection point C. The ESR signals in Figs. 6-7 and 6-8 are obtained by demodulating waveforms of the type shown in Fig. 6-6a.

If the situations depicted in Fig. 6-7 are approximated by the idealized straight-line cases shown in Fig. 6-8, then the resulting ESR signals have forms that may be easily Fourier-analyzed. The Fourier series of two half-wave and one full-wave rectified signals are given in Fig. 6-9. Using these series one may write down the expressions of the four waveforms shown in Fig. 6-8. The first

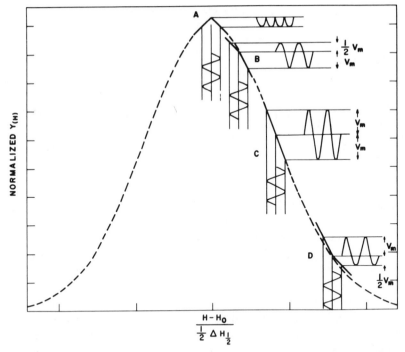

Fig. 6-8. The ESR signal produced at various points on the resonant line using the linear approximation to the lineshape. The wave forms for B', C', and D' of Fig. 6-7 are omitted for clarity.

few terms of these Fourier series are

$$A \qquad V = \frac{V_m}{\pi}\left[-\frac{1}{2} + \frac{1}{3}\cos 2\omega_m t + \frac{1}{15}\cos 4\omega_m t + \cdots\right] \tag{5}$$

$$B \qquad V = \frac{V_m}{\pi}\left[-\frac{1}{2} + \frac{3\pi}{4}\sin \omega_m t + \frac{1}{3}\cos 2\omega_m t + \frac{1}{15}\cos 4\omega_m t + \cdots\right] \tag{6}$$

$$B' \qquad V = \frac{V_m}{\pi}\left[-\frac{1}{2} - \frac{3\pi}{4}\sin \omega_m t + \frac{1}{3}\cos 2\omega_m t + \frac{1}{15}\cos 4\omega_m t + \cdots\right] \tag{7}$$

$$C \qquad V = V_m\sin \omega_m t \tag{8}$$

$$C' \qquad V = -V_m\sin \omega_m t \tag{9}$$

$$D \qquad V = \frac{V_m}{\pi}\left[\frac{1}{2} + \frac{3\pi}{4}\sin \omega_m t - \frac{1}{3}\cos 2\omega_m t - \frac{1}{15}\cos 4\omega_m t - \cdots\right] \tag{10}$$

$$D' \qquad V = \frac{V_m}{\pi}\left[\frac{1}{2} - \frac{3\pi}{4}\sin \omega_m t - \frac{1}{3}\cos 2\omega_m t - \frac{1}{15}\cos 4\omega_m t - \cdots\right] \tag{11}$$

It should be emphasized that Eqs. (5)–(11) merely approximate the true state

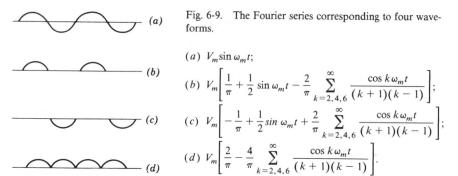

Fig. 6-9. The Fourier series corresponding to four wave-forms.

(a) $V_m \sin \omega_m t$;

(b) $V_m \left[\dfrac{1}{\pi} + \dfrac{1}{2} \sin \omega_m t - \dfrac{2}{\pi} \displaystyle\sum_{k=2,4,6}^{\infty} \dfrac{\cos k \omega_m t}{(k+1)(k-1)} \right]$;

(c) $V_m \left[-\dfrac{1}{\pi} + \dfrac{1}{2} \sin \omega_m t + \dfrac{2}{\pi} \displaystyle\sum_{k=2,4,6}^{\infty} \dfrac{\cos k \omega_m t}{(k+1)(k-1)} \right]$;

(d) $V_m \left[\dfrac{2}{\pi} - \dfrac{4}{\pi} \displaystyle\sum_{k=2,4,6}^{\infty} \dfrac{\cos k \omega_m t}{(k+1)(k-1)} \right]$.

of affairs, but nevertheless, they do furnish a mathematical insight into the actual physical situation, and so their form will be examined in detail. In Eqs. (5)–(7), (10), and (11) the sixth harmonic has 3/7 the amplitude of the fourth, and there are no odd harmonics in the ESR signals. One should pay careful attention to the signs of these seven equations. At the same point on opposite sides of a symmetrical absorption line (e.g., B and B') the fundamental term $\sin \omega_m t$ has opposite signs, while all the harmonics plus the constant term have the same sign. The constant term and all of the harmonics have one sign for amplitudes above the inflection points C and C' and the opposite sign for amplitudes below these points, while the fundamental does not obey this relation. The fundamental is absent at the top of the resonance line A, and all the other terms are absent at the inflection points C and C'.

A careful study of the first and second derivative resonance absorption curves shown in Figs. 12-1, 12-4 and 12-5 shows that they obey the above sign rules for the fundamental and second harmonic terms, respectively, in the Fourier series. This is because to first order the amplitude of the fundamental term $\sin \omega_m t$ is a measure of the slope of the resonance absorption curve, while the second harmonic term $\cos 2\omega_m t$ measures the deviation of this slope from linearity.

An analysis similar to the one given above may be made of the dispersion mode χ' of the resonance lines shown in Figs. 12-10 and 12-11, and the carrying out of this analysis is left as an *exercitatio lectori*.

Bassompierre and Pescia (1962), Bugai (1963), Doyle (1962), and Goldman (1963) discuss the effect of modulation on magnetic resonance lines. Loth et al. (1974) obtained modulated excitation ESR spectra of hydroxy phenoxy radicals. Clerj and Lambert (1971) reported a high-sensitivity field-modulated spectrometer. Blankenburg et al. (1971) designed a 125-kHz crystal-controlled field-modulation oscillator. Faulkner et al. (1967) and Wilmshurst (1967) reported transistorized power amplifiers for providing field modulation from commercial oscillators. Stesmans and van Meijel (1977) described a simple, large modulation field arrangement for use in CESR cavities. Llewellyn (1957) sent several amperes through a loop inside a cavity to produce field modulation. Edgar (1975) observed an unusual phase-reversal effect in a strongly anisotropic material (Er^{3+} in CaF_2) due to nonparallelism of the static and modulating magnetic fields.

H. Effect of Modulation Amplitude on the Resonance Line

Wahlquist (1961) studied the effect of magnetic field modulation on a Lorentzian-shaped line (see also Arndt, 1965). He combined Eq. (6-D-5) with Eq. (6-G-2) to obtain

$$Y(H) = \frac{y_m}{1 + \left(1/\tfrac{1}{2}\Delta H_{1/2}\right)^2 \left[\Delta H_0(t/t_0 - \tfrac{1}{2}) + \tfrac{1}{2}H_m\sin \omega_m t\right]^2} \tag{1}$$

This is expanded in a Fourier series

$$Y(H) = y_m\left[a_0 + \sum_{n=1}^{\infty} a_n(\Delta H_{1/2}, H_\delta, H_m)\sin n\omega_m t\right] \tag{2}$$

where

$$H_\delta = \Delta H_0\left(\frac{t}{t_0} - \frac{1}{2}\right) \tag{3}$$

as is assumed above. The Fourier amplitudes a_n

$$a_n = \frac{\omega_m}{\pi} \int_{-\pi/\omega_m}^{\pi/\omega_m} \frac{\sin n\omega_m t \, dt}{\left(\tfrac{1}{2}\Delta H_{1/2}\right)^2 + \left(H_\delta + \tfrac{1}{2}H_m\sin \omega_m t\right)^2} \tag{4}$$

were determined by contour integration, and the first three are

$$a_0 = \left(\frac{4}{H_m}\right)^2 \frac{u^{1/2}}{2(u - \gamma)(u - 2)^{1/2}} \tag{5}$$

$$a_1 = \pm a_0\left(\frac{2\gamma}{u} - 1\right)^{1/2} \tag{6}$$

$$a_2 = \left(\frac{4}{H_m}\right)^2 + (1 + 2\gamma - 2u)a_0 \tag{7}$$

where

$$u = \gamma + \left[\dot{\gamma}^2 - 16\left(\frac{H_\delta}{H_m}\right)^2\right]^{1/2} \tag{8}$$

and

$$\gamma = 1 + \left(\frac{2H_\delta}{H_m}\right)^2 + 3\left(\frac{\Delta H_{pp}}{H_m}\right)^2 \tag{9}$$

The quantity a_1 is recorded in ESR experiments when one detects at the modulation frequency ω_m, while a_2 is recorded when second harmonic detection is employed.

The properties of a_1 at the peak may be obtained by setting its derivative equal to zero

$$\frac{da_1}{dH_\delta} = -2\left(\frac{2}{H_m}\right)^3 \left(\frac{u}{u-2}\right)^{1/2} \frac{u^2 - u - 2u\gamma + 3\gamma}{(u-\gamma)^2} = 0 \tag{10}$$

which means that

$$u(1 + 2\gamma - u) = 3\gamma \tag{11}$$

From this condition one may obtain the value a_{1pp} of a_1 at the peaks of the line (inflection points of $Y(H)$):

$$a_{1pp} = \frac{\pm 2(1/\Delta H_{pp})^2 (H_m/\Delta H_{pp})}{\left\{3(H_m/\Delta H_{pp})^2 + 8 + \left[(H_m/\Delta H_{pp})^2 + 4\right]^{3/2}\right\}^{1/2}} \tag{12}$$

and the magnetic field H_δ has the value $\pm H_{\delta pp}$ at the peaks

$$H_{\delta pp} = \pm \left(\frac{\Delta H_{pp}}{2}\right) \left\{\left(\frac{H_m}{\Delta H_{pp}}\right)^2 + 5 - 2\left[4 + \left(\frac{H_m}{\Delta H_{pp}}\right)^2\right]^{1/2}\right\}^{1/2} \tag{13}$$

The modulation amplitude broadens the resonant line by the quantity under the square root sign, and the observed modulation broadened linewidth $\Delta H_{pp(obs)} = 2H_{\delta pp}$ is related to the true linewidth ΔH_{pp} by the expression

$$\Delta H_{pp(obs)} = \Delta H_{pp} \left\{\left(\frac{H_m}{\Delta H_{pp}}\right)^2 + 5 - 2\left[4 + \left(\frac{H_m}{\Delta H_{pp}}\right)^2\right]^{1/2}\right\}^{1/2} \tag{14}$$

From the condition

$$\left(\frac{d}{dH_m}\right)a_{1pp} = 0 \tag{15}$$

one finds that the amplitude a_{1pp} reaches a maximum when $H_m = 2\Delta H_{1/2}$. Substituting this into Eq. (12), one obtains

$$a_{1_{pp(\max)}} = \pm \frac{2}{(\Delta H_{1/2})^2} \tag{16}$$

The maximum that occurs at $H_m = 2\Delta H_{1/2}$ is very broad, as Fig. 6-11 indicates.

Myers and Putzer (1959) considered the modulation broadening problem from a different viewpoint than Wahlquist. They Fourier-analyzed a general lineshape function Y:

$$Y(H) = Y(H_\delta + \tfrac{1}{2}H_m\sin \omega_m t) \tag{22}$$

$$= y_m\left[a_0 + \sum_{n=1}^{\infty} (a_n\sin n\omega_m t + b_n\cos n\omega_m t)\right] \tag{23}$$

and evaluated all of the coefficients a_n and b_n explicitly for Lorentzian and Gaussian lineshapes.

The fundamental (first harmonic) results are

$$a_1 = \sum_{n=0}^{\infty} \frac{(sH_m/2)^{2n+1}}{2^{2n}(1 + sH_\delta)^{n+1}} \binom{2n+1}{n} \sin[2(n+1)\cot^{-1}(sH_\delta)] \tag{24}$$

subject to the condition

$$\frac{sH_m}{2} < \left[1 + (sH_\delta)^2\right]^{1/2} \tag{25}$$

for a Lorentzian shape, where the quantities $\binom{2n+1}{n}$ are the binomial coefficients $(2n + 1)!/[(n!)(n + 1)!]$, and

$$a_1 = e^{-(sH_\delta)^2} \sum_{n=0}^{\infty} \frac{(sH_m/2)^{2n+1}}{2^{2n}} \binom{2n+1}{n}$$

$$\times \sum_{k=n+1}^{2n+1} \frac{(-1)^k}{k!} \binom{k}{2k - 2n - 1}(2sH_\delta)^{2k-2n-1} \tag{26}$$

for a Gaussian lineshape. The scaling factor s depends upon the lineshape, and has the particular forms

$$s = \frac{2}{(3^{1/2}\,\Delta H_{pp})} = \frac{2}{\Delta H_{1/2}} \quad \text{Lorentzian} \tag{27}$$

and

$$s = \frac{2^{1/2}}{\Delta H_{pp}} = \frac{2(\ln 2)^{1/2}}{\Delta H_{1/2}} \qquad \text{Gaussian} \qquad (28)$$

for the two most important lineshapes. It may be regarded as a reciprocal

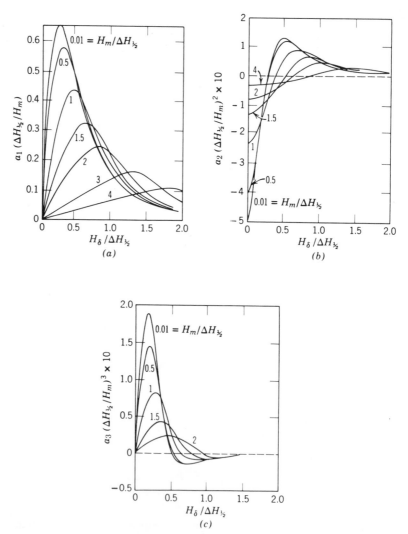

Fig. 6-10. Modulation-broadened Lorentzian lineshape. The fundamental a_1 and the first two harmonic Fourier amplitudes a_2 and a_3 of Eq. (6-H-23) are presented as a function of the normalized magnetic field $H_\delta/\Delta H_{1/2}$. Curves are drawn for various ratios ($H_m/\Delta H_{1/2}$) of the modulation amplitude to the true linewidth [Wilson (1963); Wahlquist (1959) presents a graph similar to (a)].

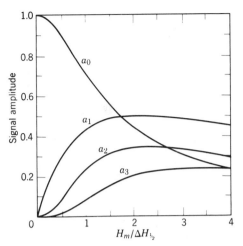

Fig. 6-11. Dependence of the maximum Lorentzian signal amplitudes on the normalized modulation amplitude $H_m/\Delta H_{1/2}$ (Wilson, 1963).

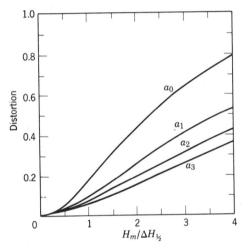

Fig. 6-12. Dependence of the distortion of the Lorentzian lineshapes on the normalized modulation amplitude $H_m/\Delta H_{1/2}$ (Wilson, 1963).

linewidth that normalizes the dimensionless quantities sH_m and sH_δ in Eqs. (24) and (26).

The Lorentzian coefficient a_1 of Eq. (26) was evaluated on a computer. The series (24) and (26) converge very slowly for high modulation amplitudes, and (24) breaks down when H_m exceeds the limits imposed by inequality (25).

The mathematical analyses of Myers and Putzer (1959) and Wahlquist (1961) that we have presented provide an insight into the manner in which the modulation amplitude broadens and distorts the resonant line. These authors

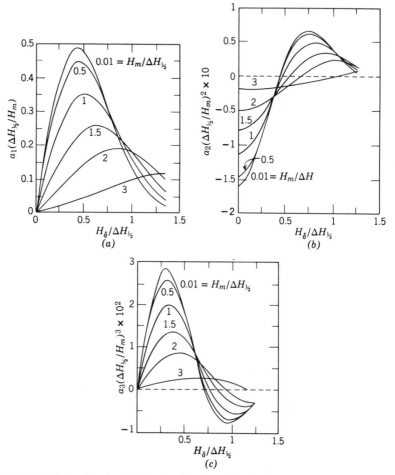

Fig. 6-13. Modulation-broadened Gaussian lineshape. The fundamental a_1 and the first two harmonic Fourier amplitudes a_2 and a_3 of Eq. 6-H-23 are presented as a function of the normalized magnetic field $H_\delta/\Delta H_{1/2}$. Curves are drawn for various ratios $(H_m/\Delta H_{1/2})$ of the modulation amplitude to the true linewidth (Wilson, 1963).

and others displayed their results graphically, and we now present some of these graphs.

Johnson (1976) discussed modulation broadening of second derivative Lorentzian and Gaussian lines. Glausinger and Sienko (1973) employed second derivative detection of ESR in metals.

Gendell et al. (1964) measured peak-to-peak linewidths from the separation of the zeros of the second derivative spectrum.

Wilson (1963) computed the lineshapes for the fundamental Fourier component a_1 and its second and third harmonics a_2 and a_3, respectively, defined by Eq. (23) for both a Lorentzian and a Gaussian lineshape, and the results are shown in Figs. 6-10 and 6-13. The peak amplitudes of the observed signal from

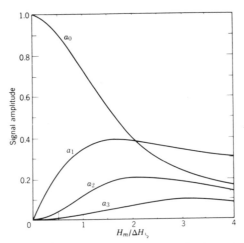

Fig. 6-14. Dependence of the maximum Gaussian signal amplitudes on the normalized modulation amplitude $H_m/\Delta H_{1/2}$ (Wilson, 1963).

the first four Fourier components a_0, a_1, a_2, and a_3 vary with the modulation amplitude H_m in the manner shown in Figs. 6-11 and 6-14. The Lorentzian signals a_1, a_2, and a_3 exhibit a linear, parabolic, and cubic dependence, respectively, on the modulation amplitude when $H_m \ll \Delta H_{1/2}$. The amplitudes a_1, a_2, and a_3 of the Gaussian lineshape are considerably reduced relative to the corresponding Lorentzian values.

The integrated areas A_1, A_2, and A_3 of the Fourier components a_1, a_2, and a_3 may be obtained by single, double, and triple integrations, respectively, of the suitably normalized function. The distortion of the nth Fourier component is defined in terms of the true area A by means of the expression

$$\text{Distortion} = \frac{|A - A_n|}{A} \tag{29}$$

Figure 6-12 shows the dependence of the distortion upon the modulation amplitude for a Lorentzian lineshape, Mohos (1975) analyzed the distortion effects of ESR spectrometers.

Smith (1964) examined the effect of modulation broadening on a Gaussian lineshape, and compared his results with Wahlquist's Lorentzian treatment. Figure 6-15 shows the variation of the observed modulation broadened peak-to-peak linewidth $\Delta H_{pp(\text{obs})}$ on the logarithm of the modulation amplitude H_m. Both quantities are normalized with respect to the true peak-to-peak linewidth ΔH_{pp}. Experimental NMR spectra of the protons in aqueous $Cr(NO_3)_3$, shown in Fig. 6-16 with $\Delta H_{pp} = 0.19$ G agreed with the theoretical curves in Figs. 6-15 and 6-17.

Figure 6-15 is not convenient to use for comparison with experimental data because the true width ΔH_{pp} is ordinarily an unknown quantity. Figure 6-17

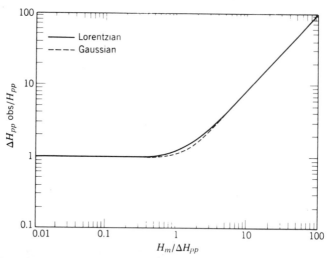

Fig. 6-15. Dependence of the amount of modulation-broadening $\Delta H_{pp(obs)}/\Delta H_{pp}$ on the normalized modulation amplitude $H_m/\Delta H_{pp}$ [adapted from Smith (1964)].

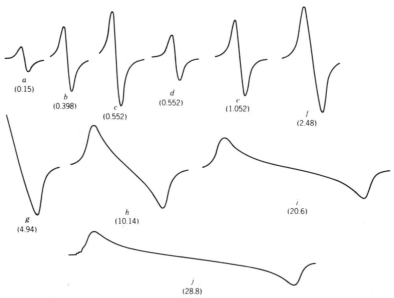

Fig. 6-16. NMR signal of protons in aqueous solution of $Cr(NO_3)_3$ ($\Delta H_{pp} = 0.19$ G) as a function of the modulation amplitude H_m. The nominal modulation frequency was 40 Hz, and the field scan rate was the same in each tracing. The gain setting used for spectra a, b, and c was twice that used to record the remaining spectra. The normalized modulation amplitude $H_m/\Delta H_{pp}$ is shown in parentheses beneath each spectrum (Smith, 1964).

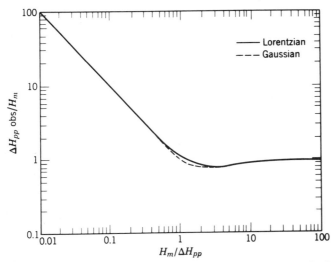

Fig. 6-17. Dependence of $(\Delta H_{pp(obs)}/H_m)$ on the normalized modulation amplitude $H_m/\Delta H_{pp})$ [adapted from Smith (1964)].

shows the experimentally measurable quantity $\Delta H_{pp(obs)}/H_m$ plotted against $H_m/(\Delta H_{pp})$, and one may use such a graph or the data in Table 6-3 to deduce the true linewidth from an overmodulated resonant line. The graph in Fig. 6-18 is useful for determining ΔH_{pp} from a slightly overmodulated resonance, which is the case most often met in practice. For extreme overmodulation ($H_m > 2\Delta H_{pp}$), one cannot accurately determine ΔH_{pp}. The Lorentzian and Gaussian curves in Figs. 6-17 and 6-18 are so close to each other that very little error is involved in using their average to determine the true width ΔH_{pp} for a lineshape intermediate between Lorentzian and Gaussian.

The peak-to-peak amplitude varies with the modulation in the manner shown in Fig. 6-19. The curves are normalized to unity at their peaks. (Figures 6-11 and 6-14 are normalized differently.) The a_{1pp} versus $H_m/\Delta H_{pp}$ plot is much more useful for distinguishing a Gaussian from a Lorentzian line than the linewidth plots.

Chapter 12 discusses lineshapes, and Fig. 12-5 defines the inner maximum slope y_1'', the outer maximum slope y_2'', and the separation between the two outer maximum slopes H_2'' of the first derivative lineshape. Wahlquist used his equations to evaluate these quantities for a Lorentzian shape, and their dependence on the normalized modulation amplitude $H_m/\Delta H_{1/2}$ is shown in Figs. 6-20 and 6-21. For comparison purposes one should recall that

$$\frac{y_1''}{y_2''} = 4 \qquad \text{Lorentzian} \qquad (30)$$

$$\frac{y_1''}{y_2''} = \frac{1}{2}e^{3/2} = 2.24 \quad \text{Gaussian} \qquad (31)$$

TABLE 6-3

Parameters for a Lorentzian (Wahlquist, 1961) and Gaussian (Smith, 1964)
Magnetic Resonance Lines as a Function of the Modulation Amplitude

$H_m/\Delta H_{1/2}$	$H_m/\Delta H_{pp}$	$\Delta H_{pp(obs)}/\Delta H_{pp}$	$\Delta H_{pp}(obs)/H_m$	a_{1pp} Normalized
Lorentzian line				
0	0	1.000	∞	0
0.1	0.173	1.006	5.815	0.13
0.2	0.346	1.029	2.973	0.248
0.4	0.694	1.114	1.610	0.478
0.8	1.388	1.432	1.035	0.784
1.2	2.08	1.903	0.907	0.930
1.6	2.78	2.387	0.873	0.987
2.0	3.46	3.000	0.866	1.000
2.4	4.16	3.564	0.869	0.992
2.8	4.86	4.221	0.876	0.974
3.2	5.56	4.884	0.883	0.952
3.6	6.24	5.537	0.890	0.929
4.0	6.94	6.288	0.897	0.905
6.0	10.40	9.55	0.922	0.800
8.0	13.84	13.0	0.938	0.721
10.0	17.34	16.4	0.949	0.659
16.0	27.72	26.5	0.967	0.541
20.0	34.64	33.7	0.973	0.488
40.0	69.4	68.2	0.986	0.353
∞	∞	∞	1.000	0
Gaussian line				
0	0	1.00	∞	0
0.12	0.141	1.00	7.095	0.148
0.24	0.282	1.00	3.573	0.291
0.48	0.564	1.007	1.842	0.551
0.96	1.128	1.039	1.044	0.887
1.44	1.692	1.178	0.859	0.993
1.68	1.974	1.454	0.834	0.995
1.92	2.26	1.645	0.826	0.983
2.40	2.82	1.862	0.831	0.943
2.88	3.38	2.343	0.844	0.898
3.36	3.94	2.856	0.858	0.857
3.84	4.52	3.384	0.870	0.819
4.32	5.08	3.922	0.880	0.785
4.80	5.64	4.465	0.889	0.755
7.20	8.46	5.013	0.921	0.639
9.60	11.28	7.786	0.939	0.564
12.00	14.10	10.6	0.956	0.497
∞	∞	13.5	1.000	0
		∞		

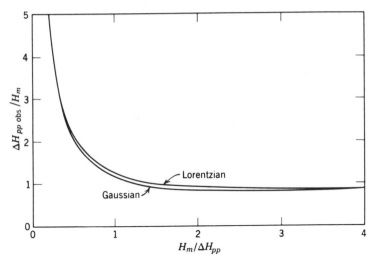

Fig. 6-18.   Linear plot of the dependence of $\Delta H_{pp(obs)}/H_m$ on $H_m/\Delta H_{pp}$ in the region of moderate modulation broadening. The data in Table 6-3 were used to construct this figure.

so we see that modulation broadening decreases the Lorentzian ratio y_1''/y_2'' toward and eventually past the theoretical Gaussian value of 2.24.

In addition,

$$H_2'' = 3^{1/2}\,\Delta H_{pp} \tag{32}$$

for both a Lorentzian and a Gaussian lineshape. It is indeed unusual to find a ratio of parameters like $H_2''/\Delta H_{pp}$ that is identical for both shapes. In terms of the half-amplitude linewidth

$$H_2'' = \Delta H_{1/2} \qquad\qquad\qquad\qquad \text{Lorentzian} \tag{33}$$

$$H_2'' = (2\ln 2)^{-1/2}\,\Delta H_{1/2} = 1.47\Delta H_{1/2} \quad \text{Gaussian} \tag{34}$$

The constants in Eqs. (30)–(32) are listed in Table 12-5.

Buckmaster (1969) et al. (1968, 1969, 1971, 1971) derived expressions for the zeros and extrema of the fourier coefficients a_0, a_1, a_2, a_3 for absorption and the analogous coefficients d_0, d_1, and d_2 for dispersion (see also Evans and Brey, 1968). Johnson (1976) showed that Lorentzian lines are more sensitive to modulation-induced distortion than Gaussian lines. To limit the distortion to below 1% requires $H_m < 0.15\Delta H_{1/2}$ for Lorentzian lines and $H_m < 0.3\Delta H_{1/2}$ for Gaussian lines.

Berger and Günthart (1962) considered the distortion of a Lorentzian resonant line that arises in a modulation-type superheterodyne spectrometer. More specifically, the magnetic field is modulated by the term $\frac{1}{2}H_m\sin\omega_m t$, and the superheterodyne detection takes place at the intermediate frequency $\omega_i \gg \omega_m$. The system that employs one lock-in detector at the frequency ω_i and

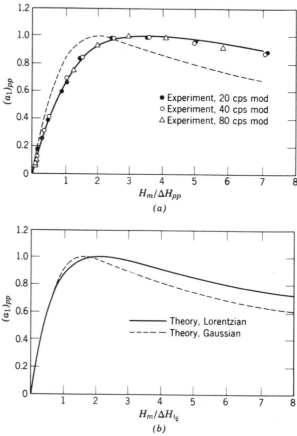

Fig. 6-19. Variation of the absorption derivative peak amplitude a_{1pp} on the modulation amplitude: (a) normalized with respect to the true peak-to-peak linewidth ΔH_{pp}, and (b) normalized with respect to the true absorption curve half-amplitude linewidth $\Delta H_{1/2}$ [adapted from Smith (1964)].

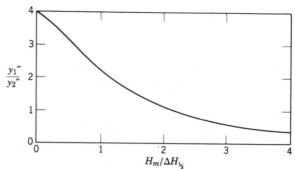

Fig. 6-20. The dependence of the ratio y_1''/y_2'' on the normalized modulation amplitude $H_m/\Delta H_{1/2}$ for a Lorentzian shape [adapted from Wahlquist (1961)].

244

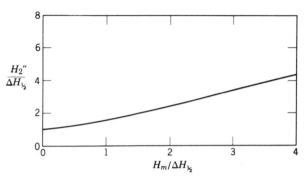

Fig. 6-21.   Variation of the quantity H_2'' on the modulation amplitude H_m for a Lorentzian shape when both are normalized relative to $\Delta H_{1/2}$ [adapted from Wahlquist (1961)].

another at ω_m is compared to the system which uses a square-law detector at the intermediate frequency, and only employs a lock-in detector for the magnetic field modulation frequency ω_m. When two phase-sensitive detectors are employed, the Lorentzian lineshape is reproduced without distortion.

When the superheterodyne spectrometer employs a quadratic first detector and only one phase-sensitive detector at ω_m, then the recorded lineshape differs from a true Lorentzian. Graphs and explicit formulas for these shapes are given by Berger and Günthart (1962). Several other workers such as Halbach (1956), Bruin and van Ladesteyn (1959), Yagi (1960), and Arndt (1965) have published graphs of the type shown in Figs. 6-10 to 6-21.

Spry (1957) used the numerical unfolding procedure of Stokes (1948) to compute the true lineshape of an arbitrary experimental recording. The "folding function" that corrects for the modulation broadening may be either calculated mathematically, or determined experimentally.

Flynn and Seymour (1960, 1961, 1962) used a series involving the moments of a generalized broadening function to derive a general formula for correcting distorted lineshapes (see also Russell and Torchia, 1962). Wilson (1963) developed a method for correcting lineshapes by relating the Fourier coefficients of the modulation-broadened line to those of the true integrated lineshape.

Andrew (1953) has shown that the true second moment of a resonance line $\langle H^2 \rangle$ is related to the second moment of a modulation broadened line $\langle H^2 \rangle_{\mathrm{obs}}$ by the expression

$$\langle H^2 \rangle_{\mathrm{obs}} = \langle H^2 \rangle + \tfrac{1}{16} H_m^2 \tag{35}$$

He further derived the following relation for the $2n$th moment

$$\langle H^{2n} \rangle_{\mathrm{obs}} = \sum_{k=0}^{n} \left(\frac{H_m^k}{2^{2k} k!} \right)^2 \frac{(2n)! \langle H^{2n-2k} \rangle}{(k+1)(2n-2k)!} \tag{36}$$

which reduces to

$$\langle H^4 \rangle_{\text{obs}} = \langle H^4 \rangle + \tfrac{3}{8}\langle H^2 \rangle H_m^2 + \tfrac{1}{128}H_m^4 \tag{37}$$

for the moment ($n = 2$). The odd moments vanish for a symmetrical lineshape.

Perlman and Bloom (1952) present experimental evidence in support of the dependence of the observed second moment on $(H_m)^2$. Verdier et al. (1961) showed that the doubly integrated area or first moment of a derivative ESR absorption signal is approximately proportional to the modulation amplitude for modulations up to twice the peak-to-peak linewidth. Wilson (1964) gives corrections for moments of lines detected at any harmonic of the modulation frequency. Rädler (1961) discusses the influence of the modulation amplitude on the Overhauser effect (see also Visweswaramurthy, 1965).

Halbach (1960) generalized the above expressions by taking into account the effect of both the modulation amplitude H_m and the modulation frequency ω'_m on the lineshape. It is assumed that there is no phase shift between the field modulation and the lock-in detector reference signal. His formulas are in terms of frequency moments $\langle \omega^n \rangle$, and using the transformation $\langle \omega^n \rangle = \gamma^n \langle H^n \rangle$, we obtain

$$\langle H^2 \rangle_{\text{obs}} = \langle H^2 \rangle + \frac{1}{3}\left(\frac{\omega_m}{\gamma}\right)^2 + \frac{1}{16}H_m^2 \tag{38}$$

$$\langle H^4 \rangle_{\text{obs}} = \langle H^4 \rangle + \langle H^2 \rangle\left[2\left(\frac{\omega_m}{\gamma}\right)^2 + \frac{3}{8}H_m^2\right] + \frac{1}{5}\left(\frac{\omega_m}{\gamma}\right)^4$$

$$+ \frac{3}{16}\left(\frac{\omega_m}{\gamma}\right)^2 H_m^2 + \frac{1}{128}H_m^4 \tag{39}$$

These expressions are valid in the absence of saturation when the linewidth, the modulation frequency, and γH_m are small compared to the resonant frequency. They are derived under the assumption that frequency and field modulation are equivalent, and this assumption is justified by Halbach (1960).

Tykarski (1974) analyzed the signal amplitude as a function of the phase shift between the modulation field and the reference signal of the lock-in detector. He showed that the width of a Lorentzien line can be calculated from the ratio of the maximum to the minimum signal amplitude. Edgar (1975) found a phase-reversal effect in Er^{3+} from the large g-factor anisotropy together with the nonparallelism of the static and modulating magnetic fields. Brotikovskii et al. (1971), Alekseev and Belondgov (1970), and Mailer et al. (1977) discuss modulation effects under saturating conditions. Galloway and Dalton (1978) gave approximation methods for calculating overmodulated spectra. Evans et al. (1978) employed fast Fourier-transform methods to simulate modulation-broadening effects. Dikanov ((1977) analyzed modulation phenomena in spin echoes to study the structure of the immediate environment of a paramagnetic species. Pasimeni (1978) studied the effect of field modulation on time-resolved ESR (see Sec. 13-E), and he found that the effective microwave field amplitude is $J_0(\alpha\gamma H_{mod}/\omega_{mod})H_1$ instead of merely H_1, where J_0 is the zero-order Bessel function and α is a dimensionless numerical constant of the order of unity. Proffitt and Gardiner (1977) discussed modulation-broadened Lorentzian lines.

In this book we have defined the modulation magnetic field by $\frac{1}{2}H_m \sin \omega_m t$. This means that H_m is the peak-to-peak modulation amplitude, so the magnetic field strength ranges between $(H_0 + H_\delta - \frac{1}{2}H_m)$, and $(H_0 + H_\delta + \frac{1}{2}H_m)$ during one modulation cycle. Some authors work in terms of the definition $H_m \sin \omega_m t$, so their expressions will differ from ours by a factor of 2. In addition, some authors express their results in terms of half of the linewidth, instead of the full linewidth. These divergent definitions can lead to considerable confusion when comparing the results reported by various workers.

I. Effect of the Modulation Frequency on the Resonance Line

In practice, the frequency $\omega_m = 2\pi f_m$ of the magnetic field modulation ordinarily satisfies the inequality

$$\frac{\omega_m}{\gamma} \ll \Delta H \tag{1}$$

where $\gamma = g\beta/\hbar$. When the modulation frequency ω_m exceeds the linewidth $\gamma\Delta H$, then sideband resonances develop that are separated by the intervals of ω_m/γ G. These sidebands are centered at the resonant field H_0 and extend over a range of H_m G, where H_m is the peak-to-peak modulation amplitude. When n, defined by

$$n = \frac{\gamma H_m}{2\omega_m} \tag{2}$$

is close to an integer, it gives the number of prominent sidebands on each side of the center line, and for certain values of n the center line itself disappears. Burgess and Brown (1952) extended the work of Smaller (1951) and Karplus (1948) to obtain the following mathematical expressions for the real and imaginary parts of the rf susceptibility χ.

$$\chi' = A \cos \omega_m t + B \sin \omega_m t \tag{3}$$

$$\chi'' = C \cos \omega_m t + D \sin \omega_m t \tag{4}$$

where

$$A = \frac{1}{2}\chi_0\omega_0 \sum_{k=-\infty}^{\infty} \frac{4\omega_m k J_k^2(n)}{\gamma H_m} \frac{[\gamma(H-H_0)+k\omega_m]}{[\gamma(H-H_0)+k\omega_m]^2 + 1/T_2^2} \tag{5}$$

$$B = \frac{1}{2}\chi_0\omega_0 \sum_{k=-\infty}^{\infty} \frac{J_k(n)[J_{k+1}(n)-J_{k-1}(n)]}{T_2}$$

$$\times \frac{1}{[\gamma(H-H_0)+k\omega_m]^2 + 1/T_2^2} \tag{6}$$

$$C = -\frac{1}{2}\chi_0\omega_0 \sum_{k=-\infty}^{\infty} \frac{4\omega_m k J_k^2(n)}{\gamma H_m T_2} \frac{1}{[\gamma(H-H_0)+k\omega_m]^2 + 1/T_2^2} \tag{7}$$

$$D = \frac{1}{2}\chi_0\omega_0 \sum_{k=-\infty}^{\infty} J_k(n)[J_{k+1}(n)-J_{k-1}(n)]$$

$$\times \frac{[\gamma(H-H_0)+k\omega_m]}{[\gamma(H-H_0)+k\omega_m]^2 + 1/T_2^2} \tag{8}$$

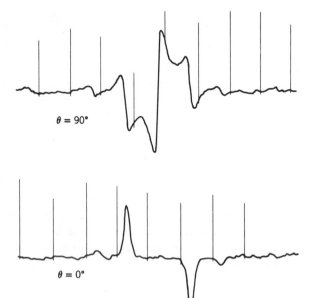

$\theta = 90°$

$\theta = 0°$

Fig. 6-22. Lock-in amplifier output versus radio frequency at a modulation frequency of 100 Hz and $(\gamma H_M/2\omega_m) = 1$ using two different phases θ of the lock-in amplifier. Vertical lines indicate 100 Hz frequency intervals (Burgess and Brown, 1952).

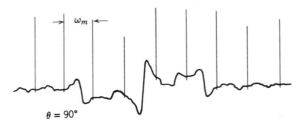

$\theta = 90°$

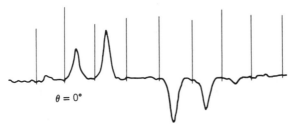

$\theta = 0°$

Fig. 6-23. Same conditions as shown in Fig. 6-22 except $(\gamma H_m/2\omega_m) = 2$ (Burgess and Brown, 1951).

and $J_k(n)$ is the kth-order Bessel function. The function χ' is symmetric and χ'' is antisymmetric about the point $H = H_0$. When $\omega_m > \gamma \Delta H$, the spectrum of both χ' and χ'' consists of a series of components that resemble absorption curves, dispersion curves, or a mixture of both, depending on the setting of the lock-in detector phase (i.e., on whether $\cos \omega_m t$, $\sin \omega_m t$, or a mixture is detected). The functions A and D are dispersion types, while B and C are absorption types.

Burgess and Brown (1952) studied the NMR of protons in water, and confirmed these equations experimentally for $n = 1$, 2, and 4 using $f_m = 100$ Hz and $H_m = 0.048$, 0.096, and 0.192 G, as shown in Figs. 6-22 to 6-24. When f_m is reduced by a factor of 4 to 25 Hz, the sidebands overlap, and a broadened line is observed, as shown in Fig. 6-25. A further reduction in f_m leads to $\omega_m \ll \gamma H_m$, and the modulation effect disappears. Gabillard and Ponchel (1962) observed sidebands in the ESR of DPPH for n in the range from 1 to 3.6 (see also Garif'yanov, 1957). Hyde and Brown (1962) observed sidebands with the tetracene positive ion. Macomber and Waugh (1965) generalized Karplus' (1948) theory. Clough and Hobson (1975) observed weak sidebands shifted from the main spectrum by a frequency equal to a methyl tunneling frequency.

Modulation frequency effects become negligible when f_m becomes much less than the linewidth ΔH expressed in frequency units

$$f_m \ll \left(\frac{\gamma}{2\pi}\right) \Delta H \tag{9}$$

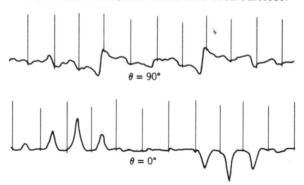

Fig. 6-24.   Same conditions as in Fig. 6-22 except $(\gamma H_m/2\omega_m) = 4$ (Burgess and Brown, 1952).

For $g = 2$, the conversion factor $\gamma/2\pi$ has the value

$$\frac{f}{H} = \frac{\gamma}{2\pi} = 2.8 \times 10^6 \text{ Hz/G} \tag{10}$$

where f is in Hz and H is in gauss. Therefore, we obtain the explicit condition

$$\frac{f_m}{\Delta H} \ll 2.8 \times 10^6 \text{ Hz/G} \tag{11}$$

which must be satisfied if modulation effects are to be neglected. A resonance line $1/10$ G wide with $g = 2$ observed with a 100-kHz modulation frequency corresponds to

$$\frac{f_m}{\Delta H} = 10^6 \text{ Hz/G} \tag{12}$$

Therefore, it is not advisable to employ 100-kHz modulation to record lines that are narrower than about a hundred milligauss. More explicitly, we may say that a 100-kHz modulation frequency corresponds to 35 mG in magnetic field units. Since lower modulation frequencies reduce the spectrometer sensitivity because of the $1/f$ crystal noise, it is recommended that a superheterodyne spectrometer be employed for such narrow lines. Hyde and Brown (1962) found that a linewidth of 100 mG recorded with a superheterodyne spectrometer broadens by about 15% with 100-kHz modulation, and their data indicate that a 25-mG linewidth will double in going from superheterodyne to 100-kHz operation. When they adjusted the lock-in detector phase away from its proper value, they found significant narrowing, in addition to the expected decrease in amplitude of a modulation-broadened resonance. Hauser (1962) has observed linewidths of 17 mG in the 1,3-bisdiphenylene allyl radical.

    Rinehart et al. (1960) discuss the effect of frequency modulation on the lineshape in a microwave spectrometer, and Acrivos (1962) and Parikh (1965) describe modulation frequency effects in NMR. Pasimeni (1978) studied the

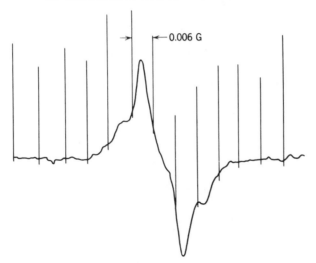

Fig. 6-25. Lock-in amplifier output versus radiofrequency at a modulation frequency of 25 Hz. Frequency intervals are 25 Hz (0.006 G) apart (Burgess and Brown, 1952).

effect of field modulation on time-resolved ESR spectra. Glarum's (1965) field-modulation techniques for resolution enhancement are discussed in Sec. 11-I.

J. Helmholtz Modulation Coils

To produce audio frequency modulation of the magnetic field, one usually mounts Helmholtz coils either on the resonant cavity itself, or on the magnet pole pieces. When they are mounted on the pole pieces, they modulate the field throughout the gap and can interfere with the use of a proton magnetometer. When mounted on the resonant cavity, they require a much lower power input for a given modulation amplitude H_m. Pole piece mounting is not feasible for frequencies above the audio range (e.g., for 100 kHz). As discussed in Sec. 9-E of the first edition the two Helmholtz coils have a radius a, and they are separated by a distance equal to this radius a. The amplitude of the magnetic field at the center between the coils with n turns on each is

$$H = \frac{B}{\mu} = \frac{8nI}{5^{3/2}a} \text{ At/m} \tag{1}$$

where a is in meters, and I is in amperes. Using the free-space permeability $\mu = 4\pi \times 10^{-7}$ H/m, expressing a in centimeters, and converting to gauss one obtains

$$B = \frac{0.64\pi nI}{5^{1/2}a} \approx 0.9\frac{nI}{a} \text{ G} \tag{2}$$

with I still in amperes. These expressions assume that the radius of the coil is much greater than the length of the two individual coils. The sample in the resonant cavity should be located at the center of the coils, as the modulating field is most homogeneous there.

The inductance L of a circular loop of wire is given by

$$L \approx \mu a \left[\ln \left(\frac{8a}{t} \right) - 2 \right] \text{ Henry} \tag{3}$$

where a is the radius of the loop, t is the radius of the wire that forms the loop, and it is assumed that $a \gg t$. A coil of n turns whose thickness and length both equal $2t \ll a$ has an inductance L given by

$$L \approx \mu a n^2 \left[\ln \left(\frac{8a}{t} \right) - 2 \right] \text{ Henry} \tag{4}$$

The quantity t may be considered as the radius of the cross section of the group of wires that form the coil. The resistance R of a piece of wire of length l and cross-sectional area A may be computed from the relationship

$$R = \frac{\rho l}{A} \tag{5}$$

where ρ is the resistivity. A loop of wire of radius a has a length $l = 2\pi a$, and therefore n turns in series have the resistance R

$$R = \frac{2\pi n a \rho}{A} \tag{6}$$

The impedance Z is

$$Z = R + j\omega L \tag{7}$$

$$= (R^2 + \omega^2 L^2)^{1/2} e^{j\phi} \tag{8}$$

where

$$\phi = \tan^{-1} \left(\frac{\omega L}{R} \right) \tag{9}$$

Usually $\omega L \gg R$, and the impedance of the coil is almost entirely inductive

$$Z \approx j\omega L \tag{10}$$

Equations (4) and (6)–(8) for the inductance, resistance, and impedance, respectively, apply to one side of a Helmholtz coil. They should be halved or doubled depending on whether or not the two sides are connected together in parallel or in series, respectively.

For Helmholtz coils to be effective, it is necessary for the modulating magnetic field to penetrate through the resonant cavity walls. An X-band waveguide has a wall thickness of 0.13 cm (0.05 in.), and a glance at Fig. 2-17 indicates that this is about the skin depth δ of copper at about 3 kHz. Therefore, modulation frequencies above ~ 1 kc should not be used with this waveguide. A frequency of 100 kHz has a skin depth in copper of 0.02 cm, and the microwave frequency of 9000 MHz has $\delta = 7 \times 10^{-5}$ cm, so if one makes a cylindrical resonant cavity out of glass and plates it inside with copper to a thickness of about 0.002 cm (0.005 in.), then it may be used with 100-kHz modulation. The requirement is that the metallic wall thickness be much greater than the skin depth at the microwave resonant frequency and much less than the skin depth at the modulation frequency. The first requirement is of somewhat greater importance than the second.

There are a number of other methods for introducing the high-frequency modulation into the resonant cavity. Several of these are: (1) One may cut a slot in the side of the cavity parallel to lines of microwave current flow (Buckmaster and Scovil, 1956). (2) A loop of wire may be located within the resonant cavity and oriented parallel to the magnetic lines of force (Llewellyn, 1957). (3) The wall of the resonant cavity adjacent to the modulation coil may be made a thickness intermediate between the skin depths of the microwave and modulation frequencies (Llewellyn, 1957; Bennett et al., 1958). (4) The magnetic field modulation may be excited by two vertical posts symmetrically placed on the base plate of a TE_{011} mode cylindrical cavity. An example of a simple push–pull modulation coil amplifier is shown in Fig. 10-22 of the first edition.

Fitzky (1958) describes modulation circuits, and Cook (1962) presents a slow-sweep generator circuit. Smith et al. (1978) show how to convert an E-3 ESR spectrometer to 1-MHz field modulation.

K. Double Modulation

Some spectrometers employ a high-frequency field modulation for use with narrow band detection, and also a simultaneous audio frequency modulation for use with video observation of the absorption line. The oscilloscope that is synchronized at the audio frequency displays the modulus of the derivative lineshape (Buckmaster and Scovil, 1956; Llewellyn, 1957). The audio frequency modulation is turned off when spectra are recorded after narrow band, high-frequency detection.

Unterberger et al. (1957) described a double modulation method that entails both amplitude-modulating the magnetic field, and frequency-modulating the klystron. As a result, the resonance observed with the klystron mode that is displayed on an oscilloscope appears much narrower than it does in the absence of the klystron frequency modulation, as illustrated in Fig. 6-26. The authors derived an equation that relates the apparent linewidth using double modulation to the true linewidth obtained with a single (field) modulation, and

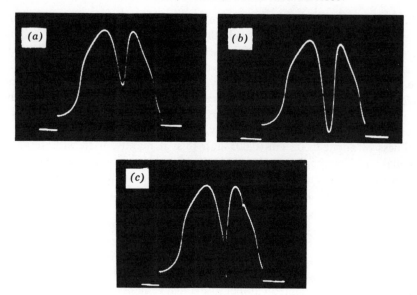

Fig. 6-26. Oscilloscopic presentation of a klystron mode and resonant cavity pip: (*a*) off resonance, (*b*) on resonance, and (*c*) on resonance plus magnetic field modulation. The cavity contains solid DPPH (Unterberger et al., 1957).

they easily obtained a reduction in width by a factor of 17. The technique may be found useful in searching for new resonances with a sillyscope.

Muromtsev et al. (1962) discuss the use of triple modulation for recording the second derivative of the resonance line. See also Rinehart and Lin (1961). Van den Boom (1971) treats double modulation in ESR.

References

Ackermann, P., J. P. Imbaud, J. P. Plumey, and G. Rives, *Rev. Phys. Appl.* **11**, 629 (1976).

Acrivos, J. V., *J. Chem. Phys.* **36**, 1097 (1962).

Alderman, D. W., *RSI* **41**, 192 (1970).

Alekseev, B. F., and A. M. Belondgov, *Izv. Vuz. Fiz.* **5**, 131 (1970).

Aleksandrov, I. V., *Theory of Nuclear Magnetic Resonance* (Tranl. Scripta Technica). Charles P. Poole, Jr., Transl., Ed., Academic Press, New York, 1966.

Andrew, E. R., *Phys. Rev.* **9**, 425 (1953).

Arndt, R., *J. Appl. Phys.* **36**, 2522 (1965).

Andriessen, W. T. M., *J. Magn. Reson.* **23**, 339 (1976).

Ault, A., and G. Dudek, *Introduction to Proton NMR Spectroscopy Halden Day Ion* Trans. 1976.

Balasurbramanian, K., L. R. Dalton, K. D. Schmalbein, and A. H. Heiss, *Chem. Phys.* **29**, 163 (1978).

Bassompierre, A., and J. Pescia, *C. R. Acad. Sci.* **254**, 4439 (1962).

Bennett, R. G., P. C. Hoell, and R. P. Schwenker, *RSI* **29**, 659 (1958).

Berger, P. A., and H. H. Günthart, *Z. Angew. Math. Phys* **13**, 310 (1962).

Blankenburg, F. J., J. E. Drumheller, and D. K. Worsencroft, *RSI* **42**, 1377 (1971).

Blume, R. J., and W. L. Williams, *RSI* **35**, 1498 (1964).

Bourdel, D., J. Pescia, and P. Lopez, *Rev. Phys. Appl.* **5**, 187 (1970).

Brandwein, L., and M. Lipsicas, *RSI* **41**, 1293 (1970).

Brotikovskii, D. I., G. M. Zhidmirov, V. B. Kazanskii, and B. N. Shelimov, *Teor. Eksp. Khim.* **7**, 245 (1971).

Bruin, F., and D. van Ladesteyn, *Appl. Sci. Res.* **7B**, 270 (1959).

Buckmaster, H. A., Relationships Between Fourier Coefficients Describing the Absorption Components a_n and the Dispersion Components d_n of a Lorentzian Line Shape, Physics Department-Univ. Calgary, Calgary, Alberta, Canada. 1969.

Buckmaster, H. A., and C. J. Dering, *J. Appl. Phys.* **39**, 4486 (1968).

Buckmaster, H. A., and H. E. D. Scovil, *Can. J. Phys.* **34**, 711 (1956).

Buckmaster, H. A., and J. D. Skirrow, *J. Appl. Phys.* **42**, 1225 (1971); *J. Magn. Reson.* **5**, 285 (1971).

Buckmaster, H. A., and J. D. Skirrow, *J. Appl. Phys.* **40**, 5386 (1969).

Bugai, A. A., *Fiz. Tverd. Tela.* **4**, 3027 (2218) (1963).

Burgess, J. H., and R. M. Brown, *RSI* **23**, 334 (1952).

Clough, S., and T. Hobson, *Proc. 18th Congr. AMPERE*, Amsterdam, 1975, p. 387.

Clough, S., and T. Hobson, *J. Phys. C* **18**, 1745 (1975).

Clerj, B., and B. Lambert, *J. Phys. E Sci. Instrum.* **48**, 619 (1971).

Collins, M. A., S. D. Devine, and W. H. Robinson, 8th Australian Spectroscopy Conference. Abstracts. Clayton, Australia. layton, Australia. 121. K6 Monash University. Australian Acad. Sci. 16–20 Aug. 1971, p. 103.

Cook, P. D., *Electron. Eng.* **34**, 320 (1962).

Coumes, A., and M. Ligeon, *J. Phys.* **25**, Suppl. No. 3, 45A (1964).

Cousins, J. E., R. Dupree, and R. L. Havill, *JSI* **40**, 407 (1963) [see *Brit. J. Appl. Phys.*, **16**, 1687 (1965)].

De Martini, F., and A. Lucchini, *Alta Frequenza* **33**, 746 (186E) (1964).

den Boef, J. H., and J. C. M. Henning, *RSI* **45**, 1199 (1974).

Dikanov, S. A., V. F. Yudanov, and Yu. D. Tsvetkov, *Zh. Strukt. Khim.* **18**, 460 (1977).

Doyle, W. T., *RSI* **33**, 118 (1962).

Dvornikov, E. V., and K. V. Grebennik, *PTE* **13**, 156 (1138) (1970).

Edgar, A., *J. Phys. E Sci. Instrum.* **8**, 179 (1975).

Eichele, H., M. Schwoerer, Ch. Kröhnke, and G. Wegner, *Chem. Phys. Lett.* **77**, 311 (1981).

Emsley, J. W., J. Feeney, and L. G. Suteliffe, *High Resolution Nuclear Magnetic Resonance Spectroscopy*, Vol. I and II, Pergamon Press, New York, 1965.

Erickson, L. E., *Phys. Rev.* **143**, 295 (1966).

Evans, J. E., P. H. Morgan, and R. H. Renaud, *Anal. Chim. Acta. Comput. Tech. Optimiz.* (*Netherlands*) **103**, 175 (1978).

Evans, T. E., and W. S. Brey, *J. Chem. Phys.* **49**, 3541 (1968).

Faulkner, E. A., and J. B. Erimbleby, *JSI* **44**, 882 (1967).

Favrot, G., R. L. Aggarwal, B. Laz, M. Balanski, R. C. C. Leite, and S. P. S. Porto, *Proceedings of the 3rd International Conference on Light Scattering in Solids. Campings, Brazil. Paris, France* **286**, 11 (1976).

Fitzky, H. G., *Z. Angew. Phys.* **10**, 489 (1958).

Flynn, C. P., *Proc. Phys. Soc.* **78**, 1546 (1961).

Flynn, C. P., and E. F. W. Seymour, *Proc. Phys. Soc.* **75**, 337 (1960); *JSI* **29**, 352 (1962).

Gabillard, R., and B. Ponchel, *C. R. Acad. Sci.* **254**, 2727 (1962).

Galloway, N. B., and L. R. Dalton, *Chem. Phys.* **32**, 189 (1978).

Galkin, A. A., O. Ya. Grinberg, A. A. Dubinskii, N. N. Kabdin, V. N. Krymov, V. I. Kurochkin, Ya. S. Lebedev, L. F. Oranskii, and V. F. Shuvalov, *Prib. Tekh. Eksp.* **20**, 284 (1977).

Ganapathy, S., *J. Inst. Electron. and Telecomm. Eng.* **23**, 152 (1977).

Garif'yanov, N. S., *Zhur. Eksper. Teor. Fiz.* **32**, 609 (503) (1957).

Garwin, R. L., and A. M. Patlach, *RSI* **36**, 741 (1965).

Gendell, J., J. H. Freed, and G. K. Fraenkel, *J. Chem. Phys.* **41**, 949 (1964).

Glarum, S. H., *RSI* **37**, 771 (1965).

Glaunsinger, W. S., and M. J. Sienko, *J. Magn. Reson.* **10**, 253 (1973).

Goldberg, I. B., and H. R. Crowe, *Anal. Chem.* **49**, 1353 (1977).

Goldberg, I. B., and H. R. Crowe, *J. Magn. Reson.* **18**, 497 (1975).

Goldman, M., *C. R. Acad. Sci.* **256**, 3643 (1963).

Gordienko, A. G. and I. O. Antonenko, *PTE* **4**, 144 (736) (1963).

Gourdon, J. C., P. Lopez, P. Ray, and J. Pescia, *C. R. Acad. Sci.* **271B**, 288 (1970).

Gourdon, J. C., C. Rey, C. Chachaty, J. C. Trombe, and J. Pescia, *C. R. Acad. Sci.* **276B**, 559 (1973).

Gourdon, J. C., B. Vigouroux, and J. Pescia, *Phys. Lett. A* **45**, 69 (1973).

Gunthard, H. H., *Ber. Bunsenges. Phys. Chem.* **78**, 1110 (1974).

Halbach, K., *Helv. Phys. Acta* **29**, 37 (1956); *Phys. Rev.* **119**, 1230 (1960).

Harris, R. K., *Nuclear Magnetic Resonance*, Vol. 1–4, American Chemical Society, New York, 1976.

Hartmann, F., *IEEE Trans. Magn.* **Mag-8(1)**, 66 (1972).

Hatch, C. F., D. Onderein, T. Sandreczki, and K. W. Kreilick, *J. Magn. Reson.* **27**, 261 (1977).

Hausser, K. H., *J. Chim. Phys.* **61**, 1610 (1964).

Hausser, K. H., *Z. Naturforsch.* **17A**, 158 (1962); *11th Colloque Ampere, Eindhoven, North Holland, Amsterdam* **1963**, p. 420.

Henning, J. C. M., and J. H. den Boef, *Phys. Rev. B* **14**, 26 (1976).

Henning, J. C. M., and J. H. den Boef, *Phys. Status Soldi B* **72**, 369 (1975).

Henning, J. C. M., J. H. den Boef, P. S. Allen, E. P. Andrew, and C. A. Bates, *Proc. 18th Ampere Congress on Magnetic Resonance and Related Phenomena*, Vol. 1, Nott. Engl., Amsterdam, 1974.

Henning, J. C. M., and J. H. den Boef, *Phys. Rev. B* **18**, 60 (1978).

Henning, J. C. M., and J. H. den Boef, *Solid State Commun.* **14**, 993 (1974).

Higgins, R. J., and Y. K. Chang, *RSI* **39**, 522 (1968).

Hikada, T., *RSI* **44**, 79 (1973).

Hill, D. A., and C. Hwang, *JSI* **43**, 581 (1966).

Hutchison, C. A., and R. C. Pastor, *J. Chem. Phys.* **21**, 1959 (1953).

Hyde, J. S., and H. W. Brown, *J. Chem. Phys.* **37**, 308 (1962).

Idoine, J. D., and J. R. Brandenberger, *RSI* **42**, 715 (1971).

Isaev-Ivanov, V. V., V. V. Lavrov, and V. N. Fomichev, *Dokl. Acad. Nauk SSSR* **229**, 70 (1976).

Ishchenko, S. S., and S. M. Okulov, *PTE*, **18**, 144 (1975).

Jackman, L. M., and F. A. Cotton, *Dynamic Nuclear Magnetic Resonance Spectroscopy*, Academic Press, New York, 1975.

Johnson, A. W., *J. Magn. Reson.* **24**, 21 (1976).

Jung, P., *12th Colloque Ampere, Bordeauz*, **1964**, p. 564.

Karplus, R., *Phys. Rev.* **73**, 1027 (1948).

Koepp, S., *Hochfrequenztech. Electroakust.* **78**, 57 (1969).

Krueger, P., and A. P. Martinelli, *Fourth International Cryogenic Engineering Conference Surrey, England*, 214 (1972).

Lancaster, G. and A. G. Smallman, *JSI* **42**, 341 (1965).

Llewellyn, P. M., *JSI* **34**, 236 (1957).

Loth, K., M. Andrist, F. Graf, and H. S. H. Güthard, *Chem. Phys. Lett.* **29**, 163 (1974).

Lynden-Bell, R. M., and R. K. Harris, *Nuclear Magnetic Resonance Spectroscopy*, Crane Russak, New York, 1975.

Macomber, J. D., and J. S. Waugh, *Phys. Rev.* **140**, A1494 (1965).

Mailer, C., T. Sarna, H. M. Swartz, and J. S. Hyde, *J. Magn. Reson.* **25**, 205 (1977).

Maxfield, B. W., and J. R. Merrill, *RSI* **36**, 1083 (1965).

Mitin, A. V., *Akust. Zh.* **21**, 86 (1975).

Mito, S., K. Okumura, and H. Mina, *Mem. Fac. Eng. Osaka City Univ.* **6**, 107 (1964).

Miyoshi, D. S. and R. M. Cotts, *RSI* **39**, 1881 (1968).

Mohos, B., *Magn. Reson. Chem. Biol. Lect. Ampere Intern. Summer Sch.*, 187 (1975).

Muha, G. M., *RSI* **36**, 551 (1965).

Muromtsev, V., A. K. Piskumov, and N. V. Verein, *Radiotekhn. Electron.* **7**, 1206 (1129) (1962).

Myers, O. E., and E. J. Putzer, *J. Appl. Phys.* **30**, 1987 (1959).

Niemelä, L., *JSI* **41**, 646 (1964).

Parikh, P., *Indian J. Pure Appl. Phys.* **3**, 34 (1965).

Pasimeni, L., *J. Magn. Reson.* **39**, 65 (1978).

Pavlovskii, A. I., N. P. Kolokolchikov, V. V. Druzhinin, O. M. Tatsenko, A. I. Bykov, and M. I. Doltenko, *Pis'Ma V Zh. Eksp. and Teor. Fiz.* **30**, 211 (1979).

Perlman, M. M., and M. Bloom, *Phys. Rev.* **88**, 1290 (1952).

Pescia, J., *Ann. Phys. (Paris)* **10**, 389 (1965).

Pierce, W. L., and J. C. Hicks, *RSI*, **36**, 202 (1965).

Poole, C. P., Jr., and H. A. Farach, *Theory of Magnetic Resonance*, Wiley, New York, 1972.

Poole, C. P., Jr., H. F. Swift, and J. F. Itzel, Jr., *J. Chem. Phys.* **42**, 2576 (1965).

Pound, R. V., and W. D. Knight, *RSI* **21**, 219 (1950).

Proffitt, M. H., and W. L. Gardiner, *J. Magn. Reson.* **25**, 423 (1977).

Rädler, K. H., *Ann. Phys.* **7**, 45 (1961).

Regel, H., and W. Regel, *Messtechnik* **80**, 328 (1972).

Rinehart, E. A., and C. C. Lin, *RSI* **32**, 562 (1961).

Rinehart, E. A., R. H. Kleen, and C. C. Lin, *J. Mol. Spectrosc.* **5**, 458 (1960).

Rinehart, E. A., and R. L. Legan, *RSI* **35**, 103 (1964).

Robinson, W. H., and S. D. Devine, *Phys. Rev. B* **17**, 3018 (1978).

Robinson, W. H., and S. D. Devine, *J. Phys. C Solid State Phys.* **10**, 1357 (1977).

Robinson, W. H., M. A. Collins and S. D. Devine, *J. Phys. E Sci. Instrum.* **8**, 139 (1975).

Robinson, W. H., and A. Edgar, *IEEE Trans. Sonics Ultrasonics* **SU-21**, 98 (1974).

Rupp, L. W., Jr., K. R. Wittig, and W. M. Walsh, Jr., *Am. J. Phys.* **44**, 655 (1976).

Russell, A. M., and D. A. Torchia, *RSI* **33**, 442 (1962).

Sari, S. O. and T. R. Carver, *RSI* **41**, 1324 (1970).

Shaffer, J. S., H. A. Farach, and C. P. Poole, Jr., *Phys. Rev. B* **13**, 1869 (1976).

Slavin, A. J., *Cryogenics* **12**, 121 (1972).

Smaller, B., *Phys. Rev.* **83**, 812 (1951).

Smith, G. E., R. E. Blankership, and M. P. Klein, *RSI* **48**, 282 (1978).

Smith, G. W., *J. Appl. Phys.* **35**, 1217 (1964).

Spry, W. J., *J. Appl. Phys.* **28**, 660 (1957).

Sthanapati, J., A. K. Ghoshal, D. K. Dey, A. K. Pal and S. N. Bhattacharyya, *J. Phys. E Sci. Instrum.* **10**, 221 (1977).

Štirand, O., *ETP* **10**, 313 (1962).

Stesmans, A., and J. van Meijel, *J. Phys. E* **10**, 339 (1977).

Stokes, A. R., *Proc. Phys. Soc.* **61**, 382 (1948).

Strandberg, M. W. P., M. Tinkham, I. H. Solt, and C. F. Davis, Jr., *RSI* **27**, 596 (1956).

Sura, P., *Acta Fac. Rerum Nat. Univ. Comenianae Phys.* **18**, 29 (1977).

Tykarski, L., *J. Phys. D Appl. Phys.* **7**, 786 (1974).

Unterberger, R. R., J. L. Garcia de Quenvedo, and A. E. Stoddard, *RSI* **28**, 616 (1957).

Van den Boom, H., *RSI* **42**, 524 (1971).

Venofrinovich, V. L., A. A. Lukhvich, S. A. Novikov, and A. A. Savitskii, *PTE* **20**, 1187 (1977).

Verdier, P. H., E. B. Whipple, and V. Schomaker, *J. Chem. Phys.* **34**, 118 (1961).

Vigouroux, B., J. C. Gordon, and J. Pescia, *J. Phys. F Metal Phys.* **6**, 1575 (1976).

Visweswaramurthy, S., *Indian J. Pure Appl. Phys.* **3**, 220, 261 (1965).

Wahlquist, H., *J. Chem. Phys.* **35**, 1708 (1961).

Wagner, R. J., and A. M. White, *Solid State Commun.* **32**, 299 (1979).

Wampler, W. R., S. Matula, B. Lengeler, and G. Durcansky, *J. Chem. Phys.* **46**, 53 (1975).

Whitford, B. G., *RSI* **32**, 919 (1961).

Wilmshurst, T. H., *JSI* **44**, 503 (1967).

Wilson, G. V. H., *J. Appl. Phys.* **34**, 3276 (1963).

Wilson, G. V. H., *JSI* **41**, 98 (1964).

Wright, J. J., and D. van der Beken, *Am. J. Phys.* **41**, 260 (1973).

Yagi, M., *Sci. Rep. Tohoku University First Ser.* (*Japan*) **44**, 5 (1960).

Yagi, Y., M. Inone, T. Naito, and T. Tatosukawa, *Japan J. Appl. Phys.* **12**, 1794 (1973).

CHAPTER 7

Detection

Electronic components cannot pass microwave frequencies, and a detector is therefore employed to convert microwave energy to a lower frequency (e.g., dc, af, or rf). The present chapter describes several detectors and detector circuits. The ultimate sensitivity that may be achieved with both bolometers and crystal detectors is discussed in Sec. 11-C. Other types of detectors and detection methods are discussed by Boivin et al. (1973), Goodwin and Jones (1961), Linev and Mochalskii (1978), Martinelli et al. (1977), Murakami et al. (1978), Paraskevopoulos et al. (1977), Rokuskika et al. (1975), Sokolov et al. (1975), Taylor and Herskovitz (1961), and Yamazaki et al. (1977). Chamberlin et al. (1979) obtained high sensitivity by using SQUID detection. Tătaru (1974) treats the sensitivity, the optimum experimental conditions, and the peculiarities of detection by either the Faraday effect or magnetic circular dichroism.

A. Calorimetric Detection

When microwave energy is absorbed in a matched load, it is converted to heat, and the temperature of the load rises. In the absence of a mechanism for heat removal the increase in temperature ΔT in degrees centigrade is related to the incident power P in watts and the time t in seconds by the relation

$$\int_0^t P \, dt = 4.186 C M \Delta T \tag{1}$$

where C is the heat capacity in calories per gram degree and M is the mass of the load in grams. Heating effects are unimportant at typical ESR power levels (e.g., 100 mW). The high power loads ordinarily employed with magnetrons, on the other hand, are equipped with cooling fins to dissipate excess power.

High microwave power levels may be measured by means of a water load with circulating water. The power in watts is computed from the rate of flow of the water (dV/dt cm^3/sec) and the temperature difference ΔT between the input and output water from the relation (RLS-11, Chap. 3)

$$P = 4.186 \left(\frac{dV}{dt} \right) \Delta T \tag{2}$$

B. Bolometer

A bolometer (or barretter) consists of a thin piece of wire that is heated by the incident microwave radiation. As a result of its positive temperature coefficient, the bolometer exhibits an increase in electrical resistance that may be detected by a Wheatstone bridge circuit such as the one shown in Fig. 7-1. From a knowledge of the increase in resistance, the thermal coefficient of resistance, and the heat capacity, one may deduce the microwave power. Commercial power meters that employ bolometer detectors incorporate more sophisticated circuits than the one shown in Fig. 7-1. Platinum is frequently employed as the thermoelectric element in the bolometer. When in use, a bolometer is ordinarily biased by passing about 8 mA of direct current through it, and thus provides it with an operating resistance R of 100 or 200 Ω. The resistance R of a bolometer with the microwave power P incident on it is related to its resistance R_0 at zero power by the expression

$$R = R_0 + kP^n \tag{1}$$

where k is a constant, and n is close to unity.

A bolometer may be employed to detect the ESR signal in a spectrometer. In this application it is most sensitive when about 20 mW of microwave power is incident upon it, and for a simple detection scheme the sensitivity decreases when the power falls below this value, as shown in Fig. 7-2. A balanced mixer detection scheme such as the one discussed in Sec. 7-D may be employed to render the sensitivity independent of power. Every bolometer has a burnout power, and if the incident microwave radiation exceeds this power the bolometer melts and "burns out."

A bolometer is ordinarily operated with a modulation frequency in the audio range of tens or hundreds of cycles per second because its response time is slow. It is advisable to keep the modulation frequency below 1 kHz. The slow response of bolometers will also distort signals that are swept through in less than a millisecond, and this severely limits their adaptability to the signal enhancement scheme of Sec. 11-I.

In theory, bolometer detection can be as sensitive as crystal detection, but in practice, it is less versatile to use. A bolometer's noise temperature is unity

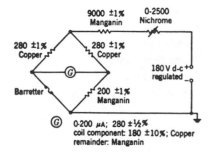

Fig. 7-1. Wheatstone bridge for use with a bolometer (RLS-11, p. 170).

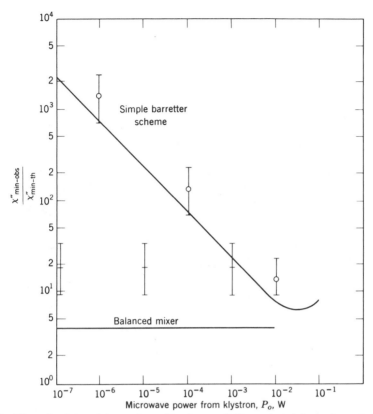

Fig. 7-2. The ratio of the minimum detectable rf susceptibility to the minimum theoretical value versus microwave power for two different bolometer (barretter) schemes. The curves correspond to the predicted sensitivity, and the vertical lines bracket the experimental values (Feher, 1957).

since it produces only Johnson noise $dN = kT\Delta f$. Several highly sensitive spectrometers that incorporate bolometers have been described in the literature. Morigaki et al. (1971) used a balanced bolometer homodyne detector. Bolometer applications have been discussed by Byrne and Cook (1963), Feher (1957), Lavelic (1962), Long (1960), Schmidt and Solomon (1966), and Urano (1955).

C. Crystals

The most commonly used detector in electron spin resonance is the crystal rectifier or diode. A common crystal conversion element consists of a "crystal" of silicon in contact with a tungsten cat whisker. The silicon contains a trace of impurity that renders it semiconducting. A typical current–voltage characteristic is shown in Fig. 7-3. A ceramic cartridge crystal is shown in Fig. 7-4.

The rectification properties of a silicon crystal are defined in terms of the equivalent circuit shown in Fig. 7-5. The incident microwave power induces the

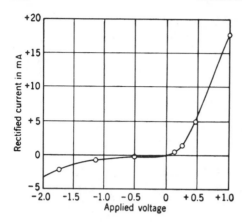

Fig. 7-3. Typical characteristic curve of a silicon rectifier (RLS-15, p. 20).

voltage drop V across the crystal and causes the current I to flow through the variable resistor R. If the resistance R is changed by a small amount ΔR, then the voltage V and current I will change by the increments ΔV and ΔI, respectively. The dynamic output impedance R_{dc} defined by

$$R_{dc} = \frac{\Delta V}{\Delta I} \tag{1}$$

is called the "dc" impedance or the "video resistance." The dc impedance of a

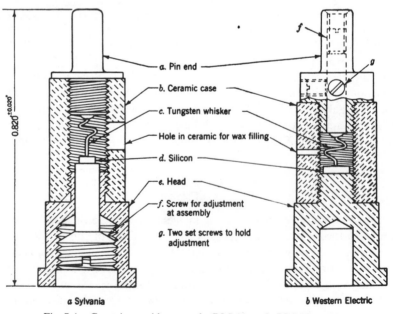

a. Pin end
b. Ceramic case
c. Tungsten whisker
Hole in ceramic for wax filling
d. Silicon
e. Head
f. Screw for adjustment at assembly
g. Two set screws to hold adjustment

a Sylvania   b Western Electric

Fig. 7-4. Ceramic cartridge crystals (RLS-11, p. 5; RLS-15, p. 16).

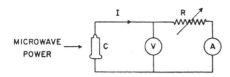

Fig. 7-5. Simplified circuit consisting of Ammeter A and voltmeter V used for measuring the rectified current I and the rectified voltage V at the crystal C. The resistor R can be adjusted between open circuit ($R \sim \infty$) and closed circuit ($R \sim 0$) conditions.

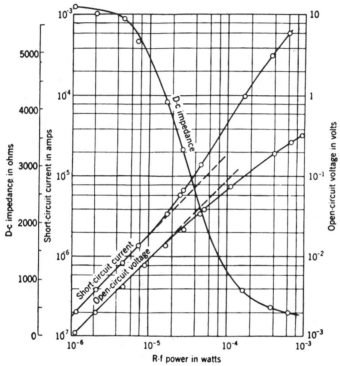

Fig. 7-6. Rectification properties of a silicon crystal rectifier at 3300 MHz (RLS-11, p. 498; RLS-15, p. 334).

typical crystal varies with the microwave power in the manner shown in Fig. 7-6. The same figure shows the power dependences of the short-circuit current obtained when $R = 0$ and the open-circuit voltage obtained when $R = \infty$. At microwatt powers, the dc impedance is fairly constant, and the rectified current is proportional to the rf power. As a result, we say that the crystal is a square-law detector. In the milliwatt region, the rectified crystal current becomes proportional to the square root of the microwave power, and the crystal is said to be a linear detector. The transition from square law to linear behavior is very gradual, and the crossover point between the two regions is typically near the power range 10^{-5} to 10^{-4} W. Crystal detectors of ESR spectrometers often operate in this transition region.

When the magnetic field is modulated at an intermediate frequency (i.f.) denoted by f_{mod}, the microwaves become amplitude-modulated at f_{mod} during the passage through resonance. At the crystal the microwave signal is demodulated, and the ESR signal enters the receiver or preamplifier as an i.f. signal. The i.f. impedance "seen" at the input terminals of the receiver is the ratio of the i.f. current to the i.f. voltage at these terminals. In many cases the i.f. impedance and dc resistance are comparable in magnitude.

The available noise power or Johnson noise dN arising from a resistor radiating into a "cold" or noiseless load at the temperature T is

$$dN = kT\Delta f \tag{2}$$

where Δf is the bandwidth. A crystal rectifier generates more noise than a resistor, and its noise power dN is given by

$$dN = tkT\Delta f \tag{3}$$

where the dimensionless quantity $t > 1$ is the so-called noise temperature or output noise ratio. Physically, t is the amount by which a resistor has to be raised in temperature to make it equivalent in noise power to a crystal maintained at room temperature. The excess noise is inversely proportional to the modulation frequency f_{mod} and for the square-law region ($P_c < 10^{-6}$ W) one may write

$$dN = \left(\frac{\gamma P_c^2}{f_{mod}} + 1 \right) kT\Delta f \tag{4}$$

while in the linear region

$$dN = \left(\frac{\gamma' P_c}{f_{mod}} + 1 \right) kT\Delta f \tag{5}$$

Feher (1957) quotes for an X-band crystal

$$\gamma \approx 5 \times 10^{14} \text{ W}^{-2} \text{ sec}^{-1} \tag{6}$$

$$\gamma' \approx 10^{11} \text{ W}^{-1} \text{ sec}^{-1} \tag{7}$$

Andrews and Bazydlo (1959) found that the noise figure in decibels from a 1N23E crystal decreased linearly with increasing frequency below 100 kHz and remained constant above this value. Bosch et al. (1961) found a $1/f_{mod}$ law obeyed from 25 Hz to 300 kHz, with excess noise still present at higher frequencies.

The conversion loss L of a crystal is the ratio of the available ESR signal power in the microwaves incident on the crystal to the available output power

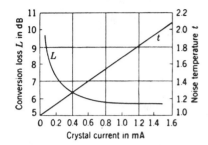

Fig. 7-7. Conversion loss L and noise temperature t as a function of the rectified current for a typical 1N23B crystal rectifier (RLS-15, p. 34).

at the frequency f_{mod} after detection. The conversion gain G is the reciprocal of L. From Feher (1957)

$$L = \frac{S'}{P_c} \tag{8}$$

in the square-law region, and

$$L = C' \tag{9}$$

in the linear region, where S' and C' are constants. For an 1N23C crystal $S' = 0.002$ W, and $C' \approx 3.3$.

It is desirable to have both a low conversion loss L and a low noise temperature t. As Fig. 7-7 indicates, the conversion loss becomes very great at low crystal currents, and the noise temperature becomes excessive at high crystal currents. The best operating point is at an intermediate current that corresponds to an intermediate power level. The leakage or amount of microwave power incident on the crystal should be adjusted to this value, but the setting is not critical since the range of power that provides a maximum signal-to-noise ratio is fairly broad. Setting the leakage in this manner is called rf bucking (see Fig. 7-8). Sometimes a dc bias is employed to improve the crystal noise figure. Long (1960) gives specific data on seven crystal diodes.

The current sensitivity β of a crystal is the number of microamperes of rectified current per microwatt of available rf power. The figure of merit M is defined by

$$M = \frac{\beta R}{(R_{\text{dc}} + R_{\text{A}})^{1/2}} \tag{10}$$

where R_{dc} is the dc or video resistance, and R_{A} is the equivalent noise resistance of a video amplifier (see RLS-15, Sec. 11-5). The JAN specifications have set R_{A} equal to 1200 Ω. It is desirable to have a high value of M.

There are two principal types of crystals. A *mixer crystal* is designed to convert two nearby microwave frequencies to an intermediate difference frequency, and is ideally suited for superheterodyne detection. It is usually

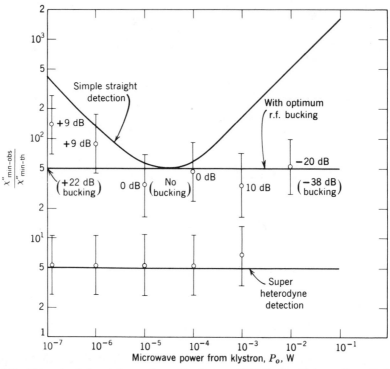

Fig. 7-8. The ratio of the minimum observable rf susceptibility to the minimum theoretical value
versus the microwave power for three crystal detection schemes. The straight detection and rf
bucking schemes employed a modulation frequency of 1 kHz. The curves correspond to the
predicted sensitivity, and the vertical lines bracket the experimental values (Feher, 1957).

characterized by its conversion loss L, output noise ratio (noise temperature) t,
receiver noise figure (see Sec. 11-C), i.f. impedance, and burnout rating. A
video crystal is designed for the detection of very low power microwave energy
and is often used in spectrometers that do not employ superheterodyne
detection. It is frequently characterized by its figure of merit M, its current
sensitivity β, its video or dc impedance R_{dc}, and its burnout rating. A video
crystal is also characterized by its minimum detectable signal, which is a signal
voltage about equal to the RMS noise, and by its tangential signal (minimum
sensitivity), which is the pulse power that raises the noise by its own width.

Buckmaster et al. (1965, 1971) made a study of the role played by the
microwave diode detector in determining the sensitivity of a spectrometer. In
their 1971 article they measured the noise temperature t, the conversion gain
G, and other significant parameters of three types of diode matched pairs,
namely, the 1N23B point contact, the HP 2627 Schottky barrier and the L4154
backward types. Figure 7-9 shows a graph of the quantity $(t/G)^{1/2}$ measured
with the diodes operated with optimum bias. We know from Eq. (11-C-25) that
when klystron and preamplifier noise are neglected the sensitivity is propor-

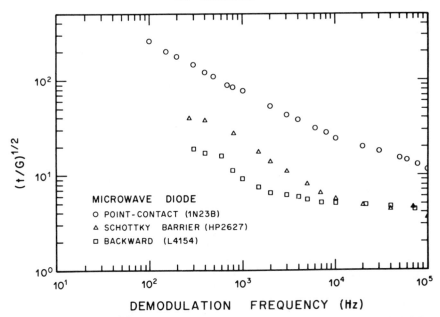

Fig. 7-9. Graph of $(t/G)^{1/2}$ as a function of the demodulation frequency for selected 1N23B, HP2627, and L4154(-4) diodes operated with optimum microwave bias (Buckmaster and Rathie, 1971).

tional to $(t/G)^{1/2}$. Thus we see that the backward diode is superior to the other two, and the Schottky barrier diode is the second best.

Crystals perform very well as detectors in ESR spectrometers, as Fig. 7-8 indicates, but their conversion characteristics are not uniform enough to render them suitable for accurate power measurements. We see from the figure that bucking improves the performance of straight or homodyne detection at both low and high powers. Den Boef (1973) designed a bucking current stabilizer to control the bias or bucking power applied to the crystal. Very high powers can damage crystals. A crystal may be checked for burnout by means of an ohmmeter. For this test the positive terminal of the ohmmeter is connected to the pointed or pin end of the crystal to measure the low resistance in the forward direction. The resistance in the forward direction should be about a thousand times that in the backward direction. For typical crystals the limiting back current obtained with a potential difference of 1 V across the crystal is between 0.1 and 0.4 mA. A simple crystal test circuit is mentioned on p. 439 of the first edition.

The first edition discusses crystals and a number of other devices such as thermal detectors, InSb photodetectors, GaAs, tunnel diodes, varactors, backward diodes, Eski diodes, and hot carrier diodes as detectors in the submillimeter range of frequencies [see first edition, Sec. 11-E, Burrus (1963), Coleman (1962, 1963), De Loach (1964)]. Kuroda et al. (1978) reported the submillimeter ESR of cobalt tutton salt. Ohtomo et al. (1975) reviewed submillimeter waves, and Geick (1977) discussed applications of submillimeter spectroscopy. Erickson (1966) used a backward wave detector with his high-frequency spectrometer, and Reinberg and Estle (1967) employed a backward wave diode in a balanced mixer.

D. Balanced Mixers

A balanced mixer employs two detectors arranged in such a manner that the ESR signal arrives at each of them in the same phase while the power coming directly from the klystron arrives out of phase. The i.f. output at the modulation frequency f_{mod} is the sum of the signals that appear at the two crystals, and so the phase relations cause the ESR signals to add and the klystron noise to cancel. Only half of the ESR signal appears at each crystal, but the two signals add so that the overall conversion loss of a balanced mixer is the same as that of a single crystal.

A balanced mixer is often constructed from a magic T in accordance with Fig. 7-10. As discussed in Sec. 4-E, the wave amplitudes B_1 and B_2 produced in the crystal arms 1 and 2 are related to the incident amplitudes A_E and A_H in the E and H arms, respectively, by the expressions [Eq. (4-E-2)]

$$B_1 = \left(\frac{j}{2^{1/2}}\right)(A_H + A_E) \tag{1}$$

$$B_2 = \left(\frac{j}{2^{1/2}}\right)(A_H - A_E) \tag{2}$$

Thus the magic T automatically causes the ESR signal to add and the klystron noise to cancel. If the klystron arm has the internal admittance Y_E and power P_E and the ESR cavity arm has the admittance Y_H and power P_H, then the power P_1 delivered to the crystal arm 1 is (RLS-16, p. 266).

$$P_1 = \frac{8g_1\left[g_H P_H(1 + Y_2 Y_E)^2 + g_E P_E(1 + Y_2 Y_H)^2\right]}{\left[(1 + Y_1 Y_H)(1 + Y_2 Y_E) + (1 + Y_1 Y_E)(1 + Y_2 Y_H)\right]^2} \tag{3}$$

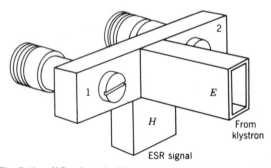

Fig. 7-10. X-Band magic-T balanced mixer (RLS-16, p. 270).

where all the admittances are normalized relative to the characteristic admittance Y_0. A similar expression may be obtained for the power P_2 in arm 2 by interchanging the subscripts 1 and 2. The quantities g_i are the conductances in the various arms of the bridge (real part of Y_1).

To estimate the liability that results from unbalanced crystals ($Y_1 \neq Y_2$; $g_1 \neq g_2$), one may simplify Eq. (3) by letting

$$Y_E = Y_H = g_E = g_H = g \tag{4}$$

As a result, one obtains the power ratio in the two arms

$$\frac{P_1}{P_2} = \frac{g_1}{g_2}\left(\frac{1 + gY_2}{1 + gY_1}\right)^2 \tag{5}$$

When $gY_1 \ll 1$ and $gY_2 \ll 1$, the power splits at the crystals in the ratio of the rf conductances, or in other words, in the inverse ratio of the rf resistances. In general, the noise suppression in a magic T balanced mixer is a complicated function of the rf resistances of the two crystals, and a simplified relation such as Eq. (5) cannot be used.

To estimate the extent to which the klystron noise suppression is dependent upon the closeness of their conversion losses to each other, let each crystal have the same incident rf power from both the ESR signal and the local oscillator. If L_1 and L_2 are the two conversion losses, then the ESR signal power is proportional to $(L_1^{1/2} + L_2^{1/2})^2$, and the klystron noise power is proportional to $(L_1^{1/2} - L_2^{1/2})^2$. The factor by which the klystron noise is suppressed relative to the signal is $(L_1^{1/2} + L_2^{1/2})^2/(L_1^{1/2} - L_2^{1/2})^2$. For a 3-dB difference in conversion loss, $L_1 = 2L_2$, and we have

$$\left(\frac{L_1^{1/2} + L_2^{1/2}}{L_1^{1/2} - L_2^{1/2}}\right)^2 = 34 \tag{8}$$

so even this small unbalance suppresses the klystron noise by 15.3 dB.

In Sec. 7-B it was mentioned that at low powers the conversion gain of a bolometer is low. A balanced mixer such as the one shown in Fig. 7-10 may feed a large amount of power P_2 to mixer bolometers or barretters. This improves the conversion gain without increasing the noise because the bolometer noise should not be power-dependent, and much of the noise from the bypassed power P_2 in Fig. 7-11 cancels in the balanced mixer. A phase shifter is provided for the bypassed power since an incorrect phase will reduce the ESR sensitivity and also produce an admixture of χ' and χ''. This system is somewhat inconvenient to operate because the phase must be reset every time either the bypassed power level P_2 is adjusted by the attenuator or the slide screw tuner setting is changed.

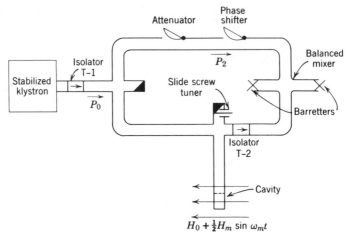

Fig. 7-11. Bolometer system with balanced mixer (Feher, 1957).

E. Superheterodyne Detection

A superheterodyne receiver employs the frequency converter shown diagrammatically in Fig. 7-12 to convert the microwave ESR signal to an intermediate frequency (i.f.) signal $\Delta\omega$ that is two or three orders of magnitude lower in frequency. In a typical case a local oscillator klystron supplies a frequency $(1/2\pi)\omega_L = 9030$ MHz, and the ESR absorption experiment is carried out at $(1/2\pi)\omega_0 = 9000$ MHz. The two frequencies ω_L and ω_0 mix in the frequency converter and produce the sum and difference frequencies $\omega_L \pm \omega_0$. The difference or i.f. frequency is detected by the mixer and later demodulated to provide the ESR resonance signal, which is either seen visually or recorded in the usual way. The magic T mixer shown in Fig. 11-10 is often employed in superheterodyne detection, and it provides for the usual cancellation of the klystron noise. Seidel (1962) and Llewellyn et al. (1962) modulate the klystron and use a sideband as the local oscillator frequency. This simplifies the superheterodyne instrumentation by eliminating the necessity of a second klystron.

The big advantage of a superheterodyne detector is that it operates at a frequency that is sufficiently high so that the $1/f$ noise in the crystal is negligible. The 30- or 60-MHz i.f. frequency of typical superheterodyne spectrometers considerably exceeds the modulation frequency of almost every

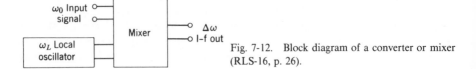

Fig. 7-12. Block diagram of a converter or mixer (RLS-16, p. 26).

field modulation system that has been described in the literature. In addition, it provides better resolution for very narrow resonant lines (e.g., tens of milligauss wide) that are broadened by very high frequency (e.g., 100 kHz) field-modulation systems.

In a superheterodyne spectrometer it is necessary to stabilize the frequency of the signal klystron, and at the same time to stabilize the frequency of the local oscillator klystron in such a manner that the difference frequency remains constant. This entails either the use of two independent AFC circuits, or the addition of a system that maintains the constant difference $\Delta\omega$ between the two klystrons. The latter is preferable.

England and Schneider (1950) describe the first superheterodyne spectrometer mentioned in the literature. Table 13-1 of the first edition lists a number of such spectrometers. Schoch et al. (1975) give a detailed description of a superheterodyne system for ESR or microwave spectroscopy.

F. Lock-In Detectors

A lock-in detector compares the ESR signal from the crystal with a reference signal and only passes the components of the former that have the proper frequency and phase. The reference voltage comes from the same oscillator that produces the field modulation voltage, and this causes the ESR signal to pass through while noise is suppressed. Thus a lock-in detector only accepts signals that "lock in" to the reference signal. Sometimes this device is given one of the more descriptive titles "phase-sensitive detector" or "coherent detector."

The operation of a lock-in detector can be described in terms of the circuit presented in Fig. 7-13, which is a simplified version of one due to Blair and Sydenham (1975). The reference oscillator produces a modulating voltage v_m

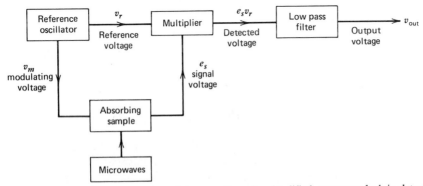

Fig. 7-13. Schematic representation of the operation of a simplified, prototype lock-in detector [adapted from Blair and Sydenham (1975)].

and a reference voltage v_r given by

$$v_m = V_m\cos(\omega t)$$

$$v_r = V_r\cos(\omega t + \phi)$$

where ϕ is a phase angle. At resonance the sample absorbs microwave energy and produces the ESR signal voltage e_s

$$e_s = E_s\cos(\omega t + \phi_s)$$

where ϕ_s is another phase angle that may be close to ϕ. The multiplier produces an output that is the product $e_s v_r$ of the reference and ESR signal voltages

$$e_s v_r = E_s V_r\cos(\omega t + \phi_s)\cos(\omega t + \phi)$$

$$= \tfrac{1}{2}E_s V_r[\cos(2\omega t + \phi + \phi_s) + \cos(\phi - \phi_s)]$$

The low pass filter removes the first term to produce the dc output voltage V_{out}

$$V_{out} = \tfrac{1}{2}E_s V_r\cos(\phi - \phi_s)$$

To maximize the sensitivity ϕ is set equal to ϕ_s, which gives

$$V_{out} = \tfrac{1}{2}E_s V_r$$

To see how the system rejects noise we consider a noise voltage e_n at the frequency ω_n and phase ϕ_n

$$e_n = E_n\cos(\omega_n t + \phi_n)$$

that enters the multiplier along with e_s. The detected noise voltage $e_n v_r$ is equal to

$$e_n v_r = E_n V_r\cos(\omega t + \phi)\cos(\omega_n t + \phi_n)$$

$$= \tfrac{1}{2}E_n V_r\{\cos[(\omega - \omega_n)t + \phi - \phi_n] - \cos[(\omega + \omega_n)t + \phi + \phi_n]\}$$

The low pass filter rejects the second term, leaving

$$V_{out} = \tfrac{1}{2}E_n V_r\cos[(\omega - \omega_n)t + \phi - \phi_n]$$

Even if ω_n is sufficiently close to ω to pass through the low pass filter, a series of noise signals will tend to arrive with random phases ϕ_n which will cause them to average to zero.

Figure 7-13 presents a simplified prototype lock-in detector. A more sophisticated version is illustrated in Fig. 7-14.

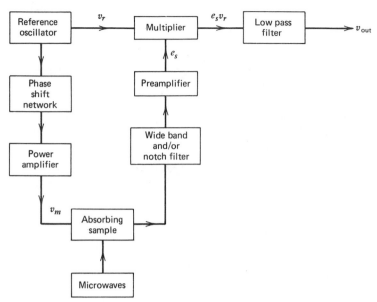

Fig. 7-14. Schematic representation of the operation of a lock-in detector [adapted from Blair and Sydenham (1975)].

The lock-in detector is essentially a narrow band filter that is phase-sensitive, and as such, it is characterized by an effective bandwidth Δf and quality factor, Q. This will be discussed in Secs. 11-C, 11-D, and 11-F.

Faulkner (1959) designed the transistorized lock-in detector shown in Fig. 7-15. The reference signal is transformer-coupled to the bases of the OC70 transistors, and the ESR signal is applied to the collectors. In the two halves of the switching cycle the transistor impedance alternates between a value very small and very large relative to the 10-kΩ load. A bandwidth of 0.2 Hz is obtained when a 100-μF capacitance is used for the integrating capacitor C. The dc drift is less than 5 μV.

The operation of a lock-in detector such as that illustrated in Fig. 7-15 may be analyzed in terms of the output voltages V_a and V_b on the two transistors. These output voltages are illustrated in Fig. 7-16 for several choices of phase. When ϕ is 0° or 180° relative to ϕ_s the output, which is the difference signal $V_b - V_a$, is a full-wave rectified voltage whose main Fourier components are a dc voltage and a double-frequency sine wave. If the input signal is 90° or 270° out of phase with the reference signal, then the average dc component of each transistor will be zero, as shown in Fig. 7-16. Of course, intermediate phase settings produce intermediate output signals.

If the input signal frequency ω_i is slightly different from the reference frequency ω_r, then it will act like an input signal of gradually changing phase, and the transistor output voltages will oscillate back and forth at the difference frequency $|\omega_i - \omega_r|$. The amplitude of this signal at $|\omega_i - \omega_r|$ becomes smaller as $|\omega_i - \omega_r|$ becomes greater. If one assumes that the input noise

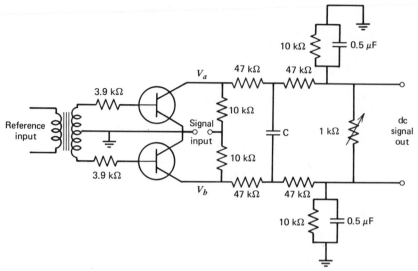

Fig. 7-15. Circuit diagram of a phase-sensitive detector using a transformer-coupled reference signal input and two matched transistors (Faulkner, 1959).

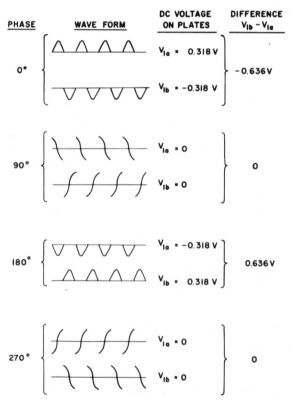

Fig. 7-16. Waveform and dc voltage on the lock-in detector plates for four values of the phase of the input signal V. The half-amplitude of the waveform on the plate is V, and the applied frequency ω_i equals the reference frequency ω_r.

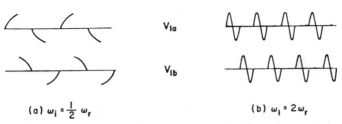

Fig. 7-17. Waveforms on the lock-in detector plates V_{1a} and V_{1b} for an input angular frequency ω_i which is (a) half and (b) twice the reference frequency ω_r. The dc voltages of both V_{1a} and V_{1b} are zero in each case.

spectrum is independent of frequency in the neighborhood of ω_r, then the output noise will peak near ω_r and decrease with $|\omega_i - \omega_r|$ for frequencies both above and below ω_r. In addition, most of the noise at the frequency $|\omega_i - \omega_r|$ will itself be rejected because it will be out of phase with the reference signal. Figure 7-17 shows the waveforms on the lock-in detector plates for $\omega_i = 2\omega_r$ and $\omega_i = \frac{1}{2}\omega_r$. This figure is drawn to the same scale as Fig. 7-16 to facilitate a comparison with the analogous zero phase (i.e., properly phased) signal for $\omega_i = \omega_r$.

To ensure phase coherence, the reference signal for the lock-in detector must be obtained from the same oscillator that produces the magnetic field (or source) modulation. Ordinarily an adjustable phase shifter is provided to ensure that the modulation and reference signals arrive in phase at the lock-in detector. A second phase shifter may be employed to synchronize the oscilloscope x-axis sweep with the lock-in detector output for visual observation of the latter. These two phases should be adjusted, respectively, to (1) maximum signal amplitude and (2) a symmetrical scope pattern.

The Faulkner circuit of Fig. 7-15 employs transistors for the switching function. Mechanical switching has also been employed (Strandberg et al., 1956). Schuster (1951) gives a lock-in detector equivalent circuit.

Sthanapati et al. (1977) designed the inexpensive phase-sensitive detector illustrated in Fig. 7-18. It utilizes the multiplication property of a commercially available current ratioing four-quadrant analog dc multiplier. This lock-in detector was designed for wide-line NMR use at 30 Hz, but with minor modification of the tuned amplifier and notch filter circuits it may be operated at 100 kHz.

The input signal to the unit is first amplified by a preamplifier with a gain that is variable between 5 and 1250 and then by a narrow-band twin T-tuned amplifier. The detected signal is multiplied by a reference signal from a phase shifter that has a range from 0° to 180°. A notch filter at twice the modulation frequency blocks the second harmonic, and a single pole, low-pass active filter of unity gain with a cutoff at one third of the modulation frequency removes low-frequency noise. The output signal is impressed on a chart recorder.

The performance of this lock-in detector is as follows: The ratio of the maximum available noncoherent input voltage to the minimum detectable

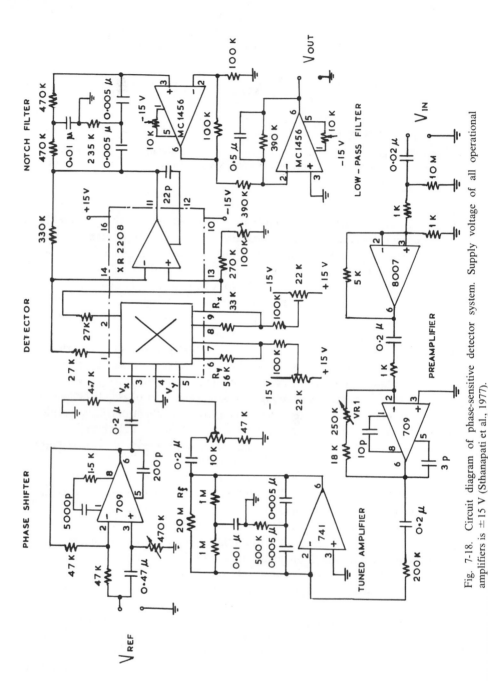

Fig. 7-18. Circuit diagram of phase-sensitive detector system. Supply voltage of all operational amplifiers is ±15 V (Sthanapati et al., 1977).

276

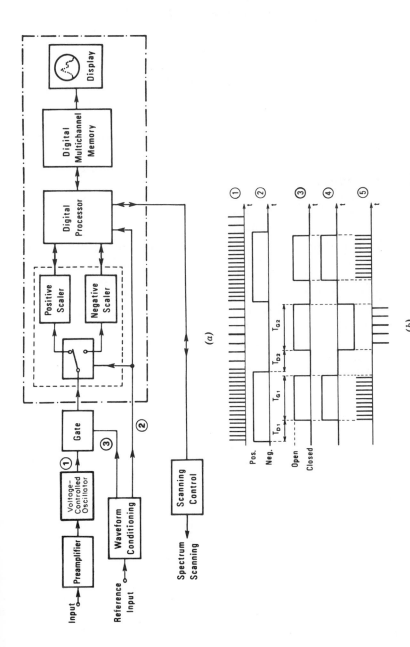

Fig. 7-19. Manner in which a square wave input signal is processed by a digital lock-in detector. (*a*) functional block diagram, (*b*) various waveforms. (1) Voltage-controlled oscillator output, (2) switch driver (3) gate opener, (4) equivalent weighting waveform of the system, and (5) circuit output (Cova et al., 1979).

277

in-phase input voltage, called the input dynamic range, is more than 90 dB between 20 Hz and 1 MHz. The ratio of the maximum noncoherent input voltage to the in-phase signal required for a full output, called the dynamic reserve, is about 36 dB. The out-of-phase rejection is 0.01%, and the dc drift is 0.1 in volts per degree Celsius.

Most lock-in detectors are of the analog type, and this limits their ability to eliminate dc and low-frequency noise. To overcome this deficiency Cova et al. (1979) designed the digital lock-in detector whose block diagram is presented in Fig. 7-19a. The precision voltage-controlled oscillator performs a voltage-to-frequency conversion, and regularly spaced pulses are generated whose spacing varies with the voltage input. These pulses are processed by the lock-in detector in the manner illustrated in Fig. 7-19b.

Halpern and Phillips (1970) used third derivative detection of the ESR signal with the circuit illustrated in Fig. 7-20. The microwave magnetic field was modulated by two sets of coils, one driven at 20 kHz and the other at 12.5 or 25 Hz, and each is detected by a PAR-120 lock-in detector. Their arrangement provides the first derivative as output of the first or 20 kHz lock-in detector, and either the second or third derivative as output of the second lock-in depending upon whether 12.5 or 25 Hz, respectively, is employed for the modulation. The use of third derivative detection provides an increase in resolution, as the spectra in Fig. 7-21 demonstrate. Others have discussed the advantages of third derivative detection (e.g., see Blair and Sydenham, 1975,

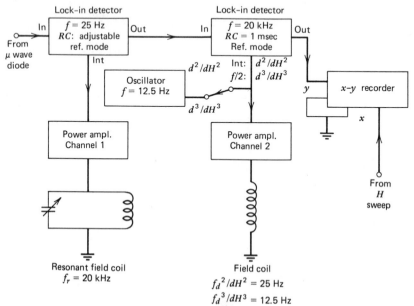

Fig. 7-20. Block diagram of a third-derivative detection system (Halpern and Phillips, 1970).

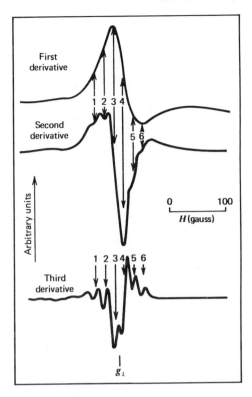

Fig. 7-21. First, second, and third derivatives of the ESR absorption signal of an Ag^{2+} nicotinate–Cd nicotinate (magnetically diluted $1:6$) polycrystalline sample at 77 K, showing the positions of the absorption lines. The third derivative was taken at 10 times the gain used for the second derivative; modulation, integration times, and all other conditions were the same in each case (Halpern and Phillips, 1970).

Cole and Duffy, 1974). Glausinger and Sienko (1973) used second derivative detection in studying spin resonance of metals.

Lock-in detector circuits of the switching type are also discussed in RLS-19, and by Baker (1954), Bösnecker (1960), Cox (1953), Dereppe (1961), Faulkner (1959), Frater (1965), Memory (1964), Nuckolls and Rueger (1952) and Rollin (1964). Current steering lock-in detectors were reported by Danby (1968) and by Grimbleby and Harding (1971), double balanced modulator ones were described by Cole and Duffy (1974), and digital ones were tested by Coles (1971) and Morris and Johnston (1968). The Popov (1960) circuit operates up to 12 MHz. Van Gerven et al. (1963) used a wattmeter as the active element in their coherent detector. Raether and Bitzer (1964) built a polarity coincidence detector. Bletzinger et al. (1965) show how an oscilloscope may be used as a lock-in detector. Beers (1965) describes a lock-in detector mode of operation that removes spurious coherent signals. Arnal and Beuchot (1964), Faulkner et al. (1964, 1966) and Williams (1965, 1970) describe transistorized phase-sensitive detectors. Edgar (1975) reported on phase reversals in ESR spectra, and Wollen (1971) discussed the perils of lock-in detection for saturated NMR lines. De Neef and Verbruggen (1974) used a static torque method in lieu of phase-sensitive detection for lines that saturate easily.

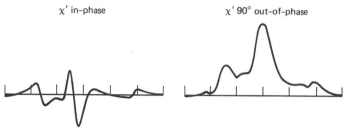

χ' in-phase   χ' 90° out-of-phase

Fig. 7-22. The χ' first harmonic in- and out-of-phase spectra of hemoglobin in glycerol, using 5-G 100-kHz field modulation and 20 mW of incident microwave power. Ticks are at 20-G intervals. (Hyde and Thomas, 1974).

In most detection schemes the phase ϕ of the lock-in detector is set equal to the phase ϕ_s of the incoming modulated ESR signal. The slow molecular motion of spin labels produces spectra whose amplitudes and shapes are sensitive to the frequency and phase setting of the lock-in detector (see Figs. 13-33 and 13-34). Such spectra are generally recorded in phase ($\phi = \phi_s$) and out of phase or in quadrature ($\phi - \phi_s = \pi/2$); Fig. 7-22 shows the dramatic change in shape that can occur when the phase is changed. In saturation transfer studies (see Sec. 13-G) in-phase and quadrature phase settings are employed. In a study of transient ESR signals from the photoexited triplet state in porphyrin molecules, Levanon and Vega (1974) found that the observed relative intensities of the component lines in the ESR signal depended upon the phase setting of the lock-in detector, and in some instances, a change in lock-in phase converted the polarization of some spectral lines from absorption to emission.

Lock-in detectors have also been discussed by Cheburkov (1976), Conradi (1977), Dvornikov and Kompankov (1976), Sastry and Ramakrishna (1977), and Uebersfeld and Erb (1956).

G. Oscilloscopic Presentation

For tuning purposes and rapid searching for unknown resonances, it is convenient to display the resonant absorption on an oscilloscope. For example, a strong absorber like α, α'-diphenyl-β-picryl hydrazyl (DPPH) may be observed visually to set the magnetic field strength dial, adjust the modulation phase, check the magnet homogeneity and modulation amplitude, and so on. A proton NMR magnetometer may be employed to calibrate the oscilloscope screen for the purpose of visually determining the linewidth, hyperfine spacings, and g factor.

If the modulation amplitude is made several times the linewidth and the signal is impressed on the oscilloscope before it reaches the narrow band amplifiers, then the scope will present the actual absorption (or dispersion) lineshape. When one employs a high-frequency plus a low-frequency modulation and narrow band detection, the oscilloscope display is the modulus of the absorption lineshape. These patterns differ from the first (or second) derivative signals that result from phase-sensitive detection with a lock-in detector.

For oscilloscopic presentation of the magnetic resonance signal, it is best to obtain the oscilloscope sweep frequency from the same oscillator that modulates the magnetic field. A phase shifter may be employed to center the pattern on the oscilloscope. When the double-frequency modulation technique is employed the low or audio frequency is used for the oscilloscope sweep. Elmore (1974) devised a method for aligning resonance peaks on a CRO screen.

H. Recorder Presentation and Response Time

When the electron spin resonance signal emerges from the lock-in detector it still contains a considerable amount of noise. A great deal of this noise may be removed by passing the signal through a low-pass filter. The filter has associated with it a time constant or response time τ_0, which is a measure of the cutoff frequency of the filter. In other words, the filter fails to pass frequencies that are much greater than the reciprocal of the time constant $1/\tau_0$; it attenuates, distorts, and retards those incoming waveforms that have frequencies in the neighborhood of τ_0, and it transmits, undisturbed, those frequencies considerably below $1/\tau_0$.

The waveform that is impressed on the ESR response filter may be considered as the derivative of the absorption (or dispersion) line, and for comparison with the above criteria, its effective frequency may be taken as the inverse of the time that it takes to scan through the resonance from one peak to the next. In other words, if the time that it takes to scan through the magnetic field range ΔH_{pp} is very short compared to the time constant τ_0, then no signal will appear on the recorder; if this time equals the time constant, then a distorted signal will result; while if one waits many time constants to complete the scan, then the recorder will faithfully reproduce the true lineshape. Of course, other criteria such as the absence of both overmodulation and saturation effects must also be met to produce an undistorted lineshape. Sometimes one encounters a drifting recorder baseline that must be recentered during a scan. Torgeson and Rhinehart (1963) have designed an automatic recorder recentering circuit.

It will be helpful to give a more quantitive treatment of the distortion that occurs at intermediate scanning rates. The time constant of a resistor R in series with a capacitor C is RC sec (ohm $\times$ farad = second). Strandberg et al. (1954) studied the distortion of an absorption signal in terms of the parameter α

$$\alpha = \frac{RC}{\tau_{1/2}} = \frac{\tau_0}{\tau_{1/2}} \tag{1}$$

where $\tau_{1/2}$ is half the time that it takes to scan through the signal between the

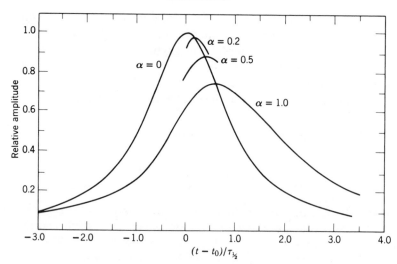

Fig. 7-23. The distortion that results from scanning too quickly through a magnetic resonance absorption line. The abscissa gives the time t relative to the time t at the center of the resonance normalized with respect to the time $\tau_{1/2}$ that it takes to scan through the width $\frac{1}{2}\Delta H_{1/2}$. The parameter α is defined by Eq. 7-H-1 (Strandberg, Johnson, and Eshbach, 1954).

half-amplitude points. They define the parameter ρ which depends on the time t as

$$\rho = \frac{(t - t_0)}{\tau_{1/2}} \tag{2}$$

where t_0 is the time at the peak of the filter's input signal. Their Lorentzian curve illustrated in Fig. 7-23 exhibits the following three distortions at $\alpha = 1$: (1) the peak amplitude decreases (2) the peak appears at a later time than it would for zero time constant, and (3) the resonance appears broader on the right than on the left, but it continues to have the same integrated area. The peak of the actual output signal always occurs where this curve intercepts the undisturbed $\alpha = 0$ curve.

If accurate linewidth and g-factor measurements are to be made, one must be careful not to employ a response time τ_0 that is too long. One can recognize response-time distortion by scanning both forward and backward through the resonance, and observing whether or not the asymmetry reverses phase since true g-factor asymmetry will always be more pronounced on the same side of the line, while response-time distortion will always broaden more on the side that appears last on the recorder. For g-factor measurements one may average the values obtained with forward and backward scans when it is not feasible to reduce α.

In theory, one may always increase the signal-to-noise ratio by increasing the response time and decreasing the rate of scan, but this always prolongs the time required to obtain data. If one carries this process too far by using scans of several hours' duration then one encounters long time drift and stability problems that can produce a drifting baseline and other deleterious effects (e.g., room temperature variations may change the magnetic field value).

It is recommended that inexperienced ESR spectroscopists practice recording the spectrum of a well-known sample like solid DPPH under a variety of scanning rates and response times. When a spectrum is being recorded to obtain quantitative data such as the g factor and lineshape, it is recommended that one employ a response time that is less than $1/10$ of the time τ_{pp} required to scan through the two peaks of the line.

The time constant may be measured by setting the magnetic field considerably off resonance, starting the recorder, and then suddenly and rapidly turning the magnet control knob to near the peak of the resonance. When this is done, the recorder will plot an exponential increase in the ESR signal I to its maximum value I_0 in accordance with the relation ($t_0 = 0$)

$$I = I_0(1 - e^{-t/RC}) = I_0(1 - e^{-t/\tau_0}) \tag{3}$$

A time-constant criterion may be established by observing that at the time

$$t = 0.693\tau_0 \tag{4}$$

one obtains

$$I = \tfrac{1}{2}I_0 \tag{5}$$

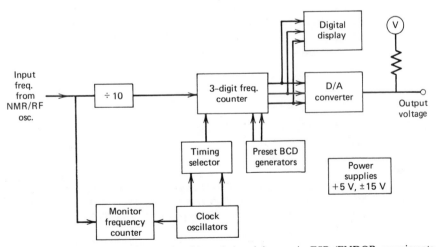

Fig. 7-24.   Block diagram of a recorder driver designed for use in ESR/ENDOR experiments (Alexander and Pugh, 1975).

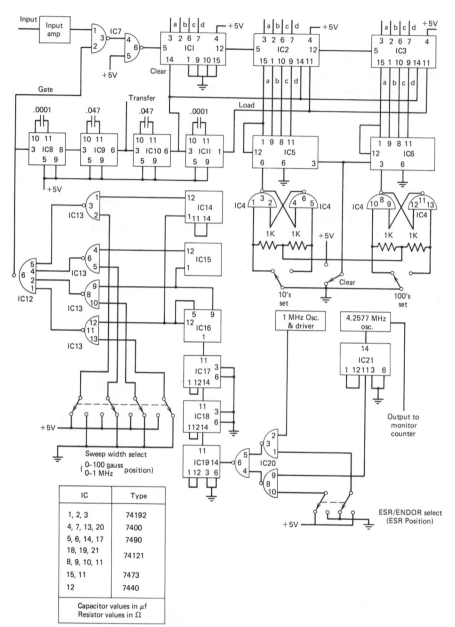

Fig. 7-25. A circuit diagram for the integrated circuits in the three-digit frequency counter, the timing selector, and the preset generators of the recorder driver illustrated in Fig. 7-24. The outputs from IC9 and ICs1, 2, and 3 go to the D/A converter, and digital display circuits shown in Fig. 7-24. The power supply connections are not shown (Alexander and Pugh, 1975).

284

and at the time

$$t = \tau_0 \tag{6}$$

one has

$$I = 0.632 I_0 \tag{7}$$

Either point may be employed to compute τ_0. The experimental technique requires that one use a strongly absorbing sample, that the resonance be set near its peak value in a time short compared to τ_0, and that the magnetic field and ESR frequency be maintained constant during the measurement. Further details are given in Sec. 11-F.

Alexander and Pugh (1975) designed a digitally controlled recorder–driver illustrated in Fig. 7-24 for ESR/ENDOR experiments. The integrated circuits of this device are presented in Fig. 7-25. Dwyer and Rodgers (1978) suggest the use of a pulse height analyzer as an x–y recorder. Vetokhin et al. (1978) reported a recording unit for field-modulated spectrometers.

I. Integration

Narrow band detection at the fundamental of the modulation frequency produces a recorder plot that is the first derivative of the absorption lineshape. Schwenker (1959) described an electromechanical analog integrator that converts the first derivative recorder pattern shown in Fig. 7-26a to the absorption

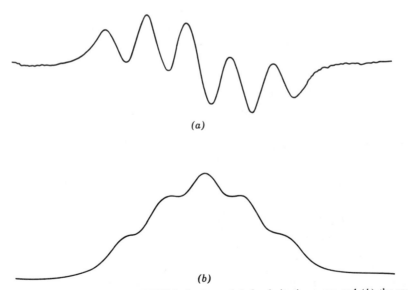

(a)

(b)

Fig. 7-26. The ESR spectrum of DPPH in benzene: (a) the derivative curve and (b) the same curve after passage through an integrator (Schwenker, 1959).

curve shown in Fig. 7-26b. This integrator is useful in applications where integrating times exceed 1 min. Kramer and Müller-Warmuth (1963) also designed an electronic integrator, and Collins (1959) and Randolph (1960) describe analog computers for use in integrating ESR spectra. Van Der Lugt and Van Overbeeke (1964) obtained a direct absorption signal on an $x-y$ recorder by driving the x axis with a voltage proportional to the instantaneous modulation amplitude. Hannula and Fee (1975) reported a simple analog integrator.

When one desires to obtain the number of unpaired spins that contributes to the recorded spectrum, it is considerably easier and quicker to perform a graphical integration of Fig. 7-26b than it is to carry out a double integration of 7-26a. This advantage is offset by the fact that the individual hyperfine lines are much better resolved in the derivative spectrum, and so derivative presentation is preferable to use for obtaining hyperfine splittings and amplitude ratios. This improvement in resolution forms the basis for the resolution enhancement of spectra, as described in Sec. 11-I.

References

Alexander, C., Jr., and H. L. Pugh, Jr., *J. Magn. Reson.* **18**, 185 (1975).

Andrews, G. B., and H. A. Bazydlo, *Proc. IRE* **47**, 2018 (1959).

Arnal, R., and G. Beuchot, *Nucl. Instrum. Methods* **28**, 277 (1964).

Baker, E. B., *RSI* **25**, 390 (1954).

Beers, Y., *RSI* **36**, 696 (1965).

Blair, D. P., and P. H. Sydenham, *J. Phys. E Sci. Instrum.* **8**, 621 (1975).

Bletzinger, P., A. Garscadden, I. Alexeff, and W. D. Jones, *JSI* **42**, 358 (1965).

Boivin, C., C. Jacolin, and J. Y. Savard, *RSI* **44**, 191 (1973).

Bosch, B. G., W. A. Gambling, and T. H. Wilmshurst, *Proc. IRE* **49**, 1226 (1961).

Bösnecker, D., *Z. Angew. Phys.* **12**, 306 (1960).

Buckmaster, H. A., and J. C. Dering, *Can. J. Phys.* **43**, 1678 (1965).

Buckmaster, H. A., and R. S. Rathie, *Can. J. Phys.* **49**, 843, 853 (1971).

Burrus, C. A., Jr., *IEEE Trans.* **MTT-11**, 357 (1963).

Byrne, J. F., and C. F. Cook, *IEEE Trans.* **MTT-11**, 379 (1963).

Chamberlin, R. V., L. A. Moberly, and O. G. Symko, *J. Low Temp. Phys.* **35**, (3–4), 337 (1979).

Cheburkov, D. I., *Meas. Tech.* **19**, 1201 (1976).

Cole, J. B., and R. M. Duffy, *J. Phys. E Sci. Instrum.* **7**, 1019 (1974).

Coles, B. A., *J. Phys. E Sci. Instrum.* **4**, 1059 (1971).

Coleman, P. D., *Proc. IRE* **50**, 1219 (1962).

Coleman, P. D., *IEEE Trans. Microwave Theory Tech.* **MTT-11**, 271 (1963).

Collins, R. L., *RSI* **30**, 492 (1959).

Conradi, M. S., *RSI* **48**, 444 (1977).

Cova, S., A. Longoni, and I. Freitas, *RSI* **50**, 296 (1979).

Cox, H. L., *RSI* **24**, 307 (1953).

Danby, P. C. G., *Electron. Eng.* **40**, 669 (1978).

DeLoach, B. C., *IEEE Trans. Microwave Theory Tech.* **MTT-12**, 15 (1964).

den Boef, J. H., *RSI* **44**, 778 (1973).

De Neef, T., and J. A. M. Verbruggen, *Phys. Lett.* **46A**, 439 (1974).

Dereppe, J. M., *RSI* **32**, 979 (1961).

Dvornikov, E. V., and B. V. Kompankov, *PTE* **19**, 1123 (1976).

Dwyer, D. J., and R. S. Rodgers, *RSI* **48**, 1678 (1978).

Edgar, A., *J. Phys. E Sci. Instrum.* **3**, 179 (1975).

Elmore, W. C., *Am. J. Phys.* **42**, 1117 (1974).

England, T. S., and E. E. Schneider, *Nature* **166**, 437 (1950).

Erickson, L. E., *Phys. Rev.* **143**, 295 (1966).

Faulkner, E. A., *JSI* **36**, 321 (1959).

Faulkner, E. A., and D. W. Harding, *JSI* **43**, 97 (1966).

Faulkner, E. A., and R. H. Stannett, *Electron. Eng.* **36**, 159 (1964).

Feher, G. *Bell Syst. Tech. J.*, **36**, 449 (1957).

Frater, R. H., *RSI* **36**, 634 (1965).

Geick, R., *IEEE Trans. Microwave Theory and Tech.* **MTT-25**, 500 (1977).

Glaunsinger, W. S., and M. J. Sienko, *J. Magn. Reson.* **10**, 253 (1973).

Goodwin, D. W., and R. H. Jones, *J. Appl. Phys.* **32**, 2056 (1961).

Grimbleby, J. B., and D. W. Harding, *J. Phys. E Sci. Instrum.* **4**, 941 (1971).

Halpern, T., and W. D. Phillips, *RSI* **41**, 1038 (1970).

Hannula, R. A., and J. A. Fee, Simple Analog Integrator, *Anal. Biochem.* **63**, 279 (1975).

Hyde, J. S., and D. D. Thomas, *Ann. N.Y. Acad. Sci.* **222**, 680 (1974).

Kramer, K. D., and W. Müller-Warmuth, *Z. Angew. Phys.* **16**, 281 (1963).

Kuroda, S., M. Motokawa, and M. Date, *J. Phys. Soc. Japan.* **44**, 1797 (1978).

Lavelic, B., *RSI* **33**, 103 (1962).

Levanon, H., and S. Vega, *J. Chem. Phys.* **61**, 2265 (1974).

Linev, V. N., and V. B. Mochalskii, *PTE* **21**, 167 (1978).

Llewellyn, P. M., P. R. Whittlestone, and J. M. Williams, *JSI* **39**, 586 (1962).

Long, M. W., *RSI* **31**, 1286 (1960).

Martinelli, M., L. Pardi, C. Pinzino, and S. Santucci, *Phys. Rev. B* **16**, 164 (1977).

Memory, J. D., *Am. J. Phys.* **32**, 83 (1964).

Morris, E. D., Jr., and H. S. Johnston, *RSI* **39**, 620 (1968).

Morigaki, K., S. Toyotomi, and Y. Toyotomi, *J. Phys. Soc. Japan* **31**, 511 (1971).

Murakami, K., S. Namba, N. Kishimoto, K. Masuda, and K. Gamo, *J. Appl. Phys.* **49**, 2401 (1978).

Nucholls, R. G., and L. J. Rueger, *Phys. Rev.* **85**, 731 (1952).

Ohtomo, I., S. Shindo, and N. Nakajima, *Denshi Tsushin Gakkaishi* **58**, 1353 (1975).

Paraskevopoulos, D., R. Meservey, and P. M. Tedrow, *Physica B and C (Neth.) Proc. of Int. Conf. on Magnetism* **86**, 1201 (1977).

Popov, Yu. V., *PTE* **3**, 77 (1960).

Raether, M., and D. Bitzer, *RSI* **35**, 837 (1964).

Randolph, M. L., *RSI* **31**, 949 (1960).

Reinberg, A. R., and T. L. Estle, *Phys. Rev.* **160**, 263 (1967).

RLS-11.

RLS-15.

RLS-16.

RLS-19, Chap. 14.

Rokuskika, S., H. Taniguchi, and J. Hatano, *Anal. Lett.* **8**, 205 (1975).

Rollin, B. V., *JSI* **41**, 239 (1964).

Sastry, V. S., and J. Ramakrishna, *Indian J. Pore Appl. Phys.* **15**, 658 (1977).

Schmidt, J., and I. Solomon, *J. Appl. Phys.* **37**, 3719 (1966).

Schoch, F., L. Prost, and Hs. H. Günthard, *J. Phys. E Sci. Instr.* **8**, 563 (1975).

Schuster, N. A., *RSI* **22**, 254 (1951).

Schwenker, R. P., *RSI* **30**, 1012 (1959).

Seidel, H., *Z. Angew. Phys.* **14**, 21 (1962).

Sokolov, Ye. A., V. P. Sazonov, and V. A. Benderskiy, *Radiotekh. Elektron.* **20**, 1314 (1975).

Sthanapati, J., A. K. Ghoshal, D. K. Dey, A. K. Pal, and S. N. Bhattacharyya, *J. Phys. E Sci. Instr.* **10**, 26 (1977).

Strandberg, M. W. P., H. R. Johnson, and J. R. Eshbach, *RSI* **25**, 776 (1954).

Strandberg, M. W. P., M. Tinkham, I. H. Solt, Jr., and C. F. Davis, Jr., *RSI* **27**, 596 (1956).

Tătaru, E., *Rev. Phys. Applique* **9**, 521 (1974).

Taylor, R. L., and S. B. Herskovitz, *Proc. IRE* **49**, 1901 (1961).

Torgeson, D. R., and W. A. Rhinehart, *RSI* **34**, 1447 (1963).

Uebersfeld, T., and E. Erb, *J. Phys. Radium* **14**, 90 (1956).

Urano, Y., *J. Phys. Soc. Japan* **10**, 864 (1955).

Van Der Lugt, W., and J. Van Overbeeke, *JSI* **41**, 702 (1964).

Van Gerven, L., A. Van Itterbeek, and L. Stals, *Paramagn. Reson.* **2**, 684 (1963).

Vetokhin, S. S., N. I. Kotovich, and V. N. Linev, *PTE* **21**, 134 (1978).

Williams, P., *JSI* **42**, 474 (1965).

Williams, P., *J. Phys. E Sci. Instrum.* **3**, 441 (1970).

Wollan, D. S., *RSI* **52**, 682 (1971).

Yamazaki, Y., H. Shimizu, and H. Hasimoto, *Tech. Rep. Osaka Univ.* **27**, 317 (1977).

CHAPTER 8

Vacuum Systems

It is frequently desired to provide samples with specialized pretreatments such as oxidation and reduction, and for this purpose a vacuum system is indispensable. This chapter briefly introduces the lector to high vacuum techniques; he should consult one of the books listed at the end of the chapter (Barr and Ankorn, 1959; Dushman, 1962; Holland et al., 1974; Roth, 1976) for further details.

A. Pretreatment of Samples

Many systems that produce strong electron spin resonance signals are unaffected by changes in the temperature, pressure, and various environmental conditions. Other systems exhibit remarkable changes in their ESR spectra when one or more of these parameters is changed, as the following examples indicate.

It is well known that α,α'-diphenyl-β-picryl hydrazyl (DPPH) produces a narrow singlet in the solid state, and a quintet with the relative intensity ratios 1–2–3–2–1 in benzene solution. When one removes all the oxygen that is dissolved in the solvent, then many additional hyperfine lines are resolved, as shown in Fig. 8-1 (see Deguchi, 1960; Ueda et al., 1962; Schmidt and Gunthard, 1967; Haniotis and Gunthard, 1968; Gubanov et al., 1973). Before the removal of the oxygen only the nitrogen hyperfine coupling constants may be determined, while after the removal of oxygen the proton hyperfine constants may also be evaluated.

Povich (1975) investigated the free radical 2,2,6,6-tetramethylpiperidinoxy.

in various air-saturated solvents and found that the linewidth increased linearly with the temperature. In addition the linewidth of DPPH in solvents without degassing increases linearly with the logarithm of the oxygen solubility. These facts suggest that ESR linewidths can be employed for determining

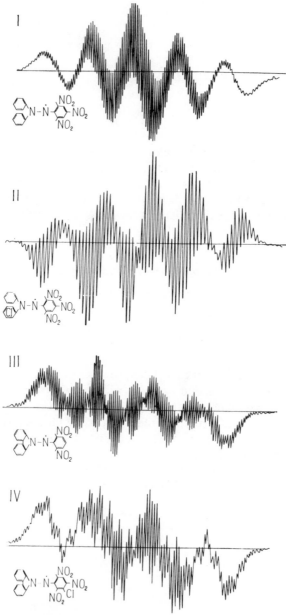

Fig. 8-1. High-resolution ESR spectra of radicals: (I) DPPH; (II) α-pentachlorophenyl, α-phenyl-β-picrylhydrzyl; (III) carbazyl-2,6-dinitrophenylnitrogen; (IV) carbazyl-5 chlor, 2,4,6-tri-nitrophenylnitrogen obtained in thoroughly degassed $5 \times 10^{-4} M$/liter solutions of a benzene–tetrahydrofuran mixture (Gubanov et al. 1973).

the concentration and diffusion coefficient of oxygen in various media. Tuttle (1975) measured the diffusion coefficient of dissolved oxygen.

In solid DPPH the linewidth depends upon the solvent from which it was precipitated (Al'tshuler and Kozyrev, 1964). Chary and Sastry (1977) discussed the solvent dependence of the ESR spectra of a copper II complex.

Various varieties of solid carbons are quite sensitive to the temperature and to the presence of certain gases such as oxygen during the thermal pretreatment that produces and destroys paramagnetic centers. Borodina et al. (1976) provide a method of pretreating coke for ESR examination. A number of pertinent articles may be found in the proceedings of the Carbon Conferences.

Many studies have been made of the absorption of various gases such as oxygen on surfaces of zeolites (e.g., Natsuko and Iwao, 1978; Kartel et al., 1978) and other materials (i.e., Cordischi et al., 1979; Gideoni and Steinberg, 1973; Miller and Haneman, 1978; Mizokawa and Nakamura, 1974; Pankratov and Prudnikov, 1971; Shimuzu et al., 1979; Szuber and Salamon, 1979, 1979; Thomas et al., 1978; Timson and Hogarth, 1971).

The low-temperature form of chromia–alumina produces ESR spectra from isolated Cr^{3+} ions, clumped Cr^{3+} ions, and Cr^{5+} ions, while Cr^{6+} is detected in this system by its near-ultraviolet charge-transfer spectrum. These three valence states may be converted reversibly into one another by heating at several hundred degrees Celsius in the presence of oxygen or hydrogen (but not a mixture of both, or an explosion may result!). For further details about this system, see the review article by Poole and MacIver (1965). Kubler et al. (1976) found that the presence of oxygen strongly effects the g factor, linewidth, and spin concentration of amorphous germanium. Bahl et al. (1973) and Connell and Pawlick (1976) studied oxygen contamination in silicon (see also Miller and Haneman, 1970).

The observed ESR spectrum of a free atom such as N, O, and P, or of a simple molecule such as NO or NO_2 in the gaseous state, is dependent upon the pressure, pumping speed, presence of buffer gases, chemical treatment of the gas container walls, and so forth. The atom or molecule under study is usually generated by dissociation in an electric discharge, and then pumped rapidly through the resonant cavity before it has time to recombine and form a diamagnetic molecule (Froben, 1974).

When an ionic crystal such as an alkali halide is heated in an atmosphere of its cation or anion, the crystal may develop an excess of either constituent, and the result is the formation of a color center such as an F center. Such color centers may also be produced by irradiation with γ rays, X rays, neutrons, and the like, as discussed in Chap. 10.

The electron spin resonance spectra of irradiated solids sometimes depend both on the gas in contact with the sample during irradiation and also on the gases introduced after the completion of the irradiation.

Thus we see that some samples require the use of very careful pretreatment and handling procedures. A vacuum system is indispensible for processing such samples. Accordingly, we devote most of this chapter to the explication of vacuum techniques. The final section discusses the electrolytic generation of radical ions, because these species require the use of careful pretreatment and handling techniques. For completeness it should also be mentioned that many samples contain paramagnetic species whose ESR spectra are not affected by changes in their physical and chemical environment. For example, the spectra of various minerals such as ruby are insensitive to most of the climatic conditions and geological upheavals to which they have been exposed for countless thousands of years.

B. Properties of Gases *in Vacuo*

Pressure

Several ranges of pressure may be distinguished:

Soft or low vacuum   760 to 10^{-3} torr

High vacuum 10^{-3} to 10^{-6} torr

Ultra-high vacuum 10^{-9} to 10^{-13} torr

One may employ mechanical pumps and rubber hoses in the region of low vacuum, while in the high-vacuum region it is necessary to use diffusion pumps and glass or metal systems. Stopcocks and ground glass joints may be used below 10^{-6} torr.

There are several units of pressure in current use: 1 atm = 760 torr = 760 mm Hg = 14.7 lb/in.^{2} = 1.0133 bar = 1.0133 × 10^5 N/m^2 = 1.0133 × 10^5 Pa. One bar = 10^6 dyn/cm^2 = 7.5 × 10^2 torr = 10^5 N/m^2 = 10^5 Pa.

Ideal Gas Law

The ideal gas law is

$$PV = nRT \tag{1}$$

where P is the pressure, V is the volume, T is the absolute temperature (273.16 + °C), n is the number of moles, and the gas constant R is given by

$$
\begin{aligned}
R &= 0.082054 & &\text{liter atm/mole deg} \\
&= 8.3144 \times 10^7 & &\text{erg/mole deg} \\
&= 8.3144 & &\text{J/mole deg} \\
&= 1.9865 & &\text{cal/mole deg}
\end{aligned}
\tag{2}
$$

At high pressures the ideal gas law usually breaks down and it is necessary to employ another relation such as the Van der Waals equation, the Beattie–

Bridgeman equation, or the virial coefficients. In the regions of high and ultrahigh vacuum, the ideal gas law usually holds true.

Maxwellian Distribution

The molecules in a gas have a Maxwellian distribution of velocities, and the probability P of finding a molecule with velocity v is given by

$$P = \left(\frac{4}{\pi^{1/2}}\right)\left(\frac{v}{v_p}\right)^2 \exp\left[-\left(\frac{v}{v_p}\right)^2\right] \tag{3}$$

where, of course,

$$\int_0^\infty P\,d\left(\frac{v}{v_p}\right) = \frac{4}{\pi^{1/2}}\int_0^\infty \left(\frac{v}{v_p}\right)^2 \exp\left[-\left(\frac{v}{v_p}\right)^2\right] d\left(\frac{v}{v_p}\right) = 1 \tag{4}$$

and v_p is the most probable velocity since

$$P = \text{maximum} \tag{5}$$

at $v = v_p$. The average velocity v_a and root-mean-square velocity v_{rms} are given by

$$v_a = \int_0^\infty vP\,d\left(\frac{v}{v_p}\right) = 1.128v_p \tag{6}$$

$$v_{rms} = \left[\int_0^\infty v^2 P\,d\left(\frac{v}{v_p}\right)\right]^{1/2} = 1.224v_p \tag{7}$$

The most probable velocity is related to the absolute temperature T and molecular weight M by

$$v_p = 1.29 \times 10^4 \left(\frac{T}{M}\right)^{1/2} \text{cm/sec} \tag{8}$$

as shown graphically in Fig. 8-2.

Mean Free Path

The average distance traversed by a molecule between successive collisions is called the mean free path l, and it is given by

$$l = (2^{1/2}\pi nd^2)^{-1} \tag{9}$$

where d is the molecular diameter, and n is the number of molecules per cubic

VACUUM SYSTEMS

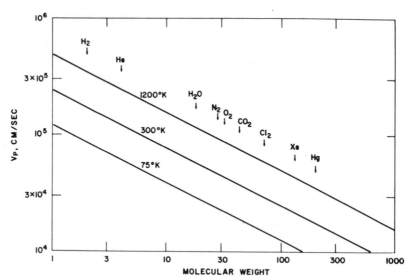

Fig. 8-2. Dependence of the most probable molecular velocity v_p on the molecular weight for three temperatures.

centimeter. The mean free paths of several common gases are listed in Table 8-1 for 1 and 760 torr. At room temperature and at a pressure of about 3×10^{-3} torr, the mean free path becomes comparable to the diameter of a vacuum system manifold so that for $P \ll 10^{-3}$ torr, the average molecule in the system will make many more collisions with the walls than with other molecules. At 10^{-6} torr and 300 K, there are about 3×10^{10} gas molecules per cubic centimeter.

TABLE 8-1

Molecular Diameters and Mean Free Paths of Several Gases at 25°C[a]

Gas	Molecular Weight	Diameter, A	Boiling Point, °C	Mean Free Path, cm	
				1 torr	760 torr
H_2	2.00	2.75	−252.8	0.93×10^{-2}	1.23×10^{-5}
He	4.00	2.18	−268.9	1.47	1.94
H_2O	18.00		100.0	0.30	0.4
Ne	20.18	2.60	−246.0	1.05	1.38
N_2	28.02	3.77	−195.8	0.52	0.69
O_2	32.00	3.64	−182.9	0.54	0.71
A	39.94	3.67	−185.8	0.53	0.67
CO_2	44.01	3.22	−78.4[b]	0.29	0.40
Hg	200.59	2.38	356.6	1.12	1.4
Air	—	3.74	—	0.51	0.67

[a] Partly from Dushman (1962, p. 32), partly from Roth (1976, Chap. 2).
[b] Sublimes.

Pumping Speed

If a large volume of V liters that is full of gas at a pressure P is connected to a perfect vacuum through a small aperture of A cm^2, then the gas will move out of the volume at the rate S (Dushman, 1962, p. 91):

$$S = \frac{V}{P - P_0} \frac{dP}{dt} = -3.64A \left(\frac{T}{M} \right)^{1/2} \text{ liter/sec} \qquad (10)$$

where T is in kelvins, M is the molecular weight, and P_0 is the limiting pressure that is attainable. For air at room temperature (300 K)

$$S = -11.7A \text{ liter/sec} \qquad (11)$$

If the gas has the pressure P_i at time $t = 0$, then at a later time t, one has the pressure P_f and

$$t = \frac{V}{S} \int_{P_i}^{P_f} \frac{dP}{P - P_0} \qquad (12)$$

$$= \frac{V}{S} \ln \left(\frac{P_f - P_0}{P_i - P_0} \right) \qquad (13)$$

At low pressures the mean free path exceeds the radius r of glass tubing, and so when gas is pumped through a piece of tubing of length l, the rate of flow (conductance) G is given approximately by

$$G \approx \frac{r^3}{l} \frac{[(T/300)(29/M)]^{1/2}}{1 + (8r/3l)} \text{ liter/sec} \qquad (14)$$

where r and l are in millimeters. Several pieces of tubing connected in series have the overall flow rate G:

$$\frac{1}{G} = \Sigma \frac{1}{G_i} \qquad (15)$$

and the composite pumping speed S_c of a pump in series with a piece of tubing is

$$\frac{1}{S_c} = \frac{1}{S} + \frac{1}{G} \qquad (16)$$

where S is the speed of the pump alone. It is important to know the pumping speed associated with a vacuum system in order to evaluate the length of time that one must wait before a desired pressure is reached. In practice, the pumping speed is often limited by other factors such as the slow desorption of gases from the surface of the sample and the walls of the vacuum system.

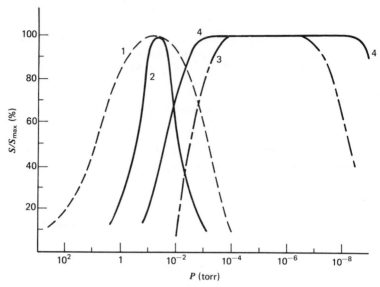

Fig. 8-3. Dependence of the pumping speed on the pressure of 1—roots mechanical pump and 2—ejector mechanical pump, 3—a diffusion pump and 4—a molecular pump [adapted from Roth (1976), p. 199].

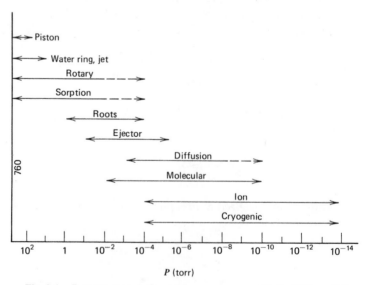

Fig. 8-4. Pressure ranges of several vacuum pumps (Roth, 1976, p. 198).

C. Pumps

Forepump

A forepump may be employed for low vacuum or for backing up a pump designed to achieve high vacuum. A mechanical pump such as Roots pump (Figs. 8-3 and 8-4) and a megavac type (compare Fig. 15-3 of first edition) can attain pressures of 10^{-4} torr. This is adequate for backing up the four high vacuum pumps listed in Fig. 8-3.

Sorption Pump

In recent years sorption pumps have been replacing mechanical pumps in some applications such as backing up ion pumps designed for operation below 10^{-2} torr. For their pumping action they make use of a zeolite molecular sieve that has a porous structure and hence a capacity to absorb a large number of gas molecules within its cavities. These cavities have entrance passages with pore diameters of about 5 Å, and hence can accommodate all the common gases listed in Table 8-1.

To effect the evacuation of a system the zeolite is cooled to the temperature of liquid nitrogen, and a valve is opened leading to the vacuum system. The air or other gases diffuse into the zeolite chamber where they are absorbed. The process continues until an equilibrium pressure of about 10^{-2} torr is reached, where no more absorption occurs. At this point the valve to the vacuum chamber is closed, and the zeolite is reactivated by allowing it to warm to room temperature, where the gases come off and are vented. Most gases are adequately released at room temperature. Water, however, is more firmly absorbed, and sometimes the zeolite must be heated to 300°C to eliminate the water vapor. The principles and operation of sorption pumps have been treated by Boers (1968), Danielson (1970), Dobrozemsky and Moraw (1971), Halama and Aggus (1974), Stern and DiPaolo (1969), Vijendran and Nair (1971), and Wheeler (1974).

Diffusion Pumps

A high vacuum is often produced by a diffusion pump, which makes use of the flow of a condensable vapor to draw along molecules from the system being evacuated and pass them on to the forepump. To accomplish this task, a diffusion pump makes use of a heater to boil the liquid and a condenser (e.g., of flowing water) to recondense it. Ordinarily a second cold trap is inserted between the diffusion pump and the vacuum system in order to hinder the vapor from diffusing into the system. From its nature, a diffusion pump will not operate unless it is backed up by a forepump.

A mercury diffusion pump such as the one shown in Fig. 8-5 is frequently employed to obtain pressures down to 10^{-6} torr. The vapor pressure of mercury is about 10^{-3} torr at 18°C as shown in Fig. 8-6 so it is desirable to employ a Dry Ice–acetone or liquid nitrogen cold trap between the pump and

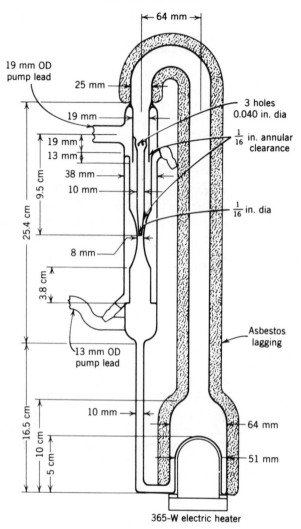

Fig. 8-5.   Mercury diffusion pump (Barr and Ankorn, 1959).

the vacuum system to prevent the diffusion of mercury into the system. A typical mercury diffusion pump can pump about 3 liter/sec.

An oil diffusion pump does not need a cold trap if one employs an oil with a vapor pressure less than the overall pressure attained. Oil diffusion pumps are considerably faster than mercury pumps, but they have the disadvantage that the oil may be decomposed by overheating, or by exposure to air when hot.

Molecular Pump

At low pressures a molecule that collides with the surface does not undergo an elastic collision but, rather, adheres to the surface for a certain time and then

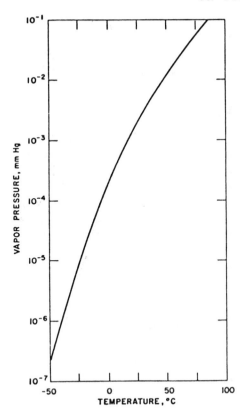

Fig. 8-6. Temperature dependence of the vapor pressure of mercury.

desorbs in a direction that does not depend upon the angle of incidence (Roth, 1976). If the surface on which the molecule was absorbed is in motion then it transmits a velocity component to the desorbing molecule. A molecular pump employs this principle of molecular drag to move molecules out of the region being evacuated. A particular type of a molecular pump called a turbo molecular pump employs stationary and rotating plates and disks to move the molecules forward. Pressures from 10^{-2} to 10^{-10} torr are achieveable with molecular pumps. The limit of 10^{-10} torr is usually due to the back diffusion of hydrogen. The principles and applications of turbomolecular pumps are described by Bachler et al. (1974), Becker and Nesseldreher (1974), Henning (1974), Mirgel (1972), and Osterstrom and Shapiro (1972).

Ion Pump

A region may be evacuated by ionizing the gas molecules and accelerating them away by an electric field. Ion pumps operate on this principle by using fast electrons from a hot filament or a cold cathode discharge to produce the ionization. Many ion pumps have getters that remove additional molecules by sorption. Ion pumps operate in the range from 10^{-4} to 10^{-14} torr, and hence an auxiliary pump must be employed to reduce the pressure to a value where the ion pump can begin to function. Two types of ion pumps in general use are the evapor ion variety, which employs a getter (Bills, 1973; Grigorov, 1973; Schwarz, 1972), and the orbitron model, which utilizes a radial electric field between concentric cylinders (Denison, 1967; Kato et al., 1974; Naik and Herb, 1968).

Cryopumps

Quite low pressures, typically in the range from 10^{-4} to 10^{-14} torr, can be obtained with cryopumps, or cryogenic pumps, which carry out their pumping action by condensing vapors at low temperatures. The subject has been reviewed by Hobson (1973), Kidnay and Hiza (1970), Moore (1962), Power (1966), and Redhead et al. (1968). In a representative case the condenser of the gases operates at 20 K, or $-253°C$, and we see from Table 8-1 that this is below the normal boiling points of all of the gases that are generally present in a vacuum system.

Cryotrapping

Condensable vapors such as water often form porous deposits with high surface areas, and these pores can trap gases through a pumping action called cryotrapping. For example, water condensed at 77 K can have an effective

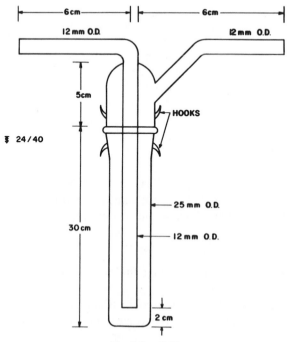

Fig. 8-7. Cold trap.

surface area as high as 600 m^2/g. This technique permits the removal of gases on surfaces that are maintained at temperatures considerably above their boiling points. Cryotrapping and cryopumps were discussed recently by Benvenuti and Calder (1972), Boissin et al. (1972), Hengevoss and Trendelenburg (1963, 1967), Muller (1966), and Templemeyer et al. (1970).

Cold Trap

Condensable vapors such as water and mercury may be removed from the vacuum system by the use of a cold trap such as the one shown in Fig. 8-7. A Dewar containing liquid nitrogen (77 K) or a Dry Ice–acetone mixture (195 K) may be employed to condense such vapors in the trap and reduce their pressure in accordance with Fig. 8-6.

D. Pressure-Measuring Devices

The pressure ranges covered by several pressure-measuring instruments are shown in Table 8-2. Each of these devices will be discussed briefly.

McLeod Gauge

The McLeod gauge shown in Fig. 8-8 is an absolute pressure-measuring device. It is operated by admitting air to the mercury reservoir. This raises the mercury level until it fills the volume V_1 and part of the two capillaries C. The pressure measurement is performed in one of two ways: (1) by lining up the mercury level in the left-hand capillary to the top of the right-hand capillary, or (2) by lining up the mercury level in the right-hand closed capillary with the level G_2. The pressure may be calculated from h or from h_1 and h_2. In practice, the pressure is usually read from a scale that is aligned with G or G_2. Since the McLeod gauge is an absolute pressure-measuring device, it may be employed

TABLE 8-2

Characteristics of Several Pressure-Measuring Devices

Name	Pressure, torr		Principle of Operation	
	Minimum	Maximum		
Monometer	10^{-1}	10^3	Height of Hg column	
Spark discharge	10^{-3}	10^2	Color of gas discharge	
Pirani gauge	10^{-6}	10^{-2}	Resistance, measured	Heating element
Thermocouple gauge	10^{-4}	1	Temperature, measured	cooled by gas
McLeod gauge	10^{-6}	10	Height of Hg column	
Ionization gauge	10^{-11}	10^{-3}	Ionization current, measured	

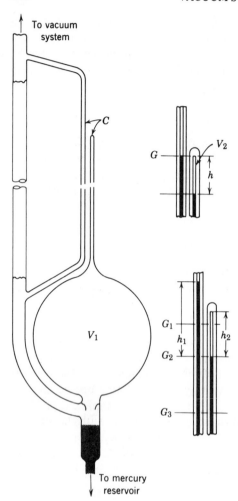

Fig. 8-8.   McLeod gauge (Barr and Ankorn, 1959).

to calibrate other gauges. It should not be employed with condensable vapors such as H_2O, or with gases that attack mercury. When not in use, it is best to keep the mercury lowered in the reservoirs. A McLeod gauge may be employed to measure pressures between 10 and 10^{-6} torr.

Spark Discharge

Pressures from 10^{-3} to 100 torr may be estimated by the color of the glow discharge in a vacuum system when a Tesla coil is applied to the glass manifold. If a leak is present the spark will jump to the leak, making the Tesla coil a useful tool in troubleshooting vacuum systems. The coil should be kept moving to avoid puncturing the glass.

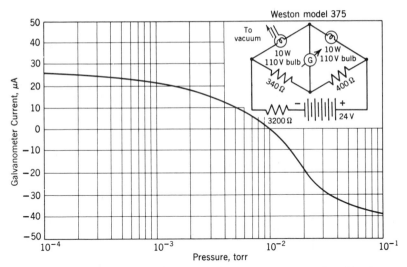

Fig. 8-9. Typical characteristic curve of galvanometer current versus pressure for a Pirani gauge (Spangenberg, 1948).

Pirani Gauge

The Pirani gauge consists of a heated filament placed in the vacuum system. When gas is present it cools the filament and lowers its resistance, as measured by a Wheatstone bridge. A typical curve of the galvamometer deflection versus the pressure is shown in Fig. 8-9.

Thermocouple Gauge

The thermocouple gauge uses a heated filament and monitors the temperature of the filament by a thermocouple. The pressure dependence of the thermocouple current follows a sigmoid curve such as the one shown in Fig. 8-10. It works in the range 10^{-4} to 1 torr.

Ionization Gauge

The ionization gauge consists of a triode with a heated filament, a positive grid to collect electrons from the filament, and a negative plate to collect positive ions formed in the gas by collisions of the electrons with gas molecules. The number of positive ions formed is a function of the pressure. The ionization gauge may be employed to measure pressures between 10^{-11} and 10^{-3} torr, and even lower pressures are reached by some gauges.

Manometer

A U-tube manometer such as the one shown in Fig. 8-11 is made from two 1-m-long glass tubes joined at the bottom to form a U and closed at one end under vacuum. The other end is attached to the vacuum system, and the

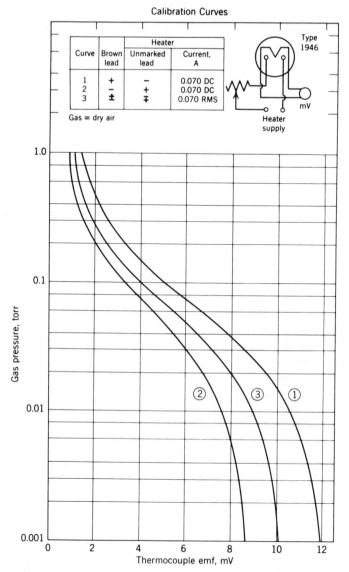

Fig. 8-10. The gas pressure versus thermocouple voltage characteristic curves of a type 1946 Thermocouple gauge (*RCA Tube Handbook*).

manometer is filled halfway with mercury so that at high vacuum the two mercury levels are equal. At atmospheric pressure the two levels will differ by 760 mm, and if the length of the tube is made longer than this, the U-tube manometer may be employed to measure pressures from about 1 torr to somewhat above 1 atm. It is convenient to provide each manometer with a meter stick for determining the height of the mercury level. A simpler version

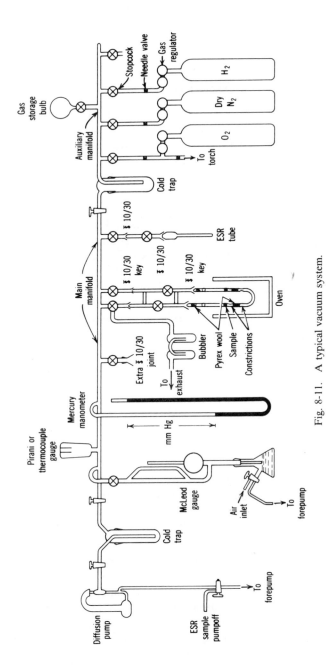

Fig. 8-11. A typical vacuum system.

305

of the manometer consists of a 1-m-long vertical tube inserted into a mercury reservoir at the lower end and attached to the vacuum system at the upper end.

E. Typical Vacuum System

A typical vacuum system for use in pretreating electron spin resonance samples is shown in Fig. 8-11. The diffusion pump at the upper left is backed up by a forepump, and they are separated by a two-position stopcock. The cold trap is used with a Dewar of liquid nitrogen (or a Dry Ice–acetone mixture) to remove condensible vapors such as water or mercury from the system. It is not advisable to use this liquid nitrogen cold trap until most of the air has been removed from the vacuum system, since otherwise gases in the air such as CO_2, H_2O and O_2 may condense in the trap and take a long time to pump out.

The vacuum system is provided with three pressure-measuring devices. In practice, it is convenient to measure the pressure from 10^{-6} to 10^{-3} torr with the McLeod gauge, from 10^{-3} to 1 torr with the thermocouple or Pirani gauge, and from 1 to 10^3 torr with the U-tube manometer. One should note that the McLeod gauge is only connected to the vacuum system through a stopcock during a pressure measurement; the thermocouple or Pirani gauge is always physically connected to the system but is only activated electrically during a pressure measurement, while the manometer is operating at all times. The McLeod gauge reservoir is preferably connected to a separate mechanical pump, but it may be operated off the same forepump that backs up the diffusion pump if care is taken to isolate the diffusion pump and the rest of the system during and for several minutes after the evacuation of the reservoir. To minimize leaks, it is advisable to make the vacuum system as simple as is feasible, with all unnecessary stopcocks eliminated. The system shown in Fig. 8-11 is a compromise between simplicity and versatility.

The two sample tube arrangements presented in Figs. 8-12 to 8-14 are shown in position on the main manifold. The gases that are employed to pretreat the samples are supplied from the auxiliary manifold, and they may be dried or purified in the cold trap before use. The two manifolds can be isolated from each other by a stopcock.

The vacuum system may be used to pretreat samples by heating or cooling them *in vacuo* or in the presence of a particular gas such as hydrogen or oxygen (but not both simultaneously or an explosion might occur!). The sample is exposed to the desired gaseous atmosphere by closing the stopcock at the left side of the main manifold in order to isolate the system from the pumps, and then opening the stopcock on the right to introduce the gas. Oxidation and reduction reactions are most efficiently carried out by adjusting the three-way stopcock in the center of the main manifold so that the gas flows into the duplex ESR pretreatment tube from the right and then leaves the system on the left via the bubbler sketched in Fig. 8-15. The rate of flow may be estimated from the number of bubbles per minute. The adapter tube is inserted between the sample and the vacuum system so that the sample may be

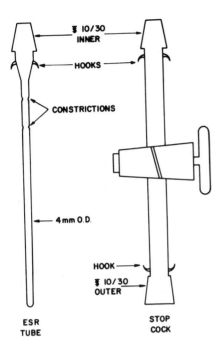

Fig. 8-12. ESR tube and stopcock (KPAH).

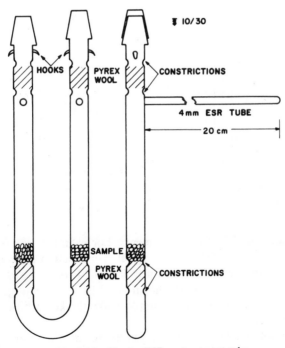

Fig. 8-13. Duplex ESR pretreatment tube.

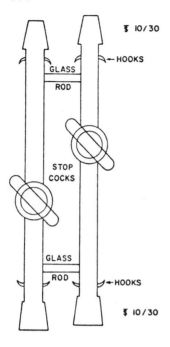

Fig. 8-14. Keyed double stopcock (KPAH) adapter.

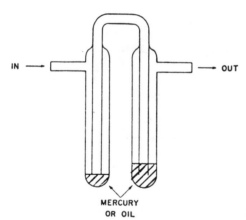

Fig. 8-15. Bubbler for monitoring the rate of flow of a gas.

removed and measured between treatments. For example, one may make use of this arrangement to study the ESR signal as a function of oxygen pressure. The pair of outer tapered joints shown in Fig. 8-14 forms a "key" that mates with the pair of inner joints as shown on Fig. 8-11. Ball joints may be substituted for the standard taper joints to make the key a little easier to use, but less vacuum tight.

After the completion of the heat or gas treatment, the sample tube may be removed from the vacuum system and the samples shaken into the ESR tubes.

One may then seal them off with a torch* and measure them in the resonant cavity. It is possible to construct a system similar to the one shown in Fig. 8-13 in which the ESR tubes face the sides rather than the front so that the sample may be placed in the resonant cavity without sealing it off, and then returned to the vacuum system for further treatment. Systems can also be built which permit the *in situ* pretreatment of samples directly in the resonant cavity. When powder samples are used, sudden pressure changes should be avoided to prevent the powder from blowing around. Samples may also be sealed off *in vacuo* sans pretreatment by using the ESR sample pumpoff three-way stopcock if care is exercised to reconnect the diffusion pump to the forepump a minute or so thereafter. McArdle (1974) described a device for sealing sample tubes under vacuum. Plachy and Wondrem (1974) designed a gas-permeable ESR tube. Ladds and Rubinson (1976) made use of an inert glass-to-glass vacuum-tight coupler.

The oven may be conveniently heated to 500°C with Pyrex sample holders, and several hundred degrees higher when quartz or Vycor is used. One may construct the entire pretreatment tube out of quartz or Vycor to eliminate the necessity of employing graded seals to the quartz ESR tubes. The oven temperature may be automatically regulated with the aid of a thermocouple and an appropriate controller. Asbestos paper may be placed around the glass at the top of the oven for thermal insulation, and sometimes a small blower or compressed air flow is used to prevent stopcock grease from melting.

The gases that are used for pretreating the sample are obtained from the auxiliary manifold. Each gas tank is supplied with a pressure regulator and is connected to the auxiliary manifold through a needle valve that provides a fine-control adjustment of the gas flow rate. An additional supply of gas may be kept in the gas storage bulb. The cold trap may be used in conjunction with a liquid nitrogen or Dry Ice–acetone bath to dry the gases, or it may be filled with a drying–purifying agent.

Cacciarelli and Stewart (1963) describe a vacuum and pressure system for NMR sample preparation, and Müller-Warmuth (1963) presents a system for free radical solutions. Wamser and Stewart (1965) describe a sample reactor and a sample transfer system for use with liquid samples. Ahn (1976) designed a cell for a capillary method of determining free-radical concentrations in solution. The cell operates under solvent vapor pressure.

F. Flow Systems

Svejda (1977) described an ultrahigh vacuum (UHV) flow system designed to follow the reactions of atoms such as H, O, and N with solid surfaces (Svejda, 1972). Their experimental arrangement permits powder samples to be cleaned, to be thermally pretreated, and finally to be measured in the spectrometer, all under high-vacuum conditions.

---

*Dark glasses should be used to protect the eyes when sealing quartz.

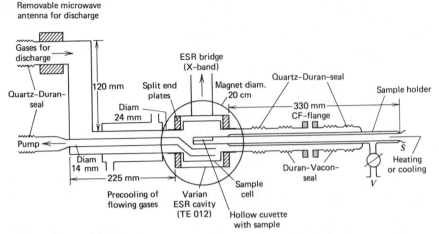

Fig. 8-16. Quartz sample cell and flow system for the study of powders in ultrahigh vacuum (Svejda, 1977).

Figure 8-16 shows the quartz sample cell and flow tube. The sample was introduced into the cell by means of a sample holder whose hollow cuvette contains a layer of powder several millimeters thick. Copper-sealed OFKC flanges permit easy sample changing and bakeability. The sample can be heated by an electrical heating element shown in Fig. 8-17a, which slides into the hollow cuvette. Cooling is accomplished with the arrangement illustrated in Fig. 8-17b. The overall vacuum system is presented in Fig. 8-18.

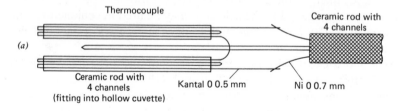

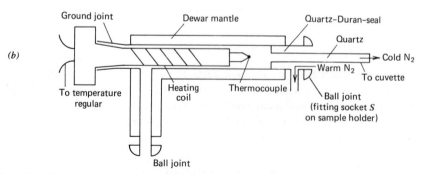

Fig. 8-17. Electrical heating element (a) and cooling device (b) for the cuvette of Fig. 8-16 (Svejda, 1977).

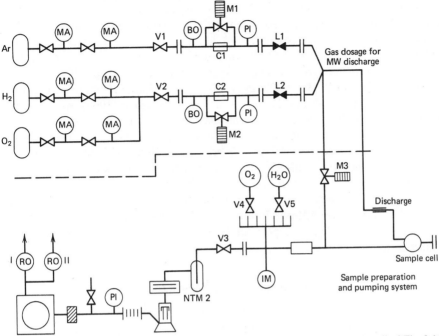

Fig. 8-18.   Block diagram of ultrahigh vacuum flow system used with the sample cell of Fig. 8-16 (Svejda, 1977).

High vacuum conditions (10^{-7}–10^{-8} torr) are achieved with the aid of an oil diffusion pump and metal tips. The pressure is measured by a Bayard–Albery gauge (1 μm). High-purity gases are introduced from cylinders at the upper left of the figure, and they are stored in glass bulbs via valves V4 and V5. Gas from the cylinders enters the system via glass capillaries C_1 and C_2, and magnet valves M_1 and M_2 are provided for rapid pumpdown. The leak valves L_1 and L_2 control the flow rate, and input pressures are measured by quartz spiral gauges (Bo) or Pirani gauges (Pl).

The quartz pieces are positioned in the large access cylindrical TE_{011} mode cavity in the manner shown in Fig. 8-16. To prevent the cavity Q from being significantly degraded the thickness of the cell wall was reduced to about 0.9 mm and that of the cuvette and return pipe was 0.5 mm. The cavity had an unloaded Q of 20,000 and a loaded Q of 2500, as measured by the absorption dip on the klystron mode (Matenaar and Schindler, 1973).

A microwave discharge unit with a demountable antenna (Fehsenfeld et al., 1964) located outside the flow system dissociated the diatomic molecules which were carried along by an argon carrier gas at 1 to 5 torr pressure. Under those experimental conditions the atoms H, O, or N arrived at the sample cell in their electronic ground states in concentrations of 10^{12} to 10^{14} cm^{-3}. Free electrons were too low in concentration to be detected (Svejda et al., 1975, 1976).

Albery et al. (1975, 1977) described tube electrodes where radical species are generated and then transported through the resonant cavity with a parabolic velocity profile v_0 under conditions of laminar flow (see also Albery et al., 1980, 1981). ESR "mix–flow" cells for use with a continuous flow method for producing short-lived paramagnetic species are described by Bennett and Thomas (1964), Borg (1964), Dixon and Norman (1963), Goldberg and Crowe (1976), and Piette et al. (1961), and reviewed by Norman (1963). Kertesz and Wolf (1973, 1975) discuss the operational characteristics of a stopped-flow system incorporating a precision solenoid valve. It provides accurate kinetic data on the disappearance of short-lived intermediates with lifetimes in the millisecond range. It can also be operated in a continuous flow mode.

Aleksandrov and Ovchinnikova (1975) designed an attachment for obtaining micropowder in a vacuum or in an inert gas atmosphere in the range from 4.2 to 300 K. The micropowder is formed near the cavity resonator.

Rokushika et al. (1975) report an ESR flow detector for the liquid chromatography of free radicals.

G. Generation of Radical Ions

Positive radical ions called radical cations may be produced by dissolving the parent compound in sulfuric or other acids (e.g., Dixon and Murphy, 1975; Eloranta et al., 1975, 1975; Holloway and Raynor, 1975; Venters et al., 1975; Zarubin et al., 1975), and negative radical ions called radical anions are generated by reduction in solution (e.g., Fessenden and Neta, 1973; Gause and Rowlands, 1976; Hirota, 1975). Most studies of radical ions involve electrochemical generation (e.g., Goldberg and Bard, 1975; Griffith et al., 1969; Gupta et al., 1971; Newton and Davis, 1975; Zagorets et al., 1975) directly in the resonant cavity, or intramuros in the terminology of Maki and Geske (1959). Koryakov and Pankratov (1971) described a polyethylene vessel for observing ESR in polar solvents such as those employed in radical ion studies. Bonnemay and Lamy (1967) and Lamy and Malaterre (1970) describe an electrochemical cell for studying reaction sites for adsorption on electrodes.

The present chapter presents the principles of the electrochemical radical ion-generation technique and will describe the original ESR electrochemical cell of Maki and Geske. The first edition may be consulted for additional cell designs and references to the literature.

The electrochemical cell that was employed by Geske and Maki (1960) is shown in Fig. 8-19. They designed a mercury-pool electrode A placed in the bottom of a 3-mm OD Pyrex tube B and centered in the microwave reflection cavity. The pool has an exposed area of 2.5 mm^2 and receives its supply voltage by means of a platinum wire C, which is sealed into the bottom of B. The tube B is surrounded by a quartz tube D that is cemented to tube E by means of epoxy resin. An aqueous saturated calomel electrode (SCE) placed in the insert tube F makes electrolytic contact with the main solution through its own soft glass–Pyrex Perley seal and a sintered glass disk bottom on the insert tube. The

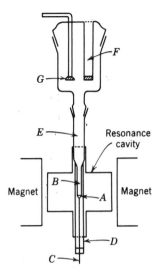

Fig. 8-19. The original electrochemical cell designed for use in an ESR spectrometer (Geske and Maki, 1960).

main solution contains the sample plus a supporting electrolyte of bulky, counterions dissolved in the solvent. It is degased with the help of a fine capillary that extends through a pinhole in the top down to within 1 mm of the mercury pool. Nitrogen gas is passed through this capillary for preliminary degassing, while a slow flow of nitrogen is maintained through the sintered glass disk G both before and during the electrolysis run. The resistance of the cell was about 20 kΩ for Geske and Maki's solutions.

The radical ions are generated electrolytically by applying the proper voltage V_A between the platinum wire and the calomel anode, and measuring the

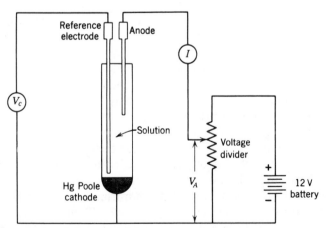

Fig. 8-20. Electrical diagram of an electrolytic cell. The 12-V power supply on the right produces the variable applied voltage drop V_A across the cell, while the reference electrode measures the cell voltage V_C, which is corrected for polarization effects. The ammeter measures the current I through the cell.

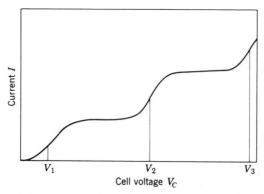

Fig. 8-21. A typical polarogram obtained with a cell of the type shown in Fig. 8-20.

voltage drop V_c across the cell from the platinum wire to the reference electrode. A typical electrical diagram is shown in Fig. 8-20. The use of a reference electrode placed much closer to the cathode than the calomel electrode helps to correct for polorization effects and the internal voltage drop within the cell. The applied voltage V_A may be varied by means of the potentiometer or voltage divider, and for a typical system the current I through the cell will vary with the cell voltage V_C in the manner shown on the polarogram in Fig. 8-21. The current rises sharply at the half-wave potentials denoted by V_1, V_2, and V_3 where there occur transformations from one ionic species to another. For example, V_1 might correspond to the formation of a mononegative hydrocarbon radical ion, while V_2 indicates the formation of the corresponding dinegative ion. The mononegative ion will be present at the plateau between V_1 and V_2 while the dinegative one is observed on the plateau between V_2 and V_3. It is best to obtain a polarogram such as the one shown in Fig. 8-21 before recording ESR spectra on an unknown system. A dinegative ion may be in either a singlet or a triplet state, with the former more probable.

The radical ions to be studied are generated electrolytically by applying a voltage V_A corresponding to a cell voltage drop V_C on the plateau between V_1 and V_2. If the gradual buildup of polarization slowly shifts the actual cell drop V_C for a constant applied voltage V_A, then V_C may be set initially toward one end of the plateau so that the drift will bring it toward the center. An electron spin resonance run may be made during the electrolysis to record the spectrum of the radical ion. Pointeau et al. (1964) describe the growth and decay of the radical ion concentration during and after the cessation of the electrolysis.

Geske and Maki used the mercury-pool electrode to circumvent the high noise level and frequency instabilities that result from allowing small drops of mercury to traverse the resonant cavity periodically. Ordinary (i.e., non-ESR) polarographic studies make use of a dropping-mercury electrode versus an aqueous saturated calomel electrode, with the half-wave potential measured relative to a saturated calomel reference electrode.

Various workers have modified the original Geske and Maki cell, and some have employed radically different designs. The first edition may be consulted for descriptions of several of these cells, and for references to some of the earlier literature.

One of the principal experimental problems encountered in setting up an electrolytic cell results from the high dielectric loss that is an intrinsic property of many of the solvents used in polarography. This loss can have a very deleterious effect on the Q of the resonant cavity, and necessitates a precise alignment of the cell at the point where the microwave electric field is a minimum. This effect on Q may be minimized by employing quartz tubing with a small inner diameter, or a flat quartz cell across a rectangular or cylindrical cavity. For the same size tubing, a cylindrical TE_{011} mode cavity is much less lossy than a rectangular TE_{102} mode cavity. The losses may be considerably reduced by working at S band (3000 Mc) instead of at X band (9500 Mc). Allendoerfer et al. (1975, 1981) improved the sensitivity by the use of the coaxial resonator described in Sec. 5-O. The improvement in sensitivity is particularly pronounced with *in situ* electrolysis when the electrodes are located inside the hollow inner conductor of the cavity.

The solutions employed in polarography contain three ingredients (1) the solvent, (2) the supporting electrolyte (bulky counter ions), and (3) the compound to be studied. The most popular solvents and supporting electrolytes may be deduced from Table 15-3 of the first edition. Dissolved oxygen produces a strong polarogram as shown in Fig. 15-22 of the first edition and has a pronounced effect on the ESR relaxation times (Hausser, 1960, 1960). In addition, a concentration of oxygen that is too small to detect polarographically is still likely to react with the radical ions and produce undesired paramagnetic by-products. Therefore, traces of dissolved oxygen must be scrupulously removed from all the solutions used in polarography. This may be most easily accomplished by degassing with nitrogen and maintaining a positive pressure of nitrogen in the system during the ESR run. A more efficient method of removing oxygen from a solution consists in freezing the solution, evacuating it, closing off the vacuum pump, and allowing the frozen solution to melt. The oxygen is more soluble in the vapor than it is in the liquid, and so several cycles of this freeze–pump–thaw technique are sufficient to remove the dissolved oxygen. As before, a positive pressure of nitrogen gas may be maintained in the system to prevent oxygen from diffusing back in. Tanaka et al. (1972) used an O-ring coupling for deoxygenation and sealing of sample tubes. Lees et al. (1961) describe a gettering technique for the removal of oxygen. Water should also be excluded from electrolytic solutions. Plachy and Wondrem (1977) controlled the concentration of oxygen, while the sample remained in the resonant cavity. They employed a special sample tube that is permeable to oxygen and nitrogen but not to most liquids, and they allowed the oxygen concentration in the sample to come to equilibrium with that in the gas stream used for maintaining the temperature control.

References

Ahn, M. K., *J. Magn. Reson.* **22**, 289 (1976).

Albery, W. J., A. T. Chadwick, B. A. Coles, and N. A. Hampson, *J. Electroanal. Chem.* **75**, 229 (1977).

Albery, W. J., B. Z. Coles, and A. M. Coupler, *J. Electroanal. Chem. Interfacial Electrochem.* **65**, 901 (1975).

Albery, W. J., R. G. Compton, A. D. Chadwick, B. A. Coles, and J. A. Lenkait, *J. Chem. Soc. Faraday Trans.* **76**, 1391 (1980).

Albery, W. J., R. G. Compton, and I. S. Kerr, *J. Chem. Soc. Perkins Trans.*, in press.

Aleksandrov, L. N., and L. V. Ovchinnikova, *PTE* **18**(5), 246 (1637) (1975).

Allendoerfer, R. D., and J. B. Carroll, Jr., *J. Magn. Reson.*, in press.

Allendoerfer, R. D., G. A. Martichek, and S. Bruckenstein, *Anal. Chem.* **47**, 890 (1975).

Al'tshuler, A. S., and B. M. Kozyrev, *Electron Paramagnetic Resonance* (transl. *Scripta Technica*, English edition edited by C. P. Poole, Jr.), Academic Press, New York, 1964.

Bachler, W., R. Frank, and E. Usselman, *Trans. 6th Int. Vac. Congr.* (Kyoto) (1974).

Bahl, S. K., S. M. Bhagat, and R. G. Glosser, *Intern, Conf. Amorphous and Liquid Semicond.*, T. Arizumi, A. Yoshida, and K. Saji, Eds., Garmisch-par-ten-kirchen, 1973, p. 1065.

Barr, W. E., and V. J. Ankorn, *Scientific and Industrial Glass Blowing and Laboratory Techniques*, Instruments Publ. Co., Pittsburgh, 1959.

Becker, W., and W. Nesseldreher, *Vak. Tech.* **23**, 12 (1974).

Bennett, J. E., and A. Thomas, *Proc. R. Soc. London A* **280**, 123 (1964).

Benvenuti, C., and R. S. Calder, *Le Vide Suppl.* **157**, 29 (1972).

Bills, D. G., *J. Vac. Sci. Technol.* **10**, 65 (1973).

Boers, A. L., Pumping Speed of Molecular Sieves in the Forevacuum Region, *Proc. 4th Internat. Vacuum Congress*, (Inst. Phys., London), 1968, p. 393.

Boissin, J. C., J. J. Thibault, and A. Richardt, *Le Vide*, Suppl. **157**, 103 (1972).

Bonnemay, M., and C. Lamy, *C. R. Acad. Sci. C* **265**, 695 (1967).

Borg, D. C., *Rapid Mixing and Sampling in Biochemistry*, B. Chance, Ed., Academic Press, New York, 1964, p. 135.

Borodina, L. V., E. I. Trif, M. M. Akhmetov, S. A. Zaitseva, and F. G. Unger, *Zavod. Lab.* **42**, 1348 (1976).

Cacciarelli, R. A., and B. B. Stewart, *RSI* **34**, 944 (1963).

Chary, M. N., and B. A. Sastry, *Indian J. Pure and Appl. Phys.* **15**, 172 (1977).

Connell, G. A. N., and J. R. Pawlick, *Phys. Rev. B* **13**, 787 (1976).

Cordischi, D., V. Indovina, M. Occhiuzzi, and A. Arieti, *J. Chem. Soc. Faraday Trans.* **75**, 533 (1979).

Danielson, P. M., *J. Vac. Sci. Technol.* **7**, 527 (1970).

Deguchi, Y., *J. Chem. Phys.* **32**, 1584 (1960).

Denison, D. R., *J. Vac. Sci. Technol.* **4**, 156 (1967).

Dixon, W. T., and D. Murphy, *J. Chem. Soc. Perkin Trans.* **2**, 850 (1975).

Dixon, W. T., and R. O. C. Norman, *J. Chem. Soc.* 3119 (1963).

Dobrozemsky, R., and R. Moraw, *Vacuum* **21**, 587 (1971).

Dushman, S., *Scientific Foundations of Vacuum Technique*, 2nd ed., Wiley, New York, 1962.

Eloranta, J., and M. Ijas, *Finn. Chem. Lett.*, 174 (1975).

Eloranta, J., and R. Koskinen, *Finn. Chem. Lett.*, 179 (1975).

Fehsenfeld, F. C., K. M. Evenson, and H. P. Broida, *RSI* **36**, 294 (1964).

Fessenden, R. W., and P. Neta, *Chem. Phys. Lett.* **18**, 14 (1973).

Froben, F. W., *Ber. Bunsen. Phys. Chem.* **78**, 184 (1974).

Gause, E. M., and J. R. Rowlands, *Spectrosc. Lett.* **9**, 219 (1976).

Gideoni, M., and M. Steinberg, *Israel J. Chem.* **11**, 543 (1973).

Geske, D. H., and A. H. Maki, *J. Am. Chem. Soc.* **82**, 2671 (1960).

Golberg, I. B., and A. J. Bard, *Magn. Reson. Chem. Biol. Lect. Ampere Int. Summer Sch.*, 225 (1975).

Goldberg, I. B., and H. R. Crowe, *J. Phys. Chem.* **80**, 247 (1976).

Griffith III, O. F., and C. P. Poole, Jr., *Spectrochim. Acta* **25A**, 1463 (1969).

Grigorov, G., *Le Vide* **28**, 14 (1973).

Gubanov, V. A., V. I. Koryakov, and A. K. Chirkov, *J. Magn. Reson.* **11**, 326 (1973).

Gupta, D. K., H. A. Farach, and C. P. Poole, Jr., *Lett. Nuovo Cimento* **2**, 20 (1971).

Halama, H. J., and J. R. Aggus, *J. Vac. Sci Technol.* **11**, 333 (1974).

Haniotis, Z., and H. S. H. Gunthard, *Helv. Chim. Acta* **51**, 561 (1968).

Hausser, K. H., *Naturwiss.* **47**, 251 (1960); *Arch. Sci.* (*Fasc. Spec.*) **13**, 239 (1960).

Hengevoss, J., and E. A. Trendelenburg, *Continuous Cryotrapping of Hydrogen and Helium by Argon at 4.2K*, Trans. 10th Am. Vac. Symp., Macmillan, New York, (1963) p.101.

Hengevoss, J., and E. A. Trendelenburg, *Vacuum* **14**, 495 (1967).

Henning, J., *New Developments in the field of High Performance Turbomolecular Pumps*, Trans. 6th Internat. Vacuum Congress (Kyoto), 1974.

Hirota, N., *Int. Rev. Sci. Phys. Chem. Ser. Two.* **4**, 85 (1975).

Hobson, J. P., B. G. Baker, and A. W. Pye, *J. Vac. Sci. Technol.* **10**, 241 (1973).

Holland, L., W. Steckelmacher, and J. Yarwood, Eds., *Vacuum Manual*, Spon, London, 1974.

Holloway, J. M., and J. B. Raynor, *J. Chem. Soc., Dalton Trans.* 737 (1975).

Kartel, N. K., N. N. Tsyba, V. V. Strelko, K. M. Nikolaev, N. S. Polyakov, and M. M. Dubinin, *Izv. Akad. Nauk USSR, Ser. Khim.* P2243 (1978).

Kato, S., M. Nojirui, and H. Oikawa, Characteristics of Orditron Pump, *Trans. 6th Internat. Vacuum Congress (Kyoto)* (1974).

Kertesz, J. C., and W. Wolf, *J. Phys. E Sci. Instrum.* **6**, 1009 (1973).

Kertesz, J. C., and W. Wolf, *Am. Lab.* **7**, 101 (1975).

Kidnay, A. J., and M. J. Hiza, *Cryogenics* **10**, 271 (1970).

Koryakov, V. I., and V. N. Pankratov, *PTE* **14**(2), 238 (1971).

Kubler, L., G. Gewinner, J. J. Koulmann, and A. Jaégl, *Phys. Status Solidi B* **78**, 149 (1976).

Ladds, M., and K. A. Rubinson, *Appl. Spectrosc.* **30** 237 (1976).

Lamy, C., and P. Malaterre, *Surf. Sci.* **22**, 325 (1970).

Lees, J., B. H. Muller, and J. D. Noble, *J. Chem. Phys.* **34**, 341 (1961).

Maki, A. H., and D. H. Geske, *Anal. Chem.* **31**, 1450 (1959).

Matenaar, H., and R. N. Schindler, *Messtechnik* **3**, 67 (1973).

McArdle, P., *J. Chem. Educ.* **51**, 124 (1974).

Miller, D. J., and D. Haneman, *Surf. Sci.* **19**, 45 (1970).

Miller, D. J., and D. Haneman, *Solid State Comm.* **27**, 91 (1978).

Mirgel, K. H., *J. Vac. Sci. Technol.* **9**, 408 (1972).

Mizokawa, Y., and S. Nakamura, *Jap. J. Appl Phys., Proc. 2nd Int. Conf. Solid Surf.*, Suppl. 2, Pt. 2, 253 (1974).

Moore, R. W., Cryopumping in the Free-Molecular Flow Regime, *Trans. 2nd Internat. Vacuum Congress*, Pergamon Press, Oxford, (1962), p. 426.

Muller, E., *Cryogenics* **6**, 242 (1966).

Muller-Warmuth, W., *A. Naturforsch.* **18A**, 1001 (1963).

Naik, P. K., and R. G. Herb, *J. Vac. Sci. Technol.* **5**, 42 (1968).

Natsuko, K., and Y. Iwao, *Bull. Chem. Soc. Japan* **51**, 991 (1978).

Newton, C. M., and D. G. Davis, *J. Magn. Reson.* **20**, 446 (1975).

Norman, R. O. C., *Lab. Pract.* **13**, 1084 (1963).

Osterstrom, G. E., and A. H. Shapiro, *J. Vacuum Sci. Tech.* **9**, 405 (1972).

Pankratov, A. A., and I. M. Prudnikov, *Teor. Eksp. Khim.* **21**, 107 (1971).

Piette, L. H., I. Yamazaki, and H. A. Mason, in *Free Radicals in Biological Systems*, M. S. Blois, Jr., H. W. Brown, R. M. Lemmon, R. O. Lindblom, and M. Weissbluth, Eds., Academic Press, New York, p. 195 (1961).

Plachy, W. Z., and D. A. Wondrem, *J. Magn. Reson.* **27**, 237 (1974).

Pointeau, R., J. Favede, and P. Delhaes, *J. Chim. Phys.* **61**, 1129 (1964).

Poole, C. P., Jr., and D. S. MacIver, *Advances Catalysis* **17**, 223 (1967).

Povich, M. J., *Anal. Chem.* **47**, 346 (1975); *J. Phys. Chem.* **79**, 1106 (1975).

Power, B. D., *High Vacuum Pumping Equipment*, Chapman & Hall, London, 1966.

RCA Tube Handbook, The Radio Corporation of America, Harrison, N.Y.

Redhead, P. A., J. P. Hobson, and E. V. Kormelsen, *The Physical Basis of Ultra-High Vacuum*, Chapman & Hall, London, 1968.

Rokushika, S., H. Taniguchi, and J. Hatano, *Anal. Lett.* **8**, 205 (1975).

Roth, A., *Vacuum Technology*, North-Holland, Amsterdam, 1976.

Schmidt, P., and Hs. H. Gunthard, *J. Appl. Math. Phys.* **17**, 404 (1967).

Schwarz, H., *J. Vac. Sci. Technol.* **9**, 373 (1972).

Shimuzu, T., M. Kumeda, I. Watanabe, and K. Kamono, *Japan J. Appl. Phys.* **18**, 1923 (1979).

Singer, L. S., *Proc. of the Fifth Carbon Cong.* Vol. 2, Pergamon, London, 1963, p. 37.

Spangenberg, K. R., *Vacuum Tubes*, McGraw-Hill, New York, 1948.

Stern, S. A., and S. F. DiPaolo, *J. Vac. Sci. Technol.* **6**, 941 (1969).

Svejda, P. S., *RSI* **48**, 1092 (1977).

Svejda, P. S., *J. Phys. Chem.* **76**, 2690 (1972).

Svejda, P., W. Hartmann, and R. Haul, *Ber. Bunsen Gesel. Phys. Chem.* **80**, 1327 (1976).

Svejda, P., R. Haul, D. Mihelcic, and R. N. Schindler, *Ber. Bunsenges*, *Phys. Chem.* **79**, 71 (1975).

Szuber, J., and B. Salamon, *Phys. Status Solidi A* **56**, 111 (1979).

Szuber, J., and B. Salamon, *Phys. Status Solidi A* **53**, 289 (1979).

Tanaka, K., R. P. Quirk, G. D. Blyholder, and D. A. Johnson, *Appl. Spectrosc.* **26**, 652 (1972).

Templemeyer, K. E., R. Dawdarn, and R. I. Young, *J. Vac. Sci. Technol.* **8**, 575 (1970).

Thomas, P. A., and M. H. Brodsky, D. Kaplan, and D. Lepine, *Phys. Rev. B* **18**(30), 9 (1978).

Timson, P. A., and C. A. Hogarth, *Thin Solid Films* **8**, 237 (1971).

Tuttle, T. R., Jr., *J. Phys. Chem.* **79**, 3071 (1975).

Ueda, H., Z. Kuri, and S. Shida, *J. Chem. Phys.* **36**, 1676 (1962).

Venters, K., R. Gavars, M. Trusule, L. M. Baider, and J. Stradins, *Latv. PSR Zinat. Akad. Vestis Khim. Wer.* 534 (1975).

Vijendran, P., and C. V. G. Nair, *Vacuum* **21**, 159 (1971).

Wamser, C. A., and B. B. Stewart, *RSI* **36**, 397 (1965).

Wheeler, W., *J. Vac. Sci. Technol.* **11**, 332 (1974).

Zagorets, P. A., V. I. Ermakov, A. G. Atanasyants, and V. V. Orlov, *Itogi. Nauki. Tekh. Rastvory. Rasplavy* **1**, 5 (1975).

Zarubin, M. Y., A. M. Kutnevich, and A. P. Rudenko, *Zh. Org. Khim.* **11**, 1284 (1975).

CHAPTER 9

Variable Temperatures and Pressures

A variable-temperature ESR study can provide a great deal of information about a spin system and its interaction with its environment. To first order, the g-factor, hyperfine interaction constants, and other terms in the Hamiltonian are independent of the temperature. The lineshape, linewidth ΔH, and relaxation times T_1 and T_2 are the principal quantities that are sensitive to the temperature, and these are discussed in detail in Chap. 13. Some spin systems have such short relaxation times that they are only detectable in the liquid-helium temperature range. Other spin systems have such long relaxation times that they saturate easily and microwatt power levels are required for their observation. Particularly dramatic changes in ΔH, T_1, T_2, the lineshape, and the intensity occur at crystallographic phase transitions such as melting, and at magnetic phase transitions such as the passage from the paramagnetic to the antiferromagnetic state (Muller, 1971; Owens et al. 1979). Plachy and Schaafsma (1969) constructed an ESR cell that has temperature gradients and fluctuations less than $10^{-3}°C$.

Several standard temperatures are shown in Table 9-1. They are easy to maintain with constant temperature baths, and may be employed to calibrate thermocouples and other types of thermometers. Figure 9-1 shows the ranges over which several thermometers operate, and Fig. 9-2 gives the temperature dependence of the thermal conductivity of various materials.

A. High Temperatures

High-temperature techniques used in pretreating samples were discussed in the previous chapter. The present section deals with electron spin resonance measurements carried out at high temperatures.

There are two general methods of maintaining a high temperature at the ESR sample. The first method consists of inserting the resonant cavity into the oven, as shown in Fig. 9-3. [Poole and O'Reilly, 1961; see also Watkins, 1959]. The gold-plated TE_{102} mode X-band rectangular resonant cavity changes its frequency by -0.17 Mc/°C over the operating range from 20 to 500°C. A section of thin-walled stainless-steel waveguide provides thermal isolation from the magic T or circulator. The heating element is made from 300 cm of No. 20 nichrome wire wound on four Transite plates, and it is activated by the power supply shown in Fig. 9-4. A rectifier is employed in this power supply to

TABLE 9-1

Several Standard Temperatures for Low-Temperature Studies

Substance	Boiling Point, K	Melting Point, K	Triple Point, K	Liquid Density, g/cm^3	Heat of Vaporization, J/cm^3
H_2O	373.1	273.1		1.0	
NH_3	239.5				
CO_2	194.5[a]			—	
Ne	165	161	83.9		
O_2	90.2		54.4	1.14	240
Ar	87.4	84	83.9	1.4	230
N_2	77.4		63.2	0.81	160
Ne	27.1	24.5	24.6	1.2	108
H_2	20.3	14	13.9	0.07	31
He^4	4.211	—	$(2.172)^b$	0.13	2.6
He^3	3.2	—	—	0.06	0.48

[a]Sublimes.
[b]Lambda point.

minimize the vibrations that may result from an ac source. It requires about 5A to produce a temperature of 400°C. The resonant cavity is flushed with nitrogen gas to prevent oxidation of its walls. The temperature is measured by a thermocouple that is not ferromagnetic (e.g., copper Constantan). Electron spin resonance at temperatures up to 1000 K may be carried out using Walsh et al.'s (1965) heated cavity assembly shown in Fig. 8-43 of the first edition. It has a water-cooled outer jacket to protect the magnet pole pieces and is powered by a Variac. The high-frequency microwaves were obtained by third harmonic generation from 10 GHz by using a GaAs point-contact diode. Persyn and Nolle (1965) used an ESR cavity in an oven up to 1000 K, Van Wieringen and Rensen (1964) operated up to 1275 K, and Chaikin (1963) attained spectrometer operation up to 1675 K.

The second method of maintaining a sample at high temperatures is to place it inside a glass or quartz tube that passes through the cavity (Singer et al., 1961; Reuveni and Maniv, 1974; Suchard and Bowers, 1972). Nitrogen gas passes through a heating element and then through the resonant cavity where it heats the sample, as shown in Fig. 9-5. The temperature may be raised by either increasing the current that flows through the heating elements, or by increasing the gas flow rate. A thermocouple is employed to monitor the sample temperature. It is best to calibrate the thermocouple reading against that of a second thermocouple placed inside an actual sample tube. The Singer (1961) arrangement shown in Fig. 9-5 makes use of water cooling coils to maintain the resonant cavity at room temperature. One may also employ a Dewar made from double-walled and evacuated quartz tubing that is silvered everywhere but inside the resonant cavity.

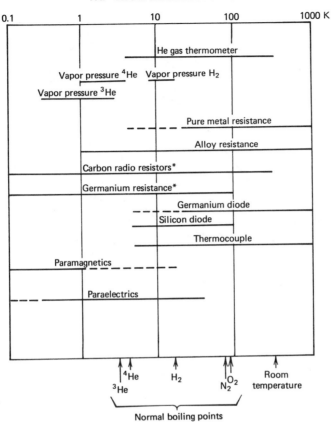

Fig. 9-1.   Thermometers for use in various temperature ranges (*indicates that a single thermometer will not cover the whole range shown with adequate sensitivity. One unit may be expected to be sensitive over about a decade in temperature (Rose-Innes, 1973).

The high-temperature cavity of Berlinger and Muller (1977) that is described in Sec. 5-P employs a Singer-type Dewar arrangement to attain temperatures up to 1300°C. Koningsberger et al. (1973) designed an ESR cavity for high-sensitivity spin concentration measurements up to 1300°C. The cavities of Giardino and Petrakis (1967), Cornell and Slichter (1969) and Dormann et al. (1976) operate up to 1100°C, 1200°C and 1400°C respectively.

Grosmaitre and Roy (1973) operated at temperatures up to 800°C using the arrangement illustrated in Fig. 9-6. It makes use of a vacuum oven placed inside of a large, inner access cavity that remains at room temperature. The heating is provided by two tungsten electrodes guided down the glass tube by an alumina tube to the point where they are welded to nichrome heating wires that are held in contact with the sample by a quartz tube.

Tanemoto and Nakamura (1978) employed a butane burner situated below the cavity as shown in Fig. 9-7 to attain temperatures as high as 1000°C.

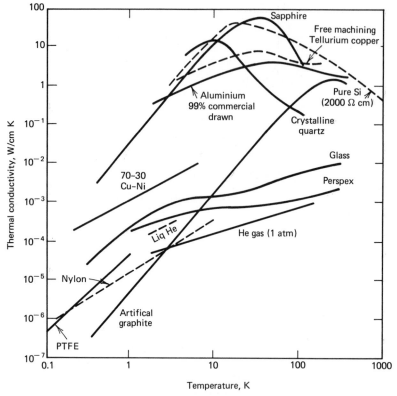

Fig. 9-2. Comparison of thermal conductivities of various substances (Rose-Innes, 1973).

If extra room is required to fit the Dewar in the cavity, then one may increase the small dimension of a TE_{102} rectangular resonant cavity without appreciably effecting the resonant frequency, and use a larger hole. Increasing this dimension will increase Q somewhat, but the losses in the additional glassware will result in a net decrease in Q. The principal electrical effect of such a dimensional change will be to lower the energy density in the cavity. A large X-band rectangular resonant cavity may be constructed from oversize RG 51/U waveguide (2.84 × 1.28 cm ID). An easier way to obtain more room for a Dewar is to employ a TE_{011} mode cylindrical cavity such as the one discussed in Sec. 5-D.

A temperature-control unit may be designed to stabilize the ESR sample temperature to a predetermined value. This is accomplished by taking the difference between a standard preset voltage and the thermocouple voltage, and using this difference as an error signal to turn on the heating element (or the gas flow). If the temperature exceeds the desired temperature, then the error voltage will be opposite in polarity to its low temperature value, and the heater will not turn on. When the temperature is too low, the heating element is activated (compare Fig. 16-7 of the first edition). The tendency to overshoot

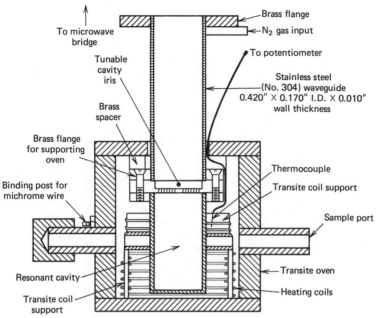

Fig. 9-3. High-temperature resonant cavity with Transite oven and Nichrome wire heating element (Poole and O'Reilly, 1961).

the desired temperature during the initial warmup may be minimized by a periodic application of the power until the operating temperature is reached. A control system of this type is discussed in Sec. 16-C of the first edition. Kertesz and Wolf (1975) designed a solenoid valve for a stopped-flow system.

Shaulov et al. (1973) employed the arrangement illustrated in Fig. 9-8 to control the temperature at the sample in a Dewar to a reported accuracy of 0.003°C. It achieves this with the aid of a gas stream stablized in velocity and in temperature and a stirrer that equalizes the temperature over the sample

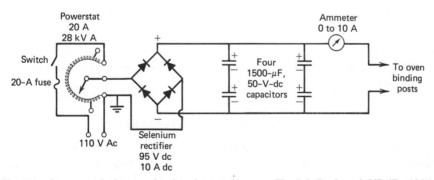

Fig. 9-4. Power supply for oven heating element shown on Fig. 9-3 (Poole and O'Reilly, 1961).

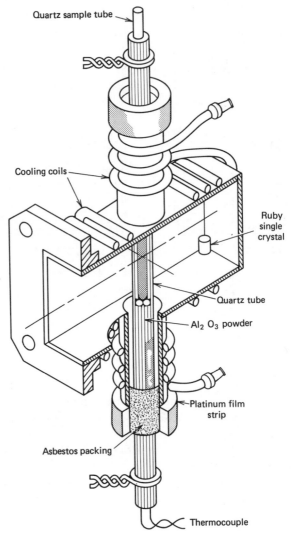

Fig. 9-5. A microwave resonant cavity for electron spin resonance measurements at high and low temperatures (Singer et al., 1961).

volume. Their apparatus was used to investigate the formation of local concentrations near the liquid–gas critical point in a binary gas system containing free-radical probes. Measurements were made over the range −50° to 240°C.

B. Low Temperatures (77–300 K)

The instrumentation for the low-temperature region is similar to that employed above room temperature. Again, there are two general methods of attaining the desired temperature. In the first method the entire resonant cavity is inserted

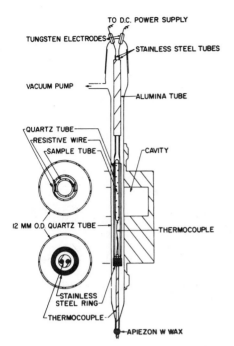

Fig. 9-6. Sketch of a vacuum oven located inside a cavity for high-temperature measurements (Grosmaitre and Roy, 1973).

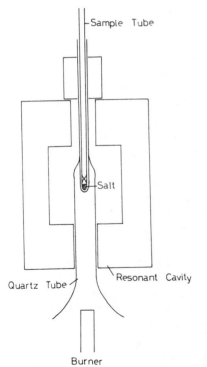

Fig. 9-7. Experimental arrangement for heating a sample up to 1000°C with a butane–oxygen gas flame (Tanemoto and Nakamura, 1978).

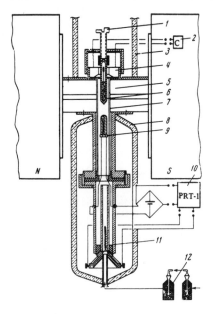

Fig. 9-8. Thermostat system for the sample in the resonator of the spectrometer: 1—regulator for the vertical displacement adjustment, 2—chopper for the solenoid used to produce stirring, 3—"parolon" heat insulation, 4—solenoid, 5—ESR cavity resonator, 6—ampule with investigated substance, 7—quartz tube, 8—"parolon" heat insulation, 9—platinum resistance thermoter, 10—temperature regulator, 11—heater, 12—stabilizer for the flow of the heat transfer gas (Shaulov et al., 1973).

into the Dewar and cooled by the refrigerant, as shown in Fig. 5-6. The resonant cavity shown in this figure may be disassembled at the plane of zero rf current flow for irradiation purposes. A hydrocarbon glass, such as 3-methyl pentane, was employed to seal the cavity by applying it when the lower half of the cavity was immersed in liquid nitrogen. While obtaining ESR data, the entire cavity must be submerged in liquid nitrogen to minimize the noise that results from the bubbling of the refrigerant around the cavity walls. Chasteen (1977) minimizes bubbling in Dewar inserts by a coating of glycerine on the bottom of the sample tube. The bubbling can be eliminated by blowing helium gas into the liquid nitrogen. Nelson and Villa (1978) solve the bubbling problem by pumping on the liquid nitrogen until it freezes. The cavity is evacuated to prevent oxygen or nitrogen from condensing within. The lead gasket and mica window form a vacuum seal above the resonant cavity. An inch or two of thin-walled stainless-steel waveguide will provide adequate thermal isolation for the resonator. A small air gap in the waveguide may also be used to achieve isolation (Poole, 1958).

The second method of achieving a low temperature is to make use of an arrangement such as that shown in Fig. 9-5 (Singer et al., 1961). The nitrogen gas is precooled by flowing it through a coil of copper tubing immersed in a liquid nitrogen or a Dry Ice–acetone bath. The nitrogen should be passed through a drying tube (e.g., calcium chloride or Drierite) before traversing the coil in order to prevent moisture condensation from clogging the coil. A typical arrangement will make use of a pressure head of 500 torr (~ 10 lb/in^2) stepped down by a needle valve to the desired flow rate. A flow meter may be employed to monitor the gas flow rate. The needle valve is employed as a coarse control on the temperature, and the current in the heating element constitutes a fine control. Reuveni and Maniv (1974) and Yanagi et al. (1978) describe continuous gas flow low-temperature systems. Windle (1976) discusses temperature control in an E-3 spectrometer.

If a double-walled evacuated Dewar such as that described in the last section is employed, then the "warming coils" that replace the cooling coils of Fig. 9-5 may be eliminated. For the most efficient operation, double-walled and evacuated Dewar tubing should be employed to carry the cold gas from the cooling coils to the resonant cavity. Ball joints or tapered joints may be used for joining several sections of such Dewar tubing. A temperature-stabilization circuit such as the one described in the preceding and following sections may also be employed at low temperatures.

Another method of maintaining the resonant cavity at temperatures between 77 K and room temperature is to place the bottom surface of the resonant cavity in intimate thermal contact with a copper rod that is several inches long. The lower end of the rod dips into liquid nitrogen, and a heating coil is placed near the top. The temperature of the cavity is determined by the level of the refrigerant and also by the current through the heating coil (see, for example, Ure, 1957). Von der Weid (1976) described a liquid-nitrogen cryostat for optical and ESR use. Yudin (1975) carried out ESR studies in liquid oxygen mixed with liquid nitrogen, and Sharrock (1979) described an inexpensive technique for obtaining low-temperature spectra by maintaining a minimum flow rate of liquid-nitrogen coolant past the sample.

Dalal and Miller (1981) studied lossy samples down to 77 K by using two melting-point capillaries side by side in a standard 4 mm OD, 3 mm ID sample tube and a conventional cooled-nitrogen flow system. The pair of capillaries were oriented in the plane of maximum H_1 field the same way a flat quartz cell is oriented.

A Dewar containing liquid nitrogen can be inserted directly into the resonant cavity, but this is not recommended since the boiling of the nitrogen produces a high noise level.

Homemade containers for liquid nitrogen may be fabricated from styrofoam (polyfoam or polystyrene foam) (see Froelich and Kenty, 1951; Marshall, 1955). The evaporation rate from polyfoam is greater than that from silvered vacuum vessels. For example, Nelson (1956) quotes an evaporation rate of 3 g/min for polyfoam versus 0.6 g/min for a silvered glass Dewar.

Several constant level-controllers for liquid nitrogen have been described in the literature (e.g., see Roizen and Gannus, 1961; Phillips and Owens, 1963; Nelson, 1963). It is usually best to shield the tops of Dewars so that moisture will not condense within.

C. Very Low Temperature (Below 77 K)

Most very-low-temperature studies are carried out at the temperature of liquid helium (4.2 K). Temperatures somewhat below 4.2 K such as the λ point of helium (2.172 K) are obtained by reducing the pressure above the helium. Very-low-temperature experiments are much more difficult to carry out than those at 77 K and above. The books listed at the end of the chapter may be referred to for general background material on cryogenics (Lounasmaa, 1974; MacKinnon, 1966; Mendelssohn, 1960; Vance and Duke, 1962; Sittig and Kidd, 1963; Vance, 1964; Rose-Innes, 1973; Johnson, 1961; Scott, 1959). The series *Progress in Low Temperature Physics* may be consulted for recent advances in experimental techniques.

Refrigerants

Liquid hydrogen and liquid helium are available commercially, but some institutions that consume large quantities build their own liquifiers. Many large-scale consumers find it economical to install a helium recovery system.

Storage vessels

Both liquid hydrogen and liquid helium are ordinarily stored in the inner chamber of a double-walled Dewar. Liquid nitrogen fills the outer chamber, and it is necessary to maintain a minimum level of liquid nitrogen to prevent excessive evaporation of the helium. Whitehouse et al. (1965) describe an economical homemade cryostat. Fradkov (1961, 1961) has described a liquid-helium cryostat design in which the escaping helium vapor cools a shield to between 72 and 80 K, and thereby obviates the necessity of using liquid nitrogen. The inner sections of Dewars are ordinarily evacuated to minimize heat loss, although powder insulation has also been used. Liquid-helium cryostats are available commercially. A special transfer tube is employed to transfer the liquid helium to the cryostat where the low-temperature experiment will be carried out. The efficiency of a cryostat depends on the evaporation rate of the liquid helium. Bewilogua and Lippold (1962) describe a simple apparatus for checking the evaporation rate from Dewars.

Sometimes it is desirable to perform irradiation experiments at liquid-helium temperatures. Rieckhoff and Weissbach (1962) describe an X-band cavity for such optical studies. A large number of literature articles have been written that describe Dewars provided with transparent windows for ultraviolet, visible, and infrared irradiation (e.g., see Schoen and Broida, 1962; Lotkova and Fradkov, 1961). Such Dewars are commercially available.

Temperatures below 4.2 K may be reached by several methods such as by pumping on liquid helium and by adiabatic demagnetization. Ginodman et al. (1973) pumped on ^{3}He to cover the range from 0.5 to 1.3 K.

Thermometers

A number of different thermometric principles have been employed for measuring temperatures in the liquid-helium region. The helium temperature scales are based on the temperature dependence of the vapor pressure of the main

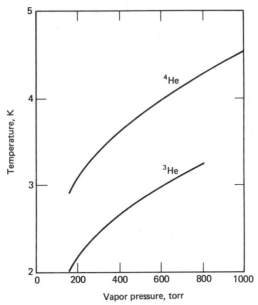

Fig. 9-9.   Temperature dependence of the vapor pressure of the two helium isotopes ^{4}He and ^{3}He.

isotope ^{4}He and also of the lighter isotope ^{3}He, as shown in Fig. 9-9. Kagarakis (1972) proposed a nuclear quadrupole resonance thermometer.

Thermocouples such as copper–constantan, chromel–constantan, and chromel–iron are useful at low temperatures as the curves plotted in Fig. 9-10 indicate. The sensitivity of a thermocouple, which is measured by its thermoelectric power, that is, by its rate of change of thermoelectric emf with temperature, decreases in the manner indicated in Fig. 9-10 as the temperature is lowered. It is best to calibrate a thermocouple at two or more of the fixed points shown in Table 9-1.

Pure metals at high temperatures normally exhibit an electrical resistivity that is approximately proportional to the temperature. As a result, the change in resistivity per degree is constant. At low temperatures the rate of change of resistivity with temperature becomes much less, and the metal approaches a constant resistivity at the lowest temperatures, as shown for platinum in Fig. 9-11. Therefore the resistivity of a metal is a poorer thermometer at liquid-helium temperature than it is above liquid-nitrogen temperature. Alloys have also been used as resistance thermometers.

In contrast to the behavior of metals, as the temperature T is lowered carbon resistors increase their resistance in the manner indicated in Fig. 9-11. These curves correspond to a resistance that varies with the temperature in accordance with the expression (Hoare et al., 1961)

$$\log R + \frac{K}{\log R} = A + \frac{B}{T} \tag{1}$$

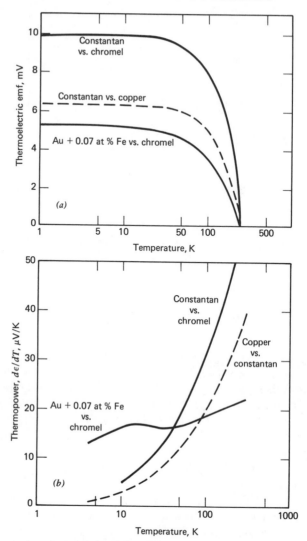

Fig. 9-10.   Temperature characteristics of three thermocouples. (*a*) Thermoelectric emf (reference junction at 273 K). (*b*) Thermoelectric power (Rose-Innes, 1973).

where *K*, *A*, and *B* are experimentally determined constants. 1-W Allen Bradley resistors between 2.7 and 270 Ω are typically used. Carbon resistance thermometers are very useful for measuring the temperature inside Dewars, and usually only require the use of an inexpensive ohmmeter instead of the expensive precision potentiometer that is ordinarily used with thermocouples and metallic thermometers. Pearce et al. (1956) used carbon resistors down to 0.3 K. Blake et al. (1958) describe a resistance thermometer bridge for measuring temperatures in the liquid-helium range. Ambler and Plumb (1960)

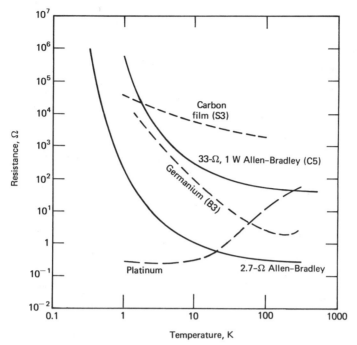

Fig. 9-11. Typical resistance versus temperature relations for several resistance thermometers (Rose-Innes, 1973).

have discussed the effect of stray rf fields on carbon resistor thermometers. Strips of carbon have also been employed as thermometers.

The ranges of other resistance thermometers suitable for low-temperature work are shown in Fig. 9-1.

A resistance thermometer may be conveniently employed for indicating the level of liquid helium in a Dewar. Nechaev (1961) and Meiboom and O'Brien (1963) employ a capacitor for this purpose.

ESR Apparatus

Now that some background has been presented on cryogenic techniques, several specific experimental arrangements will be described. Casini et al. (1976) designed an X-band TE_{102} mode brass rectangular resonant cavity with the smaller side reduced to 4 mm and the broad sidewall thickness decreased to 0.3 mm to permit the 100-kHz modulation to penetrate, since the skin depth δ is 0.5 mm for this case. The modulation coils are mounted on the outside of this wall. There is a small decrease in sensitivity due to the lower Q value of 2500 but a large increase arising from the 100-kHz modulation relative to a lower-frequency design.

Perlson and Weil (1975) designed the low-temperature X-band cavity illustrated in Fig. 9-12, which operates in the range from 12 to 300 K. Sample

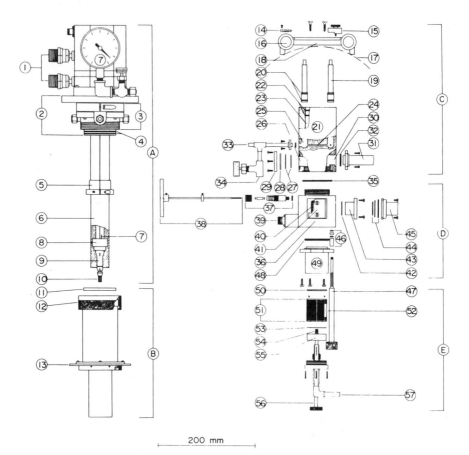

Fig. 9-12. Unit assembly drawing of low-temperature ESR cavity system. Subassemblies: A—Displex refrigerator; B—Displex–vacuum-shroud interface; C—vacuum shroud; D—modulation coil support; E—cavity support. Components: 1—self-sealing gas couplings; 2—displex instrument flange; 3—split-ring adaptor; 4—O-rings; 5—first stage of Displex refrigerator; 6—heat shield; 7—hydrogen-vapor thermometer; 8—cold end of refrigerator; 9—sample holder; 10—sample; 11—spacer ring; 12—union; 13—360° dial; 14—vernier scale; 15—clamp; 16—turnbuckles; 17—magnet pole caps; 18—support plate; 19—pistons; 20—O-rings; 21—brass block; 22—O-rings; 23—main cylinder; 24—interconnecting drill holes; 25—piston holes; 26—irradiation port; 27—O-rings; 28—fused quartz window; 29—window sash; 30—pump-out port; 31—flanged pipe; 32—O-ring; 33—compressed gas inlet system; 34—compressed gas outlet system; 35—O-ring; 36—Delrin plastic block; 37—vacuum-tight guiding assembly for screwdriver; 38—stainless-steel tipped screwdriver; 39—twin connector; 40—modulation coils; 41—plastic caps; 42—O-ring; 43—waveguide; 44—O-rings and glass window; 45—waveguide; 46—vacuum-tight guiding assembly for iris coupling screw; 47—iris coupling screw; 48—O-ring; 49—cylindrical spacer; 50—EPR cavity top plate; 51—EPR cavity cup; 52—iris coupling hole; 53—indium gasket; 54—frequency tuning rod; 55—cooling labyrinth; 56—threaded shaft; 57—feed and return lines for cold gas (Perlson and Weil, 1975).

332

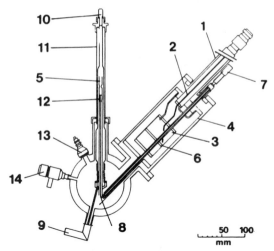

Fig. 9-13.   Cooling unit for use at X band over the temperature range 1.8–300 K: 1—transfer tube arm; 2—helium delivery tube; 3—helium chamber; 4—gas-cooled radiation shield; 5—ESR sample region; 6—sintered bronze disk; 7—needle valve; 8—capillary tube; 9—gas exhaust port; 10—access for samples; 11—quartz tube; 12—thermocouple and heater; 13—electrical lead-through; 14—evacuation valve (Campbell et al., 1976).

cooling is accomplished by a Displex closed-cycle, two-stage helium gas refrigerator operated electrically. The refrigerator capacity falls off from typical values of 5 W at 77 K to 1 W at 17 K. A TE_{011} mode rectangular cavity is equipped with 100-kHz field modulation. Some features of the system are the ability to rotate and manipulate single crystals throughout the operating range of temperature, the capability of *in situ* irradiation and visual sample observation, and the facility to transfer samples to the Displex refrigerator whose first stage is at approximately 80 K.

The low-temperature system illustrated in Fig. 9-12 has five main subassemblies: (A) the refrigerator, (B) the Displex–vacuum shroud interface, (C) the vacuum shroud, (D) the modulation coils, and (E) the cavity. The present system is superior to the one employing the Joule–Thompson effect, which was reported earlier by Weil et al. (1967).

Campbell et al. (1976, 1976) designed the X-band helium cooling unit illustrated in Fig. 9-13, which operates in the range from 1.8 to 300 K, and Fig. 9-14 shows the gas flow circuit associated with this cooling unit. The liquid helium is delivered to transfer tube arm 1 of the cooling unit via a flexible transfer tube, and it passes through helium delivery capillary tube 2 to a reservoir helium chamber 3 near the sample. Needle valve 7 controls the flow of helium through capillary tube 8 to the sample region 5. To cool the sample needle valve 7 is opened carefully and gas is pulled slowly through the system via port 9. A similar cooling unit illustrated in Fig. 9-15 operates at Q band over the temperature range from 4.8 to 300 K.

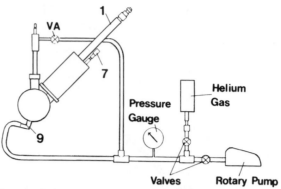

Fig. 9-14. Gas flow circuit for X-band sample cooling unit illustrated in Fig. 9-13. The valve is denoted by VA, and numbers 1, 7, and 9 are the same as in Fig. 9-13 (Campbell et al., 1976).

A number of studies have been made using a matrix isolation technique in which paramagnetic species and inert gas molecules are deposited on a refrigerated cold finger of sapphire or other material that is inserted into the resonant cavity (e.g., Delannoy et al., 1978; Herring et al., 1972; Jen et al., 1958, 1960; Jinguji et al., 1976; Kasai and McLeod, 1971; Knight and Wise, 1979; Lindsay et al., 1976; McDowell and Shimokoshi, 1974). The experimental arrangement of Dillon et al. (1959) for carrying out matrix isolation studies at liquid helium temperature is described in Sec. 16-C of the first edition. The apparatus of Zhitnikov and Kolesnikov (1964, 1965) mentioned in that section is designed to trap atoms in inert matrices at 77 K. In more recent articles Descamps and Gianese (1972) and Rettori et al. (1974) presented designs for cold finger crystals, the latter operating with ^{3}He as the refrigerant.

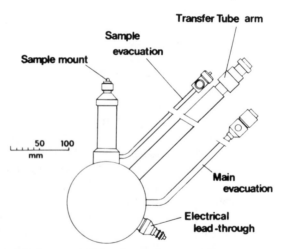

Fig. 9-15. Cooling unit for use at Q band over the temperature range from 4.8 to 300 K (Campbell et al., 1976).

McDowell et al. (1970) published the arrangement illustrated in Fig. 9-16 for carrying out matrix isolation studies. It employs an X-band superheterodyne spectrometer utilizing 400-Hz modulation and operating at power levels below 0.2 mW. A rotatable thin copper target was employed in lieu of the usual sapphire rod. The matrix and paramagnetic gases were premixed before being sprayed on the cold surface by means of a thin-walled quartz nozzle. In a typical case spraying rates of 2×10^{18} molecules/minute and total spraying times of 20 min were found to provide strong signals (see Bennett et al., 1964).

Two liquid-helium-temperature ESR systems were described in Sec. 16-C of the first edition. In the Flournoy et al. (1960) arrangement the resonant cavity was located in a glass Dewar, and a temperature controller for the heater provided any temperature between 4.2 and 77 K. In the Dillon et al. (1959) system a Dewar tip inserted into the resonant cavity contains the sample mounted on a sapphire rod.

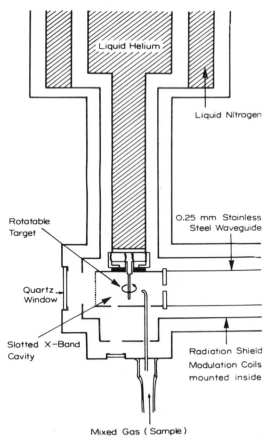

Fig. 9-16.   Low-temperature apparatus for mixed-gas deposition on and ESR measurements of a sample on a rotatable copper target cold finger located in a TE_{102} resonant cavity (McDowell et al., 1970).

A number of articles have appeared in the literature that describe methods of maintaining temperatures at any value between the boiling points of liquid helium and liquid nitrogen. For example, see Adkins (1961), Bigeleisen et al. (1968), Blake and Chase (1963), Bratashevskii et al. (1974), Campbell et al. (1976, 1976), Cataland et al. (1961), Hausmann et al (1970), Laquer and Decker (1966), Perlson and Weil (1975), and Shimashek (1961). The MacKinnon (1972) system operates from 1.6 to 300 K. The low-cost cooling device of Albracht (1974) operated at 35 GHz down to 4.8 K. The $\lambda = 8$ mm liquid-helium cavity of Abdulsabirov et al. (1971, 1972) permits low-temperature sample changing. Lutes (1962) and Vetchinkin (1961) discussed the attainment of controlled temperatures below 4.2 K. Matsumura et al. (1974) designed an X-band insertion-type liquid-helium Dewar. Henry and Dolecek (1950) employed a stainless-steel Dewar; Tolparov (1975) reported a helium cryostat for an RE-1301 spectrometer. Nistor (1970) described a low-temperature cavity. Ranon and Stamires (1970) presented a simple variable coupler for a Varian liquid-helium cavity. The plastic ENDOR cavity of Van Camp et al. (1976) operates down to 2 K. Rothwarf et al. (1972) described variable temperature probes for making magnetic susceptibility measurements in the range 4.2 to 300 K.

Abkowitz and Honig (1962) describe a K-band ESR apparatus that operates at a steady-state temperature of 0.48 K using continuously recycled ^{3}He. The upper part of the waveguide contains two aperiodic bends to prevent infrared radiation from passing down the waveguide. The system has been operated at 0.3 K by pumping on the liquid ^{3}He. Cowen et al. (1964), Rettori et al. (1974), and Svare and Seidel (1964), also describe ^{3}He systems.

Seven of the spectrometers described in Chap. 13 of the first edition have been operated at liquid-helium temperature: (1) Feher and Kip (1955) (Sec. 13-G); (2) Buckmaster and Scovil (1956) (Sec. 13-H); (3) Feher (1957) (Sec. 13-M); (4) Rose-Innes (1957) (Sec. 13-N); (5) Mock (1960) (Sec. 13-P); (6) Teaney et al. (1961) (Sec. 13-Q); (7) Llewellyn et al. (1962) (Sec. 13-R) as have others listed in Table 13-1 of the first edition. Zverev (1961) describes an X-band spectrometer for measuring spin-lattice relaxation times between 2 and 60 K, and Federov (1963, 1965) discusses a radio frequency (42 Mc) ESR spectrometer that operates between 0.15 and 4.2 K using adiabatic demagnetization. A stub tuner for impedance matching coaxial cables to a resonant cavity in liquid helium is described by de Klerk (1963). Maxwell and Schmidt (1958) described a superconducting cavity resonator made by electroplating lead on brass. This cavity operates at several hundred megahertz, and has a Q of 400,000. It incorporates a demountable cryostat for operation at liquid-helium temperatures.

The mechanical properties of metals at low temperatures are discussed by Klyavin and Stepanov (1959) and Parker and Sullivan (1963). Plastics are treated by Giauque et al. (1952) and epoxy resin cements by Netzel and Dillinger (1961). Wheatley et al. (1956) cover thermal contact and insulation below 1 K. Salinger and Wheatley (1961) measured the magnetic susceptibilities of materials that are commonly incorporated into cryogenic apparatus. Horwitz and Bohm (1961) recommend an indium O-ring seal that remains leak-tight when immersed in liquid helium. Wexler et al. (1950) had previously employed gold gaskets for this purpose. Schulte (1965) recommends a Teflon ribbon as a self-adhesive tape from liquid-helium temperature to 350 K. Mathes (1963) discusses the low-temperature properties of dielectrics. Lundin and Aasa (1972) discuss the use of self-pressurization for maintaining a low temperature at the sample. Sadreev (1979) described anomalies of dynamics of a paramagnetic crystal resonator. See also Galkin et al. (1976).

D. High Pressures

The lattice parameters of solids and fluids depend upon both the temperature and the pressure. Since it is easier to vary the temperature in an experiment, most investigators follow this procedure. The present section will describe some experimental techniques that may be used to carry out electron spin resonance studies at high pressures (see Benedek, 1963; Smith, 1963). It should be mentioned in passing that the lattice parameters may be varied over an even wider range by the use of isomorphous solid solutions (Poole et al., 1962; Poole and Itzel, 1964). Ulmer (1971) discusses researach techniques at high pressures.

Birnbaum (1950), Birnbaum and Maryott (1951), and Philips (1955) describe microwave devices for studying the complex dielectric constant of compressed gases up to 100 bars. Lawson and Smith (1959) employed two conically tapered single crystals of alumina as a high-pressure microwave window that is capable of supporting a pressure of 10 or more kilobars in a circular waveguide and its terminating high-Q resonator. At the same time, the window provides an impedance match from a standard 1-cm circular waveguide at atmospheric pressure to the cavity that sustains an internal hydrostatic pressure of up to 10 kilobars. Vallauri and Forsbergh (1957) describe a wide-band dielectric cell for use at S band at pressures up to 1500 atm.

Walsh and Bloembergen (1957) studied nickel fluorosilicate ($NiSiF_6 \cdot 6H_2O$) under high pressure and observed the pressure dependence of the zero-field parameter D (see also Andreev and Sugakov, 1975; Hurren et al. 1969; Kasatochin et al. 1975; Lukin and Tsintsadze, 1975, 1975. They employed a high-pressure cell good up to 10 kbars enclosed in a nonmagnetic Be–Cu cylinder with a pressure feed plug at one end. The cavity was immersed in the pressure-transmitting fluid petroleum ether. The unaxial stress equipment and spectrometer illustrated in the first edition were used by Walsh (1959, 1961) in conjunction with this high-pressure cell. Gardner et al. (1963) describe a high-pressure spectrometer. The low-temperature cavity of Berlinger and Muller (1977) that is described in Sec. 5-P has been used by the authors for uniaxial stress studies.

Rupp et al. (1977) describe a 35-GHz moderate-pressure, low-temperature spectrometer that generates ~ 3.5 kbars of hydrostatic pressure in the range from ambient temperature to liquid-helium temperature. A more recent design

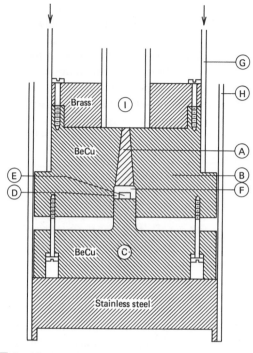

ZZ Sapphire

Fig. 9-17. Cross-sectional view of the piston and cylinder for high-pressure ESR showing: A—sapphire cone, B—Be-Cu cylinder; C—Be-Cu piston; D—Everdur washer; E—sample; F—silicone grease; G—"inner" stainless steel tube; H—fixed "outer" stainless steel tube; I—waveguide and flange (Rupp et al., 1977).

(Rupp et al., 1977) operates up to 8 kbars. A conical sapphire resonant cavity of the type described by Lawson and Smith (1959) was employed in a piston–cylinder configuration, as shown in Fig. 9-17. This assembly makes use of a single crystal sapphire cone and a Be–Cu piston containing the sample surrounded by silicone grease. Figure 9-18 shows the primary drive with a sliding vacuum seal between the waveguide and inner tube. The authors believe that silicone grease used as a pressure fluid remains hydrostatic even at 4.2 K over the pressure range 0 to 3.5 kbars. On occasion they have used the apparatus up to 8 kbars. Sample changing is easy because only eight screws need to be removed.

Berlinger (1982) designed the K-band system illustrated on Fig. 9-19 for applying uniaxial stresses to a sample up to 35 kp and a maximum temperature of 1300 K. Figure 9-20 gives details of the resonant cavity.

Henning and den Boef (1978) describe a method for applying periodic uniaxial stress to a sample in a technique called strain-modulated electron spin resonance (SMESR). The strain modulates the internal crystal fields in the sample, and the detection is carried out at their modulation frequency, as was

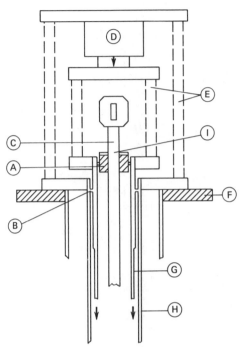

Fig. 9-18. Schematic view of the primary drive for high-pressure cylinder shown in Fig. 9-17 showing: A—waveguide-"inner" tube sliding vacuum seal; B—"inner"-"outer" tube sliding vacuum seal; C—waveguide bend; D—ram; E—support tubes; F—Dewar flange; G—"inner" stainless-steel tube; H—fixed "outer" stainless-steel tube; I—waveguide vacuum seal (Rupp et al., 1977).

mentioned in Sec. 6-F (Collins et al. 1971; den Boef and Henning, 1974, 1975; Henning and den Boef, 1974, 1975, 1976, 1978; Robinson et al., 1975, 1977, 1977, 1978). Henning and den Boef (1978) used the transducer strain modulator illustrated in Fig. 9-21 to apply periodic stress to the MgO crystal, and they mounted the transducer on the microwave cavity in the manner illustrated in Fig. 9-22. Two Phillips PX-5 bimorph elements produce the strain and two others sense it, as explained in the figure caption. The driving voltage is provided by a tuned oscillator. The sensor bar is connected to the probe circuitry illustrated in Fig. 9-23 to improve the accuracy of the strain measurements. This experimental arrangement was used to measure the strain dependance of the g tensor of cubic Cr^{3+} ions in MgO.

Clark and Wait (1964) describe an ESR spectrometer for studying pressure effects up to 10 kbars, Goodrich et al. (1964) designed a K-band high-pressure resonant cavity, Barnett et al. (1978) report a system operating up to 60 kbars, Kozhukhar et al. (1975, 1975, 1976, 1976) designed a high-pressure chamber with a resonator for low-temperature work, Assmus and Silber (1975) and Holmberg et al. (1974) describe high-pressure cryostats for studying single crystals at low temperature [see also Galkin et al. (1975)]. The Assmus and

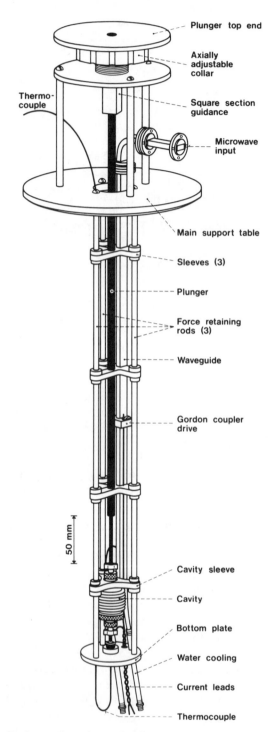

Fig. 9-19. Compressional stress system for operating with pressures up to 35 kp at temperatures up to 1300 K. Figure 9-20 shows the cavity details (Berlinger, 1982).

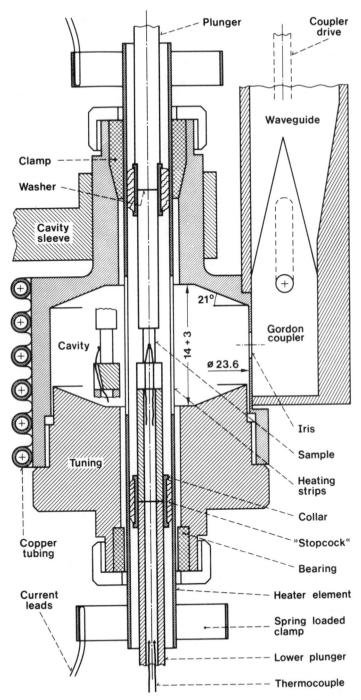

Fig. 9-20. Details of the resonant cavity used with the system of Fig. 9-19. The two figures are related by 90° with respect to each other (Berlinger, 1982).

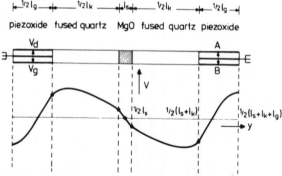

Fig. 9-21. Symmetrical longitudinal strain modulator. Strain is excited in two Piezoxide driver bars (voltage V_d) and detected in two sensor bars (V_g); A and B may correspond either to V_d or to V_g. The polarization is indicated by arrows. Lower trace: displacement waveform for a $\frac{3}{2}\lambda$ mode. Note that the strain ($e_{yy} = dv/dy$) is discontinuous at the interfaces Piezoxide–quartz and quartz–sample (Henning and den Boef, 1978).

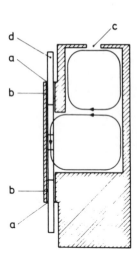

Fig. 9-22. Sketch of TE_{102} microwave cavity with strain modulator showing (a) Teflon slabs; (b) microwave choke; (c) coupling iris to waveguide; (d) transducer. The rf magnetic field lines are indicated (Henning and den Boef, 1978).

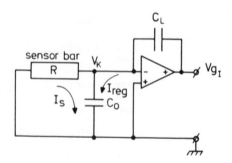

Fig. 9-23. Electronic probe circuit connected to sensor bar associated with the strain modulator of Fig. 9-21 (Henning and den Boef, 1978).

342

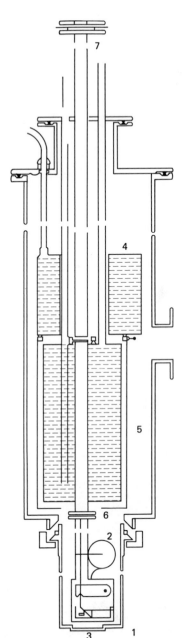

Fig. 9-24. High-pressure cryostat cavity system: 1—cavity with stress lever system; 2—100-kHz transformer for modulation; 3—quartz window for irradiation; 4—liquid-nitrogen tank; 5—liquid-hydrogen tank; 6—copper waveguide; 7—stainless-steel waveguide. The lever system of the cryostat is not shown. The design details of the cavity are presented in Fig. 9-25 (Assmus and Silber, 1975).

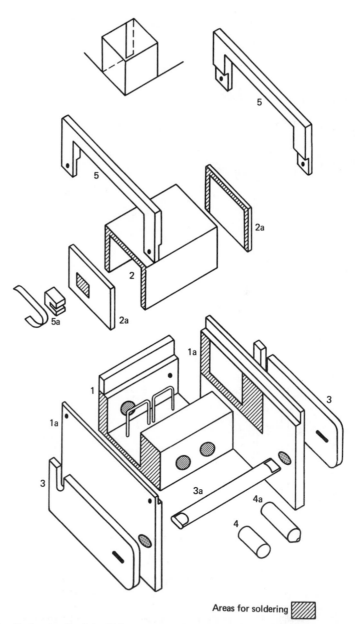

Fig. 9-25. Design details of the high-pressure cavity shown in Fig. 9-24 1, 1a—Upper part of the cavity; 2, 2a—lower (movable) part of the cavity; 3, 3a—cavity lever system (the lever acting on the sapphire is not shown); 4, 4a—pistons (sapphire and stainless steel); 5, 5a—sliding system for the movable part of the cavity (Assmus and Silber, 1975).

Silber high-pressure cryostat and cavity are illustrated in Figs. 9-24 and 9-25. Hwang et al. (1976) and Neilo et al. (1977) studied the pressure dependance of ordering and relaxation, Nagamura (1980) et al. (1974, 1975, 1979) investigated radicals in stressed fibers using the equipment illustrated in Fig. 9-26, and Hale et al. (1975) reported on strain effects in the ESR of Sb donors in Ge. Cevec and Srinivasan (1978) designed a simple pressure-locked cavity for use at high hydrostatic pressures and temperatures down to liquid nitrogen. Barnett et al. (1978) constructed a high-pressure system using a single crystal sapphire that serves both as a solid X-band microwave cavity and as an anvil in a Bridgeman pressure geometry. Pressures up to 60 kbars are attainable. Plachy and Schaafsma (1969) made use of a microwave helix in their high-pressure apparatus. High-pressure cavities were designed by Alaeva et al. (1971, 1972), Stankowski et al. (1976), Timofeev et al. (1971), Zaitov (1970), and Zhurkin et al. (1970). The Alaeva et al. (1971) cell worked to 100 kbars. High-pressure

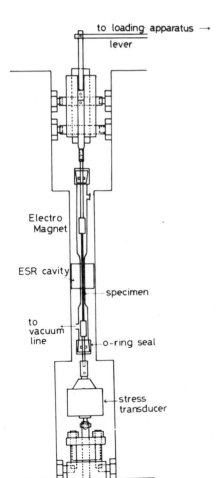

Fig. 9-26. Apparatus for observing ESR of specimens held under constant stress, constant strain, or constant elongation. At one end is a load-transmitting brass rod connected to a stress transducer, and at the other end is a rod attached via a lever to a motor-driven loading framework. Stretching is carried out at 1.5×10^{-5} torr (Nagamura and Takayanagi, 1974).

ESR systems have been reported by Fainstein and Oseroff (1971, 35 GHz), Filippov (1975), and Kobayashi (1972). The system of Szumowski and Falkowski (1976) permits angular dependence studies between 80 and 400 K at pressures up to 3500 kg/cm^2. Wolbarst (1976) reported a high-pressure, low-temperature ESR–ENDOR cavity, and Fainstein and Oseroff (1971) and Kasatochkin et al. (1974, 1977) employed ENDOR at high pressure.

Some references to additional ESR work at high pressure are Barberis et al. (1977), Bendik et al. (1971), Berthier et al. (1976), Bowden et al. (1971), Brekhovskikh et al. (1972, 1973), Davidov et al. (1976), Filippov and Ryzhmanov (1977), Grochulski et al. (1978), Henning and den Boef (1978), Hwang et al. (1973), Larys et al. (1975), Mamola and Wu (1975), Müller et al. (1970), Nelson et al. (1967), Rupp et al. (1977), Schwab and Hillmer (1975), Serway et al. (1973), Shaulov et al. (1973), Shoemaker (1974), Shoemaker and Lagendijk (1977), Suassuna et al. (1977), Vial and Buisson (1975), von Waldkirch et al. (1972, 1973), Waplak (1975), Watkins and Ham (1970).

One confusing aspect of high-pressure research is the fact that various investigators use different units of pressure. Benedek (1963) gives the conversion factors 1 kg/cm^2 = 0.9698 atm = 0.9807 bar = 14.22 $lb/in.^2$ = 0.9807 × 10^6 dyn/cm^2 = 10^4 kg/m^2 = 10^4 Pa.

References

Abdulsabirov, R. Yu., V. S. Kropotov, and V. G. Stepanov, *PTE* **14**, 232 (1971).

Abdulsabirov, R. Yu., V. S. Kropotov, and V. G. Stepanov, *Cryogen.* **12**, 393 (1972).

Abkowitz, M., and A. Honig, *RSI* **33**, 568 (1962).

Adkins, C. J., *JSI* **38**, 305 (1961).

Alaeva, T. I., L. F. Vereshchagin, S. V. Kasatochkin, Yu A. Timofeev, and E. N. Yakovlev, *PTE* **14**(1), 223 (264) (1971).

Alaeva, T. I., L. F. Vereshchagin, V. V. Gvozdev, Yu. A. Timofeev, V.S. Shanditsev, and E. N. Yakovlev, *PTE* **15**(5), 206 (1533) (1972).

Albracht, S. P. J., *J. Magn. Resonance* **13**, 299 (1974).

Ambler, E., and H. Plumb, *RSI* **31**, 656 (1960).

Andreev, A. A., and V. I. Sugakov, *Ukr. fiz. Zh.* **12**, 1998 (1975).

Assmus, W., and D. Silber, *RSI* **46**, 299 (1975).

Barberis, G. E., J. F. Suassuna, C. Rettori, and C. A. Pela, *Solid State Commun.* **23**, 603 (1977).

Barnett, J. D., S. D. Tyagi and H. M. Nelson, *RSI* **49**, 348 (1978).

Bendik, N. T., L. S. Milevskii, and E. G. Smirnov, *Sov. Phys. Semicond.* **5**, 749 (1971).

Benedek, G. B., *Magnetic Resonance at High Pressure*, Interscience, New York, 1963.

Bennett, J. E., and A. Thomas, *Proc. R. Soc. London A* **280**, 123 (1964).

Berlinger, W., *RSI* **53**, 338 (1982).

Berlinger, W., and K. A. Muller, *RSI* **48**, 1161 (1977).

Berthier, C., J. R. Cooper, D. Jerome, G. Soda, C. Weyl, J. M. Fabre, and L. Giral, *Mol. Cryst. Liq. Cryst.* **32**, 267 (1976).

Bewilogua, L., and H. Lippold, *ETP* **10**, 373 (1962).

Bigeleisen, J., F. P. Brooks, T. Ishida, and S. V. Ribnikar, *RSI* **39**, 353 (1968).

Birnbaum, G., *RSI* **21**, 169 (1950).

Birnbaum, G., and A. A. Maryott, *J. Appl. Phys.* **22**, 95 (1951).

Blake, C., and C. E. Chase, *RSI* **34**, 984 (1963).

Blake, C. B., C. E. Chase, and E. Maxwell, *RSI* **29**, 715 (1958).

Bowden, C. M., H. C. Meyer, and P. L. Donoho, *Phys. Rev. B* **3**, 645 (1971).

Bratashevskii, Yu. A., F. N. Bukhan'ko, and O. Z. Shapiro, *PTE* **17**, 241 280 (1974).

Brekhovskikh, S. M., V. A. Tyulkin, and I. N. Palandov, *Zh. Prikl. Spektrosk.* **17**, 318 (1972).

Brekhovskikh, S. M., V. A. Tyulkin, and I. N. Polandov, *Izv. Akad. Nauk SSSR Neorg. Mater.* **9**, 1021 (1973).

Buckmaster, H. A., and H. E. D. Scovil, *Can. J. Phys.* **34**, 711 (1956).

Campbell, S. J., I. R. Herbert, C. B. Warwick, and J. M. Woodgate, *J. Phys. E Sci. Instrum.* **9**, 443 (1976); *RSI* **47**, 1172 (1976).

Casini, G., P. Graziani, R. Linari, and A. Tronconi, *RSI* **47**, 318 (1976).

Cataland, G. M. H. Edlow, and H. H. Plumb, *RSI* **32**, 980 (1961).

Cevec, P., and R. Srinivasan, *RSI* **49**, 1282 (1978).

Chaikin, A. M., *PTE* **6**, 178 (1185) (1963).

Chasteen, N. D., *J. Magn. Reson.* **27**, 349 (1977).

Clark, A. F., and D. F. Wait, *RSI* **35**, 863 (1964).

Collins, M. A., S. D. Devine, W. H. Robinson, *8th Australian Spectroscopy Conference. Clayton, Australia*, Australian Acad. Sci., 16–20 Aug. 1971.

Cornell, E. K., and C. P. Slichter, *Phys. Rev.* **180**, 358 (1969).

Cowen, J. A., R. D. Spence, H. Van Till, and H. Weinstock, *RSI* **35**, 914 (1964).

Dalal, N. S., and J. M. Miller, *J. Magn. Reson.* **42**, 496 (1981).

Davidov, D., P. Urban, and L. D. Longinotti, *Solid State Commun.* **9**, 249 (1976).

de Klerk, J., *RSI* **34**, 183 (1963).

Delannoy, D. D., B. Tribollet, B. Valadier, and A. Erbeia, *J. Chem. Phys.* **68**, 2252 (1978).

den Boef, J. H., and J. C. M. Henning, *RSI* **45**, 1199 (1974).

den Boef, J. H., and J. C. M. Henning, *Proc. 18th Colloque Ampere Nottingham 1974*, North Holland, Amsterdam, 1975.

Descamps, D. P., and P. Gianese, *Cryogenics* **12**, 466 (1972).

Dillon, J. F., Jr., S. Geschwind, V. Jaccarino, and A. Machalett, *RSI* **30**, 559 (1959).

Dormann, E., D. Hone, and V. Jaccarino, *Phys. Rev. B* **14**, 2715 (1976).

Fainstein, C., and S. B. Oseroff, *RSI* **42**, 547 (1971).

Federov, B. V., *PTE* **4**, 98 (686) (1963); *Cryogenics* **5**, 12 (1965).

Feher, G., *Bell Syst. Tech. J.* **36**, 449 (1957).

Feher, G., and A. F. Kip, *Phys. Rev.* **98**, 337 (1955).

Filippov, A. I., *PTE* **18**, 249 (1975).

Filippov, A. I., and Yu. M. Ryzhmanov, *Fiz. Tverdogo Tela.* **19**, 1267 (1977).

Flournoy, J. M., L. H. Baum, and S. Siegel, *RSI* **31**, 1133 (1960).

Fradkov, A. B., *Sov. Phys. Dokl.* **5**, 888 (1961).

Fradkov, A. B., *PTE* **4**, 170 (796) (1961).

Froelich, H. C., and C. Kenty, *RSI* **22**, 214 (1951).

Galkin, A. A., G. G. Grinev, V. I. Kurochkin, and E. D. Nemchenko, *Defektoskopiya* **12**, 121 (1976).

Galkin, A. A., A. Yu. Kozhukhar, S. N. Lumin, and G. A. Tsintsadze, *High Temp.–High Pressures, 5th International Conference on High Pressure Physics and Technology* (summaries only), **8**, 612,6. 26–31 May (1975).

Gardner, J. H., N. W. Hill, C. Johansen, D. Larson, W. Murri, and N. Nelson, *RSI* **34**, 1043 (1963).

Giardino, D. A., and L. Petrakis, *RSI* **38**, 1180 (1967).

Giauque, W. F., T. H. Geballe, D. N. Lyon, and J. J. Fritz, *RSI* **23**, 169 (1952).

Ginodman, V. B., V. G. Zhurkin, V. F. Troitskii, and A. B. Fradkov, *PTE* **16**, 659 (1973).

Goodrich, R. G., G. E. Everett, and A. W. Lawson, *RSI* **35**, 1596 (1964).

Grochulski, T., K. Leibler, and A. Sienkiewicz, *Phys. Status Solidi A* **47**, K169 (1978).

Grosmaitre, C. S., and J. C. Roy, *RSI* **44**, 652 (1973).

Hale, E. B., J. R. Dennis, and Shih-hua Pan, *Phys. Rev. B* **12**, 2553 (1975).

Hausmann, A., W. Sardes and P. Schreiber, *Cryogenics* **10**, 70 (1970).

Henning, J. C. M., and J. H. den Boef, *Solid State Comm.* **14**, 993 (1974), *Phys. Status Sol.* **72b**, 369 (1975), *Phys. Rev. B* **14**, 26 (1976); **18**, 60 (1978); *Appl. Phys.* **16**, 353 (1978).

Henry, W. E., and R. L. Dolecek, *RSI* **21**, 496 (1950).

Herring, F. G., C. A. McDowell, and J. C. Tait, *J. Chem. Phys.* **57**, 4564 (1972).

Hoare, F. E., L. C. Jackson, and N. Kurt, *Experimental Cryophysics*, Butterworths, London, 1961.

Holmberg, G. E., W. P. Unruh, and R. J. Friauf, *International Conference on Colour Centres in Ionic Crystals.* (1974).

Horwitz, N. H., and H. V. Bohm, *RSI* **32**, 857 (1961).

Hurren, W. R., H. M. Nelson, E. G. Larson, and J. H. Gardner, *Phys. Rev.* **185**, 624 (1969).

Hwang, J., D. Kivelson, and W. Plachy, *J. Chem. Phys.* **58**, 1753 (1973).

Hwang, J. S., K. V. S. Rao, and J. H. Freed, *J. Phys. Chem.* **80**, 1490 (1976).

Jen, C. K., S. N. Foner, E. L. Cochran, and V. A. Bowers, *Phys. Rev.* **112**, 1169 (1958); *J. Chem. Phys.* **32**, 963 (1960).

Jinguji, M., K. C. Lin, C. A. McDowell, and P. Raghunathan, *J. Chem. Phys.* **65**(10), 3910 (1976).

Johnson, V. J., Ed., *Properties of Materials at Low Temperatures*, Pergamon, New York, 1961.

Kagarakis, C. A., *Tech. Chron.* **7**, 631 (1972).

Kasai, P. H., and D. McLeod, Jr., *Chem. Phys.* **55**, 1566 (1971).

Kasatochkin, S. V., and E. N. Yakovlev, *High Temp.–High Pressures, 5th International Conference on High Pressure Physics and Technology* (summaries only), **8**, 615-16, May 1975 (1977).

Kasatochkin, S. V., Yu. A. Timofeev, T. I. Alaeva, L. F. Vereshchagin, and E. N. Yakovlev, *PTE* **17**(6), 140 (1717) (1974).

Kertesz, J. C., and W. Wolf, *Am. Lab.* **7**, 101 (1975).

Klyavin, O. V., and A. V. Stepanov, *Fiz. Tverd. Tela. Suppl.*, *Sbornik* **1**, 241 (1959).

Knight, L. B., Jr., and M. B. Wise, *J. Chem. Phys.* **71**, 1578 (1979).

Kobayashi, T., *Acta Crystallogr. A* (*Denmark*). 9th International Congress of Crystallography of the International Union of Crystallography. Abstracts only. Vol. A28, Pt. 4, Suppl. Kyoto, Japan. S242. Internat. Union of Crystallography. 26 Aug.–7 Sept. 1972 (1972).

Koningsberger, D. C., G. J. Muller, and B. Pelupessy, *J. Phys. E Sci. Instrum.* **6**, 306 (1973).

Kozhukhar, A. Yu., S. N. Lukin, and G. A. Tsintsadze, *Fiz. Tverdogo Tela* **17**, 1870 (1975).

Kozhukhar, A. Yu., S. N. Lukin, G. A. Tsintsadze, and V. A. Shapovalov, *Prib. Tekh. Eksp.* **18**, 198 (1975).

Kozhukhar, A. Yu., S. N. Lukin, G. A. Tsintsadze, and V. Shapovalov, *Cryogenics* **16**, 441 (1976).

Kozhukhar, A. Yu., and G. A. Tsintsadze, *Sov. Phys. Solid State* **17**, 2234 (1976).

Laquer, H. L., and D. L. Decker, *Cryogenics* **6**, 109 (1966).

Larys, L., J. Stankowski, and M. Krupski, *Acta Phys. Pol. A* **A50**, 351 (1975).

Lawson, A. W., and G. E. Smith, *RSI* **30**, 989 (1959).

Lindsay, D. M., D. R. Herschbach, and A. L. Kwiram, *Mol. Phys.* **32**, 1199 (1976).

Llewellyn, P. M., P. R. Whittlestone, and J. M. Williams, *JSI* **39**, 586 (1962).

Lotkova, E. N., and A. B. Fradkov, *PTE* **1**, 188 (1961).

Lounasmaa, O. V., *Experimental Principles and Methods Below 1K*, Academic Press, New York, 1974.

Lukin, S. N., and G. A. Tsintsadze, *Zh. Eksp. Teor. Fiz.* **69**, 250 (1975).

Lukin, A. N., and G. A. Tsintsadze, *Fiz. Tverdogo Tela* **17**, 1872 (1975).

Lundin, A., and R. Aasa, *J. Magn. Reson.* **8**, 70 (1972).

Lutes, O. S., *RSI* **33**, 1008 (1962).

MacKinnon, J. A., *RSI* **43**, 1847 (1972).

Mackinnon, L., *Experimental Physics at Low Temperatures*, Wayne State University Press, Detroit, Mich., 1966.

Mamola, K. and R. Wu, *J. Phys. Chem. Solids* **36**, 1323 (1975).

Marshall, L., *RSI* **26**, 614 (1955).

Mathes, K. N., *Dielectrics in Space Symposium*, Westinghouse Electric Corp., Pittsburgh, 1963, p. 14.

Matsumura, Z., M. Chikira, S. Kubota, and T. Isobe, *RSI* **45**, 596 (1974).

Maxwell, E., and A. F. Schmidt, *Bull. Inst. Froid, Annexe* **95** (1958–1961).

McDowell, C. A., H. Nakajima, and P. Raghunathan, *Can. J. Chem.* **48**, 805 (1970).

McDowell, C. A., K. Shimokoshi, *J. Chem. Phys.* **60**, 1619 (1974).

Meiboom, S., and J. P. O'Brien, *RSI* **34**, 811 (1963).

Mendelssohn, K., *Cryophysics*, Interscience, New York, 1960.

Mock, J. B., *RSI* **31**, 551 (1960).

Müller, K. A., in *Structural Phase Transitions and Soft Modes* S. J.Samuelsen, E. A. Anderson, and J. Feder, Eds., (Universitetes Forlaget, Oslo, 1971.

Müller, K. A., W. Berlinger, and J. C. Slonczewski, *Phys. Rev. Lett.* **25**, 734 (1970).

Nagamura, T., Chap 14, in *Polymers*, Part C, R. A. Fava, Academic Press, New York, 1980.

Nagamura, T. and Devries, K. L., *Polymer Engrg. Sci.* **19**(2), p 89 (1979).

Nagamura, T., and M. Takayanagi, *J. Polym. Sci. Polym. Phys. Ed.* **12**, 2019 (1974); **13**, 567 (1975).

Nechaev, Yu. I., *PTE* **4**, 174 (801) (1961).

Neilo, G. N., Prokhorov, A. D. and G. A. Tsintsadze, *Zh. Eksp. Teor. Fiz.* **72**, 1081 (1977).

Nelson, H. C., and J. F. Villa, *J. Magn. Reson.* **31**, 515 (1978).

Nelson, H. M., D. B. Larson, and J. H. Gardner, *J. Chem. Phys.* **47**, 1994 (1967).

Nelson, L. S., *RSI* **27**, 655 (1956).

Nelson, L. C., *JSI* **40**, 428 (1963).

Netzel, R. G., and J. R. Dillinger, *RSI* **32**, 855 (1961).

Nistor, S. V., *Rev. Roumaine Phys.* **15**, 603 (1970).

Owens, F., C. P. Poole, Jr., and H. A. Farach, *Magnetic Resonance of Phase Transitions*, Academic Press, New York, 1979.

Parker, C. M., and J. W. W. Sullivan, *Ind. Eng. Chem.* **55**, 18 (1963).

Pearce, D. C., A. H. Markham, and J. R. Dillinger, *RSI* **27**, 240 (1956).

Perlson, B. D., and J. A. Weil, *RSI* **46**, 874 (1975).

Persyn, G. A., and A. W. Nolle, *Phys. Rev. A* **140**, 1610 (1965).

Philips, C. S. E., *J. Chem Phys.* **23**, 2388 (1955).

Phillips, T. R., and D. R. Owens, *JSI* **40**, 426 (1963).

Plachy, W. Z., and T. J. Schaafsma, *RSI* **40**, 1590 (1969).

Poole, C. P., Jr., Thesis, University of Maryland, 1958.

Poole, C. P., Jr., and J. F. Itzel, Jr., *J. Chem. Phys.* **41**, 287 (1964).

Poole, C. P., Jr., W. L. Kehl, and D. S. MacIver, *J. Catalysis* **1**, 407 (1962).

Poole, C. P., Jr., and D. E. O'Reilly, *RSI* **32**, 460 (1961).

Progr. Low Temperature Physics, D. F. Brewer, Ed., North-Holland Publ., a continuing series beginning in 1955.

Ranon, U., and D. N. Stamires, *RSI* **41**, 147 (1970).

Rettori, C., H. M. Kim, and D. Davidov, *Cryogenics* **14**, 285 (1974).

Reuveni, A., and S. Maniv, *RSI* **45**, 1290 (1974).

Rieckhoff, K. E., and R. Weissbach, *RSI* **33**, 1393 (1962).

Robinson, W. H., M. A. Collins, and S. D. Devine, *J. Phys. E Sci. Instrum.* **8**, 139 (1975).

Robinson, W. H., and S. D. Devine, *IEEE Trans. on Sonics and Ultrasonics* **SU-2198** (1974); *J. Phys. E Sci. Instrum.* **8**, 139 (1975); *J. Phys. C Solid State Phys.* **10**, 1357 (1977).

Robinson, W. H., and S. D. Devine, *Phys. C Solid State Phys.* **10**, 1357 (1977).

Robinson, W. H., and S. D. Devine, *Phys. Rev. B* **17**, 3018 (1978).

Roizen, L. I., and V. K. Gannus, *PTE* **2**, 191, (399) (1961).

Rose-Innes, A. C., *JSI* **34**, 276 (1957).

Rose-Innes, A. C., *Low Temperature Laboratory Techniques*, English Univ. Press, London, 1973.

Rothwarf, F., D. Ford, and L. W. Dubeck, *RSI* **43**, 317 (1972).

Rupp, L. W., Jr., P. S. Peercy, and W. M. Walsh, Jr., *RSI* **48**, 877 (1977); in press.

Sadreev, A. F., *Fiz. Nizk. Temp.* **5**, 883 (1979).

Salinger, G. L., and J. C. Wheatley, *RSI* **32**, 872 (1961).

Schoen, L. H., and H. P. Broida, *RSI* **33**, 470 (1962).

Schulte, *RSI* **36**, 706 (1965).

Schwab, G., and W. Hillmer, *Phys. Stat. Sol. B* **70**, 237 (1975).

Scott, R. B., *Cryogenic Engineering*, Van Nostrand, New York, 1959.

Serway, R. A., S. A. Marshall, and R. B. Robinson, *Phys. Stat. Sol. B* **56**, 319 (1973).

Shaulov, A. Yu., N. I. Andreeva, A. G. Sklyarova, A. L. Buchachenko, N. S. Enikolopyan, and Yu. Kh. Shaulov, *Sov. Phys. JETP* **36**, 82 (1973).

Sharrock, P., *J. Magn. Reson.* **33**, 465 (1979).

Shimashek, E., *PTE*, **4**, 173 (800) (1961).

Shoemaker, D., *Phys. Rev. B* **9**, 1804 (1974); **15** 115 (1977).

Shoemaker, D., and A. Lagendijk, *Phys. Rev. B* **15**, 115 (1977).

Singer, L. S., W. H. Smith, and G. Wagoner, *RSI* **32**, 213 (1961); see also L. S. Singer, *J. Appl. Phys.* **30**, 1463 (1959).

Sittig, M. and S. Kidd, *Cryogenic Research and Applications*, Van Nostrand, New York, 1963.

Smith, J. A. S., *High Pressure Physics and Chemistry*, Vol. 2, Academic Press, New York, 1963, p. 293.

Stankowski, J., A. Galezewski, Krupski, M., S. Waplak, and A. Gierszal, *RSI* **47**, 128 (1976).

Suassuna, J. A., G. E. Barberis, C. Rettori, and C. A. Pela, *Solid State Comm.* **22**, 347 (1977).

Suchard, S. N., and K. W. Bowers, *J. Chem. Phys.* **56**, 5540 (1972).

Svare, I., and G. Seidel, *Phys. Rev. A* **134**, 172 (1964).

Szumowski, J., and K. Falkowski, *RSI* **47**, 252 (1976).

Tanemoto, K., and T. Nakamura, *Japan J. Appl. Phys.* **17**, 1561 (1978).

Teaney, D. T., M. P. Klein, and A. M. Portis, *RSI* **32**, 721 (1961).

Timofeev, Yu. A., S. V. Kasatochkin, T. I. Alaeva, L. F. Vereschchaginm, and E. N. Yakovlev, *PTE* **14**, 164 (1776) (1971).

Tolparov, Yu. N., *PTE* **18**, 260 (1975).

Ulmer, G. C., Ed., *Research Techniques for High Pressure and Temperature*, Springer Verlag, Berlin, 1971.

Ure, R. W., Jr., *RSI* **28**, 836 (1957).

Vallauri, M. G., and P. W. Forsbergh, Jr., *RSI* **28**, 198 (1957).

van Camp, H. L., C. P. Scholes, ad R. A. Isaacson, **47**, 516 (1976).

Vance, R. W., Ed., *Cryogenic Technology*, Wiley, New York, 1964.

Vance, R. W., and W. M. Duke, Eds., *Applied Cryogenic Engineering*, Wiley, New York, 1962.

van Wieringen, J. S., and J. G. Rensen, *Proceedings of the Twelfth Colloque Ampere, Bordeaux*, North-Holland, Amsterdam, 1964, p. 229.

Vetchinkin, A. N., *PTE* **1**, 192 (1961).

Vial, J. C., and R. Buisson, *Phys. Rev. B* **12**, 405 (1975).

Von der Weid, J. P., *Rev. Bras. Fis.* **6**, 1 (1976).

von Waldkirch, Th., and K. A. Muller, *Helv. Phys. Acta* **46**, 331 (1973).

von Waldkirch, Th., K. A. Muller, and W. Berlinger, *Phys. Rev. B* **5**, 4324 (1972).

Walsh, W. M., *Phys. Rev.* **114**, 1473, 1485 (1959); **122**, 762 (1961).

Walsh, W. M., and N. Bloembergen, *Phys. Rev.* **107**, 904 (1957).

Walsh, W. M., J. Jeener, and N. Bloembergen, *Phys. Rev. A* **139**, 1338 (1965).

Waplak, S., *Radiospektrosk., Ciala Stalega*, 327 (1975).

Watkins, G. D., *Phys. Rev.* **113**, 79 (1959).

Watkins, G. D., and F. S. Ham, *Phys. Rev. B* **1**, 4071 (1970).

Weil, J. A., P. Schindler, and P. M. Wright, *RSI* **38**, 659 (1967).

Wexler, A., W. S. Corak, and G. T. Cunningham, *RSI* **21**, 259 (1950).

Wheatley, W. C., D. F. Griffing, and T. L. Estle, *RSI* **27**, 1070 (1956).

Whitehouse, J. E., T. A. Callcott, J. A. Naber, and J. S. Raby, *RSI* **36**, 768 (1965).

Windle, J. J., *J. Mag. Reson.* **22**, 487 (1976).

Wolbarst, A. B., *RSI* **47**, 255 (1976).

Yanagi, H., T. Tamura, Y. Tabata, *J. Fac. Eng. Univ. Tokyo Ser. B* **34**, 639 (1978).

Yudin, E. P., *Dokl. Akad. Nauk SSSR* **217**, 63 (1975).

Zaitov, M. M., *Cryogenics* **10**, 254 (1970).

Zhitnikov, P. A., and N. V. Kolesnikov, *Cryogenics* **5**, 129 (1965); *PTE* **3**, 189 (682) (1964).

Zhurkin, B. G., V. I. Novikov, N. A. Penin, and N. N. Sibeldin, *PTE* 166 (1970).

Zverev, G. M., *PTE* **5**, 109 (930) (1961).

CHAPTER 10

Irradiation

A. Types of Irradiation

When a fluid or solid is irradiated, the type of radiation damage that can occur is dependent on the energy of the incoming photons or other particles. Table 10-1 gives the energy associated with various bombarding photons and particles, and Table 10-2 lists the energies associated with several crystallographic and molecular quantities. The energy units used in these two tables are electron volts (eV) and kilocalories per mole (kcal/mole) since these are the two units generally employed by physicists and chemists, respectively, for most of the quantities listed in Table 10-2. Appendix A and Sec. 10-J discuss the relations between the various energy units. The energy of thermal neutrons (0.025 eV) corresponds to kT at room temperature.

The first thing to notice about Table 10-1 is that the energies associated with γ rays, X rays, and the bombarding particles (except thermal neutrons) are many orders of magnitude greater than a typical bond energy (50–100 kcal/mole). When these high-energy particles pass through matter, they gradually lose their energy by producing radiation damage such as displacing atoms from their lattice positions, ionizing atoms, and electronically exciting atoms. The last two effects are dominant in dielectrics such as ionic crystals, polymers, and glasses, while bombarding charged particles usually produce atomic displacements in metals. In liquids and noncrystalline solids, it is usually the whole molecule that is ionized or electronically excited.

Most radiation damage studies employ an external source for the irradiation of the substance under study. It is also possible to synthesize compounds that contain short-lived radioactive isotopes which spontaneously decay into different nuclei by emitting neutrons, electrons, protons, α particles, or γ rays. Such *in situ* radiation products produce radiation damage, while the new nuclei that are formed constitute impurities in the lattice. For example, Kroh et al. (1962) studied the tritium-produced radiation damage in THO. Regulla and Deffner (1976) studied dosimetry in the megarad range by ESR. Recent ESR reviews have appeared of radiation damage in ionic crystals (Kabler, 1972; Radhakrishna and Chowdari, 1977) and in organic crystals (Box and Segre, 1972; Shields, 1974; Hadley, 1980). Many radiation damage conferences have been held.

TABLE 10-1

Typical Energies per Photon or per Particle of Several Types
of Irradiation Sources

Radiation	Energy		Typical Source
	kcal/mole	eV	
γ Rays	10^6–10^8	10^5–10^7	^{60}Co
X Rays	10^3–10^6	40–40,000	X-Ray tube
Ultraviolet	70–350	3.2–15	Arc lamp
Visible	35–70	1.6–3.2	Incandescent lamp
Infrared	1–35	0.04–1.6	Incandescent lamp
Electrons	$\sim 2 \times 10^7$	$\sim 1 \times 10^6$	Van de Graaff
Protons	$\sim 4 \times 10^8$	$\sim 2 \times 10^7$	Cyclotron
Thermal neutrons	~ 0.6	~ 0.025	Nuclear reactor
Fast neutrons	$\sim 10^8$	$\sim 5 \times 10^6$	Nuclear reactor
α Particles	4×10^8	$\sim 2 \times 10^7$	Cyclotron

TABLE 10-2

Typical Energies Associated with Several Molecular and Lattice Characteristics

System	Energy		Comments
	kcal/mole	eV	
Bonding energy per nucleon in nucleus	2×10^8	8×10^6	Nuclear excited states are in this energy range
Activation energy of diffusion in ionic lattice	20	1	
Lattice energy of ionic solid	~ 200	~ 9	Approximate energy to completely remove atom from ionic lattice
F-center energy in alkali halide lattice	50–100	2–4	From optical spectra
Covalent bond energy	50–100	2–4	Organic compounds
Activation energy of thermal conductivity	$\begin{cases} 20\text{–}70 \\ 7\text{–}20 \end{cases}$	1–3 0.3–0.9	Intrinsic $\rbrace$ Alkali halide Extrinsic $\rbrace$ lattice defects
Electronic transitions of organic molecules	50–150	2–6	
Characteristic group vibrational frequencies	1–10	0.04–0.4	Vibrational frequencies of organic chemical groups
Lattice vibrations of ionic solid	0.2	10^{-2}	
Rotations of diatomic molecules	10^{-3}–10^{-2}	10^{-4}–10^{-3}	Region of microwave spectroscopy

B. Neutrons

Neutrons are produced in a nuclear reactor, and most neutron irradiation experiments are carried out at such a reactor. Neutrons may also be obtained in other ways such as by bombarding certain materials with charged particles from high-energy accelerators, and from natural radioactive sources such as the α emitter beryllium. Fast neutrons have energies of several million electron volts (MeV), while thermal neutrons are in thermal equilibrium with their environment and have energies of the order of 1/40 of an electron volt (eV).

Since neutrons are not charged particles they do not experience a Coulombic attraction or repulsion when they approach atoms or ions, but rather their collisions are of the hard sphere type at energies where the nucleus itself is not excited. Fast neutrons may collide with and knock ionized atoms out of their lattice sites. When a fast neutron passes through a solid, it leaves a trail of radiation damage behind it. The displaced atoms have a considerable amount of kinetic energy, which they may dissipate by ionization and by production of localized high-frequency lattice vibrations. The latter corresponds to a thermal spike or a localized heating of the neighborhood of the displaced atom that lasts for a very small fraction of a microsecond.

A neutron may penetrate a nucleus in its path and produce a nuclear disintegration with the emission of one or more γ rays, protons, additional neutrons, and so on. The fission fragments will go on to produce additional damage, and the transformed nucleus will become a foreign atom in the host lattice. Thermal neutrons are particularly effective in producing nuclear reactions. Fast neutrons are easily thermalized by collisions with protons in hydrogen-rich materials such as hydrocarbons, since the maximum energy transfer from a bombarding particle takes place when it strikes a particle of equal mass, as will be explained in the next section.

C. Charged Particles

The following charged particles listed in the order of increasing mass are often used for irradiation projectiles: electrons (e), positrons (e^+), protons (p or $^1H^+$), deuterons (d, D^+, or $^2H^+$), tritons (T or $^3H^+$), and α particles (α or He^{2+}). The masses of both an electron and a positron are several thousand times less than that of the other particles listed. Heavier nuclei may also be accelerated and used for irradiation sources.

The energy E_c required to displace an atom from its lattice site permanently (perhaps to an intersticial site) is about 25 eV, and for such a displacement to take place, it is necessary to transfer at least this threshold energy E_c to the atom. If the incident particle of mass M_1 possesses the energy E, then the maximum energy E_{max} transferred by a collision to an atom of mass M_2 is given by

$$E_{max} = \frac{4M_1M_2}{(M_1 + M_2)^2} E \tag{1}$$

An atomic displacement can take place when $E_{max} > E_c$. Relativistic effects must be taken into account in the case of electron bombardment, which leads to the following expression:

$$E_{max} = \frac{2(E + 2mc^2)}{M_2c^2} E \qquad (2)$$

where $M_1 = m$ is the electronic mass and c is the velocity of light. The most important rest energies are

$$mc^2 = 0.511 \text{ MeV} \quad \text{electrons} \qquad (3)$$

$$M_pc^2 = 938 \text{ MeV} \quad \text{protons} \qquad (4)$$

and for an atom of atomic weight N, one has to a high approximation

$$M_2c^2 = 931N \text{ MeV} \qquad (5)$$

Both the rate at which a charged particle loses energy in a solid, and the range of the particle are strongly dependent upon the particle's mass, as shown in Figs. 10-1 and 10-2. It is of interest to know the ranges of various particles in ESR sample tubes and samples. Ranges are usually expressed in the units mg/cm^2 or g/cm^2, and the latter may be converted to centimeters by dividing by the density ρ g/cm^3. Protons in aluminum have a range of 0.5 g/cm^2 at 17.5 MeV and 10 g/cm^2 at 100 MeV. Electrons in aluminum have a range of 1 g/cm^2 at 2 MeV and 10 g/cm^2 at 20 MeV. Since aluminum has a density of 2.7 g/cm^3, both 48-MeV protons and 5-MeV electrons have a range of 1 cm. These

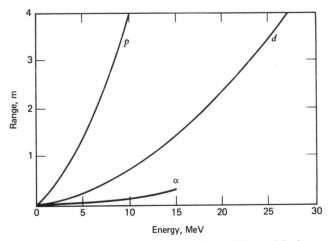

Fig. 10-1. Mean ranges of (p) protons, (d) deuterons, and (α) alpha particles in standard air as a function of their energy.

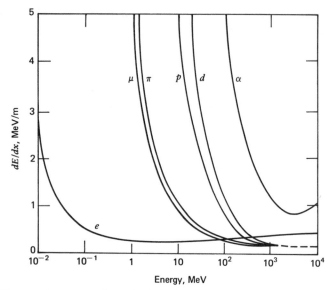

Fig. 10-2. The stopping power of standard air for (e) electrons, (μ) muons, (π) pions, (p) protons, (d) deuterons, and (α) alpha particles as a function of their energy.

data may be found in Sec. 8 of the *American Institute of Physics Handbook*. Ranges expressed in the units g/cm² are approximately independent of the particular absorber, so the above data may be employed to estimate the extent to which charged particles will penetrate shields and samples. For example, if a sample is irradiated within an ESR tube, the irradiation must be energetic enough to penetrate the sample tube and reach the specimen within.

The wavelength λ associated with a particle of mass M and energy E is

$$\lambda = \frac{h}{(2ME)^{1/2}} \tag{6}$$

where h is Planck's constant. In terms of the Compton wavelength λ_0

$$\lambda_0 = \frac{h}{mc} = 0.0242621 \text{ Å} \tag{7}$$

this becomes

$$\lambda = \lambda_0 \left(\frac{mc^2}{2E} \right)^{1/2} \left(\frac{m}{M} \right)^{1/2} \tag{8}$$

where mc^2 is given by Eq. (3). The ratio of the proton mass M_p to the electron mass m is given by

$$\frac{M_p}{m} = 1837 \tag{9}$$

Putting numerical values into Eq. (8) one obtains for $M = M_p$

$$\lambda = \frac{2.86 \times 10^{-4}}{E^{1/2}} \quad \text{for protons} \tag{10}$$

and for $M = m$

$$\lambda = \frac{1.23 \times 10^{-2}}{E^{1/2}} \quad \text{for electrons} \tag{11}$$

where the particle energy E is in MeV, and λ is in Å (10^{-8} cm). It is interesting to note that the circumference $2\pi a_0$ of the first Bohr orbit in a hydrogen atom is

$$2\pi a_0 = \frac{h^2}{2\pi mc^2} = 3.32 \text{ Å} \tag{12}$$

which is identical with the wavelength λ calculated from Eq. (11) using the energy of the first Bohr orbit or hydrogen ground state

$$E = 1 \text{ Rydberg} = \frac{e^2}{2a_0} = 13.54 \text{ eV} \tag{13}$$

Electron diffraction studies are made with wavelengths of the order of typical crystallographic spacings or several angstrom units, and the energies employed are about 100 eV. Such low-energy electrons only penetrate the first few atomic surface layers of the solid under investigation, and so electron diffraction studies are useful for the determination of the crystallographic structure of surfaces. The structure of the entire lattice is provided by X-ray diffraction since X rays of the same wavelength are so much more penetrating.

Some recent ESR studies of centers in solids irradiated with high-energy electrons are by Andrei et al. (1976), Chambaudet et al. (1976), Muller (1971), Radhakrishna and Chowdari (1976), Sagstuen and Alexander (1978), Sargent and Grady (1976), Shields (1974), Taylor et al. (1978), Shiraishi et al. (1976), and Tsuji et al. (1970). Von der Weid and Rebeiro (1974) used proton irradiation, and a number of recent articles have been published on neutron radiation damage (e.g., Radhakrishna and Chowdari, 1976; Shields, 1974; Yugo, 1977). Gupta et al. (1975) investigated pile-irradiated barite. Nakata and Morishima (1976) produced paramagnetic species by bombardment with ^{31}P ions. Gerasimenko et al. (1976) bombarded Si single crystals with ions, Pavlov et al. (1975) studied the ESR of neutron and alpha-particle irradiated SiC, Fujiwara et al. (1974) produced hot ions by α irradiation. Chambaudet et al. (1976) and Henriksen et al. (1970) employed heavy ion bombardment, Adachi and Machi (1978), Botvin et al. (1973), Brosius et al. (1974), Brower (1971), and Lee et al. (1978) used the ion implantation technique, and Zimmermann et al. (1977) studied negative ions field-emitted into liquid helium. Iwasaki et al. (1976) produced hot atoms by X irradiation.

D. Gamma Rays and X Rays

It may be seen from Table 10-1 that γ rays are photons whose wavelength λ

$$\lambda = \frac{hc}{E} = \frac{c}{f} \tag{1}$$

is usually much less than atomic spacings (0.1 to 0.4 μm), while X rays are photons whose wavelengths are often comparable to atomic spacings. The wavelengths of γ rays are much greater than nuclear radii, which lie between 2×10^{-13} and 9×10^{-13} cm. The dielectric constant ϵ and permeability μ of most materials for γ rays and X rays are close to the free-space values ϵ_0 and μ_0, respectively, so these photons will traverse matter at close to the speed of light *in vacuo* $[1/(\mu_0\epsilon_0)^{1/2}]$. Lower-frequency photons such as those in the visible region travel through matter slower than light in free space since in this case $\epsilon > \epsilon_0$.

When high-energy photons traverse matter, they become scattered (i.e., change their direction of propagation), lose intensity (i.e., the number of photons decreases), and loose energy (i.e., the photon frequency decreases) by the photoelectric effect, Compton effect, and pair production.

Gamma rays possess nuclear energies, and are produced in some types of nuclear transformations when one nucleus is converted to another. A convenient source of γ rays is ^{60}Co, which emits an electron and two γ rays of 1.332 and 1.172 MeV, and thereby decays to ^{60}Ni with the suitable half-life of 5.27 yr. The nuclear transformation follows the scheme

$$^{60}\text{Co} \rightarrow {}^{60}\text{Ni}^* + e \tag{2}$$

$$^{60}\text{Ni}^* \rightarrow {}^{60}\text{Ni} + \gamma \tag{3}$$

where the asterisk (*) indicates a ^{60}Ni nucleus in an excited nuclear energy state.

When fast-moving electrons strike matter, they interact with the matter to produce X rays with energies up to their own energy. Such bombarding electrons produce a broad, continuous X-ray spectrum known as Bremsstrahlung. This spectrum will have superimposed upon it fairly sharp regions of greater intensity corresponding to the characteristic X-ray wavelengths of the target. X Rays of a particular wavelength, called monochromatic X rays, may be produced by preferentially suppressing the continuous background relative to the sharp lines.

In practice, monochromatic X rays are produced in a Coolidge tube by thermionically emitting electrons from a heated cathode, accelerating them to a high velocity, and focusing them on to a heavy metal target. The X rays emitted by the target have energies (wavelengths) that are dependent upon the particular target element employed. X-Ray tubes are available commercially. Naryadchikov et al. (1962) present an experimental arrangement for *in situ* ESR measurements during X irradiation.

Most irradiations have been carried out at room temperature, although a number of investigators employed low-temperature radiation techniques (e.g., Ching and Box, 1977; Hesse et al., 1971; Madden and Bernhard, 1980; Truesdale et al., 1980). This can elucidate radical mobility and formation and decay kinetics (e.g., Edwards et al., 1974; Hama et al., 1979; Lyons et al., 1975; Mamedov et al., 1978; Smith et al., 1981; and Zagorets et al., 1976) and also identify primary radiation products (e.g., Adams et al., 1976; Box et al., 1973, 1976, and Morishima, 1970).

Irradiation damage studies are carried out with inorganic materials (e.g., Boate et al., 1978; da Silva et al., 1976; Hariharan and Sobhanadri; 1969, Jiang and Jui, 1976; Morton et al., 1979; Morton-Blake, 1970; Radhakrishna et al., 1976; and Yu and Chang, 1976), biological materials (e.g., Bernhard et al., 1977; Eaton and Keighley, 1971; Shimada et al., 1974; and Usmanov et al., 1978), and also in particular amino acids (e.g., Box et al., 1974; Crippa et al., 1974, 1975; Lassmann and Damerau, 1971; Minegishi et al., 1972; Rezk and Johnsen, 1978; Shields and Hamrick, 1976; and Sinclair, 1971). Some irradiations are carried out using X-ray sources (Anderson, 1977; Box et al. 1978; Claridge and Greenaway, 1972; Flossmann and Westhof, 1978; Goeffroy, 1973; Geoffroy et al., 1979; Hamrick et al., 1971; Hoffmann and Pöss, 1978; Lin and Nickel, 1972; Morton-Blake, 1970) and others using γ rays (e.g., Boate and Preston, 1978; Dodelet et al., 1970; Ermakovich et al., 1977; Faucitano et al., 1976; Griscom et al., 1976; Kiss et al., 1977; Menczel et al., 1978; Nagai and Gillbro, 1977; Wong, 1978; and Yasukawa and Matsuzaki 1969). A number of irradiated organic compounds including various amino acids have been investigated by ENDOR [compare Sec. 14-B; see, e.g., Bernhard et al. (1976), Box et al. (1975, 1975, 1976), Budzinski et al. (1975), Castleman and Moulton (1972), Ching et al. (1978), Edlund et al. (1973), Kotake and Miyagawa (1976), Kou et al. (1976), Lamotte and Gloux (1973), Muto et al. (1977), Wells and Ko (1978), and Whelan (1969)] and ELDOR [compare Sec. 14-D; see Lund et al. (1975)].

Almost all the work just discussed involved irradiated solids in the powder or single crystal form. Studies have also been carried out with irradiated solutions [e.g., see Ershov et al. (1971), Joshi et al. (1976), Kominami et al. (1976), and Mao and Kevan (1974)].

E. Ultraviolet, Visible, and Infrared Radiation

The photons in the visible and ultraviolet spectral regions, called optical photons, are treated separately from the γ rays and X rays for several reasons. In the first place, they traverse matter at velocities below the velocity of light *in vacuo* c because the relative dielectric constant ϵ/ϵ_0 exceeds unity. They may be produced without the use of specialized equipment, they may be diffracted by prisms and gratings, and they are easily separated into monochromatic beams. In addition, they may be plane-polarized or elliptically polarized. One important characteristic of optical photons that causes them to merit a separate classification is the fact that their energies are comparable to the electronic and bond energies of molecules, as shown in Table 10-3 to 10-5.

An optical photon may be absorbed by a transition from a bonding level G to an antibonding level E_A of a diatomic molecule as shown in Fig. 10-3a, and in this case, spontaneous bond rupture will result since the antibonding energy state has its lowest energy when the atoms are seperated at infinity. Another predissociation mechanism entails the absorption of a photon in a transition from the ground-state bonding level G to an excited bonding level E_B that intersects an antibonding level E_A. As the diatomic molecule in the level E_B vibrates back and forth between the internuclear separations r_0 and r_3, as shown in Fig. 10-3b, it passes through the intersection with the antibonding level E_A. At this point the molecule may cross over from the state E_B to the state E_A, and spontaneously dissociate. The process of breaking up a molecule into its component free radicals or ions by optical irradiation is called photolysis.

Figure 10-3 is drawn in accordance with the Franck–Condon principle, which states that electronic transitions occur so much more rapidly than the motions of the nuclei during molecular vibrations that the nuclei may be considered as fixed during the electronic transition. In the ground vibrational

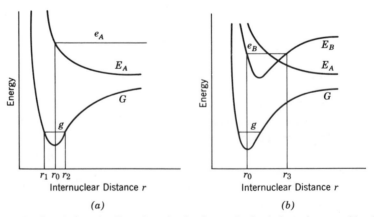

Fig. 10-3. Predissociation of a diatomic molecule when excitation is from the ground level to (a) an antibonding level and (b) a bonding level that intersects an antibonding level (Poole, 1958).

level g of the ground electronic level G, the molecule vibrates between the internuclear separations r_1 and r_2 shown in Fig. 10-3a. The electronic transitions shown in Fig. 10-3 are assumed to occur when the molecule is at the midpoint r_0.

A comparison of Tables 10-3 to 10-5 shows that in many cases the longest wavelength optical absorption maximum listed in Table 10-3 exceeds typical C—H and C—C bond energies, and when it does not do so, there are shorter-wavelength optical absorptions that do have this property. In aromatics such as benzene and naphthalene, the shorter-wavelength absorption is much more intense; that is, a much greater percentage of the shorter wavelength incident photons is absorbed. Infrared photons are less energetic than bond dissociation energies, and thus they do not break chemical bonds directly. The amount of absorption is expressed in terms of the extinction coefficient ϵ (liter/mole cm), which relates the intensity of the incident light I_0 to the transmitted I light in terms of the concentration C (mole/liter) and sample thickness d (cm) by the relation

$$I = I_0 10^{-\epsilon C d} \tag{1}$$

TABLE 10-3

Optical Absorption Maxima λ_m and Extinction Coefficients ϵ_m
for the Longest Wavelength Band
of Various Organic Compounds

Compounds	λ_m, μm	ϵ_m
Ethylene	162.5	15,000
Octene-2-*cis*	183	13,000
Cyclohexene	182	6,600
1,3-butadiene	217	21,000
2,4,6-Octatriene-1-ol	265	53,000
Octyne-1 and octyne-2	222.5	4,800
Benzene	{ 254	204 }
	{ ~ 195	6,900 }
Naphthalene	{ ~ 297.5	230 }
	{ ~ 275	9,300 }
Anthracene	~ 365	9,000
Mesitylene	270	300
1,3-Pentadiene	224	24,000
Acetaldehyde	285	15
Acetone	279.5	15
Benzaldehyde	330	20
p-Benzoquinone	{ 445	20
	{ 276	300
Nitromethane	270	15
Nitrobenzene	350	200
Phenol	270	1,450
Aniline	280	1,430

TABLE 10-4
Bond Energies E and Total Atomization Energies
of Representative Organic Compounds[a]

Compound	Bond	E_c	E_i	E	Number of Bonds	Total	Atomization Energy (kcal/mole) Calc.	Exp.
CH$_3$Cl	C—H			98.7	3	296.1	380	377
	C—Cl	60.9	22.5	83.4	1	83.4		
CH$_3$COCH$_3$	C—H			98.7	6	592.2	935	938
	C—C	84.3	0	84.3	2	168.6		
	C = O	114.3	60.2	174.5	1	174.5		
CH$_3$CH$_2$OH	C—H			98.7	5	493.5	770	771
	C—C	82.7	0	82.7	1	82.7		
	C—O	46.1	34.8	80.9	1	80.9		
	O'—H	50.2	62.3	112.5	1	112.5		
(CH$_3$CO)$_2$O	C—H			98.7	6	592.2	1330	1325
	C—C	85.4	0	85.4	2	170.8		
	C—O	72.1	38.0	110.1	2	220.2		
	C = O	113.4	59.8	173.2	2	346.4		
C$_6$H$_5$CHO	C—H			97.3	5	486.5	1582	1581
	C—H	87.6	9.1	96.7	1	96.7		
	C—C	122.6	0	122.6	6	735.6		
	C—C	85.4	0	85.4	1	85.4		
	C = O	116.1	61.2	177.3	1	177.3		
CH$_3$CH$_2$SH	C—H			98.7	5	493.5	737	733
	C—C	83.2	0	83.2	1	83.2		
	C—S	64.5	7.3	71.8	1	71.8		
	S—H	70.9	17.5	88.4	1	88.4		
(CH$_3$)$_3$N	C—H			98.7	9	888.3	1105	1100
	C—N	54.1	18.1	72.2	3	216.6		
C$_6$H$_5$CH = CH$_2$	C—H			98.7	3	296.1	1748	1753
	C—H			97.3	5	486.5		
	C—C	122.6	0	122.6	6	735.6		
	C—C	85.4	0	85.4	1	85.4		
	C = C	144.5	0	144.5	1	144.5		
CH$_3$COOCH$_2$CH$_3$	C—H			98.7	8	789.6	1319	1323
	C—C	83.4	0	83.4	2	166.8		
	C—O	45.2	34.1	79.3	1	79.3		
	C—O	72.1	38.0	110.1	1	110.1		
	C = O	113.4	59.8	173.2	1	173.2		
C$_6$H$_5$Br	C—H			97.3	5	486.5	1293	1290
	C—C	122.6	0	122.6	6	735.6		
	C—Br	57.2	14.0	71.2	1	71.2		

[a] The covalent (E_c) and ionic (E_i) contributions to the bond energies are shown. The energies are given in kcal/mole (Sanderson, 1976).

TABLE 10-5

Standard Bond Energies—Experimental
and Calculated (Sanderson, 1976)

Bond	Experimental (kcal/mole)	Calculated (kcal/mole)
C—H	99	98
N—H	93	93
P—H	76	76
O—H	111	113
S—H	82	87
C—C	83	83
C = C	146	144
C—O'	84	84
C = O''	192	192
C—S	65	71
C—Cl	78–81	84
C—Br	66–68	71
C—I	51–57	55
Si—Cl'''	86	94
N—F	64–65	67
P—Cl''	79	79
P—Br'	65	63
P—I'	51	47
O—F	44–45	48
O—Cl	48–52	49
S—Cl	60	66

Some authors define an absorption coefficient α in terms of the equation

$$I = I_0 e^{-\alpha d} \tag{2}$$

and, of course,

$$\alpha = 2.303 \epsilon C \tag{3}$$

Table 10-3 does not include the shorter wavelength absorption maxima, and it is important to emphasize that in many cases they have much greater extinction coefficients than the longest wavelength transition that is tabulated.

Ultraviolet light sometimes produces free radicals through a photosensitization mechanism whereby one molecular species absorbs the photons and passes the energy on to another species that dissociates into two free radicals. For example the photolysis of H_2O_2 resulted in (1) the absorption of a photon by H_2O_2 with the production of OH radicals, and (2) the extraction of a hydrogen atom from the surrounding rigid solvent with the formation of H_2O and the free radical (Gibson et al., 1957; Smith and Wyard, 1961; Rustigi and Riesz, 1978).

After dissociation occurs, the "cage effect" causes the photolysis products to become trapped by the surrounding medium at the site of the photodecomposition, and thereby to recombine rapidly. This is the main quantum yield limiting mechanism since as a result only a very small fraction of the incident photons actually produces stabilized free radicals. This cage effect is less important with higher-energy radiation, which is more capable of transferring to the free-radical fragment sufficient kinetic energy to permit its escape from the "cage."

Von der Weid (1974) described a cryostat for optical and X-band measurements, and microwave cavities for *in situ* irradiation were designed by Chamel et al. (1976), Imbusch et al. (1967), and Piette and Landgraf (1960). The modular cavity resonators of Berlinger and Müller (1977) that were discussed in Sec. 5-P may be employed for optical irradiation studies. Representative recent work on photolytically produced free radicals was published by Cook and Vincent (1977), Gazzinelli and Schirmer (1977), and Nemec and Monig (1975). Möhwald and Sackmann (1973) found 0.1-G-wide lines in triplet excitons and used optical emission spectra to aid in the interpretation. Richter et al. (1976) formed radicals by two-photon absorption, and Hutchison and Scott (1974) determined the zero-field levels of the photoexcited triplet state in naphthalene. Ayscough et al. (1976) studied ESR kinetics using pulse photolysis, Trifunac et al. (1975) reported submicrosecond pulse radiolysis by time-resolved ESR, and Yamashita et al. (1975, 1975, 1976) employed laser radiation. McLachlan and Sealy (1976) published *Photochemistry in a Spin*, Wosinski et al. (1976, 1977) investigated spin-dependant photoconductivity in silicones, and Shimada et al. (1975) caused the ESR spectrum arising from trapped β-cyclopentanyl radicals to narrow by irradiating it with the infrared band corresponding to the C—H stretching frequency. Sprague and Willard (1975) sharpened the spectra from radicals by uv irradiation; Kaufmann and Koschel (1978) studied the ESR and photoluminescence of Cr^{3+} in GaP, and Muramoto (1973) optically detected the excited $\bar{E}(^2E)$ state of Cr^{3+} in Al_2O_3.

Some typical ultraviolet radiation damage studies were carried out by Anpo et al. (1980), Atkins et al. (1975), Bluhn and Weinstein (1976), Bowers et al. (1976), Carstensen (1970), Devolder and Goudmand (1975), Hellecbrand and Wuensche (1976), Hudson et al. (1976), Ikegamt and Watanabe (1976), Lind and Loeliger (1976), Muto and Iwasaki (1973), Owens and Vogel (1976), Sakurai et al. (1976), and Smith et al. (1978); and related experiments on biological samples were published by Azizova et al. (1979), Blum et al. (1978), Islomov et al. (1979), Kipnis et al. (1972), Kruglyakova and Zhil'tsova (1979), Meybeck (1979, 1979), Odinokova et al. (1974, 1979), Yamanashi and Zuclich (1978), and Zhil'tsova et al. (1979).

High-power infrared (ir) radiation is capable of heating solids and fluids to high temperatures, and as a result many free radicals and lattice defects will anneal out by migration, recombination, and neutralization processes. Ultraviolet (uv) and visible light sources frequently emit a considerable amount of infrared radiation, and it is best to interpose a filter between the lamp and the sample to remove the ir. For high powers, it may be necessary to water-cool the filter. A convenient method of filtering out infrared radiation is to employ two quartz plates cemented to a Pyrex cylinder by means of black wax or some other cement. Such a cylinder filled with distilled water forms an effective filter, and it may be prevented from overheating by the use of flowing water.

F. Gas Discharges

A commonly used method of producing free radicals in the gaseous state is by means of a gas discharge (Howatson, 1976). Such a discharge may be produced by a high voltage (e.g., 2500 V) between two electrodes in a discharge tube. An electrodeless discharge may be produced by winding a coil around the glass tubing that carries the gas and connecting it to a high-power transmitter with a frequency of several megahertz. A higher-frequency electrodeless discharge may be produced by placing the evacuated quartz tube in an S-band resonant cavity and supplying several hundred watts of microwave power to the cavity from a diathermy unit, or from the magnetron of a microwave oven. Experimental details pertaining to these methods of producing gas discharges and references to the literature are given in Sec. 17-F and Table 17-8 of the first edition.

Most discharge-tube experimental setups pump the gas through both the discharge tube and the resonant cavity. It is important to allow sufficient space between the discharge and the resonant cavity so that the discharge is not actually in the cavity. Sometimes a very broad cyclotron resonance may be observed in the discharge itself. On the other hand, it is also important not to place the resonant cavity too far from the discharge since this would allow most of the atoms to recombine before reaching the cavity. An inert carrier gas such as argon may be employed to prevent recombination of the radicals in the gas, and the wall of the tube between the discharge and the resonant cavity may be coated with a substance to prevent the destruction of free radicals during surface collisions, as described in the first edition. The products of electric discharges may be condensed on cold surfaces at liquid-helium temperature and studies may be made of the ESR spectra of the trapped atoms and free radicals. Dewar–cavity systems for this purpose are described in Sec. 9-C.

Molecules may be dissociated thermally by heating them in an oven to a high temperature (e.g., 2500 K). This dissociation method has been used extensively in atomic and molecular beam experiments.

G. Time and Temperature Effects

When a material is irradiated, the rate of production of radiation damage is greatest at the start of the irradiation, and gradually decreases with time. The accumulated or integrated radiation damage, on the other hand, monotonically or continuously increases with irradiation time and eventually approaches an asymptotic value for prolonged irradiations. An example of this behavior is shown in Fig. 17-5 of the first edition. See also Adrian et al. (1975), Alger et al. (1959), Chambaudet et al. (1976), Iwasaki et al. (1976), Lee et al. (1976), Nakata and Morishima (1976), Parrot et al. (1977), Von der Weid and Ribeiro (1974), and Yamashita and Kashiwagi (1975). Yamaga et al. (1978) found that holes self-trapped at low temperature became mobile at higher temperatures and moved to new sites where they were retrapped.

When a particular system is under investigation, it is best to obtain a graph of the spin concentration as a function of time in order to decide upon the best exposure time. The approach to an asymptotic number of spins at long irradiation times is sometimes referred to as a saturation phenomenon, and this should not be confused with the ESR saturation that occurs at high microwave powers, as discussed in Sec. 13-C. Sometimes one observes a more complex dependence of the spin concentration on irradiation time as illustrated by Fig. 17-6 of the first edition. For example, in phosphorous ion-bombarded silicon (Nakata and Morishima, 1976) a change in dosage by four orders of magnitude increased the total number of spins by a factor of 10, and a plot of relative number of spins versus dose exhibited structure. Von der Weid and Ribeiro (1974) found that the saturation level of spins produced in CaF_2 and SrF_2 by proton bombardment increased with the incident proton energy. One radical species can predominate after short irradiation times and another after long times. A number of studies have been made of short-lived free radicals using techniques such as flash photolysis or pulse radioloysis and time-resolved ESR (Ayscough et al., 1976; Fessenden et al., 1973, 1977; Kominami et al., 1976; Sargent and Grady, 1976; Trifunac et al., 1975; Yamashita, 1976). Hsi et al. (1973) constructed a rapid scan apparatus for flash photolysis ESR. Earlier flash photolysis ESR apparatus has been reported by Atkins et al. (1970, 1975), Dutton et al. (1972), Hales and Bolton (1970, 1972), Hirasawa et al. (1968), Ohno and Sohma (1969), Sohma et al. (1968), and Warden and Bolton (1972). Photofragmentations (Steenken et al., 1975, 1975; Yasukawa et al., 1976) redox reactions (Bolton et al., 1975; Bukowka and Zecki, 1975; Forster et al., 1976; Voznyak et al., 1976; Vyas et al., 1976) and other photoinduced changes (Fauguenoit and Claes, 1971; Fujimoto and Saito, 1972; Fujiwara et al., 1973, 1974; Hama and Shinohara, 1970; Lin, 1973; Moan, 1975) have been investigated using ultraviolet radiation.

The rate of production of paramagnetic centers may be expressed in terms of the quantum yield or number of centers produced per incident photon (or particle). Optical irradiation rates are ordinarily expressed in terms of quantum yield, and this is generally very much less than one since a given optical photon usually does not have enough energy to produce more than one bond rupture, and the cage effect makes immediate recombination much more likely than radical stabilization. One should recall that the 253.7-μm mercury resonance line corresponds to 113 kcal/mole or 4.9 eV if one wishes to compare the quantum yield unit to the G value defined in the next paragraph.

High-energy photons (X rays and γ rays) and high-energy particles (electrons, protons, etc.) are capable of undergoing many ionizing and replacement collisions as they traverse matter, and the primary radiation products usually are energetic enough to produce secondary radiation damage, with the result that quantum yields may greatly exceed unity. For such high-energy radiations it is customary to express the rate of radiation damage in terms of the G value or number of free radicals formed per 100 eV of energy absorbed in a material. It was mentioned earlier that the threshold energy per atomic displacement is about 25 eV, so one expects the G value to be of the order of unity. Such G values are often observed.

After termination of the irradiation, the ESR signal will often decay with time. In most cases it merely decays, while in other cases it is converted to an ESR signal from a new paramagnetic species. The decay is referred to as annealing, and usually one may accelerate the annealing process by heating the specimen, (Botvin et al., 1973; Close et al., 1977; Frick and Siebert, 1977; Gewinner et al., 1975; Halliburton et al., 1977; Hori and Kispert, 1978; Kawabata, 1976; Lee et al., 1976; Livingston et al., 1955; Shimizu et al., 1977; Taylor et al., 1978; Vakhidov et al., 1974; Yugo, 1977) as illustrated in Fig. 17-7 of the first edition. This figure demonstrates that a small temperature change is often capable of changing the rate of decay by more than an order of magnitude. Sometimes one can produce annealing by exposure to ultraviolet or visible light, a process referred to as bleaching since it is often employed to decolorize alkali halide crystals by the removal of color centers.

There is also a spacial dimension to radiation damage, and sometimes the paramagnetic centers are not uniformly distributed throughout the sample. Gerasimenko et al. (1976) presented plots of the number of defects produced in silicon monocrystals versus the depth in the sample and also versus the radiation dose. Iwasaki et al. (1976) found that X irradiation of decane single crystals produced alkyl radicals in pairs with a spatial distribution that depends upon the irradiation temperature. Deuteration was found to reduce the spacial distribution. Zaikin et al. (1976) studied surface state traps in PbS monocrystals. Kodzhespirov et al. (1974) determined the optical and thermal depths of photosensitive paramagnetic centers.

ESR is useful for studying the reaction kinetics involved in the formation and decay of free radicals. For first-order kinetics the concentration of free radicals decays exponentially with time. Thus if the sample contains N_0 atoms at the termination of the irradiation, then the number that remains after a time t is given by

$$N = N_0 e^{-kt} \tag{1}$$

where k is a measure of the decay rate. In typical cases the rate constant k decreases with increasing temperature in accordance with the relation

$$k = k_0 \exp\left(-\frac{\Delta E}{K_B T}\right) \tag{2}$$

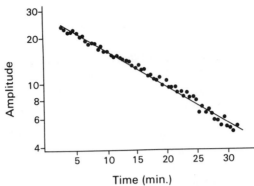

Fig. 10-4. Kinetics of the free-radical decay in irradiated α-alanine at 162 K showing the linear decrease in the logarithm of the ESR amplitude with time. The slope of the line gives the rate constant k in accordance with Eq. (10-G-3) (Smith et al., 1981).

where ΔE is the activation energy for the recombination or decay process. A two-step procedure is followed to determine the activation energy. First $k(T)$ is found for various temperatures by plotting the logarithm of the relative ESR signal amplitude Y_0/Y versus the time as shown in Fig. 10-4:

$$\log\left(\frac{Y}{Y_0}\right) = -kt \tag{3}$$

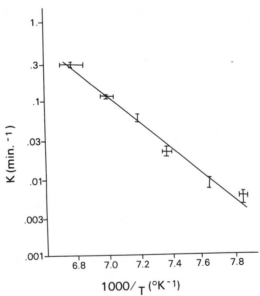

Fig. 10-5. Arrhenius plot of the logarithm of the rate constant k for free-radical decay in irradiated glycine versus the reciprocal temperature. The slope of the straight line gives the activation energy ΔE for the radical decay process in accordance with Eq. (10-G-4). (Smith et al., 1981).

since Y_0/Y is proportional to N_0/N in Eq. (1). The slope of each straight line gives the rate constant k at a particular temperature. Then the logarithms of k are plotted versus $1/T$ as illustrated in Fig. 10-5:

$$\log k = \log k_0 - \frac{\Delta E}{k_B T} \qquad (4)$$

to determine the frequency factor k_0 and the activation energy ΔE from the intercept and slope, respectively, of the resultant straight line.

A number of workers have studied the kinetics of radical formation and decay (e.g., Andrew et al., 1974; Che and Tench, 1976; Chong and Itoh, 1973; Eliav and Levanon, 1976; Geoffroy et al., 1976; Ikeya, 1976; Nadzhafova and Sharpatyi, 1978; Nemec and Monig, 1974, 1977; Sanner, 1970; Smith et al., 1981; and Yamashita and Kashiwagi, 1975). Some studies of this type have examined the reaction kinetics involving the decay of one radical into another (Clark et al., 1976; Dulcic and Herak, 1977; Gregoli et al., 1977; Haak and Benson, 1976; Saxebol and Sagstuen, 1974; Shields and Hamrick, 1976). To describe the overall kinetics it is necessary to write coupled equations for the two radicals.

Sometimes selective bleaching can be employed to remove one radical species that masks the presence of another (e.g., Halliburton et al., 1977; Kroh et al., 1975, 1977; Nauwelaerts et al., 1977; and Suryanarayana and Sobhanadri, 1974). Bleaching can also transform one radical species into another (e.g., Flossmann et al., 1976; Iacona et al., 1977; Zehner et al., 1976). Pung and Lushchik (1976) optically bleached nonparamagnetic centers to produce paramagnetic photodissociation products. Other bleaching data are reported by Cheema and Smith (1971, 1977), Kroh and Plonka (1975), and Watterich and Raksanyi (1977). Weiter and Finch (1975) studied free radical effects in the photodamage of human lenses that leads to cataract formation.

H. Experimental Techniques

If one is not worried about the effect of air, a sample may be irradiated first, and then afterwards placed in an ESR tube. One often prefers to carry out the irradiation *in vacuo* or in a particular atmosphere, and in this case the sample may be sealed in an ESR tube before beginning the irradiation. Quartz ESR tubes are transparent in the near uv from the visible to about 220 μm. When high-energy radiation is employed, it will produce paramagnetic centers in the ESR tube, as is evidenced by the pronounced coloration of the tube. One may anneal away these centers by shaking the sample to the upper end of the ESR tube and heating the lower end with a Bunsen burner or torch, taking care not to warm the sample. The annealing process will decolorize the tube and render it completely transparent. The efficacy of the annealing technique may be tested by recording and comparing ESR spectra of unirradiated, irradiated, and annealed ESR tubes. At the end of this chapter are listed several references (Anderson and Weil, 1959; Castle et al., 1963; Dienes, 1960; Friebele et al.,

1974; Fröman et al., 1959; Griscom et al., 1976; Hines and Arndt, 1960; Mackey, 1963; Shamfarov and Smirnova, 1963; Silsbee, 1961; van Wieringen and Kats, 1963; Weeks, 1963) to the spectra of irradiated quartz since quartz is the material customarily employed for constructing ESR tubes. Brown (1975) studied gamma-irradiated pyrex.

Low-temperature irradiation may be carried out by placing ESR tubes in a Dewar under an ice–water mixture, a Dry Ice–acetone bath, or liquid nitrogen. Care should be exercised not to have too thick a layer of refrigerant above the sample when employing irradiation with low penetrating power. The transmittance of distilled water is 85% down to 200 μm. Near-ultraviolet radiation can also penetrate several centimeters of liquid nitrogen.

When silica ESR sample tubes are irradiated with 2-MeV electrons from a van de Graaff generator at 77 K and then removed from the liquid-nitrogen bath, they gradually warm up, and during this warmup process they go through a certain temperature range where they glow. This glowing is an example of thermoluminescence. The ESR tubes are still darkly colored when they reach room temperature and must be heated in a Bunsen burner to completely anneal away the coloration. Examples of the correlation between ESR and thermluminescence studies in other materials are given by Doi (1977), Gromov et al. (1975), Ikeya (1976), Kalder et al. (1976), Lopez et al. (1977), Mori (1977), Moskalonov (1977), Nambi et al. (1974), Riggin et al. (1976), and Vainer et al. (1976).

Some workers have recorded ESR spectra while simultaneously irradiating the sample with electrons (Fessenden and Schuler, 1963; Molin et al., 1960; Sargent and Grady, 1976; Shiraishi et al., 1976), gamma rays (Murabayashi et al., 1977), X rays (Naryadchikov et al., 1962), ultraviolet light (Piette and Landgraf, 1960), and other irradiations. A light pipe may be employed to conduct visible and ultraviolet radiation to the sample position for such studies. Slots in the cavity endplate may also be used for this purpose as discussed in Sec. 5-C.

I. Effect of Radiation on Materials

High-energy irradiation produces lattice damage that alters a number of physical and chemical properties of the irradiated material. This section briefly mentions some of the principal effects of irradiation. The comments made are very general, and are subject to numerous exceptions. Different results are often obtained with neutrons, charged particles, γ rays, and X rays.

Electrical Properties

The electrical resistivity of most materials is strongly depended upon the irradiation time.

Optical Properties

Many of the centers produced by irradiation have ultraviolet, visible, and infrared absorption bands, and a study of these bands helps to differentiate and identify the various centers. It is often possible to selectively anneal a specific center by irradiation at a particular band of optical wavelengths. The various optical bands often anneal at different temperatures. One sometimes observes luminescence (optical irradiation at one wavelength with emission at another), phosphorescence (delayed luminescence), electroluminescence (luminescence induced by the application of an electric field), and other forms of luminescence. Many investigators have studied the optical absorption spectra and ESR simultaneously, such as Bartram et al. (1976), Bogomolova et al. (1978), Chong and Itoh (1973), Clark et al. (1976), Coufal et al. (1975), da Silva et al. (1976), Hahn et al. (1975), Hattori and Muto (1974), Hiratsuka et al. (1976), Mehendru and Kumar (1975), Moan and Kaalhus (1974), Murti et al. (1976), Radhakrishna et al. (1974, 1976), Slifkin et al. (1977), Sportelli et al. (1977), Suryanarayana and Sobhanadri (1974). Von der Weid and Ribeiro (1974) correlated ESR and optical dichroism measurements on free radicals.

Dimensional Changes

The bulk density tends to decrease with irradiation. Part of this may be due to lattice expansion in the neighborhood of interstitial atoms. Anisotropy effects often occur; for example, during the irradiation period, graphite expands considerably along the crystallographic (interplanar) c direction and contracts slightly perpendicular to it.

Stored Energy

Energy is stored in the lattice during irradiation and released during annealing.

Mechanical Properties

Mechanical properties such as strength, ductibility, hardness, brittleness, and so on are strongly influenced by irradiation. The effect of irradiation on the mechanical properties is of considerable practical importance in reactor design, and other applications.

Magnetic Properties

Many radiation-produced centers are paramagnetic and contribute to the overall magnetic susceptibility. Structure-sensitive properties such as the permeability are more affected by irradiation than nonstructure-sensitive properties such as saturation magnetization.

Crystal Structure

Radiation tends to render a crystal lattice more disordered, and sometimes gross changes in structure are produced.

Surface Properties

The catalytic and adsorption properties, as well as corrosion resistance, are sensitive to irradiation.

Chemical Properties

Polymeric and catalytic properties are sensitive to irradiation.

J. Units and Definitions of Terms

There are several systems of units currently employed in the irradiation literature. This section defines the more common units and relates them to each other.

Quantum Yield

The number of free radicals or other irradiation products produced per incident photon.

G Value

The number of free radicals or other irradiation products produced per 100 eV of energy absorbed from the incident beam of radiation.

Extinction Coefficient ϵ

This is defined by Eq. (10-E-1). It has the units liter/mole cm, where C is in moles per liter and d is in centimeters. This unit is often employed in optical irradiation studies.

Absorption Coefficient α

This is defined by Eq. (10-E-2) and has the units cm^{-1}. It is often referred to as the linear absorption coefficient. One should note that $\alpha = 2.303\epsilon C$.

Optical Density ϵCd

This is defined by Eq. (10-E-1).

Mass Absorption Coefficient α_m

This is also employed for high-energy radiation, and is defined in terms of the density ρ g/cm^3 by the relation

$$\alpha_m = \frac{\alpha}{\rho} \tag{1}$$

It has the units cm^2/g, and tends to be roughly independent of the nature of the absorbing material. The areal density ρd has the units g/cm^2, and for thin films it is easily measured by weighing. Sometimes α_m is referred to simply as an absorption coefficient.

Rad

This corresponds to the absorption of 100 erg/g in a material. A megarad equals 10 J/g. This is the unit of absorbed dose.

Roentgen (R)

This corresponds to the release of 2.58 C/kg of electrical charge in air under standard conditions. This is the unit of exposure.

Rutherford (rd)

This is defined as a disintegration rate of 10^6 disintegrations/sec.

Curie (Ci)

The curie was originally defined as the amount of radon in equilibrium with 1 g of radium, but its current definition is 3.7×10^{10} disintegrations/sec. It has its submultiples the millicurie (mCi) and microcurie (μCi).

Relative Biological Effectiveness (RBE)

A dimensionless number is assigned to each type of radiation; namely, 1 for electrons, X rays, and γ rays, 10 for neutrons, and 20 for α particles.

References

Adachi, S., and Y. Machi, *Japan J. Appl. Phys.* **17**, 229 (1978).

Adams, S. M., E. E. Budzinski, and H. C. Box, *J. Chem. Phys.* **65**, 998 (1976).

Adrian, F. J., V. A. Bowers, and E. L. Cochran, *J. Chem. Phys.* **63**, 919 (1975).

Alger, R. S., T. H. Anderson, and L. A. Webb, *J. Chem. Phys.* **20**, 695 (1959).

Anderson, R. S., *J. Chem. Phys.* **66**, 5610 (1977).

Anderson, J. H., and J. A. Weil, *J. Chem. Phys.* **31**, 427 (1959).

Andrei, E. Y., A. Katzir, and J. T. Suss, *Phys. Rev. B* **13**, 2831 (1976).

Andrew, E. R., H. J. Gale, W. S. Hinshaw, W. Vennart, P. S. Allen, E. P. Andrew, and C. A. Bates, *Proc. of the 18th Ampere Congress on Magnetic Resonance and Related Phenomena* **1**, 271 (1974).

Anpo, M., C. Yun, and Y. Kubokawa, *J. Catal.* **61**, 267 (1980).

Atkins, P. W., A. J. Dobbs, and K. A. McLauchlan, *J. Chem. Soc. Faraday Trans. 2* **71**, 1269 (1975).

Atkins, P. W., K. A. McLachlan, and A. F. Simpson, *J. Phys. E* **3**, 347 (1970).

Ayscough, P. B., T. H. English, and D. A. Tong, *J. Phys. E* **9**, (1976).

Azizova, O. A., A. I. Islomov, D. I. Roshchupkin, D. A. Predvoditelev, A. N. Remizov, and Yu. A. Vladimirov, *Biofizika* **24**, P396 (1979).

Bartram, R. H., L. A. Kappers, and G. G. DeLeo, *Phys. Rev. B* **14**, 5482 (1976).

Bernhard, W. A., J. Huttermann, A. Muller, D. M. Close, and G. W. Fouse, *Radiat. Res.* **68**, 390 (1976).

Bernhard, W. A., D. M. Close, J. Huettermann, and G. Zehner, *J. Chem. Phys.* **67**, P1211 (1977).

Berlinger, W., and K. A. Müller, *RSI* **48**, 1161 (1977).

Bluhn, A. L., and J. Weinstein, *Spectrosc. Lett.* **8**, 43 (1976).

Blum, H., J. C. Salerno, and J. S. Leigh, Jr., *J. Magn. Reson., Proc. 6th Int. Symp. Magn. Reson.* **30**, 385 (1978).

Boate, A. R., J. R. Morton, and K. F. Preston, *J. Phys. Chem.* **82**, 718 (1978).

Boate, A. R., and K. F. Preston, *Inorg. Chem.* **17**, 1669 (1978).

Bogomolova, L. D., V. A. Jachkin, V. N. Lazukin, T. K. Pavlushkina, and V. A. Shmuckler, *J. Non-Crystalline Solids* **28**, 375 (1978).

Bolton, J. R., K. S. Chen, A. H. Lawrence, and P. Demayo, *J. Am. Chem. Soc.* **97**, 1832 (1975).

Botvin, V. A., Yu. V. Gorelkinskii, V. A. Kudryashev, and V. O. Sigle, *Fiz. Tekh. Poluprovodn.* **8**, 1614 (1973).

Bowers, P. R., K. A. McLachlan, and R. C. Sealy, *J. Chem. Soc. Perkin Trans.* **2**, 915 (1976).

Box, H. C., and E. E. Budzinski, *J. Chem. Phys.* **59**, 1588 (1973).

Box, H. C., and E. E. Budzinski, *J. Chem. Phys.* **62**, 197 (1975); **64**, 1593 (1976).

Box, H. C., E. E. Budzinski, and H. C. Freund, *J. Chem. Phys.* **61**, 2222 (1974).

Box, H. C., E. E. Budzinski, and K. T. Lilga, *J. Chem. Phys.* **64**, 4495 (1976).

Box, H. C., W. R. Potter, and E. E. Budzinski, *J. Chem. Phys.* **62**, 3476 (1975).

Box, H. C., and E. Segre, *Ann. Rev. Nucl. Sci.* **22**, 355 (1972).

Box, H. C., E. E. Budzinski, and G. Potlienko, *J. Chem. Phys.* **69**, 1966 (1978).

Brosious, P. R., J. W. Corbett, and J. C. Bourgoin, *Phys. Status Solidi A* **21**, 677 (1974).

Brower, K. L., *Phys. Rev. B* **4**, 1968 (1971).

Brown, G., *J. Mater. Sci.* **10**, 1481 (1975).

Budzinski, E. E., and H. C. Box, *J. Chem. Phys.* **63**, 4927 (1975).

Bukowka, J., and Z. Zecki, *Rocz. Chem.* **49**, 11 (1975).

Carstensen, P., *Acta Polytech. Scandinavia* (*Sweden*) *Ch.* **97**, 3 (1970).

Castle, J. G., Jr., D. W. Feldman, P. G. Klemens, and R. A. Weeks, *Phys. Rev.* **130**, 577 (1963).

Castleman, B. W., and G. C. Moulton, *J. Chem. Phys.* **57**, 2762 (1972).

Chambaudet, A., A. Bernas, and J. Roncin, *Radiat. Eff. 9th Int. Conf. Solid State Nucl. Track Detectors* **34**, 57 (1976).

Chamel, M., R. Chicault, and Y. Merle Baubigne, *J. Phys. E* **9**, 87 (1976).

Che, M., and A. J. Tench, *J. Chem. Phys.* **64**, 2370 (1976).

Cheema, S. V., and M. J. A. Smith, *J. Phys. C Solid State* **4**, 1231 (1971).

Cheema, S. V., and M. J. A. Smith, *Solid State Comm.* **24**, 686 (1977).

Ching, L. K., and H. C. Box, *J. Chem. Phys.* **68**, 5357 (1978); **67**, 2811 (1977).

Chong, T., and N. Itoh, *J. Phys. Soc. Japan* **35**, 518 (1973).

Claridge, R. F. C., and F. T. Greenaway, *J. Magn. Reson.* **8**, 316 (1972).

Clark, C. D., E. W. J. Mitchell, N. B. Urli, and J. W. Corbett, *Int Conf. Radiat. Effects Semicond.* **12**, 45 (1976).

Close, D. M., R. A. Farley, and W. A. Bernhard, *Radiat. Res.* **73**, 212 (1977).

Cook, W. T., and J. S. Vincent, *J. Chem. Phys.* **67**, 5766 (1977).

Coufal, H. J., E. Luscher, M. Krusius, and M. Vuorio, *Proceedings of the International Conference on Low Temperature Physics*, 14 (1975).

Cottrell, T. L., *The Strengths of Chemical Bonds*, 2nd ed., Butterworths, London, 1958.

Crippa, P. R., and E. Loh, *Radiat. Eff.* **25**, 13 (1975).

Crippa, P. R., R. A. Tedeschi, and A. Veoli, *Int. J. Radiat. Biol.* **25**, 497 (1974).

Da Silva, E. C., G. M. Gaulberto, H. Vargas, and C. Rettori, *Chem. Phys. Lett.* **37**, 138 (1976).

Da Silva, E. C., G. M. Gualberto, C. Rettori, H. Vargas, M. E. Foglio, and G. E. Barberis, *J. Chem. Phys.* **65**, 3461 (1976).

Devolder, P., and P. Goudmand, *C. R. Hebd. Seances Acad. Ser. C* **280**, 1281 (1975).

Dienes, G. J., *J. Phys. Chem. Solids* **13**, 272 (1960).

Dodelet, J. P., C. Fauguendit, M. Siquet, and P. Claes, *Ann. Soc. Sci. Bruxelles, Ser. 1* **84**, 107 (1970).

Doi, A., *Japan J. Appl. Phys.* **16**, 925 (1977).

Dulcic, A., and J. N. Herak, *Radiat. Res.* **71**, 75 (1977).

Dutton, P. L., J. S. Leigh, and M. Seibert, *Biochem. Biophys. Res. Comm.* **46**, 406 (1972).

Eaton, W. C., and J. H. Keighley, *Appl. Polym. Symp.* **18**, 263 (1971).

Edlund, O., A. Lund, and A. Graslund, *J. Magn. Reson.* **10**, 7 (1973).

Edwards, J., S. P. Mishra, and C. R. Martyn Symons, *J. Chem. Soc. Chem. Commun.*, 556 (1974).

Eliav, U., and H. Levanon, *Chem. Phys. Lett.* **55**, 369 (1976).

Ermakovich, K. K., V. N. Lazukin, V. M. Tatarintsev, and I. V. Chapeleva, *Sov. Phys. Solid State* **19**, 11 (1977).

Ershov, B. G., A. I. Mustafaev, and A. K. Pikaev, *Int. J. Radiat. Phys. Chem.* **3**, 71 (1971).

Fauguenoit, C., and P. Claes, *Bull. Soc. Chim. Belg.* **80**, 323 (1971).

Faucitano, A., A. Buttafav, and F. F. Martinotti, *J. Chem. Soc. Perkin Trans.* **2**, 1014 (1976).

Fessenden, R. W., G. E. Adams, E. M. Fieldem, and B. D. Michael, *Proc. of the 5th L. H. Gray Conf. on E. Proc. in Radiat. Chem. and Biol.* **6**, 60 (1973).

Fessenden, R. W., and R. H. Schuler, *J. Chem. Phys.* **39**, 2147 (1963).

Fessenden, R. W., L. T. Muus, P. W. Atkins, D. A. Mclauchlan, and J. B. Pederson, *Chem. Induced Magn. Polarization* **12**, 119 (1977).

Flossmann, W., and E. Westhof, *Radiat. Res.* **73**, 75 (1978).

Flossmann, W., E. Westhof, and A. Muller, *J. Chem. Phys.* **64**, 1688 (1976).

Forster, M., K. Loth, M. Andrist, U. P. Fringeli, and H. H. Gurnthard, *Chem. Phys.* **17**, 59 (1976).

Frick, B., and D. Siebert, *Phys. Status Solidi A* **41**, (1977).

Friebele, E. J., D. L. Griscom, and G. H. Siger, Jr., *J. Appl. Phys.* **5**, 3424 (1974).

Fröman, P. O., R. Petter-son, and T. Vanngard, *Ark. Fys.* **15**, 559 (1959).

Fujimoto, M., and S. Saito, *J. Chem. Phys.* **57**, 3704 (1972).

Fujiwara, M., N. Tamura, and H. Hirai, *Bull. Chem. Soc. Japan* **46**, 701 (1973).

Fujiwara, S., P. S. Allen, E. R. Andrew, and C. A. Bates, *Proc. of the 18th Ampere Congress on Magnetic Resonance and Related Phenomena* **1**, 9 (1974).

Gazzinelli, R., and O. F. Schirmer, *J. Phys. C* **10**, 145 (1977).

Geoffroy, M., *Helv. Chim. Acta* **56**, 1552 (1973).

Geoffroy, M., L. Ginet, and E. A. C. Luche, *Chem. Phys. Lett.* **38**, 327 (1976).

Geoffroy, M., and A. Llinares, *Helv. Chim. Acta* **62**, 1605 (1979).

Gerasimenko, N. N., G. A. Gudaev, A. V. Dvurechenskii, L. S. Smirnov, and S. A. Sokolov, *Fiz. Tekh. Poluprovodn.* **10**, 1237 (1976).

Gewinner, G., L. Kubler, J. J. Koulmann, and A. Jaegle, *Phys. Stat. Sol. B* **70**, 595 (1975).

Gibson, J. F., D. J. E. Ingram, M. C. R. Symons, and M. G. Townsend, *Trans. Faraday Soc.* **53**, 914 (1957).

Gregoli, S., M. Olast, and A. Bertinchamps, *Radiat. Res.* **70**, 255 (1977).

Griscom, D. L., G. H. Sigel, Jr., and R. J. Ginther, *J. Appl. Phys.* **47**, 960 (1976).

Gromov, V. V., L. G. Karaseva, and E. I. Saunin, *J. Radioanal. Chem. 8th Radiochemical Conference* **30**, 441 (1975).

Gupta, N. M., J. M. Luthra, and M. D. Sastry, *J. Lumin.* **10**, 305 (1975).

Haak, R. A., and B. W. Benson, *Radiat. Res.* **68**, 381 (1976).

Hadley, J. H. Jr., *Magn. Reson. Rev.* **6**, 56 (1980)

Hahn, A., W. Lohmann, M. Hillerbrand, and U. Deffner, *Radiat. and Environ. Biophys.* **11**, 265 (1975).

Hales, B. J., and J. R. Bolton, *Photochem. Photobrol.* **12**, 239 (1970); *J. Amer. Chem. Soc.* **94**, 3314 (1972).

Halliburton, L. E., C. D. Norman, K. Saha, and L. A. Kappers, *Phys. Rev. B* **16**, 884 (1977).

Hama, Y., K. Hamanaka, and T. Horiuchi, *Radiat. Phys. Chem.* **13**, 13 (1979).

Hama, Y., and K. Shinohara, *Mol. Phys.* **18**, 279 (1970).

Hamrick, P. J., Jr., H. Shields, and C. C. Whisnant, *Radiat. Res.* **48**, 234 (1971).

Hariharan, N., J. Sobhanadri, *Mol. Phys.* **17**, 507 (1969).

Hattori, S., and Y. Muto, *Bull. Nagoya Inst. Technol.* **26**, 143 (1974).

Hellecbrand, J., and P. Wuensche, *Faserforsch. Textiltech.* **27**, 589 (1976).

Henriksen, T., P. K. Horan, and W. Snipes, *Radiat. Res.* **43**, 1 (1970).

Hesse, C., N. Leray, and J. Ronoin, *Mol. Phys.* **22**, 137 (1971).

Hines, R. L., and R. Arndt, *Phys. Rev.* **119**, 623 (1960).

Hirasawa, R., T. Mukaiba, and H. Hasegawa, *RSI* **39**, 935 (1968).

Hiratsuka, H., K. Sekiguchi, Y. Hatano, and Y. Tanizaki, *Chem. Phys. Lett.* **55**, 358 (1976).

Hoffmann, K., and D. Pöss, *Phys. Status Solidi A* **45**, 263 (1978).

Hori, Y., and L. D. Kispert, *J. Chem. Phys.* **69**, 3826 (1978).

Howatson, A. M., *Introduction to Gas Discharges*, Pergamon, New York, 1976.

Hsi, E. S. P., L. Fabes, and J. R. Bolton, *Rev. Sci. Instrum.* **44**, 197 (1973).

Hudson, A., M. F. Lappert, and P. W. Lednor, *J. Chem. Soc. Dalton Trans.*, 2369 (1976).

Hutchison, C. A., Jr., and G. W. Scott, *J. Chem. Phys.* **61**, 2240 (1974).

Iacona, C., J. P. Michaut, and J. Roncin, *J. Chem. Phys.* **67**, 5658 (1977).

Ikegamt, Y., Watanabe, *Chem. Lett.*, 1007 (1976).

Ikeya, M., *Solid State Phys. Japan* **11**, 591 (1976).

Imbusch, G. F., S. R. Chinn, and S. Geschwind, *Phys. Rev.* **161**, 295 (1967).

Islomov, A. I., O. A. Azizova, G. G. Odinokova, and D. I. Roschchupkin, *Biofizika.* **24**, 403 (1979).

Iwasaki, M., K. Toriyama, H. Muto, and K. Numome, *Chem. Phys. Lett.* **39**, 90 (1976); *J. Chem. Phys.* **65**, 596 (1976).

Jiang, T. Yu., and H. C. Jui, *Chin. J. Phys.* **14**, 68 (1976).

Joshi, A., S. Rustgi, and P. Riesz, *Int. J. Radiat. Biol. Relat. Stud. Phys. Chem. Med.* **30**, 151 (1976).

Kabler, M. N., in *Point Defects in Solids*, J. A. Crawford and L. M. Slifkin, Eds., Academic Press, New York, 1972.

Kalder, K. A., T. N. Kyarner, Ch. B. Lushchik, A. F. Malysheva, R. V. Mileniana, *Izv. Akad. Nauk SSSR Ser. Fiz.* **40**, 2313 (1976).

Kaufmann, U., and W. H. Koschel, *Phys. Rev. B* **17**, 2081 (1978).

Kawabata, K., *J. Chem. Phys.* **65**, 2235 (1976).

Kipnis, A. B., O. A. Azizova, P. I. Levenko, I. G. Shifrin, and E. M. Shesheev, *Radiobiologiya* **12**, 375 (1972).

Kiss, L., M. Irie, and K. Hayashi, *Proc. Tihany Symp. Radiat. Chem.* **4**, 419 (1977).

Kodzhespirov, F. F., M. F. Bulanyi, and I. A. Terev, *Fiz. Tverdogo Tela* **16**, 3159 (1974).

Kominami, S., S. Rokushika, and H. Hatano, *Int. J. Radiat. Biol.* **30**, 525 (1976).

Kotake, Y., and I. Miyagawa, *J. Chem. Phys.* **64**, 3169 (1976).

Kou, W. W. H., and H. C. Box, *J. Chem. Phys.* **64**, 3060 (1976).

Kroh, J., and A. Plonka, *Chem. Phys. Lett.* **52**, 371 (1977).

Kroh, J., and A. Plonka, *Nukleonika* **20**, 841 (1975).

Kroh, J., B. C. Green, and J. W. T. Spinks, *Can. J. Chem.* **40**, 413 (1962).

Kruglyakova, K. E., and V. M. Zhil'tsova, *Fotobiol. Zhivotn. Kletki, Dokl. Vses. Simp., Ist.*, 64 (1979).

Lamotte, B., and P. Gloux, *J. Chem. Phys.* **59**, 3365 (1973).

Lassmann, G., and W. Damerau, *Mol. Phys.* **21**, 551, 555 (1971).

Lee, Y. H., P. R. Brosious, and J. W. Corbett, *Phys. Status A* **50**, 237 (1978).

Lee, K. H., G. E. Holmberg, and J. H. Crawford, Jr., *Solid State Commun.* **20**, 283 (1976).

Lin, W. C., and J. M. Nickel, *J. Chem. Phys.* **57**, 3581 (1972).

Lin, T. S., *Chem. Phys. Lett.* **19**, 410 (1973).

Lind, H., and H. Loeliger, *Tetrahedron Lett.*, **2569** (1976).

Livingston, R., H. Zeldes, and E. H. Taylor, *Disc. Faraday Soc.* **19**, 166 (1955).

Lopez, F. J., F. Jaque, A. J. Fort, F. Agullo-Lopez, *J. Phys. and Chem. Solids* **38**, 1101 (1977).

Lund, A., T. Gillbro, D. F. Feng, and L. Kevan, *Chem. Phys.* **7**, 414 (1975).

Lyons, A. R., G. W. Neilson, S. P. Mishra, and M. C. R. Symons, *J. Chem. Soc. Faraday Trans. 2* **(71)**, 363 (1975).

Mackey, J. H., Jr., *J. Chem. Phys.* **39**, 74 (1963).

McLachlan, K., and R. Sealy, *New Sci.* **69**, 433 (1976).

Madden, K. P., and W. A. Bernhard, *J. Chem. Phys.* **72**, 31 (1980).

Mao, S. W., and L. Kevan, *Chem. Phys. Lett.* **24**, 505 (1974).

Mamedov, A. P., V. R. Mardukhaev, and E. Yu. Salaev, *Khim. Vys. Energ.* **12**, 418 (1978).

Mehendru, P. C., and N. Kumar, *Phys. Status Solid B* **72**, 143 (1975).

Menczel, J., J. Varga, and G. Hardy, *Period. Polytech. Chem. Eng.* **22**, 231 (1978).

Meybeck, A., *Int. J. Cosmet. Sci.* **1**, 199 (1979).

Meybeck, A., *Parfums. Cosmet. Aromes.* **28**, 73 (1979).

Minegishi, A., Y. Shinozaki, and G. Meshitsuka, *J. Chem. Phys.* **56**, 2481 (1972).

Moan, J., Kaalhus, *J. Chem. Phys.* **61**, 3556 (1974).

Moan, J., *Photochem. Photobiol.* **22**, 111 (1975).

Molin, Yu. N., A. T. Koritskii, A. G. Semenov, N. Ya. Buben, and V. N. Shamshev, *PTE* **6**, 73 (931) (1960).

Möhwald, H., and E. Sackmann, *Chem. Phys. Lett.* **21**, 43 (1973).

Mori, K., *Phys. Status Solidi A* **42**, 375 (1977).

Morishima, H., *Radiat. Res.* **44**, 605 (1970).

Morton-Blake, D. A., *J. Phys. Chem.* **73**, 2964 (1969); **74**, 1508 (1970).

Morton, J. R., K. F. Preston, and S. J. Strach, *J. Phys. Chem.* **83**, 3418 (1979).

Moskalonov, A. V., *Fiz. Tverdogo Tela* **19**, 1440 (1977).

Muller, O., *Z. Angew. Phys.* **31**, 275 (1971).

Murabaysahi, S., M. Shiotani, and J. Sohma, *Chem. Phys. Lett.* **48**, 80 (1977).

Muramoto, T., *J. Phys. Soc. Japan* **35**, 921 (1973).

Murti, Y. V. G. S., G. Samuel, and S. Sivaraman, *Phys. Status Solidi A* **34**, 615 (1976).

Muto, H., and M. Iwasaki, *J. Chem. Phys.* **58**, 2454 (1973).

Muto, H., M. Iwasaki, and Y. Takahashi, *J. Chem. Phys.* **66**, 1943 (1977).

Nadzhafova, N. A., V. A. Sharpatyi, *Dokl. Akad. Nauk. Az. SSSR* **34**, 35 (1978).

Nagai, S., and T. J. Gillbro, *Phys. Chem.* **81**, 1793 (1977).

Nakata, H., H. Morishima, *Phys. Status Solidi A* **33**, 633 (1976).

Nambi, K. S. V., A. M. Ghose, D. V. Gopinath, and J. H. Hubbell, *Proceedings of the International Symposium on Radiation Physics* **8**, 234 (1974).

Naryadchikov, A. D., A. D. Grishina, and N. A. Bakh, *PTE* **3**, 192 (1962).

Nauwelaerts, F., M. Lemahieu, and J. Ceulemans, *J. Chem. Phys.* **66**, 140 (1977).

Nemec, H. W., and H. Monig, *Radiat. and Environ. Biophys.* **11**, 227 (1975); **14**, 137 (1977).

Odinokova, G. G., O. A. Azizova, Yu. A. Vladimirov, A. N. Remizov, and D. I. Roshchupkin, *Biofizika.* **24**, 202 (1979).

Odinakova, G. G., O. A. Azizova, A. G. Kotov, A. N. Remizov, and D. I. Roshchupkin, *Fiz.-Khim. Osn. Funkts. Nadmolekulyarnykh Strukt. Kletki, Mater. Vses. Simp.* **2**, 44 (1974).

Ohno, K., and J. Sohma, *Chem. Instrum.* **2**, 121 (1969).

Owens, F. J., and V. L. Vogel, *J. Chem. Phys.* **64**, 851 (1976).

Parrot, R., C. Naud, C. J. Delbeck, and P. H. Yuster, *Phys. Rev. B* **15**, 137 (1977).

Pavlov, N. M., M. I. Iglitsyn, M. G. Kosaganova, and and V. N. Solmatin, *Fiz. Tekh. Poluprovodn.* **9**, 1279 (1975).

Piette, L. H., and W. C. Landgraf, *J. Chem. Phys.* **32**, 1107 (1960).

Poole, C. P., Jr., Thesis, University of Maryland, 1958, p. 9.

Pung, L. A., and A. Ch. Lushchik, *Fiz. Tverdogo Tela* **18**, 1176 (1976).

Radhakrishna, S., and T. Bhask Ar Rao, *Phys. Status Solidi A* **35**, 715 (1976).

Radhakrishna, S., B. V. R. Chowdari, M. V. Seshachala, *1974 International Conference on Colour Centres in Ionic Crystals* **12**, 19 (1974).

Radhakrishna, S., B. V. R. Chowdari, and A. K. Viswanath, *Chem. Phys. Lett.* **40**, 134 (1976).

Radhakrishna, S., and B. V. R. Chowdari, *Fortschr. Phys.* **25**, 511 (1977).

Regulla, D. F., and U. Deffner, *Radiat. Pollut. Abatement, Proc. Int. Conf. Esna Work Group "Waste Irradiat."* **I**, 21 (1976).

Rezk, A. M. H., and R. H. Johnsen, *Int. J. Radiat. Biol. Relat. Stud. Phys. Chem. Med.* **34**, 337 (1978).

Richter, P. I., J. E. Johnson, L. L. Finch, G. C. Moulton, *J. Chem. Phys.* **65**, 5527 (1976).

Riggin, M., S. Radhakrishna, and P. W. Whippey, *Phys. Status Solidi A* **37**, 51 (1976).

Rustigi, S., and P. Riesz, *Int. J. Radiat. Biol.* **33**, 21 (1978).

Sagstuen, E., and C. Alexander, Jr., *J. Chem. Phys.* **68**, 762 (1978).

Sakurai, H., T. Uchida, and M. Kira, *J. Organomet. Chem.* **197**, 15 (1976).

Sanderson, R. T., *Chemical Bonds and Bond Energy*, Academic Press, New York, 1976.

Sanner, T., *Radiat. Res.* **44**, 13 (1970).

Sargent, F. P., and E. M. Grady, *Chem. Phys. Lett.* **38**, 130 (1976); **39**, 188 (1976).

Saxebol, G., and E. Sagstuen, *Int. J. Radiat. Biol.* **26**, 373 (1974).

Shamfarov, Ya. L., and T. A. Smirnova, *Fiz. Tverd. Tela* **5**, 1046 (1963).

Shields, H. S., and P. J. Hamrick, Jr., *J. Chem. Phys.* **64**, 263 (1976).

Shields, H., *Magn. Resonance Rev.* **3**, 375 (1974).

Shimada, S., H. Kashiwabara, and J. Sohma, *Mol. Phys.* **30**, 1195 (1975).

Shimada, M., Y. Nakamura, Y. Kusama, O. Matsuda, N. Tamura, and E. Kageyams, *J. Appl. Polym. Sci.* **18**, 3379 (1974).

Shimizu, T., M. Kumeda, and M. Suzuki, *Oyo Buturi* **46**, 444 (1977).

Shiraishi, H., H. Kadoi, Y. Katsumura, Y. Tabata, K. Oshima, *J. Phys. Chem.* **80**, 2400 (1976).

Silsbee, R. H., *J. Appl. Phys.* **32**, 1459 (1961).

Sinclair, J., *J. Chem. Phys.* **55**, 245 (1971).

Slifkin, M. A., Y. M. Suleiman, and B. G. Phillips, *Radiat. Eff.* **32**, 33 (1977).

Smith, P., L. B. Gilman, R. D. Stevens, C. Vignola de Hargrave, *J. Magn. Res.* **29**, 545, 20 (1978).

Smith, R. C., and S. J. Wyard, *Nature* **191**, 897 (1961).

Smith, C., C. P. Poole, Jr., and H. A. Farach, *J. Chem. Phys.* **74**, 993 (1981).

Sohma, J., T. Komatsu, and Y. Kanda, *Japan J. Appl. Phys.* **7**, 298 (1968).

Sportelli, L., H. Neubacher, and W. Lohmann, *Biophys. Struct. and Mech.* **3**, 317 (1977).

Sprague, E. D., and J. E. Willard, *J. Chem. Phys.* **63**, 2603 (1975).

Steenken, S., *Photochem. Photobiol.* **22**, 157 (1975).

Steenken, S., W. Jaenicke-Zauner, and D. Schulte-Frohlinde, *Photochem. Photobiol.* **21**, 21 (1975).

Steenken, S., E. D. Sprague, and D. Schulte-Frohlinde, *Photochem. Photobiol.* **22**, 19 (1975).

Suryanarayana, D., and J. Sobhanadri, *J. Chem. Phys.* **61**, 2827 (1974).

Suryanarayana, D., and J. Sobhanadri, *J. Nonmet.* **2**, 113 (1974).

Takeuchi, N., T. Mizuno, H. Sasakura, and M. Ishiguro, *J. Phys. Soc. Japan* **18**, 743 (1963).

Taylor, P. C., U. Strom, and S. G. Bishop, *Phys. Rev. B* **18**, 511 (1978).

Trifunac, A. D., K. W. Johnson, B. E. Clifft, and R. H. Lowers, *Chem. Phys. Lett.* **35**, 566 (1975).

Truesdale, R., H. A. Farach, and C. P. Poole, Jr., *Phys. Rev. B* **22**, 365 (1980).

Tsuji, K., K. Hayashi, and S. Okamura, *J. Polym. Sci. A-1 Polym. Chem.* **8**, 583 (1970).

Usmanov, Kh. U., U. A. Azizov, D. S. Khamidov, and K. Sultanov, *Uzv. Khim. Zh.*, 34 (1978).

Vainer, V. S., A. I. Veinger, and Yu. A. Polonskii, *Fiz. Tverdogo Tela* **18**, 409 (1976).

Vakhidov, Sh., M. A. Vakhidova, A. N. Lobachev, O. K., Melnikov, and N. S. Triodina, *Dokl. Akad. Nauk SSSR* **217**, 1310 (1974).

van Wieringen, J. S., and A. Kats, *Philips Res. Rept.* **12**, 432 (1957); van Wieringen, J. S., Y. Haven, and A. Kats, *11th Ampere Colloquium*, Eindhoven, 1963, p. 403.

von der Weid, J. P., *Rev. Bras. Fis* **6**, 1 (1974).

von der Weid, J. P., and S. C. Ribeiro, *1974 Int'l. Conf. on Colour Centres in Ionic Crystals* **12**, 19 (1974).

Voznyak, V. M., I. I. Proskuryakov, E. I. Eifimov, V. A. Kim, and V. B. Evstigneev, *Itofi Issled. Mekh. Fotosint. Tezisy Dokl. Vsws. Knof. 9th*, 25 (1976).

Vyas, H. M., B. B. Adeleke, and J. K. S. Wan, *Spectrosc. Lett.* **9**, 663 (1976).

Warden, J. T., and J. R. Bolton, *J. Am. Chem. Soc.* **94**, 4351 (1972).

Watterich, A., and K. Raksanyi, *Phys. Status Solidi B* **80**, 105 (1977).

Weeks, R. A., *Phys. Rev.* **130**, 570 (1963).

Weiter, J. J., and E. D. Finch, *Nature* **254**, 536 (1975).

Wells, J. W., and Ko, Ching-Lung, *J. Chem. Phys.* **69**, 1848 (1978).

Whelan, D. J., *J. Chem. Phys.* **49**, 4734 (1969).

Wong, P. K., *Polymer* **19**, 785 (1978).

Wosinski, T., and T. Figielski, *Phys. Stat. Sol. B* **83**, 93 (1977).

Wosinski, T., T. Figielski, and A. Makosa, *Phys. Stat. Sol. A* **37**, K57 (1976).

Yamaga, M., Y. Hayashi, and H. Yoshioka, *J. Phys. Soc. Japan* **44**, 154 (1978).

Yamanashi, B. S., and J. A. Zuclich, *Ophthalmic Res.* **10**, 140 (1978).

Yamashita, M., and H. Kashiwagi, *Japan J. Appl. Phys.* **14**, 421 (1975).

Yamashita, M., and H. Kashiwagi, *IEEE J. Quantum Electron 5th Biannual Conf. on Laser Engineering and Applications* **12**, 90 (1975).

Yamashita, M., *Denshi Gijutsu SOGO Kenkyusho Kenkyu* **758**, 1 (1976).

Yasukawa, T., and K. Matsuzaki, *Tohoku Daigaku Hisuiyoeki Kagaku Kenkyusho Hokoku* **19**, 257 (1969).

Yasukawa, T., H. Kubota, and Y. Ogiwara, *J. Polym. Sci. Polym. Chem. Ed.* **14**, 1617 (1976).

Yu, J -T., and J -H. Chang, *Chin. J. Phys.* (*Taipei*) **14**, 68 (1976).

Yugo, S., *Rep. Univ. Electro-Commun.* **27**, 89 (303) (1977).

Zagorets, P. A., A. G. Shostenko, A. I. Krivonosov, A. M. Dodonov, and B. A. Gorelik, and M. V. Postrigan, *Khim. Vys. Energ.* **10**, 280 (1976).

Zaikin, Yu., S. A. Karyagin, Yu. A. Zarifyants, V. F. Kiselev, *Fiz. Tekh. Poluprovodn.* **10**, 2221 (1976).

Zehner, H., W. Flossman, E. Westhof, and A. Muller, *Mol. Phys.* **32**, 869 (1976).

Zhil'tsova, V. M., Z. P. Geibova, O. A. Azizova, K. E. Kruglyakova, and Ya. A. Vladimirov, *Fotobiol. Zhivotn. Kletki, Dokl. Vses. Simp.* **135** (1979).

Zimmermann, P. H., J. F. Reichert, and A. J. Dahm, *Phys. Rev.* **B15**, 2630 (1977).

CHAPTER 11

Sensitivity

This chapter will begin by presenting Feher's (1957) classical analysis of the sensitivity of electron spin resonance spectrometers and will then discuss several other approaches to the same subject. The latter part of the chapter will elaborate upon the manner in which various instrumental parameters and experimental conditions affect the sensitivity, and will recommend ways of selecting optimum conditions. We assume throughout that allowed transitions are being observed with the microwave magnetic field H_1 perpendicular to the constant applied field. Forbidden transitions with lower intensity are observed with these fields parallel to each other, and Brogden and Butterworth (1965) even reported the detection of doubly forbidden satellites. Chiarini et al. (1975) and Martinelli et al. (1977) discussed longitudual detection in double resonance experiments.

Many authors reported observing forbidden transitions (e.g., De Groot et al., 1973; Draeger and Wolfmeier 1976; Grivet and Mispelter, 1974; Kispert and Chang, 1973; Royaud and Sardos, 1981; Schlick and Kevan, 1976). Poole and Farach (1971, 1972) gave a systematic exposition of forbidden transitions. Bales and Lesin (1976) used spin-flip satellite lines as a tool in identifying paramagnetic centers in disordered solids.

A. Q and Frequency Changes at Resonance

When a paramagnetic sample in a resonant cavity is tuned through the resonance condition $\hbar\omega = g\beta H$, the unpaired spins interact with the rf field H_1 at the sample. This interaction manifests itself as a frequency change (dispersion) or a Q change (absorption) in the resonant cavity. The absorption mode will be discussed in detail, and then a few words will be said about dispersion. Most ESR spectrometers detect only absorption, but some (e.g., Ryter et al., 1955; Faulkner, 1962) operate in both modes.

A paramagnetic sample in a resonant cavity of frequency $\omega_0 = 2\pi\nu_0$ will absorb incident microwave energy at the resonant magnetic field H_0 given by

$$\hbar\omega_0 = g\beta H_0 \tag{1}$$

where $\hbar$ is Planck's constant divided by 2π, β is the Bohr magneton, and g is the dimensionless g factor, which is often close to 2. The average power absorbed per unit volume is given by

$$P = \tfrac{1}{2}\omega_0 H_1^2 \chi''$$
(2)

where H_1 is the amplitude of the microwave magnetic field at the sample and χ'' is the microwave or rf susceptibility. The power absorbed at resonance will manifest itself as a change in the quality factor or Q of the cavity,

$$\frac{1}{Q} = \frac{1}{Q_u} + \frac{1}{Q_\chi}$$
(3)

where Q_u is the unloaded Q of the cavity in the absence of resonant absorption and Q_χ is the contribution of the resonant absorption to the Q. This quantity Q_u includes cavity losses plus the dielectric and conductivity losses in the sample. At resonance, the Q changes by an amount ΔQ:

$$\Delta Q = -Q^2 \Delta\left(\frac{1}{Q}\right) = -\frac{Q_u^2}{Q_\chi}$$
(4)

where we assume

$$Q_\chi \gg Q_u$$
(5)

The loaded Q is defined in Sec. 5-B, and Q_χ has the form

$$Q_\chi = \frac{(\mu_0/2)\int_{\text{cavity}} H_1^2 \, dV}{(\mu_0/2)\int_{\text{sample}} H_1^2 \chi'' \, dV}$$
(6)

If χ'' is homogeneous over the sample volume, then one has

$$Q_\chi = \frac{\int_{\text{cavity}} H_1^2 \, dV}{\chi'' \int_{\text{sample}} H_1^2 \, dV} = \frac{1}{\chi'' \eta}$$
(7)

where η is the filling factor discussed in Sec. 5-H, and χ'' is considered a constant for each sample. The filling factor is a measure of the fraction of the microwave energy that interacts with the sample, and explicit values of η are given in Sec. 5-H for several commonly used sample arrangements. Equations (4) and (7) furnish the change in Q:

$$\Delta Q = \chi'' \eta Q_u^2$$
(8)

where the negative sign is omitted for simplicity. To convert this equation from MKS to cgs units, multiply the right-hand side by 4π.

The equivalent circuits for reflection and transmission cavities are shown in Figs. 5-24. The coupling coefficients to the waveguide are discussed in Sec. 5-F, and the same terminology will be used here.

Reflection Cavity and Square Law Detector

The power P_c into a resonant cavity is given by (Feher, 1957)

$$P_c = \frac{\frac{1}{2}(nE)^2 R_c}{\left(R_c + R_g n^2\right)^2} = \frac{2 P_w \left(R_c / R_g n^2\right)}{\left(1 + R_c / R_g n^2\right)^2} \tag{9}$$

and the maximum power P_w available from the source is

$$P_w = \frac{1}{4}\left(\frac{E^2}{R_g}\right) \tag{10}$$

where R_g is the generator's internal resistance, R_c is the resonant cavity resistance, and n is the turns ratio of the equivalent circuit transformer.

At match $R_c = R_g n^2$. At resonance, the change in reflected power ΔP_r equals the change in the power ΔP_c absorbed by the sample in the cavity as a result of the change ΔR_c in the equivalent cavity resistance R_c:

$$\frac{\Delta P_c}{P_w} = \frac{1}{P_w} \frac{\partial P_c}{\partial R_c} \Delta R_c = \frac{2 \Delta R_c}{R_g n^2} \frac{1 - \left(R_c / R_g n^2\right)}{\left(1 + R_c / R_g n^2\right)^3} \tag{11}$$

For a square-law detector whose output is proportional to the incident power (e.g., crystals in the microwatt region, or bolometers), we optimize ΔP_c with respect to the coupling parameter $R_g n^2$ by setting

$$\frac{\partial \Delta P_c}{\partial \left(R_g n^2\right)} = 0 \tag{12}$$

with the result that

$$\frac{R_g n^2}{R_c} = 2 \pm 3^{1/2} = \begin{cases} \text{VSWR} & R_g n^2 \geqslant R_c \\ 1/\text{VSWR} & R_g n^2 \leqslant R_c \end{cases} \tag{13}$$

The positive sign is for an overcoupled cavity and the negative sign for an undercoupled cavity. Each case gives a VSWR of $2 + 3^{1/2}$, and from Eq. (11), we have for the two cases

$$\frac{\Delta P_c}{P_w} = \pm 0.193 \frac{\Delta R_c}{R_c} = \mp 0.193 \frac{\Delta Q}{Q} = \mp 0.193 \chi'' \eta Q_u \tag{14}$$

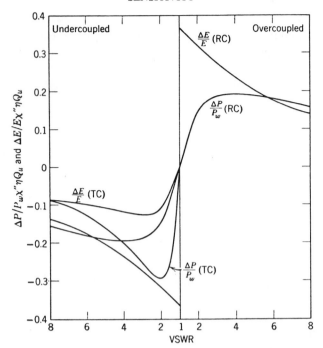

Fig. 11-1. The ESR signal $\Delta P/P_w$ and $\Delta E/E$ normalized relative to $\chi''\eta Q_u$ as a function of VSWR for square-law ($\Delta P/P_w$) and linear ($\Delta E/E$) detectors used with a reflection (RC) and transmission (TC) cavity (Feher, 1957).

for the maximum ESR signal. An overcoupled cavity produces less noise, as will be shown below. Equation (11) is plotted in Fig. 11-1, and the symmetry of the graph shows that for a given VSWR the same sensitivity may be obtained with both the overcoupled and the undercoupled cases. Goldberg and Crowe (1977) discussed the effect of cavity loading on analytical ESR.

Transmission Cavity with Square-Law Detector

The power into the load P_l is

$$P_l = \frac{E^2 n_1^2 n_2^2 R_g}{\left[R_g n_1^2 + R_c + R_l n_2^2\right]^2} \tag{15}$$

The change in power ΔP_l at resonance is related to the maximum available power $P_w = E^2/(4R_g)$ by the expression

$$\frac{\Delta P_l}{P_w} = \frac{1}{P_w}\frac{\partial P_l}{\partial R_c}\Delta R_c = \frac{8 n_1^2 n_2^2 R_g^2 \Delta R_c}{\left(R_g n_1^2 + R_c + R_l n_2^2\right)^3} \tag{16}$$

and at match $R_g = R_l$. If we maximize ΔP_l with respect to both $R_g n_1^2$ and $R_l n_2^2$, we obtain

$$R_g n_1^2 = R_l n_2^2 = R_g n^2 = R_c \qquad (17)$$

Since the cavity input impedance equals the sum of R_c and $R_l n_2^2$, it follows that the cavity is overcoupled, and the VSWR has the form

$$\text{VSWR} = \frac{\left(R_l n_2^2 + R_c\right)}{R_g n_1^2} \qquad (18)$$

It is not possible to undercouple a transmission cavity with equal input and output coupling constants. Using condition (17) and setting $n_1 = n_2$ we obtain

$$\frac{\Delta P_l}{P_w} = \frac{1}{P_w} \frac{\partial P_l}{\partial R_c} \Delta R_c = \frac{8}{27} \frac{\Delta R_c}{R_c} = -\frac{8}{27} \frac{\Delta Q}{Q} = -\frac{8}{27} \chi'' \eta Q_u \qquad (19)$$

Reflection Cavity and Linear Detector

In this case, one detects the reflected voltage E_r, which is related to the incident voltage E by

$$E_r = E\Gamma = -E\left(1 - \frac{2\,\text{VSWR}}{1 + \text{VSWR}}\right) \qquad (20)$$

where the reflection coefficient Γ is defined by Eq. (2-C-7). The VSWR for a reflection cavity is given by

$$\text{VSWR} = \left(R_g n^2 / R_c\right)^{\pm 1} \qquad (21)$$

where we shall adopt the convention that the upper sign is for the overcoupled case and the lower sign for the undercoupled case. At resonance, the change in the reflected voltage ΔE_r arises from a change ΔR_c in the equivalent cavity resistance R_c

$$\frac{\Delta E_r}{E} = \frac{1}{E} \frac{\partial E_r}{\partial R_c} \Delta R_c = \pm 2\,\Delta R_c \frac{R_g n^2}{\left(R_g n^2 + R_c\right)^2} \qquad (22)$$

using the same sign convention. For a linear detector whose output is proportional to the input voltage (e.g., crystals at milliwatt or watt power levels), the optimum coupling occurs when

$$\frac{\partial \Delta E_r}{\partial R_g n^2} = 0 \qquad (23)$$

which corresponds to the condition of match

$$R_g n^2 = R_c \tag{24}$$

As a result, Eq. (22) gives

$$\frac{\Delta E_r}{E} = \pm \frac{\Delta R_c}{2R_c} = \mp \frac{\Delta Q}{2Q} = \mp \frac{\chi'' \eta Q_u}{2} \tag{25}$$

Thus we should work near match for maximum sensitivity, but to avoid distortion, it is best to detune the cavity sufficiently so that it will remain on one side of match while sweeping through the resonant line. The observed sign of the signal will indicate the type of coupling, since during resonance adsorption the reflected voltage E_r will increase for an undercoupled cavity and decrease for an overcoupled cavity.

Transmission Cavity and Linear Detector

The voltage E_t observed at a linear detector after passing through a transmission cavity is

$$E_l = \frac{E n_1 n_2 R_g}{R_g n_1^2 + r + R_l n_2^2} \tag{26}$$

where again for maximum sensitivity both couplings are the same ($R_g n_1^2 = R_l n_2^2 = R_g n^2$). At resonance, one obtains

$$\frac{\Delta E_l}{E} = \frac{1}{E} \frac{\partial E_l}{\partial R_c} \Delta R_c = -\frac{\Delta R_c}{R_c} \frac{R_c R_g n^2}{\left(R_c + 2R_g n^2\right)^2} \tag{27}$$

When

$$\frac{\partial E_l}{\partial R_g n^2} = 0 \quad \text{then} \quad R_g n^2 = \frac{1}{2} R_c \tag{28}$$

and we arrive at the condition

$$\frac{\Delta E_l}{E} = -\frac{1}{8} \chi'' \eta Q_u \tag{29}$$

Figure 11-1 shows $\Delta P/P_w$ and $\Delta E/E$ from Eqs. (11), (16), (22), and (27) plotted as a function of the VSWR. Crystal detectors used with reflection cavities are ordinarily operated somewhere between their linear and square-law regions, and so in this case one may work at an intermediate VSWR, perhaps at a value near 2. When noise is taken into account (see Sec. 7-C), then we see that in the milliwatt and watt range it is advisable to operate crystal detectors with a constant amount of incident power or "leakage" P_d that is independent of the power P_c that enters the resonant cavity. At high powers, $P_c \gg P_d$, this corresponds to tuning the cavity close to match.

Most ESR studies are carried out with reflection cavities, although transmission cavities can be useful with conducting samples (Domdey and Voitlander, 1974). Flesner et al. (1976) discussed transmission ESR as a probe of a metallic interface, and Janossy (1975) reviewed transmission ESR (compare Orbach et al., 1975).

This section has discussed the change in Q that results from the behaviour of the absorption or the imaginary part of the magnetic susceptibility χ'' at resonance. During resonance there will also be a dispersion mode that corresponds to a change in the frequency of the resonant cavity as a result of the real part of the rf magnetic susceptibility χ'. The reason for this frequency change may be deduced by considering the velocity v of electromagnetic radiation in a dielectric medium

$$v = \frac{1}{(\mu\epsilon)^{1/2}} = \lambda f \tag{30}$$

with a dielectric constant ϵ and a permeability $\mu = \mu_0(1 + \chi)$, where ϵ and μ are considered real. In a resonant cavity the wavelength λ (or the guide wavelength λ_g) is fixed by the dimensions of the resonator. As a result, if the susceptibility or dielectric constant is changed by the amount $\Delta\chi$ and losses are neglected to (i.e., $\chi'' = 0$), then the frequency will change by

$$\Delta f = \frac{1}{\lambda(\mu_0\epsilon)^{1/2}} \left[\frac{1}{(\chi')^{1/2}} - \frac{1}{(\chi' + \Delta\chi)^{1/2}} \right] \tag{31}$$

At resonance, $\chi = \chi' + j\chi''$ changes both its real and imaginary parts in conformity with the Kramers–Kronig relations

$$\chi'(\omega) = \chi'(\infty) + \frac{2}{\pi} \int_0^\infty \frac{\omega'\chi''(\omega')}{\omega'^2 - \omega^2} d\omega' \tag{32}$$

$$\chi''(\omega) = -\frac{2\omega}{\pi} \int_0^\infty \frac{\chi'(\omega') - \chi'(\infty)}{\omega'^2 - \omega^2} d\omega' \tag{33}$$

These relations are very important physically because they ensure that the amplitude of the observed signal from χ' will be comparable to that observed from χ'' (Rajan, 1962). If a phase shifter or slide screw tuner is employed to vary the effect length of line in front of the resonant cavity (Sec. 4-N), then the shape of the observed first derivative signal will change with the phase in the

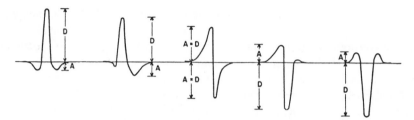

Fig. 11-2. Change in shape of a spectrum from solid DPPH when the bridge is tuned from dispersion to absorption and then back to dispersion.

manner shown in Fig. 11-2. If the klystron is stabilized on the sample resonant cavity, then it will "follow" the frequency variations, and the dispersion mode χ' will be stabilized out. This is why most ESR spectrometers record only the absorption mode χ''.

When recording broad lines Kiel (1975) employed a low Q (~ 1000) undercoupled cavity to minimize interference between absorption and dispersion signals.

If a mixed mode is recorded, then the percentage of absorption $\%\chi''$ may be approximated from the expression

$$\%\chi'' = \frac{100\chi''}{(\chi' + \chi'')} \tag{34}$$

$$\approx 100 \left(\frac{2A/(A+D) - [2A/(A+D)]_{\chi'}}{1 - [2A/(A+D)]_{\chi'}} \right) \tag{35}$$

where the parameters A and D are defined in Fig. 11-2. For pure absorption, $D = A$, and for pure dispersion, D/A has the values [Sec. 12-H; Table 12-5; Pake and Purcell (1948)]

$$\frac{D}{A} = \begin{cases} 8 & \text{Lorentzian} \\ 3.5 & \text{Gaussian} \end{cases} \tag{36}$$

which lead to

$$\left(\frac{2A}{A+D} \right)_{\chi'} = \begin{cases} \frac{2}{9} & \text{Lorentzian} \\ \frac{4}{9} & \text{Gaussian} \end{cases} \tag{37}$$

Peter et al. (1962) also discuss a technique for obtaining the ratio of χ'' to χ'.

If the klystron is stabilized at the frequency f_k on a reference cavity, and recordings are made of a DPPH sample in a second resonant cavity at the frequency f_c, then the $\%\chi''$ was found experimentally to vary with the frequency deviation $\Delta f = f_k - f_c$ in the manner shown in Fig. 11-3.

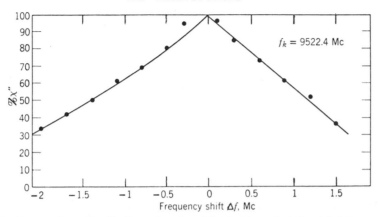

Fig. 11-3. Percent absorption ($\%\chi''$) as a function of the frequency deviation Δf of the resonant cavity frequency f_c from the klystron frequency f_k, where $\Delta f = f_k - f_c$. The measurements were made using $f_k = 9522.4$ MHz. The data were obtained by former Pittsburgh steeler fullback J. F. Itzel, Jr.

Paramagnetic species in aqueous solution cells that are thin compared to the wavelength, for example, 0.2 mm at X band, produce lineshapes that are absorptive. In thick aqueous samples dielectric losses due to the imaginary part ϵ' of the dielectric constant

$$\epsilon = \epsilon' - j\epsilon' \tag{38}$$

modulate the magnetic losses of the paramagnetic spins and produce a lineshape that is a mixture of absorption and dispersion (Gillberg and LaForce, 1973).

Accou et al. (1972) discussed dispersion in weak and strong rf fields. Roch et al. (1971) carried out absorption and dispersion measurements at 10 to 16 GHz using interferometers [see also Marzouk et al. (1975), Sardos (1969, 1972)]. Bonori et al. (1981b) and Pristupa and Kadyrov (1974) designed spectrometers for the observation of both absorption and dispersion.

B. Noise Sources

In the previous section, the magnitude of the ESR signal detected by the crystal (or bolometer) was discussed for several experimental arrangements. The factor that determines the sensitivity of a spectrometer is the signal-to-noise ratio, and so the present section will discuss the various types of noise that are capable of effecting this ratio (see RLS-24). Obvious noise sources such as the lack of voltage regulation and mechanical transients will not be mentioned. The next section will discuss the signal-to-noise ratio itself. Buckmaster et al. (1971, 1972) discussed noise measurements and spectrometer sensitivity. Meijer (1975) analyzed noise in a homodyne spectrometer.

It might be mentioned in passing that cigarette smoke is paramagnetic and should be excluded from resonant cavities as well as from lungs (Menzel et al., 1976; Pryor et al., 1976; Scott et al., 1975).

Johnson Noise

The available noise power of Johnson noise dN arising from a resistor radiating into a "cold" or noiseless load at a temperature T is

$$dN = kT\Delta f \tag{1}$$

where k is Boltzmann's constant and Δf is the bandwidth in cycles per second. A crystal rectifier generates more noise than a resistor, and so its noise power is characterized by the noise temperature t as discussed in Sec. 7-C.

Vacuum Tube Noise

Shot effect noise arises in vacuum tubes as a result of randomness in the emission of electrons by the filament. Partition noise results from randomness in the partition of the total current among the electrodes. Other noise sources include the flicker effect, induced grid noise, and positive ion fluctuations.

Detector Noise

A bolometer detector produces Johnson noise $dN = kT\Delta f$ while a crystal detector produces the noise power $dN = tkT\Delta f$, where the dimensionless "noise temperature" t exceeds 1. The excess noise in a crystal is inversely proportional to the modulation frequency f_{mod}, and this may be expressed analytically by writing

$$t - 1 = \frac{F(P)}{f_{\mathrm{mod}}} \tag{2}$$

The term $F(P)$ is a quadratic function of the power in the microwatt region and a linear function of the power in the milliwatt and watt region, as discussed in Sec. 7-C [e.g., Eqs. (7-C-3), (7-C-4), and (7-C-5)]. In a superheterodyne spectrometer the second detector sometimes introduces appreciable noise. Klystron noise suppression in a balanced mixer is discussed in Sec. 7-D.

Klystron Noise

The noise figure F of a klystron is related to noise power dN in the frequency interval Δf by (RLS-7, p. 472; RLS-24, p. 113; Feher, 1957)

$$dN = FkT\Delta f \tag{3}$$

and this may be rearranged to

$$\frac{1}{2}F = \frac{1}{2}\left(\frac{dN}{P_w}\right)\left(\frac{P_w}{kT\Delta f}\right) = sP_w \tag{4}$$

in terms of the klystron output power P_w.

The relative klystron noise decreases in going to higher modes. FM klystron noise is considerably more troublesome than AM noise. There is an appreciable decrease in the klystron noise output in going from an intermediate frequency of 30 MHz to one of 90 MHz. At 90 MHz the klystron noise was sufficiently low so as not to increase the effective crystal noise temperature, while the reverse was true at 30 MHz (RLS-24, p. 113).

A microwave generator such as a klystron will produce a great deal of noise if its supply voltages are not maintained at constant values. The klystron output is most sensitive to fluctuations in the reflector voltage, and least sensitive to variations in the filament supply, with variations of the beam or accelerator voltage in an intermediate category. Sections 3-E and 3-F discuss klystron power supplies and stabilization schemes.

Feher (1957) points out that, for some common operating conditions, noise voltages that result from frequency instabilities enter as a first-order effect when one is tuned to the dispersion mode χ', and as a second-order effect when one is tuned to the absorption mode χ''. An overcoupled cavity is less sensitive than an undercoupled one to this source of noise.

Amplifier Noise

An amplifier will amplify the impressed signal plus its accompanying noise, and in addition it will add some noise of its own. Sometimes noise is generated in the i.f. receiver. This will be discussed in the next section.

Cavity Vibrations

One should be sure of having a sufficiently rigid mounting of the resonant cavity to avoid vibrations that will increase the noise level. It is particularly important to make sure that all of the screws are tight, and so forth. Another noise source arises from eddy currents generated in the cavity walls by a high-frequency modulation signal. Mechanical vibrations can be produced by an interaction of these currents with the applied magnetic field. The amplitude of this interaction is approximately proportional to the applied field strength, and the frequency of vibration is the modulation frequency or a harmonic thereof. It will affect the recorder trace by producing a continuous shift of the baseline. Amplitude and phase shifts may also occur. This interference may be minimized in a rectangular cavity by properly orienting it relative to the magnetic field, or by using a glass cavity coated with silver to a thickness that is much greater than the microwave skin depth but must less than the modulation skin depth. Vibrations in unsupported thin cavity walls may be damped out by a thick coat of cement or other dielectric material.

At high modulation amplitudes, the eddy currents in the cavity walls will heat the cavity and shift its frequency. Sometimes, it is necessary to await the establishment of thermal equilibrium at the high modulation amplitude to obviate a sloping background signal. In low-temperature experiments, this heat dissipation effect may boil off refrigerant excessively.

C. Signal-to-Noise Ratio

The sensitivity of an ESR spectrometer is proportional to the ratio of the signal amplitude to the noise amplitude on a recorder or oscilloscope. The results of the last section will be applied toward a determination of the minimum detectable rf susceptibility with various types of detecting systems. This will provide us with the tools needed for deriving a general formula for computing the numer of spins in a sample. The derivation itself will be postponed until the next section. In this section and the remainder of the chapter, we assume that the resonant cavity is properly matched to the transmission line.

As the noise and the signal proceed from the generator through various networks, the ratio of the incremental noise power dN to the signal power S increases because of the extra noise generated by the networks themselves. The noise power is written as a differential dN because it corresponds to noise in the bandwidth Δf centered at the signal frequency f_m, while the signal S is ordinarily at only one frequency (e.g., the modulation frequency). The ratio of dN to S at the output of a network dN_1/S_1 is related to the corresponding ratio dN_0/S_0 at the input by (RLS-16, Sec. 1.4)

$$\frac{dN_1}{S_1} = F_1\left(\frac{dN_0}{S_0}\right) \tag{1}$$

where $F_1 \geq 1$ is the noise figure of the network. For a network without any internal sources of noise, $F_1 = 1$ since both the noise and the signal undergo equal amplification at the gain G_1:

$$G_1 = \frac{S_1}{S_0} \tag{2}$$

As a result, at a detection and amplification temperature T_d,

$$dN_1 = F_1 G_1 dN_0 = F_1 G_1 kT_d \Delta f \tag{3}$$

$$= G_1 kT_d \Delta f + (F_1 - 1)G_1 kT_d \Delta f \tag{4}$$

where we equated the input noise power dN_0 to the Johnson noise $kT_d \Delta f$. This equation means that the network amplifies the input noise $kT_d \Delta f$ by its gain G_1, and in addition it adds the noise $(F_1 - 1)G_1 kT_d \Delta f$ to the signal.

The overall or integrated noise figure N_1 of the network is obtained by integrating over all frequencies

$$N_1 = \int_0^\infty F_1 G_1 kT_d \, df \tag{5}$$

The noise power that would have appeared at the output of an ideal or noiseless network is

$$\int_0^\infty G_1 kT_d \, df$$

so we write

$$\langle F_1 \rangle = \frac{\int_0^\infty FG_1 kT_d \, df}{\int_0^\infty G_1 kT_d \, df} \tag{6}$$

The effective noise bandwidth B is defined by

$$B = \frac{1}{G_{max}} \int_0^\infty G \, df \tag{7}$$

For two networks in series, the second network with the noise figure F_2 and gain G_2 will amplify its input noise dN_1 by the gain G_2, and in addition, it will add the noise $(F_2 - 1)G_2 kT_d \, df$ to the signal. As a result, the output noise power dN_2 from the second network has the form

$$dN_2 = G_2 \, dN_1 + (F_2 - 1)G_2 kT_d \Delta f \tag{8}$$

and using Eq. (3) we may write

$$dN_2 = \left[F_1 + \frac{(F_2 - 1)}{G_1} \right] G_1 G_2 kT_d \Delta f \tag{9}$$

$$= F_{12} G_1 G_2 kT_d \Delta f \tag{10}$$

where F_{12} is the overall noise figure

$$F_{12} = F_1 + \frac{(F_2 - 1)}{G_1} \tag{11}$$

For n cascaded sections

$$dN_n = F_{12 \cdots n} G_1 G_2 \cdots G_n kT_d \Delta f \tag{12}$$

and Eq. (11) can be generalized to

$$F_{12\cdots n} = F_1 + \frac{(F_2 - 1)}{G_1} + \frac{(F_3 - 1)}{G_1 G_2} + \cdots + \frac{(F_n - 1)}{G_1 G_2 \cdots G_{n-1}} \qquad (13)$$

Thus after any high gain network, the overall noise figure is not appreciably influenced by additional networks, even though their individual noise figures are relatively high. We can immediately see the advantage of using a maser preamplifier because it introduces a gain G_1 with a noise figure F_1 close to 1.

In an ordinary spectrometer random fluctuations in the source and microwave components result in an input noise figure $F_K > 1$ before the signal reaches the detector. Hence the noise power dN_K incident on the detector is given by ($F_1 = F_K$; $G_1 = G_K \cong 1$)

$$dN_K = F_K k T_d \Delta f \qquad (14)$$

from Eq. (3) since the gain G_K before reaching the detector is assumed to be close to unity. The output noise power from the detector dN_d depends upon the detector noise figure F_d and upon the detector gain G_d. These quantities may be used to replace F_2 and G_2 in Eq. (10), but it is more customary to characterize detectors by their noise temperature t defined by

$$t = G_d F_d = G_2 F_2 \qquad (15)$$

and the insertion loss L, which is the reciprocal of the conversion gain G_d:

$$L = \frac{1}{G_d} \qquad (16)$$

Upon placing these relations in Eq. (9), we obtain for the noise power at the output of the detector

$$dN_d = \left(\frac{F_K}{L} + t - \frac{1}{L} \right) k T_d \Delta f \qquad (17)$$

From Eqs. (12) and (13) we obtain for the noise power dN_{amp} at the output of the preamplifier ($F_{amp} = F_3$; $G_{amp} = G_3$)

$$dN_{amp} = \left(\frac{F_K}{L} + t - \frac{1}{L} + F_{amp} - 1 \right) G_{amp} k T_d \Delta f \qquad (18)$$

Now that the noise output from the preamplifier has been evaluated, we consider the ESR signal under ideal conditions. For the linear detector with the reflection cavity, we have the magnitude

$$\Delta E_r = \tfrac{1}{2} \chi'' \eta Q_u E \qquad (19)$$

where the incident power is

$$P_w = \frac{E^2}{4R_0} \tag{20}$$

For the limit of detectability χ''_{min}, we equate the ESR signal ΔE_r with the RMS voltage across a terminating resistance R_0 on a transmission line of characteristic impedance R_0

$$\Delta E_r = (4R_0 k T_d \Delta f)^{1/2} \tag{21}$$

with the result that

$$\chi''_{\text{min}} = \left(\frac{2}{\eta Q_u}\right)\left(\frac{kT_d \Delta f}{P_w}\right)^{1/2} = \chi''_{\text{min(TH)}} \tag{22}$$

One should bear in mind that χ'' is in MKS units, and the right-hand side of Eq. (22) must be divided by 4π to convert it to the usual cgs units. Feher's expression for Eq. (22) has the factor $2^{1/2}$ in the numerator. Throughout this chapter we freely quote Feher's results without bothering to correct for this factor. We shall see in the next section that it is more convenient to replace P_w in Eq. (22) and the subsequent ones by an expression containing the energy density $(1/2\mu)\langle H_1^2 \rangle_w$ immediately outside the cavity iris. At X band we have the typical values $Q_u = 5000$, $\Delta f = 0.1$ Hz, $P_w = 10^{-2}$ W, $V_c \approx 10$ cm³, and $\eta = 2V_s/V_c$ [Eq. (5-H-20)]. Hence (Feher, 1957)

$$\chi''_{\text{min}} V_s \approx \frac{8\pi \times 10^{-15}}{P_w^{1/2}} \tag{23}$$

This expression is plotted in Fig. 11-4 as a function of the microwave power P_w in watts. Similar expressions are obtained for the other cavity arrangements.

The minimum voltage ΔE_R will become $(G_{\text{amp}}/L)\Delta E_R$ at the output of the preamplifier, and so we may take into account the noise introduced by the detector and preamplifier by replacing $(G_{\text{amp}}/L)\Delta E_R$ from Eq. (21) by dN_{amp} of Eq. (18) to obtain

$$\chi''_{\text{min(OB)}} = \frac{2}{Q_u \eta}\left[(F_k - 1 + (t + F_{\text{amp}} - 1)L)\frac{kT_d \Delta f}{P_w}\right]^{1/2} \tag{24}$$

We denote χ''_{min} defined by Eq. (22) as $\chi''_{\text{min(TH)}}$ and χ''_{min} defined by Eq. (24) as $\chi''_{\text{min(OB)}}$, and consider the ratio $\chi''_{\text{min(OB)}}/\chi''_{\text{min(TH)}}$:

$$\frac{\chi''_{\text{min(OB)}}}{\chi''_{\text{min(TH)}}} = \left[F_k - 1 + (t + F_{\text{amp}} - 1)L\right]^{1/2} \geq 1 \tag{25}$$

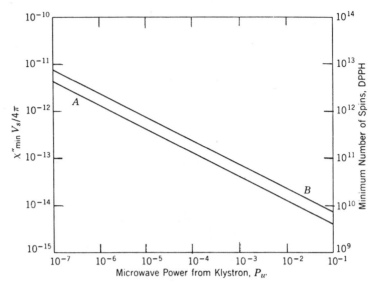

Fig. 11-4. (*A*) The minimum detectable number of DPPH spins; (*B*) the minimum detectable rf susceptibility $\chi''_{min}V_s/4\pi$ as a function of the applied microwave power P_w in watts. The curves correspond to the conditions $Q_u = 5000$, $\Delta f = 0.1$ sec^{-1}, $T = 300$ K, $\omega/\Delta\omega = 1500$, and no saturation (Feher, 1957).

as a measure of the extent to which a spectrometer approaches the ideal sensitivity where

$$L = F_k = F_{amp} = t = 1 \quad \text{and} \quad \chi''_{min(OB)} = \chi''_{min(TH)}$$

The response laws and noise temperature relations found in Secs. 7-B and 7-C may be used to evaluate these last two expressions, and this was done by Feher (1957). For bolometer detection employing a balanced mixer

$$\frac{\chi''_{min(OB)}}{\chi''_{min(TH)}} \sim 4 \tag{26}$$

as may be seen by consulting Fig. 7-2. A complicated expression is required for the simple bolometer curve on this figure.

Figure 7-8 shows the ratio of the observed to the theoretical minimum susceptibility for three arrangements of the crystal detector. In the simple straight detection scheme, the bridge is adjusted so that the microwave power P_d that is incident on the crystal equals 10% of the microwave power P_w that is incident on the cavity. We assume $F_{amp} \sim 1$ which approximates a step-up transformer. For this arrangement at a modulation frequency of 1000 Hz we obtain (Feher, 1957)

$$\frac{\chi''_{min(OB)}}{\chi''_{min(TH)}} = \left(\frac{1 + 5 \times 10^9 P_w^2}{50 P_w}\right)^{1/2} \qquad \text{quadratic}$$

$$= (3 \times 10^7 P_w)^{1/2} \qquad \text{linear} \tag{27}$$

for the square-law (quadratic) and linear regions, respectively, of the IN23C crystalline response. If optimum bucking is employed, then the actual power level at the crystal is maintained equal to the minimum value in the simple detection scheme of Fig. 7-8, and this entails biasing the crystal with a power $P_B \sim 1.4 \times 10^{-6}$ W. Using the relations of Sec. 7-C, one may show that

$$\frac{\chi''_{min(OB)}}{\chi''_{min(TH)}} = \left[2S'\left(\frac{\gamma F_{amp}}{f}\right)^{1/2}\right]^{1/2} \approx \frac{300}{f^{1/4}} = 55 \tag{28}$$

for the typical modulation frequency of 1 kHz. The optimum bucking power is proportional to the square root of the modulation frequency. Negative bucking or the reduction of the power ratio P_d/P_w when $P_w > P_B$ is easily accomplished by using the slide screw turner to bring the bridge closer to balance. When $P_B > P_w$, then it is necessary to employ an arrangement like that shown in Fig. 7-11 to obtain the necessary power at the crystal. When making saturation curve measurements at very low powers (e.g., $P_w < P_B$) in order to determine the relaxation times T_1 and T_2, it is often more convenient to work with $P_d < P_B$ on an ordinary ESR spectrometer, and thereby sacrifice some sensitivity. This avoids the necessity of setting up an arrangement such as that shown in Fig. 7-11. An examination of Fig. 7-8 reveals that when $P_d = 0.1P_B$, the sensitivity $\chi''_{min(OB)}/\chi''_{min(TH)}$ decreases by less than a factor of 2, but this allows the power measurements to be conveniently made an order of magnitude lower than is possible with the optimum leakage condition $P_d = P_B$. Indeed, by sacrificing a factor of 10 in sensitivity one can work conveniently at a power level 1/200th of the value that can be reached with the setting $P_d = P_B$.

A superheterodyne spectrometer ordinarily operates at 30 or 60 MHz and thereby renders negligible the generation of crystal noise. For this case, Feher estimates for a balanced mixer operating in the linear region of the crystal

$$\frac{\chi''_{min(OB)}}{\chi''_{min(TH)}} \sim [L(t + F_{i.f.} - 1)]^{1/2} \sim 5 \tag{29}$$

which is independent of the microwave power, where $F_{i.f.}$ is the noise figure of the 30- or 60-MHz receiver. The klystron noise tends to cancel, and the ESR signal adds in the balanced mixer. Note that the sensitivity obtained with superheterodyne detection is comparable to that obtained by using a 100-KHz field-modulation scheme. Bolometer detection also gives about the same sensitivity. The 100-kc field modulation arrangement is easier to set up and operate than the superheterodyne one, but it is not useful for detecting resonant lines that are much less than 100 mG wide. Hausser (1962) has observed lines as narrow as 17 mG.

Frater (1965) designed a synchronous demodulator that subtracts a modulated microwave signal from a calibrated microwave source and thereby increases the signal-to-noise ratio. A balanced mixer produces the same result and is more useful in ESR applications.

Teaney et al. (1961) compared the sensitivity of a superheterodyne spectrometer to that of the bimodal spectrometer discussed in Sec. 5-K. They studied several superheterodyne receivers and found the noise figures shown in Fig. 11-5 for video (homodyne) and lock-in detection. At low powers the conventional superheterodyne spectrometer produced a signal-to-noise ratio proportional to the square root of the microwave power, as expected, but the sensitivity fell off above a klystron power of 10 mW in accordance with Fig. 11-6. The decrease in sensitivity at high power levels is the result of random frequency modulation that disturbs the bridge balance. This source of noise is more troublesome when detecting dispersion than it is with absorption. It increases with the microwave power because of the bridge being brought closer to match when it is tuned to maintain the optimum power P_d at the detector. The bimodal spectrometer employs a bimodal cavity (Portis and Teaney, 1958) whose balance is completely independent of the frequency. As a result, this source of noise is minimized, and the bimodal spectrometer maintains its

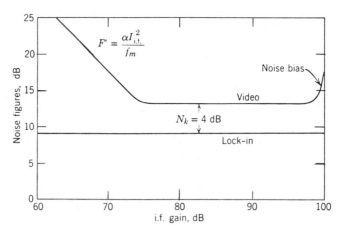

Fig. 11-5. Noise figures for superheterodyne receivers employing video and homodyne (phase-sensitive) detection of the intermediate frequency. The improved noise figure of the lock-in scheme results from the suppression of carrier noise. In the video system, the excess noise F at low i.f. gain results from flicker noise generated by the i.f. current in the mixer diodes, while that at high i.f. gain arises from the audio-frequency beat between noise components at the second detector (Teaney et al., 1961).

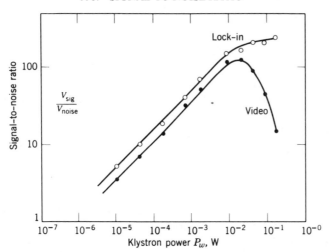

Fig. 11-6.   Signal-to-noise ratios for a bridge spectrometer. The linear parts of the curves correspond to a noise figure of 9 dB for lock-in detection and 13 dB for video detection as shown on Fig. 15-5 (Teaney et al., 1961).

expected sensitivity at high powers, as shown in Fig. 11-7. Brodwin and Burgess (1963) discuss an induction spectrometer that uses an H-plane microwave junction in lieu of a bimodal cavity. Bonori et al. (1981a, 1981b) built induction spectrometers. Tataru (1973, 1974) reports the sensitivity of ESR detection using the Faraday effect and magnetic circular dichroism, respectively. Mahendru and Parshad (1964) discuss the merit of using several parallel

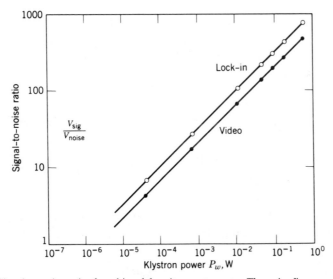

Fig. 11-7.   Signal-to-noise ratios for a bimodal cavity spectrometer. The noise figures are the same as those noted in Fig. 11-6 for the bridge spectrometer (Teaney et al., 1961).

crystal detectors when operating at high microwave power levels. Bresler et al. (1971) improved the sensitivity by employing balanced mixers.

Hedvig (1959) used a dielectric slab in a rectangular resonant cavity to appreciably increase the filling factor η and thereby enhance the sensitivity. Section 5-J gives the theory behind this technique. An enormous increase in sensitivity is obtained by making a properly shaped single crystal into a dielectric resonant cavity (compare Sec. 5-N).

Several detailed theoretical discussions of the sensitivity of ESR spectrometers have appeared (e.g., Atsarkin et al., 1962; Bresler et al., 1958; Buckmaster and Dering, 1965; Goldsborough and Mandel, 1960; Misra, 1963; Muller, 1960; Muromtsev et al., 1962; Payne, 1964; Smidt, 1960; Schneider, 1963; Tsung-tang et al., 1965; Valeriu and Pascaru, 1963; Wilmshurst et al., 1962). Various spectrometer arrangements are discussed, and their sensitivities are compared by Faulkner (1964). Korablev (1969) treats the sensitivity of a superheterodyne spectrometer using the AFC of a master oscillator. Moore and Yalcin (1973) report the experimental measurement of exponential time constants in the presence of noise. Stockhausen and Strassman (1975) compared ESR and dielectric measurements on broad background lines of manganese in solution. Mollier et al. (1973) discussed the absolute limit of sensitivity in microwave spectroscopy. Kloza (1971) treated sensitivity problems in spectrometers. Koningsberger et al. (1973) designed a cavity for ultimate sensitivity measurements up to 1300 K. Goldberg (1978) discussed improvements in ESR analytical accuracy. Chang (1974) proposed a simple setup for quantitative ESR measurements.

D. Number of Spins

The results of the last section may be combined with the Curie–Weiss law to provide a general formula for the number of spins in the sample. The dependence of the minimum detectable number of spins on various parameters such as the instrumental conditions (e.g., frequency), the spin system's environment (e.g., temperature), and its characteristics (e.g., linewidth) will now be discussed in detail.

In the previous section we derived the expression [Eq. (11-C-24)]

$$\chi_{min} = \frac{2}{Q_u \eta} \left[\frac{F_k - 1 + (t + F_{amp} - 1)L}{P_w} (kT_d \Delta f) \right]^{1/2} \tag{1}$$

for the minimum detectable rf susceptibility. The rf susceptibility $\chi''(\omega)$ is related to the static susceptibility χ_0 by the Bloch equation solution

$$\chi''(\omega) = \frac{\frac{1}{2}\chi_0 \omega_0 T_2}{1 + (\omega - \omega_0)^2 T_2^2} \tag{2}$$

This expression reaches its maximum value at $\omega = \omega_0$

$$\chi''(\omega_0) = \frac{\chi_0 \omega_0}{3^{1/2} \Delta\omega_{pp}} \tag{3}$$

where we have replaced the spin–spin relaxation time T_2 by its equivalent reciprocal linewidth $\Delta\omega$

$$T_2 = \frac{2}{\Delta\omega_{1/2}} = \frac{2}{3^{1/2} \Delta\omega_{pp}} \tag{4}$$

The static susceptibility χ_0 is related to the g factor, spin S, and sample temperature T_s by the Curie–Weiss law* (1-G-2)

$$\chi_0 = \frac{Ng^2\beta^2 S(S+1)}{3V_s k(T_s + \Delta)} \tag{5}$$

where N is the number of spins in the sample of volume V_s. When the Weiss constant Δ vanishes, we have the Curie law*

$$\chi_0 = \frac{Ng^2\beta^2 S(S+1)}{3V_s kT_s} \tag{6}$$

From Eqs. (1), (3), and (6), the minimum detectable number of spins for systems obeying the Bloch equations and Curie law is (Poole, 1958)

$$N_{min} = \frac{6V_s kT_s}{Q_u \eta g^2 \beta^2 S(S+1)} \left(\frac{\Delta\omega_{pp}}{\omega_0} \right)$$

$$\times \left[\frac{F_k - 1 + (t + F_{amp} - 1)L}{P_w} (3kT_d \,\Delta f) \right]^{1/2} \tag{7}$$

To work in magnetic field units one merely replaces the ratio $\Delta\omega_{pp}/\omega_0$ by $\Delta H_{pp}/H_0$. If all the constants are combined into one constant K, and P_w is replaced by $P_w \omega_0^2$ for reasons discussed below, then Eq. (7) simplifies to

$$N_{min} = \frac{V_s T_s K}{Q_u \eta g^2 S(S+1)\omega_0} \left(\frac{\Delta H_{pp}}{H_0} \right)$$

$$\times \left[\frac{T_d \,\Delta f}{P_w} [F_k - 1 + (t + F_{amp} - 1)L] \right]^{1/2} \tag{8}$$

One wishes to make N_{min} as small as possible for maximum sensitivity. The

*This is for free radicals and the first transition series. For rare earths, replace S by J.

number of spins in the sample N_{spin} is given by

$$N_{spin} = y'_m N_{min} \tag{9}$$

where y'_m is the signal-to-noise ratio.

For the three typical detection schemes discussed in Sec. 11-C we have

$$F_k - 1 + (t + F_{amp} - 1)L = \begin{cases} 16 & \text{bolometer} \\ \left(\dfrac{1}{500P_d} + 10^9 P_d \right) & \text{square law} \\ 3 \times 10^8 P_d & \text{linear} \end{cases} \left. \begin{matrix} \\ \\ \\ \end{matrix} \right\} \begin{matrix} \text{IN23C} \\ \text{crystal} \end{matrix} \tag{10}$$

where P_d is the "leakage" power incident on the crystal. In Sec. 5-H we found that

$$\frac{V_s}{\eta} = (\text{const})V_c \tag{11}$$

where the dimensionless constant is of the order of unity. The factor $g^2 S(S + 1)$ must be taken into account when one compares spectra from two spin systems that differ in their g factors and spins. Decreasing the sample temperature T_s increases the Q so that there are two reasons why the sensitivity increases with decreasing temperature. Decreasing the temperature of the detector also increases the sensitivity, but this is not ordinarily done in practice. Decreasing the detection–amplification bandwidth Δf by means of a narrow band amplifier and a lock-in detector can produce a pronounced improvement in sensitivity. This is elaborated upon in Secs. 11-F and 11-I. For unsaturated lines, N_{min} is inversely proportional to the square root of the microwave power, while above saturation N_{min} increases with increasing power. Thus oversaturation decreases the sensitivity.

The dependence of the sensitivity upon the frequency can be clarified by grouping into a constant K all of the terms in Eq. (7) that are independent of the frequency. It is assumed that the detection and amplification stages are in this category. In addition, the resonant cavity has its dimensions scaled in proportion to the wavelength, and the same sample is in the same relative position in the cavity. As a result Eq. (8) with $\Delta\omega_{pp}/\omega_0$ simplifies to

$$N_{min} = \frac{KV_s}{Q_u \eta \omega_0^2 (P_w)^{1/2}} \tag{12}$$

This expression has the extra ω_0 that we added to Eq. (8) in order to take into account the dependence of the energy density (i.e., $\langle H^2 \rangle_w$) in the waveguide outside the cavity iris on the total microwave power P_w. Before proceeding further, we shall justify this modification of Feher's results.

It is the average value of H_1^2 at the sample $\langle H_1^2 \rangle_s$ that produces ESR absorption, and this is related to the root-mean-square value of H_1^2 in the cavity $\langle H_1^2 \rangle_c$ through the filling factor η. Equation (5-H-74) relates $\langle H_1^2 \rangle_c$ to the rms rf magnetic field outside the cavity $\langle H^2 \rangle_w$ that results from Poynting's vector $\frac{1}{2} \mathbf{E} \times \mathbf{H}$

$$\langle H_1^2 \rangle_c = \frac{2}{\pi} Q_L \left(\frac{V_w}{V_c} \right) \langle H^2 \rangle_w \tag{13}$$

We will, of course, maintain the ratio V_w/V_c constant when we vary the frequency. For a rectangular waveguide with the large dimension $a = \frac{1}{2}\lambda_g$ ($2a = d$) and small dimension b we obtain from Eqs. (5-H-66) and (5-H-70) the following expression for the microwave power P_w incident on the cavity

$$P_w = abZ_0 \langle H^2 \rangle_w \tag{14}$$

Since the waveguide dimensions a and b are each inversely proportional to the frequency it follows that

$$\langle H^2 \rangle_w \propto \omega_0^2 P_w \tag{15}$$

which justifies the extra ω_0 that we inserted into Eq. (8).

The physical significance of Eq. (15) may be illustrated from another point of view. The total power incident on the cavity per cycle is given by $(2\pi/\omega_0)P_w$

$$\left(\frac{2\pi}{\omega_0} \right) P_w = 2\pi \left(\frac{ab}{\omega_0} \right) Z_0 \langle H^2 \rangle_w \text{ J/cycle} \tag{16}$$

and at match it all enters and is dissipated in the cavity. Since Q is defined as 2π times the ratio of the total stored energy U to the total energy dissipated per cycle [Eq. (5-A-13)], we have

$$U = \left(\frac{Q_L}{\omega_0} \right) P_w \tag{17}$$

By definition

$$U = \frac{1}{2} \mu \int_{\text{cavity}} H_1^2 \, dV = \frac{1}{2} \mu V_c \langle H_1^2 \rangle_c \tag{18}$$

and therefore

$$\langle H_1^2 \rangle_c = \left(\frac{2Q_L}{\mu V_c \omega_0} \right) P_w \tag{19}$$

in agreement with a similar expression of Zverov (1961). The use of Eq. (13) allows us to write

$$\langle H^2 \rangle_w = \left(\frac{2}{\mu V_w \omega_0} \right) P_w \tag{20}$$

which has the same frequency dependence as Eq. (14) since both ab and $V_w \omega_0$ are inversely proportional to ω_0^2.

Another way to arrive at this proportionality is to say that $\langle H^2 \rangle_w$ is proportional to Poynting's vector $\frac{1}{2} \mathbf{E} \times \mathbf{H}$, and that P_w is proportional to Poynting's vector times the waveguide cross-sectional area. Section 5-I explains how to measure H_1.

Constant Incident Power P_w

Now we return to the discussion of the frequency dependence of $N_{\min}$ in accordance with Eq. (12). If the sample size is scaled to the same extent as the cavity dimensions, then the filling factor η remains constant. We saw in Secs. 5-C and 5-D that Q_u equals a geometric factor times λ/δ. Scaling the cavity dimensions produces a Q_u proportional to λ/δ, which is inversely proportional to $\omega_0^{1/2}$, so we obtain for the minimum detectable spin concentration $N_{\min}/V_s$ (spins/cm^3 or moles/l) at a constant microwave power P_w

$$\frac{N_{\min}}{V_s} \propto \frac{1}{\omega_0^{3/2}} \tag{21}$$

Therefore for a constant filling factor η and microwave power level P_w, a factor of 2 increase in frequency produces about a factor of 3 increase [i.e., $2(2^{1/2})$] in sensitivity. This case corresponds to the usual experimental situation in which there is plenty of sample available, and the limiting factor is the amount that can be put into the cavity without appreciably lowering the Q. In a practical case, the sample dielectric loss may be frequency-sensitive, and the same filling factor may not be optimum at both frequencies.

If the sample is small and cannot conveniently be varied in size (e.g., a small single crystal), then we assume that the filling factor η is proportional to V_s/V_c independent of the frequency, which gives

$$N_{\min} = \frac{K V_c}{Q_u \omega_0^2} \tag{22}$$

Since the cavity volume V_c is proportional to ω_0^{-3} we obtain

$$N_{\min} \propto \frac{1}{\omega_0^{9/2}} \tag{23}$$

Thus for a constant sample size and a constant microwave power, it is very advantageous to go to higher frequencies. The physical reason for this is that at the higher frequency the energy density in the cavity increases and in addition the sample fills a much greater percentage of the cavity volume, which results in a greatly enhanced filling factor. A factor of 2 increase in frequency provides a factor of 22 increase in sensitivity.

Constant rf Field at Sample

It is shown in Sec. 13-C that there is an optimum value of $\langle H_1^2 \rangle_s$ at the onset of saturation that produces the greatest signal-to-noise ratio. If there is enough power available to saturate the spin system at two microwave frequencies, then the maximum sensitivity for each frequency will correspond to the same $\langle H_1^2 \rangle_s$ value. To attain this condition, each microwave generator or klystron will be set at the appropriate power level P_w. This may be deduced by noting that for the same sample–cavity relationship (except for a scaling factor) the ratio $\langle H_1^2 \rangle_s / \langle H_1^2 \rangle_c$ will be independent of frequency. From Eq. (19), we see that it is necessary to maintain a constant ratio $(Q_l P_w / V_c \omega_0)$ in order to preserve the same $\langle H_1^2 \rangle_s$ at the sample. As a result, Eq. (12) gives

$$N_{\min} = \frac{KV_s}{\eta \omega_0^2 (V_c Q_u \omega_0)^{1/2}} \tag{24}$$

and for the two cases of a constant ratio of V_s/V_c and a constant V_s, respectively, we obtain

$$\frac{N_{\min}}{V_s} \propto \frac{1}{\omega_0^{3/4}} \tag{25}$$

$$N_{\min} \propto \frac{1}{\omega_0^{15/4}} \tag{26}$$

Constant Incident Poynting Vector

Feher (1957) compared sensitivities for the case in which $\langle H^2 \rangle_w$ is kept constant, and he obtained the resulting equations for $V_s/V_c = \text{const}$ and $V_s = \text{const}$, respectively [see Eq. (15)]:

$$\frac{N_{\min}}{V_s} \propto \frac{1}{\omega_0^{1/2}} \tag{27}$$

$$N_{\min} \propto \frac{1}{\omega_0^{7/2}} \tag{28}$$

TABLE 11-1

Dependence of the Minimum Detectable Concentration (N_{min}/V_s) and the Minimum Detectable Number of Spins (N_{min}) on the Microwave Frequency ω_0

Spectrometer Conditions	N_{min}/V_s (η = const)	N_{min} (V_s = const)
Constant microwave power P_w	$1/\omega_0^{3/2}$	$1/\omega_0^{9/2}$
Constant energy density at sample ($\langle H_1^2 \rangle_s$ = const) ($\langle H_1^2 \rangle_c$ = const)	$1/\omega_0^{3/4}$	$1/\omega_0^{15/4}$
Constant energy density in waveguide outside sample ($\frac{1}{2}\mathbf{E} \times \mathbf{H}$ = const) ($\langle H_\omega^2 \rangle$ = const)	$1/\omega_0^{1/2}$	$1/\omega_0^{7/2}$

The results obtained for these six cases are summarized in Table 11-1. These expressions are useful for comparing results obtained at different frequencies. Bhagat and Venkataraman (1976) describe a simple microwave kit to convert an X-band spectrometer for K-band use. Müller et al. (1976) discussed variable frequency spectrometers using slow wave helices. Newman and Urban (1975) described a variable frequency ESR technique.

The effect of modulating the magnetic field at an amplitude H_{mod} much less than the linewidth ΔH_{pp} is to produce a resonant line whose amplitude is proportional to H_{mod}. When H_{mod} greatly exceeds ΔH_{pp}, then further increases produce a weaker ESR line. One may approximate the effects of H_{mod} on the sensitivity by multiplying Eq. (7) or (8) by $(\Delta H_{pp}/H_{mod})$ when $H_{mod} \ll \Delta H_{pp}$ as is done explicitly in Eq. (32). Section 6-H gives further details on this effect.

The area A of an ESR resonance absorption is given by (Table 12-5)

$$A = \begin{cases} 1.06 y_m \Delta H_{1/2} & \text{Gaussian} & (29a) \\ 1.57 y_m \Delta H_{1/2} & \text{Lorentzian} & (29b) \\ \Lambda y_m \Delta H_{1/2} & \text{general} & (29c) \end{cases}$$

where y_m is the peak amplitude, and Λ is a general shape factor. For a first derivative spectrum, Table 12-5 gives

$$A = \begin{cases} 1.03 y_m' (\Delta H_{pp})^2 & \text{Gaussian} & (30a) \\ 3.63 y_m' (\Delta H_{pp})^2 & \text{Lorentzian} & (30b) \\ \Lambda' y_m' (\Delta H_{pp})^2 & \text{general} & (30c) \end{cases}$$

where y'_m is the derivative amplitude and Λ' is a function characteristic of the lineshape. The sensitivity relations such as Eq. (8) developed earlier in this section were for a Lorentzian lineshape. They may be generalized by multiplying the right-hand side by $\Lambda/1.57$ for absorption lines and $\Lambda'/3.63$ for first derivative lines as is done in Eq. (30). This is important when comparing two resonances with different lineshapes.

Eaton and Eaton (1980) reported that the area can be measured to within a few percent if care is taken to hold constant or to correct for various spectrometer and sample features. Spectrometer features included linearity of the magnetic field scan, amplifier gain and modulation amplitude, deviation from square law detector response, distribution of microwave and modulation fields over the sample, and distortions arising from sample tubes, Dewars and background signals. Sample features include sample holder material and wall thickness, solvent or matrix type, relaxation times, peak width, lineshape, g value, and spin multiplicity.

When the resonance consists of several hyperfine components, we may take this into account by a summing over the signal-to-noise ratios y_i of the component lines $\Sigma y'_i$. Let

$$D_m = \frac{1}{y'_m} \Sigma y'_i \qquad (31)$$

where y'_m is the signal-to-noise ratio of the strongest line, and D_m is the corresponding multiplicity factor. The parameter D is the amount by which the signal-to-noise ratio of the strongest hyperfine component must be multiplied to give the equivalent amplitude of the singlet that would result if all of the hfs collapsed to a single line of the same width. For n equally coupled protons or other spin-$\frac{1}{2}$ nuclei, we have the multiplicity factors shown in Table 11-2. These data assume fully resolved hyperfine components with a binomial intensity ratio. The experimental multiplicity factor will decrease as the resolution decreases, as can be seen from an examination of Fig. 1-3.

Hydrazyl (DPPH) in benzene solution has two equally coupled nitrogen nuclei with spin $I = 1$, and an intensity ratio $1:2:3:2:1$. Therefore its D_m value is 3. However, the first derivative lines overlap so that the minimum of

TABLE 11-2

Multiplicity Factors D_m for n Fully Resolved
Equally Coupled Nuclei with Nuclear Spin $I = \frac{1}{2}$ (Poole, 1958)]

n	D_m	n	D_n
1	2.00	5	3.20
2	2.00	6	3.20
3	2.67	7	3.66
4	2.67	8	3.66

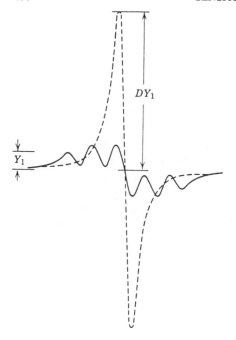

Fig. 11-8. Quintet Spectrum of DPPH (—) showing the hypothetical singlet (---) that would be obtained if $I = 0$ for ^{14}N. The figure was prepared by O. F. Griffith III.

one line is near the maximum of the next, and as a result all but the two outside components are considerably decreased in intensity. Consequently the measured D_m value is much greater than the calculated value. To circumvent this difficulty, one may use the outermost peak (line number 1) of the hfs since its amplitude is not effected by the overlap of the components. Its multiplicity factor D_1 is 9, which means that the amplitude of this first component is $1/9$ of the singlet amplitude that would result if I equaled zero for nitrogen. Knowing this enables the outside DPPH line to be used in checking ESR spectrometer sensitivity, as shown in Fig. 11-8.

The appropriate D value may be placed in the numerator of the right-hand of the sensitivity relations (χ''_{min} and N_{min}) to deduce N_{min} in the presence of hfs.

If the modulation amplitude, lineshape, and hyperfine structure (multiplicity factor) are taken into account, then Eq. (8) will have the form

$$N_{min} = \frac{V_s T_s DK}{Q_u \eta g^2 S(S + 1)\omega_0} \left(\frac{\Delta H_{pp}}{H_0} \right) \left(\frac{\Delta H_{pp}}{H_{mod}} \right)$$

$$\times \left(\frac{\Lambda'}{3.63} \right) \left[\frac{T_d \Delta f}{P_w} [F_k - 1 + (t + F_{amp} - 1)L] \right]^{1/2} \quad (32)$$

for first derivative lines. If the sample is lossy, then Q_u in a sensitivity formula such as this should be replaced by Q'_u defined in Sec. 11-J. The number of spins in a sample N_{spin} is related to the minimum detectable number N_{min} by the signal-to-noise ratio y'_m:

$$N_{spin} = N_{min} y'_m \tag{33}$$

In practice, the number of spins in a sample is often determined by comparing the sample under study with a standard sample. It is customary to carry out this comparison with the same spectrometer, so most of the terms in Eq. (32) will be identical for both samples. Therefore one may write

$$\frac{N_{spin}^A}{N_{spin}^B} = \frac{V_s^A D^A / Q_u^A \eta^A (g^A)^2 S^A (S^A + 1)(P_w^A)^{1/2}}{V_s^B D^B / Q_u^B \eta^B (g^B)^2 S^B (S^B + 1)(P_w^B)^{1/2}}$$

$$\times \frac{(\Delta H_{pp}^A / H_0)(\Delta H_{pp}^A / H_{mod}^A)(\Lambda'^A / 3.63) y'^A_m}{(\Delta H_{pp}^B / H_0)(\Delta H_{pp}^B / H_{mod}^B)(\Lambda'^B / 3.63) y'^B_m} \tag{34}$$

where the superscripts A and B refer to the two types of spins. A comparison of this type is ordinarily carried out in the same sample tube ($\eta = $ const, $V_s = $ const), with a negligible change in Q ($Q_l = $ const), and at the same power level ($P_w = $ const) and modulation amplitude ($H_{mod} = $ const), so that one may simplify Eq. (34) to

$$\frac{N_{spin}^A}{N_{spin}^B} = \left(\frac{D^A}{D_B}\right)\left(\frac{g^B}{g^A}\right)^2 \left[\frac{S^B(S^B + 1)}{S^A(S^A + 1)}\right]$$

$$\times \left(\frac{\Delta H_{pp}^A}{\Delta H_{pp}^B}\right)^2 \left(\frac{\Lambda'^A}{\lambda'^B}\right)\left(\frac{H_{mod}^B}{H_{mod}^A}\right)\left(\frac{y'^A_m}{Y'^B_m}\right) \tag{35}$$

This expression is simplified even further when both types of spins have the same spin S and g factor (e.g., when both are free radicals). The ratio (y'^A_m / y'^B_m) may be looked upon as the ratio of the amplitudes on the recorder.

E. Temperature

When a spin system is studied at a series of temperatures and microwave frequencies, it is often desirable to compare the resulting data with each other. In this section the ordinary ESR doublet is discussed in terms of its Boltzmann populations, and two graphs are presented to assist in comparing experimental data. The less common triplet state is treated from the same viewpoint, and the results are compared graphically to those obtained with the doublet.

Doublet State

For a spin system with $S = \frac{1}{2}$ the population of spins in the upper energy level n_2 will be related to the number in the lower energy level n_1 by the Boltzmann distribution

$$n_2 = n_1\exp\left(-\frac{\Delta E}{kT}\right) \tag{1}$$

where the energy separation ΔE for the Zeeman term in the Hamiltonian for $g = 2$ has the value

$$\Delta E = \hbar\omega = g\beta H \sim \tfrac{1}{3}\ cm^{-1} \sim 6.3 \times 10^{-24}\ J \tag{2}$$

at X band (9.6 GHz). Since $k = 1.38 \times 10^{-23}$ J/deg, we obtain

$$n_2 = n_1\exp\left(-\frac{0.46}{T}\right) \sim n_1\left(1 - \frac{0.46}{T}\right) \tag{3}$$

where the binomial expansion is valid above liquid-helium temperature. The ESR signal Y_D is proportional to the population difference $n_1 - n_2$, and this has the form

$$Y_D = Y_0\left(\frac{n_1 - n_2}{n_1 + n_2}\right) = Y_0\left[\frac{1 - \exp(-\hbar\omega/kT)}{1 + \exp(-\hbar\omega/kT)}\right] \tag{4}$$

where Y_0 is the signal amplitude at $T \to 0$. The temperature dependence of the ESR signal Y_D is shown in Figs. 11-9 and 11-10 for several typical microwave frequencies and for several typical magnetic field strengths (assuming $g = 2.0$ in the latter case). Note that at X band (10 GHz), it is necessary to lower the temperature to the liquid helium range to appreciably populate the lower Zeeman level and render the ratio Y_D/Y_0 greater than 10%.

At high temperatures ($\hbar\omega \ll kT$), the exponential term may be expanded in a power series, and the sensitivity becomes inversely proportional to the temperature. At low temperature ($\hbar\omega \gg kT$), the exponentials may be neglected, and the sensitivity becomes independent of temperature. These limiting cases may be summarized by the relations

$$Y_D = \left(\frac{\hbar\omega}{2kT}\right)Y_0 = \left(\frac{g\beta H}{2kT}\right)Y_0 \qquad \hbar\omega \ll kT \tag{5}$$

$$Y_D = Y_0 \qquad\qquad\qquad\qquad\qquad \hbar\omega \gg kT \tag{6}$$

Equation (5) corresponds to almost every case discussed elsewhere in this book. In particular, the discussion in the last section is based upon the Curie law (11-D-6).

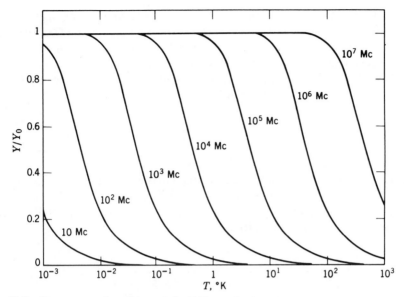

Fig. 11-9. Temperature dependence of the ESR amplitude $Y = Y_D$ relative to its value Y_0 at $T = 0$ for several microwave frequencies.

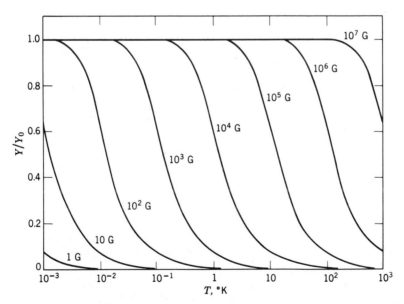

Fig. 11-10. Temperature dependence of the ESR amplitude $Y = Y_D$ relative to its value Y_0 at $T = 0$ at several magnetic field strengths for a g factor of 2.0.

411

Variations in the temperature will result in other more complicated effect that influence the sensitivity. As the temperature is lowered, the Q of the cavity will increase to render the system more sensitive. For liquid-helium studies the spin-lattice relaxation time of solids often becomes very long (Poole and Farach, 1971a). This produces resonant lines that saturate at relatively low (e.g., microwatt) power levels [compare Eqs. (13-C-3) and (13-C-4)].

Triplet State

The paramagnetic absorption associated with a triplet state has been discussed by Bijl et al. (1959). It corresponds to the energy level diagram (b) or (c) of Fig. 11-11. In the case of a *low lying triplet* (Fig. 11-11b), the ESR signal arises from the two transitions $1 \to 2$ and $2 \to 3$, and its amplitude Y_{LT} is determined by the average populations as follows

$$Y_{LT} = Y_0 \left[\frac{n_1 - n_2}{n_1 + n_2 + n_3 + n_s} + \frac{n_2 - n_3}{n_1 + n_2 + n_3 + n_s} \right] \tag{7}$$

$$\approx \frac{1 - \exp(-2\hbar\omega/kT)}{1 + \exp(-\hbar\omega/kT) + \exp(-2\hbar\omega/kT) + \exp(-\Delta E_T/kT)} Y_0 \tag{8}$$

where the singlet population $n_s \ll n_1$ is neglected since $\Delta E_T \gg \hbar\omega$. Above the

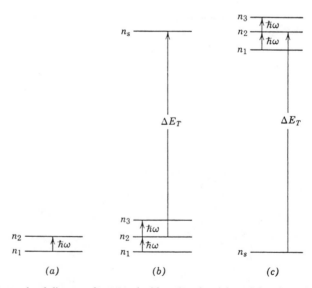

Fig. 11-11.   Energy level diagrams for (a) a doublet, (b) a low-lying triplet plus a singlet, and (c) a singlet plus a high-lying triplet state. The symbol n_i refers to the population of the ith energy level.

liquid-helium temperature range $\hbar\omega \ll kT$ and

$$Y_{LT} \approx \frac{2\hbar\omega Y_0}{kT} \frac{1}{[3 + \exp(-\Delta E_T/kT)]} \tag{9}$$

For a *high lying triplet* such as the one shown in Fig. 11-11c, one has from Eq. (7)

$$Y_{HT} \approx \frac{[1 - \exp(-2\hbar\omega/kT)]}{1 + [1 + \exp(-\hbar\omega/kT) + \exp(-2\hbar\omega/kT)]}$$

$$\times \frac{\exp[(\hbar\omega - \Delta E_T)/kT]}{\exp[(\hbar\omega - \Delta E_T)/kT]} Y_0 \tag{10}$$

$$\approx \frac{2\hbar\omega_0 Y_0}{kT} \frac{1}{[3 + \exp(\Delta E_T/kT)]} \tag{11}$$

since we assume $\Delta E_T \gg \hbar\omega$.

The ratio of the sensitivity (i.e., amplitude ratio) for each triplet case relative to the doublet one for $\hbar\omega \ll kT$ is given by

$$\frac{Y_{LT}}{Y_D} = \frac{4}{3 + \exp(-\Delta E_T/kT)} \tag{12}$$

$$\frac{Y_{HT}}{Y_D} = \frac{4}{3 + \exp(\Delta E_T/kT)} \tag{13}$$

The temperature dependence of these ratios is shown graphically in Fig. 11-12. The low-lying triplet provides greater ESR sensitivity than an ordinary doublet at all temperatures, while the ESR signal from the high-lying triplet is relatively strong only for temperatures that correspond to $kT \gg \Delta E_T$. The ratios Y_{LT}/Y_0 and Y_{HT}/Y_0 may be obtained by comparing Figs. 11-9, 11-10, and 11-12. In this section we did not take into account the $S(S + 1)$ term in Eqs. (11-D-7) and (11-D-8), which will also affect the sensitivity ratio of triplet-to-doublet spin states.

Burland et al. (1976), Dietrich and Schmid (1976), Reineker (1975, 1976, 1976, 1978), and Zieger and Wolf (1978) discuss lineshapes of triplet excitons, and Andreev and Sugakov (1975), Konzelmann et al. (1975), Kottis and Lefebvre (1963, 1964), Krebs and Sackmann (1976), Schouler et al. (1976), Vergragt et al. (1979), and Zinsli and El-Sayed (1973) treated triplet states. Dobryakov et al. (1976), Genoud and Decorps (1979), Glarum and Marshall (1975), Henling and McPherson (1979), Hinkel et al. (1978), and Lemaistre and Kottis (1978) reported on radical pairs, and Corvaja and Pasimeni (1976) and Mukai and Sakamoto (1978) studied biradicals.

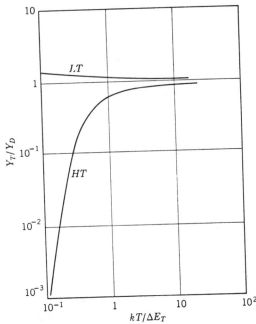

Fig. 11-12. Ratio of the ESR amplitude for a triplet state Y_T relative to a doublet state Y_D for a low-lying triplet LT and a high-lying triplet HT. The abscissa kT is normalized relative to the triplet energy ΔE_T defined on Fig. 11-11. [adapted from Bijl et al. (1959)].

F. Narrow Band Amplification and Detection

The ESR signal appears at a single frequency (e.g., 100 kHz or 30 MHz) while the noise that accompanies the signal has components extending over the entire spectrum of frequencies that can be passed by the detector and preamplifier. The signal-to-noise ratio is considerably increased by narrow band amplification and detection (compare Chap. 7). The narrow band amplifiers remove noise components that lie outside of their passband of width Δf centered at the modulation frequency f_m.

The lock-in detector (synchronous detector or coherent detector) is more efficient because it removes noise components that differ from the signal in either frequency or phase. The time-constant network between the lock-in detector and the recorder removes the noise whose frequency exceeds the reciprocal of the time constant. The enhancement scheme discussed below in Sec. 11-I removes noise at frequencies both above and below the reciprocal of the effective time constant. Since the Johnson noise power discussed earlier in this chapter is proportional to the bandwidth, the noise-reducing circuits under discussion may be looked upon as decreasing the effective bandwidth of the system. Figure 11-13 shows schematically the relationship between the effective bandwidth Δf_2, of the tuned amplifier and that of the lock-in detector RC filter combination. A 100-kc narrow band amplifier might have a Q of 20 to

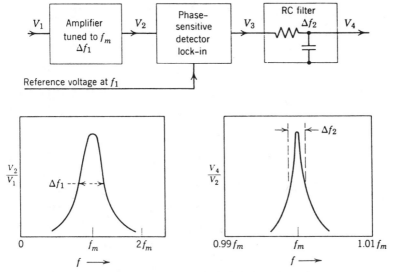

Fig. 11-13.   The decrease in the effective bandwidth Δf when the voltage V_2 traverses the phase-sensitive detector and RC filter [adapted from Feher (1957)].

give it a bandwidth of 5 kHz while a 100-kHz coherent detector followed by a 10-sec time constant might have an effective bandwidth of 0.1 Hz.

Štirand (1962) studied the effects of a linear network of time constant τ on the position and shape of a first derivative signal. He assumed a Gaussian lineshape:

$$Y'(H) = 1.65 y'_m \left(\frac{H - H_0}{\frac{1}{2} \Delta H_{pp}} \right) \exp\left[-\frac{1}{2} \left(\frac{H - H_0}{\frac{1}{2} \Delta H_{pp}} \right)^2 \right] \tag{1}$$

with a linear scan

$$\frac{(H - H_0)}{\Delta H_{pp}} = \frac{t}{b_0} \tag{2}$$

where b_0 is the peak-to-peak linewidth ($b_0 = \Delta t_{pp}$) of the undistorted line in temporal units. The undistorted lineshape $Y'(t)$ considered as a function of time t is

$$Y'(t) = 1.65 y'_m \left(\frac{t}{b_0/2} \right) \exp\left[-\frac{1}{2} \left(\frac{t}{b_0/2} \right)^2 \right] \tag{3}$$

When the first derivative signal (3) emerges from the lock-in detector, it frequently traverses a time constant (τ) network that converts it to a new

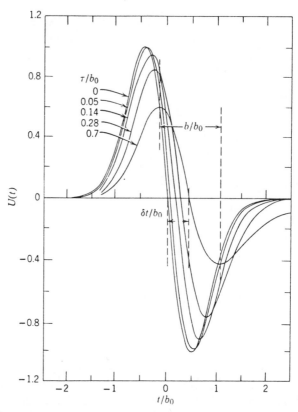

Fig. 11-14. Variation in the shape and position of the ESR line $U(t)$ after it emerges from a filter of time constant τ. Values of b/b_0 and $\delta t/b_0$ are shown for the broadest line where $\tau/b_0 = 0.7$ [adapted from Štirand (1962)].

distorted shape function $U(t)$. When $\tau \ll b_0$, we have no distortion, and so

$$\lim_{\tau/b_0 \to 0} U(t) = Y'(t) \qquad (4)$$

When the time constant τ is not negligible relative to the undistorted temporal linewidth b_0, then the distorted lineshape $U(t)$ has the form shown in Fig. 11-14. Note that for increasing values of the t/b_0, the following three changes take place in $U(t)$: (1) the amplitude decreases, (2) the linewidth b broadens ($b = \Delta t_{pp(\text{OB})}$ increases), and (3) it takes longer to reach the crossover point in the center of the line where $U(t) = 0$. The normalized increase in the peak-to-peak linewidth b/b_0 is shown in Fig. 11-15, and the time delay $\delta t/b$ in reaching the crossover point is depicted in Fig. 11-16 as a function of the normalized time constant τ/b_0.

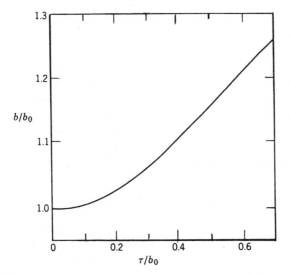

Fig. 11-15. The variation in the linewidth b/b_0 as a function of the normalized filter time constant τ/b_0 [adapted from Štirand (1962)].

These figures show that when $\tau \ll b_0$, the time delay effect is much greater than the increase in width or distortion in shape. Fortunately we are ordinarily not bothered by the time delay, while the amplitude–width–shape errors can be serious. One should note that the lineshape of the first half of the spectrum (the part that traverses the filter first) corresponding to $t/b_0 < 0$ is much less distorted than the second half where $t/b_0 > 0$. Figure 7-23 exhibits the same

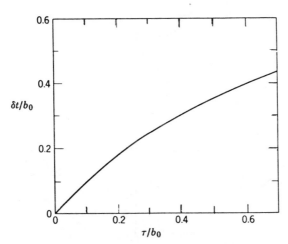

Fig. 11-16. The variation of the shift in position $\delta t/b_0$ of the ESR line as a result of the passage through the filter of normalized time constant τ/b_0 [adapted from Štirand (1962)].

characteristics as the above figures. For best reproducibility one should work with $\tau \ll b_0$, while for best sensitivity it is advantageous to use $\tau \sim b_0$. A useful *lex pollicis* is to select τ equal to one-tenth b_0.

In some cases the optimum setting of the lock-in detector differs from the standard setting. This occurs, for example, when saturation transfer effects exist (compare Sec. 13-G) and in the intermediate motion region of liquids (compare Sec. 12-O) characteristic of spin labels (compare Sec. 12-S). Under these conditions the phase setting becomes an adjustable paramater. Alquie et al. (1971) separated two spectral lines by varying the phase, and Edgar (1975) discussed phase reversals in spectra.

G. Resonant Cavities and Helices

Wilmshurst et al. (1962) analyzed the sensitivity of various ESR spectrometers in terms of their optimum operating points on a Smith chart (Sec. 2-C of the first edition). They examined the transmission cavity, absorption cavity, and reflection cavity schemes shown in Fig. 11-17 and found the corresponding

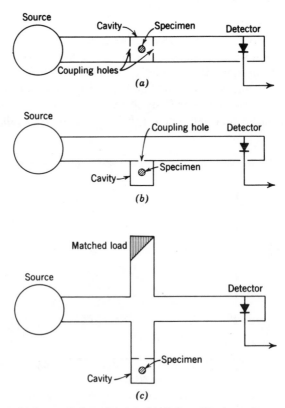

Fig. 11-17. Schematic representation of (*a*) transmission, (*b*) absorption, and (*c*) reflection cavity spectrometers (Wilmshurst et al., 1962).

relative sensitivities Y_{min} to vary with the tuning conditions in the manner shown in Fig. 11-18. The data on these three figures fit the equations

$$Y_{min} = \begin{cases} \dfrac{4(\alpha - 1)}{\alpha^2} & \begin{cases} \text{transmission } (Y_T) \\ \text{absorption } (Y_A) \end{cases} \\[2ex] \dfrac{1 + \Gamma}{1 - \Gamma} & \text{reflection } (Y_R) \end{cases} \tag{1}$$

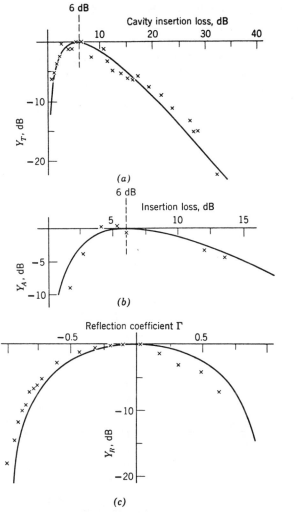

Fig. 11-18.   Variation of the relative sensitivity Y with (a) the insertion loss of a transmission cavity, (b) the insertion loss of an absorption cavity, and (c) the reflection coefficient of a reflection cavity. The calculated solid curves are fitted to the data at their respective maxima (Wilmshurst et al., 1962).

where α is the voltage insertion ratio, Γ is the voltage reflection coefficient, and of course, $Y_R =$ VSWR for the reflection cavity. In terms of Q, the optimum coupling in each case corresponds to a loaded Q equal to one half of the unloaded Q.

Under the conditions of maximum sensitivity the dispersion mode χ' is repressed with the transmission and absorption cavity arrangements. The reflection cavity is considerably more flexible since it may be conveniently tuned to provide either absorption χ'' or dispersion χ' under optimum coupling conditions. The absorption cavity is useful for generating circular polarization to determine the sign of the g factor (see Sec. 5-O).

The diagram of a reflection helix or traveling wave tube spectrometer is given in Fig. 11-19, and the equivalent circuits of three helix spectrometers are shown in Fig. 11-20. Wilmshurst et al. (1962) showed that for the transmission case the rate of change of the detected voltage E_l with the resistivity of the helix r is given by

$$\frac{1}{E_g}\frac{\partial E_l}{\partial r} = -\frac{1}{2}j\omega cx\frac{2\sinh\gamma x + \left(n^2 + 1/n^2\right)\cosh\gamma x}{\left[2\cosh\gamma x + \left(n^2 + 1/n^2\right)\sinh\gamma x\right]^2} \tag{2}$$

where γ is the propagation constant in the helix [compare Eq. (2-D-13) of the first edition]:

$$\gamma = \alpha + j\beta = \left[j\omega c(r + j\omega l)\right]^{1/2} \tag{3}$$

and normally $r \ll \omega l$. In these relations E_g is the generator voltage, n is the

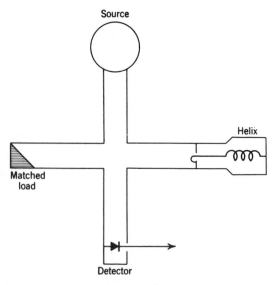

Fig. 11-19.   The essential elements of a reflection helical spectrometer (Wilmshurst et al., 1962).

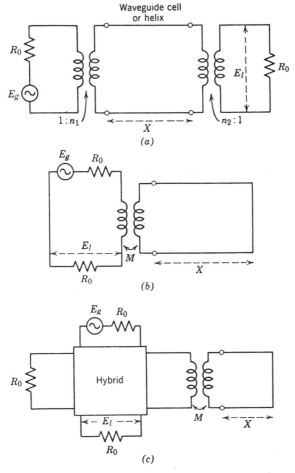

Fig. 11-20.   The equivalent circuits for (*a*) transmission, (*b*) absorption, and (*c*) reflection helical spectrometers (Wilmshurst et al., 1962).

transformer ratio, and x is the length of the helix. Equation (2) has a maximum with respect to n^2 when

$$n^2 = 1 \qquad\qquad \text{traveling waves} \qquad\qquad (4)$$

$$n^2 + \frac{1}{n^2} = \frac{2(1 - \sinh^2 \gamma x)}{\sinh \gamma x \cosh \gamma x} \quad \text{standing waves} \qquad (5)$$

The traveling-wave mode leads to the expression

$$\frac{1}{E_g} \frac{\partial E_l}{\partial r} = -\frac{j\omega cx}{4\gamma} \exp(-\gamma x) \qquad\qquad (6)$$

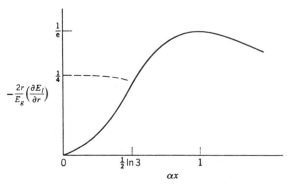

Fig. 11-21. Variation in the relative sensitivity of the helix with length x when operating in (—) the traveling wave mode and (---) the standing wave mode (Wilmshurst et al., 1962).

and its solution

$$E_l = \tfrac{1}{2} E_g \exp(-\gamma x) \tag{7}$$

The sensitivity for absorption $(2r/E_g)(\partial E_l/\partial r)$ is a maximum when $x = 1/\alpha$, and varies with the length of the helix, as shown in Fig. 11-21. At the point of maximum sensitivity we have

$$\frac{2r}{E_g} \frac{\partial E_l}{\partial r} = \frac{1}{e} \tag{8}$$

where e is the natural logarithm base (compare Table 12-5). The standing wave mode leads to maximum sensitivity when x is an integral number of half-wavelengths ($\beta x = m\pi$) in the range $0 < \alpha x < (\tfrac{1}{2})\ln 3$, with the explicit values at its endpoints:

$$
\begin{aligned}
\frac{2}{E_g} \frac{\partial E_l}{\partial r} &= -\frac{j\omega c}{8\gamma\alpha} & \alpha x &= 0 \\
&= -\frac{0.63 j\omega c}{8\gamma\alpha} & \alpha x &= \frac{1}{2}\ln 3
\end{aligned}
\tag{9}
$$

It is best to keep x as small as possible to maximize the sensitivity $(2/E_g)(\partial E_l/\partial r)$. From Eq. (9) with $\alpha x = 0$ we can show that [compare first edition, p. 567]

$$\frac{2r}{E_g} \frac{\partial E_l}{\partial r} = -\frac{1}{4} \tag{10}$$

The sensitivities of the two modes are compared in Fig. 11-21 as function of αx. Equation (10) leads to the same result for the minimum detectable signal as

(11-C-19). Wilmshurst made a similar analysis of the absorption and reflection traveling-wave spectrometers and found that they all produce about the same sensitivity as the cavity spectrometers, provided the cavities are no more than a few half-wavelengths long.

The advantages of a slow wave structure or traveling-wave tube are that the effective length is increased by the slowing factor S defined by

$$S = \frac{\text{velocity of light } in \ vacuo}{\text{effective axial velocity of microwave energy}} \tag{11}$$

This is another way of saying that the microwave energy is concentrated by the factor S. Modulating coils and optical irradiation may easily be introduced into the helix, and in double resonance experiments, the helix can also serve as the NMR coil. Since the helix is broadbanded, it is not necessary to stabilize the source as precisely as in the cavity case, and frequency modulation noise (Bosch and Gambling 1961, 1962) becomes unimportant. The helix can be used over a wide range of frequencies. It may be matched at room temperature and then immersed in liquid nitrogen or helium without appreciably changing its properties (Webb, 1962). A slight disadvantage is the fact that the axis of polarization varies periodically along the helix. Slowing factors of 100 can be achieved by helices, and other types of slow wave structures are able to produce $S \sim 1000$ with smaller bandwidths (DeGrasse et al., 1959). The effective Q of a slow wave structure of length x is given by

$$Q = 4\pi S\left(\frac{x}{\lambda}\right) \tag{12}$$

where λ is the free-space wavelength. Mittra (1963) discusses wave propagation on helices.

Webb (1962) discusses the matching of a helix to a transmission line, and three of his arrangements are presented in Fig. 11-22. In this figure, screw A compensates for the frequency dependence of the match and screw B adjusts for different sample loadings. Screw A may be adjusted to provide either absorption or dispersion modes. A complete double resonance spectrometer incorporating a helix is shown in Fig. 11-23. Hausser and Reinhold (1961) used a helical waveguide for an Overhauser effect experiment. ESR double resonance spectrometers that employ helices were designed by Werner (1964) and Kenworthy and Richards (1965). Muller (1976) and Volino (1969) applied helices to ESR.

Before concluding this section a few words of practical importance will be added about resonators. It is very important to keep clean the inside surface of the resonant cavity and everything that enters it such as sample tubes and Dewars. After a spectrum is recorded, it is always wise to repeat the scan with an empty sample tube. Sometimes quartz produces a weak "blank" absorption near $g = 2$, but it is much superior to Pyrex in this respect (Brown, 1975, Amanis et al., 1975). The inner surface of a resonant cavity is often gold-plated

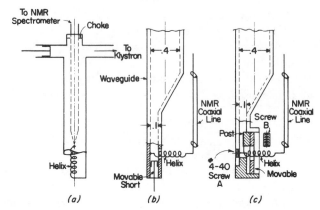

Fig. 11-22. Methods of matching a helix (a) to a coaxial line and, (b) and (c), to a waveguide. The latter two cases show the NMR input for double-resonance studies (Webb, 1962).

to minimize spurious resonances. Solvents such as phene (benzene) or dimethylketone (acetone) may be employed to remove organic deposits from cavity walls.

During low-temperature studies, the condensation of moisture within the resonant cavity can be troublesome owing to its high dielectric loss. The losses due to water vapor and oxygen are discussed in RLS-23 (p. 663).

H. Masers and Parametric Amplifiers

It was shown in Secs. 11-B and 11-C that the detector and preamplifier not only amplify the incoming signal with its accompanying noise, but in addition,

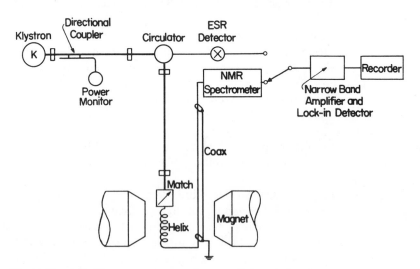

Fig. 11-23. A double-resonance spectrometer employing a helix (Webb, 1962).

they introduce additional noise of their own. It is possible to minimize the introduction of such additional noise by the use of a maser, parametric amplifier, or traveling-wave amplifier. The first two will be discussed here. Section 3-B treats traveling-wave tubes.

Masers or microwave amplifiers for stimulated emission of radiation have many applications as low noise amplifiers (Kollberg, 1973; Reid, 1973; Siegman, 1964), and their use in ESR was reviewed in Sec. 14-H of the first edition. A detailed discussion of masers as such is beyond the scope of this book, so we will limit ourselves to their application to ESR spectrometers. The advantage of employing a maser for a preamplifier is its very low noise figure. It provides a stage of amplification before detector and transistor noise enters to obscure the signal.

The first edition of this work contained a description of a maser (Sec. 13-O) and of a superheterodyne K-band spectrometer that incorporated an ammonia maser with a 30-Hz bandwidth as a preamplifier [Sec. 14-H, Gambling and Wilmshurst (1963)]. This maser and preamplifier operated at liquid-nitrogen temperature and provided up to 17 dB improvement to signal-to-noise ratio. Mollier et al. (1972) discussed the sensitivity of microwave spectrometers using masers.

Lecar and Okaya (1963) made use of a ruby laser as a preamplifier in a spectrometer. The laser operated at liquid-helium temperature. Flesner et al. (1977) employed the two-stage reflection ruby maser illustrated in Fig. 11-24 to improve the sensitivity of a transmission spectrometer (Schultz, 1972). This maser has an electronic tuning range from 9 to 10 GHz, a gain in excess of 20 dB, and an instantaneous bandwidth exceeding 10 MHz. All microwave components except the masing section are at room temperature. The maser improves the system signal-to-noise ratio by a factor of 5, as illustrated by the recording presented in Fig. 11-25. The maser design of Flesner et al. has been scaled down to provide a K-band device (Clauss, 1975). Hardin and Uebersfeld (1971) constructed a spectrometer in which the beam of active molecules of an ammonia maser enters a transmission cavity resonator, and the waveguide from one side of this transmission cavity is terminated by a reflection sample cavity located between the polepieces of a magnet. This spectrometer can detect ESR signals at power levels as low as 10^{-12} W. Herman (1969) discussed the influence of the lineshape on the power amplification characteristics of a multicavity reflection maser.

There are other types of microwave amplifiers besides masers. Hoentzsch et al. (1978) introduced a GaAs low-noise microwave preamplifier in a homodyne spectrometer and this increased the signal-to-noise ratio by more than a factor of 10 at low modulation frequencies to render it superior to a superheterodyne spectrometer. The block diagram of the spectrometer with the GaAs amplifier (No. 10) in series with a uniline isolator is presented in Fig. 11-26. The noise figure plots of Fig. 11-27 indicate the superiority of the present arrangement over spectrometers without preamplification.

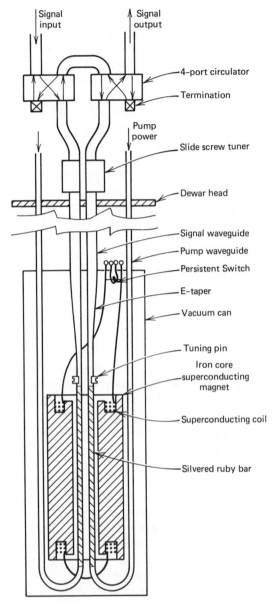

Fig. 11-24. Schematic diagram of a two-stage reflection ruby maser for use in a transmission spectrometer (Flesner et al., 1971).

426

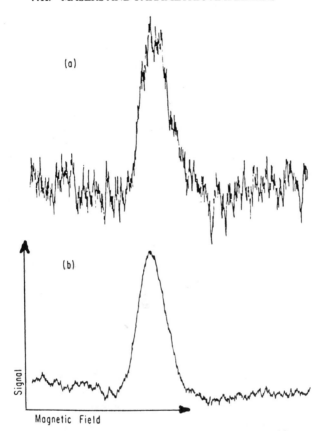

Fig. 11-25. Transmission ESR spectra of 60 ppm Er in Ag at 1.4 K (*a*) without maser and (*b*) with maser (Flesner et al., 1977).

Another device that can provide low-noise, high-frequency amplification is the parametric amplifier. This also goes under the name of junction diode, mavar (modulation amplification through variable resistance), reactance amplifier, varactor, and variable resistance amplifier. The principle of operation is explained in Sec. 14-H of the first edition.

Jelenski (1962) designed a parametric ESR spectrometer that employed a pump frequency f_p of ~ 700 MHz and a signal frequency f_s of ~ 220 MHz. The block diagram of the spectrometer presented in Fig. 11-28 shows the sample located in the coil of the parametric amplifier. The 1-kc oscillator is employed for recorder presentation after passage through the amplifier, filter, and synchronous detector, while 100 Hz is used for visual presentation. From Fig. 11-29 we see that in this application parametric amplifiers are still low enough in frequency to eliminate noise effectively, while at X band, masers are theoretically much superior. Hollocher et al. (1964) employed a commercial

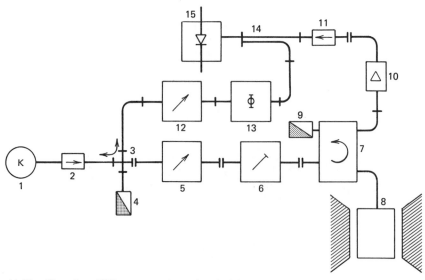

Fig. 11-26. Homodyne ESR spectrometer equipped with a low-noise GaAs microwave preamplifier: 1—klystron; 2, 11—isolater; 3—cross coupler; 4, 9—matched load; 5, 6, 12—attenuator; 7
—circulater; 8— cavity; 10— microwave preamplifier; 13—phase shifter; 14—directional coupler; 15—detector (Hoentzsch et al., 1978).

varactor diode parametric amplifier to reduce the noise figure of their ESR
spectrometer from 15 to 4.1 dB. Deloach (1963) described a 54 GHz parametric amplifier. Electron spin oscillator spectrometers employing a regenerative
microwave feedback loop are described by Payne (1964) and Sloan et al.
(1964).

Takao and Hayashi (1969) used a parametron preamplifier (i.e., a phase-
coherent degenerate parametric amplifier) to improve the sensitivity of an

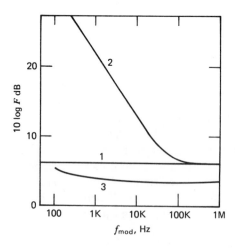

Fig. 11-27. Modulation frequency dependence of the calculated noise figure of a
spectrometer (1) with superheterodyne detection sans microwave preamplification,
(2) with homodyne detection sans microwave preamplification, and (3) with homodyne detection and microwave preamplification (Hoentzsch et al., 1978). Note how
the preamplification overcomes the low
modulation frequency inadequacies of homodyne detection (Hoentzsch et al., 1978).

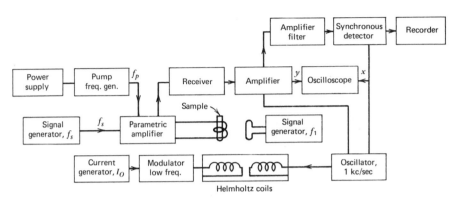

Fig. 11-28.   Block diagram of a spectrometer employing a parametric amplifier (Jelenski, 1962).

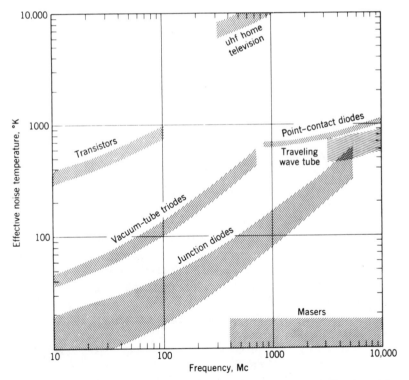

Fig. 11-29.   The frequency dependence of the effective noise temperature as a function of the type of detector (Uhlir, 1959).

429

X-band superheterodyne spectrometer. The signal-to-noise ratio improved by about 6 dB at low microwave powers ($P < 2$ mW) and by about 20 dB for higher powers ($P > 10$ mW). The greater improvement at higher powers was ascribed to the elimination of coherent FM klystron noise that is out of phase with the ESR signal because of the dispersive effect of the sample cavity [see also Takao and Watanabe (1971)].

The noise temperature or relative noisiness of various types of preamplification and detection devices varies with the frequency in the manner illustrated in Fig. 11-29 (Uhlir, 1959).

I. Enhancement of Spectra

The earlier sections in this chapter discussed the sensitivity attainable using standard electronic components, including a long time-constant filter preceding the recorder. This filter is very efficient in removing high-frequency noise, but it does not affect the low-frequency noise. An enhancement technique of continuous averaging may be employed to remove both low- and high-frequency noise, and thereby increase the signal-to-noise ratio. The theory and application of this technique are presented in this section.

The section concludes with another type of enhancement technique designed to increase the resolution of an ESR spectrum.

Signal-to-Noise Enhancement

Klein and Barton (1963, 1963) describe a digital computer technique that makes use of a multichannel pulse-height analyzer to enhance the signal-to-noise ratio of ESR spectra by continuous averaging.

In conventional lock-in detector techniques, the ESR signal is sent through a long time-constant filter and then recorded. The signal-to-noise ratio is enhanced by increasing the time constant τ and the time T required for the scan, while keeping the ratio τ/T constant. The continuous average or time domain filtering technique (Klein and Barton, 1963) requires the same overall time T for the measurement, but divides this time into T/n intervals and makes a total of n scans while using a time constant of τ/n. The n scans are summed, and this causes the signals to add and produce a resultant with n times the amplitude of a single scan. The noise components, on the other hand, will add randomly and give a resultant that is $n^{1/2}$ times the noise of a single traverse. As a result the signal-to-noise ratio increases as the square root of the time constant τ (or total scanning time T). This is the same result that is obtained for the high-frequency components of noise when a single long scan is used. However, the long scan does not filter out very low-frequency noise, and it tends to be upset by very strong noise impulses or spikes. The prominent low-frequency noise arises from the crystal $1/f$ noise power and other sources. The continuous averaging technique achieves its remarkable enhancement by its efficient removal of both high- and low-frequency noise.

The multichannel pulse-height analyzer and its associated equipment are shown in Fig. 11-30. It employs a voltage-to-frequency converter to convert the output voltage of the lock-in detector to a form that is suitable for storage in the memory of the analyzer. This unit produces a pulse train whose rate is proportional to its instantaneous input voltage. The magnetic field H is driven from the address in order to set up a one-to-one correspondence between the channel number and H. A field sweeping regulator is employed to ensure that the field rather than the current follows the address analog. To avoid magnet time constant effects, the address advance is programed in a triangular manner. The data are displayed on either a recorder or a sillyscope.

The apparatus shown in Fig. 11-30 is arranged to operate as an average response computer. The ferrite core memory unit consists of 400 channels, and each has a capacity of 10^6 counts. The clock that advances the address by one channel for each pulse is continuously adjustable from a rate of 1 msec/channel upward. The arithmetic counting register and control circuits constitute the logic units that add (or subtract) the pulses entering the input circuit to (or from) the count stored in the memory.

Figure 11-31 shows the enhancement that has been obtained with ESR spectra. Signal-to-noise ratios have been enhanced by one to two orders of magnitude without sacrifice of bandwidth. When hyperfine structure is present, there should be enough channels to allow more than 25 for each hfs component. Sampling techniques can be used with transient signals. The first edition gives additional references to the literature.

Sensitivity enhancement is sometimes a compromise tactic in which the signal-to-noise ratio is increased at the cost of distorting the spectral shape (Caprini et al., 1970). Posner (1974) employed an integration technique that converts first derivative spectra to adsorption signals of greater signal-to-noise ratio without introducing any distortion of the spectrum.

Resolution Enhancement

Many ESR spectra containing partially resolved hyperfine structure are composed of individual resonant lines whose width and shape are either known or may be approximated. In this case the spectrum may be sharpened by the method (Allen et al., 1964) shown in Fig. 11-32. The second derivative of the lineshape $(d^2/dt^2)f(t)$ shown in the center is multiplied by a constant C and then subtracted from the upper lineshape $f(t)$ [denoted by $Y'(t)$ or $U(t)$ in Sec. 11-F] to produce the narrowed lower lineshape. A Gaussian line narrowed by subtracting some of its second derivative and adding some of its fourth derivative is shown in Fig. 11-33. The success of this method depends upon the fact that the even derivatives of absorption lines are narrower in the center than the original lineshape, and similarly higher-order odd derivatives are narrower than the first one (compare Sec. 11-K and Fig. 11-38). Allen et al. (1964) discuss filters that may be employed to perform the differentiations, additions, and subtractions. They made use of both digital and analog com-

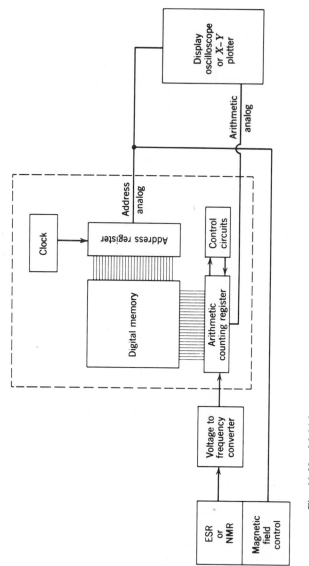

Fig. 11-30. Multichannel pulse-height analyzer (within dotted line) and associated electronic equipment. The system is arranged to operate as an average response computer (Klein and Barton, 1963).

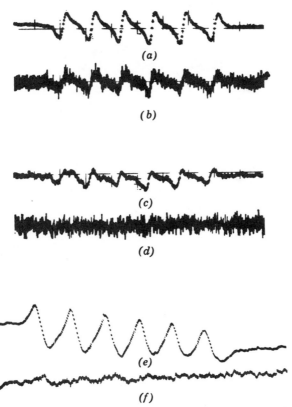

Fig. 11-31.   First-derivative ESR spectra of Mn^{2+} in H_2O at the output of the pulse height analyzer. The time constant was 10^{-2} sec and each traverse lasted for 5 sec. (a) 100 traverses, 5×10^{14} spins; (b) 1 traverse, 5×10^{14} spins; (c) 500 traverses, 5×10^{13} spins; (d) 1 traverse, 5×10^{13} spins; (e) 5000 traverses, 5×10^{13} spins; (f) 1 traverse, 5×10^{13} spins (Klein and Barton, 1963).

puters in their experimental arrangements. Filtered and unfiltered ESR spectra are presented in Fig. 11-34.

The digital equipment required for the above resolution enhancement scheme is rather expensive, and the substitution of analog techniques introduces noise and makes sharpening adjustments critically dependent on the rate of scan. Glarum (1965) presented an alternative technique that avoids these difficulties. He showed that modulating the magnetic field with a complex (nonsinusoidal) waveform and using phase-sensitive detection can reduce the observed width of the spectral lines. This resolving technique has several advantages: (1) the sharpening is independent of the rate of scan; (2) the sharpened spectrum is obtained directly on the recorder; (3) distortion due to ringing is eliminated; (4) derivatives higher than the third can be included; and (5) the required equipment is inexpensive and readily integrated into existing spectrometer systems.

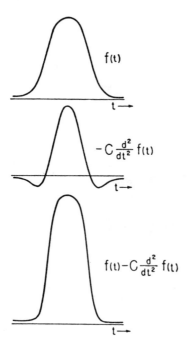

$f(t)$

$-C\dfrac{d^2}{dt^2}\,f(t)$

$f(t)-C\dfrac{d^2}{dt^2}\,f(t)$

Fig. 11-32. Qualitative representation of the principle of resolution enhancement, where $f(t)$ = the lineshape function; C = a constant dependent on linewidth (Allen et al., 1964).

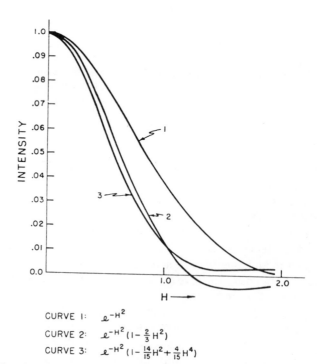

CURVE 1: ϱ^{-H^2}

CURVE 2: $\varrho^{-H^2}\left(1-\tfrac{2}{3}H^2\right)$

CURVE 3: $\varrho^{-H^2}\left(1-\tfrac{14}{15}H^2+\tfrac{4}{15}H^4\right)$

Fig. 11-33. Gaussian absorption line: 1—undistorted, 2—sharpened to its second derivative, 3—sharpened to its fourth derivative (Allen et al., 1964).

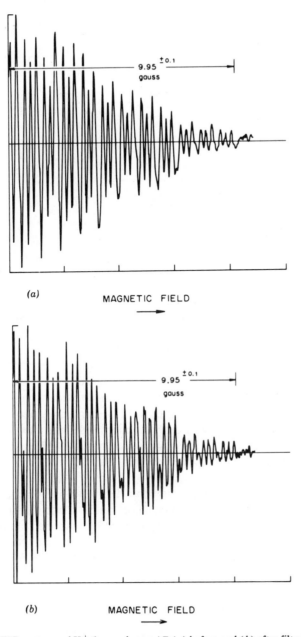

$9.95 \, {}^{\pm\,0.1}$
gauss

(a)

MAGNETIC FIELD

$9.95 \, {}^{\pm\,0.1}$
gauss

(b)

MAGNETIC FIELD

Fig. 11-34. ESR patterns of K^+ (benzophenone)$^-$ (a) before and (b) after filtering by the analog line sharpening circuit (Allen et al., 1964).

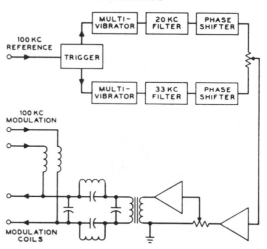

Fig. 11-35.   Block diagram of field-modulation line sharpener (Glarum, 1965).

In the Glarum technique, the magnetic field is modulated at several frequencies $f_{\text{mod}}, f_{\text{mod}}/3, f_{\text{mod}}/5, \ldots$, while retaining a single receiver and detector at f_{mod}. For small amplitude modulation ($H_{\text{mod}} \ll \Delta H_{pp}$ for all frequencies), this method is mathematically equivalent to using detector circuits at $f_{\text{mod}}, f_{\text{mod}}/3, f_{\text{mod}}/5, \ldots$, and recombining terms later.

Figure 11-35 shows a transistorized line-sharpening field modulator. The 100-kHz reference signal from a conventional modulation circuit triggers two multivibrators to produce the 33- and 20-kHz sharpening frequencies. Each frequency is filtered, adjusted in phase and amplitude, and then impressed on the modulation coils. Figure 11-36 shows a spectrum of DPPH in benzene after sharpening with one (33 kHz) and two (33 plus 20 kHz) modulation subharmonics.

Wood and Yen (1969) designed a second derivative line-sharpening device. The third derivative detection technique of Halpern and Phillips (1970) is described in Sect. 7-F. Bieber and Gough (1976) applied autocorrelation techniques to the recording and analysis of spectra, and Nishikawa and Someno (1975) employed discrete Fourier-transform methods for the analysis

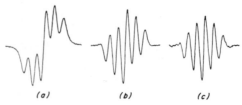

(a)   (b)   (c)

Fig. 11-36.   First-derivative spectrum of DPPH in benzene with (a) zero, (b) one, and (c) two sharpening terms (Glarum, 1965).

of ESR data. Ernst and Benz (1970) discussed a least squares analysis of overlapping spectra.

Overlapping spectra arising from separate species may be better resolved by working at a higher frequency band or by employing ENDOR-induced ESR [see Chap. 14; Niklas and Spaeth, 1980; Schweiger et al. 1981). Schweiger and Günthard (1982) proposed a technique for resolving such overlapping spectra by the use of a circularly polarized modulating field which causes each species to be in phase at a different phase angle. A particular component may be suppressed by detecting out of phase relative to it. Andersson (1976) separated spectra by the difference in their 90° out-of-phase components under saturation conditions.

J. Choice of Sample Size

The discussion of Sec. 5-H and the calculations of the type that are summarized in Table 5-2 show that, in general, the ESR signal increases with an increase in the size of the paramagnetic sample, and also with an increase in Q. If a sample is lossy, then an increase in its size will (1) increase the number of paramagnetic centers and (2) decrease Q, and these two factors will have opposite effects on the sensitivity. Feher (1957) distinguishes two limiting cases: (1) insulators and semiconductors that exhibit a dielectric loss proportional to E^2 arising from the imaginary part ϵ'' of the dielectric constant ϵ:

$$\frac{\epsilon}{\epsilon_0} = \epsilon' + j\epsilon'' \tag{1}$$

and (2) low resistivity samples that produce losses that are proportional to H^2 and that, for example, may arise from surface currents. Each will be considered in turn.

We showed in Sec. 11-D that the ESR signal y_m' is proportional to the produce ηQ_u:

$$\eta Q_u \propto V_s Q_u \tag{2}$$

where the filling factor η, which is proportional to the sample volume V_s, has been evaluated explicitly for several experimental arrangements in Sec. 5-H. In the presence of a lossy sample, the factor Q_u in the sensitivity formulas [e.g., (11-D-1) to (11-D-34)] should be replaced by Q_u' defined by

$$\eta Q_u' = \frac{\eta}{1/Q_u + 1/Q_\chi + 1/Q_\epsilon + 1/Q_\mu} \tag{3}$$

where Q_ϵ and Q_μ, respectively, arise from the dielectric and H^2 losses in the sample as mentioned above. These are defined by

$$\frac{1}{Q_\epsilon} = \frac{\int_{\text{sample}} \epsilon_0 \epsilon'' |E|^2 \, dV}{\int_{\text{cavity}} \mu H_1^2 \, dV} = \frac{\epsilon_0}{\mu_0 V_c \langle H_1^2 \rangle_c} \int_{\text{sample}} \epsilon'' |E|^2 \, dV \tag{4}$$

$$\frac{1}{Q_\mu} = \frac{\int_{\text{sample}} \mu_0 \mu'' H_1^2 \, dV}{\int_{\text{cavity}} \mu H_1^2 \, dV} = \frac{1}{V_c \langle H_1^2 \rangle_c} \int_{\text{sample}} \mu'' H_1^2 \, dV \tag{5}$$

where ϵ'' and μ'' are the losses under consideration and $\mu/\mu_0 = \mu' + j\mu''$. We consider the optimum sample size V_s for two specific sample shapes and orientations.

First let us consider a sample of the type shown in Fig. 5-5 placed in a "square" TE_{102} mode with $2a = d$. From Eq. (5-H-31) we have

$$\eta = \frac{V_s}{V_c} = \left(\frac{\pi d}{V_c}\right) r^2 \tag{6}$$

for a sample tube of radius r. Using Eqs. (5-H-8) to (5-H-10), Eq. (5-H-18), and the coordinate system of Fig. 5-26 in Eq. (4), we obtain

$$\frac{1}{Q_\epsilon} = \frac{1}{abd} (4\epsilon'') \int_0^a \sin^2 \frac{\pi x}{a} \, dx \int_{-r}^r \sin^2 \frac{2\pi z}{d} \, dz \int_{-r\sin\varphi}^{r\sin\varphi} dy \tag{7}$$

$$= \frac{4r\epsilon''}{bd} \int_{-r}^r \left[1 - \left(\frac{z}{r}\right)^2\right]^{1/2} \sin^2 \frac{2\pi z}{d} \, dz \tag{8}$$

Ordinarily the tube radius r is sufficiently small relative to the cavity dimension a, so that we may approximate $\sin 2\pi z/d$ by its argument, and ergo, we obtain, using $d = 2a$

$$\frac{1}{Q_\epsilon} = \frac{4\pi^2 r^4}{a^2 bd} \epsilon'' \int_{-1}^1 \left[1 - \left(\frac{z}{r}\right)^2\right]^{1/2} \left(\frac{z}{r}\right)^2 d\left(\frac{z}{r}\right) \tag{9}$$

$$= \frac{\pi^3 r^4}{2a^2 bd} \epsilon'' = \frac{\pi^3 r^4}{V_c d} \epsilon'' \tag{10}$$

where ϵ'' is the ratio of the imaginary part of the dielectric constant of the sample to that of free space. [More properly, E^2 should be replaced by a value closer to $\epsilon' E^2$ (compare Sec. 2-I).] For the dielectric loss case, $1/Q_\mu = 0$, and we neglect the resonance absorption $1/Q_X$, which gives

$$\eta Q_u' = \frac{(\pi d/V_c)r^2}{1/Q_u + (\pi^3/2V_c d)\epsilon'' r^4} \tag{11}$$

This has a maximum when

$$\frac{\partial(\eta Q_u')}{\partial r} = 0 \tag{12}$$

which leads to the condition

$$\frac{\pi^3}{2V_c d}\epsilon'' r^4 = \frac{1}{Q_u} = \frac{1}{Q_\epsilon} \tag{13}$$

or in other words

$$(Q_u' \text{ with sample}) = \tfrac{1}{2}(Q_u \text{ sans sample}) \tag{14}$$

This result is only valid for $r \ll a$. Thus to obtain the highest sensitivity, one should use a sample that reduces the loaded Q of the cavity to one-half its initial value.

For the case of magnetic losses, using Eq. 5-H-13 with $d = 2a$

$$\frac{1}{Q_\mu} = \frac{\int_{\text{sample}} \mu'' H_1^2 \, dV}{V_c \langle H_1^2 \rangle_c} = \frac{\tfrac{1}{2}(\pi r^2 d)H_0^2}{V_c \langle H_1^2 \rangle_c}\mu'' \tag{15}$$

$$= \mu'' \frac{2\pi r^2}{ab} \tag{16}$$

where the integration is similar to Eq. (5-H-27)ff. As a result

$$\eta Q_u' = \frac{2(\pi d/V_c)r^2}{(1/Q_u) + [\mu''(\pi/ab)]r^2} \tag{17}$$

which has no maximum, so the larger the sample tube, the higher the sensitivity.

These calculations were presented in detail to illustrate the method. Feher (1957) considered the case of a planar sample of dimensions $a \times b \times \delta$ with $\delta \ll d$ placed at the bottom of a rectangular TE_{102} mode cavity. He obtained the condition

$$(Q_u' \text{ with sample}) = \tfrac{2}{3}(Q_u \text{ sans sample}) \tag{18}$$

for dielectric losses and found that the larger the sample size the greater the sensitivity for H_1^2 losses.

The physical reason that an optimum sample size exists for the two cases with dielectric loss is as follows. When the size increases by an increment Δr or Δz, the additional sample is located in a position of lower H_1^2 and higher E^2, and so it contributes less to the signal and more to the losses than the rest of the sample. This process becomes more pronounced as the size increases until a point is reached where the loss term predominates, and a further increase in size lowers the sensitivity. When the losses are proportional to H_1^2, on the other hand, both the signal and the losses increase at the same rate, and the constant term $1/Q_u$ that is in the denominator of Eq. (3) prevents an optimum size from being reached.

Goldberg and Crowe (1977) studied the effect of cavity loading on ESR spectra and used the results to quantify spectra for analytical purposes. The results obtained with intense signals from large samples vary with the parameter measured. Figures 11-37 and 11-38 compare the results obtained with first derivative amplitudes, double integration over a range of four and also over a range of ten linewidths, and the product of the amplitude times the square of the linewidth. Both the Lorentzian and the Gaussian cases are shown. Similar results were obtained with feromagnetic samples (Goldberg et al., 1979).

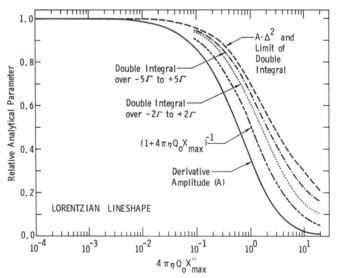

Fig. 11-37. Plot of the relative peak-to-peak amplitude (A), peak-to-peak amplitude multiplied by the square of the linewidth ($A\Delta^2$) and double integrals over different magnetic field regions versus the cavity loading parameter $4\pi\eta Q_0 \chi''_{max}$ for a Lorentzian lineshape. The parameter $(1 + 4\pi\eta Q_0 \chi''_{max})$ is included for reference. Values of the relative parameter less than unity reflect deviations from linearity in the particular ESR signal versus concentration or quantity of material. The symbol Γ denotes half of a linewidth (Goldberg and Crowe, 1977).

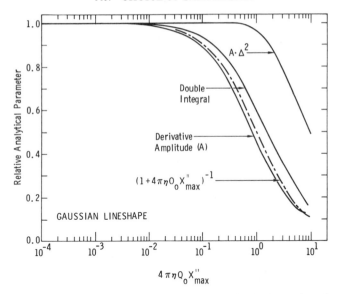

Fig. 11-38. Plot of the relative peak-to-peak amplitude, double integral, and peak-to-peak amplitude multiplied by the square of the linewidth for a Gaussian lineshape. Deviations from unity in the relative values reflect a nonlinearity of the ESR signal versus concentration or quantity of material. The notation is the same as on Fig. 11-37 (Goldberg and Crowe, 1977.)

Overloading the sample cavity causes a spurious broadening of the ESR signal (Burgardt and Seehra, 1977; Gourdon et al., 1970; Gupta et al., 1972; Seehra et al., 1968, 1974; Vigouroux et al., 1973). Bass (1973) reports on optimal sample sizes.

For analytical applications the detector should be biased at the same operating point for each sample. For powers above 10 mW enough power is generally reflected from the cavity to provide good detection sensitivity. Below 10 mW sensitivity falls off, and external biasing is required to restore it, as indicated in Fig. 11-39.

One of the most frequently employed lossy sample materials is water (aqueous solutions). Several authors (Cook and Mallard, 1963; Stoodley, 1963, 1963; Wilmshurst, 1963) have discussed the feasibility of employing long cavities to increase the sensitivity with such samples. For the reasons advanced by Stoodley and Wilmshurst and from arguments similar to those presented in Sec. 5-H, one may conclude that it is best to keep the resonant cavity as short as possible. Lossy samples may be analyzed in capillary tubes, and their cylindrical shape is most adaptable for use with cylindrical resonators. The most efficient arrangement for a rectangular TE_{102} cavity is a thin flat quartz sample cell located at the region of maximum magnetic field strength and minimum electric field strength, and oriented parallel to the smallest cross section of the cavity. Various workers have made use of flat cells with X-band thicknesses between 0.25 and 1 mm (Cook and Mallard, 1963; Dixon and

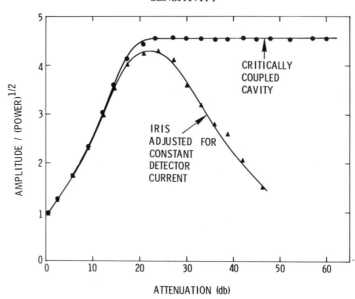

Fig. 11-39. Plot of the signal amplitude divided by the square root of the microwave power versus the attenuation of the microwave power where 0 dB of attenuation corresponds to 580 mW incident power. The curve labeled "critically coupled cavity" corresponds to the use of external biasing with an optimum cavity coupling setting. The other curve is obtained by using the cavity iris setting to adjust the detector current (Goldberg, 1978).

Norman, 1963; Piette et al., 1961; Storey et al., 1962). Electrolytic or polarographic studies (compare Sec. 8-G) are often carried out with flat cells. Lagercrantz and Persson (1964) used two flat cells in a dual cavity. A thin flat cell also works well in the center of a TE_{011} mode cylindrical cavity. Dielectric losses may be reduced by operating at S band (3000 Mc) or lower in frequency. Kent and Mallard (1965) describe a vhf (80 MHz) spectrometer designed for use with wet biological tissue samples. Decorps and Fric (1972) describe a high-resolution, low-frequency spectrometer. Brown (1974) constructed a 1-GHz spectrometer with a triplate line cavity, and Allendoerfer et al. (1975, 1980) employed a coaxial cavity, both described in Sec. 5-P, for studying aqueous samples.

Roberts and Derbyshire (1961) describe a universal crystal mount designed for use with a rectangular TE_{102} resonant cavity. It provides for rotation of a crystal about two orthogonal axes, and gives a reproducibility of 2°. Similar crystal mounts are described by Bil'dyukevich (1963, 1964, 1965), Engel (1964), Danilyuk et al. (1965), Manenkov and Prokhorov (1965), May and White (1969), and Russell (1965). Weil and Claridge (1970) designed an internal electromagnetic goniometer that uses two orthogonal pickup coils to monitor the rf field orientation relative to the sample position. Yo et al. (1978) describe a liquid crystal spinning sample, and Krebs and Sackmann (1976) report a liquid crystal orientation cell.

K. Standard Samples and Number of Spins

To obtain quantitative spin concentrations and g factors by electron spin resonance, it is necessary to compare an "unknown" sample to a known one. The calibration sample may be "run" before and after the unknown, or it may be placed in the cavity simultaneously and recorded superimposed on the unknown. A dual cavity (compare Sec. 5-M) provides the best features of both methods. Several standard samples useful for calibration purposes are discussed in this section.

One of the most frequently used calibration samples is α,α'-diphenyl-β-picryl hydrazyl (DPPH) (e.g., Slifkin and Suleiman, (1976). It has been used in the solid state as a g-factor standard ($g = 2.0036 \pm 0.0003$), and in benzene solution its hyperfine structure has been used to calibrate the number of gauss per recorder division. It may be employed in both phases as an intensity standard, and it does not saturate easily (Lloyd and Pake, 1953; Bloembergen and Wang, 1954; Berthet, 1955). A series of standard solutions diluted successively by factors of $10^{1/2}$ or 10 can be very useful.

Solid DPPH exhibits a g-factor anisotropy (Cohen et al., 1952; Kikuchi and Cohen, 1954) that is more pronounced at low temperatures (Singer and Kikuchi, 1955). The narrow Lorentzian-like singlet linewidth of solid DPPH is due to exchange narrowing, and the Lorentzian-shaped quintet observed in dilute solutions arises from the Brownian motion or molecular tumbling. If one could remove the exchange interaction in the solid, then one would observe a dipolar broadened Gaussian resonance. The linewidth in solid DPPH varies from 1.5 to 4.7 G depending on the solvent from which it is crystallized (Bodi and Goara, 1964; see Table 7.1 of Al'tschuler and Kozyrev, 1964). In more recent work Servant and Palangie (1975) investigated the hyperfine evolution of DPPH solutions with concentration. Yashchenko et al. (1975, 1976) discussed hydroxyl biradicals, and Verlinden et al. (1974) reported on the low-temperature magnetic properties of DPPH.

Despite its advantages, DPPH also has certain undesirable features (e.g., Feher, 1957; Uebersfeld, 1959). As a result, several other standards have been proposed. L. S. Singer (1959) and Singer et al. (1961) suggested the use of a small single crystal of ruby (0.5% Cr/Al_2O_3) as a secondary standard because it is chemically stable, and has an ESR signal that is strongly orientation-dependent. The crystal may be permanently situated at the edge of the resonator and oriented to produce a signal at a g factor that will not interfere with the spectra under study. (e.g., $g = 1.5$ or 2.5 is often convenient). The ruby spectrum may be recorded under the same instrumental conditions immediately before or after the unknown, and since both are in the same resonator, there will be an automatic compensation for such effects as the cavity Q and the modulation amplitude. Figure 9-5 shows Singer's experimental arrangement. Thompson and Waugh (1965) describe an adjustable ruby standard in which the amplitude and effective g factor may be varied at will.

Chang et al. (1978) have developed a synthetic ruby intensity standard that is available from the National Bureau of Standards as Standard Reference

Material No. 2601. This standard is chemically and physically stable under laboratory conditions. It is annealed and chemically etched after cutting to remove strains and surface damage. The Cr^{3+} concentrations were determined by static susceptibility measurements. The linewidth is 14 G for the $\frac{1}{2} \leftrightarrow -\frac{1}{2}$ transition and 16 G for the others: At X band (9.5 GHz) with the applied field parallel to the c axis there are three transitions at the magnetic field strengths 718.7, 3425.0, and 7568.88 G with the relative intensity ratios $3:4:3$. The angular dependances of the line positions and intensities were calculated for frequencies 6, 9.5, 14.5, 18, 25, 35, and 50 GHz. Small angular misalignments do not appreciably affect the measurements. The Bureau provides samples doped with $(3.694 \pm 0.011) \times 10^{15}$ Cr^{3+} ions/mg Al_2O_3 in square plate samples (nominally $1.5 \times 1.5 \times 0.5$ mm), and bar samples (nominally $0.5 \times 0.5 \times 4$ mm).

Guilbault and co-workers (1968, 1969, 1970) and Moyer and McCarthy (1969) evaluated the potentiality of ESR as a quantitative analytical tool. More recently Warren and Fitzgerald (1977) assessed the precision of quantitative ESR using copper(II) bound to ion exchange resins. They measured signal integrated areas from the expressions

$$A = \frac{3.63 ym'(\Delta H_{pp})^2}{GH_{\text{mod}}} \tag{1}$$

and obtained more precise results for A when the modulation amplitude H_{mod} was kept constant and only the receiver gain G was varied. Their standard deviation D:

$$D = \left[D_{\text{instr}}^2 + D_{\text{samp}}^2 \right]^{1/2} \tag{2}$$

arose from instrumental uncertainties D_{instr} coming from the width D_w, amplitude D_y, rescanning D_{scan}, receiver gain charge D_g, and modulation amplitude D_{mod}:

$$D_{\text{instr}} = \left[D_w^2 + D_y^2 + D_{\text{scan}}^2 + D_g + D_{\text{mod}}^2 \right]^{1/2} \tag{3}$$

and the uncertainties D_{samp} in sample preparation D_{prep}, in repacking the sample tube D_{repack}, and in reinserting the tube into the cavity D_{insert}:

$$D_{\text{samp}} = \left[D_{\text{prep}}^2 + D_{\text{repack}}^2 + D_{\text{insert}}^2 \right]^{1/2} \tag{4}$$

Their measured relative standard deviations were 4.3% for D_{instr} and 3.5% for D_{samp}. The lower limit of detection was 10^{-6} M and 10^{-7} M for the ammonia and orthophenanthoroline complexes of Cu^{2+}, respectively, and the corresponding upper limits of linearity were 10^{-3} M and 10^{-4} M, which provides three decades for quantitative concentration measurements. The upper limit of linearity was attributed to spin–spin exchange contributions to the relaxation. In earlier work Guilbault and Lubrano (1968) used the fourth peak from downfield of a Mn^{2+} complex, and Brinkman and Fresser (1971) employed a Ni^{2+} complex for quantitative ESR studies.

Tinkham and Strandberg (1955), Krongelb and Strandberg (1959), Westenberg and DeHaas (1964), and Evenson and Burch (1966) used a molecular oxygen standard. Filipovich and Sanders (1959) proposed the use of gaseous $^{16}O^{18}O$ as an intensity standard, while Westenberg and DeHaas (1965) used gaseous NO. Hoskins and Pastor (1960) recommend the use of charred dextrose as a frequency marker because the linewidth (10.6 G), g value, and spin-lattice relaxation time are constant between liquid helium and room temperature. This material is easily prepared and is stable when stored in a sealed container. Miner et al. (1972) suggest the Mn^{2+} hextet spectrum ($g = 1.9992 \pm 0.003$, $A = 173.77 \pm 0.04$ G) in the powdered mineral forsterite, $MgSi_2O_4$. Baican et al. (1976) proposed the stable free radical 1-cyano-2,2-diphenylpyrazyl as a standard. Leskov et al. (1976) reported a primary standard for ESR. The Varian standard is $3.3 \times 10^{-4}\%$ pitch on KCl. It contains 10^{13} spins per centimeter of length and has $g = 2.0028$, $\Delta H_{pp} = 1.7 G$ and $A/(y'_m \Delta H_{pp}^2) = 5.46$ (see Sec. 12-B).

Dalal et al. (1981, 1981) proposed the use of the paramagnetic compound K_3CrO_8, potassium perchromate, as a standard. It is easy to prepare (Riesenfeld, 1905) and check for purity (McGarvey, 1962; Stromberg, 1963). The solid can be stored for several years in a refrigerator and strongly basic (pH $\geqslant$ 13) solutions are stable for weeks for making g-value and magnetic field calibrations, but accurate spin concentration determinations require fresh solutions. The isotropic g value of 1.97120 ± 0.00005 for the main line ($^{52}Cr^{5+}$) does not interfere with signals at the free electron position, and the 20.0-G spacing between the two lowest $^{53}Cr^{5+}$ hyperfine components provides a good calibration of the scan. The linewidth of 0.75 ± 0.05 G ensures greater accuracy than more commonly used standards. The g value of a K_3CrO_8 crystal is constant to within ±0.0005 over the temperature range from 77 to 400 K. One liability of potassium perchromate is that it is not recommended for use above 100°C, because it decomposes at 170°C and may explode at 178°C (Mellor, 1931).

Fremy's salt or peroxylamine disulfonate dissolved in saturated sodium carbonate forms the $NO(SO_3)_3^{2-}$ ion with three hyperfine components separated by 13.00 ± 0.07 G, and has $\Delta H_{pp} = 0.26 \pm 0.02$ G, and $g = 2.0057 \pm 0.0001$ (Wertz et al., 1961). The solution may be stored in a refrigerator for several months but decays if stored at room temperature (see Lloyd and Pake, 1954; Burgess, 1958; Cram and Reeves, 1958). These may be used for calibrating the magnet scan. McBrierty and Cook (1965) recommend a nitric oxide complex for this purpose.

Yariv and Gordon (1961) and Burgess (1961) suggest the use of $CuSO_4 \cdot 5H_2O$ crystals as an intensity standard (see also Feher, 1957). Copper sulfate is readily available commercially. Foerster (1960) uses magnesium arsenate.

In Chapter 12 we discuss methods of integrating ESR spectra in order to determine the number of spins, the moments of the line, and so on. Wyard

(1965) doubly integrates ESR spectra on a desk calculator and obtains a 5% accuracy using intervals equal to a quarter of the linewidth.

Burgess (1961) describes an analog method for rapidly determining the integrated area of a spectrum from its first derivative curve. The technique consists of tracing the recorded curve on cardboard and ascertaining its moment on a balance. Burgess recommends copper sulfate pentahydrate ($CuSO_4 \cdot 5H_2O$) as a standard to calibrate the recordings, and thereby determine the absolute number of spins. Köhnlein and Müller (1962) discussed the errors of this method arising from the finite length of the spectrum. Aasa and Vänngärd (1964) obtained 2% agreement between calculated and measured intensity ratios of copper and cobalt complexes (effective spin $\frac{1}{2}$, axial symmetry, and small hfs energy.) Chang (1974) reported 5% reproducibility using a single cavity.

Randolph (1960) made use of the analog computer shown in Fig. 11-40 to determine the total absorption of ESR spectra. Other details on analog integrators are found in Sec. 7-I and in a paper by Collins (1959). The direct valuation of the area by his method is shown in the lower part of Fig. 11-41. The data on Fig. 11-42 demonstrate that a moderate amount of overmodulation does not affect the area, and its effect on the first derivative curve is considerably more pronounced that it is on the straight absorption curve. This is important to know because for barely detectable resonances, it is necessary to overmodulate. Figure 11-41 shows that structure is better resolved with first derivative presentation.

To obtain accurate concentrations Tomkiewicz and Taranko (1978) double-integrated first derivative curves over a range of nine linewidths. Hall (1972) showed that the integration errors are much greater for Lorentzian first derivative lines than they are for Gaussian ones. Johnson and Chang (1965) and Halpern and Phillips (1970) recorded the third derivatives for even greater resolution. Wessel and Schwarz (1978) describe a small computer system for the collection and analysis of spectra.

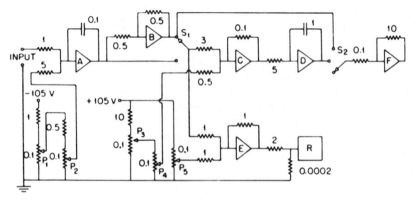

Fig. 11-40. Schematic circuit for the analog computation of ESR spectra. The figure gives typical resistances in megaohms and typical capacitances in microfarads (Randolph, 1960).

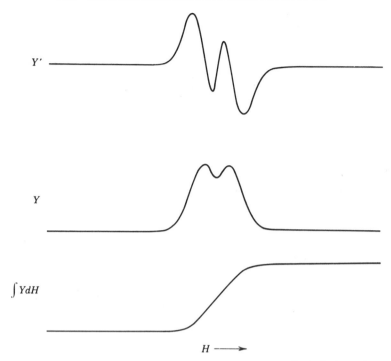

Fig. 11-41.   Simultaneous plots of the ESR absorption derivative Y', the ESR absorption Y, and the integral of the ESR absorption $\int Y\,dH$ of gamma-irradiated glycylglycine versus the magnetic field H (H increases from left to right). The peaks are separated by ~ 17 G on Y, and the modulation amplitude is about 1.2 G (Randolph, 1960).

Zuckermann (1973) prepared an automatic compensation method for the absolute determination of the change in magnetic susceptibility at resonance. The method is a further development of an antimodulation technique suggested by Dymanus (1969). It employs a compensating circuit with an electronically controlled element (diode ferroelectric or electroacoustic element) placed in the cavity and controlled by the modulation voltage so that it compensates for the modulation of the Q due to the oscillation of χ'' with the magnetic field at resonance. Figure 11-43 shows the compensator scheme connected to a standard spectrometer. The device is calibrated by supplying the element with a testing voltage at the frequency $2 f_{\mathrm{mod}}$. The resulting signal is compared to one from the microwave frequency modulation at f_{mod}. The measured susceptibility is independent of the power, field modulation amplitude, amplifier gain, and so forth and is accurate to 0.1%.

A dual sample cavity for use in comparing an unknown spectrum to a standard is described in Sec. 5-M. One advantage of this arrangement is that it allows one to use a standard sample whose amplitude and linewidth are comparable to the unknown. In addition, it obviates the necessity of working with overlapping spectra.

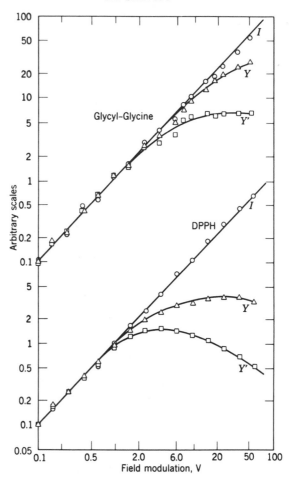

Fig. 11-42. The peak-to-peak derivative amplitude Y', absorption amplitude Y, and integrated area $I = \int Y\, dH$ versus the peak-to-peak modulation amplitude. The DPPH sample contained about 2×10^{16} spins. One volt corresponds to a peak-to-peak modulation amplitude of 1.2 G (Randolph, 1960).

Smith and Wilmshurst (1963) provided a rectangular TE_{103} mode dual cavity analogous to the TE_{105} one shown in Fig. 5-38 with separate 465-kHz modulation for each sample by means of single-turn hairpin loops. The modulation phase is adjusted to be opposite at each sample, and the spectrometer records the difference between the two signals, as shown in Fig. 11-44. This arrangement is useful for identifying spectra and for separating individual spectra from a mixed material. Incomplete cancellation can reveal small differences in the width, shape, or position of two resonances. Differential techniques have been extensively applied to ultraviolet, visible, and infrared spectroscopy.

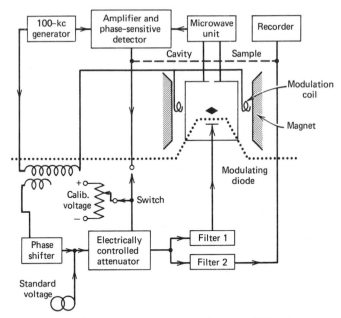

Fig. 11-43. Compensating circuit for absolute magnetic susceptibility determination (below dotted line) shown connected to a standard homodyne spectrometer (above dotted line) (Zuckermann, 1973).

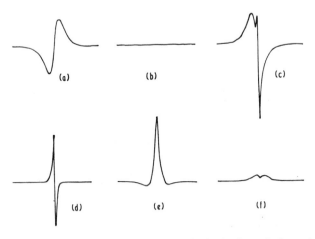

Fig. 11-44. Electron spin resonance difference spectra obtained using a dual sample cavity from (a) a single sample of powdered coal, (b) two identical samples of coal under exact balance; (c) a single sample of coal to which one crystal of DPPH has been added; (d) the two samples used for (a) and (c) balanced to give the DPPH spectrum; (e) two identical samples of coal that are subjected to a difference in magnetic field; (f) samples as in (e) with equal magnetic fields. (a) to (d) were obtained using the same spectrometer gain, and (c) and (f) were obtained with the gain increased 7 times (Smith and Wilmshurst, 1963).

449

In this chapter we have discussed the measurement of intensities and g factors in electron spin resonance. Yariv and Gordon (1961) describe a procedure for determining the number of spins by a measurement of the cavity reflection coefficient Γ. Scheffler and Stegmann (1963) present a method for determining g factors to a precision of about 5×10^{-6}, and Segal et al. (1965) report values to within two or three parts per million. See also Blois et al. (1961).

The ESR absorption in an unknown sample may be measured by an NMR spectrometer in a field of about a dozen gauss, and then calibrated against a standard NMR sample under the same instrumental conditions except for an increase in the magnetic field strength by three orders of magnitude.

Belson (1964) describes a method for producing spherical polished samples for ferrimagnetic studies, Howling and Hoskins (1965) present a high-speed method for obtaining thin flakes of metal samples, and Schone and Olson (1965) show how to make a hollow metal single crystal.

L. Computer Data Processing

In recent years electron spin resonance, as well as other branches of spectroscopy, has moved in the direction of computer control of spectrometer settings and computer processing of output data. Several authors who have written about data handling are Denisov (1975), Gruber et al. (1977), Plato and Möbius (1972), Wessel (1975, 1976, 1978), and Schwarz (1976). These articles treat such topics as digital data processing, transposing data to punched tape, smoothing and time averaging, single and double integration, subtraction of spectra, and oscilloscope monitoring of results as data processing proceeds. Bieber and Gough (1976) wrote a computer program for applying autocorrelation techniques to ESR. Nishikawa and Someno (1975) employed Fourier-transform techniques to reduce the computer storage and improve the computer speed. Section 12-V contains more details on this topic.

M. Spectrometer Systems

Chapter 13 of the first edition contained a table listing the characteristics of the spectrometers described in the literature up to the end of 1965, and it also provided descriptions of some of the more important or novel systems. This chapter has been eliminated from the present work, and as a substitute for it the present section comments upon some spectrometers that have been more recently reported in the literature.

Rupp et al. (1970) described a simplified spectrometer that employed a Gunn diode. It is suitable for instructional laboratory use. Hatch and Kreilick (1972) reported a broad-band low-field spectrometer operating over the range from 10 to 20 MHz.

Bernstein and Dobbs (1973) designed a zero-field spectrometer that operates either with a resonant cavity or a broad-band traveling-wave tube helix as a sample holder. A circuit is given for producing square-wave magnetic field modulation with a fast rise time. Erickson (1966) built a zero-field spectrometer equipped with a regenerative microwave oscillater and a reentrant cavity. He canceled the earth's field to within 10 mG.

High-frequency spectrometers have been reported by Geifman et al. (1978) for $\lambda = 8$ mm, by Van Meijel (1977) for $\lambda = 5$ mm, by Gershenzon and Putilov (1970) for $\lambda = 3$ mm, by Galkin et al. (1977) for $\lambda = 2$ mm and by Gershenzon et al. (1975).

Spectrometers have been designed by den Boef and Henning (1974) that use strain modulation (compare Sec. 14-J) and by Clerjaud and Lambert (1971) and Koepp (1969) that employ source modulation.

In additional articles Kinoshita (1970) reported a high-power spectrometer, Pascaru (1973) described a transisterized system, and Schoch et al. (1975) designed an X-band superheterodyne system for ESR or microwave spectroscopy. Slezak et al. (1976) gave the general operating principles for spectrometers, and Muller (1977) discussed recent developments in ESR techniques. Belonogov et al. (1978) report a spectrometer.

References

Aasa, R., and T. Vänngärd, *Z. Naturforsch.* **19A**, 1425 (1964).

Accou, J., P. Van Hecke, and L. Van Gerven, 17th Congress Ampere Turku, Finland, 1972.

Allen, L. C., H. M. Gladney, and S. H. Glarum, *J. Chem. Phys.* **40**, 3135 (1964).

Allendoerfer, R. D., and J. B. Carroll, *J. Magn. Reson.* **37**, 497 (1980).

Allendoerfer, R. D., G. A. Martinchek, and S. Bruckenstein, *Anal. Chem.* **47**, 890 (1975).

Alquie, A. M., C. Taupin, and A. P. Legrand, *C.R. Hebd. Sean. Acad. Sci. B* **172**, 973 (1971).

Al'tshuler, S. A., and B. M. Kozyrev, *Electron Paramagnetic Resonance* (trans. from the Russian by *Scripta Technica*, C. P. Poole, Jr., Ed.), Academic Press, New York, 1964.

Amanis, I. K., J. G. Kliava, J. J. Purans, and A. N. Truhin, *Phys. Status Solidi A (Germany)* **31**, 165 (1975).

Andersson, L. O., *Proc. XIXth Congress Ampere*, Heidelberg, 1976, H. Brunner, K. H. Hausser, and D. Schweitzer, Eds.,) p. 535.

Andreev, V. A., and V. I. Sugakov, *Fiz. Tverdogo Tela* **17**, 1963 (1975).

Atsarkin, V. A., M. E. Zhabotinskii, and A. V. Frantsesson, *Radiotekhn. Electron.* **7**, 866 (820) (1962).

Biacan, R., N. Negoita, and A. T. Balaban, *Rev. Roum. Phys.* **21**, 213 (1976).

Bales, B. L., and E. S. Lesin, *J. Chem. Phys.* **65**, 1299 (1976).

Bass, I. A., *Izmer. Tekh.* **16**, 48 (1973).

Belonogov, A. M., D. P. Volnyagin, V. Z. Drapkin, V. V. Zyats, V. V. Ovcharov, V. A. Pashinskii, A. S. Serdyuk, and I. I. Ugolev, *PTE* **21**(2), 244 (1978).

Belson, H. S., *RSI* **35**, 234 (1964).

Bernstein, E. R., and G. M. Dobbs, *RSI* **44**, 1314 (1973).

Berthet, G., *C.R. Acad. Sci.* **241**, 1730 (1955); *Ann. Phys. (Paris)* **3**, 629 (1958).

Bhagat, V. R., and B. Venkataraman, *Pramana (India)* **9**, 605 (1976).

Bieber, K. D., and T. E. Gough, *J. Magn. Reson.* **21**, 285 (1976).

Bijl, D., H. Kainer, and A. C. Rose-Innes, *J. Chem. Phys.* **30**, 765 (1959).

Bil'dyukevich, A. L., *PTE* **6**, 186 (1194) (1963); **2**, 185 (453) (1964); *Cryogenics* **5**, 205, 277 (1965).

Bloembergen, N., and S. Wang, *Phys. Rev.* **93**, 72 (1954).

Blois, M. S., Jr., H. W. Brown, and J. E. Maling, *Free Radicals in Biological Systems*, Academic Press, New York, 1961, p. 121.

Bodi, A., and P. Goara, *Stud. Cercetari Fiz.* (*Rumania*) **15**, 385 (1964).

Bonori, M., C. Franconi, P. Galuppi, and C. A. Tiberio, submitted for publication (1981a).

Bonori, M., C. Franconi, P. Galuppi, C. A. Tiberio, and W. Cecchetti, submitted for publication (1981b).

Bosch, B. G., and W. A. Gambling, *J. Brit. Inst. Radio Eng.* **24**, 389 (1962); **21**, 503 (1961).

Bresler, S. E., E. M. Saminskii, and E. N. Kazbekov, *Zhur. Tekh. Fiz.* **27**, 2535 (2357) (1958).

Bresler, S. E., E. N. Kazbekov, and V. N. Fomichev, *Zh. Tekh. Fiz.* **41**, 1237 (1971).

Brinkman, W. J., and H. Freiser, *Anal. Lett.* **4**, 513 (1971).

Brodwin, M. W., and T. J. Burgess, *IEEE Trans. Instrum. Meas.* **IM-12**, 7 (1963).

Brodgen, T. W. P., and J. Butterworth, *Proc. Phys. Soc.* **86**, 877 (1965).

Brown, G., *J. Phys. E Sci. Instrum.* **7**, 635 (1974).

Brown, G., *J. Mater. Sci.* **10**, 1481 (1975).

Buckmaster, H. A., and S. Cohn-Sfetcu, *17th Congress Ampere, Turku, Finland*, p. 31 (1972).

Buckmaster, H. A., and R. S. Rathie, *Can. J. Phys.* **49**, 8, 853 (1971).

Buckmaster, H. A., and J. C. Dering, *Can. J. Phys.* **43**, 1088 (1965).

Burgardt, P., and M. S. Seehra, *Phys. Rev. B* **16**, 1802 (1977).

Burgess, J. H., *J. Phys. Radium* **19**, 845 (1958).

Burgess, V. R., *JSI* **38**, 98 (1961).

Burland, D. M., D. E. Cooper, M. D. Fayer, and C. R. Gochandur, *Chem. Phys. Lett.* **42**, 279 (1976).

Caprini, M., S. Cohn-Sfetco, and A. M. Manov, *IEEE Trans. Audio Electroacoust.* **AU-18**, 389 (1970).

Chang, R., *Anal. Chem.* **46**, 1360 (1974).

Chang, T., D. Foster, and A. H. Kahn. *J. Res. Natl. Bur. Stand.* **83**, 133 (1978).

Chang, T., and A. H. Kahn, *Electron Paramagnetic Resonance Intensity Standard*: *SRM 2601*, *Description and USE NBS Publication*, 260 (1978).

Chiarini, F., M. Martinelli, L. Pardi, and S. Santucci, *Phys. Rev. B* **12**, 847 (1975).

Clauss, R. C., *The Deep Space*, Progr. Rep. 42, Jet Prepulsion Laboratory, Pasadena, Calif, pp. 65–68, 1975.

Clerjaud, B., and B. Lambert, *J. Phys. E* **4**, 619 (1971).

Cohen, V. W., C. Kikuchi, and J. Turkevich, *Phys. Rev.* **85**, 379 (1952).

Collins, R. L., *RSI* **39**, 492 (1959).

Cook, P., and J. R. Mallard, *Nature* **198**, 145 (1963).

Corvaja, C., and L. Pasimeni, *Chem. Phys. Lett.* **39**, 261 (1976).

Cram, D. J., and R. A. Reeves, *J. Am. Chem. Soc.* **80**, 3094 (1958).

Dalal, N. S., J. M. Millar, M. S. Jagadeesh, and M. S. Seehra, *J. Chem. Phys.* **74**, 1916 (1981).

Dalal, N. S., M. M. Suryan, and M. S. Seehra, *Anal. Chem.* **53**, 938 (1981).

Danilyuk, Yu. L., P. L. Pakhol'chik, and F. A. Koleda, *PTE* **1**, 213 (222) (1965).

Decorps, M., and C. Fric, *J. Phys. E* **5**, 337 (1972).

DeGrasse, R. W., E. O. Schulz-Du Bois, and H. E. D. Scovil, *Bell Syst. Tech. J.* **38**, 305 (1959).

De Groot, M. S., C. A. De Lange, and A. A. Monster, *J. Magn. Reson.* **10**, 51 (1973).

Deloach, B. C., *Proc. Inst. Elect. Electron. Eng.* **51**, 1153 (1963).

Den Boef, J. H., and J. C. M. Henning, *RSI* **45**, 1199 (1974).

Denisov, S. A., *PTE* **18**(2), 240 (1975).

Dietrich, W., and D. Schmid, *Phys. Status Solidi B* **74**, 609 (1976).

Dixon, W. T., and R. O. C. Norman, *J. Chem. Soc.* 3119 (1963).

Dobryakov, S. N., G. G. Lazarev, Ya. S. Lebedev, and M. V. Serdobov, *Zh. Strukt. Khim.* **17**, 814 (1976).

Domdey, H., and J. Voitlander, *Z. Naturforsch.* **29a**, 949 (1974).

Draeger, K. and Wolfmeier, *Z. Naturforsch. A* **31a**, 1057 (1976).

Dymanus, A., *Physica* **25**, 859 (1969).

Eaton, S. S., and G. R. Eaton, *Bull. Mag. Res.* **1**, 130 (1980).

Edgar, A., *J. Phys. E. Sci. Instr.* **8**, 179 (1975).

Engel, J., *ETP* **12**, 253 (1964).

Erickson, L. E., *Phys. Rev.* **143**, 295 (1966).

Ernst, R. R., and H. Benz, *IEEE Trans. Audio Electroacoust.* **AU-18**, 380 (1970).

Evenson, K. M., and D. S. Burch, *J. Chem. Phys.* **44**, 1714 (1966).

Faulkner, E. A., *JSI* **39**, 135 (1962); *Lab. Pract.* **13**, 1065 (1964).

Feher, G., *Bell Syst. Tech. J.* **36**, 449 (1957).

Filipovich, G., and T. M. Sanders, Jr., *RSI* **29**, 293 (1959).

Flesner, L. D., D. R. Fredkin, and S. Schultz, *Solid State Commun.* **18**, 207 (1976).

Flesner, L. D., S. Schultz, and R. Clauss, *RSI* **48**, 1104 (1977).

Foerster, G. V., *Z. Naturforsch.* **15a**, 1079 (1960).

Frater, R. H., *RSI* **36**, 634 (1965).

Galkin, A. A., O. Ya. Grinberg, A. A. Dubinskii, N. N. Kabdin, V. N. Krymov, V. I. Kurochkin, Ya. S. Lebedev, L. F. Oranskii, and V. F. Shuvalov, *PTE* **20**(4), 284 (1229) (1977).

Gambling, W. A., and T. H. Wilmshurst, *Phys. Lett.* **5**, 228 (1963); *Proc. 12th Colloq. Ampere*, Bordeaux, 1963, p. 171.

Geifman, I. N., M. D. Glinchouk, and B. K. Krulikovsky, *Zh. Eks. Teor. Fiz.* **75**, 1468 (1978).

Genoud, F., and M. Decorps, *Mol. Phys.* **34**, 1583 (1979).

Gershenzon, E. M., and P. A. Putilov, *PTE* **13**, 144 (477) (1970).

Gershenzon, E. M., E. M. Kuleshov, A. A. Negirev, L. A. Orlov, and B. N. Tumanov, *PTE* **18**, 845 (1975).

Gillberg-La Force, G., and R. C. La Force, *J. Chem. Phys.* **58**, 5402 (1973).

Glarum, S. H., *RSI* **36**, 771 (1965).

Glarum, S. H., and J. H. Marshall, *J. Chem. Phys.* **62**, 956 (1975).

Goldberg, I. B., and H. R. Crowe, *Anal. Chem.* **49**, 1353 (1977).

Goldberg, I. B., H. R. Crowe, R. M. Housley, and E. H. Cirlin, *J. Geophys. Res.* **84**, 1025 (1979).

Goldberg, I. B., *J. Magn. Reson.* **32**, 233 (1978).

Goldsborough, J. P., and M. Mandel, *RSI* **31**, 1044 (1960).

Gourdon, J. C., P. Lopez, Ch. Roy, and J. Pescia, *C. R. Acad. Sci.* **B271**, 288 (1970).

Grivet, J. P., and J. Mispelter, *Mol. Phys.* **27**, 15 (1974).

Gruber, H., G. Junge, and M. Vogel, *Mess. Steuern Regeln* **20**, 158 (1977).

Guilbault, G. G., and G. T. Lubrano, *Anal. Lett.* **1**, 725 (1968).

Guilbault, G. G., and T. Meisel, *Anal. Chem.* **41**, 1100 (1969); *Anal. Chim. Acta* **50**, 151 (1970).

Guilbault, G. G., and E. S. Moyer, *Anal. Chem.* **42**, 441 (1970); *Anal. Chim. Acta* **50**, 151 (1970); *Anal. Lett.* **3**, 563 (1970).

Gupta, R. P., M. S. Seehra, and W. E. Vehse, *Phys. Rev. B* **5**, 92 (1972).

Hall, P. L., *J. Phys. D Appl. Phys.* **5**, 673 (1972).

Halpern, T., and W. D. Phillips, *RSI* **41**, 1038 (1970).

Hardin, J., and J. Uebersfeld, *Rev. Phys. Appl.* **6**, 169 (1971).

Hatch, G. F., and R. W. Kreilick, *J. Magn. Reson.* **8**, 126 (1972).

Hausser, K. H., *Ampére Colloquium*, Eindhoven, 1962, p. 420.

Hausser, K. H., and R. Reinhold, *Z. Naturforsch.* **16A**, 1114 (1961).

Hedvig, P., *Acta Phys. Hung.* **10**, 115 (1959).

Henling, L. M., and G. L. McPherson, *Phys. Rev. B* **16**, 4756 (1977).

Herman, M. A., *Arch. Elektrotech (Poland)* **18**, 635 (1969).

Hinkel, H., H. Port, H. Sixl, M. Schwoerer, P. Reineker, and D. Richardt, *Chem. Phys.* **31**, 101 (1978).

Hoentzsch, C., J. R. Niklas, and J. M. Spaeth, *RSI* **49**, 1100 (1978).

Hollocher, T. C., W. H. From, and N. S. Bromberg, *Phys. Med. Biol.* **9**, 65 (1964).

Hoskins, R. H., and R. C. Pastor, *J. Appl. Phys.* **31**, 1506 (1960).

Howling, D. H., and J. M. Hoskins, *RSI* **36**, 400 (1965).

Janossy, A., *Magy. Fiz. Foly.* **23**, 213 (1975).

Jelenski, A., *Ampere Colloquium*, Eindhoven, 1962, p. 734.

Johnson, C. S., Jr., and R. Chang, *J. Chem. Phys.* **43**, 3183 (1965).

Kent, M., and J. R. Mallard, *JSI* **42**, 505 (1965).

Kent, M., and J. R. Mallard, *Nature* **207**, 1195 (1965).

Kenworthy, J. G., and R. E. Richards, *JSI* **42**, 675 (1965).

Kiel, A., *Phys. Rev. B* **12**, 1868 (1975).

Kikuchi, C., and V. W. Cohen, *Phys. Rev.* **93**, 394 (1954).

Kinoshita, N., *Bull. Electrotech. Lab. (Japan)* **32**, 639 (1970).

Kispert, L., and K. Chang, *J. Magn. Reson.* **10**, 162 (1973).

Klein, M. P., and G. W. Barton, Jr., *RSI* **34**, 754 (1963); *Paramagnetic Res.* **2**, 698 (1963).

Kloza, M., *Pomiary Autom. Konir. (Poland)* **17**, 337 (1971).

Koepp, S., *Hochfrequenztech. Elecktroakust.* **78**, 57 (1969).

Kohnlein, W., and A. Müller, *Phys. Med. Biol.* **6**, 599 (1962).

Kollberg, E. L., *Proc. IEEE* **61**, 1323 (1973).

Koningsberger, D. C., G. J. Mulder, and B. Pelupessy, *J. Phys. E* **6**, 306 (1973).

Konzelmann, U., D. Kilpper, and M. Schwoerer, *Z. Naturforsch.* **30a**, 754 (1975).

Korablev, I. V., *Radiotekh. Elektron.* **14**, 1292 (1969).

Kottis, P., and R. Lefebvre, *J. Chem. Phys.* **39**, 393 (1963); **41**, 379 (1964).

Krebs, P., and E. Sackmann, *J. Magn. Reson.* **22**, 359 (1976).

Krongelb, S., and M. W. P. Strandberg, *J. Chem. Phys.* **31**, 1196 (1959).

Lagercrantz, C., and L. Persson, *RSI* **35**, 1605 (1964).

Lecar, H., and A. Okaya, *Paramagn. Reson.* **2**, 675 (1963).

Lemaistre, J. P., and P. Kottis, *J. Chem. Phys.* **68**, 2730 (1978).

Leskov, A. S., V. N. Zinchenko, N. P. Ilgasova, O. V. Ogarenko, and G. V. Shmeleva, *Izmer. Tekh.* **19**, 48 (1976).

Lloyd, J. P., and G. E. Pake, *Phys. Rev.* **92**, 1576 (1953); **94**, 579 (1954).

Marzouk, M., J. Royaud, and R. Sardos, *C. R. Acad. Sci.* **B281**, 501 (1975).

Mahendru, P. C., and R. Parshad, *RSI* **35**, 1618 (1964).

Manenkov, A. A., and A. M. Prokhorov, *Radiotekh. Elektron.* **1**, 469 (1956).

Martinelli, M., L. Pardi, C. Pinzino and S. Santucci, *Phys. Rev.* **16**, 164 (1977).

May, I. W., and K. J. White, *RSI* **40**, 1639 (1969).

McBrierty, V. J., and P. D. Cook, *Nature* **205**, 1197 (1965).

McGarvey, B. R., *J. Chem. Phys.* **37**, 2001 (1962).

Meijer, G. C. M., *J. Appl. Sci. Eng. A* **1**, 129 (1975).

Mellor, J. W., *A Treatise on Inorganic and Theoretical Chemistry*, Longmans, Green, & Co., New York, 1931, Vol. XI, p. 356.

Menzel, E. R., W. R. Vincent, and J. R. Wasson, *J. Magn. Reson.* **21**, 321 (1976).

Miner, G. K., T. P. Graham, and G. T. Johnston, *RSI* **43**, 1297 (1972).

Misra, M., *Indian J. Pure Appl. Phys.* **1**, 37 (1963); **3**, 54 (1965).

Mittra, R., *IEEE Trans. Antennas Propag.* **AP-11**, 585 (1963).

Mollier, J. C., J. Hardin, and J. Uebersfeld, *17th Congress Ampere Nuclear Magnetic Resonance and Related Phenomena, Turku, Finland*, 1972.

Mollier, J. C., J. Hardin, and J. Uebersfeld, *17th International Colloquium. Microwave Spectroscopy, Florence, Italy*, 296 (1973).

Moore, W. S., and T. Yalcin, *J. Magn. Reson.* **11**, 50 (1973).

Moyer, E. S., and W. J. McCarthy, *Anal. Chim. Acta* **48**, 79 (1969).

Mukai, K., and J. Sakamoto, *J. Chem. Phys.* **68**, 1432 (1978).

Müller, K. A., *Arch. Sci. (Fasc. Spec.)* **13**, 342 (1960).

Müller, K. A., *Proc. of the IVth Ampere International Summer School, Pula, Yugoslavia*, 637 (1976). Ampere Group Europ. Phys. Soc. Magn. Reson., 1977.

Muromtsev, V., A. K. Piskunov, and N. V. Verein, *Radiotekh. Elektron.* **7**, 1206 (1129) (1962).

Newman, D. J., and W. Urban, *Adv. Phys.* **24**, 793 (1975).

Niklas, J. R., and J. M. Spaeth, *Phys. Status Solidi* **101b**, 221 (1980).

Nishikawa, T., and K. Someno, *Anal. Chem.* **47**, 1290 (1975).

Orbach, R., M. Krusius, and M. Vucrio, *Proceedings of the 14th International Conference on Low Temperature Physics, Otaniemi, Finland*, 375 (1975).

Pake, G. E., and E. M. Purcell, *Phys. Rev.* **74**, 1184 (1948).

Pascaru, I., *Stud. Cercet. Fiz. (Rumania)* **25**, 1007 (1973).

Payne III, J. B., *IEEE Trans. Microwave Theory Tech.* **MTT-12**(1), 48 (1964).

Peter, M., D. Shaltiel, J. H. Wernick, H. J. Williams, J. B. Mock, and R. C. Sherwood, *Phys. Rev.* **126**, 1395 (1962).

Piette, L. H., I. Yamazaki, and H. S. Mason, *Free Radicals in Biological Systems*, Academic Press, New York, 1961, p. 195.

Plato, M., and K. Möbius, *Messtechnik* **80**, 224 (1972).

Poole, C. P., Jr., Thesis, Dept of Physics, University of Maryland, 1958.

Poole, C. P., Jr., and H. A. Farach, *Relaxation in Magnetic Resonance*, Academic Press, New York, 1971a.

Poole, C. P., Jr., and H. A. Farach, *The Theory of Magnetic Resonance*, Wiley-Interscience, New York, 1972.

Poole, C. P., Jr., and H. A. Farach, *J. Magn. Reson.* **4**, 312 (1971); **5**, 305 (1971).

Portis, A. M., and D. T. Teaney, *J. Appl. Phys.* **29**, 1692 (1958).

Posner, D. W., *J. Magn. Res.* **14**, 129 (1974).

Pristupa, A. I., and D. I. Kadyrov, *PTE* **17**(1), 114 (1974).

Pryor, W. A., K. Terauchi, and W. H. Davis, Jr., *Environ. Health Perspect.* **16**, 161 (1976).

Rajan, R., *J. Sci. Ind. Res. (India)* **21B**, 445 (1962).

Randolph, M. L., *RSI* **31**, 949 (1960).

Reid, M. S., et al., *Proc. IEEE* **61**, 1330 (1973).

Reineker, P., *Phys. Status Solidi B* **74**, 121 (1976).

Reineker, P., *Solid State Commun.* **25**, 859 (1978).

Reineker, P., *Phys. Status Solidi B* **70**, 189 (1975).

Reineker, P., *Chem. Phys.* **16**, 425 (1976).

Riesenfeld, E. H., *Chem. Ber.* **38**, 4068 (1905).

RLS-6.

RLS-7.

RLS-11.

RLS-13, Chap. 8.

RLS-16.

RLS-22.

RLS-23.

RLS-24.

Roberts, G., and W. Derbyshire, *JSI* **38**, 511 (1961).

Roch, J., R. Sardos, K. Haye, and R. Chastanet, *Revue Phys. Appl.* (*France*) **6**, 247 (1971).

Royaud, J., and R. Sardos, in preparation, 1981.

Rupp, L. W., Jr., and W. M. Walsh, Jr., *Am. J. Phys.* **38**, 238 (1970).

Russell, D. B., *J. Chem. Phys.* **43**, 1996 (1965).

Ryter, Ch., R. Lacroix, and R. Extermann, *Onde Elect.* **35**, 490 (1955).

Sardos, R., *Revue de Phys. Appl.* **4**, 29 (1969).

Sardos, R., *RSI* **43**, 1726 (1972).

Scheffler, K., and H. B. Stegmann, *Ber. Brunsengesell. Phys. Chem.* **67**, 864 (1963).

Schlick, S., and L. Kevan, *J. Magn. Reson.* **21**, 129 (1976).

Schneider, F., *Z. Instrumentenkd.* **71**(12), 315 (1963).

Schoch, F., L. Prost, and H. H. Gunthard, *J. Phys. E* **8**, 563 (1975).

Schone, H. E., and P. W. Olson, *RSI* **36**, 843 (1965).

Schouler, M. C., M. Decorps, and F. Genoud, *Mol. Phys.* **32**, 1671 (1976).

Schultz, S., in *Measurements of Physical Properties*, Vol. IV, E. Passaglia, Ed., Wiley, New York, 1972, Chap. VB.

Schwarz, V. D., and V. R. Wessel, *ETP* **24**, 531 (1976).

Schweiger, A., and H. H. Günthard, submitted for publication (1982).

Schweiger, A., M. Rudin, and J. Forrer, *J. Chem. Phys. Lett.* **80**, 376 (1981).

Scott, T. W., K. C. Chu, and M. Venugopalan, *Naturwiss.* **62**, 532 (1975).

Seehra, M. S., *RSI* **39**, 1044 (1968).

Seehra, M. S., and R. P. Gupta, *Phys. Rev. B* **9**, 197 (1974).

Segal, B. G., M. Kaplan, and G. K. Fraenkel, *J. Chem. Phys.* **43**, 4191 (1965).

Servant, R., and E. Palangie, *C.R. Hebd. Seances Acad. Sci. Ser. B* **280**, 239 (1975).

Siegman, A. E., *Microwave Solid State Masers*, McGraw-Hill, New York, 1964.

Singer, L. S., *J. Appl. Phys.* **30**, 1463 (1959).

Singer, L. S., and C. Kikuchi, *J. Chem. Phys.* **23**, 1738 (1955).

Singer, L. S., W. H. Smith, and G. Wagoner, *RSI* **32**, 213 (1961).

Slezak, A., and R. Hrabanski, *Zesz. Nauk. Politech. Czestochow, Nauki Podstawowe* **18**, 69 (1976).

Slifkin, M. A., and Y. M. Suleiman, *Radiat. Eff.* **27**, 111 (1976).

Sloan, III, E. L., A. Ganssen, and E. C. LaVier, *Appl. Phys. Lett.* **4**, 109 (1964).

Smidt, J., *Arch. Sci. (Fasc. Spec.)* **13**, 337 (1960).

Smith, R. C., and T. H. Wilmshurst, *JSI* **40**, 371 (1963).

Štirand, O., *ETP* **10**, 313 (1962).

Stockhausen, M., and M. Strassman, *Magn. Reson. Relat. Phenom. Proc. Congr. Ampere, 18th* **2**, 535 (1975).

Stoodley, L. G., *Nature* **198**, 1077 (1963).

Stoodley, L. G., *J. Electron. Control* **14**, 531 (1963).

Storey, W. H., C. M. Moneta, and D. G. Cadena, *Nature* **195**, 963 (1962).

Stromberg, R., *Acta Chem. Scan.* **17**, 1563 (1963).

Takao, I., and T. Hayashi, *IEEE Trans. Microwave Theory Tech.* **MTT-17**, 107 (1969).

Takao, I., and K. Watanabe, *RSI* **42**, 788 (1971).

Tataru, E., *Stud. Cercet. Fiz. (Rumania)* **25**, 901 (1973).

Tataru, E., *Rev. Phys. Appl. (France)* **9**, 521 (1974).

Teaney, D. T., M. P. Klein, and A. M. Portis, *RSI* **32**, 721 (1961).

Thompson, D. S., and J. S. Waugh, *RSI* **36**, 552 (1965).

Tinkham, M., and M. W. P. Strandberg, *Phys. Rev.* **97**, 941 (1955).

Tsung-tang, Sun, Chaing Wen-chia, Shun Lien-fong, and Mao Tin-fong, *Acta Phys. Sinica* **21**, 866 (1965).

Tomkiewicz, Y., and A. R. Taranko, *Phys. Rev.* **B18**, 723 (1978).

Uebersfeld, J., *J. Chim. Phys.* **56**, 805 (1959).

Uhlir, A., Jr., *Sci. Am.* **200**, 118 (1959).

Valeriu, A., and I. Pascaru, *Rev. Phys. (Rumania)* **8**, 481 (1963).

Van Meijel, J., A. Stesmans, and J. Witters, *Proc. Meeting on Characterization Noncrystalline Solids, Rev. Phys. Appl. (France)* **12**, 905 (1977).

Vergragt, P. J., J. A. Kooter, and J. H. Van der Waals, *Mol. Phys.* **33**, 1523 (1979).

Verlinden, R., P. Grobet, and L. Van Gerven, *Chem. Phys. Lett.* **27**, 535 (1974).

Vigouroux, B., J. C. Gourdon, P. Lopez, and J. Pescia, *J. Phys. E Sci. Instrum.* **6**, 557 (1973).

Volino, F., CEA-R-3706, 1969. (*Phys. Abs.* A7060121, 1970).

Volodin, A. P., M. S. Khaikin, and V. S. Edel'man, *PTE*, **17**, 1719 (1974).

Warren, D. C., and J. M. Fitzgerald, *Anal. Chem.* **49**, 250 (1977).

Webb, R. H., *RSI* **33**, 732 (1962).

Weil, J. A., and R. F. C. Claridge, *RSI* **41**, 140 (1970).

Werner, K., *Hochfrequenztech. Elektrokust.* **73**, 116 (1964).

Wertz, J. E., D. C. Reitz, and F. Dravnieks, *Free Radicals in Biologicals Systems*, M. S. Blois, Jr., et al., Ed., Academic Press, New York, 1961, p. 183.

Wessel, R., and D. Schwarz, *Exp. Tech. Phys.* **23**, 641 (1975); **24**, 195 (1976); **26**, 195 (1978).

Westenberg, A. A., *J. Chem. Phys.* **43**, 1544 (1965).

Westenberg, A. A., and N. DeHaas, *J. Chem. Phys.* **40**, 3087 (1964); **43**, 1550 (1965).

Wilmshurst, T. H., *Nature* **199**, 477 (1963).

Wilmshurst, T. H., W. A. Gambling, and D. J. E. Ingram, *J. Electron. Control* **13**, 339 (1962).

Wood, D. E., and T. F. Yen, *Proc. Symp. Electron Spin Resonance of Metal Complexes*, Cleveland, Ohio, 135 (1969).

Wyard, S. J., *JSI* **42**, 769 (1965).

Yariv, A., and J. P. Gordon, *RSI* **32**, 462 (1961).

Yashchenko, G. I., V. I. Koryakov, and A. K. Chirkov, *Zh. Obshch. Khim.* **46**, 409 (1976).

Yashchenko, G. N., V. I. Koryakov, and A. K. Chirkov, *Tr. Inst. Khim. Ural. Nauchn. Tsentr, Akad. Nauk SSSR* **34**, 99 (1975).

Yo, C. H., R. Poupko, and R. M. Hornreich, *Chem. Phys. Lett.* **54**, 142 (1978).

Zieger, J., and H. C. Wolf, *Chem. Phys.* **29**, 209 (1978).

Zinsli, P. E., and M. A. El-Sayed, *Chem. Phys. Lett.* **20**, 171 (1973).

Zuckermann, B., *RSI* **44**, 1118 (1973).

Zverov, G. M., *PTE* **5**, 109 (930) (1961).

CHAPTER 12

Lineshapes

A great deal of information can be obtained from a careful analysis of the width and shape of a resonant absorption line. The first few sections of this chapter analyze in detail the two most common lineshapes, the Lorentzian and the Gaussian, when they are detected and recorded under various types of instrumental conditions that faithfully reproduce their true shape. The remainder of the chapter discusses various mechanisms that broaden spectral lines. The broadening (and narrowing) that arises under the influence of the measuring arrangement is discussed in Chaps. 6, 11, and 13. Homogeneous and inhomogeneous broadening is discussed in Sec. 13-B.

In this chapter, unless otherwise stated, we assume that the microwave power level is sufficiently low so that saturation is avoided. As a result the integrated area A of a resonance absorption line is proportional to the number of spins, and the measured moments of the lines are physically significant in terms of the theories of Van Vleck (1948) and others. We also assume that the modulation amplitude is much less than the linewidth. If the recorded spectrum is distorted by either saturation or overmodulation, then the methods described in Secs. 13-C and 6-H, respectively, may be used to convert the observed area and moments to the physically meaningful ones discussed in this chapter. Mohos (1975) analyzed the distortion effects of ESR spectrometers.

In this chapter we are concerned with lineshapes of paramagnetic liquids and solids in their optical ground states. [Some work has been carried out near solidification temperatures, e.g., Dormann et al. (1977) and Morantz and Thompson (1970), and Kiel (1962, 1963) discussed lineshapes of ESR in optically excited states.]

Lineshape and linewidth studies have also made of a number of gases (e.g., Carrington, 1968; Cook and Miller, 1973; Fehsenfeld et al., 1965; Fisanick-Englot and Miller, 1976; Freed, 1964; Gordon and McGinnis, 1968; Lin and Kevan, 1977; Takagi and Kojima, 1974; Uehara et al., 1971; Zegarski et al., 1975).

A. Resonance Absorption Line

The area A under the resonance absorption curve Y shown in Figs. 12-1a and 12-1c is given by

$$A = \sum_{j=1}^{m} y_j \left(H_j - H_{j-1} \right) \tag{1}$$

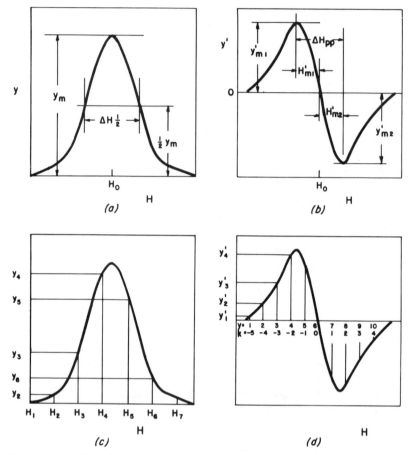

Fig. 12-1. (a), (c) absorption and (b), (d) absorption derivative lineshapes. The lineshape parameters are defined in (a) and (b), and the method of integration is illustrated in (c) and (d).

where y_j is the amplitude of the resonance absorption line at the magnetic field H_j, as shown in Fig. 12-1c. In practice it is convenient to select equal intervals $(H_j - H_{j-1})$ along the field direction so that

$$A = (H_j - H_{j-1}) \sum_{j=1}^{m} y_j \tag{2}$$

where there are m intervals in all. This area is usually proportional to the number of unpaired spins in the sample. The precision of the area determination is increased by increasing m and decreasing the interval $(H_j - H_{j-1})$.

The nth moment $\langle H^n \rangle$ of a resonance absorption is defined by

$$\langle H^n \rangle = \frac{H_j - H_{j-1}}{A} \sum_{j=1}^{m} (H_j - H_0)^n y_j \tag{3}$$

where the summation limits are the same as in Eq. (2), and in general, these moments are functions of the field H_0. If a resonance absorption line is symmetrical, then H_0 is the magnetic field that divides the line into two halves, each of which contains half the area A:

$$\int_{-\infty}^{H_0} Y(H)\, dH = \frac{1}{2} A = \int_{H_0}^{\infty} Y(H)\, dH \tag{4}$$

where Y is the lineshape function. This criterion may be used to determine H_0, and we shall use this definition even for unsymmetrical lines. Another method for determining H_0 of a symmetric line entails the computation of the first moment:

$$\langle H^1 \rangle = \frac{H_j - H_{j-1}}{A} \sum_{j-1}^{m} (H_j - H_0) y_j \tag{5}$$

which vanishes for the proper choice of H_0. If this expression is not zero, then let

$$\langle H^1 \rangle = B = \frac{\displaystyle\sum_{j=1}^{m} (H_j - H_0) y_j}{\displaystyle\sum_{j=1}^{m} y_j} \tag{6}$$

and when B is inserted into Eq. (5), the corrected value of $\langle H^1 \rangle$ will vanish since

$$\langle H^1 \rangle_{\text{corr}} = \frac{H_j - H_{j-1}}{A} \sum_{j=1}^{m} (H_j - H_0 - B) y_j = 0 \tag{7}$$

The correction factor B is the displacement shown in Fig. 12-2. As a result, Eq. (3) may be rewritten as

$$\langle H^n \rangle_{\text{corr}} = \frac{H_j - H_{j-1}}{A} \sum_{j=1}^{m} (H_j - H_0 - B)^n y_j \tag{8}$$

In practice one can usually select H_0 so that $B = 0$, and use Eq. (3), but Eqs. (6) plus (8) may be employed if a more refined computation of A or $\langle H^n \rangle$ is desired. It should be emphasized that the vanishing of the first moment and all other odd moments is a property of symmetric absorption lines, so B cannot be defined in a physically meaningful way for an asymmetric absorption line.

Hyde and Pilbrow (1980) showed that the first moment of a powder pattern vanishes to a good approximation when H_0 of Eq. (5) is chosen as the magnetic field $H_0 = h\upsilon/g_0 B$ corresponding to the average $g_0 = \frac{1}{3}(g_1 + g_2 + g_3)$ of the three principal g factors g_1, g_2, and g_3 (see Sec. 12-K).

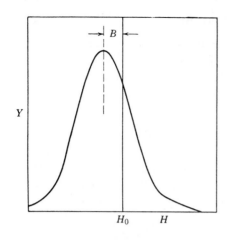

Fig. 12-2. Correction factor B that results from an improper choice of the resonant field H_0. When the resonant line is unsymmetrical, the proper choice of H_0 is not as obvious as it is in this case.

TABLE 12-1

Computation of Correction Factor B, Area A and Second and Fourth Moments of a Gaussian Lineshape using Eqs. (6), (2), and (8) of Sec. 12-A

j	y_j	$\left(\dfrac{H_j - H_0}{\frac{1}{2}\Delta H_{1/2}}\right)$	$y_j\left(\dfrac{H_j - H_0}{\frac{1}{2}\Delta H_{1/2}}\right)$	$\left(\dfrac{H_j - H_0 - B}{\frac{1}{2}\Delta H_{1/2}}\right)$	$y_j\left(\dfrac{H_j - H_0 - B}{\frac{1}{2}\Delta H_{1/2}}\right)^2$	$y_j\left(\dfrac{H_j - H_0 - B}{\frac{1}{2}\Delta H_{1/2}}\right)^4$
1	0.000015	−3.0	−0.0000	−4.0	0.0002	0.0038
2	0.00020	−2.5	−0.0005	−3.5	0.0015	0.0300
3	0.00195	−2.0	−0.0039	−3.0	0.0178	0.1600
4	0.0131	−1.5	−0.0197	−2.5	0.0819	0.5117
5	0.0626	−1.0	−0.0626	−2.0	0.2508	1.0032
6	0.2101	−0.5	−0.1051	−1.5	0.4727	1.0630
7	0.5000	0.0	0.0000	−1.0	0.5000	0.5000
8	0.8380	0.5	0.4190	−0.5	0.2095	0.0524
9	1.0000	1.0	1.0000	0	1.5353	3.3241
10	0.8380	1.5	1.2570	0.5	×2	×2
11	0.5000	2.0	1.0000	1.0	3.0706	6.6482
12	0.2101	2.5	0.5253	1.5		
13	0.0626	3.0	0.1881	2.0		
14	0.0131	3.5	0.0459	2.5		
15	0.00195	4.0	0.0078	3.0		
16	0.00020	4.5	0.0018	3.5		
17	0.000015	5.0	0.0001	4.0		
	4.259		5.4450			
			−0.1919			
			4.2531			

$$B = \frac{\frac{1}{2}\Delta H_{1/2}}{\sum\limits_{j=1}^{m} y_j} \sum_{j=1}^{m} y_j\left(\frac{H_j - H_0}{\frac{1}{2}\Delta H_{1/2}}\right) = \frac{4.2531}{4.259}\left(\tfrac{1}{2}\Delta H_{1/2}\right) = 0.5\,\Delta H_{1/2}$$

$$H_j - H_{j-1} = \tfrac{1}{2}\left(\tfrac{1}{2}\Delta H_{1/2}\right) = \tfrac{1}{4}\Delta H_{1/2}$$

$$A = |H_j - H_{j-1}| \sum_{j=1}^{m} y_j = \left(\tfrac{1}{4}\Delta H_{1/2}\right)(4.259) = 1.065\,\Delta H_{1/2}$$

$$\langle H^2 \rangle = \frac{H_j - H_{j-1}}{A}\left(\tfrac{1}{2}\Delta H_{1/2}\right)^2 \sum_{j=1}^{m}\left(\frac{H_j - H_0 - B}{\frac{1}{2}\Delta H_{1/2}}\right)^2 y_j = \frac{\left(\tfrac{1}{2}\Delta H_{1/2}\right)^2}{4.259}(3.0706) = 0.18(\Delta H_{1/2})^2$$

$$\langle H^4 \rangle = \frac{H_j - H_{j-1}}{A}\left(\tfrac{1}{2}\Delta H_{1/2}\right)^4 \sum_{j=1}^{m}\left(\frac{H_j - H_0 - B}{\frac{1}{2}\Delta H_{1/2}}\right)^4 y_j = \frac{\left(\tfrac{1}{2}\Delta H_{1/2}\right)^4}{4.259}(6.6482) = 0.097(\Delta H_{1/2})^4$$

The use of Eqs. (2), (7), and (8) is illustrated by the calculation with a Gaussian lineshape ^GY that is shown in Table 12-1.

$$^GY(H_j) = {}^Gy_j = y_m\exp\left[-0.693\left(\frac{H_j - H_0}{\frac{1}{2}\Delta H_{1/2}}\right)^2\right] \tag{9}$$

or more generally

$$^GY(H) = y_m\exp\left[-0.693\left(\frac{H - H_0}{\frac{1}{2}\Delta H_{1/2}}\right)^2\right] \tag{10}$$

where $\ln 2 = 0.693$. We let $y_m = 1$, and compute the correction B, obtaining from Table 12-1

$$\frac{B}{\frac{1}{2}\Delta H_{1/2}} = \frac{\sum\limits_{j=1}^{m}(H_j - H_0)y_j}{\sum\limits_{j=1}^{m}y_j} = \frac{4.253}{4.259} \sim 1 \tag{11}$$

For this calculation we selected the convenient increments

$$H_j - H_{j-1} = \tfrac{1}{4}\Delta H_{1/2} \tag{12}$$

where $\Delta H_{1/2}$ is the full linewidth between half-amplitude points (i.e., the half-amplitude linewidth) defined in Fig. 12-1a. The use of smaller increments will considerably improve the accuracy of the results. Since the lineshape is symmetrical, the calculation of A, $\langle H^2\rangle$, and $\langle H^4\rangle$ can be limited to a calculation with half the line if one doubles the result. (It is important to use only half the amplitude of y at the center point in this abbreviated calculation.)
From Table 12-1 and Eq. (2), we obtain

$$A = 1.065\,\Delta H_{1/2} \tag{13}$$

and from Eq. (8)

$$\langle H^2\rangle = \frac{H_j - H_{j-1}}{A}\left(\tfrac{1}{2}\Delta H_{1/2}\right)^2\sum\limits_{j=1}^{m}\left(\frac{H_j - H_0 - B}{\frac{1}{2}\Delta H_{1/2}}\right)^2 y_j \tag{14}$$

$$= 0.721\left(\tfrac{1}{2}\Delta H_{1/2}\right)^2 \tag{15}$$

$$\langle H^4\rangle = \frac{H_j - H_{j-1}}{A}\left(\tfrac{1}{2}\Delta H_{1/2}\right)^4\sum\limits_{j=1}^{m}\left(\frac{H_j - H_0 - B}{\frac{1}{2}\Delta H_{1/2}}\right)^4 y_j \tag{16}$$

$$= 1.56\left(\tfrac{1}{2}\Delta H_{1/2}\right)^4 \tag{17}$$

The preceding summations may be represented more elegantly by means of the following two integrals:

$$A = \int_{-\infty}^{\infty} Y\,dH = 2\int_{0}^{\infty} Y\,dH \tag{18}$$

and

$$\langle H^n \rangle = \begin{cases} 0 & n \text{ odd} \\ \dfrac{2}{A}\displaystyle\int_{0}^{\infty} (H - H_0)^n Y\,dH & n \text{ even} \end{cases} \tag{19}$$

The change in the lower limit $(-\infty \to 0)$ is only valid for symmetrical lines. For a Gaussian lineshape (10), the change of variable $x = [(H - H_0)/(\frac{1}{2}\Delta H_{1/2})](\ln 2)^{1/2}$ yields

$$A = \frac{2y_m\left(\frac{1}{2}\Delta H_{1/2}\right)}{(\ln 2)^{1/2}} \int_{0}^{\infty} \exp(-x^2)\,dx \tag{20}$$

$$= \left[\frac{\frac{1}{2}\sqrt{\pi}}{(\ln 2)^{1/2}}\right] y_m\,\Delta H_{1/2} \tag{21}$$

$$= 1.0643\, y_m\,\Delta H_{1/2} \tag{22}$$

To evaluate the moments, one may employ the expression

$$\int_{-\infty}^{\infty} x^n \exp\left[-\left(\frac{x}{a}\right)^{1/2}\right] dx = a^{n+1}\Gamma\left[\frac{(n+1)}{2}\right] \tag{23}$$

where the gamma function $\Gamma[(n+1)/2]$ is defined by

$$\Gamma\left[\frac{(n+1)}{2}\right] = \frac{1\cdot3\cdot5\cdots(n-1)}{(2)^{n/2}}\pi^{1/2} = \frac{(n-1)!!}{2^{n/2}}\pi^{1/2} \tag{24}$$

for the even moments (even n), while from symmetry the odd moments (odd n) vanish. Since $dH = d(H - H_0)$, we have

$$\langle H^n \rangle = \frac{\left(\frac{1}{2}\Delta H_{1/2}\right)^n}{\pi^{1/2}(\ln 2)^{n/2}}\Gamma\left[\frac{(n+1)}{2}\right] = \frac{\left(\frac{1}{2}\Delta H_{1/2}\right)^n (n-1)!!}{(2\ln 2)^{n/2}} \tag{25}$$

with the result that

$$\langle H^2 \rangle = 0.721\left(\frac{1}{2}\Delta H_{1/2}\right)^2 \tag{15}$$

and

$$\langle H^4 \rangle = 1.56 \left(\tfrac{1}{2} \Delta H_{1/2} \right)^4 \tag{17}$$

as found above.

For a Lorentzian lineshape

$$^L Y(H) = \frac{y_m}{1 + \left[(H - H_0)/\tfrac{1}{2}\Delta H_{1/2} \right]^2} \tag{26}$$

one may let $x = 2(H - H_0)/\Delta H_{1/2}$, and obtain from Eq. (18)

$$A = y_m \Delta H_{1/2} \int_0^\infty \frac{dx}{1 + x^2} \tag{27}$$

and letting $x = \tan \theta$, and $dx = \sec^2 \theta \, d\theta$ the integral becomes

$$A = y_m \Delta H_{1/2} \int_0^{\pi/2} d\theta \tag{28}$$

$$= \left(\frac{\pi}{2} \right) y_m \Delta H_{1/2} \tag{29}$$

$$= 1.57 y_m \Delta H_{1/2} \tag{30}$$

Since the Lorentzian lineshape is symmetrical, all odd moments vanish; and for the second moment we have

$$\langle H^2 \rangle = \frac{y_m}{A} \int_{-\infty}^\infty \frac{(H - H_0)^2 \, dH}{1 + \left[(H - H_0)/\tfrac{1}{2}\Delta H_{1/2} \right]^2} \tag{31}$$

$$= \frac{2}{\pi} \left(\tfrac{1}{2}\Delta H_{1/2} \right)^2 \int_0^\infty \frac{x^2 \, dx}{1 + x^2} \tag{32}$$

which diverges to an infinite result. This is evident if one considers that for large x, the integral behaves like $\int_0^\infty dx$. Thus, for a Lorentzian lineshape all odd moments vanish, and all even moments are infinitely large.

It is important to note that the area of an absorption line is proportional to the amplitude times the linewidth $\Delta H_{1/2}$, and the constant of proportionality depends on the lineshape. The units of the area are amplitude units times gauss, or number of unpaired spins. Of particular interest is the fact that for the same amplitude y_m and the same linewidth $\Delta H_{1/2}$, a Lorentzian line corresponds to $(1.56/1.06)$ or 1.48 times as many spins as a Gaussian line. One should bear in mind that this factor of 1.48 applies only to absorption lines themselves, since the derivative lineshape requires a different normalization factor, $4(2\pi/3e)^{1/2} = 3.51$, from Table 12-5.

We have emphasized the vanishing of all the odd moments of symmetrical absorption lines such as those with Gaussian and Lorentzian shapes. It is the even moments that vanish for the corresponding dispersion mode lineshapes. The moments of the first derivative of an absorption line are defined to agree with the corresponding moments of the absorption line itself, as discussed in the next section.

To record the actual absorption line one may either perform a dc experiment (without modulation) as discussed in Sec. 6-E, or an electronic integrator may be employed to integrate the ESR signal before it is recorded (see Sec. 7-I). Divers methods of integrating spectra are discussed in Sec. 11-K. A quick integration method is given by Charlier et al. (1964).

The extent to which an unsymmetrical resonant line is asymmetric may be expressed quantitatively by giving the dimensionless ratio

$$\text{Skewness} = \frac{\langle H^3 \rangle}{\langle H^2 \rangle^{3/2}} \tag{33}$$

Svare (1965) calculated third moments for $g\beta H \geqslant kT$ and showed that paramagnetic lines are asymmetric at low temperatures. Equation (7) states that experimental lineshapes are defined with vanishing first moments. If instead the line center is selected as an unperturbed resonant field, then first moments may exist (e.g., see Svare and Seidel, 1964).

B. First-Derivative Absorption Line

If the narrow band amplifier and phase-sensitive detector are tuned to the modulation frequency, and the modulation amplitude H_{mod} is much less than the linewidth, then the recorded lineshape becomes the first derivative Y' of the absorption line Y (see Sec. 6-G):

$$Y' = \left(\frac{d}{dH} \right) Y \tag{1}$$

shown in Fig. 12-1b. Again we use equal intervals $(H_j - H_{j-1})$ as shown in Fig. 12-1d, so we may write

$$y_j = \left(H_j - H_{j-1} \right) \sum_{i=j}^{m} y_i' \tag{2}$$

From Eq. (12-A-2), the area A becomes

$$A = \left(H_j - H_{j-1} \right)^2 \sum_{j=1}^{m} \sum_{i=j}^{m} y_i' \tag{3}$$

$$= \left(H_j - H_{j-1} \right)^2 \sum_{j=1}^{m} j y_j' \tag{4}$$

The second and fourth moments are given by

$$\langle H^2 \rangle = \frac{(H_j - H_{j-1})^2}{A} \sum_{j=1}^{m} \sum_{i=j}^{m} (H_j - H_0)^2 y_i' \tag{5}$$

$$\langle H^4 \rangle = \frac{(H_j - H_{j-1})^2}{A} \sum_{j=1}^{m} \sum_{i=j}^{m} (H_j - H_0)^4 y_i' \tag{6}$$

Equations (2)–(6) presuppose that the baseline is chosen properly, so that

$$\sum_{j=1}^{m} y_j' = 0 \tag{7}$$

If this is not the case, then we redefine Y_j' by subtracting a correction term B_1 defined by the displacement shown in Fig. 12-3a

$$\sum_{j=1}^{m} (y_j' - B_1) = 0 \tag{8}$$

whence

$$B_1 = \frac{1}{m} \sum_{j=1}^{m} y_j' \tag{9}$$

Using this, the area is given by

$$A = (H_j - H_{j-1})^2 \sum_{j=1}^{m} \sum_{i=j}^{m} (y_i' - B_1) \tag{10}$$

$$= (H_j - H_{j-1})^2 \sum_{j=1}^{m} j(y_j' - B_1) \tag{11}$$

Equations (5) and (6) presuppose the proper choice of crossover point, and if it is not properly chosen, then a second correction factor B_2 defined by Fig. 12-3a and the relation

$$\sum_{j=1}^{m} (H_j - H_0 - B_2)^2 (y_j' - B_1) = 0 \tag{12}$$

or more explicitly

$$B_2 = \frac{1}{2} \frac{\displaystyle\sum_{j=1}^{m} (H_j - H_0)^2 (y_j' - B_1)}{\displaystyle\sum_{j=1}^{m} (H_j - H_0)(y_j' - B_1)} \tag{13}$$

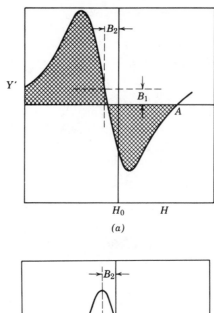

(a)

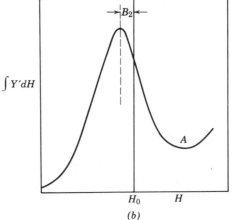

(b)

Fig. 12-3. (a) The derivative line for an improper choice of baseline and resonant field H_0, and (b) the shape of the resonant line (a) after integration. A is the point where Y' crosses the incorrect baseline for the second time, and thereby causes the integrated curve to begin to rise again. The shaded areas above and below the baseline of (a) will be equal for the proper choice of baseline (i.e., when $B_1 = 0$).

may be included, to give

$$\langle H^2 \rangle = \frac{(H_j - H_{j-1})^2}{A} \sum_{j=1}^{m} \sum_{i=j}^{m} (H_j - H_0 - B_2)^2 (y_i' - B_1) \qquad (14)$$

The fourth moment has the form

$$\langle H^4 \rangle = \frac{(H_j - H_{j-1})^2}{A} \sum_{j=1}^{m} \sum_{i=j}^{m} (H_j - H_0 - B_2)^4 (y_i' - B_1) \qquad (15)$$

An improper choice of the baseline (i.e., $B_1 \neq 0$) will produce an integrated shape of the type shown in Fig. 12-3b. The baseline of the integrated shape will be the same at both ends of the line only if the derivative area (shaded on Fig. 12-3a) is equal above and below the baseline. An improper choice of H_0 (i.e., $B_2 \neq 0$) will not affect the area or integrated shape, but will produce erroneous values of the moments. Loveland and Tozer (1972) devised a numerical method for the double integration of a first-derivative spectrum in the presence of baseline drift.

The use of Eqs. (10)–(14) is illustrated by the calculations shown on Table 12-2. They are performed with a Gaussian lineshape $^G Y'(H)$

$$^G Y'(H_j) = {}^G y_j'$$

$$= 1.649 y_m' \left(\frac{H_j - H_0}{\frac{1}{2}\Delta H_{pp}} \right) \exp\left[-\frac{1}{2}\left(\frac{H_j - H_0}{\frac{1}{2}\Delta H_{pp}} \right)^2 \right] \qquad (16)$$

TABLE 12-2

Computation of the Correction Factors B_1 and B_2 illustrated for a Gaussian Derivative Lineshape, Using Eqs. (9) and (13) of Sec. 12-B

j	y_j'	$(y_j' - B_1)$	$\left(\dfrac{H_j - H_0}{\frac{1}{2}\Delta H_{pp}}\right)$	$\left(\dfrac{H_j - H_0}{\frac{1}{2}\Delta H_{pp}}\right)(y_j' - B_1)$	$\left(\dfrac{H_j - H_0}{\frac{1}{2}\Delta H_{pp}}\right)^2 (y_j' - B_1)$	$\left(\dfrac{H_j - H_0 - B_2}{\frac{1}{2}\Delta H_{pp}}\right)$
1	0.1022	0.0022	−3.0	−0.0066	0.020	−4.0
2	0.1126	0.0126	−2.5	−0.032	0.079	−3.5
3	0.155	0.055	−2.0	−0.110	0.220	−3.0
4	0.280	0.180	−1.5	−0.270	0.405	−2.5
5	0.546	0.446	−1.0	−0.446	0.446	−2.0
6	0.903	0.803	−0.5	−0.401	0.200	−1.5
7	1.100	1.000	0.0	0	0	−1.0
8	0.828	0.728	0.5	0.303	0.152	−0.5
9	0.100	0.000	1.0	0	0	0
10	−0.628	−0.728	1.5	−1.058	−1.587	0.5
11	−0.900	−1.000	2.0	−2.000	−4.000	1.0
12	−0.703	−0.803	2.5	−2.008	−5.020	1.5
13	−0.346	−0.446	3.0	−1.338	−4.014	2.0
14	−0.080	−0.180	3.5	−0.630	−2.205	2.5
15	0.045	−0.055	4.0	−0.220	−0.880	3.0
16	0.0874	−0.0126	4.5	−0.057	−0.255	3.5
17	0.0978	−0.0022	5.0	−0.011	−0.055	4.0
	1.700			−8.285	−16.494	

$$B_1 \frac{\displaystyle\sum_{j=1}^{m} y_j'}{m} = \frac{1.700}{17} = 0.1$$

$$B_2 = \frac{1}{2}\frac{\displaystyle\sum_{j=1}^{m}(H_j - H_0)^2(y_j' - B_1)}{\displaystyle\sum_{j=1}^{m}(H_j - H_0)(y_j' - B_1)} = \frac{1}{4}(\Delta H_{pp})\left(\frac{16.494}{8.285}\right) = 0.996\left(\frac{1}{2}\Delta H_{pp}\right) \approx 1.00\left(\frac{1}{2}\Delta H_{pp}\right)$$

where $1.649 = e^{1/2}$, and the quantities y'_m and ΔH_{pp} are the derivative half-amplitude and the peak-to-peak full linewidth (compare Fig. 12-1 with $y'_{m1} = y'_{m2} = y'_m$. For the sample calculation given in Table 12-2, we let $y'_m = 1$ and $|H_j - H_0| = \frac{1}{4}\Delta H_{pp}$ so that $y'_j = \pm 1$ at the peaks of the line. We deliberately displace the baseline and crossover point from their true values as shown in Fig. 12-3 to illustrate the calculation of the correction factors B_1 and B_2.

From the sum of the second column in Table 12-2 and Eq. (9), we find that the correction factor B_1 equals $0.1\Delta H_{pp}$, and we subtract 0.1 from each number in the second column to obtain the third column. The fourth column lists the values of $2(H_j - H_0)/\Delta H_{pp}$ obtained with the incorrect choice of crossover point, and the sums of the fifth and sixth columns are put in Eq. (13) to obtain $B_2 = 1.00$. This is subtracted from the fourth column to give the seventh column. A proper choice of baseline and crossover point would have given the third and seventh columns directly. The calculations themselves are given in the lower part of the table.

The next step is to compute the area, second moment, and fourth moment, and this is done in Table 12-3 after adjusting the values of y'_j and $(H_j - H_0)$ so that the correction factors B_1 and B_2 vanish. The computations carried out in the table are self-explanatory. The use of Eqs. (3), (5), and (6) to calculate A, $\langle H^2 \rangle$, and $\langle H^4 \rangle$ is illustrated at the lower part of the table. Note that the subtotals in the seventh and eighth columns differ considerably for the first half $(1 \leqslant j \leqslant 8)$ and the second half $(9 \leqslant j \leqslant 17)$ of the line. If one chooses much smaller magnetic field increments so that $|H_j - H_{j-1}| \ll \Delta H_{pp}$, then the subtotals for each half of the line will become equal, instead of differing by a factor of 2 or 3. It is interesting that the calculated moments are so close to the theoretical ones shown in Table 12-5, despite the large size of the increments $|H_j - H_{j-1}|$. This occurs because the errors on the two halves of the resonance line are in opposite directions and tend to cancel. If the increments are made sufficiently small, then it is only necessary to carry out the calculation shown in Table 12-3 with half of the lineshape, as was done in the previous section (compare Table 12-1). When $(H_j - H_{j-1})$ is not much less than ΔH_{pp}, then it is important to select one point at each peak in the resonant line [i.e., the ratio $\frac{1}{2}\Delta H_{pp}/(H_j - H_{j-1})$ should be an integer]. More care is required in calculations with the first derivative Y' because errors can accumulate in the double summation more easily than in the single summation of the straight absorption line Y.

The calculation presented in Table 12-3 may be called the even-moment first-derivative method, since it entails summations of the type

$$\sum_{j=1}^{m} \sum_{i=1}^{j} (H_j - H_0)^n y'_i$$

with even n. The odd moment method of calculating the area and moments

TABLE 12-3

Sample Calculation of the Area and Second and Fourth Moments of a Gaussian First-Derivative Lineshape by the Method of Even Moments, using Eqs. (3), (5), and (6) of Sec. 12-B[a]

j	y'_j	$\sum\limits_{i=1}^{j} y'_i$	$\left(\dfrac{H_j - H_0}{\frac{1}{2}\Delta H_{pp}}\right)$	$\left(\dfrac{H_j - H_0}{\frac{1}{2}\Delta H_{pp}}\right)^2$	$\left(\dfrac{H_j - H_0}{\frac{1}{2}\Delta H_{pp}}\right)^4$	$\left(\dfrac{H_j - H_0}{\frac{1}{2}\Delta H_{pp}}\right)^2 \sum\limits_{i=1}^{j} y'_i$	$\left(\dfrac{H_j - H_0}{\frac{1}{2}\Delta H_{pp}}\right)^j \sum\limits_{i=1}^{j} y'_i$
1	0.002	0.002	−4.0	16.0	256.0	0.03	0.51
2	0.013	0.015	−3.5	12.25	150.0	0.18	2.25
3	0.055	0.070	−3.0	9.0	81.0	0.63	5.67
4	0.180	0.250	−2.5	6.25	39.1	1.56	9.78
5	0.446	0.696	−2.0	4.0	16.0	2.74	11.36
6	0.803	1.499	−1.5	2.25	5.06	3.37	7.59
7	1.000	2.499	−1.0	1.00	1.00	2.50	2.50
8	0.728	3.227	−0.5	0.25	0.063	0.81	0.20
		(8.258*)				(11.82*)	(39.86*)
9	0	3.227	0	0	0	0	0
10	−0.728	2.499	0.5	0.25	0.063	0.62	0.16
11	−1.000	1.499	1.0	1.00	1.00	1.50	1.50
12	−0.803	0.696	1.5	2.25	5.06	1.57	352
13	−0.446	0.250	2.0	4.0	16.0	1.00	4.00
14	−0.180	0.070	2.5	6.25	39.1	0.44	2.74
15	−0.055	0.015	3.0	9.0	81.0	0.14	1.22
16	−0.013	0.002	3.5	12.25	150.0	0.02	0.30
17	−0.002	0	4.0	16.0	256.0	0	0
		(8.258*)				(5.29*)	(13.44*)
		16.516**				17.11**	53.30**

$$H_j - H_{j-1} = \frac{1}{4}\Delta H_{pp}$$

$$A = |H_j - H_{j-1}|^2 \sum_{j=1}^{m} \sum_{i=1}^{j} y'_i = \left(\frac{1}{4}\Delta H_{pp}\right)^2 (16.52) = 1.03(\Delta H_{pp})^2$$

$$\langle H^2 \rangle = \frac{|H_j - H_{j-1}|^2}{A} \sum_{j=1}^{m} \sum_{i=1}^{j} (H_j - H_0)^2 y'_i = \frac{17.11}{16.52}\left(\frac{1}{2}\Delta H_{pp}\right)^2 = 0.259(\Delta H_{pp})^2$$

$$\langle H^4 \rangle = \frac{|H_j - H_{j-1}|^2}{A} \sum_{j=1}^{m} \sum_{i=1}^{j} (H_j - H_0)^4 y'_i = \frac{53.30}{16.52}\left(\frac{1}{2}\Delta H_{pp}\right)^4 = 0.202(\Delta H_{pp})^4$$

[a] Subtotals are indicated by asterisks (*), and grand totals by (**).

makes use of the following equations, and is illustrated in Table 12-4.

$$A = |H_j - H_{j-1}| \sum_{j=1}^{m} (H_j - H_0)y'_j \tag{17}$$

$$\langle H^2 \rangle = \frac{|H_j - H_{j-1}|}{3A} \sum_{j=1}^{m} (H_j - H_0)^3 y'_j \tag{18}$$

TABLE 12-4

Sample Calculation of the Area and Second and Fourth Moments of a Gaussian First-Derivative
Lineshape by the Method of Odd Moments, Using Eqs. (17)–(19) of Sec. 12-B[a]

j	y'_j	$\left(\dfrac{H_j - H_0}{\frac{1}{2}\Delta H_{pp}}\right)$	$\left(\dfrac{H_j - H_0}{\frac{1}{2}\Delta H_{pp}}\right)^3$	$\left(\dfrac{H_j - H_0}{\frac{1}{2}\Delta H_{pp}}\right)^5$	$y'_j\left(\dfrac{H_j - H_0}{\frac{1}{2}\Delta H_{pp}}\right)$	$y'_j\left(\dfrac{H_j - H_0}{\frac{1}{2}\Delta H_{pp}}\right)^3$	$y'_j\left(\dfrac{H_j - H_0}{\frac{1}{2}\Delta H_{pp}}\right)^5$
1	0.002	4.0	64.0	102	~ 0	0.13	2.05
2	0.013	3.5	42.9	525	0.05	0.56	6.83
3	0.055	3.0	27.0	243	0.17	1.49	13.37
4	0.180	2.5	15.6	98	0.45	2.81	17.64
5	0.446	2.0	8.0	32	0.89	3.57	14.27
6	0.803	1.5	3.4	7.6	1.20	2.73	6.10
7	1.000	1.0	1.0	1	1.00	1.00	1.00
8	0.728	0.5	0.1	~ 0	0.36	0.07	~ 0
9	0	0	0	0	0	0	0
					4.12	12.36	61.26
					$\times 2$	$\times 2$	$\times 2$
					8.24	24.72	122.5

$$H_j - H_{j-1} = \frac{1}{4}\Delta H_{pp}$$

$$A = |H_j - H_{j-1}| \sum_{j=1}^{m}(H_j - H_0)y'_j = \left(\frac{1}{4}\Delta H_{pp}\right)(8.24)\left(\frac{1}{2}\Delta H_{pp}\right) = 1.03(\Delta H_{pp})^2$$

$$\langle H^2\rangle = \left(\frac{H_j - H_{j-1}}{3A}\right)\sum_{j=1}^{m}(H_j - H_0)^3 y'_j = \frac{\frac{1}{4}\Delta H_{pp}(24.72)}{3 \times 1.04(\Delta H_{pp})^2}\left(\frac{1}{2}\Delta H_{pp}\right)^3 = 0.250(\Delta H_{pp})^2$$

$$\langle H^4\rangle = \left(\frac{H_j - H_{j-1}}{5A}\right)\sum_{j=1}^{m}(H_j - H_0)^5 y'_j = \frac{\frac{1}{4}\Delta H_{pp}(122.5)}{5 \times 1.04(\Delta H_{pp})^2}\left(\frac{1}{2}\Delta H_{pp}\right)^5 = 0.185(\Delta H_{pp})^4$$

[a] Negative signs are omitted.

$$\langle H^4\rangle = \frac{|H_j - H_{j-1}|}{5A}\sum_{j=1}^{m}(H_j - H_0)^5 y'_j \qquad (19)$$

$$\langle H^n\rangle = \frac{|H_j - H_{j-1}|}{(n+1)A}\sum_{j=1}^{m}(H_j - H_0)^{n+1} y'_j \qquad (20)$$

These relations are the summation analogues to their integral counterparts, which will be discussed next. Since Eqs. (17)–(20) are completely symmetrical, it is only necessary to calculate for half of the resonance line, as is done in Table 12-4.

If the lineshape is unsymmetrical, then one may treat each half separately with the numbering convention shown in Fig. 12-4. It is best to use smaller intervals ($|H_j - H_{j-1}| \ll \Delta H_{pp}$) for asymmetric lines. A quantitative measure

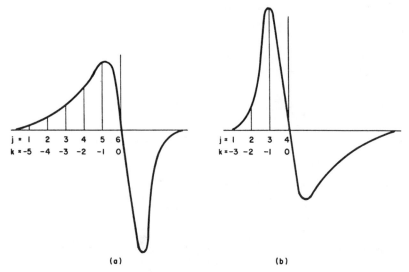

Fig. 12-4. Method of integrating an unsymmetrical absorption derivative lineshape: (a) on the low field side, and (b) on the high field side.

of the amount of asymmetry of the lineshape is given by the skewness:

$$\text{Skewness} = \frac{\langle H^3 \rangle}{\langle H^2 \rangle^{3/2}} \tag{21}$$

and for lines that are close to Gaussian the curtosis defined by

$$\text{Curtosis} = \left(\frac{\langle H^4 \rangle}{\langle H^2 \rangle^2} \right) - 3 \tag{22}$$

measures the deviation from Gaussian, or, from another viewpoint, it measures the center-to-wings balance of the line (Abramowitz and Stegun, 1972; Sýkora and Vogt, 1979).

The first-derivative lineshape formulas may also be put in integral form. Recall that the absorption amplitude Y is related to the derivative amplitude Y' by the expression

$$Y(H) = \int_{-\infty}^{H} Y'(H)\, dH \tag{23}$$

and the area A and nth moment $\langle H^n \rangle$ are obtained through integrating by parts

$$A = \int_{-\infty}^{\infty} Y\, dH \tag{24}$$

$$= -\int_{-\infty}^{\infty} (H - H_0) Y'\, dH \tag{25}$$

since for all lineshapes of finite area

$$\lim_{H \to \pm \infty} (H - H_0)Y = 0 \tag{26}$$

The nth moment is given by

$$\langle H^n \rangle = \int_{-\infty}^{\infty} (H - H_0)^n Y \, dH \tag{27}$$

$$= -\frac{1}{n+1} \int_{-\infty}^{\infty} (H - H_0)^{n+1} Y' \, dH \tag{28}$$

which is valid if

$$\lim_{H \to \pm \infty} \left[(H - H_0)^{n+1} Y \right] = 0 \tag{29}$$

This condition (29) is satisfied by a Gaussian lineshape for all n, but not by a Lorentzian lineshape (except for $n = 0$).

There are limitations on the functional dependence of Y' for physically acceptable lineshapes. For example, from Eq. (25), Y must vanish at $H \to \pm \infty$ if A is to be finite, and in addition,

$$\int_{-\infty}^{\infty} Y' \, dH = 0 \tag{30}$$

if A is to be physically meaningful.

As mentioned above, the value of the nth moment is dependent on the choice of H_0, and to make this choice unambiguous, one may further require that

$$\int_{-\infty}^{\infty} (H - H_0)^2 Y' \, dH = 0 \tag{31}$$

If these two requirements are not satisfied, then one may define correction factors B_3 and B_4 so that

$$\int_{-\infty}^{\infty} (Y' - B_3) \, dH = 0 \tag{32}$$

$$\int_{-\infty}^{\infty} (H - H_0 - B_4)^2 (Y' - B_3) \, dH = 0 \tag{33}$$

as was done in the summation case above [Eqs. (8)–(13)]. In the ensuing discussion it will be assumed that the correction factors B_3 and B_4 are both zero. Recall that a Gaussian line has the shape

$$^G Y(H) = y_m \exp\left[-\left(\frac{H - H_0}{\frac{1}{2}\Delta H_{1/2}} \right)^2 \ln 2 \right] \tag{34}$$

and differentiation yields

$$^{G}Y'(H) = \frac{2y_m \ln 2}{\frac{1}{2}\Delta H_{1/2}} \left(\frac{H - H_0}{\frac{1}{2}\Delta H_{1/2}} \right) \exp\left[-\left(\frac{H - H_0}{\frac{1}{2}\Delta H_{1/2}} \right)^2 \ln 2 \right] \qquad (35a)$$

Equation (16) gave the Gaussian line in terms of its maximum amplitude y_m, which occurs at

$$H - H_0 = \pm\tfrac{1}{2}\Delta H_{pp} \qquad (36)$$

and hence Y' has the form

$$^{G}Y'(H) = e^{1/2}y_m' \left(\frac{H - H_0}{\frac{1}{2}\Delta H_{pp}} \right) \exp\left[-\frac{1}{2} \left(\frac{H - H_0}{\frac{1}{2}\Delta H_{pp}} \right)^2 \right] \qquad (35b)$$

Therefore the maximum absorption amplitude y_m and the maximum first derivative amplitude y_m' are related by the expression

$$y_m = e^{1/2}y_m' \left(\tfrac{1}{2}\Delta H_{pp} \right) \qquad (37)$$

where $e^{1/2} = 1.6487$. The Gaussian lineshape of Eq. (16) may also be written in the form

$$^{G}Y'(H) = y_m' \left(\frac{H - H_0}{\frac{1}{2}\Delta H_{pp}} \right) \exp\left\{ -\frac{1}{2} \left[\left(\frac{H - H_0}{\frac{1}{2}\Delta H_{pp}} \right)^2 - 1 \right] \right\} \qquad (35c)$$

which is sometimes more convenient for calculations.

The integrations that are required for using the Gaussian first-derivative lineshape to calculate the even moments are fairly easy using Eqs. (23) and (24) of the last section, and they lead to the same results obtained with the absorption line itself, so they will not be repeated here. Since the lineshape in Eq. (35) is in terms of ΔH_{pp} and y_m' rather than $\Delta H_{1/2}$ and y_m, the answers obtained in the derivative case have a slightly different form from those calculated with the absorption line $Y(H)$ itself. The characteristics of a Gaussian lineshape are summarized in the following expressions and in Table 12-5:

$$\Delta H_{1/2} = (2 \ln 2)^{1/2} \Delta H_{pp} \qquad (38)$$

$$A = (2\pi e)^{1/2} \left(\tfrac{1}{2}\Delta H_{pp} \right)^2 y_m' \qquad (39)$$

and

$$\langle H^n \rangle = (n - 1)!! \left(\tfrac{1}{2}\Delta H_{pp} \right)^n \quad \text{(for even } n\text{)} \qquad (40)$$

TABLE 12-5

Comparison of Gaussian and Lorentzian Lineshapes

Parameter	Gaussian Shape	Lorentzian Shape
$\Delta H_{1/2}/\Delta H_{pp}$	$(2\ln 2)^{1/2} = 1.1776$	$3^{1/2} = 1.7321$
$y_m/(y_m' \, \Delta H_{pp})$	$e^{1/2}/2 = 0.8244$	$4/3 = 1.3333$
$y_m/(y_m'' \, \Delta H_{pp}^2)$	$1/4 = 0.2500$	$3/8 = 0.3750$
$A/(y_m \, \Delta H_{1/2})$	$\frac{1}{2}(\pi/\ln 2)^{1/2} = 1.0643$	$\pi/2 = 1.5708$
$A/(y_m' \, \Delta H_{pp}^2)$	$\frac{1}{2}(\pi e/2)^{1/2} = 1.0332$	$2\pi/3^{1/2} = 3.6276$
$\langle H^2 \rangle/(\Delta H_{1/2})^2$	$1/(8\ln 2) = 0.1803$	∞
$\langle H^4 \rangle/(\Delta H_{1/2})^4$	$3/[64(\ln 2)^2] = 0.0974$	∞
$\langle H^2 \rangle/(\Delta H_{pp})^2$	$1/4 = 0.2500$	∞
$\langle H^4 \rangle/(\Delta H_{pp})^4$	$3/16 = 0.1875$	∞
y_1''/y_2''	$\frac{1}{2}e^{3/2} = 2.2409$	$64^{1/3} = 4$
$H_1''/\Delta H_{pp}$	0.626	0.567
$H_2''/\Delta H_{pp}$	$3^{1/2} = 1.7321$	$3^{1/2} = 1.7321$
$H_3''/\Delta H_{pp}$	2.52	$81^{1/4} = 3$
$d'(0)/d'(3^{1/2}) = A/B$	$7/2 = 3.50$	$2^3 = 8$

$$e = 2.718282 \qquad \left. \begin{array}{l} \ln 2 \\ \log_e 2 \end{array} \right\} = 0.693147$$

$$\pi = 3.141593$$

$$2^{1/2} = 1.414214 \qquad \pi^{1/2} = 1.772454$$

$$3^{1/2} = 1.732051 \qquad (\ln 2)^{1/2} = 0.832555$$

$$e^{1/2} = 1.648722$$

where $(n-1)!! = (n-1)(n-3) \cdots (5)(3)(1)$. It is noteworthy that the nth even moment has such a simple relationship to $(\frac{1}{2}\Delta H_{pp})^n$.

The Lorentzian derivative lineshape has the normalized form

$$^L Y'(H) = \frac{16 y_m'\left[(H - H_0)/\frac{1}{2}\Delta H_{pp}\right]}{\left\{3 + \left[(H - H_0)/\frac{1}{2}\Delta H_{pp}\right]^2\right\}^2} \tag{41}$$

where

$$y_m = \tfrac{4}{3}(\Delta H_{pp}) y_m' \tag{42}$$

It is straightforward to compute the following relations for this lineshape:

$$A = \frac{2\pi}{3^{1/2}} y_m'(\Delta H_{pp})^2 \tag{43}$$

$$\Delta H_{1/2} = 3^{1/2} \Delta H_{pp} \tag{44}$$

The properties of the Lorentzian lineshape are given in Table 12-5.

The lineshape formulas are important to take into account when comparing the areas of different curves. For example, when a Lorentzian and a Gaussian spectrum have the same derivative amplitude y'_m and width ΔH_{pp}, the former corresponds to $(3.63/1.03) = 3.51$ times as many spins as the latter. This may be seen from the fifth entry of Table 12-5, where for this case,

$$\frac{\text{Lorentzian area}}{\text{Gaussian area}} = 4\left(\frac{2\pi}{3e}\right)^{1/2} = 3.51 \tag{45}$$

The corresponding figure for the ratio of the areas of a Lorentzian absorption line to a Gaussian one with the same amplitude y_m and width $\Delta H_{1/2}$ is $(1.57/1.06) = 1.48$. McClintock and Orr (1968) published a short tabulation related to Table 12-5.

In this and the preceding section we discussed the summation forms for computing the area and moments of spectral lines. These summations are readily adapted for computer calculations. The correction factors B, B_1, and B_2 are easily calculated by a computer before proceeding to the determination of A and the $\langle H^n \rangle$. Collins (1959), Lebedev et al. (1963), Murthy (1963), and Young (1964) discuss such computer calculations. Tolkachev and Mikhailov (1964) describe a nomogram for the double integration of an ESR signal.

C. Second-Derivative Absorption Line

If the narrow band amplifier and lock-in detector are tuned to the second harmonic of the modulation frequency, then the recorded signal will be the second derivative of the resonance absorption. The second derivative may also be recorded with a double modulation arrangement. These second derivative signals are not customarily employed for computing areas and moments, and so this section is confined to a discussion of the integral formulation of the lineshape.

The second derivative of the lineshape Y'' is defined by

$$Y' = \int_{-\infty}^{H} Y'' \, dH \tag{1}$$

so that

$$A = \int_{-\infty}^{\infty} dH \int_{-\infty}^{H} dH' \int_{-\infty}^{H'} Y'' \, dH'' \tag{2}$$

$$= \frac{1}{2} \int_{-\infty}^{\infty} (H - H_0)^2 Y'' \, dH \tag{3}$$

and

$$\langle H^n \rangle = \frac{1}{(n+1)(n+2)} \int_{-\infty}^{\infty} (H - H_0)^{n+2} Y''(H) \, dH \tag{4}$$

where the formulas are derived by integrating by parts twice [compare Eqs. (23)–(31) of the last section].

The explicit form of Y'' for a normalized Gaussian lineshape may be obtained by differentiating Eq. (16) or (35c) of the last section:

$$
{}^G Y''(H) = y''_m \left[\left(\frac{H - H_0}{\frac{1}{2}\Delta H_{pp}} \right)^2 - 1 \right] \exp\left[-\frac{1}{2} \left(\frac{H - H_0}{\frac{1}{2}\Delta H_{pp}} \right)^2 \right] \tag{5}
$$

where $y_m = y''_m(\frac{1}{2}\Delta H_{pp})^2$ defines the normalization constant y''_m. The maxima may be found by differentiating ${}^G Y''$, and setting the result equal to zero:

$$
\left(\frac{d}{dH} \right) {}^G Y''(H) = 0 \tag{6}
$$

to obtain the expression

$$
\frac{H - H_0}{\frac{1}{2}\Delta H_{pp}} \left[\left(\frac{H - H_0}{\frac{1}{2}\Delta H_{pp}} \right)^2 - 3 \right] \exp\left[-\frac{1}{2} \left(\frac{H - H_0}{\frac{1}{2}\Delta H_{pp}} \right)^2 \right] = 0 \tag{7}
$$

which has the five solutions

$$
\frac{(H - H_0)}{\frac{1}{2}\Delta H_{pp}} = 0, \quad \pm 3^{1/2} \pm \infty \tag{8}
$$

The quantities H''_1 and H''_3 shown in Fig. 12-5 correspond to the points where Y'' equals $\frac{1}{2}y''_1$ and $\frac{1}{2}y''_2$, respectively, and H''_2 is the separation of the peaks. For a Gaussian lineshape these parameters have the explicit values

$$
H''_1 = 0.626\,\Delta H_{pp} \tag{9}
$$

$$
H''_2 = 3^{1/2}\,\Delta H_{pp} \tag{10}
$$

$$
H''_3 = 2.52\,\Delta H_{pp} \tag{11}
$$

and the ratio y''_1/y''_2 is given by

$$
\frac{y''_1}{y''_2} = \frac{1}{2}e^{3/2} = 2.241 \tag{12}
$$

For a normalized Lorentzian line ($y'_m = \frac{9}{32}y''_m \Delta H_{pp}$),

$$
{}^L Y'' = 27 y''_m \left[\frac{1 - [(H - H_0)/\frac{1}{2}\Delta H_{pp}]^2}{\{3 + [(H - H_0)/\frac{1}{2}\Delta H_{pp}]^2\}^3} \right] \tag{13}
$$

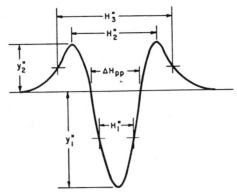

Fig. 12-5. Definitions of parameters used to characterize the second derivative of a resonant absorption line. The lineshape shown is Gaussian.

and its maxima are easily found to be at the same numerical values as in the Gaussian case:

$$\frac{(H - H_0)}{\frac{1}{2}\Delta H_{pp}} = 0, \quad \pm 3^{1/2}, \quad \pm \infty \tag{14}$$

As a result, for a Lorentzian lineshape we obtain

$$H_1'' = 0.567 \Delta H_{pp} \tag{15}$$

$$H_2'' = 3^{1/2} \, \Delta H_{pp} \tag{16}$$

$$H_3'' = 3 \, \Delta H_{pp} \tag{17}$$

$$\frac{y_1''}{y_2''} = 4 \tag{18}$$

Some researchers such as those in Walter Gordy's group routinely record second-derivative spectra [see also, e.g., Hori and Kispert (1978)].

D. Comparison of Lorentzian and Gaussian Lineshapes

The Lorentzian shape is slightly sharper in the center and decreases much more slowly in the wings beyond the half-amplitude or first-derivative points. The two shape functions may be compared in the manner shown in Fig. 12-6, where the two absorption curves are drawn with the same half-amplitude linewidths $\Delta H_{1/2}$, and the two first- and second-derivative curves are plotted with the same peak-to-peak linewidths ΔH_{pp}. The lineshapes of experimental spectra are frequently compared to these two theoretical shapes by matching them at three points to the Lorentzian and Gaussian shapes in the manner shown in Fig. 12-7. (See also Poole and Farach, 1974).

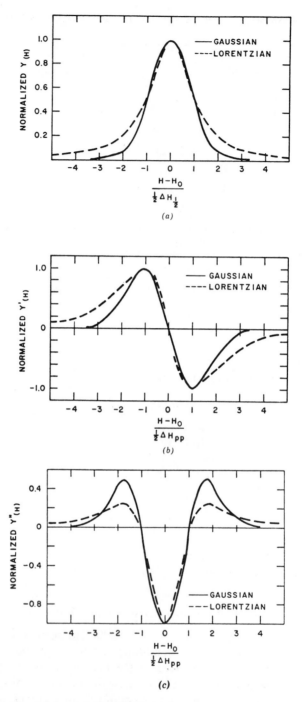

Fig. 12-6. (*a*) Lorentzian and Gaussian absorption curves with the same half-amplitude linewidth. (*b*) Lorentzian and Gaussian absorption first-derivative curves with the same peak-to-peak linewidth. (*c*) Lorentzian and Gaussian absorption second-derivative curves with the same peak-to-peak linewidth. [Judeikis (1964) plots (*a*) on a log–log scale graph.]

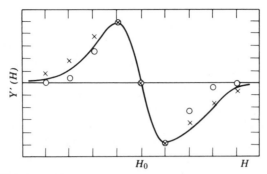

Fig. 12-7. An asymmetric experimental spectrum $Y'(H)$ matched to (×) Lorentzian and (○) Gaussian shapes with the same amplitude y'_m and width ΔH_{pp}.

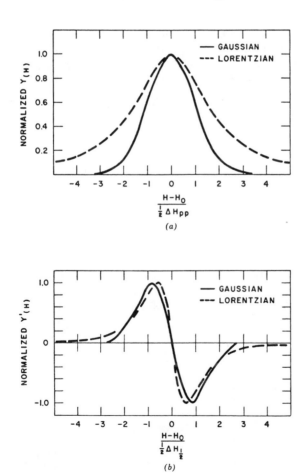

Fig. 12-8. (a) Lorentzian and Gaussian absorption curves with the same peak-to-peak linewidth ΔH_{pp}. (b) Lorentzian and Gaussian absorption first-derivative curves with the same half-amplitude linewidth $\Delta H_{1/2}$.

Figure 12-8 shows the dramatic difference between the two lineshapes when the absorption curves are matched at their inflection points, and when the first-derivative curves are matched at the half-amplitude points.

The difference between Lorentzian and Gaussian lineshapes is illustrated by the data in Tables 12-6 to 12-10. The dispersion lineshape is discussed in Sec. 12-H. Van Gerven and Van Itterbeek (1961) discuss Lorentzian and Gaussian lineshapes arising in weak magnetic field experiments.

The lineshape of an observed spectrum $Y'(H)$ may be deduced from its peak amplitude y'_m and linewidth ΔH_{pp} by using the data of Table 12-7 to calculate the expected values for Lorentzian and Gaussian shapes. These points are placed on the experimental spectrum in the manner shown in Fig. 12-7. The illustrated spectrum has a shape between the two at low fields, and close to the former at high fields. Tables 12-6, 12-8, 12-9, and 12-10 may be used for checking other types of spectra. Marguardt et al. (1961) and Ibers and Swalen (1962) describe how to obtain a least-squares fit between a calculated and an observed spectrum, and Johnston and Hecht (1965) treat this problem (see also Weidner and Whitmer, 1953; Turoff et al., 1967). One may use different techniques to subtract a known spectrum from an unknown to reveal hidden structure, as discussed in Sec. 11-K and illustrated in Fig. 11-44 (see also Smith and Pieroni, 1964).

TABLE 12-6

Amplitudes of Gaussian and Lorentzian Absorption Lineshapes at Multiples of $\frac{1}{2}\Delta H_{1/2}$ from the Center as Shown in Figure 12-6a Using Data Computed from Eqs. (9) and (26) of Sec. 12-A

$\left(\dfrac{H - H_0}{\frac{1}{2}\Delta H_{1/2}} \right)$	Amplitude $Y(H)$	
	Gaussian	Lorentzian[a]
0	1.0000	1.0000
0.5	0.8409	0.8000
1	0.5000	0.5000
1.5	0.2101	0.3077
2	0.0626	0.2000
3	0.00195	0.1000
4	1.5×10^{-5}	0.0588
5	—	0.0384
6	—	0.0270
7	—	0.0200
8	—	0.0154
9	—	0.0122
10	—	0.0099

[a]When $|H - H_0| > 10\,\Delta H_{1/2}$, one may approximate $Y(H) \sim y_m[(H - H_0)/\frac{1}{2}\,\Delta H_{1/2}]^{-2}$ for the Lorentzian shape.

TABLE 12-7

Amplitudes of Gaussian and Lorentzian Absorption
First-Derivative Lineshapes at Multiples of $\frac{1}{2}\Delta H_{pp}$ from
the Center, as Shown in Fig. 12-6b Using Data Computed
from Eqs. (16) and (41) of Sec. 12-B

$\left(\dfrac{H - H_0}{\frac{1}{2}\Delta H_{pp}}\right)$	Amplitude $Y'(H)$	
	Gaussian[a]	Lorentzian[b]
0	0.0000	0.0000
0.5	0.7275	0.7574
1	1.0000	1.0000
1.5	0.8029	0.8701
2	0.4461	0.6531
3	0.0549	0.3333
4	0.0022	0.1773
5	3×10^{-5}	0.1020
6	1.5×10^{-7}	0.0631
7	—	0.0414
8	—	0.0285
9	—	0.0204
10	—	0.0151

[a]Note that $1.649\, x \exp(-x^2/2) = x \exp[-(x^2 - 1)/2]$, where
$x = [(H - H_0)/(\frac{1}{2}\Delta H_{pp})]^2$.
[b]When $|H - H_0| > 10\,\Delta H_{pp}$, we may set $Y'(H) \sim 16 y'_m[(H - H_0)/\frac{1}{2}\Delta H_{pp}]^{-3}$ for the Lorentzian shape.

Singer et al. (1957) and Tikhomirova and Voevodskii (1959) present graphical methods for determining whether or not an experimental spectrum is Gaussian or Lorentzian in shape. Young (1964) describes a Fortran II program for calculating and displaying Gaussian and Lorentzian absorption and derivative lines.

Swarup (1959) observed a gradual transformation of the ESR lineshape of Cr^{3+} in potassium cobalticyanide and potassium aluminum alum single crystals from Lorentzian to Gaussian as the chromium concentration changed.

Royaud and Treville (1974) observed the same gradual transformation in solutions of the free radical galvinoxyl in carefully deoxygenated dimethoxyethane as a function of concentration. They found that the experimental integrated area A given by the expression:

$$A = \Lambda' y'_m (\Delta H_{pp})^2 \tag{1}$$

varied from a low concentration value $\Lambda' = 1.033$ characteristic of Gaussian lines to a high concentration value $\Lambda' = 3.63$ characteristic of Lorentzian lines, where the values of Λ' are obtained from Table 12-5. A figure in their article shows the dependence on the integrated area A of galvinoxyl radicals on the product $y_m''(H_{pp})^2$ for increasing radical concentrations. Grant and Strandberg (1964) developed a statistical theory of spin interactions according to which a resonance line is Lorentzian in the center and Gaussian in the wings (see also Anderson and Welling, 1965). Burland et al. (1977) recorded triplet exciton optical spectral lines that were Gaussian on the high-frequency side and Lorentzian on the low-frequency side. Deigen et al. (1970) recorded spectra of Fe^{3+} in Al_2O_3 that were between Gaussian and Lorentzian in shape. Zimmerman et al. (1972) found the spectrum of Cu^{2+} in $Cu\,(C_6H_5SO_3)_2 \cdot 6H_2O$ to be Lorentzian for the magnetic field along one axis and Gaussian along the other. Kawamori et al. (1970) found the lineshape of zinc-doped copper acetate monohydrate to be close to Gaussian below 50 K and nearly Lorentzian above 60 K.

TABLE 12-8

Amplitudes of Gaussian and Lorentzian Absorption Second-Derivative Lineshapes at Multiples of $\frac{1}{2}\Delta H_{pp}$ from the Center as Shown on Figure 12-6c Using Data Computed from Eqs. (5) and (13) of Sec. 12-C[a]

$\left(\dfrac{H - H_0}{\frac{1}{2}\Delta H_{pp}}\right)$	Amplitude $Y''(H)$	
	Gaussian	Lorentzian
0	−1.0000	−1.0000
0.5	−0.6619	−0.5899
1	0.0000	0.0000
1.5	0.4058	0.2332
$3^{1/2}$	0.4463	0.2500
2	0.4060	0.2362
3	0.0889	0.1250
4	0.0050	0.0590
5	10^{-4}	0.0295
6	—	0.0159
7	—	0.0092
8	—	0.0057
9	—	0.0036
10	—	0.0025

[a]The maximum amplitude occurs at $(H - H_0)/\frac{1}{2}\Delta H_{pp} = 3^{1/2}$ for both lineshapes. It has the value $2e^{-3/2} = 0.44626$ for the Gaussian shape and $\frac{1}{4}$ for the Lorentzian shape.
[b]When $|H - H_0| > 10\,\Delta H_{pp}$, we may use the approximation $Y''(H) \sim 27 y_m''[(H - H_0)/\frac{1}{2}\,\Delta H_{pp}]^{-4}$ for Lorentzian lines.

TABLE 12-9

Amplitudes of Lorentzian Dispersion Lineshape
at Multiples of $\frac{1}{2}\Delta H_{1/2}$ from the Center
Using Data Calculated from Eq. (1) of Sec. 12-H

$\left(\dfrac{H - H_0}{\frac{1}{2}\Delta H_{1/2}}\right)$	Amplitude, d^a
0	0.0000
0.5	0.8000
1	1.0000
1.5	0.9231
2	0.8000
3	0.6000
4	0.4706
5	0.3846
6	0.3243
7	0.2800
8	0.2462
9	0.2195
10	0.1980

[a] When $|H - H_0|/\frac{1}{2}\Delta H_{1/2}$ exceeds 10 we may
use the approximation $d \sim \Delta H_{1/2}/(H - H_0)$.

TABLE 12-10

Amplitudes of Lorentzian Dispersion First-Derivative
Lineshape at Multiples of $\frac{1}{2}\Delta H_{pp}$ from the Center
Using Data Calculated from Eq. (2) of Sec. 12-H

$\left(\dfrac{H - H_0}{\frac{1}{2}\Delta H_{pp}}\right)$	Amplitude, d'^a
0	-1.0000
0.5	-0.7811
1	-0.3750
1.5	0.0816
$3^{1/2}$	0.0000
2	0.0612
3	0.1250
4	0.1080
5	0.0842
6	0.0651
7	0.0510
8	0.0408
9	0.0332
10	0.0274

[a] $d' \sim 3[(H - H_0)/\frac{1}{2}\Delta H_{pp}]^{-2}$ when $|H - H_0| >$
$10\,\Delta H_{pp}$.

E. Overlapping Resonances

The lineshapes that result from the overlapping of two to six hyperfine components with a Gaussian lineshape, and Gaussian intensity distributions were reconstructed manually by Poole and Anderson (1959) and are shown in Fig. 1-3. A more extensive set of overlapping spectra were worked out on a computer and published in the book by Lebedev and Voevodksii (1963) (see also Rotaru et al., 1964). Kummer (1963) discusses whether or not more than one set of parameters (coupling constants) is compatible with an experimental spectrum. Garrett et al. (1966) synthesized unresolved Mn II structure.

First-derivative spectral lines are narrower and more resolved than the absorption spectrum itself, and second-derivative spectra are even better in this respect. A glance at Fig. 11-41 and a careful comparison of the curves in Figs. 12-6 and 12-7 will convince one of this fact. The difficulty with second-derivative presentation is that the peaks shown in Fig. 12-6c at $H - H_0 = \pm 3^{1/2} \Delta H_{pp}$ above the baseline tend to interfere with the adjacent hyperfine components to produce misleading spectra. This liability is considerably more pronounced with Gaussian than it is with Lorentzian shapes. Johnson and Chang (1965) and Halpern and Phillips (1970) recorded third-derivative spectra for even greater resolution (see Sec. 7-F).

Ageno and Frontali (1963) give a graphical method for separating overlapping Gaussian curves, and Allen (1962) discusses several methods for unravelling complex spectra. Allen et al. (1964) and Glarum (1965) show how to enhance considerably the resolving power of an ESR spectrometer (compare Sec. 11-I). Kaplan (1965) discusses linewidths of inhomogeneously broadened lines.

Robinson et al. (1976) resolved overlapping spectra from different radical species by determining the differences between an ordinary partially saturated ESR spectrum and the spectrum recorded in the presence of a resonant radio frequency field. They described an instrumental arrangement for rapidly attaining this difference spectrum.

It is possible to synthetically produce a complex resonance line on an oscilloscope by means of electronic Gaussian or Lorentzian function generators. This may be compared to an observed spectrum, and the hyperfine intensity ratio may be deduced from the individual output of each function generator (Giardino and Wisert, 1965).

F. Finite-Moment Lorentzian-Type Lineshapes

The Lorentzian lineshape has an infinite second moment, as was mentioned above. One frequently encounters lineshapes that are Lorentzian near the center but fall off faster than a Lorentzian in the wings, and consequently give a finite second moment. A simple lineshape of this type is the cutoff Lorentzian (Kittel and Abrahams, 1953). It is assumed that the amplitude is identical to that corresponding to a Lorentzian line from $-a \leq H - H_0 \leq a$, and zero beyond these limits, as shown in Fig. 12-9. The first-derivative curve is cut off at $H - H_0 = \pm a$ without taking account of the discontinuity there.

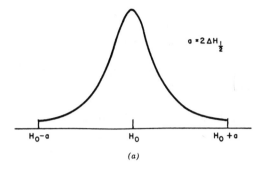

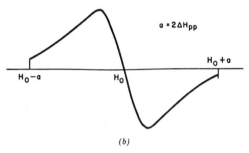

Fig. 12-9.   Cutoff Lorentzian lineshape: (a) absorption curve and (b) first-derivative curve.

The area of a cut-off Lorentzian line is given by

$$A = y_m \int_{H_0-a}^{H_0+a} \frac{dH}{1 + \left[(H - H_0)/\tfrac{1}{2}\Delta H_{1/2}\right]^2} \tag{1}$$

$$= y_m \Delta H_{1/2} \int_0^{\tan^{-1}(2a/\Delta H_{1/2})} d\theta \tag{2}$$

$$= y_m \Delta H_{1/2} \left[\frac{\pi}{2} - \cot^{-1}\left(\frac{2a}{\Delta H_{1/2}}\right)\right] \tag{3}$$

where $\tan\theta = 2(H - H_0)/\Delta H_{1/2} = \cot(\pi/2 - 0)$. When

$$\frac{2a}{\Delta H_{1/2}} \gg 1 \tag{4}$$

one may expand the arc cotangent in a power series, and neglect all but the first term, giving

$$\cot^{-1}\left(\frac{2a}{\Delta H_{1/2}}\right) \sim \frac{\Delta H_{1/2}}{2a} \tag{5}$$

so that

$$A \sim y_m \, \Delta H_{1/2} \left(\frac{\pi}{2} - \frac{\Delta H_{1/2}}{2a} \right) \tag{6}$$

Thus the area of a Lorentzian line that is cut off considerably beyond the half power point is equal to the area of the corresponding Lorentzian curve less a small correction term. In terms of the first derivative lineshape one has, of course (neglecting the discontinuity at $H = H_0 \pm a$),

$$A = \left(\frac{4}{3^{1/2}} \right) y_m' (\Delta H_{pp})^2 \left(\frac{\pi}{2} - \frac{3^{1/2} \, \Delta H_{pp}}{2a} \right) \tag{7}$$

The nth even moment of a cutoff Lorentzian lineshape is given by

$$\langle H^n \rangle = \frac{y_m}{A} \int_{H_0-a}^{H_0+a} \frac{(H - H_0)^n \, dH}{1 + \left[(H - H_0)/\frac{1}{2}\Delta H_{1/2} \right]^2} \tag{8}$$

$$= \frac{2 \left(\frac{1}{2}\Delta H_{1/2} \right)^{n+1} y_m}{A} \left[\frac{1}{n-1} \left(\frac{2a}{\Delta H_{1/2}} \right)^{n-1} - \int_0^{\tan^{-1}(2a/\Delta H_{1/2})} \tan^{n-2} \theta \, d\theta \right] \tag{9}$$

and the second moment has the explicit form

$$\langle H^2 \rangle = \frac{\left(\frac{1}{2}\Delta H_{1/2} \right)^2 \left[2a/\Delta H_{1/2} - \pi/2 + \cot^{-1}(2a/\Delta H_{1/2}) \right]}{\left[\pi/2 - \cot^{-1}(2a/\Delta H_{1/2}) \right]} \tag{10}$$

In the limit $a \gg \Delta H_{1/2}$ only the first term contributes in Eqs. (9) and (10), so we have for every nth even moment

$$\langle H^n \rangle \approx \frac{a^{n-1} \, \Delta H_{1/2}}{(n-1)\pi} \tag{11}$$

This gives the following particular expressions for the second and fourth moments, respectively (Kittel and Abrahams, 1953):

$$\langle H^2 \rangle \approx a \frac{\Delta H_{1/2}}{\pi}$$

$$\langle H^4 \rangle \approx \left(\frac{1}{3\pi} \right) a^3 \, \Delta H_{1/2} \tag{12}$$

The treatment just presented may also be applied to a cutoff Gaussian lineshape, and the results will be in terms of incomplete gamma functions.

Sharply cutoff lines do not, of course, appear in nature; a more realistic model that produces finite moments is a line that is Lorentzian in the center and Gaussian in the wings. Resonant lines that approximate this shape have been observed in, for example, exchange-narrowed systems. (See p. 484).

Menzel and Wasson (1975) employed the square Lorentzian lineshape

$$^{L\,2}Y(H - H_0) = \frac{y_m}{\left[1 + (H - H_0)^2/\Gamma^2\right]^2} \tag{13}$$

to approximate an ordinary Lorentzian $^{L}Y(H)$ in the center and provide the following finite second moment:

$$\langle H^2 \rangle = \Gamma^2 \tag{14}$$

and the linewidths

$$\Delta H_{1/2} = 2\Gamma\left(\sqrt{2} - 1\right)^{1/2} \tag{15}$$

and

$$\Delta H_{pp} = \frac{2\Gamma}{\sqrt{5}} \tag{16}$$

Fourth- and higher-order moments are infinite for this shape. Menzel and Wasson also treated the square Gaussian lineshape that is itself Gaussian with a change in width.

The finite-moment lineshapes treated in this section are useful for computational purposes only. They do not correspond to shapes that actually exist in nature, and they do not have any theoretical justification.

G. Convolution Shape

When the spectral line is broadened independently by both Gaussian and Lorentzian effects, the absorption amplitude is given by the following convolution (Posener, 1959):

$$Y_{(v, b)} = \frac{b}{\pi} \int_{-\infty}^{\infty} \frac{e^{-x^2}\, dx}{b^2 + (v - x)^2} \tag{1}$$

where b is a measure of the ratio of the Lorentzian linewidth $^{L}\Delta H_{1/2}$ to the Gaussian width $^{G}\Delta H_{1/2}$

$$b = (\ln 2)^{1/2} \frac{^{L}\Delta H_{1/2}}{^{G}\Delta H_{1/2}} \tag{2}$$

and v is the distance from the line center H_0 in units of $^{G}\Delta H_{1/2}/2\ln 2$, as follows:

$$v = 2(\ln 2)^{1/2} \frac{(H - H_0)}{{}^G\!\Delta H_{1/2}} \tag{3}$$

This Voigt profile function (1) has been tabulated for various values of b and v. For a particular resonance line ${}^G\!\Delta H_{1/2}$ and ${}^L\!\Delta H_{1/2}$ will be constant, so that the shape Y becomes a function only of $(H - H_0)$:

$$Y(H - H_0) = \frac{(\ln 2)^{1/2}}{\pi} \left(\frac{{}^L\!\Delta H_{1/2}}{{}^G\!\Delta H_{1/2}} \right)$$

$$\times \int_{-\infty}^{\infty} \frac{e^{-x^2}\, dx}{\left({}^L\!\Delta H_{1/2}/{}^G\!\Delta H_{1/2}\right)^2 \ln 2 + \left(2(\ln 2)^{1/2}\left[(H - H_0)/{}^G\!\Delta H_{1/2}\right] - x\right)^2} \tag{4}$$

Unfortunately this cannot be integrated in closed form, but Posener (1959) and Ganapathy and Parthasarathi (1977) have published tables of it. Castner (1952) evaluated Eq. (4) in terms of the error function integral and described how to determine the ratio ${}^L\!\Delta H_{1/2}/{}^G\!\Delta H_{1/2}$ (see also Farach and Teitlebaum, 1967).

The convolution lineshape is applicable to the description of a Gaussian-shaped envelope formed from narrow Lorentzian spin packets (see Sec. 13-B). Stoneham (1972) derived an expression relating the observed peak-to-peak linewidth to the Gaussian and Lorentzian widths. Strandberg (1973) discussed Gaussian–Lorenzian convolutions, and Al'tshuler et al. (1973) and Kemple and Stapleton (1972) applied the shape to specific systems.

H. Dispersion Line

A Lorentzian dispersion signal d has the form

$$d = \frac{2\left[(H - H_0)/\tfrac{1}{2}\Delta H_{1/2}\right]}{1 + \left[(H - H_0)/\tfrac{1}{2}\Delta H_{1/2}\right]^2} \tag{1}$$

and its first derivative d' is

$$d' = 3 \frac{3 - \left[(H - H_0)/\tfrac{1}{2}\Delta H_{pp}\right]^2}{\left\{3 + \left[(H - H_0)/\tfrac{1}{2}\Delta H_{pp}\right]^2\right\}^2} \tag{2}$$

These two expressions are individually normalized so that each has a maximum amplitude of 1, and as usual $\Delta H_{1/2} = 3^{1/2} \Delta H_{pp}$. The derivative d' vanishes at $H - H_0 = \pm \frac{1}{2} \Delta H_{1/2}$ and reaches extrema at

$$\frac{(H - H_0)}{\frac{1}{2}\Delta H_{pp}} = 0, \pm 3 \tag{3}$$

Numerical values for Eqs. (1) and (2) are given in Tables 12-9 and 12-10. The ratio of $d'(0)/d'(3^{1/2})$ equals 8 for a Lorentzian line and 7/2 for a Gaussian line. This ratio is called D/A in Sec. 11-A and corresponds to y_1''/y_2'' in Fig. 12-5 for the absorption derivative. Nagasawa (1965) discusses the variation of the ratio D/A and other dispersion-mode parameters on the modulation frequency and relaxation time T_1.

Pake and Purcell (1948, 1949) analyzed the absorption χ' and dispersion χ'' of the magnetic resonance susceptibility χ:

$$\chi = \chi' - i\chi'' \tag{4}$$

for both Lorentzian and Gaussian lineshapes. They used the simplified notation

$$\chi' = \frac{x}{(1 + x^2)} \qquad\qquad \text{Lorentzian} \tag{5a}$$

$$= \frac{2}{\pi^{1/2}} e^{-x^2/\pi} \int_0^{x/\pi^{1/2}} \exp y^2 \, dy \quad \text{Gaussian} \tag{5b}$$

and

$$\chi'' = \begin{cases} \dfrac{1}{1 + x^2} & \text{Lorentzian} \\[2ex] e^{-x^2/\pi} & \text{Gaussian} \end{cases} \tag{6a} \tag{6b}$$

where the dimensionless variable x corresponds to a normalized value of $(H - H_0)$, and their normalization constants differ from ours. Their lineshapes for the real and imaginary parts of the susceptibility χ are plotted in Fig. 12-10, and the corresponding first derivative lineshapes are shown in Fig. 12-11.

When the ESR sample is held in a high-Q resonant cavity, the dispersion mode χ' corresponds to a change in frequency, and the absorption mode χ'' corresponds to a change in Q as one scans through the resonance condition. When the klystron frequency is stabilized on the sample resonant cavity, the dispersion signal is "stabilized out" and only χ'' is observed. When the klystron frequency is stabilized on an auxiliary cavity, then one observes either

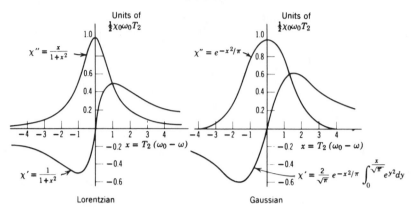

Fig. 12-10. A comparison of the dispersion χ' and absorption χ'' curves for Lorentzian and Gaussian lineshapes (Pake and Purcell, 1948, 1949).

dispersion, absorption, or a mixture of the two, depending on the phase adjustment on the slide screw tuner or phase shifter in the microwave bridge. If an experimental spectrum contains a mixture of χ' and χ'', then one may use the technique discussed at the end of Sec. 11-A, or that of Peter et al. (1962) to obtain the ratio of χ' to χ''.

Chovino et al. (1971) used an interferometric technique (Sardos, 1969; Bonis and Sardos, 1973; Chastenet et al., 1970, 1971, 1981) to measure the variations in relative amplitude $-\Delta I/I$ and phase $\Delta\phi$ of the ESR signal, as illustrated on Fig. 12-12a and 12-13a. The quantities $-\Delta I/I$ and $\Delta\phi$ corresponding to absorption and dispersion are, respectively, proportional to $\chi''_{\perp} - \chi''_{\perp_0}$ and

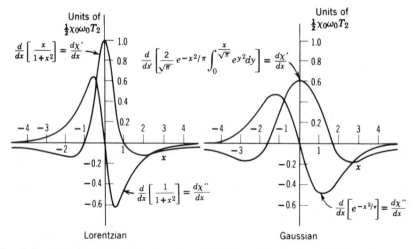

Fig. 12-11. A comparison of the dispersion $(d\chi'/dx)$ and absorption $(d\chi''/dx)$ derivatives for Lorentzian and Gaussian lineshapes (Pake and Purcell, 1948, 1949).

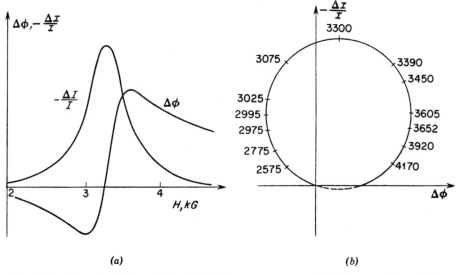

Fig. 12-12. Plot of $-I/\Delta I$ and $\Delta\phi$ of MnO (a) versus the magnetic field strength H and (b) against each other. The circular shape of the latter plot indicates a Lorentzian line (Chovino et al., 1971).

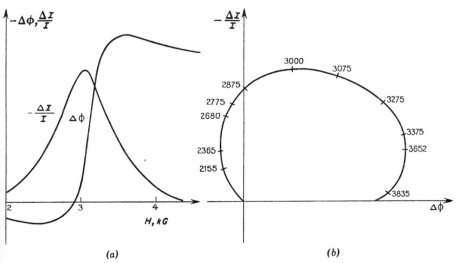

Fig. 12-13. Plot of $-\Delta I/I$ and $\Delta\phi$ of NiO (a) versus the magnetic field strength H and (b) against each other. The oval shape of the latter plot is suggestive of a Gaussian line. (Chovino et al., 1971).

$\chi'_\perp - \chi'_{\perp_0}$. The factor $-\Delta I/I$ can be plotted against $\Delta\phi$ and the result called a Cole–Cole plot is a circle for a Lorentzian line and an oval shape for a Gaussian line. The Lorentzian plot presented in Fig. 12-12b was obtained from MnO and the Gaussian plot of Fig. 12-13b corresponds to NiO.

Accou et al. (1973) studied dispersion of DPPH in the range from 5 to 30 MHz using rf magnetic field amplitudes between several milligauss and 2 G. In these ranges the static magnetic field value for the center of the resonant line became comparable to the linewidth, and the rf magnetic field became comparable to the static field at resonance. They describe the block diagram of the spectrometer employed for making the measurements.

Roch et al. (1971) reported experiments on ESR absorption and dispersion of paramagnetic salts at 10 and 16 GHz carried out with the aid of three wave-recording interferometers. Anderson and Ullman (1967) employed Cole–Cole plots in molecular relaxation studies, and Servant et al. (1976) used Cole–Cole plots in X-band Faraday and Cotton–Mouton–Voigt effect studies. Mailer et al. (1980) used a bimodal cavity for dispersion.

I. Dysonian Shape

The theory of the ESR lineshapes obtained from conduction electrons in metals was worked out by Dyson (1955) and confirmed experimentally by Feher and Kip (1955) in their extension of the original work of Griswold et al. (1952) (see Cousins and Dupree, 1965). Dyson showed that the lineshape depends upon the time T_D that it takes an electron to diffuse through the skin depth δ, the time T_T that it takes for the electron to traverse the sample, the electron spin-lattice relaxation time T_1, and the electron spin–spin relaxation time T_2 (for metals $T_1 = T_2$). In the region of the normal skin depth where the electron mean free path Λ is small in comparison with the skin depth δ, Feher and Kip (1955) give the following formulas for the lineshape Y and its derivative Y' in the units of absorbed power P and power absorbed per unit angular frequency $Y'_\omega = dP/d\omega$ (or $Y'_H = dP/dH$).

(1) For $T_T \ll T_D$ one has

$$Y = \frac{\omega H_1^2}{4}(V)\omega_0\chi_0 T_2 \frac{1}{1 + (\omega - \omega_0)^2 T_2^2} \tag{1}$$

$$Y'_\omega = -\frac{\omega H_1^2}{4}(V)\omega_0\chi_0 T_2 \frac{2(\omega - \omega_0)T_2^2}{\left[1 + (\omega - \omega_0)^2 T_2^2\right]^2} \tag{2}$$

where V is the volume of the sample, H_1 is the rf magnetic field amplitude, χ_0 is the paramagnetic part of the static susceptibility, and $\omega_0 = g\beta H_0/\hbar$ is the resonant frequency. This is a Lorentzian lineshape that is independent of diffusion.

(2) For sufficiently thick samples one has $T_T \gg T_D$ and $T_T \gg T_2$; and for an arbitrary ratio of $T_D/T_2 = R^2$ the lineshape has the form

$$Y = -\left[\frac{\omega H_1^2}{4}(A\delta)\omega_0\chi_0 T_2\right]\left(\frac{T_D}{2T_2}\right)$$

$$\times \left\{ \frac{R^4(\chi^2 - 1) + 1 - 2R^2\chi}{\left[(R^2\chi - 1)^2 + R^4\right]^2}\left[\frac{2\xi}{R(1+\chi^2)^{1/2}} + R^2(\chi + 1) - 3\right]\right.$$

$$\left. + \frac{2R^2(1 - \chi R^2)}{\left[(R^2\chi - 1)^2 + R^4\right]^2}\left[\frac{2\eta}{R(1+\chi^2)^{1/2}} + R^2(\chi - 1) - 3\right]\right\} \quad (3)$$

where $\chi = (\omega - \omega_0)T_2$, $\xi = [(\omega - \omega_0)/(|\omega - \omega_0|)][(1 + \chi^2)^{1/2} - 1]^{1/2}$, $\eta = [(1 + \chi^2)^{1/2} + 1]^{1/2}$, and $A = $ the surface area. Equation (3) is plotted in Fig. 12-14, and its graphically determined first derivative y' is shown in Fig. 12-15 for the cases $T_D/T_2 \ll 1$ and $T_D/T_2 \to \infty$. Equation (3) and its first derivative have simplified forms for the following two limiting cases:

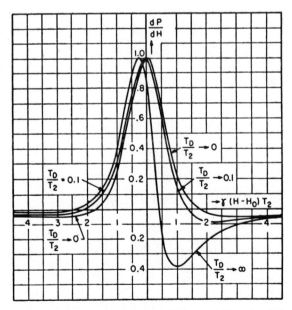

Fig. 12-14. The ESR power absorption $Y = P$ in thick metal films for different ratios of the diffusion time T_D to the spin–spin relaxation time T_2 (Feher and Kip, 1955).

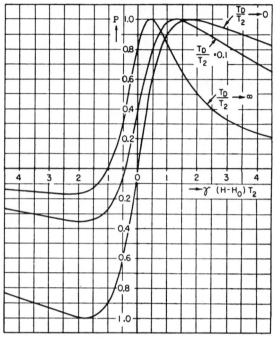

Fig. 12-15. Derivative of the ESR power absorption $Y' = dP/d\omega$ in thick metal films for different ratios of the diffusion time T_D to the spin–spin relaxation time T_2 (Feher and Kip, 1955).

(2a) In this case we have $T_T \gg T_2 \gg T_D$, which corresponds to metals of high conductivity (low temperature):

$$Y \cong \left[\frac{\omega H_1^2}{4}(A\delta)\omega_0 \chi_0 T_2 \right]$$

$$\times \left[\frac{\omega - \omega_0}{|\omega - \omega_0|} \left\{ \frac{T_D}{T_2} \left[\frac{(1 + \chi^2)^{1/2} - 1}{1 + \chi^2} \right] \right\}^{1/2} \right] \quad (4)$$

$$Y_\omega' \cong -\left[\frac{\omega H_1^2}{4}(A\delta)\omega_0 \chi_0 T_2^2 \left(\frac{T_D}{T_2} \right)^{1/2} \right]$$

$$\times \left\{ \frac{\left[2 - (1 + \chi^2)^{1/2} \right]\left[1 + (1 + \chi^2)^{1/2} \right]^{1/2}}{2(1 + \chi^2)^{3/2}} \right.$$

$$\left. + \left(\frac{T_D}{T_2} \right)\left[\frac{(1 + \chi^2)^{1/2} - 1}{1 + \chi^2} \right]^{1/2} \right\} \quad (5)$$

When $T_D/T_2 \to 0$, the second term within the curly brackets becomes negligible relative to the first.

(2b) Here we have $T_T \gg T_D \gg T_2$ corresponding to thick films with slowly diffusing magnetic dipoles. This condition applies to the case of paramagnetic impurities distributed throughout the volume of the metal. This condition also corresponds to the nuclear magnetic resonance of metals (Bloembergen, 1952) since the nuclei are almost completely stationary. The lineshapes are

$$Y = \left[\frac{\omega H_1^2}{4}(A\delta)\omega_0\chi_0 T_2 \right]\left[\frac{1}{2}\frac{1 - T_2(\omega - \omega_0)}{1 + T_2^2(\omega - \omega_0)^2} \right] \tag{6}$$

$$Y_\omega' = \left[\frac{\omega H_1^2}{4}(A\delta)\omega_0\chi_0 T_2 \right]\left[\frac{T_2}{4}\frac{T_2^2(\omega - \omega_0)^2 - 2T_2(\omega - \omega_0) - 1}{\left[1 + T_2^2(\omega - \omega_0)^2\right]^2} \right] \tag{7}$$

using Wagoner's (1960) correction for Eq. (7).

Equations (1)–(7) all omit terms in Dyson's formulas that replace $(\omega - \omega_0)$ by $(\omega + \omega_0)$, since near resonance the latter terms are negligible for sufficiently narrow lines. The above expressions also neglect the effect of surface relaxations that are important whenever there exist strong spin-dependent forces during a collision of the electron with the surface. Surface relaxation effects are more important in the cases of thin films and small particles than they are in the thick cases. Dyson's original paper (1955) should be consulted for further details.

Figures 12-16 and 12-17 show the dependence of two lineshape parameters on the ratio $(T_D/T_2)^{1/2}$. They may be employed to determine values of T_2 and the g factor. Figures 12-18 and 12-19 reveal how the asymmetry of the line varies with the ratio $(T_D/T_2)^{1/2}$. Feher and Kip (1955) discuss the physical significance of the ratio A/B. Wagoner (1960) applied these figures to graphite. Glausinger and Sienko (1973) published similar $(T_D/T_1)^{1/2}$ graphs for second-derivative absorption curves.

Feher and Kip (1955) found satisfactory agreement between Dyson's theory and their experimental results. They assumed that the mean free path Λ is small compared to the skin depth δ, and that all of the electrons have the velocity v in accordance with the expression

$$T_D = \frac{3(\delta^2/v\Lambda)}{2} \tag{8}$$

One may use the well-known conductivity formula

$$\sigma = \frac{Ne^2\Lambda}{m^*v} \tag{9}$$

where N is the number of electrons per cubic centimeter and m^* is the effective

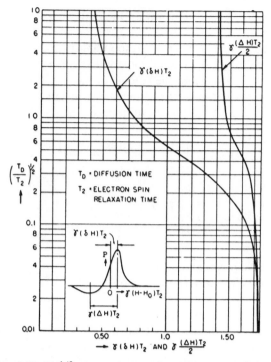

Fig. 12-16.   Plots of $(T_D/T_2)^{1/2}$ versus $\gamma(\delta H)T_2$ and $\gamma\Delta HT_2/2$ for the ESR power absorption in thick metal films (Feher and Kip, 1955).

mass, to obtain

$$T_D = \frac{3}{2}\left(\frac{\delta^2 Ne^2}{\sigma m^* v^2}\right) \tag{10}$$

When the electron mean free path Λ exceeds the skin depth δ, then the skin effect is anomalous, and the exact theory of Reuter and Sondheimer (1948) must be utilized. Kittel extended the Dyson theory to the case of the anomalous skin effect $\delta < \Lambda$ under the assumption that the diffusion time T_D is short in relation to the relaxation time. The lineshape in terms of the complex impedance $Z = A + iB$ is

$$Y = \left[\frac{\omega H_1^2}{4}(\Lambda A)\omega_0\chi_0\left(\frac{c^2}{4\pi\omega\Lambda}\right)^2(6T_2\tau)^{1/2}\right]$$

$$\times\left\{(B^2 - A^2)\left[\frac{(1+\chi^2)^{1/2}+1}{1+\chi^2}\right]^{1/2} + 2AB\frac{\omega - \omega_0}{|\omega - \omega_0|}\left[\frac{(1+\chi^2)^{1/2}-1}{1+\chi^2}\right]^{1/2}\right\}$$

$$\tag{11}$$

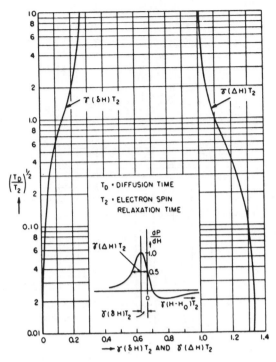

Fig. 12-17. Plots of $(T_D/T_2)^{1/2}$ versus $\gamma(\delta H)T_2$ and $\gamma(\Delta H)T_2$ for the derivative of the ESR power absorption in thick metal films (Feher and Kip, 1955).

$$Y'_H = \left[\frac{\omega H_1^2 \gamma T_2}{4} (\Lambda A) \omega_0 \chi_0 \left(\frac{c^2}{4\pi\omega\Lambda} \right)^2 (6T_2\tau)^{1/2} \right]$$

$$\times \left\{ \frac{B^2 - A^2}{2} \left(\frac{2 + (1+\chi^2)^{1/2}}{1+\chi^2} \right) \left[\frac{(1+\chi^2)^{1/2} - 1}{1+\chi^2} \right]^{1/2} \right.$$

$$\left. + AB \left(\frac{\omega - \omega_0}{|\omega - \omega_0|} \right) \frac{\left[(1+\chi^2)^{1/2} - 2 \right]\left[(1+\chi^2)^{1/2} + 1 \right]}{(1+\chi^2)^{3/2}} \right\} \quad (12)$$

The ratio B/A may be obtained from Fig. 1 of Reuter and Sondheimer (1948). Equations (11) and (12) are plotted in Fig. 12-20 for the region of the completely anomalous skin effect where $\delta \ll \Lambda$ and $B/A = 3^{1/2}$. Webb (1967) extended Dyson's theory to the case of spherical particles with the normal skin effect. Kahn (1977) discussed the theory of microwave eddy currents and ESR in materials of intermediate conductivity.

Dresselhaus et al. (1955) give expressions for the lineshape obtained from plasma resonance in crystals in terms of the parameters ν and ν', which are

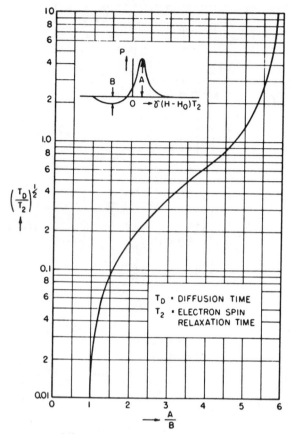

Fig. 12-18. The ratio $(T_D/T_2)^{1/2}$ versus A/B for the ESR power absorption in thick metal films (Feher and Kip, 1955).

defined by $(\omega\tau = \nu)$

$$\nu_c = \omega_c\tau = \left(\frac{eH}{m^*c}\right)\tau \tag{13}$$

$$\nu' = \omega'\tau = \nu - \frac{\nu_p^2}{\nu} \tag{14}$$

$$\nu_p = \omega_p\tau = \tau\left(\frac{L_iNe^2}{m^*}\right)^{1/2} \tag{15}$$

where H is the applied magnetic field, L_i is the depolarization factor, N is the conduction electron concentration, ω is the applied frequency, ω_c is the cyclotron frequency, ω_p is the plasma frequency, and τ is the relaxation time. The theoretical lineshape is shown in Figs. 20-19 and 20-20 of the first edition

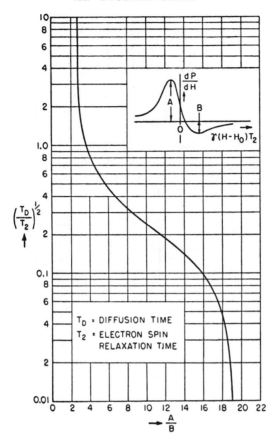

Fig. 12-19. The ratio $(T_D/T_2)^{1/2}$ versus A/B for the derivative of the ESR power absorption in thick metal films (Feher and Kip, 1955).

for two types of modulation in the limiting case $\nu \ll \nu'$ so that $\nu' = -\nu_p^2/\nu$. Rukhadze and Silin (1962) discuss cyclotron resonance lineshapes in nonrelativistic plasmas.

Jansen and Sperlich (1975), Jarrett et al. (1974), Monod (1978), Sperlich and Muller (1975), Urban et al. (1976), and Vigouroux et al. (1978) studied lineshapes in metals. Some authors (e.g., Hauser and Hutton, 1976; Mizokawa and Nakamura, 1975; Movahgar and Schweitzer, 1977; Voget-Grote et al., 1976) compared ESR results to the data from resistivity measurements to gain more insight into the mechanisms involved in conduction ESR relaxation and lineshapes.

Vigouroux et al. (1976) studied conduction ESR using amplitude modulation of the microwave magnetic field and monitoring of $\partial Mz/\partial t$ by a coil. They employed a magnetron oscillating at 9.3 GHz amplitude modulated by a pin diode fed by a sine wave generator. The sample was placed along the axis of a cylindrical TE_{011} mode resonator and surrounded by a pickup coil that fed the

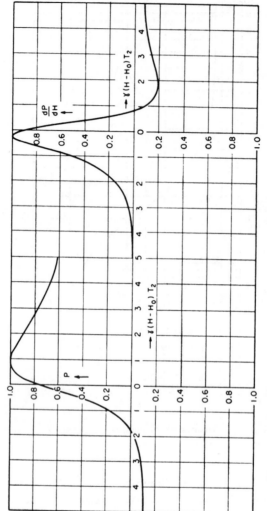

Fig. 12-20. Theoretical curves for the ESR absorption $Y = P$ and its derivative $Y' = dP/dH$ in the region of the completely anomalous skin effect (Feher and Kip, 1955).

detected signal to a microammeter. Details of the apparatus were given by Gourdon et al. (1970, 1973). Janssens and Witters (1972) described the detection of broad CESR lines by a reflection spectrometer. Stesmans and Van Meijel (1977) designed a simple, large modulation field setup for CESR cavities.

A number of authors have reported measurements of conduction ESR in recent years (see, for example, Antonenko et al., 1977; Berthault et al., 1977; Beuneu and Monod, 1977; Berkovits et al., 1979; Chazalviel, 1975; Damay and Sienko, 1974; Dobson, 1973; Dokter et al., 1977; Domdey and Voitlander, 1974; Dumas, 1978; Dvurechensky et al., 1975; Engel et al., 1974; Gerasimenko et al., 1975; Hasegawa et al., 1977; Janossy, 1975; Kushida, 1974; Knight, 1976; Kodera, 1970; Menard and Walker, 1974; Meservey and Tedrow, 1978; Pifer and Magno, 1971; Rettori et al., 1978; Seipler and Elschner, 1976; Sharp-Dent et al., 1976; Witters, 1972).

J. Small Clusters of Spins

The local magnetic field $\mathbf{H}$ produced by a magnetic dipole of spin $\mathbf{S}_1$ at a distance $\mathbf{r}$ from its center is given by

$$\mathbf{H} = \left(\frac{g\beta}{r^3} \right) \left[\mathbf{S}_1 - 3\mathbf{r} \left(\frac{\mathbf{r} \cdot \mathbf{S}_1}{r^2} \right) \right] \tag{1}$$

This local magnetic field alters the total value of $\mathbf{H}$ at the positions of neighboring unpaired spins, and thereby broadens their resonant lines. The dipolar interaction energy $\mathcal{K}_{dd}$ between $\mathbf{S}_1$ and an identical magnetic moment with spin $\mathbf{S}_2$ is given by

$$\mathcal{K}_{dd} = \left(\frac{g^2\beta^2}{r^3} \right) \left[\mathbf{S}_1 \cdot \mathbf{S}_2 - \frac{3(\mathbf{r} \cdot \mathbf{S}_1)(\mathbf{r} \cdot \mathbf{S}_2)}{r^2} \right] \tag{2}$$

In a strong applied magnetic field $\mathbf{H}_0$ oriented at an angle relative to the vector $\mathbf{r}$, the two spins will be quantized along H_0, and for like spins $S_1 = S_2$, Eq. (2) becomes

$$\mathcal{K}_{dd} = S(S + 1) \left(\frac{g^2\beta^2}{r^3} \right) (1 - 3\cos^2\theta) \tag{3}$$

If the spins are not identical, then $\mathcal{K}_{dd}$ is reduced by the factor $2/3$. In other words, a like neighbor is $3/2$ as effective in broadening a line as an unlike neighbor.

By averaging the dipolar interaction over all angles corresponding to a random distribution of directions $\mathbf{r}$, the lineshape for a pair of interacting dipoles was found by Pake (1948) to have the form shown in Fig. 12-21c. This lineshape was observed experimentally by Pake (1948) and by Gutowsky et al.

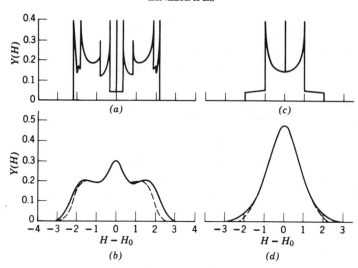

Fig. 12-21. Powder pattern lineshapes obtained for (a) three nuclei of spin $\frac{1}{2}$ at the apices of an equilateral triangle and (c) two nuclei when the component (or single crystal) linewidth in each case is a delta function. When the component lines have a finite width the observed lineshapes change from (a) and (c) to the solid lines of (b) and (d), respectively. The dashed lines of (b) and (d) correspond to experimental data (Andrew and Bersohn, 1950). The abscissa is in the units $3\mu/2R^3$, where μ is the magnetic moment, and R is the distance between the nuclei.

(1949). The lineshape collapses to half its width when the nuclei rotate rapidly about a line perpendicular to the internuclear axis.

A number of workers have studied radical pairs, such as Box (1971), Carr et al. (1974), Flossman et al. (1975, 1975), Gooijer et al. (1972), Hutchison and King (1973), Leigh (1970), Nelsen et al. (1977), Pekcan and Keyman (1980), Trifunac and Avery (1974), and van Gerven and Accou (1975). H–D exchange has been studied by ENDOR (Teslenko, 1975).

The treatment just discussed has been generalized to three magnetic moments forming the apices of an equilateral triangle by Andrew and Bersohn (1950), (See also Doremieux-Morin, 1979). The powder pattern lineshape has nine points that "blow up" to infinity, as shown in Fig. 12-21a. In practice these infinities are rendered finite by other broadening mechanisms. When this three-spin system rotates about the threefold symmetry axis normal to its plane, it produces the same lineshape as two coupled spins [Fig. 12-21c and d]. The case of a three-spin system with one nucleus between a pair of another species was analyzed by Waugh et al. (1953). Tsukerblat et al. (1976) determined the ESR line profile of randomly oriented trigonal exchange clusters of three nuclei. Blinc et al. (1966) analyzed a nearly linear five-spin system.

A number of workers have reported lineshape narrowing by rotating groups, hindered rotation, and tunneling (Adrian et al., 1973; Alexander et al., 1974; Beram et al., 1976; Berclaz and Geoffroy, 1976; Bernhard and Ezra, 1974; Bogan and Kispert, 1972; Chase, 1970; Clough and Poldy, 1973; Clough et al.,

1969, 1972, 1974, 1974; Dattagupta and Blume, 1974, 1974; Davidson and Miyagawa, 1970, 1972; Freed, 1965; Isoya et al., 1978; Kaminski, 1976; Jinguji et al. (1976); Krishnamurthy, 1972; Krusic et al., 1971; Lee and Rogers, 1976; Maniv et al., 1977; Maruani et al., 1968; McDowell and Shimokoshi, 1974; McDowell et al., 1973; Mooij and de Boer, 1976; Moriuchi and Sohma, 1971, 1972; Mottley et al., 1975; Ohkura et al., 1975; Reuveni and Luz, 1976; Miller, 1972; Stankowski et al., 1973; Subramanian and Narasimhan, 1972; Teslenko and Shanina, 1975; Vugman et al., 1975; Welter and Windsch, 1971; Zamaraev et al., 1978).

Kothe (1977) studied symmetrical three-spin systems in nematic liquids, Burlamacchi (1976) examined phosphate complexes in solution, Berthault et al. (1977) developed a statistical model for the ESR lineshape of lithium colloids, and Heeger and Foner (1976) investigated dilute magnetic alloys. Martinelli et al. (1977) give approximate expressions for the linewidth and intensity of longitudinally detected ESR spectra. Waller and Rogers (1974) proposed methods for the determination of axial g tensors. Porumb and Slade (1976) found a computational model for systematizing spectra of axially symmetric systems. Weill et al. (1977) and Ovchinnikov and Konstantinov (1976) discuss anisotropic systems. Suassuna et al. (1977) reported linewidth anisotropies in thin films, and Gesmundo (1975) studied exchange coupled Cr^{3+} clusters.

Ovchinnikov and Konstantinov (1976) treated the ESR spectra of axially symmetric systems. Porumb and Slade (1971) developed a computational model for synthesizing spectra from axially symmetric sites, Hentschel et al. (1978) analyzed the spectra in partially ordered sites, Servant et al. (1976) carried out computer simulation for complete g factor anisotropy in the Faraday and Cotton–Mouton–Voigt effects, Ovchinnikov and Konstantinov (1976) treated singularities in ESR spectra from a partially oriented matrix, Martinelli (1977) discussed the lineshape in longitudinally detected ESR, Benner (1977) reported on the influence of radiation damping on the lineshape, and anisotropies were treated by a number of authors (e.g., Jinguji et al., 1976; Suassuna et al., 1977; Waller and Rogers, 1974; and Weill et al., 1977).

Some linewidth and lineshape studies have been made of spectra from ordered magnetic systems [e.g., Akhiezer et al., 1976; Glass and Elliott, 1976; Shumilkina and Obraztsov, 1975; Taylor and Coles, 1972; Tomashpolskii, 1974) such as ferromagnetic samples (e.g., Angelov et al., 1977; Babushkin et al., 1975; Bazuev et al., 1977; Janossy et al., 1974; Miyadai et al., 1971; Shrivastava, 1975; Shumilkina and Obraztsov, 1975; Tomashpolskii, 1974) and antiferromagnetic samples (e.g., Anders and Zvyagin, 1973; Gulley and Jaccarino, 1972; Gupta et al., 1972; Seehra, 1972; and Seehra and Huber 1975; Tanaka and Kondo, 1973, 1975). Line broadening of low dimensional exchange-coupled spins with uniaxial anisotropy was calculated by Tanaka (1969, 1970, 1971, 1975), Tanaka et al. (1971, 1973, 1975, 1976), Benner (1977), and Hughes et al. (1975). Pomerantz et al. (1978), Richards (1973), and Shia and Kokoszka (1974) present experimental data on two-dimensional magnetic structures. Berim et al. (1976) calculated the temperature dependence of the first and second moments of a linear chain of spins coupled by the Ising exchange interaction, and they compared the influences of the Ising and Heisenberg exchange interactions on the line profile.

K. Anisotropic g Factors

Many paramagnetic single crystals have electron spin resonance absorptions that occur at magnetic field strengths that depend on the orientation of the crystal with respect to the magnetic field direction. This phenomenon can be explained by noting that for paramagnetic substances the fine structure Hamiltonian $\mathcal{H}$, omitting the terms that depend on the nuclear spin is given by (Poole and Farach, 1972)

$$\mathcal{H} = \mathbf{S} \cdot \boldsymbol{D} \cdot \mathbf{S} + \beta \mathbf{S} \cdot \boldsymbol{g} \cdot \mathbf{H}$$

where $\boldsymbol{D}$ and $\boldsymbol{g}$ are tensors that are constants for a given orbital state, β is the Bohr magneton, and $\mathbf{H}$ is the strength of the applied magnetic field. If the total electronic spin S is $\frac{1}{2}$, the first term, which represents the zero field splitting, vanishes (Kramers' degeneracy). Then the resonance magnetic field H' is

$$H' = \frac{h\nu}{g'\beta} \tag{1}$$

where the value of H' depends on the orientation of the crystal in accordance with the expression

$$g' = \left(g_1^2\cos^2\theta_1 + g_2^2\cos^2\theta_2 + g_3^2\cos^2\theta_3 \right)^{1/2} \tag{2}$$

Here the θ_i are the angles between the direction of the magnetic field and the coordinate axes in which the g tensor is diagonal, and the g_i are the diagonal components of $\boldsymbol{g}$. This coordinate system need not be the same as the symmetry axes of the crystal, but it corresponds to the symmetry axes for the local internal electric field. The g factor can also be expressed in spherical coordinates θ and ϕ:

$$g(\theta, \phi) = \left(g_1^2\sin^2\theta \sin^2\phi + g_2^2\sin^2\theta \cos^2\phi + g_3^2\cos^2\theta \right)^{1/2} \tag{3}$$

and if H_0 as defined as the magnetic field strength corresponding to the free electron g factor g_0:

$$H_0 = \frac{h\nu}{g_0\beta} \tag{4}$$

then Eq. (1) can be written

$$H'(\theta, \phi) = \frac{H_0 g_0}{g(\theta, \phi)} \tag{5}$$

where $H'(\theta, \phi)$ is the magnetic field strength that produces resonant absorption when its direction is at the orientation θ, ϕ relative to the g-factor principal axes.

If a study is made of a single crystal with one angularly dependent spectral line recorded with the magnetic field in the direction θ, ϕ relative to the crystal symmetry axes then the resonant absorption lineshape is given by the product $P(\theta, \phi)Y\{[H - H'(\theta, \phi)]/\Delta H(\theta, \phi)\}$, where in general the linewidth can be angularly dependent and the factor $P(\theta, \phi)$ takes into account the angular dependence of the probability that the microwave field will induce a transition. For a powder sample in which the individual microcrystallites are randomly oriented the powder pattern lineshape $I(H)$ is obtained by integrating this monocrystal lineshape over a unit sphere [Poole and Farach (1980)]

$$I(H) = \int_0^{2\pi} \int_0^{\pi} P(\theta, \phi) Y\left(\frac{H - H'(\theta, \phi)}{\Delta H(\theta, \phi)}\right) \sin\theta \, d\theta \, d\phi$$

In general this is not a convenient expression to work with, because H' is a complex function of the position and is not easy to integrate.

A number of authors have published graphs of the experimentally observed linewidth anisotropy in various systems [Huang et al., 1972 (FR); Hinkel et al., 1978 (FR); Dattagupta and Blume, 1974 (CO_2^-); Hughes and Soos, 1970 (CO_2^-); Byberg et al., 1974 (BrO_3^{2-}); Sperlich et al., 1974 (VO_6); Henning and den Boef, 1978 (Cr^{3+}); Bugai et al., 1975 (Cr^{3+}); Bugai et al., 1971, 1973 (Fe^{3+}); Shia and Kokoszka, 1974 (Mn^{2+}); Sastry and Sastry, 1974 (Cu^{2+}); Hoffmann and Goslar, 1975 (Cu^{2+}); Arbilly et al., 1975 (Er); Urban et al., 1976 (Gd^{3+}); Elschner and Weimann, 1971 (Gd^{3+}); Dietrich, 1976 (Yb^{3+}); Wolfe and Jeffries, 1971 (Yt); Ermakovich et al., 1977 (centers)].

We will illustrate the calculation of $I(H)$ for a completely anisotropic g factor (Bloembergen and Rowland, 1953; Kohin and Poole, 1958; Kneubuhl, 1960; Poole and Farach, 1980) when both the linewidth and the transition probability $[P(\theta, \phi) = 1]$ are independent of orientation. Some authors (Searl et al., 1959, 1961; Kneubuhl and Natterer, 1961; Ibers and Swalen, 1962; Kottis and Lefebvre, 1963; Lardon and Günthard, 1966; Iwasaki, 1966; Aasa and Vänngärd, 1970; Shteinshneider and Zhidomirov, 1970) have taken the angular dependence of $P(\theta, \phi)$ into account in calculating powder patterns. Hauser (1972) and Hauser and Renaud (1972) calculated the powder pattern for a completely anisotropic g factor and an angularly dependent linewidth.

The orientation of a particular molecule in the sample may be specified by the two angles θ_1 and θ_2 since all the θ_i are not independent. It is more convenient, however, to choose an alternate pair of independent variables u and ψ to specify the orientation. These are defined by

$$u = \cos\theta_1$$

$$\cos^2\theta_2 = (1 - u^2)\sin^2\psi \tag{6}$$

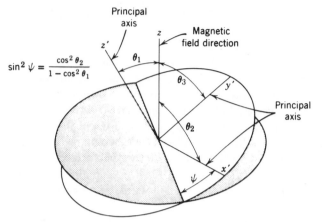

Fig. 12-22. Illustration of the angles used to specify the orientation of a completely anisotropic crystal (Kohin and Poole, 1958).

where θ_1 and ψ are two of the Euler angles describing the orientation of the molecular axes. Specification of the third Euler angle ϕ is unnecessary since rotations about the direction of the applied magnetic field are immaterial. These parameters are illustrated in Fig. 12-22. We then obtain

$$g' = \left[\left(g_1^2 - g_3^2\right)u^2 + \left(g_2^2 - g_3^2\right)(1 - u^2)\sin^2\psi + g_3^2\right]^{1/2} \tag{7}$$

We now calculate the probability $p(u_1, \psi_1)$ that a molecule will assume an orientation u_1, ψ_1 such that $u \leqslant u_1 \leqslant u + du$ and $\psi \leqslant \psi_1 \leqslant \psi + d\psi$. Let $p(u_1, \psi_1) = p(u_1)p(u_1 \mid \psi_1)$, where $p(u_1)$ is the probability that $u \leqslant u_1 \leqslant u + du$ and $p(u_1 \mid \psi_1)$ is the probability that, for a given u_1, $\psi \leqslant \psi_1 \leqslant \psi + d\psi$. Assuming a random orientation, both $p(u_1)$ and $p(u_1 \mid \psi_1)$ are constants times du and $d\psi$, respectively. Then the normalized probability is

$$4\pi p(u, \psi) = du\, d\psi \tag{8}$$

It is now necessary to integrate this probability distribution over the lineshape of the component lines. We take $Y(H - H')$ to be the lineshape factor for the absorption in a single crystal, where H is the applied magnetic field and $H' = H'(u, \psi)$ is the field for the resonance maximum of the single crystal as given by Eqs. (1) and (2). Here Y is assumed not to depend explicitly on the orientation (see Livingston and Zeldes, 1956). It is normalized such that

$$\int_{-\infty}^{\infty} Y(H - H')\, dH = 1.$$

Then, for a randomly oriented group of crystals, the normalized intensity $I(H)$

is given by

$$I(H) = \frac{1}{4\pi} \int_{-1}^{1} \int_{0}^{2\pi} Y(H - H') \, du \, d\psi$$

$$= \frac{2}{\pi} \int_{0}^{1} \int_{0}^{\pi/2} Y(H - H') \, du \, d\psi \qquad (9)$$

To perform this integration we use Eq. (7) to change the independent variables from u, ψ to u, g', and we then integrate first over the variable u. This may be accomplished by putting $d\psi = (\partial\psi/\partial g) \, dg$. For a fixed value of g', the limits of integration for u may be evaluated by mapping the area of integration onto the u, g' plane (Fig. 12-23). The integral thereby decomposes into two integrals depending on whether g' is less than or greater than g_2. We write

$$I_A(g') = \left(\frac{1}{g'} \right) \left(\frac{\partial\psi}{\partial g'} \right) du \quad \text{for } g_2 \leqslant g' \leqslant g_1$$

and

$$I_B(g') = \left(\frac{1}{g'} \right) \left(\frac{\partial\psi}{\partial g'} \right) du \quad \text{for } g_3 \leqslant g' \leqslant g_2$$

Then in region A, $u_{(\psi=\pi/2)} \leqslant u \leqslant u_{(\psi=0)}$, and in region B, $0 \leqslant u \leqslant u_{(\psi=0)}$. By means of suitable transformations both of these integrals may be converted to complete ellipitic integrals of the first kind K (see Jahnke and Emde, 1945). We

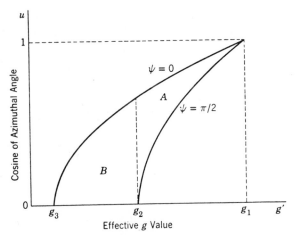

Fig. 12-23. Mapping of the u, g' plane showing regions A and B over which the integration for the intensity of a powder pattern is to be performed (Kohin and Poole, 1958).

then obtain

$$I_A(g') = \left(\frac{1}{b_g}\right) K\left(\frac{a_g}{b_g}\right) \tag{10a}$$

$$I_B(g') = \left(\frac{1}{a_g}\right) K\left(\frac{b_g}{a_g}\right) \tag{10b}$$

where

$$a_g = \left[(g_2^2 - g_3^2)(g_1^2 - g^2)\right]^{1/2} \tag{11a}$$

$$b_g = \left[(g_1^2 - g_2^2)(g^2 - g_3^2)\right]^{1/2} \tag{11b}$$

Equation (10) can be used to obtain the probability distribution as a function of g.

If we replace all g values in Eq. (10) by the appropriate H values, the normalized intensity is given by

$$I(H) = \frac{2H_1 H_2 H_3}{\pi}\left\{ \int_{H_1}^{H_2} \frac{I_A(H')}{(H')^2} Y(H - H')\, dH' \right.$$

$$\left. + \int_{H_2}^{H_3} \frac{I_B(H')}{(H')^2} Y(H - H')\, dH' \right\} \tag{12}$$

where the factor $H_1 H_2 H_3/(H')^2$ appears after making this replacement. Now if we assume that the component lines are extremely narrow, we can put $Y(H - H') = \delta(H - H')$, where $\delta(H - H')$ is the Dirac delta function. Then

$$I(H) = \frac{2H_1 H_2 H_3}{\pi H^2}\begin{cases} \dfrac{1}{b_H} K\left(\dfrac{a_H}{b_H}\right) & H_1 \leqslant H \leqslant H_2 \\[2mm] \dfrac{1}{a_H} K\left(\dfrac{b_H}{a_H}\right) & H_2 \leqslant H \leqslant H_3 \\[2mm] 0 & \text{otherwise} \end{cases} \tag{13}$$

where

$$a_H = \left[(H_3^2 - H_2^2)(H^2 - H_1^2)\right]^{1/2} \tag{14a}$$

$$b_H = \left[(H_2^2 - H_1^2)(H_3^2 - H^2)\right]^{1/2} \tag{14b}$$

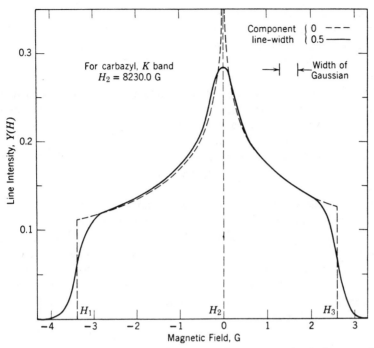

Fig. 12-24. Calculated powder pattern lineshape for carbazyl assuming both zero and nonzero component linewidths (Kohin and Poole, 1958).

This lineshape has been plotted in Fig. 12-24 taking carbazyl (N-picryl-9-amino-carbazyl) as an example. The values assumed for the parameters are taken from Kikuchi and Cohen (1954). The three g factors g_1, g_2, and g_3 are related to the magnetic field strengths H_1, H_2, and H_3 by the relation

$$H_1 = \frac{h\nu}{g_1 \beta} \tag{15a}$$

$$H_2 = \frac{h\nu}{g_2 \beta} \tag{15b}$$

$$H_3 = \frac{h\nu}{g_3 \beta} \tag{15c}$$

where $h\nu$ is the quantum of microwave energy. These relations may be used to deduce g factors from powder pattern spectra. Bloembergen and Rowland (1953) calculated a similar lineshape using the approximation $(g_1 + g_2)/(g_2 + g_3) = 1$.

If the component lineshape cannot be assumed to be extremely narrow, then the integrals in Eq. (9) may be evaluated numerically without difficulty. The infinite peak is so narrow that the area lost by cutting it off after $K = \pi$ is negligible. The general effect of integrating over a line of nonzero width is to lower and widen the peak, and to round off the shoulders. This is illustrated in Fig. 12-24 for carbazyl, where a Gaussian lineshape having a width of 0.5 G is assumed.

This description of the lineshape for a paramagnetic substance is particularly useful for the determination of the principle g values of a substance when it is impossible to prepare the substance in the form of a single crystal. This is also true in the case of many free radicals formed directly in solids by ultraviolet rays, X rays, or gamma rays, or condensed in solids after their formation in an electric discharge. In either case, the free radicals would be expected to see randomly oriented crystalline fields. The three g tensor components can be most easily obtained by examining the derivative of the absorption, as is normally done in electron spin resonance spectrometers (Fig. 12-25). If the component linewidth is smaller than the separation between the g tensor components, the derivative will have maxima very near to g_1 and g_3, and will cross the axis sharply very close to g_2. The g tensor components may then be obtained quite accurately. As the component linewidth becomes wider, however, the peaks at g_1 and g_3 tend to become obscured, as shown in Fig. 12-26 for the case of axial symmetry where $g_1 = g_2 = g_\perp$ and $g_3 = g_\parallel$, with $g_\perp < g_\parallel$.

If the anisotropic molecule rotates about randomly with a correlation time that is much shorter than the reciprocal of the spread of the spectrum in frequency units $\hbar/(|g_1 - g_3|\beta H)$, then a single resonant line is obtained at the average g factor g_0 given by

$$g_0 = \tfrac{1}{3}(g_1 + g_2 + g_3) \tag{16}$$

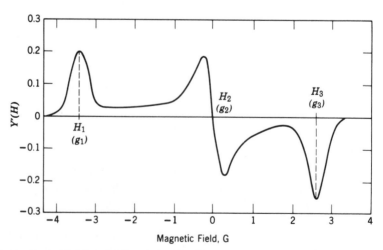

Fig. 12-25. Derivative of the ESR powder pattern lineshape for carbazyl (Kohin and Poole, 1958).

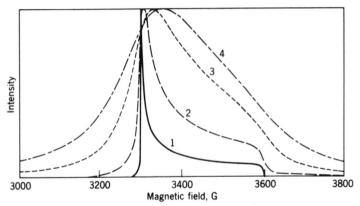

Fig. 12-26.   The effect of increasing the component linewidth on the calculated spectrum for an axially symmetric powder pattern. Lorentzian linewidths: 1—1 G; 2—10 G; 3—50 G; 3—50 G; 4 —100 G (Ibers and Swalen, 1962).

The correlation time is the average time that it takes a molecule to move through an angle of one radian or a distance equal to its diameter.

If there is an axis of magnetic symmetry, then the spin will be characterized by $g_\parallel$ along the axis and $g_\perp$ at right angles to the axis. With axial symmetry the three g factors of the general case may be related to $g_\parallel$ and $g_\perp$ as follows:

$$\left.\begin{array}{l} g_1 = g_\parallel \\ g_2 = g_3 = g_\perp \end{array}\right\} g_\perp < g_\parallel \qquad (17a)$$

$$\left.\begin{array}{l} g_1 = g_2 = g_\perp \\ g_3 = g_\parallel \end{array}\right\} g_\parallel < g_\perp \qquad (17b)$$

To obtain the lineshape we follow the development of Poole and Farach (1980). For the case (17b) Eqs. (3) and (5) reduce to

$$g = \left(g_\parallel^2 \cos^2\theta + g_\perp^2 \sin^2\theta\right)^{1/2} \qquad (18)$$

$$H' = \frac{g_0 H_0}{\left(g_\perp^2 + \left(g_\parallel^2 - g_\perp^2\right)\cos^2\theta\right)^{1/2}} \qquad (19)$$

where H' varies with the angle θ between the limiting values $H_\parallel$ and $H_\perp$ given by

$$g_\parallel H_\parallel = g_\perp H_\perp = g_0 H_0 \qquad (20)$$

where $H = H_\parallel$ for $\theta = 0$ and $H = H_\perp$ for $\theta = \pi/2$. The transition probability $P(\theta)$ for a linearly polarized oscillatory microwave field is given by (Bleaney,

1950, 1960)

$$P(\theta) = \frac{1}{2}\left[\left(\frac{H'(\theta)}{H_\parallel}\right)^2 + 1\right] \tag{21}$$

normalized to $P(\theta) = 1$ for $\theta = 0$.

The powder pattern lineshape is

$$I(H) = \frac{H_\parallel\left(g_\parallel^2 - g_\perp^2\right)^{1/2}}{2} \int_{H_\parallel}^{H_\perp} \frac{\left(H'^2 + H_\parallel^2\right)Y(H')\,dH'}{\left(g_\parallel^2 H_\parallel^2 - g_\perp^2 H'^2\right)^{1/2} H'^2} \tag{22}$$

where the limits of integration are selected for the case $H_\parallel < H_\perp$. If the individual component linewidth is infinitely narrow, then we obtain the delta function lineshape

$$I(H) = \frac{H_\parallel\left(H^2 + H_\parallel^2\right)}{2H^2}\left(\frac{g_\parallel^2 - g_\perp^2}{g_\parallel^2 H_\parallel^2 - g_\perp^2 H^2}\right)^{1/2} \tag{23}$$

which has the limiting $I(H) = 1$ for $H = H_\parallel$ and becomes infinite $H = H_\perp$. The lineshape for $g_\perp \sim 1.1 g_\parallel$ has the form shown in Fig. 12-26. Mathematically it corresponds to the merging of the infinity found at g_2 in Fig. 12-24 with the discontinuity at the edge of the lineshape. The case of axial symmetry is frequently met in practice. Again, rapid rotation will reduce the spectrum to a symmetrical resonance centered at g_0:

$$g_0 = \tfrac{1}{3}\left(g_\parallel + 2g_\perp\right) \tag{24}$$

Waller and Rogers (1975) developed methods for calculating axial **g** tensors that require less single crystal data than had been hitherto needed. The input data may include g values from measurements on single crystals, polycrystalline powders, or solutions. They considered 22 cases classified into six groups with the following input data: (1) rotations about two axes, (2) one rotation and the value of $g_\parallel$, $g_\perp$, or g_0, (3) one rotation and the g value from an arbitrary orientation, (4) two g values, $g_\parallel$ and $g_\perp$, (5) three g values and $g_\parallel$, $g_\perp$, or g_0, and (6) four g values. A computer program was written that covers these cases.

L. Powder Patterns

Until now we have dealt mainly with powder patterns of angularly dependent singlets. When more than one resonance line is present then the observed spectrum is a summation of component hyperfine patterns of the type (Poole

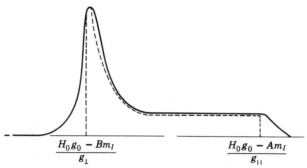

Fig. 12-27. Powder pattern lineshape for one hyperfine component m_I when $g_0 H_0 \gg A$, B for the case of axial symmetry (O'Reilly, 1958).

and Farach, 1980)

$$I(H) = \sum_i \int_{H_{i1}}^{H_{i2}} P_{i(H')} Y_i \left(\frac{H - H'}{\Delta H_i(H')} \right) G_{i(H')} \, dH' \tag{1}$$

The lineshape illustrated in Fig. 12-27 is an example of a component from the above integral. It is quite common to find overlapping hyperfine component powder patterns in polycrystalline samples containing Cu^{2+} and overlapping fine structure component spectra from Fe^{3+}.

The effect of axially symmetric hyperfine structure on the lineshape of a sample with an axially symmetric g factor was solved by O'Reilly (1958) with the aid of the Hamiltonian $\mathcal{H}$

$$\left(\frac{1}{\beta} \right) \mathcal{H} = g_\| H_z S_z + g_\perp \left(H_x S_x + H_y S_y \right) + A I_z S_z$$

$$+ B \left(I_x S_x + I_y S_y \right) \tag{2}$$

where I is the nuclear spin, and the hyperfine constants A and B are in gauss. Each hyperfine component is characterized by its projection M_I along the z axis, which lies in the magnetic field direction. There are components corresponding to M_I that assume integer or half-integer values ranging from I to $-I$. As shown in Fig. 12-27 each component hyperfine line will stretch from $(H_0 g_0 - Bm_I)/g_\perp$ to $(H_0 g_0 - Am_I)/g_\|$ on the magnetic field scale, where $H_0 = h\nu/g_0\beta$, and g_0 is defined by Eq. (24). One often finds experimentally that the powder patterns of the various hyperfine components overlap each other. Bales et al. (1975) separated isotropic and anisotropic hyperfine components by making measurements at two microwave frequencies.

A number of investigators have performed calculations and obtained theoretical lineshape functions for polycrystalline substances, and these calculations are summarized in Table 12-11. Taylor et al. (1974) reviewed ESR spectra

TABLE 12-11
List of Articles Reporting Powder Pattern Lineshapes (Poole and Farach, 1980)

Symmetries			Lineshape, δ—delta function L—Lorentzian G—Gaussian	Comments	References
g Factor g	Hyperfine A	Zero Field D			
Axial	—	—	δ		Bleaney (1950, 1951, 1960)
Asymmetric	—	—	δ		Bloembergen and Rowland (1953)
Axial	—	—	δ		Sands (1955)
Isotropic	—	Axial	δ	NMR	Singer (1955)
Asymmetric	—	—	δ	Spin $= \frac{3}{2}$	Kohin and Poole (1958)
Axial	Axial	—	δ		O'Reilly (1958)
Axial	—	—	L, G	Orientation-dependent transition probability	Searl et al. (1959, 1960, 1961)
Asymmetric	—	—	δ		Kneubuhl (1960)
Isotropic	Asymmetric	—	δ, G	Good graphs	Blinder (1960)
Isotropic	Asymmetric	—	G		Lefebvre (1960)
Isotropic	Asymmetric	—	δ	Spin $= \frac{1}{2}$, nuclear Zeeman $I = \frac{1}{2}$	Sternlicht (1960)
Asymmetric	—	—	L		Chirkov and Kokin (1960)
Axial	Axial	—	G		Neiman and Kivelson (1961)
Asymmetric	—	—	—		Lefebvre (1961)
Axial	—	—	L	Spin $> \frac{1}{2}$, orientation-dependent transition probability	Kneubuhl and Natterer (1961)
Isotropic	Anisotropic	—	δ, G		Lord and Blinder (1961)
Axial	Axial	—	δ, G		Kneubuhl et al. (1961)
Isotropic	Asymmetric	—	G		Cochran et al. (1961)
Axial	—	—	—		Heuer (1961)
Axial	—	—	L	Orientation-dependent transition probability	Ibers and Swalen (1962)
Axial	—	—	δ		Gersmann and Swalen (1962)
Isotropic	Isotropic	—	δ, L		Wakim and Nolle (1962)

Axial	Axial	—	L, G	Orientation-dependent transition probability	Vänngård and Aasa (1963)
Isotropic	—	—	G		Kottis and Lefebvre (1963)
Asymmetric	Asymmetric	—	L, G	Computer Calculation	Weil and Hecht (1963)
Asymmetric	—	—	L, G		Lebedev (1962)
Axial	Axial	—	G		Hughes and Rowland (1964)
Axial	Axial	—	Arbitrary		Korolkov and Potapovich (1964)
Asymmetric	—	—	L, G	Valid for small anisotropy	Johnston and Hecht (1965)
Asymmetric	Asymmetric	—	L	Large ^{19}F anisotropy	Malley (1965)
Asymmetric	Asymmetric	—	—	Nuclear Zeeman $I = \frac{1}{2}$	Iwasaki et al. (1965)
Asymmetric	—	—	—		Lefebvre and Maruani (1965)
Axial	Anisotropic	—	L	Computer simulation	Schoffa and Burk (1965)
Axial	Axial	—	—	Orientation-dependent transition probability	Chachaty and Maruani (1966)
Axial	Axial	—	L		Lardon and Günthard (1966)
Isotropic	—	—	δ	Spin $= \frac{1}{2}$, nuclear Zeeman $I = \frac{1}{2}$, orientation-dependent transition probability	Iwasaki (1966)
Axial	Asymmetric	—	L, G	Shape-transform graph	Che et al. (1969)
—	—	—	—	Discusses singularities	Coope (1969); Maruani et al. (1970)
Axial	—	Asymmetric	—	Spin $= \frac{3}{2}$, no orientation-dependent transition probability, central powder pattern folded back	Mohrmann et al. (1970)
Axial	Axial	—	—		Garrett et al. (1966)
Asymmetric	—	—	—	g_z/g_x vs. g_x/g_y, orientation-dependent transition probability	Aasa and Vänngard (1970)
Axial	Axial	—	L	Different principal directions, orientation-dependent transition probability	Shteinshneider and Zhidomirov (1970)
Axial	Axial	—	L, G	Quadrupole, nuclear Zeeman	Tsay et al. (1971)
Isotropic	Isotropic	Axial	—	Spin $= \frac{5}{2}$, doubling gives signs	Mialhe et al. (1971)

TABLE 12-11
Continued

g Factor g	Hyperfine A	Zero Field D	Lineshape, δ—delta function L—Lorentzian G—Gaussian	Comments	References
Isotropic	Axial	—	G	^{14}N hfs axial; rotating methyl group	Moriuchi and Sohma (1971, 1972)
Isotropic	—	Axial	δ	Spin $> \frac{1}{2}$	Poole and Farach (1972), Shaffer et al. (1976)
Asymmetric	—	—	—	ΔH angular dependence	Hauser (1972); Hauser and Renaud (1972)
Isotropic	—	—	δ	$H = \beta I g H + \alpha \Sigma I_i H_j E_k$ ($i \neq j \neq k$) Stark term	Blazha and Roitsin (1972)
Axial	Axial	—	L, G	Spin $= \frac{1}{2}$, vanadyl	Narayana and Sastry (1972)
Asymmetric	Asymmetric	—	G	Nuclear Zeeman + asymmetric quadrupole	Herring et al. (1972)
Isotropic	—	Asymmetric	G	spin $= \frac{5}{2}$; includes B_4^M zero field	Sweeney et al. (1973)
Isotropic	—	Axial	—	Spin $= 1$, O_2 libration	Hirokawa (1973, 1974, 1978)
Asymmetric	—	Asymmetric	—	Spin $= 1$, different principal directions	Gaponenko et al. (1973)
Asymmetric	—	Asymmetric	δ	Spin $> \frac{1}{2}$	Blazha and Roitsin (1974)
Asymmetric	—	Asymmetric	G	Spin $= \frac{5}{2}$, $g\beta H < D$	Shcherbakova and Istamin (1975)
Asymmetric	—	—	G	Spin $= \frac{1}{2}$; $l_1 = l_2 = \frac{1}{2}$; torsional-oscillation axis distribution	Jinguji et al. (1976)
Axial	Axial	—	δ	$l_1 = l_2 = l_3 = \frac{1}{2}$; extra singularities	Tsukerblat and Chobanu (1976) Belinskii et al. (1974), Ovchinnikov and Konstantinov (1976)
Axial	Axial	—	—		Clark et al. (1978)
Isotropic	—	Asymmetric	—	Spin $= \frac{5}{2}$; g vs. 3 E/D plot	Niarchos et al. (1978)

in polycrystalline solids. Draghicescu (1963) has reviewed anisotropy in ESR studies, but the contents of his paper are not readily available linguistically. Hughes and Rowland (1964) present numerical integrations of anisotropic Gaussian lineshapes. Lefebvre and Maruani (1965) developed a computer program for interpreting the ESR spectra of dilute free radicals in amorphous solids. Taylor and Bray (1970) have discussed the use of perturbation techniques in simulating powder spectra. Aasa and Vänngård (1964) obtained a correction factor that may be used to convert integrated areas to spin concentrations by taking into account the dependence of the transition probability on the g factor (Bleaney, 1960).

Burns (1961) and Weil and Hecht (1963) discuss the experimental determination of anisotropic g factors and zero-field splitting parameters from powdered samples. Weil and Anderson (1958) show how to evaluate completely anisotropic g factors from ESR data obtained at a series of orientations in the magnetic field. Scheffler and Stegmann (1963) determined g factors to a precision of $\sim 5 \times 10^{-6}$ (see also Bronstein and Volterra, 1965). Kneubühl and Natterer (1961) discuss the complex ESR lineshapes in anisotropic substances with fine structure ($S > \frac{1}{2}$, $D \neq 0$).

Shaltiel and Low (1961) discuss the effects of mosaic structure on anisotropic broadening in dilute single crystals. This broadening results from the various crystallites of the single crystal making slightly different angles with respect to the applied magnetic field direction. Shaltiel and Low derived a general formula for the lineshape and width, and estimated the average deviation of the crystallites from the crystal symmetry axis. See also Curtis et al. (1965). Lünd and Vänngård (1965) show how to evaluate fine and hyperfine constants in a single crystal with hyperfine structure and several molecules per unit cell. Schonland (1959) used a similar method to determine principal g values. Scott et al. (1965) present a study of rare earth double nitrates whose linewidths are strongly orientation dependent. Erdös (1964) published a theoretical discussion of anisotropic g factors. Lee (1978) discussed the powder pattern behavior of the electron spin echo modulation effect.

When hyperfine structure is present, a delta-function powder pattern can have extra singularities in addition to the ones discussed above. For example Miahle et al. (1971) showed that Mn^{2+} in an axially symmetric environment has the usual infinity at $\theta = \pi/2$ and another at the angle θ corresponding to the condition

$$\tan^2\theta = \frac{8\left[1 + (HA/H_0)M_I + (3A^2/4DH)M_I\right]}{1 - (GA^2/DH_0)M_I} \tag{3}$$

This permits the determination of the relative signs of the hyperfine (A) and zero-field (D) tensors. Other authors (Neiman and Kivelson, 1961; Gersmann and Swalen, 1962); Larin, 1968; Shteinschneider et al., 1970, 1972; Ovchinnikov and Konstantinov, 1978; Zhidomirov et al., 1975) have treated the presence of anomalous lines or singularities in powder patterns, and the latter article contains analytical solutions for such lines.

Mialhe et al. (1973, 1976) used an intensity spin operator to describe the line intensity angular dependence when crystal field and hyperfine interactions are comparable, and they applied their theoretical results to explain the spectra of Mn^{2+} in Al_2O_3.

M. Triplet States

Kottis and Lefebvre (1963) calculated the lineshape for randomly oriented molecules in a triplet state (see also Hudson and McLachlan, 1965). Lemaistre and Kottis (1978) carried out calculations of triplet lineshapes for molecular aggregates embedded in ordered lattices. They computed the temperature dependence of the lineshape for a dimer that is shown in Fig. 12-28. The dimer exhibits well-resolved structure at low temperatures due to quantum exchange narrowing, poorly resolved structure at intermediate temperatures arising from homogeneous and inhomogeneous broadening, and well-resolved structure at high temperatures due to classical exchange from hopping between sites. Lemaistre and Kottis (1978) also reported calculations for linear chains of various lengths, and they present a figure similar to Fig. 12-28 in which a linear chain of 10 molecules produces a sharp line spectrum at low temperature (5 K), a broad asymmetric line at intermediate temperatures (25 K), and a sharp singlet at high temperatures (50 K). Lineshape changes for shorter chain lengths are also shown.

Schouler et al. (1976) investigated dimerization of the free-radical nitrodisulphonate ion. Henling and McPherson (1977) studied coupled pairs of Gd^{3+} ions in the linear chain $CsMX_3$ lattice, where M is Mg or Cd and X is Cl or Br. Biradicals were studied by Corvaja and Pasimeni (1976), Mukai and Sakamoto (1978), and Parmon and Zhidomirov (1976); and Toy and Smith (1970) investigated a dimeric cupric complex.

Some recent articles on triplet states are by Bramwell et al. (1975), Dietrich et al. (1974, 1976), Estes et al. (1978), Freed et al. (1971), Furrer and Stehlik (1978), Grechishkin (1976), Hori and Kispert (1978), Dennis et al. (1975), Hudson (1976), Hutchison et al. (1970, 1974), Konzelmann et al. (1975), Kottis (1967), Mukai and Sakamoto (1978), Sharnof and Aiteb (1973), Villa and Hatfield (1970), Zieger and Wolf (1978), and Brenner et al. (1974).

Triplet state excitons have been investigated by Andreev and Sugakov (1976), Berim and Kessel (1977), Berk et al. (1981), Bizzaro et al. (1981), Botter et al. (1976), Burland et al. (1977), Dietrich and Schmid (1976), Frankevich et al. (1978), Hibma and Kommandeur (1975), Hinkel et al. (1978), Kawakubo (1979), Lemaistre and Zewail (1980), Marrone et al. (1973), Mavroyannis (1977), Mosina and Yablokov (1974), Nöthe et al. (1978), Reineker (1974, 1974, 1975, 1975, 1976, 1976, 1978), Shain (1972), Schmidberger and Wolf (1972), Schmidt et al. (1977), and Wasiela et al. (1974).

In a number of the articles that are cited above Reineker et al. expressed the Hamiltonian for triplet-state excitons in terms of creation and annihilation

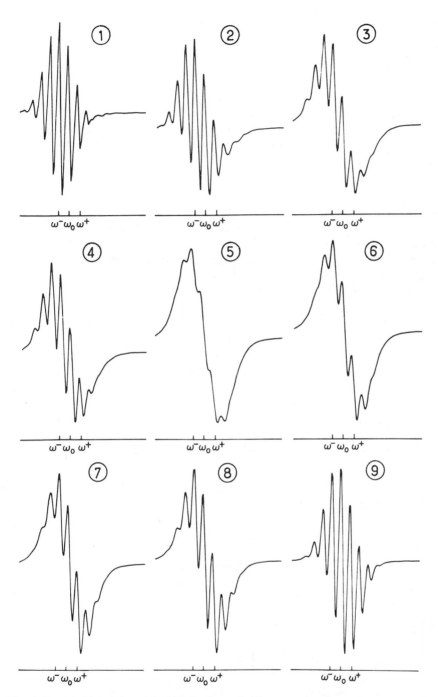

Fig. 12-28. Model variation of the ESR lineshape of a dimer as a function of the temperature T and a high-temperature limit relaxation rate U^0. The parameters used for these calculations are: $V = 1.25$ cm^{-1}; $\Delta = 5$ G and nine couples of values T and U^0 with $T_1 = 2$ K, $U_1^0 = 1$ G; $T_2 = 4$ K, $U_2^0 = 2$ G; $T_3 = 6$ K, $U_3^0 = 3$ G; $T_4 = 8$ K, $U_4^0 = 4$ G; $T_5 = 15$ K, $U_5^0 = 6$ G; $T_6 = 20$ K, $U_6^0 = 20$ G; $T_7 = 30$ K, $U_7^0 = 8$ G; $T_8 = 40$ K, $U_8^0 = 9$ G; $T_9 = 50$ K, $U_9^0 = 20$ G (Lemaistre and Kottis, 1978).

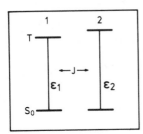

Fig. 12-29. Energy level scheme for exchange-coupled naphthalene pairs with different singlet–triplet excitation energies E_1
and E_2 (Hinkel et al., 1978).

operators b_1^+ and b_1, respectively, for a localized electron hole pair. As an
example of this work we consider the case of triplet naphthalene guest pairs in
a host crystal of perdeuterated naphthalene between 0.7 and 4.2 K. The two
naphthalenes have the ground-state singlet S_0 to triplet T excitation energies E_1
and E_2, respectively, and interact via the exchange interaction $J = 1.2$ cm^{-1},
as indicated in Fig. 12-29, and they have the Zeeman splitting between
Davydov levels illustrated in Fig. 12-30.

The temperature dependence of the ESR spectrum arising from the level pair
$E_{46} = E_6 - E_4$ and $E_{35} = E_3 - E_5$ was calculated in terms of the Haken
Strobol (1967, 1973) model for the coupled coherent and incoherent motion of
Frenkel excitations influenced by phonons in a stochastic manner (Haken and
Reineker, 1972). The ESR lineshapes illustrated in Fig. 12-31 are determined
by the parameter γ_0, which is a measure of the strength of the phonon-induced
energy fluctuations and increases with increasing temperature. The figure
presents plots of the imaginary part of the susceptibility χ'' as a function of the
frequency ω expressed in the units of gauss for various values of γ_0 expressed

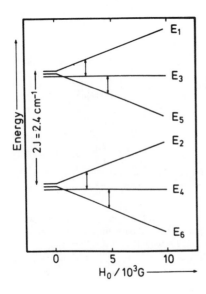

Fig. 12-30. Zeeman energy scheme of upper and
lower Davydov components. $\updownarrow$ indicates ESR transitions (Hinkel et al., 1978).

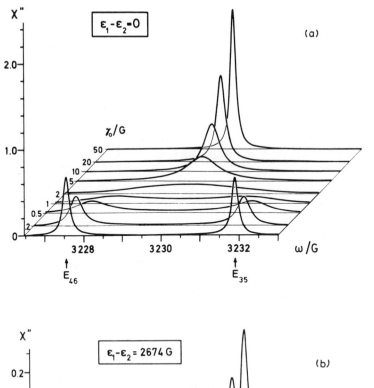

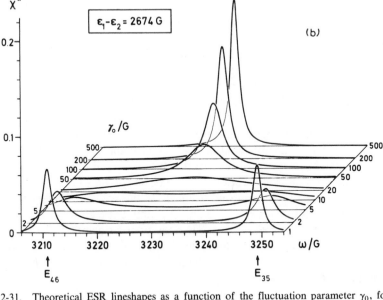

Fig. 12-31.   Theoretical ESR lineshapes as a function of the fluctuation parameter γ_0, for the (E_4-E_6) and (E_3-E_5) transitions of Fig. 12-30 (Hinkel et al., 1978).

in gauss, where

Frequency in GHz $= 1.4g$ (magnetic field in kilogauss)

and the g factor is 2.003.

The cases for equal ($\epsilon_1 = \epsilon_2$) and unequal ($\epsilon_1 \geqslant \epsilon_2$) triplet excitation energies are shown. We see from the figure that the spectrum is a doublet at lower temperature, where γ_0 is less than the energy separation ($E_{35}-E_{46}$), and a singlet at higher temperatures, where γ_0 is greater than ($E_{35}-E_{46}$).

Several authors have studied quartet states in three spin systems (Brickmann and Kothe, 1973; Kothe et al. 1974, 1978; Luckhurst and Zannoni, 1976; Reibisch et al., 1972), quintet states in four spin systems (Mackowiak, 1975), and coupled triplet states (Benk and Sixl, 1981; Huber et al., 1980, 1981; Schwoerer et al. 1981).

N. Amorphous and Semirandom Distributions

In several previous sections, we discussed the lineshapes found with randomly oriented microcrystallites of spin systems with definite Hamiltonian parameters. These situations occur with an ionic or molecular crystal in which the paramagnetic species has a definite site symmetry. There is another type of material, namely, a glassy or amorphous substance, in which the paramagnetic species is randomly oriented and in addition exhibits a range of Hamiltonian parameters. The powder spectrum of such a material is then an average over both angles and Hamiltonian parameters. As a result, different techniques are often applied for the calculation of the powder spectra of amorphous solids.

A simple example will illustrate how glassy samples produce spectra that differ from those of powders. Consider the case of an axially symmetric g factor in which the strongest absorption in the powder pattern occurs at the field value $H = H_\perp$, as indicated in Fig. 12-26. If the paramagnetic species is at a site of axial symmetry with a fixed value of $g_{\parallel}$ and a range of values of $g_\perp$, then the resonance absorption at $g_\perp$ will be spread over a range of field strengths, whereas the absorptions from all of the spins at the $g_{\parallel}$ point will superimpose and add. As a result, the most prominent feature on the spectrum will correspond to $g_{\parallel}$ instead of $g_\perp$. Thus the basic principle in the interpretation of spectra from glassy samples is the determination of those magnetic field values called stationary points where there is resonant absorption by spins with a wide range of site symmetries.

As an example of how the spectra from amorphous samples can be interpreted, consider the case of a ^6S transition ion such as Fe^{3+}, which has the Hamiltonian (Nicklin et al., 1976)

$$\mathcal{H} = g\beta \mathbf{H} \cdot \mathbf{S} + D\left[\mathbf{S}_z^2 - \tfrac{1}{3}S(S+1)\right] + E\left(S_x^2 - S_y^2\right) \qquad (1)$$

where g is isotropic, D ranges over values greater than the Zeeman term:

$$g\beta H < |D| < D_{max} \tag{2}$$

and the asymmetry parameter $\eta = 3E/D$ can range between 0 and 1:

$$0 \leqslant \eta \leqslant 1. \tag{3}$$

Since $S = 5/2$ for a ^6S state there are in general six energy levels, and the observed spectrum arises from various transitions between these levels. Energy absorption will occur at those magnetic field strengths H where the distance between two energy levels equals the microwave energy $h\nu$, and this provides an effective g factor

$$H = H_0 \left(\frac{g_0}{g} \right) \tag{4}$$

where the magnetic field H_0 corresponds to the free electron value g_0 through the expression

$$H_0 = \frac{h\nu}{g_0 \beta} \tag{5}$$

Associated with each value of D and η and with each orientation of the magnetic field at a fixed microwave frequency there is a series of transitions, each of which is characterized by an effective g value. After solving the Hamiltonian problem by diagonalizing the matrix associated with H of Eq. (1), a series of graphs can be made in which g_0/g is plotted against $D/h\nu$ for a fixed value of η and a particular orientation. Figure 12-32 illustrates such a graph for $\eta = 0.6$, the individual curves corresponding to the magnetic field aligned along the x, y, and z principle directions. A characteristic feature of this graph is the fact that most of the curves are fairly horizontal for values of D/h greater than unity. The g factors corresponding to such flat regions are called stationary values. Crystallites with a wide range of zero-field splittings will absorb microwaves at close to the same magnetic field positions corresponding to these stationary values. Plots such as that shown in Fig. 12-32 can be constructed for various values of η and for a series of orientations, and the regions where the curves are flat can be compared to experimentally observed g values to deduce the range of D and E values in the sample. Aasa (1970) has given graphs similar to Fig. 12-32 but of g/g_0 versus $D/h\nu$ on logarithmic scales.

A number of articles have been published that report the ESR spectra of amorphous samples, and these have been reviewed by Poole et al. (1981) and Poole and Farach (1980). Bales et al. (1975) used proton satellite lines (Livingston and Zeldes, 1956; Poole and Farach, 1971, 1971, 1972) measured

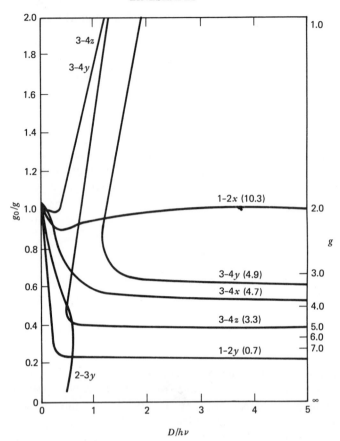

Fig. 12-32. Graph of some effective g values (right ordinate) for $\eta = 0.6$ plotted as g_0/g (left ordinate) versus $D/h\nu$. These graphs were constructed for the 1–2 and 3–4 transitions with the magnetic field H in the x principal direction, for the 1–2, 2–3, and 3–4 transitions with H in the y principal direction and for the 3–4 transition with H in the z principal direction. Each curve is labeled with its transition and the corresponding relative transition probabilities are given in parentheses for the region $3 < D/h\nu < 4$ (Nicklin et al., 1976).

at two microwave frequencies to separate the isotropic and anisotropic hyperfine constants in glassy systems.

Until now we have treated cases where paramagnetic species have their axes randomly oriented in direction. Sometimes the microcrystallites are not randomly oriented but, rather, have a preferred alignment or distribution $\rho(\delta)$ relative to a direction in the sample, where δ is the angle between the microcrystalline axis and the preferred direction. For this situation, ESR measurements can be carried out with the magnetic field set at various angles ψ relative to this preferred direction, and each spectrum $I(H, \psi)$ will be an

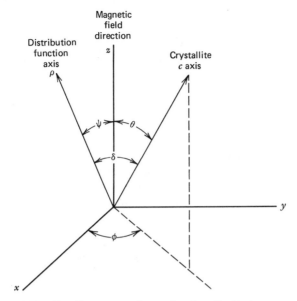

Fig. 12-33. Coordinate system for semirandom distribution calculation.

average over microcrystallite distributions, and hence powderlike in appearance. An analysis of a series of such spectra can provide the distribution function $\rho(\psi)$ (Poole and Farach, 1980). More complex distributions in terms of polar angles $\rho(\theta, \phi)$ or Euler angles $\rho(\alpha, \beta, \gamma)$ (Hentschel et al., 1978) can also be treated.

To describe the axial or cylindrically symmetric distribution case mathematically we follow Poole and Farach (1980) and select a Cartesian coordinate system with the magnetic field in the z direction, the preferred axis of the distribution function ρ in the xz plane with the polar angle ψ and the crystallite symmetry axis c in a general θ, ϕ direction, as shown in Fig. 12-33. We treat the specific case in which the g factor is $g_\parallel$ for the c-axis direction and $g_\perp$ at right angles to c. In this coordinate system the crystallite distribution function $\rho(\delta)$ is a function of the angle δ between the crystallite axis c and the preferred direction of ρ. The angle δ may be written in terms of θ, ϕ, and ψ:

$$\cos \delta = \cos \theta \cos \psi + \sin \theta \sin \psi \cos \phi \tag{6}$$

Eqs. (12K-19) and (12K-20) are combined, and if expressed in the form

$$\cos^2\theta = \frac{g_\parallel^2 H_\parallel^2 - g_\perp^2 H'^2}{H'^2\left(g_\parallel^2 - g_\perp^2\right)} \tag{7}$$

then $\cos \delta$ can be written

$$\cos \delta = \frac{\left(g_{\parallel}^2 H_{\parallel}^2 - g_{\perp}^2 H'^2 \right)^{1/2} \cos \psi + g_{\parallel} \left(H'^2 - H_{\parallel}^2 \right)^{1/2} \sin \psi \cos \phi}{H' \left(g_{\parallel}^2 - g_{\perp}^2 \right)^{1/2}} \qquad (8)$$

$$= A(H') + B(H') \cos \phi \qquad (9)$$

As a result the distribution function $\rho(\delta)$ may be treated as a function $\rho(H', \phi)$ of H' and ϕ since H' is a function of θ and ψ is a constant for each spectrum. Combining these various expressions and recalling Eq. (12K-23) gives the lineshape $I(H, \psi)$ for the magnetic field at an angle ψ relative to the preferred direction of the distribution function $\rho(\delta)$

$$I_{(H,\psi)} = \frac{H_{\parallel} \left(g_{\parallel}^2 - g_{\perp}^2 \right)^{1/2}}{2} \int_{H_{\parallel}}^{H_{\perp}} \frac{\left(H'^2 + H_{\parallel}^2 \right) Y_{(H')} \, dH'}{H'^2 \left(g_{\parallel}^2 H_{\parallel}^2 - g_{\perp}^2 H'^2 \right)^{1/2}} \int_0^{2\pi} \rho(H', \phi) \, d\phi$$

$$(10)$$

The ϕ dependence conveniently factors from the integral over H'.

A number of examples of the magnetic resonance study of semirandom distributions exist in the literature. For example, McDowell et al. (1973) reported the ESR of ClO_2 molecules preferentially oriented in a rare gas matrix at 4.2° with the molecular plane parallel to the deposition surface, and other paramagnetic species are similarly oriented (Kasai et al., 1971; Kasai et al., 1966; Knight et al., 1971; Korst et al., 1972). Panepucci and Farach (1977) found the apatite microcrystallites in irradiated bone to be preferentially aligned along the bone axis. Blazha et al. (1973) were able to reconstruct the semirandom defect distribution around a paramagnetic center from an analysis of the lineshape. See also Hentschel et al. (1978), Korst et al. (1972), Libertini et al. (1969, 1974), and McDowell et al. (1974).

O. Intermediate Motion Regime in Liquids

High-viscosity liquids constitute a transition region in which the spectral features are intermediate between the rigid lattice and fluid cases, and they vary continuously from one to the other as the viscosity decreases. A number of authors have proposed theoretical treatments of this intermediate motion region (Anderson and Ullman, 1967; Atkins and Kivelson, 1966; Dalton et al., 1975; Kivelson, 1957; Kivelson et al., 1970; Korst et al., 1964, 1978; Moriuchi and Sohma, 1974; Reiter, 1974; Sames, 1967; Sillescu and Kivelson, 1968; Thomas and McConnell, 1974; Wilson and Kivelson, 1966), and we shall be content to discuss some general features of it following the treatment in Poole and Farach (1980). Bolton (1972), Sullivan (1974, 1978), Allendoerfer (1980), and Stevenson (1980) have reviewed solution ESR. See also Kubo and Tomita (1954).

The Brownian motion of a liquid is characterized by a correlation time τ that is a measure of the rapidity of the motion. In classical dielectric relaxation experiments, the molecules are treated as spheres of radius a undergoing isotropic rotational reorientation in a liquid of viscosity η, and the resulting Debye correlation time τ_D varies inversely with the temperature in accordance with the Einstein–Stokes relationship

$$\tau_D = \frac{1}{2D} \tag{1}$$

$$= \frac{4\pi\eta a^3}{kT} \tag{2}$$

where D is the rotational diffusion constant (Poole and Farach, 1971). The corresponding correlation time τ_c for the motion of the magnetic moments in an ESR experiment is one third of the classical Debye value

$$\tau_D = 3\tau_c \tag{3}$$

corresponding to

$$\tau_c = \frac{1}{6D} = \frac{4\pi\eta a^3}{3kT} \tag{4}$$

From a more general viewpoint, if we consider τ_c as the correlation time for random perturbations involved in ESR measurements and τ_D as its dielectric relaxation counterpart, then τ_c is related to τ_D through the expression

$$\tau_D = f\tau_c \tag{5}$$

where f varies between 1 and 3 and assumes the limiting value of 3 for the isotropic Brownian motion model mentioned above. If the random motion is described by a jump model wherein the molecules reorient via large angular shifts of random length, then τ_c and τ_D are equal, that is, $f = 1$ (see also Alexander et al., 1977). Figure 12-34 shows the ranges over which various techniques may be employed to measure rotational correlation times for different sizes of molecules (Thomas, 1978).

To see the significance of the correlation time τ_c, consider the situation wherein a random field $H(t)$ that perturbs the motion of a spin at a time t is exponentially relaxed via the expression $\exp(-t/\tau_c)$ to its initial value at $t = 0$. The situation just described may be expressed analytically by writing the following autocorrelation function $\langle H_\alpha(t + \tau)H_\alpha(t)\rangle$, which is a time average of one of the components H_x, H_y, or H_z of $H(t)$.

$$\langle H_\alpha(t + \tau)H_\alpha(t)\rangle = \langle H_\alpha^2\rangle\exp\left(-\frac{|t|}{\tau_c}\right). \tag{6}$$

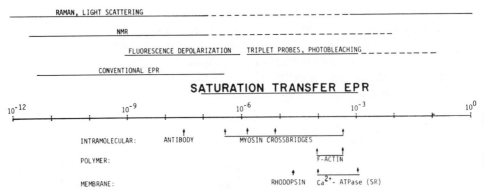

Fig. 12-34. Approximate ranges of rotational correlation times τ_c in seconds where various techniques can be employed to measure them. Several biological molecules are indicated (Thomas, 1978).

In this simple case, the spin-lattice (T_1) and spin–spin (T_2) relaxation times are

$$\frac{1}{T_1} = \gamma^2 \left[\langle H_x^2 \rangle + \langle H_y^2 \rangle \right] \frac{\tau_c}{1 + \omega_0^2 \tau_c^2} \tag{7}$$

$$\frac{1}{T_2} = \gamma^2 \left[\langle H_z^2 \rangle \tau_c + \tfrac{1}{2} \left(\langle H_x^2 \rangle + \langle H_y^2 \rangle \right) \frac{\tau_c}{1 + \omega_0^2 \tau_c^2} \right] \tag{8}$$

This gives for the linewidth

$$\Delta \omega_{1/2} = \gamma \Delta H_{1/2} = \frac{2}{T_2} \tag{9}$$

where the perturbing fields generally are isotropic:

$$\langle H_x^2 \rangle = \langle H_y^2 \rangle = \langle H_z^2 \rangle$$

There are two limiting cases:

$$\begin{aligned} T_1 &= T_2 \quad \omega_0 \tau_c \ll 1 \\ T_1 &\gg T_2 \quad \omega_0 \tau_c \gg 1 \end{aligned} \tag{10}$$

the former being typical of low-viscosity liquids and the latter of solids.

The discussion thus far has been for a single symmetrical resonant line whose width varies in accordance with the factor $\omega_0 \tau_c$. Of greater interest is a spectrum arising from asymmetric Hamiltonian tensors, and the shape of such a spectrum is characterized by the value of the parameter $2\tau_c / T_2$ or $\tau_c \Delta \omega_{1/2}$. In the slow-motion regime corresponding to $\tau_c \Delta \omega_{1/2} \gg 1$, a rigid lattice powder pattern is obtained, while in the opposite or rapid motion limit

$\tau_c \Delta\omega_{1/2} \ll 1$, the powder pattern collapses to a singlet characteristic of a liquid. At intermediate rates of reorientation where $\tau_c \Delta\omega_{1/2} \simeq 1$, the lineshape is a poorly resolved or partially collapsed powder pattern.

Sillescu (1971) examined these various cases in terms of a fluctuating time-dependent axially symmetric Zeeman Hamiltonian:

$$H(t) = \beta\left[g_\perp \left(S_x H_x + S_y H_y \right) + g_{\parallel} S_z H_z \right] \tag{11}$$

If one adopts the notation

$$g_0 = \tfrac{1}{3}\left(g_{\parallel} + 2g_\perp \right) \tag{12}$$

$$\Delta g = \left(g_{\parallel} - g_\perp \right) \tag{13}$$

then in the limit of $|\Delta g| \ll g_0$ the Hamiltonian may be written in the form

$$H(t) = g_0 \beta H S_z + \tfrac{1}{3}\Delta g\left[3\cos^2\theta(t) - 1\right]\beta H S_z \tag{14}$$

where the time dependence arises from random local fluctuations of the angle $\theta(t)$ at a rate characterized by the correlation time τ_c. If $\tau_c \Delta\omega_{1/2}$ greatly exceeds unity, then very little averaging occurs and the observed spectrum is a powder pattern of the type illustrated in Figs. 12-27 and 12-35. In the opposite limit when $\tau_c \Delta\omega_{1/2}$ is much less than unity, then the factor $[3\cos^2\theta(t) - 1]$ almost completely averages out to produce the slightly asymmetric singlet illustrated in Fig. 12-36. The intermediate case where $\tau_c \Delta\omega_{1/2} \simeq 1$ corresponds

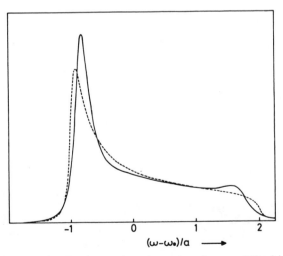

$(\omega - \omega_0)/a \longrightarrow$

Fig. 12-35.   ESR spectrum in the slow reorientation region $\tau\Delta\omega_{1/2} = 200$ arising from a slightly averaged, fluctuating, axially symmetric Zeeman Hamiltonian: (——) Brownian-motion model, (---) jump model (Sillescu, 1971).

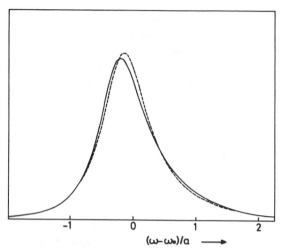

Fig. 12-36. ESR spectrum in the rapid reorientation region $\tau_c \Delta\omega_{1/2} = 0.6$ arising from an almost completely averaged fluctuating axially symmetric Zeeman Hamiltonian: (——) Brownian-motion model, (– – –) jump model [from Sillescu (1971)].

to a partially averaged but still recognizable powder pattern as presented in Fig. 12-37. These figures show the effect that the random fluctuation mechanism has on the lineshape by presenting curves for the rotational Brownian motion model ($\tau_c = \tau_D/3$) and also for the rotational random jump model ($\tau_c = \tau_D$) mentioned above. We see from the figures that for the same value of $\tau_c \Delta\omega_{1/2}$ the Brownian motion model leaves the powder pattern features somewhat more prominent.

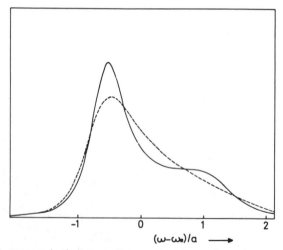

Fig. 12-37. ESR spectrum in the intermediate reorientation region $\tau_c \Delta\omega_{1/2} = 4$ arising from a partially averaged, fluctuating axially symmetric Zeeman Hamiltonian: (——) Brownian-motion model, (– – –) jump model [from Sillescu (1971)].

In the slow-motion region, which is generally taken as the range of correlation times 10^{-9} sec $< \tau_c < 10^{-3}$ sec, partially averaged powder pattern spectra are observed. For shorter times, τ_c becomes of the order of the period of the microwaves ($\omega_0 \tau_c \simeq 1$), and for even shorter times, $\omega_0 \tau_c \ll 1$, as explained above. For longer correlation times $\tau_c > 10^{-3}$ sec, the rigid lattice limit is approached.

A number of authors have treated the completely anisotropic and axially symmetric g-factor lineshapes at intermediate correlation times. Some of the calculations used specialized mathematical techniques such as a continued fraction expansion (Misra et al., 1975), graphical methods (Friedman et al., 1976), projection operators (Atkins and Hills, 1975), Monte Carlo (Pederson, 1972) and Markov (Freed 1977; Sillescu and Kivelson, 1968) procedures, and in some cases hyperfine interactions were included (Antsiferova et al., 1970, 1972; Alexander et al., 1974; Beram et al., 1976; Goldman et al., 1972, 1972, 1973, see also Baram, 1979).

A considerable amount of experimental work has been carried out studying ESR in solution, and recent investigations are listed in the reviews mentioned at the beginning of this section. Some typical work has been done, for example, with vanadyl complexes (Eagles and McClung, 1975; Hoel and Kivelson, 1975, 1975; Kowert and Kivelson, 1976; Stoklosa and Wasson, 1976) and with solutions of $3d^5$ ions such as Mn^{2+} (Burlamacchi, 1971, 1976; Burlamacchi and Romanelli, 1971; Culvahouse and Francis, 1977; Reed et al., 1971; Reichert et al., 1979; Strother et al., 1971; Suchard and Bowers, 1972). Bruno and Freed (1974) used ELDOR to study lineshapes in the slow motional region. Very slow motions such as in plastic cyclohexane (Volino and Rousseau, 1972) have also been investigated.

P. Concentration and Temperature Dependence

Free radicals and transition ions in solution have linewidths that tend to increase linearly with concentration C, from a residual value ΔH_0 at zero concentration in accordance with the expression

$$\Delta H_{1/2} = \Delta H_0 + KC \tag{1}$$

where ΔH_0 and K are functions of the temperature. When the concentration becomes large enough, exchange effects become appreciable, and at high concentrations they cause the lines to narrow.

Ayant (1975, 1976) studied the deuterated tanone radical (tetramethyl 2,2,6,6,-peperidone-4, oxide), which gives a triplet hyperfine pattern owing to the nuclear spin $I = 1$ of the nitrogen. He found that the concentration-independent width ΔH_0 has the following η/T dependence:

$$\Delta H_0 = \frac{aT}{\eta} + b + (A + BM + FM^2)\frac{\eta}{T} \tag{2}$$

where η is the viscosity and T is the temperature. The first term aT/η arises from the spin rotational interaction, the second term b is due to unresolved hyperfine interactions and instrumental effects, and the third one, which depends upon the hyperfine component line through M, will be discussed in the next section. The concentration-dependent width factor K of Eq. (1):

$$K = d\frac{\eta}{T} + e\frac{T}{\eta}\phi\left(\frac{T}{\eta}\right) \qquad (3)$$

has a dipolar part $d\eta/T$ and an exchange part that depends upon the temperature through a complicated function $\phi(T/\eta)$, which will not be repeated here. Ayant (1976) and Ayant et al. (1975) found satisfactory agreement with experiment from 183 to 363 K for concentrations from 0.2×10^{-3} to $4 \times 10^{-3} M$/liter. Anisimov et al. (1971) found that Eq. (1) was satisfied for two piperidine-type radicals when the lines were broadened by the paramagnetic nitrates of iron-group transition ions, with cadmium nitrate used as a blank to eliminate nonparamagnetic effects.

A number of investigations have been made of the concentration dependence of the linewidth, and we quote a few recent representative articles. Measurements have been made in aqueous solution (Anisimov et al., 1971; Hwang et al., 1975; Levanon et al., 1970; Parmon et al., 1977; Strother et al., 1971; Urban et al., 1975; Vishnevskaya et al., 1975), in glycerol solution (Anisimov et al., 1971; Hwang et al., 1975; Levanon et al., 1970; Parmon et al., 1977), in methanol solution (Ayant, 1976; Ayant et al., 1975; Burlamacchi et al., 1970, 1973), in other organic solvents (Ayant, 1976; Ayant et al., 1975; Burlamacchi et al., 1970, 1973; Haupt et al., 1962; Hwang et al., 1975; Kowert and Kivelson, 1976), and in an inorganic liquid melt (Tanemoto, 1978). Studies in solids have included glasses (Bales et al., 1974; Friebele et al., 1972; Hwang et al., 1975; Kumeda et al., 1977; Onda and Morigaki, 1973; Zamaraev et al., 1978), layers and surfaces (Muha and Sakamoto, 1978; Murakami et al., 1978; Nagai et al., 1972; Poole et al., 1962, 1964, 1967; Shiga et al., 1971), monocrystals (Bugai et al., 1975; Gillbro et al., 1975; Morigaki et al., 1971; Muramoto, 1973; Richards, 1976; Shul'man et al., 1971, 1972; Teodorescu et al., 1977), polycrystals (Burzo et al., 1978; Poole et al., 1964, 1967; Shiraishi et al., 1976), and miscellaneous solids (Chovino et al., 1971; Koopmann et al., 1973; Nguyen, 1977; Ochiai and Matsuura, 1976; Rettori et al., 1973). The concentration dependence of ESR lineshapes has been reviewed by Poole and Farach (1980).

Variable-temperature linewidth studies have been carried out with unpaired spins, free radicals, and transition ions in various solids such as monocrystals (Ajiro et al., 1975; Byberg et al., 1974; Clerjaud et al., 1970; Henner et al., 1978; Kemple and Stapleton, 1972; Lahiry et al., 1978; McMillan and Marshall, 1968; Misra et al., 1973; Muramoto, 1973; Owens, 1972; Raoux, 1975; Stoneham et al., 1972; Teodorescu, 1977), polycrystals and powders (Sperlich et al., 1974; Stasz, 1977), amorphous solids (Gewinner et al., 1975; Kubler et al., 1976; Misra and Sharma, 1975; Watanabe et al., 1976), semiconductors (Arizumi et al., 1972; Brodsky and Title, 1976; Feichtinger et al., 1978; Kirillov et al., 1976; Kishimoto et al., 1977; Kumeda et al., 1977; Lepine, 1970; Ochiai and Matsuura, 1978; Sugihara, 1975; Thomas and Kaplan, 1976; Voget-Grote et al., 1976), and surfaces of solids (Hasegawa and Yazaki, 1978; Muha and Sakamoto, 1978; Nagai et al., 1972; Poole et al., 1962, 1967), in ordered solids (Angelov et al., 1977; Huber and Seehra, 1975, 1976; Chicault et al., 1977; Nazarian et al., 1972; Owens, 1977, Seehra and Gupta, 1974; Sugano and Kuroda, 1977; Svare and Seidel, 1964; Tanaka and Kondo, 1976) and in other solids (Baranov et al., 1976; Boate et al., 1976; Dormann et al., 1977; Ehrenfreund et al., 1977; Koopmann et al., 1973; Korolev et al., 1975; Servant, 1975; Stoppels et al., 1976; Ursu and Burzo, 1972; Vittoria et al., 1978). Poole and Farach (1980) have tabulated the ESR results of recent temperature-dependent studies.

Some pressure-dependent linewidth studies have been reported. For example, Filippov and Donskaya (1972, 1973) studied the spectra of VO^{2+}, Cr^{3+}, and Mn^{2+} in aqueous solutions up to 6000 atm, and Grouchulski et al. (1978) investigated the temperature dependence of MnTe at 1, 2, and 4 kbars. Kozhukhar and Tsintsadze (1975) measured the pressure dependence of the linewidth of $NiSiF_6 \cdot 6H_2O$ up to 12 kbars at 7 K, and they found a sharp reduction in the width at the pressure at which the trigonal component of the crystal field vanishes. Kozhurkhar et al. (1978) studied the influence of hydrostatic compression on Tb^{3+} in lanthanium ethylsulphate crystals. Olson and Castner (1978) determined the effect of unaxial stress on both absorption and dispersion signals.

In recent years a number of experimental studies have been made of temperature-dependent linewidths and lineshapes in fluid systems; we mention some typical ones. Some linewidth studies have been reported with gases (e.g., Oluwole et al., 1973). Investigations have been carried out in various solvents such as water (Strother et al., 1971; Wilson et al., 1976; Antsiferova et al., 1970; Vishnevskaya et al., 1975), alcohol (Burlamacchi et al., 1970, 1971; Hausser, 1961; Hoel and Kivelson, 1975; Kowert and Kivelson, 1976; Aasa, 1970; Abdrakhmanov et al., 1978; Leblond et al., 1971, 1971; Poupko, 1973; Rengan et al., 1972; Romanelli and Burlamacchi, 1976; Sargent and Grady, 1976), dimethoxysulphoxide (Burlamacchi et al., 1970; Garrett et al., 1966; Romanelli and Burlamacchi, 1976), dimethyl formamide (Burlamacchi et al., 1970, 1971; Kaminski, 1976; Romanelli and Burlamacchi, 1976; Subramanian and Narasimhan, 1972), tetrahydrofuran (Haupt et al., 1962; Kooser et al., 1969), other organic solvents (Burlamacchi et al., 1970, 1971, 1971; Garrett et al., 1966; Kooser et al., 1969; Misra et al., 1976, 1976; Romanelli and Burlamacchi, 1976; Van et al., 1974), and the melt of the sample itself (Dormann et al., 1977; Tanemoto et al., 1978). The subject has been reviewed by Poole and Farach (1980).

Q. Hyperfine Components in Solution

Hyperfine multiplets arising from radicals or transition ions in solution often exhibit a linewidth that varies with the magnetic quantum number M from one component to the next across the spectrum in accordance with the expression:

$$\Delta H_{1/2} = \Delta H_0 + A + BM + FM^2 \qquad (1)$$

which was derived theoretically and checked experimentally by Kivelson (1960, 1966), Atkins and Kivelson (1966), and Wilson and Kivelson (1966). The residual linewidth ΔH_0 generally arises from spin–rotational interactions (Hoel and Kivelson, 1975).

When the line broadening arises from incomplete averaging of the g factor and hyperfine coupling interactions in the limit of rapid tumbling in strong magnetic fields (e.g., 10^3 or 10^4 G), the terms A, B, and F are given by (Poole and Farach, 1972)

$$A = \frac{2}{15}(\Delta g)^2 \left(\frac{\beta H_0}{h}\right)^2 \tau_c + \frac{1}{20}(\Delta a)^2 I(I+1)\tau_c \qquad (2)$$

$$B = -\frac{4}{15}\left(\frac{\beta H_0}{h}\right)(\Delta g \, \Delta a)\tau_c \qquad (3)$$

$$F = \frac{1}{12}(\Delta a)^2 \tau_c \qquad (4)$$

where the factors containing g and a arise from the g-factor and hyperfine tensor principal values relative to their respective isotopic parts g_0 and a_0:

$$(\Delta a)^2 = (a_{xx} - a_0)^2 + (a_{yy} - a_0)^2 + (a_{zz} - a_0)^2 \qquad (5)$$

$$(\Delta g \, \Delta a) = (a_{xx} - a_0)(g_{xx} - g_0) + (a_{yy} - a_0)(g_{yy} - g_0)$$

$$+ (a_{zz} - a_0)(g_{zz} - g_0) \qquad (6)$$

Other interactions can also contribute to A, B, and F. The factor $(\beta H_0/h)$ causes the widths to depend upon the frequency.

The term BM gives the spectrum an unsymmetrical appearance. A more refined theory adds the terms GM^3 and JM^4 to Eq. (1), but these are usually small. For rotational diffusion in the Debye approximation τ_c is given by Eq. (12-O-4). Eagles and McClung (1975) studied vanadyl diketonate complexes and found deviations from the Kivelson linewidth theory at low temperatures and high viscosities.

Equations (1) to (6) were derived for the high-field case typical of X-band (3400 G, 10^{10} MHz) spectra. Barbarin and Germain (1975) derived analogous expressions for low fields (30 G, 80 MHz) and carried out measurements at the high- and low-frequency bands in order to evaluate the contribution to the linewidth arising from the g-factor and hyperfine anisotropies.

Poole and Farach (1980) have tabulated the results of a number of studies involving hyperfine component-dependent linewidths.

For the case of an electronic spin S greater than $\frac{1}{2}$ the linewidths and intensities of the various fine structure transitions will differ as a result of the M-dependent incomplete averaging of the zero field anisotropy tensor D. The widths and intensities depend upon the inner product Δd

$$(\Delta d)^2 = \tfrac{2}{3}(D^2 + 3E^2) \tag{7}$$

and the correlation time τ_c. Burlamacchi (1971) calculated $\Delta H/(\Delta d)^2$ and the intensity as a function of $\omega_0 \tau_c$ for these transitions, and the resulting plots are presented in Fig. 12-38. We see from these plots that for $\omega \tau_c \ll 1$ the total intensity converges to the central $|\frac{1}{2}\rangle \leftrightarrow |-\frac{1}{2}\rangle$ transition. For $\omega_0 \tau_c \gg 1$ the other transitions broaden beyond detectability, and only the $|\frac{1}{2}\rangle \leftrightarrow |-\frac{1}{2}\rangle$

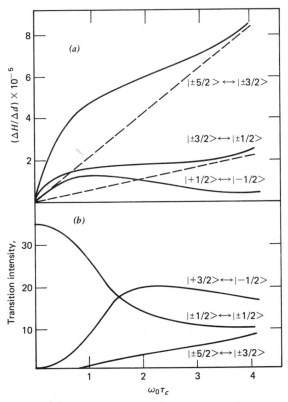

Fig. 12-38. Theoretical curves for an $S = 5/2$ sextet electron spin state as a function of $\omega \tau_c$: (a) linewidth ΔH normalized relative to zero-field D term, (b) signal intensity. The transition corresponding to each curve is indicated (Burlamacchi, 1971).

transition is observed. Near $\omega_0 \tau_c \simeq 1$ the central transition reaches a maximum in width, and near $\omega_0 \tau_c \simeq 1.5$–2 the $| + \frac{3}{2} \rangle \leftrightarrow | \pm \frac{1}{2} \rangle$ transition makes a partial contribution to the observed spectra. Burlamacchi (1971), Burlamacchi et al. (1971, 1978), and Romanelli and Burlamacchi (1976) checked this theory experimentally with Fe^{3+} and Mn^{2+} in various solvents over a wide range of temperature and found satisfactory agreement (see also Garrett et al., 1966; Misra et al., 1975).

Reed et al. (1971) calculated the linewidths and transition probabilities of the fine-structure transitions as eigenvalues and eigenvectors, respectively, of a relaxation matrix and obtained curves that resembled those of Fig. 12-38. They found that an individual Mn^{2+} solution spectrum could be fit with a range of τ_c and $(D^2 + 3E^2)$ values, but that the simultaneous fit of X-band (9.1 GHz) and Q-band (35 GHz) spectra from the same solution provide unique values of τ_c and $(D^2 + 3E^2)$.

R. Alternating Linewidths

Freed and Fraenkel (1962) considered linewidths arising from hyperfine patterns in which nuclei are symmetrically equivalent but not completely equivalent. By incompletely equivalent they mean equivalent with respect to the zero-order Hamiltonian so that they have the same gyromagnetic ratios ($\gamma_i = \gamma_j$), spins ($I_i = I_j$), and hyperfine interaction constants ($A_i = A_j$), but inequivalent with respect to a perturbing Hamiltonian such as the dipolar interaction ($H_{ddi} \neq H_{ddj}$). Completely equivalent nuclei also have the same perturbing Hamiltonian (e.g., $H_{ddi} = H_{ddj}$).

The linewidth $\Delta H_{1/2}$ for the general or incompletely equivalent case is given by

$$\Delta H_{1/2} = \sum_{i,j} J_{ij} M_i M_j \tag{1}$$

$$= J_{11}(M_1^2 + M_2^2) + 2 J_{12} M_1 M_2 \tag{2}$$

since $J_{11} = J_{22}$ and $J_{12} = J_{21}$ from the symmetrical equivalence of the nuclei. Each hyperfine line has a degeneracy of 1 when $M_1 = M_2$ and 2 when $M_1 \neq M_2$, since in the latter case (M_1, M_2) and (M_2, M_1) give superimposed lines.

There are three special cases, namely, the in-phase case ($J_{12} = J_{11}$) when the nuclei are completely equivalent, the out-of-phase case ($J_{12} = -J_{11}$), and the uncorrelated case ($J_{12} = 0$) with the following respective linewidths:

$$\Delta H_{1/2} = J_{11}(M_1 + M_2)^2 \quad J_{12} = J_{11} \text{ in phase}$$

$$\Delta H_{1/2} = J_{11}(M_1 - M_2)^2 \quad J_{12} = -J_{11} \text{ out of phase} \tag{3}$$

$$\Delta H_{1/2} = J_{11}(M_1^2 + M_2^2) \quad J_{12} = 0 \text{ uncorrelated}$$

Thus the general case produces linewidths that vary in a complex manner from one hyperfine component to another. For some choices of the ratio J_{12}/J_{11} certain hyperfine components have very narrow widths and others are broad. The theory explains some observed spectra in which the widths tend to alternate wide and narrow between successive components, and other observed spectra in which the various amplitudes deviate from the binomial case owing to the broadening of particular components beyond detection.

Examples of articles that reported alternating linewidth effects are Bolton et Carrington (1962), DeBoer (1964), Freed and Fraenkel (1962), Freed et al. (1962), Iwaizumi and Isobe (1965), and Pijpers et al. (1971). Froncisz and Hyde (1980) explained the M_I dependence of the linewidths of Cu^{2+} in glassy spectra on the microwave frequency for the three frequencies 2.6, 3.8, and 8.9 GHz.

S. Spin Labels

Nitroxides and other stable free radicals serve as probes called spin labels or spin probes, which attach themselves to or occupy positions near active sites in proteins or other molecules. They produce spectra whose linewidths and lineshapes monitor the extent to which molecular motion averages anisotropies in the local environment. In addition the g-factor and hyperfine coupling constant are sensitive to the hydrophobic or hydrophilic nature of the environment.

The most commonly used spin labels are nitroxides that produce a three-line hyperfine pattern due to the unpaired electron on the nitrogen atom of the $\cdot NO$ group. An example of a $I = 1$ spin label is 2, 1, 2, 1, 6, 6,-tetramethyl-4-piperidone-1-oxyl:

In general both g and A are anisotropic with the following typical principal values:

$$g_x = 2.008 \quad A_x = 5.5G$$

$$g_y = 2.006 \quad A_y = 4.0G$$

$$g_z = 2.003 \quad A_z = 30.0G \tag{1}$$

Motional effects on nitroxide spectra were analyzed by Freed (1965) in the slow tumbling region where the correlation time τ_c is in the range 10^{-9} sec $< \tau_c < 10^{-6}$ sec as was discussed above in Sec. 12-O. When the rotational diffusion constant D is isotropic, τ_c may be approximated by the Einstein–Stokes relationship (12-O-2). For peroxylamine disulphonate D is axially symmetric with the value $D_\parallel$ along the y direction and $D_\perp$ in the xz plane, and the rotational correlation time τ_R, which is analogous to τ_c in Eq. (12-O-4), is given by (Freed, 1965, 1976)

$$\tau_R = \frac{1}{6\left(D_\parallel D_\perp\right)^{1/2}} \tag{2}$$

There is quite an extensive literature on this subject, and some general and review articles on spin labels, their theory and practice, have been written: Axel (1976), Berlinger (1976), Dugas (1973), Swartz (1972), Libertini et al. (1974), Luckhurst (1976), McConnell and McFarland (1970), Parmon et al. (1977), and Thomas and McConnell (1974). Some typical articles were written by Ahn (1976), Balaban et al. (1973), Benton et al. (1975), Berger (1978), Berlinger (1976), Bullock and Cameron (1976), Coffey et al. (1975, 1978), Danna and Wharton (1970), Freed (1976, 1976), Gaffney (1976), Gaffney and Lin (1976), Grant (1975), Hemminga (1975), Hensen et al. (1974), Hyde et al. (1972, 1974, 1975), Kirillov et al. (1975, 1976), Kolt'tover (1975), Kuznetsov et al. (1974), Lazjzerowicz-Bonneteau (1976), Leyko et al. (1978), Libertini et al. (1969), Lin and Freed (1979), Livshits (1976), Luckhurst et al. (1976, 1976), Mason et al. (1974), McConnell and McFarland (1970), Nagamura et al. (1976), Nordio (1975), Parmon et al. (1975), Pilo-Veloso et al. (1976), Poggi and Johnson (1970), Polnaszek and Freed (1975), Poole and Farach (1980), Püsnik and Schara (1976), Reddoch and Konishi (1979), Schaafsma (1975), Schindler and Seelig (1974), Seelig (1976), Shiotani and Sohma (1974), Smigel et al. (1974), Smith (1972) Smith et al. (1976), Strykov et al. (1971, 1974), Taylor et al. (1978), Tigyi and Hideg (1976), Tormala et al. (1976), Van et al. (1974), and Vignais and Devaux (1976). Particularly interesting studies can be carried out near phase transitions (e.g., Marušič and Schwerdtfeger, 1973). The linewidths and lineshapes in the ESR spectra of various spin labels were analyzed by several authors (e.g., Freed, 1977; Kothe, 1977; Kothe et al., 1976; Nordio, 1970; Nordio et al., 1972). The technique of ENDOR (compare Sec. 14-A) has also been used with spin labels (e.g., see Mottley et al., 1975; Thomann et al., 1980; Vieth and Hausser, 1972, 1974). Popp and Hyde (1981) investigated the effect of oxygen on spin probe work, and they recommend the use of deoxygenated samples.

T. Phase Transitions

Electron spin resonance is a favorable method for studying phase transitions because ESR spectra often undergo pronounced changes during passage through the transition temperature (Owens et al., 1979). The parameters that can change near the transition point are the lineshape, the linewidth, the intensity, and the relaxation time. Seehra and Sturm (1975) mention three experimental factors to be avoided in measuring intrinsic linewidths near magnetic transition temperatures: (1) cavity overloading (compare Sec. 11-J), (2) irregularities in sample broadening due to a nonuniform demagnetization factor, and (3) temperature dependence of the demagnetization factor near the transition. Some recent typical articles will be quoted on phase transitions involving ferromagnets: Angelov et al. (1977), Babushkin et al. (1975), Bazuev et al. (1977), Cochrane et al. (1978), Huber and Seehra (1975, 1976), Janossy et al. (1974), Miyadai et al. (1971), Scheithe et al. (1978), Seehra et al. (1975, 1977), Shrivastava (1975). Shumilkina and Obraztsov (1975), Stasz (1977, 1978), Sugano and Kuroda (1977), Taylor and Coles (1972), and Tomash-polskii (1974); antiferromagnetic specimens: Anders and Zvyagin (1973), Gulley and Jaccarino (1972), Gupta et al. (1970, 1972), Huber et al. (1974), Kunii and Kasuya (1979), Maekawa (1972), Poole and Itzel (1964), Seehra (1972), Seehra et al. (1970, 1974, 1975, 1976), Tanaka and Kondo (1973, 1975), and Tornero et al. (1975); ferroelectric materials: Buzare and Fayet (1977), Kool and Glasbeek (1977), Lamotte and Gaillard (1972), Lippe et al. (1976), Müller et al. (1972, 1974), Owens (1978), Truesdale et al. (1980, 1982), Von Waldkirch et al. (1972, 1973), Windsch (1976), review by Müller and von Waldkirch (1975); solid-to-liquid transitions: Dormann et al. (1977), Morantz and Thompson (1970); semiconductor-to-metal or metal-to-insulator transitions: Houlihan et al. (1974, 1975, 1976), Sugano and Kuroda (1977), Schlenker et al. (1972); transitions at high pressure: Larys et al. (1975); and others: Friederich et al. (1978), Hirokawa (1978), Krupski and Stankowski (1976), Lang et al. (1977), Möhwald and Sackmann (1974).

Some of the above studies were carried out in the neighborhood of the transition temperature were pronounced, discontinuous variations in width and shape can occur [see also Campbell and Lawson (1973), Charles et al. (1973), Gupta and Seehra (1970), Houlihan and Mulay (1974), Huber et al. (1974), Krupski and Stankowski (1976), Lang et al. (1977), Möhwald and Sackmann (1974), Hirokawa (1978), Seehra (1971, 1972), Stasz et al. (1976), and Velichko and Gusev (1976)].

ESR studies in the neighborhood of phase transitions have provided critical exponents through parameters that depend upon the temperature with laws of the type $| T - T_c |^p$. For example, near T_c one can write

$$\Delta H \sim K \, | T - T_0 |^p$$

for magnetic transitions, and various authors have predicted values of p (e.g., $= \frac{3}{2}$) for three-dimensional systems (Kawasaki, 1968; Mori, 1963; Mori and Kawasaki, 1962; Seehra, 1971) and low-dimensional systems [Richards (1973), $p = -1.5$, Huber and Seehra, (1973), $p = -2.5$]. Experimental measurements have provided values ranging from -0.4 to -2.5 for three-dimensional systems (Angelov 1972; Battles, 1971; Burgiel and Strandberg, 1965; Dillon and Remeika, 1967; Le Craw et al., 1967; Petrov and Kizhaev, 1970; Seehra and Castner, 1970; Toyata and Hirakawa, 1971), two-dimensional systems (Boesch et al., 1971; deWijn et al., 1972; Nagata and Date, 1964; Reidel and Willet, 1975; Yokozawa, 1971), and one-dimensional systems Ajiro et al., 1975; Gulley et al., 1970; Morimoto and Date, 1970; Nagata, 1976; Oshina et al., 1976; Tazuke and Nagata, 1971, 1975). In one- and two-dimensional materials the linewidth is often anisotropic (Boesch et al., 1971; Chao and Englesberg 1976; Dietz et al., 1971; Ferrieu and Boucher, 1976; McGregor et al., 1976; Morimoto and Date, 1970; Nagata and Date, 1964; Richards, 1974; Richards et al., 1973, 1974; Willet and Extine, 1973; Yokozawa, 1971). A typical such dependence is

$$\Delta H = a + b(3\cos^2\theta - 1) + c(3\cos^2\theta - 1)^2$$

where θ is the angle between the dc magnetic field and the plane of the two-dimensional magnetic system. Lagendijk (1976) investigated magic angle linewidths in a one-dimensional Heisenberg magnet.

Line broadening of low-dimensional exchange-coupled spins with uniaxial anisotropy was calculated by Tanaka (1969, 1970, 1971, 1975), Tanaka et al. (1969, 1973, 1975), Benner (1977), and Hughes et al. (1975). Pomerantz et al. (1978), Shia et al. (1974) and Yamada et al. (1972) present some experimental data on two-dimensional magnetic structures. Berim et al. (1976) calculated the temperature dependence of the first and second moments of a linear chain of spins coupled by the Ising exchange interaction, and they compared the influences of the Ising and Heisenberg exchange interactions on the line profile. Kondal and Seehra (1981) measured the linewidth of volume modes in EuS and found it to vary inversely as the mode quantum number $1/n_2$, where $n_2 = 5, 7, 9, \ldots, 41$.

Another parameter that can give an exponent is the zero field D term (Owens, 1979), which has the following temperature dependence near T_c:

$$D = aT + b\,|\,T_c - T\,|^{1/2}$$

and if the site lacks inversion symmetry the temperature dependence of D is

$$D = aT + b\,|\,T_c - T\,|$$

Mooy and Bolhuis (1976) used changes in the D and E zero-field parameters to obtain a critical exponent.

The magnetic resonance of phase transitions was recently treated in a volume by Owens et al. (1979).

U. Liquid Crystals

As the temperature of a normal liquid approaches the melting point, orientational order is lost and then, at the melting point, long-range translational order is lost. Thus a liquid has neither orientational order nor long-range

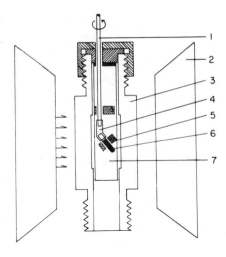

Fig. 12-39. Schematic diagram of a liquid crystal rotating apparatus. 1—Rotating rod, 2—magnet, 3—ESR cavity, 4—flexible connection, 5—sample holder, 6—sample, 7—supporting plate [from Yo et al. (1978)].

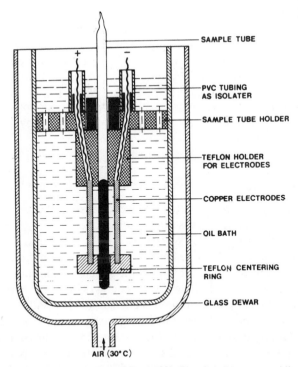

Fig. 12-40. Experimental assembly for liquid–crystal orientation. The glass Dewar containing the isolating paraffin oil has an inner diameter of 55 mm, an outer diameter of 80 mm, and a length of 120 mm. The condenser is composed of copper electrodes with dimension $60 \times 10 \times 1$ mm^3. Degassing the oil permits electric fields, up to 40,000 V/cm to be applied. The dc high-voltage power supply consists of a Tesla transformer and rectifier [from Krebs and Sackmann (1976)].

translational order. Some materials do not pass directly from the solid to the liquid state, but, rather, they pass through one or more intermediate, partly ordered states called liquid-crystal states. For example, at a particular temperature a solid may lose some of its translational order while preserving its orientational order to form what is called a smectic liquid-crystal phase, at a higher temperature it may lose the remainder of its translational order while retaining it orientational order to form what is called a nematic liquid crystal phase, and at a still higher temperature it may lose its orientational order and form a normal liquid. Liquid crystals have some properties that are liquidlike and some that are crystallike. For example, a nematic phase looks like and flows like a low-viscosity liquid, while optical birefringence and NMR measurements are indicative of a crystalline solid.

Yo et al. (1978) studied the lag angle between the director or axis that characterizes the molecular order and the external magnetic field for a nematic liquid crystal with the aid of the experimental arrangement shown in Fig. 12-39. Krebs and Sackmann (1976) prepared electric field-oriented mixtures of cholesteryl chloride and cholesteryl laurate using the apparatus illustrated in Fig. 12-40. This permits the application of electric fields up to 30,000 V/cm to orient the sample.

There is an extensive literature on the NMR of liquid crystals, and the subject has been reviewed by Doane (1979). ESR can be employed to study liquid crystals by dissolving paramagnetic compounds in them. A number of typical ESR studies of liquid crystals have been carried out on various systems (Brooks et al., 1976; Fryburg and Gelerinter, 1970; Glarum and Marshall, 1966, 1967; Hemminga, 1975; Kothe, 1977; Rao et al., 1977; Polnaszek and Freed, 1975; Schara, 1975; Seelig, 1976; Singstad, 1975; Lin and Freed, 1979), especially with nitroxide spin labels (Bales et al., 1974; Barbarin et al., 1978).

V. Computer Applications

A number of recent articles have discussed the use of computers in ESR spectroscopy; we discuss a few typical ones.

In a series of articles Wessel (1975, 1978) and also Schwarz (1976) discussed the use of a small computer for data sampling and analysis in ESR spectroscopy. They treated double integration, computer subtraction of spectra, time averaging, and smoothing. Evans et al. (1978), Gouch and Hacker (1972), Ling (1974), Mackey et al. (1969), Pedersen (1972), Pilbrow and Winfield (1973), and Sagstuen (1977) described computer programs for simulating ESR spectra. DuVarney and Anderson (1970) designed an analog computer for the rapid analysis of poorly resolved spectra. Vollmer and Caspary (1977) developed one computer program for determining spin concentrations by double integration together with the linewidth and g factor. Their second program determines the fractions of individual component spectra in a composite spectrum. Servant et al. (1976) carried out computer simulation for completely asymmetric g factors taking into account the transition probability. Farach and Poole (1968) published a program that calculates complex hyperfine multiplets. Coffey et al. (1975) simulated the spectra from axially symmetric nitroxide spin labels, and in another article (1976) they programmed the saturation transfer spectra of axially symmetric spin labels. Galloway and Dalton (1978) presented approximate methods for computing saturation transfer spectra. Porumb and Slade (1976) devised a computational model for systems with axial symmetry. Robinson et al. (1974) calculated spectral shapes from the fast-motion regime to the rigid-lattice limit, and Gregoli et al. (1977) provided a computer-assisted analysis of temperature-dependent ESR spectra.

A number of authors have used computer analysis techniques for the interpretation of particular experimental spectra. Friebele et al. (1974, 1976) assumed gaussian distributions in excited-state energy-level splittings to explain the spectra in Ge-doped silica core optical fiber. Griscom et al. (1976) elucidated the detailed nature of boron defects in irradiated silica core optic fibers. Gregoli et al. (1977) carried out a computer-assisted analysis of DNA nucleotide spectra arising from gamma radiation. Hartig and Dertinger (1971) computer-analyzed free-radical formation after irradiation. Hoffman and Poss (1978) employed a computer-fit program to decompose a complex spectrum into subspectra that permit the analysis of superhyperfine structures. Kawano et al. (1975) computer-simulated the angular dependence of Eu^{2+} spectra from alkali halide hosts. Mengeot et al. (1975) tested a theoretical model of radiation-produced spectra in hydroxyapatite using computer simulation. Jain and Upreti (1977) computed spin Hamiltonian parameters for $S = \frac{1}{2}$ ions in crystals. Vivien et al. (1976) fit a spectrum with an 0.131-cm^{-1} zero-field term by computer methods. Gubanov et al. (1972), Owen and Vincow (1971), Taylor and Bray (1970, 1972), Tynan and Yen (1970), and Vitko and Huddleston (1976) described computer programs for glass and powder spectra. The program of Pomponiu and Balaban (1972) interprets free-radical spectra. Lefebvre and Maruani (1963, 1965) presented computer programs for dilute radicals in amorphous solids. Robinson et al. (1974) calculated lineshapes from the fast-motion to the rigid-lattice limits. DeNeef et al. (1970) used computer methods for choosing lineshape and reliability parameters. Posener (1973) discussed fast ESR matrix diagonalization, and Goldberg et al. (1975) explains how to control an ESR spectrometer by a computer. Nishikawa and Someno (1975) used Fourier-transform computer storage to reduce computer memory requirements and increase computer speed. Bieber and Gough (1976) applied autocorrelation techniques to the analysis of ESR spectra. Kotomin et al. (1975) used the method of successive excursions to expand complex spectra. Warner and McPhate (1975) combined polynomial regression and a nonderivative search technique for spectral analysis.

Misra (1976) evaluated spin Hamiltonians by a least-squares fitting method, and Misra and Sharp (1976) presented a technique for computer-analyzing ESR data when the spin Hamiltonian has large off-diagonal elements. Smith et al. (1976) prepared a general method for determining spin Hamiltonian parameters and applied it to a Hamiltonian including only Zeeman and zero-field terms. The general formalism of Waller and Rogers (1973) converts the overspecification of tensor elements into a determination of the three rotational misalignments. Complex spectra can be unraveled with the aid of autocorrelation functions (Allen, 1962) since such functions provide maxima corresponding to splitting constants (Bieber and Gough, 1976; Ziegler and Hoffman, 1968). In a related approach Brumby (1979, 1979) employed significance plots to determine hyperfine coupling constants. Tkach et al. (1975) developed a system for the automatic mathematical processing of electronic paramagnetic data. McGregor et al. (1975) carried out ESR calculations by the eigenfield method. Fischer (1971) and Kertesz et al. (1974) developed programs for spectral analysis. Dalton and Kwiram (1972) computer-analyzed the ENDOR spectra of low-symmetry materials. A number of laboratories have developed a variety of computer data-processing procedures (e.g., Carr et al., 1977; Gruber et al., 1974; Herring et al., 1979; Kloza, 1971; Plato and Möbius, 1972). Computers are also widely used in NMR (Bricker and Martin, 1972; Carr et al., 1977; Diehl et al., 1972; Sherman et al., 1974; Storek, 1976; Zupan et al., 1977). The use of computers in magnetic resonance was reviewed by Van Camp and Heiss (1981), Swalen and Lusebrink (1973), and Thyer (1971).

Computer are particularly useful for carrying out literature searches. The most commonly employed technique involves the use of key words, such as asking for all articles that contain the pair "lineshape" and "ESR," or synonyms thereof, in the title or abstract. For example, one might include "line profile" and perhaps "second moment," "envelope," "Lorentzian," and "Gaussian" with "lineshape" in the first group and the key words "ESR," "EPR," "spin resonance," and "paramagnetic resonance" in the other group, and ask for all articles that contain one word or phrase from each group in either the title or the abstract. Both *Chemistry Abstracts* and *Physics Abstracts* may be searched to provide article references, titles, and author addresses. In the case of *Physics Abstracts* the abstract itself can be printed out.

These computer literature searches can provide lists of references that are more than 80% or 90% pertinent when the key words are chosen properly. Occasionally, however, one is surprised to obtain unexcepted articles in a computer search. For example, EPR and ESR are abbreviations for "east Pacific rise" (Anderson and Hobart, 1976; Rea, 1978), "electroslag refining" (Ritter and deMorton, 1977), "electroslag remelting" (Lipiarz, 1975), "ethylene propylene rubber" (Philbrick and Portinari, 1976; Steiner, 1977), "Exxon Production research" (Ledgerwood, 1976), and "experimental power reactor" (Fasolo et al., 1977; McHargue and Scott, 1976; Usher and Powell, 1976).

W. Origin of the Moments of Spectral Lines

The first few sections of this chapter explained how to determine the moments of experimentally observed resonance lines. This section explains how to compute moments from the positions of paramagnetic centers and nuclei in the crystal lattice for comparison with the measured values. The basic article on this subject was written by Van Vleck (1948). Judeikis (1964) analyzed the errors that result from neglecting the "wings" of the line in calculating moments from observed spectra.

A system of electron spins S_j interacting with an applied magnetic field H (Zeeman energy) produces a resonance line whose width is determined by the dipolar interaction and the exchange interaction with other spins in accordance with the Hamiltonian

$$\mathcal{K} = \underbrace{g\beta H \Sigma S_{zj}}_{j} + \underbrace{\sum_{k>j} \tilde{A}_{jk} \mathbf{S}_j \cdot \mathbf{S}_k}_{\text{Exchange energy}}$$

$$\underbrace{+ g^2\beta^2 \sum_{k>j} \left[\frac{\mathbf{S}_j \cdot \mathbf{S}_k}{r_{jk}^3} - 3\frac{(\mathbf{r}_{jk} \cdot \mathbf{S}_j)(\mathbf{r}_{jk} \cdot \mathbf{S}_k)}{r_{jk}^5} \right]}_{\text{Dipole–dipole (dipolar) energy}} \tag{1}$$

where r_{jk} is the distance between the jth and kth spins, and $\tilde{A}_{jk}$ is proportional to the exchange integral J_{jk}. One should note that the first term in the dipolar energy is of the same form as the exchange interaction.

When a single crystal is considered, the various moments of the spectral line depend upon the direction cosines of the applied magnetic field H relative to the crystalline axes as a result of the scalar products $(\mathbf{r}_{jk} \cdot \mathbf{S}_j)$. For example, an array of spins arranged on a simple cubic lattice has a second moment $\langle H^2 \rangle$ given by

$$\langle H^2 \rangle = \left(\frac{36.8}{d^6} \right) g^2\beta^2 [\tfrac{1}{3}S(S+1)][(\lambda_1^4 + \lambda_2^4 + \lambda_3^4) - 0.187] \tag{2}$$

where λ_1, λ_2, and λ_3 are the direction cosines of the applied magnetic field H relative to the principal cubic axes, and d is the lattice constant or nearest-neighbor spin–spin separation.

A number of experimental studies have been made of systems with orientation dependent linewidths and lineshapes (e.g., Abraham et al., 1971; Arbilly et al. 1975; Bendt, 1970; Bogan et al., 1972; Boscaino et al., 1973, 1974, 1976; Bugai et al., 1971, 1973, 1975; Byberg et al., 1974; Dattagupta et al., 1974; Davidov et al., 1975; Dietrich et al., 1976; Elschner and Weimann, 1971; Ermakovich et al., 1976, 1977; Henning and den Boef, 1978; Hinkel et al., 1978; Hoffmann et al., 1975; Huang et al., 1972; Hughes et al., 1970; Janecka et al., 1971; Krishnamurthy, 1972; Kondal and Seehra, 1981; Lee and Brodbeck, 1977; Marshall et al., 1978; Nordio and Segre, 1976; Panepucci and Farach, 1977; Petrova et al., 1971; Reineker, 1976; Rettori et al., 1974; Richards et al., 1974; Saraswat and Upreti, 1977; Sastry and Sastry, 1971, 1974; Schoemaker et al., 1970, 1973, Shia et al., 1974; Sperlich et al., 1974; Takagi et al., 1976, 1976; Teodorescu, 1977; Upreti, 1974; Urban et al., 1976; Wolfe and Jeffries, 1971; Zevin, 1961). The linewidth is especially anisotropic in one-dimensional systems (Bartkowski et al., 1972, 1972; Duffy et al., 1974; Hughes et al., 1975; Ikebe and Date, 1971; Kumano and Ikegami, 1978; McGregor and Soos, 1976; Nagata et al., 1978; Okuda et al., 1972; Soos et al., 1973; Yamada et al., 1975; see Sec. 12-T). Poole and Farach (1980) reviewed orientation-dependent linewidths.

For a powder sample with cubic symmetry the second moment (Aasa, 1970) is averaged over a sphere to give

$$\langle H^2 \rangle = \frac{3}{5} g^2 \beta^2 S(S + 1) \sum_j \left(\frac{1}{r_{jk}} \right)^6 \tag{3}$$

where, of course, the term $j = k$ is excluded from the summation. It should be emphasized that Eq. (3) is for a general type of cubic symmetry, while Eq. (2) is for the more particular simple cubic structure. These two expressions are for identical spins. If one type of spin is broadened by another type (e.g., a free-radical electron spin broadened by protons), the second moment is decreased by the factor $4/9$ to give

$$\langle H^2 \rangle = \frac{4}{15} g^2 \beta^2 S(S + 1) \sum_j \left(\frac{1}{r_{jk}} \right)^6 \tag{4}$$

In these formulas the g factor corresponds to the spin that causes the broadening, not the one being measured. Van Vleck (1948) also gives formulas for fourth moments, and Glebashev (1957) computed sixth moments by the Van Vleck method. The two Eqs. (3) and (4) do not include subsidiary lines at ~ 0, $2g\beta H$, and $3g\beta H$ in the moment calculation since they are not observed at high fields. They were removed by truncating the Hamiltonian. Their inclusion would increase $\langle H^2 \rangle$ by the factor $10/3$ [for the "$10/3$ effect," see Van Vleck (1948)]. The subsidiary lines and the $10/3$ effect have been observed experimentally (e.g., see Castner and Seehra, 1971; Seehra, 1971; Kennedy et al., 1970; McGavin and Tennant, 1976; Murakami and Kawamori, 1977; Sastry and Sastry, 1974; Soos et al., 1973).

Half-field lines are especially prominent in low-dimensional systems (Hamashima and Ajiro, 1978; Komatsubara et al., 1978, 1978; Lagendijk and Schoemaker, 1977; Murakami and Kawamori, 1977; Nagata et al., 1977; Yamada and Nishizaki, 1977). Davidov et al. (1975, 1977) and Zevin et al. (1975, 1976) discussed Van Vleck paramagnets.

When the broadening arises from protons then deuteration will decrease the linewidth by the factor $g_p/g_d = 6.514$ and reduce the second moment by the factor $(g_p/g_d)^2 = 42.43$. Some authors have employed deuteration to narrow lines and increase the resolution (e.g., Bales et al., 1974; Castleman and Moulton, 1972; Dimney and Gordy, 1980; Finch, 1979; Gillbro et al., 1975;

Graslund et al., 1975; Johnson and Moulton, 1978; Kalyanaraman and Kispert, 1978). Selective deuteration is also useful for identifying radicals (e.g., Box et al., 1975; Clough and Poldy, 1973; Finch et al., 1979; Flossman and Westhof, 1978; Lamotte and Gloux, 1973). Silver (1975) studied ^{17}O labeled compounds.

The exchange interaction does not influence the second moment, but it does affect the fourth moment. In the absence of exchange, dipolar broadening produces a Gaussian-shaped absorption curve. The exchange interaction between like spins causes the absorption line to be more peaked in the center, and to taper more gradually in the wings, and thereby renders the lineshape closer to Lorentzian than Gaussian. On the other hand, the exchange interaction between unlike spins tends to broaden absorption lines, but in practice this effect is much less prevalent than exchange narrowing by like spins. Ayant (1976) and Gurak et al. (1976) observed exchange broadening. Kittel and Abrahams (1953) showed that in the case of magnetically dilute crystals, dipolar broadening produces a Lorentzian shape (see also Grant and Strandberg, 1964; McMillian and Opechowski, 1960; Niemark and Vugmeister, 1975; Wyard, 1965). Dipolar broadening due to solvent segregation in frozen aqueous solutions is discussed by Ross (1965).

When the exchange interaction energy is very large compared to the dipolar energy, the linewidth ΔH is proportional to the square of the linewidth ΔH_{dd} that arises from dipole–dipole broadening divided by the rate of exchange ω_e expressed in frequency units (Anderson and Weiss, 1953; Van Vleck, 1948)

$$\Delta H \approx \frac{\gamma(\Delta H_{dd})^2}{\omega_e} \tag{5}$$

$$\approx \frac{(\Delta H_{dd})^2}{H_e} \tag{6}$$

The contributions from the exchange H_c and the dipolar H_{dd} interactions are given by

$$H_e = \left(\frac{1.7J}{g\beta}\right)[S(S+1)]^{1/2} \tag{7}$$

$$\Delta H_{dd} = 2.3(g\beta\rho)[S(S+1)]^{1/2} \tag{8}$$

where S is the spin, J is the exchange integral, and ρ is the density of spins per cubic centimeter. The last two equations are for a simple cubic lattice, and are given by Anderson and Weiss (1953) to illustrate the use of Eqs. (5) and (6) for extreme exchange narrowing. Fedders (1971) presented a microscopic view of exchange narrowing. In solutions exchange can occur intramolecularly (e.g., Koryakov et al., 1975; Shimada and Szwarc, 1975, 1975) or between different radical species (e.g., Box and Budzinski, 1974; Castellano et al., 1976; Shimozato et al., 1975), see also Ayant (1976), Doba et al. (1976), Odgaard et al. (1975), and Stillman and Schwartz (1976). Goyal and Seth (1977), Mackowiak (1975) Mao and Kreilick (1976), Samokhvalov et al. (1972, 1975, 1976), and Wieckowaski (1975, 1975) recently discussed exchange, Cusumano and Troup (1973) treated the lineshape in the wings of an exchange-narrowed line, Zaspel and Drumheller (1977) and Gulley et al. (1976) discussed the temperature dependence of the exchange interaction, and Heinzer (1974) provided a least-squares analysis of exchange-broadened spectra.

Molecular rotation has the effect of narrowing a magnetic resonance line. Andrew et al., (1958, 1958, 1959) showed that molecular rotation produces weak sidebands or satellite lines that are difficult to observe, but that contribute to the overall second moment in such a way that the second moment remains invariant in magnitude Clough et al. (1972) observed the tunneling sidebands of methyl group hyperfine structure. A number of authors studied hindered rotation or tunneling (Ban and Chachaty, 1973; Birkner and Schattke, 1975; Meservey et al., 1977, 1978, 1978, 1979; Michalik and Kevan, 1978; Rius et al., 1976), especially of methyl groups (Clough et al., 1969, 1972, 1973, 1974, 1974; Chen and Kochi, 1974; Beckmann and Clough 1978). Polak (1976) deduced the libration amplitude from the second moment. Bloch (1958) and Sarles and Cotts (1958) discuss line narrowing by double frequency irradiation. Köhler (1965) treats the lineshape and second moment with superimposed Overhauser and solid-state effects. Tumanov (1962) and Grant (1964) discuss cross relaxation and moments.

Kambe and Ollom (1956) modified Van Vleck's theory to include quadrupole coupling or the crystalline field effect (see Volkoff, 1953). Kopvillem (1958) extended the work of Van Vleck (1948) to the case where the g factor is a tensor. Yokota (1952) showed that Van Vleck's moment method is equivalent to the Fourier transform technique, which is often used in the theory of pressure broadening. O'Reilly and Tsang (1962, 1963) employed lattice harmonics of the appropriate crystal field group to compute second and fourth moments, and to make vibrational corrections to moments. McMillan and Opechowski (1960, 1961) discuss the temperature dependence of moments, and Englman (1976) presents a motionally narrowing diagnostic for ESR lines in vibronic systems. Lin and Kevan (1977) applied second moment determinations to the lineshapes of trapped electrons. Korst et al. (1963) averaged Van Vleck's moment formulas over the zero-point vibrations in solids. Thermal motion effects have been discussed by Shmueli et al. (1976) and by Sjoblom (1979). Murthy (1963) presents a method for calculating moments by means of an analog computer. Verdier et al. (1961) used a generalized convolution transform to analyze moments.

Sjoblom (1976) showed that the second moment of a dipolar solid can be written as a sum of products of real second-order spherical harmonics. Ding-Ping and Kevan (1976) used second moment analyses for ENDOR spectra. ENDOR linewidths are also discussed by Derouane and Vedrine (1974), Helbert et al. (1972), and Kevan et al. (1979).

We see from Eq. (8) that the linewidth is expected to be dependent on the concentration or density of paramagnetic ions. A number of workers have carried out studies of this, and in Sec. 12-P we quote some recent representative articles.

One may deduce from Eqs. (3) and (4) that the calculation of moments requires the computation of lattice sums such as those that Mayer (1933), Gutowsky and McGarvey (1952), Kaplan (1961), and Cheng (1961) evaluated for several lattices.

Divers articles such as Hervé (1957, 1960), Kopvillem (1960), Koloskova and Kopvillem (1960), and Vincow and Johnson (1963) may be consulted for experimental measurements of moments. Vincow and Johnson (1963) derived expressions for the effects of finite linewidth, satellite lines, carbon-13 in natural abundance, g-factor anisotropy, and π electron-spin density on the second moments of π electron radicals. Loudon (1960). Meyer et al. (1966) and Stoneham (1965) discuss the relationship between lineshapes obtained during ordinary ESR studies and those obtained in acoustic resonant absorption experiments.

Most of the theories of resonance linewidths and lineshapes, such as those discussed in this chapter, are only valid for cases in which the linewidth in gauss is small compared to the applied magnetic field. Garstens (1954) modified the original Bloch theory (1946) for the case in which this assumption is not valid, and Garstens et al. (1954) and Garstens and Kaplan (1955) obtained experimental data at low magnetic fields and low resonant frequencies ($f_0 < 15$ Mc) on DPPH that agreed with their modified theory. Rogers et al. (1959) explained the narrowing of the linewidth of solid DPPH in low fields by the $10/3$ effect discussed earlier in this section, and Tchao (1960) attributes the residual width of 2.1 G at zero field to anisotropic diamagnetism of the benzene rings (see also Walter et al., 1956).

We conclude this section with some parenthetical remarks about summations. Suppose a set of four spins has the simplified (but unrealistic) Hamiltonian $\mathcal{H}$

$$\mathcal{H} = g\beta H \sum_j S_{zj} + g^2\beta^2(1 - 3\cos^2\theta) \sum_{k>j} \left(\frac{1}{r_{jk}}\right)^3 \tag{9}$$

One may write the right-hand side explicitly as follows:

$$\mathcal{H} = g\beta H[S_{z1} + S_{z2} + S_{z3} + S_{z4}] + g^2\beta^2(1 - 3\cos^2\theta)$$

$$\times \left[\left(\frac{1}{r_{12}}\right)^3 + \left(\frac{1}{r_{13}}\right)^3 + \left(\frac{1}{r_{14}}\right)^3 + \left(\frac{1}{r_{23}}\right)^3 + \left(\frac{1}{r_{24}}\right)^3 + \left(\frac{1}{r_{34}}\right)^3\right] \tag{10}$$

This is expected because the interaction energy term $g^2\beta^2(1 - 3\cos^2\theta)/r_{ij}^3$

should only be counted once for each pair of spins. The summation

$$\sum_{k \neq j} \left(\frac{1}{r_{jk}}\right)^3 = \left[\left(\frac{1}{r_{12}}\right)^3 + \left(\frac{1}{r_{13}}\right)^3 + \left(\frac{1}{r_{14}}\right)^3 + \left(\frac{1}{r_{21}}\right)^3 + \left(\frac{1}{r_{23}}\right)^3 + \left(\frac{1}{r_{24}}\right)^3 \right.$$

$$\left. + \left(\frac{1}{r_{31}}\right)^3 + \left(\frac{1}{r_{32}}\right)^3 + \left(\frac{1}{r_{34}}\right)^3 + \left(\frac{1}{r_{41}}\right)^3 + \left(\frac{1}{r_{42}}\right)^3 + \left(\frac{1}{r_{43}}\right)^3 \right]$$

(11)

has twice as many terms as $\Sigma_{k>j} r_{gh}^{-3}$ so that

$$\sum_{k>j} \left(\frac{1}{r_{jk}}\right)^3 = \frac{1}{2} \sum_{k \neq j} \left(\frac{1}{r_{jk}}\right)^3 \tag{12}$$

since of course $r_{jk} = r_{kj}$. Some authors use the summation on the right-hand side of Eq. (12) instead of that on the left-hand side.

In the second-moment calculation, we select a typical spin and calculate the broadening interaction with all of the others in the sample. To illustrate the method by our four-spin case, we choose $k = 2$, and assume (erroneously) that Eq. (3) is valid for such a small cluster. Ergo,

$$\langle H^2 \rangle = \frac{3}{5} g^2 \beta^2 S(S + 1) \left[\left(\frac{1}{r_{12}}\right)^6 + \left(\frac{1}{r_{32}}\right)^6 + \left(\frac{1}{r_{42}}\right)^6 \right] \tag{13}$$

In practice, the second-moment sum is over about 10^{24} atoms, but only the nearby ones make an appreciable contribution to the result.

One may approximate the second moment by summing directly over the nearest neighbors (nn), and then integrating over the remaining spins in the sample. If the density or number of spins per unit volume is ρ, then the average number in a thin spherical layer of thickness dr at a distance r from the origin will be $4\pi r^2 \rho \, dr$. Therefore

$$\langle H^2 \rangle \approx \frac{3}{5} g^2 \beta^2 S(S + 1) \left[\sum_{nn} \left(\frac{1}{r_{jk}}\right)^6 + 4\pi\rho \int_{a_0}^{\infty} \frac{r^2 \, dr}{r^6} \right] \tag{14}$$

$$\approx \frac{3}{5} g^2 \beta^2 S(S + 1) \left[\sum_{nn} \left(\frac{1}{r_{jk}}\right)^6 + \frac{4}{3}\pi\rho \left(\frac{1}{a_0}\right)^3 \right] \tag{15}$$

where a_0 is the distance to the nearest spin that is not included in the summation. The same units of length must be used for a_0, r_{jk}, and ρ; for example, ρ might be the number of spins per cubic angstrom unit. For this approximation to be valid, the magnitude of the summation term must greatly exceed that of the integral term. If there are several species of spins in the sample, then each type will contribute to the moments, and should be summed separately. The numerical factor of $3/5$ in Eqs. (14) and (15) for like spins must be changed to $4/15$ for unlike spins, as explained above in this section.

Van Vleck's theory (1948) of moments is a high-temperature approximation that is valid in most experimental arrangements. Below liquid-helium temperature and at high frequencies one often attains the low-temperature condition $g\beta H > kT$ (compare Sec. 11-E, especially Figs. 11-9 and 11-10). In this region the theory of moments must be modified by weighting the spin levels with their proper Boltzmann factors, as discussed by Pryce and Stevens (1950), Kambe and Usui (1952), McMillian and Opechowski (1960, 1961), Svare and Seidel (1964), and Svare (1965). Grant (1964) discusses moment theory from a mathematical viewpoint. See also Kaplan (1976) and Lundin and Sobov (1977).

In this section we discussed the molecular origin of the moments of spectral lines. There are also many instrumental factors that broaden resonances such as magnetic field inhomogeneities (Sec. 6-A), modulation effects (e.g., Tykarski, 1974; Secs. 6-H and 6-I), variation in cavity Q across the resonance line (Vigouroux et al., 1973), and AFC effects (Dumais and D'Ausigne, 1980) that are treated elsewhere in this monograph.

References

Aasa, R., *J. Chem. Phys.* **52**, 3919 (1970).

Aasa, R., and T. Vänngärd, *Z. Naturforsch.* **19a**, 1425 (1964).

Aasa, R., and T. Vänngärd, *J. Chem. Phys.* **52**, 1612 (1970).

Abdrakhmanov, R. S., and T. A. Ivanova, *J. Mol. Struct.* **46**, 229 (1978).

Abraham, M. M., L. A. Boatner, C. B. Finch, and R. W. Reynolds, *Phys. Rev.* **3**, 2864 (1971).

Abramowitz, M., and I. A. Stegun, Eds., *Handbook of Mathematical Functions*, Dover, New York, 1972.

Accou, J., P. Van Hecke, and L. Van Gerven, *17th Colloque Ampere Turku, Finland* (1973), p. 379.

Adrian, F. J., E. L. Cochran, and V. A. Bowers, *J. Chem. Phys.* **59**, 56 (1973).

Ageno, M., and C. Frontali, *Nature* **198**, 1294 (1963).

Ahn, M. K., *J. Chem. Phys.* **64**, 134 (1976).

Ajiro, Y., S. A. Friedberg, and N. S. Vander-Ven, *Phys. Rev. B* **12**(1), 29 (1975).

Ajiro, Y., S. Matsukawa, S. Yawada, and T. Haseda, *J. Phys. Soc. Japan* **39**, 259 (1975).

Akhiezer, I. A., D. P. Belozorov, and E. M. Chudnovskii, *Fiz. Nizk. Temp.* **2**, 47 (1976).

Alexander, S., A. Beram, and Z. Luz, *Mol. Phys.* **27**, 441 (1974).

Alexander, S., Z. Luz, Y. Naor, and R. Poupko, *Mol. Phys.* **33**, 1119 (1977).

Allen, L. C., *Nature* **196**, 663 (1962).

Allen, L. C., H. M. Gladney, and S. H. Glarum, *J. Chem. Phys.* **40**, 3235 (1964).

Allendoerfer, R. D., *Magn. Reson. Rev.* **5**, 175 (1980).

Al'tshuler, T. S., I. A. Garifullin, E. G. Kharakhash'yan, and L. F. Shatrukov, *Sov. Phys. Solid State* **14**, 2213 (1973).

Anders, A. G., and A. I. Zvyagin, *Sov. Phys. JETP* **36**(6), 1151 (1973).

Anderson, R. N., and M. A. Hobart, *J. Geophys. Res.* **81**, 2968 (1976).

Anderson, J. E., and R. Ullman, *J. Chem. Phys.* **47**, 2178 (1967).

Anderson, H. G., and H. Welling, *Phys. Rev. A* **139**, 321 (1965).

Anderson, P. W., and P. R. Weiss, *Rev. Mod. Phys.* **25**, 269 (1953).

Andreev, V. A., and V. I. Sugakov, *Sov. Phys. JETP* **36**, 1151 (1976).

Andrew, E. R., *Phys. Rev.* **51**, 425 (1953); *Bulletin du Groupement Ampere, 8th Colloque Ampere*, 1959, p. 103.

Andrew, E. R., and R. Bersohn, *J. Chem. Phys.* **18**, 159 (1950).

Andrew, E. R., A. Bradbury, and R. G. Eades, *Arch. Sci. Spec. Publ.* **11**, 223 (1958).

Andrew, E. R., and R. A. Newing, *Proc. Phys. Soc.* **72**, 959 (1958).

Angelov, S., *Nauki Bulg. Akad.* **5**, 101 (1972).

Angelov, S., J. P. Mercurio, B. Chevalier, and J. Etourneau, *Solid State Commun.* **23**, 531 (1977).

Anisimov, O. A., A. T. Nikitaev, K. I. Zamaraev, and Yu. N. Molin, *Theor. Exper. Chem.* **7**, 556 (1971).

Antonenko, A. Kh., N. N. Gerasimenko, A. V. Dvurechenskii, *Fiz. Tekh. Poluprovodn.* **11**, 559 (1977).

Antsiferova, L., and N. Korst, *Chem. Phys. Lett.* **15**, 439 (1972).

Antsiferova, L. I., A. V. Lazarev, and V. B. Siryukov, *Zh. ETF Pis. Red.* **12**, 108 (1970).

Arbilly, D., G. Deutscher, E. Grunbaum, R. Orbach, and J. T. Suss, *Phys. Rev. B* **12**, 5068 (1975).

Arizumi, T., A. Yoshida, K. Saji, J. Stuke, and W. Brenig, *Proc. Int. Conf. Amorpous and Liquid Semiconductors* **2** (1972).

Atkins, P. W., and B. P. Hills, *Mol. Phys.* **29**, 761 (1975).

Atkins, P. W., and D. Kivelson, *J. Chem. Phys.* **44**, 169 (1966).

Axel, F. S., *Biophys. Struct. Mech.* **2**, 181 (1976).

Ayant, Y., *J. Phys.* **37**, 219 (1976).

Ayant, Y., R. Besson, and A. Salvi, *J. Phys.* **36**, 571 (1975).

Babushkin, V. S., A. A. Samokhvalov, S. P. Tataurov, and M. I. Simonova, *Fiz. Tverd. Tela* **17**, 1527 (1975).

Balaban, A. T., N. Negoita, and R. Baican, *J. Magn. Res.* **9**, 18 (1973).

Bales, B. L., M. K. Bowman, L. Kevan, and R. N. Schwartz, *J. Chem. Phys.* **63**, 3008 (1975).

Bales, B. L., J. Helbert, and L. Kevan, *J. Phys. Chem.* **78**, 221 (1974).

Bales, B. L., J. A. Swenson, and R. N. Schwartz, *Mol. Cryst. Liq. Cryst.* **28**, 143 (1974).

Ban, B., and C. Chachaty, *Can. J. Chem.* **51**, 3889 (1973).

Baranov, P. G., R. A. Zhitnokov, and N. G. Romanov, *Fiz. Tverd. Tela* **18**, 1742 (1976).

Barbarin, F., and G. P. Germain, *J. Phys.* **36**, 475 (1975).

Barbarin, F., B. Chevarin, J. P. Germain, C. Fabre, and D. Cabaret, *Mol. Cryst. Liq. Cryst.* **46**, 181, 195 (1978).

Bartkowski, R. R., M. J. Hennessy, B. Morosin, and P. M. Richards, *Solid State. Comm.* **11**, 405 (1972).

Bartkowski, R. R., and B. Morosin, *Phys. Rev. B* **6**, 4209 (1972).

Battles, J. W., *J. Appl. Phys.* **42**, 1286 (1971).

Bazuev, G. V., A. A. Samokhvalov, Yu. N. Morozov, I. I. Matveenko, V. S. Babushkin, T. I. Arbuzova, and G. P. Shveikin, *Sov. Phys. Solid State* **19**, 1913 (1977).

Beckmann, P., and S. Clough, *J. Phys. C* **11**, 4055 (1978).

Belinskii, M. I., B. S. Tsukerblat, and A. V. Ablov, *Sov. Phys. Solid State* **16**, 639 (1974).

Bendt, P. J., *Phys. Rev. B* **2**, 4366, 4375 (1970).

Benk, H., and H. Sixl, *Molecular Physics* **42**, 77 (1981).

Benner, H., *Appl. Phys.* **13**, 141 (1977).

Benton, J. E., and R. M. Lynden-Bell, *J. Chem. Soc. Faraday Trans. II* **71**, 807 (1975).

Beram, A., *J. Chem. Phys.* **71**, 2503 (1979).

Beram, A., Z. Luz, and S. Alexander, *J. Chem. Phys.* **64**, 4321 (1976).

Berclaz, T., and M. Geoffroy, *Mol. Phys.* **32**, 815 (1976).

Berger, R., *Verres Refract.* **32**, 172 (1978).

Berim, G. O., M. M. Zaripov, and A. R. Kessel, *Sov. Phys. Solid State* **17**, 1744 (1976).

Berim, G. O., and A. R. Kessel, *Phys. Status Solidi B* **79**, 489 (1977).

Berk, N. F., W. Bizzaro, J. Rosenthal, and L. Yarmus, *Phys. Rev. B* **23**, 5661 (1981).

Berkovits, V. I., C. Hermann, G. Lampel, A. Nakamura, and V. I. Safarov, *Phys. Rev. B*, 1767 (1979).

Berlinger, L. J., *Spin Labeling: Theory and Practice*, Academic Press, New York, 1976.

Berman, J. A., R. N. Schwartz, and B. L. Bales, *Mol. Cryst. Liq. Cryst. (GB)* **28**, 51 (1974).

Bernhard, W. A., and F. S. Ezra, *J. Chem. Phys.* **60**, 1707 (1974).

Berthault, T. A., S. Bedere, and J. Matricon, *J. Phys. Chem. Solids* **38**, 913 (1977).

Beuneu, F., and P. Monod, *Physica B and C* **86-88B and C**, 256 (1977).

Bieber, K. D., and T. E. Gough, *J. Magn. Reson.* **21**, 285 (1976).

Birkner, G. K., and W. Schattke, *J. Phys. C* **8**, 1205 (1975).

Bizzaro, W., L. Yarmus, J. Rosenthal, and N. F. Berk, *Phys. Rev. B*, **23**, 5673 (1981).

Blazha, M. G., A. A. Bugai, V. Maksimenko, and A. B. Roitsin, *Sov. Phys. Solid State* **14**, 1779 (1973).

Blazha, M. G., and A. B. Roitsin, *Sov. Phys. Solid State* **14**, 416 (1972); **16**, 327 (1974).

Bleaney, B., *Proc. Phys. Soc.* **A63**, 407 (1950); **75**, 621 (1960); *Phil. Mag.* **42**, 441 (1951).

Blinc, R., Z. Trontelj, and B. Volavsek, *J. Chem. Phys.* **44**, 1028 (1966).

Blinder, S. M., *J. Chem. Phys.* **33**, 748 (1960).

Bloch, F., *Phys. Rev.* **70**, 460 (1946); **111**, 841 (1958).

Bloembergen, N., *J. Appl. Phys.* **23**, 1383 (1952).

Bloembergen, N., E. M. Purcell, and R. V. Pound, *Phys. Rev.* **73**, 679 (1948).

Bloembergen, N., and T. J. Rowland, *Acta Met.* **1**, 731 (1953); *Phys. Rev.* **97**, 1679 (1955).

Boate, A. R., J. R. Morton, and K. F. Preston, *J. Phys. Chem.* **80**, 2954 (1976).

Boesch, H. P., V. Schmocker, E. Wladner, E. Emerson, and J. Drumheller, *Phys. Lett. A* **36**, 461 (1971).

Bogan, C. M., and L. D. Kispert, *J. Chem. Phys.* **57**, 3109 (1972).

Bolton, J. R., *Magn. Reson. Rev.* **1**, 195 (1972).

Bolton, J. R., and A. Carrington, *Mol. Phys.* **5**, 161 (1962).

Bonis, B., and R. Sardos, *C.R. Acad. Sci.* **276**, 689 (1973).

Boscaino, R., M. Brai, and I. Ciccarello, *Phys. Rev. B* **13**, 2798 (1976).

Boscaino, R., M. Brai, and I. Ciccarello, *18th Ampere Congress Nottingham*, 139 (1974).

Boscaino, R., M. Brai, I. Ciccarello, and G. Contrino, *Phys. Lett.* **46A**, 190 (1973).

Botter, B. J., C. J. Nonhof, J. Schmidt, and J. H. van der Waals, *Chem. Phys. Lett.* **43**, 210 (1976).

Box, H. C., Second Symp. Electron Spin Resonance Spectroscopy, *J. Phys. Chem.* **75**, 3426 (1971).

Box, H. C., and Budzinski, E. E. *J. Chem. Phys.* **60**, 3337 (1974).

Box, H. C., H. G. Freund, K. T. Lilqa, and E. E. Budzinski, *J. Chem. Phys.* **63**, 2059 (1975).

Bramwell, F. R., M. E. Laterza, M. L. Spinner, *J. Chem. Phys.* **62**, 4184 (1975).

Brenner, H. C., C. A. Hutchinson, Jr., and M. D. Kemple, *J. Chem. Phys.* **60**, 2180 (1974).

Bricker, J. L., and W. L. Martin, *Digest of Papers of the Six Annual IEEE Computer Society International Conference*, 333 (1972).

Brickmann, J., and G. Kothe, *J. Chem. Phys.* **59**, 2807 (1973).

Brodsky, M. H., and R. S. Title, *AIP Conf. Proc.* (1976).

Brooks, S. A., G. R. Luckhurst, G. F. Pedulli, and J. Roberts, *J. Chem. Soc. Faraday. Trans.* 2, **72**, 651 (1976).

Bronstein, J., and V. Volterra, *Phys. Lett.* **16**, 211 (1965).

Brumby, S., *J. Magn. Reson.* **34**, 317 (1979); **35**, 357 (1979).

Bruno, G. V., and J. H. Freed, *Chem. Phys. Lett.* **25**, 328 (1974).

Bugai, A. A., D. G. Duliv, B. K. Krulinkovskii, and A. B. Roitsin, *Phys. Status Solidi* **B57**, K15 (1973).

Bugai, A. A., M. D. Glinchuk, M. F. Delgen, and V. M. Maksimenko, *Phys. Status Solidi* **B44**, 1999 (1971).

Bugai, A. A., P. T. Levkovskii, and V. M. Maksimenko, *Ukr. Fiz. Zh.* **15**, 508 (1971).

Bugai, A. A., V. M. Maksimenko, and L. A. Suslin, *Phys. Status Solidi* **19**, K149 (1973).

Bugai, A. A., B. K. Krulikovskii, V. M. Maksimenko, and A. B. Roitsin, *Sov. Phys. JETP* **40**, 377 (1975).

Bullock, A. T., and G. G. Cameron, *Struct. Stud. Macromol. Spectrosc. Meth.* **272**, 273 (1976).

Burgiel, J. C., and M. W. P. Strandberg, *J. Phys. Chem. Solids* **26**, 865 (1965).

Burlamacchi, L., *J. Chem. Phys.* **55**, 1205 (1971).

Burlamacchi, L., *Gazz. Chim. Ital.* **106**, 347 (1976).

Burlamacchi, L., G. Martini, M. F. Ottaviani, and M. Romanelli, *Adv. Mol. Relax. Inter. Proc.* **12**, 145 (1978).

Burlamacchi, L., G. Martini, and M. Romanelli, *J. Chem. Phys.* **59**, 3008 (1973).

Burlamacchi, L., and M. Romanelli, *J. Chem. Phys.* **55**, 979 (1971).

Burlamacchi, L., G. Martini, and E. Tiezzi, *J. Phys. Chem.* **74**, 3980 (1970).

Burland, D. J., D. E. Cooper, M. D. Fayer, and C. R. Gochanour, *Chem. Phys. Lett.* **52**, 279 (1977).

Burns, G., *J. Appl. Phys.* **32**, 2048 (1961).

Burzo, E., and M. Balanescu, *Solid State Commun.* **28**, 693 (1978).

Buzare, J. Y., and J. C. Fayet, *Solid State Commun.* **21**, 1097 (1977).

Byberg, J. R., S. J. K. Jensen, and B. S. Kirkegaard, *J. Chem. Phys.* **61**, 138 (1974).

Campbell, T. G., and A. W. Lawson, *AIP Conf. Proc., 19th Annual Conference on Magnetism and Magnetic Materials* **18**, 759 (1973).

Carr, S., A. Fritzson, A., L. Hull, T. C. Werner, G. Williams, and S. C. Lee, *Microcomputer Design and Application*, 213 (1977).

Carr, S. G., T. D. Smith, and J. R. Pilbrow, *J. Chem. Soc. Faraday Trans. II* **70**, 497 (1974).

Carrington, A., *Proc. R. Soc. London A* **302**, 291 (1968).

Castellano, A., J. P. Catteau, and A. Lablache-Combier, *J. Phys. Chem.* **80**, 2614 (1976).

Castleman, B. W., and G. C. Moulton, *J. Chem. Phys.* **57**, 2762 (1972).

Castner, T. G., Jr., *Phys. Rev.* **115**, 1506 (1952).

Castner, T. G., Jr., and M. S. Seehra, *Phys. Rev. B* **4**, 38 (1971).

Chachaty, C., and J. Maruani, *Can. J. Chem.* **44**, 2681 (1966).

Charles, S. W., J. Popplewell, and P. A. Bates, *J. Phys. F Metal Phys.* **3**, 664 (1973).

Chao, N. C., and M. Englesberg, *Phys. Rev. B* **14**, 271 (1976).

Charlier, A., H. Danan, and P. Taglang, *J. Phys. (France)* **25**, *Suppl.* **11**, 183A (1964).

Chase, L. L., *Phys. Rev. B* **2**, 2308 (1970).

Chastanet, R., and K. Haye, *Physica* **103B**, 301 (1981).

Chastanet, R., K. Haye, and R. Sardos, *C. R. Acad. Sci.* **273**, 785 (1971).

Chastanet, R., and R. Sardos, *C. R. Acad. Sci.* **271**, 571 (1970).

Chazalviel, J. N., *Phys. Rev. B* **11**, 1555 (1975).

Che, M., J. Demarquay, and C. Naccache, *J. Chem. Phys.* **51**, 5177 (1969).

Chen, K. S., and J. K. Kochi, *J. Am. Chem. Soc.* **96**, 794 (1974).

Cheng, H., *Phys. Rev.* **124**, 1359 (1961).

Chicault, R., and R. Buisson, *J. Phys.* **38**, 795 (1977).

Chirkov, A. K., and A. A. Kokin, *Zh. Eksper. Teor. Fiz.* **29**, 1381 (964) (1960).

Chovino, E., R. Sardos, and R. Chastanet, *C. R. Acad. Sci.* **273**, 557 (1971).

Clark, C. O., C. P. Poole, Jr., and H. A. Farach, *J. Phys. C* **11**, 769 (1978).

Clerjaud, B., A. Kuhn, and B. Lambert, *16th Congr. Ampere Bucharest*, 1009 (1970).

Clough, S., J. Hill, and F. Poldy, *J. Phys. C* **5**, 518, 1739 (1972).

Clough, S., J. R. Hill, and M. Punkkinen, *J. Phys. C* **7**, 3413, 3779 (1974).

Clough, S., T. Hobson, P. S. Allen, E. R. Andrew, and C. A. Bates, *Proc. 18th Ampere Congr. Magn. Reson. Relat. Phenom.* **11**, 387 (1974).

Clough, S., and F. Poldy, *J. Chem. Phys.* **51**, 1953, 2076 (1969).

Clough, S., and F. Poldy, *J. Phys. C* **6**, 1953, 2357 (1973).

Cochran, E. L., F. J. Adrian, and V. A. Bowers, *J. Chem. Phys.* **34**, 1161 (1961).

Cochrane, R. W., F. T. Hedgcock, and A. W. Lightsone, *Can. J. Phys.* **56**, 68 (1978).

Coffey, R., B. H. Robinson, and L. R. Dalton, *Mol. Phys.* **31**, 1703 (1978).

Coffey, R., B. H. Robinson, and L. R. Dalton, *Chem. Phys. Lett.* **35**, 360 (1975); *Mol. Phys.* **31**, 1703 (1976).

Collins, R. C., *RSI* **30**, 492 (1959).

Cook, T. J., and T. A. Miller, *J. Chem. Phys.* **59**, 1342 (1973).

Cook, T. J., and T. A. Miller, *J. Chem. Phys.* **59**, 1352 (1973).

Coope, J. A. R., *Chem. Phys. Lett.* **3**, 539 (1969).

Corvaja, C., and L. Pasimeni, *Chem. Phys. Lett.* **39**, 261 (1976).

Cousins, J. E., and R. Dupree, *Phys. Lett.* **14**, 177 (1965).

Curtis, D. A., C. J. Kirkby, and J. S. Thorp, *Brit. J. Appl. Phys.* **16**, 1681 (1965).

Culvahouse, J. W., and C. L. Francis, *J. Chem. Phys.* **66**, 1079 (1977).

Cusumano, C., and G. J. Troup, *Phys. Lett. A* **44A**, 441 (1973).

Dalton, L. R., P. Coffey, L. A. Dalton, B. H. Robinson, and A. D. Keith, *Phys. Rev. A* **11**, 488 (1975).

Dalton, L. R., and A. L. Kwiram, *J. Chem. Phys.* **57**, 1132 (1972).

Damay, P., M. J. Sienko, *Phys. Rev. B* **13**, 603 (1974).

Danna, J. A., and J. H. Wharton, *J. Chem. Phys.* **53**, 4047 (1970).

Dattagupta, S., and M. Blume, *Phys. Rev. B* **10**, 4540, 4551 (1974).

Davidov, D., C. Rettori, and V. Zevin, *Solid State Commun.* **16**, 247 (1975).

Davidov, D., V. Zevin, R. Levin, and D. Shattiel, *Phys. Rev. B* **15**, 2771 (1977).

Davidov, D., V. Zevin, J. M. Bloch, and C. Rettori, *Solid State Commun.* **17**, 1279 (1975).

Davidson, R. B., and I. Miyagawa, *J. Chem. Phys.* **52**, 1727 (1970); **57**, 1815 (1972).

DeBoer, E., and A. P. Praat, *Mol. Phys.* **8**, 291 (1964).

Deigen, M. F., I. N. Geifman, and M. D. Glinchyk, *Fiz. Tverd. Tela* **12**, 1469 (1970).

De Neef, T., D. C. Koningsberger, and P. Van Der Leeden, *Appl. Sci. Res.* **22**, 251 (1970).

Dennis, L. W., and D. S. Tinti, *J. Chem. Phys.* **62**, 2015 (1975).

Derouane, E. G., and J. C. Vedrine, *Chem. Phys. Lett.* **29**, 222 (1974).

DeWijn, H. W., L. R. Walker, J. L. Davis, and H. J. Guggenhiem, *Solid State Commun.* **11**, 803 (1972).

Diehl, P., E. Fluck, and R. Kosheld, *NMR Basic Principles and Practice*, Springer-Verlag, Berlin, 1972.

Dietrich, J., *Stat. Phys. Sol.* **73**, K57 (1976).

Dietrich, W., and D. Schmid, *Phys. Status Solidi* **74b**, 609 (1976).

Dietrich, W., L. Schmidt, and D. Schmid, *Phys. Lett.* **28**, 249 (1974).

Dietz, R. F., F. R. Merrit, P. Dingle, D. Hone, B. G. Silbernagel, and P. M. Richards, *Phys. Rev. Lett.* **26**, 1186 (1971).

Dillon, J. F., and J. P. Remeika, in *Magnetic Resonance and Relaxation*, North-Holland, Amsterdam, 1967, p. 480.

Dimney, L. J., and W. Gordy, *Proc. Natl. Acad. Sci.* **77**, 343 (1980).

Ding-Ping, L., and L. Kevan, *Chem. Phys. Lett.* **40**, 517 (1976).

Doane, J. W., NMR of Liquid Crystals, Chap. 4 in *Magnetic Resonance of Phase Transitions*, F. J. Owens, C. P. Poole, Jr., and H. A. Farach, Eds., Academic Press, New York, 1979.

Doba, T., M. Ogasaware, and H. Yoshida, *Mem. Fac. Eng. Hokkaido Univ.* 14 (1976).

Dobson, J. F., *Phys. Lett. A* **44A**, 171 (1973).

Dokter, H. C., D. Davidov, J. M. Bloch, I. Felner, and S. Shaltiel, Proceedings of the International Conf. on Magn. Alloys and Oxides, *J. Magn. Magn. Mater.*, 7 (1977).

Domdey, H., and J. Voitlander, *Z. Naturforsch. A* **29a**, 949 (1974).

Doremieux-Morin, C., *J. Magn. Reson.* **33**, 505 (1979).

Dormann, E., R. D. Hogg, D. Hone, and V. Jaccarino, *Physica* **86-88B**, 1183 (1977).

Draghicescu, P., *Stud. Cercelari Fiz.* **14**, 201 (1963).

Dresselhaus, G., A. F. Kip, and C. Kittel, *Phys. Rev.* **100**, 618 (1955).

Duffy, W., Jr., J. E. Venneman, D. L. Strandberg, and P. M. Richards, *Phys. Rev. B* **9**, 2220 (1974).

Dugas, H., *Can. J. Spectrosc.* **18**, 110 (1973).

Dumais, J. C., and V. M. D'Ausigne, private communication, 1980.

Dumas, J., *Phys. Status Solidi B* **86**, K75 (1978).

DuVarney, R. C. and R. E. Anderson, *J. Magn. Reson.* **3**, 235 (1970).

Dvurechensky, A. V., N. N. Gerasimenko, V. B. Glazman, G. Carter, J. S. Collingon, and W. A. Grant, *Applications of Ion. Beams of Materials*, Coventry, Warwicks, England, 1975.

Dyson, F. J., *Phys. Rev.* **98**, 349 (1955).

Eagles, T. E., and R. E. D. McClung, *Can. J. Phys.* **53**, 1492 (1975).

Ehrenfreund, E., S. K. Khanna, A. F. Garito, and A. J. Heeger, *Solid State Commun.* **22**, 139 (1977).

Elschner, B., and G. Weimann, *Solid State Commun.* **9**, 1935 (1971).

Engel, U., K. Baberschke, G. Koopman, P. S. Allen, E. R. Andrew, and C. A. Bates, *Proc. of the 18th Ampere Congress on Magnetic Resonance and Related Phenomena*, **11**, Nottingham, England, Amsterdam, Netherlands (1974).

Englman, R., *J. Phys. Lett.* (*France*) **37**, 1261 (1976).

Erdos, P., *Helv. Phys. Acta* **37**, 493 (1964).

Ermakovich, K. K., V. N. Lazukin, V. M. Talarintsev, and I. V. Chepeleva, *Fiz. Tverd. Tela* **19**, 3488 (1977) (tr, p. 2040).

Ermakovich, K. K., V. M. Lazukin, I. V. Chepeleva, and V. I. Aleksandrov, *Fiz. Tverd. Tela*. **18**, 1755 (1976).

Estes, W. E., J. R. Wasson, J. W. Hall, and W. E. Hatfield, *Inorganic Chem*. **17**, 3657 (1978).

Evans, J. C., P. H. Morgan, and R. H. Renaud, *Anal. Chim. Acta Comput. Tech. Optimiz*. **103**, 175 (1978).

Farach, H. A., and H. Teitelbanm, *Can. J. Phys*. **45**, 2913 (1967).

Farach, H. A., and C. P. Poole, Jr., *Quantum Chem. Prog. Exch. Newsletter* **23**, 14 (1968), Program No. 129.

Farach, H. A., and H. Teitelbaum, *Can. J. Phys*. **45**, 2913 (1967).

Fasolo, J. A., M. S. Lubell, and C. Whitmire, *Proc. of the Seventh Symposium on Engineering Problems of Fusion Research*, Knoxville, Tenn., 1977.

Fedders, P. A., *Phys. Rev. B* **3**, 2352 (1971).

Feher, G., and A. Kip, *Phys. Rev*. **98**, 337 (1955).

Fehsenfeld, F. C., L. R. Megill, and L. K. Droppleman, *J. Chem. Phys*. **43**, 3618 (1965).

Feichtinger, H., J. Welti, and A. Gechwantner, *Solid State Commun*. **27**, 867 (1978).

Ferrieu, F., and J. P. Boucher, *Magn. Reson. Relat. Phenom. Proc. Congr. Ampere*, *19th*, 457 (1976).

Fessenden, R. W., *J. Chem. Phys*. **48**, 3735 (1968). (^{17}O labeling, see p. 549).

Filippov, A. I., and I. S. Donskaya, *Dokl. Akad. Nauk. CCCP* **205**, 138 (1972); **214**, 1124 (1974); *Zhur. Fiz. Khim*. **47**, 1271 (1973).

Finch, L. L., J. E. Johnson, and G. C. Moulton, *J. Chem. Phys*. **70**, 3662 (1979).

Fisanick-Englot, G. J., and T. A. Miller, *J. Chem. Phys*. **64**, 786 (1976).

Fischer, L., *J. Mol. Spectrosc*. **40**, 414 (1971).

Flossmann, W., A. Mueller, and E. Westhof, *Mol. Phys*. **29**, 703 (1975).

Flossmann, W., E. Westhof, and A. Mueller, *Phys. Rev. Lett*. **34**, 959 (1975).

Flossman, W., and E. Westhof, *Radiat. Res*. **73**, 75 (1978).

Frankevich, E. L., V. I. Lesin, and A. I. Pristupa, *Zh. Eksp. Teor. Fiz*. **75**, 415 (1978).

Freed, J. H., Chap. 3 in *Spin Labeling, Theory and Practice*, Academic Press, New York, 1976, p. 53.

Freed, J. H., *ACS Symp. Ser*. **34**, 1 (1976).

Freed, J. H., *J. Chem. Phys*. **41**, 7 (1964).

Freed, J. H., *J. Chem. Phys*. **43**, 1710 (1965).

Freed, J. H., *J. Chem. Phys*. **66**, 4183 (1977).

Freed, J. H., G. V. Bruno, and C. Polnaszek, *J. Chem. Phys*. **55**, 5270 (1971).

Freed, J. H., G. K. Fraenkel, *J. Chem. Phys*. **37**, 1156 (1962); **39**, 326 (1963); **40**, 1815 (1964); **41**, 699, 2077, 3623 (1964).

Freed, J. H., P. H. Rieger, and G. K. Fraenkel, *J. Chem. Phys*. **37**, 1881 (1962).

Friebele, E. J., D. L. Griscom, G. H. Sigel, Jr., *4th Int. Conf. Phys. Non-Crystalline Solids* (1976).

Friebele, E. J., D. O. Griscom, and G. H. Sigel, Jr., *J. Appl. Phys*. **45**, 3424 (1974).

Friebele, E. J., L. K. Wilson, and D. L. Kisner, *J. Am. Ceram. Soc*. **55**, 164 (1972).

Friederich, A., D. Kaplan, N. Sol, and R. H. Wallis, *J. Phys. Lett*. (*France*) **39**, 343 (1978).

Friedman, H. L., L. Blum, and G. Yue, *J. Chem. Phys*. **65**, 4396 (1976).

Froncisz, W., and J. S. Hyde, *J. Chem. Phys*. **73**, 3123 (1980).

Fryburg, G. C., and E. Gelerinter, *J. Chem. Phys*. **52**, 3378 (1970).

Furrer, R., and D. Stehlik, *Semicond. Insul. Proc. 3 Spec. Colloq. Ampere Opt. Tech. Magn. Reson. Spectrosc.* **4**, 295 (1978).

Gaffney, B. J., *Spin Labeling*, 183 (1976).

Gaffney, B. J., and D. C. Lin, *Enzymes Biol. Membr.* **1**, 71 (1976).

Galloway, N. B., and L. R. Dalton, *Chem. Phys.* **30**, 445 (1978); **32**, 189 (1978).

Ganapathy, S., and V. Parthasarathi, *Indian J. Pure Appl. Phys.* **15**, 63 (1977).

Gaponenko, V. A., L. V. Mosina, and Yu. V. Yablokov, *J. Struct. Chem.* **14**, 442 (1973).

Garrett, B. B., K. De Armond, and H. S. Gutowski, *J. Chem. Phys.* **44**, 3393 (1966).

Garrett, B. B., and L. O. Morgan, *J. Chem. Phys.* **44**, 890 (1966); C. C. Hinckley and L. O. Morgan, *J. Chem. Phys.* **44**, 898 (1966).

Garstens, M. A., *Phys. Rev.* **93**, 1229 (1954).

Garstens, M. A., L. S. Singer, and A. H. Ryan, *Phys. Rev.* **96**, 53 (1954).

Garstens, M. A., and J. I. Kaplan, *Phys. Rev.* **99**, 459 (1955).

Gerasimenko, N. N., V. B. Glazman, and A. V. Dvurechenskii, *Fiz. Tekh. Poluprovodn.* **9**, 1734 (1975).

Gersmann, H. K., and S. K. Swalen, *J. Chem. Phys.* **36**, 3221 (1962).

Gesmundo, F., *Nuovo Cimento B* **258**, 795 (1975).

Gewinner, G., L. Kubler, J. J. Koulmann, and A. Jaegle, *Phys. Status Solidi* **B59**, 395 (1975): **70**, 595 (1975).

Giardino, D. A., and C. O. Wisert, private communication, 1965.

Gillbro, T., A. Lund, and Y. Shimoyama, *Chem. Phys. Lett.* **32**, 529 (1975).

Glarum, S. H., and J. H. Marshall, *J. Chem. Phys.* **62**, 956 (1975).

Glarum, S. H., *RSI* **36**, 771 (1965).

Glarum, S. H., and J. H. Marshall, *J. Chem. Phys.* **44**, 2884 (1966).

Glarum, S. H., and J. H. Marshall, *J. Chem. Phys.* **46**, 55 (1967).

Glass, H. L., and M. T. Elliott, *J. Cryst. Growth* **34**, 285 (1976).

Glaunsinger, W. S., and M. J. Sienko, *J. Magn. Reson.* **10**, 253 (1973).

Glebashev, G. Ya., *Zh. Eksper. Teor. Fiz.* **32**, 82 (1957).

Goldberg, I. B., H. R. Crowe, and R. S. Carpenter II, *J. Magn. Reson.* **18**, 84 (1975).

Goldman, S. A., G. V. Bruno, and J. H. Freed, *J. Phys. Chem.* **76**, 1858 (1972); *J. Chem. Phys.* **59**, 3071 (1973).

Goldman, S. A., G. V. Bruno, C. F. Polnaszek, and J. H. Freed, *J. Chem. Phys.* **56**, 716 (1972).

Gooijer, C., N. H. Velthorst, and C. Maclean, *Mol. Phys.* **24**, 1361 (1972).

Gordon, R. G., and R. P. McGinnis, *J. Chem. Phys.* **49**, 2455 (1968).

Gouch, T. E., and R. G. Hacker, *J. Magn. Res.* **6**, 129 (1972).

Gourdon, J. C., P. Lopez, P. Rey, and J. Pescia, *C.R. Acad. Sci.* **271B**, 288 (1970).

Gourdon, J. C., C. Rey, C. Chachaty, J. C. Trombe, and J. Pescia, *C. R. Acad. Sci.* **276B**, 559 (1973).

Goyal, M. L., and B. M. Seth, *Acta Phys. Pol. A* **A52**, 787 (1977).

Grant, C. W. M., *Biophys. J.* **15**, 949 (1975).

Grant, W. J. C., *Physica* **30**, 1433 (1964); *Phys. Rev. A* **134**, 1554, 1564, 1574 (1964); **135**, 1265 (1964).

Grant, W. J. C., and M. W. P. Strandberg, *Phys. Rev. A* **135**, 715, 727 (1964).

Graslund, A., A. Ehrenberg, A. Rupprecht, G. Strom, and H. Crespi, *Int. J. Radiat. Biol.* **28**, 313 (1975).

Grechishkin, V. S., *Fiz. Tverd. Tela.* **18**, 2116 (1976).

Gregoli, S., M. Olast, and A. Bertinchamps, *Radiat. Res.* **70**, 255 (1977); **72**, 201 (1977).

Griscom, D. L., G. H. Sigel, Jr., and R. J. Ginther, *J. Appl. Phys.* **47**, 960 (1976).

Griswold, T. W., A. F. Kip, and C. Kittel, *Phys. Rev.* **88**, 951 (1952).

Grouchulski, T., K. Leiber, and Sienkiewicz, *Phys. Stat. Sol.* **47**, K169 (1978).

Gruber, K., J. Forrer, A. Schweiger, and Hs. H. Günthard, *J. Phys. E. Sci. Instr.* **7**, 569 (1974).

Gubanov, V. A., V. D. Inishev, and A. K. Chirkov, *Zh. Struk. Khim.* **13**, 349 (1972).

Gulley, J. E., and V. Jaccarino, *Phys. Rev. B* **6**, 58 (1972).

Gulley, J. E., D. Hone, D. Scalapino, and B. G. Silbernagel, *Phys. Rev. B* **1**, 1020 (1970).

Gulley, J. E., V. Jaccarino, and S. Foner, *Magnetism. Selected Topics*, London, England, 665 (1976).

Gupta, R. P., and M. S. Seehra, *Phys. Lett.* **33A**, 347 (1970).

Gupta, R. P., M. S. Seehra, and W. E. Vehse, *Phys. Rev. B* **5**, 92 (1972).

Gurak, J., M. Lynch, and G. Kokoszka, *Inorg. Nucl. Chem. Lett.* **12**, 927 (1976).

Gutowsky, H. S., G. B. Kistiakowsky, G. E. Pake, and E. M. Purcell, *J. Chem. Phys.* **17**, 972 (1949).

Gutowsky, H. S., and B. R. McGarvey, *J. Chem. Phys.* **21**, 1423 (1952).

Haken, H., and G. Strobl, in *The Triplet State*, A. B. Zahlen, Ed., Cambridge Univ. Press, Oxford, 1967.

Haken, H., and P. Reineker, *Z. Phys.* **249**, 253 (1972).

Haken, H., and G. Strobl, *Z. Phys.* **262**, 135 (1973).

Halpern, T., and W. D. Phillips, *RSI* **41**, 1038 (1970).

Hamashima, M., and Y. Ajiro, *J. Phys. Soc. Japan* **44**, 1743 (1978).

Hartig, G., and H. Dertinger, *Int. J. Radiat. Biol.* **20**, 577 (1971).

Hasegawa, S., and S. Yazaki, *Thin Solid Films* **55**, 15 (1978).

Hasegawa, S., S. Yazaki, and T. Shimizu, *J. Noncryst. Solids* **27**, 215 (1977).

Haupt, J., K. Kramer, and W. Muller Warmuth, *XIth Colloque Ampere*, p. 709, 1962.

Hauser, J. J., and R. S. Hutton, *Phys. Rev. Lett.* **37**, 868 (1976).

Hauser, C., *Helv. Phys. Acta* **45**, 683 (1972).

Hauser, C., and B. Renaud, *Phys. Status Solidi A* **10**, 161 (1972).

Hausser, K. H., *Z. Elek.* **65**, 636 (1961).

Heeger, A. J., in *Magnetism Selected Topics*, S. Foner, Ed., Gordon & Breach, London, England, 1976, p. 345.

Heinzer, J., *J. Magn. Res.* **13**, 124 (1974).

Helbert, J., L. Kevan, and B. L. Bales, *J. Chem. Phys.* **57**, 723 (1972).

Hemminga, M. A., *Chem. Phys.* **6**, 87 (1975).

Henling, L. M., and G. L. McPherson, *Phys. Rev. B* **16**, 4756 (1977).

Henner, E., I. Shaposhnikov, B. Bonis, and R. Sardos, *J. Magn. Reson.* **32**, 107 (1978).

Hensen, K., W. O. Riede, H. Sillescu, A. V. Wittgenstein, *J. Chem. Phys.* **61**, 4365 (1974).

Hentschel, R., J. Schlitter, H. Sillescu, and H. Spiess, *J. Chem. Phys.* **68**, 56 (1978).

Henning, J. C. M., and J. H. den Boef, *Phys. Rev. B* **18**, 60 (1978).

Hervé, J., *C.R. Acad. Sci.* **244**, 1182 (1957); **245**, 653 (1957); *Ann. Phys.* (*Paris*) **5**, 321 (1960).

Herring, F. C., C. A. McDowell, and J. C. Tait, *J. Chem. Phys.* **57**, 4564 (1972).

Herring, F. G., J. Mayo, and P. S. Phillips, *J. Magn. Reson.* **35**, 413 (1979).

Heuer, K., *Jenaer. Jahrbuch* **I**, 233 (1961).

Hibma, T., and J. Kommandeur, *Phys. Rev. B* **12**, 2608 (1975).

Hinkel, H., H. Port, H. Sixl, M. Schwoerer, P. Reineker, and D. Richardt, *Chem. Phys.* **31**, 101 (1978).

Hirokawa, S., *Mol. Phys.* **36**, 29 (1978).

Hirokawa, S., *J. Phys. Soc. Japan* **35**, 12 (1973); **37**, 897 (1974).

Hoel, D., and D. Kivelson, *J. Chem. Phys.* **62**, 1323 (1975); **62**, 4535 (1975).

Hoffman, K., and D. Poss, *Phys. Status Solidi A* **45**, 263 (1978).

Hoffmann, S. K., and J. R. Goslar, *Acta Phys. Polon.* **A48**, 707 (1975).

Hori, Y., and L. D. Kispert, *J. Chem. Phys.* **69**, 3826 (1978).

Houlihan, J. F., and L. N. Mulay, *Phys. Stat. Sol. B* **61**, 647 (1974).

Houlihan, J. F., W. J. Danley, and L. N. Mulay, *J. Solid State Chem.* **12**, 265 (1975).

Houlihan, J. F., D. P. Madaxsi, and L. N. Mylay, *Mat. Res. Bull.* **11**, 307 (1976).

Huang, T. Z., R. P. Taylor, and Z. G. Soos, *Phys. Rev. Lett.* **28**, 1054 (1972).

Huber, R. A., M. Schwoerer, H. Benk, and H. Sixl, *Chem. Phys. Lett.* **78**, 416 (1981).

Huber, H., and M. S. Seehra, *Phys. Lett.* **43A**, 311 (1973).

Huber, R. A., and M. Schwoerer, *Chem. Phys. Lett.* **72**, 10 (1980).

Huber, D. L., and M. S. Seehra, *Phys. Status Solidi B* **74**, 145 (1976).

Huber, D. L., and M. S. Seehra, *J. Phys. Solids* **36**, 723 (1975).

Huber, D. L., M. S. Seehra, and P. W. Verbeek, *Phys. Rev. B* **9**, 4988 (1974).

Hudson, A., and A. D. McLachlan, *J. Chem. Phys.* **43**, 1518 (1965).

Hudson, A., *Electron Spin Reson.* **3**, 62 (1976).

Hughes, D. G., and T. J. Rowland, *Can. J. Phys.* **42**, 209 (1964).

Hughes, R. C., B. Morosin, P. M. Richards, and W. Duffy, Jr., *Phys. Rev. B* **11**, 1795 (1975).

Hughes, R. C., and Z. G. Soos, *J. Chem. Phys.* **52**, 6302 (1970).

Hutchison, C. A., Jr., J. W. Nicholas, and G. W. Scott, *J. Chem. Phys.* **53**, 1906 (1970).

Hutchison, C. A., Jr., and J. S. King, Jr., *J. Chem. Phys.* **58**, 392 (1973).

Hutchison, C. A., Jr., and G. W. Scott, *J. Chem. Phys.* **61**, 2240 (1974).

Hwang, J. S., R. P. Mason, L. P. Hwang, and J. H. Freed, *J. Phys. Chem.* **79**, 489 (1975).

Hyde, J. S., and J. R. Pilbrow, *J. Magn. Reson.* **41**, 447 (1980).

Hyde, J. S., and L. Dalton, *Chem. Phys. Lett.* **16**, 568 (1972).

Hyde, J. S., and D. D. Thomas, *Ann. N.Y. Acad. Sci.* **222**, 689 (1974).

Hyde, J. S., M. D. Smigel, L. R. Dalton, and L. A. Dalton, *J. Chem. Phys.* **62**, 1655 (1975).

Ibers, J. A., and J. D. Swalen, *Phys. Rev.* **127**, 1914 (1962).

Ikebe, M., and M. Date, *J. Phys. Soc. Japan* **30**, 93 (1971).

Isoya, J., J. A. Weil, and R. F. C. Claridge, *J. Chem. Phys.* **69**, 4876 (1978).

Iwaizumi, M., and T. Isobe, *Bull. Chem. Soc.* **38**, 1547 (1965).

Iwasaki, M., *J. Chem. Phys.* **45**, 990 (1966).

Iwasaki, M., K. Toriyama, and B. Eda, *J. Chem. Phys.* **42**, 63 (1965).

Jahnke, E., and F. Emde, *Tables of Functions*, Dover, New York, 1945, p. 73.

Jain, A. K., and G. C. Upreti, *Mol. Phys.* **34**, 273 (1977).

Janecka, J., H. M. Vyas, and M. Fujimoto, *J. Chem. Phys.* **54**, 3229 (1971).

Janossy, A., *Magy. Fiz. Poly.* **23**, 213 (1975).

Janossy, A., P. Monod, P. S. Allen, E. R. Andrew, and C. A. Bates, *Proc. 18th Ampere Congr. Magn. Reson. Relat. Phenom.* **2**, 313 (1974).

Jansen, K., and G. Sperlich, *Solid State Commun.* **17**, 1179 (1975).

Janssens, L., and J. Witters, *17th Cong. Ampere Nucl. Magn. Reson. Relat. Phenom. Turku, Finland* (1972).

Jarrett, H. S., J. E. Gulley, and P. C. Hoell, *AIP Conf. Proc. 20th Annu. Conf. Magn. Magn. Mater.* **24** (1974).

Jinguji, M., K. C. Lin, C. A. McDowell, and P. Raghunathan, *J. Chem. Phys.* **65**, 3910 (1976).

Johnson, C. S., Jr., and R. Chang, *J. Chem. Phys.* **43**, 3183 (1965).

Johnson, J. E., and G. C. Moulton, *J. Chem. Phys.* **69**, 3108 (1978).

Johnston, T. S., and H. G. Hecht, *J. Mol. Spectrosc.* **17**, 98 (1965).

Judeikis, H. S., *J. Appl. Phys.* **35**, 2615 (1964).

Kahn, A. H., *Phys. Rev. B* **16**, 64 (1977).

Kalyanaraman, B., and L. D. Kispert, *J. Chem. Phys.* **68**, 5219 (1978).

Kambe, K., and J. F. Ollom, *J. Phys. Soc. Japan* **11**, 50 (1956).

Kambe, K., and T. Usui, *Progr. Theor. Phys.* **8**, 3029 (1952).

Kaminski, W., *Z. Naturforsch.* **25a**, 639 (1976).

Kaplan, J. I., *J. Chem. Phys.* **28**, 278 (1958); **34**, 2205 (1961); **42**, 3789 (1965); *J. Phys. Soc. Japan* **19**, 1994 (1964), *J. Magn. Reson.* **21**, 153 (1976).

Kasai, P. H., and D. McLeod, *J. Chem. Phys.* **55**, 1566 (1971).

Kasai, P. H., W. Weltner, and E. B. Whipple, *J. Chem. Phys.* **42**, 1120 (1965); **44**, 5281 (1966).

Kawakubo, I., *Mol. Cryst. Liq. Cryst.* **55**, 163 (1979).

Kawamori, A., and S. Matsuura, *J. Phys. Soc. Japan* **29**, 1173 (1970).

Kawano, K., R. Nakata, and M. Sumita, *Rep. Univ. Electro-Commun.* (*Japan*) **26**, 19 (1975).

Kawasaki, K., *Prog. Theor. Phys.* **39**, 285 (1968).

Kemple, M. D., and H. J. Stapleton, *Phys. Rev. B* **5**, 1668 (1972).

Kennedy, T. A., S. H. Choh, and G. Seidel, *Phys. Rev. B* **2**, 3645 (1970).

Kertesz, J. C., M. B. Wolf, and W. Wolf, *Comput. Programs Biomed.* **4**, 21 (1974).

Kevan, L., P. A. Narayana, K. Toriyama, and M. Iwasaki, *J. Chem. Phys.* **70**, 5006 (1979).

Kiel, A., *Phys. Rev.* **126**, 1292 (1962); **2**, 525 (1963).

Kikuchi, C., and W. V. Cohen, *Phys. Rev.* **93**, 394 (1954).

Kirillov, S. T., M. A. Kozhushner, and V. B. Stryukov, *Zh. Eksp. Teor. Fiz.* **68**, 2249 (1975).

Kirillov, V. I., V. S. Postnikov, S. I. Rembeza, and A. I. Spirin, *Fiz. Tverd. Tela* **18**, 1108 (1976).

Kirillov, S. T., and V. B. Stryukov, *Zh. Fiz. Khim.* **50**, 2938 (1976).

Kishimoto, N., K. Morigaki, K. Murakami, A. Shimizu, and A. Hiraki, *Phys. Status Solidi B* **80**, K113 (1977).

Kittel, C., and E. Abrahams, *Phys. Rev.* **90**, 238 (1953).

Kivelson, D., *J. Chem. Phys.* **27**, 1087 (1957); **33**, 1094 (1960); **45**, 1324 (1966).

Kivelson, D., M. G. Kivelson, and I. Oppenheim, *J. Chem. Phys.* **52**, 1810 (1970).

Kloza, M., *Pomiary Autom. Kontr.* **17**, 337 (1971).

Kneubuhl, F. K., W. S. Koski, and W. S. Caughey, *J. Am. Chem. Soc.* **83**, 1607 (1961).

Kneubuhl, F. K., *J. Chem. Phys.* **33**, 1074 (1960); *Helv. Phys. Acta* **35**, 259 (1962).

Kneubuhl, F. K., and B. Natterer, *Helv. Phys. Acta* **35**, 710 (1961).

Knight, L. B., W. C. Easley, and W. Weltner, *J. Chem. Phys.* **54**, 1610 (1971).

Knight, W. D., *AIP Conf. Proc.*, 88 (1976).

Kodera, H., *J. Phys. Soc. Japan* **28**, 89 (1970).

Kohin, R. P., and C. P. Poole, Jr., *Bull. Am. Phys. Soc.* **11**, 3, 8 (1958).

Köhler, R., *Ann. Phys.* (*Germany*) **15**, 389 (1965).

Koloskova, N. G., and U. Kh. Kopvillem, *Fiz. Tverd. Tela* **2**, 1368 (1960).

Kol'tover, V. K., *Itogi Nauki Tekh. Biofiz* **4**, 16 (1975).

Komatsubara, T., and K. Nagata, *J. Phys. Soc. Japan* **45**, 826 (1978).

Komatsubara, T., K. Iio, and K. Nagata, *J. Phys. Soc. Japan* **45**, 351 (1978).

Kondal, S. C., and M. S. Seehra, *Phys. Rev. B.* **22**, 5482 (1980); *J. Phys. C. Solid State*, to be published.

Konzelmann, U., D. Kilpper, and M. Schwoerer, *Z. Naturforsch A* **30a**, 754 (1975).

Kool, Th. W., and M. Glasbeek, *Solid State Commun.* **22**, 103 (1977).

Koopmann, G., K. Babarschke, and U. Engel, *Solid State Commun.* **12**, 997 (1973).

Kooser, R. G., W. V. Volland, and J. H. Freed, *J. Chem. Phys.* **50**, 5243 (1969).

Kopvillem, U. Kh., *Zh. Eksper. Teor. Fiz.* **34**, 1040 (719) (1958); **38**, 151 (109) (1960).

Korolev, V. D., S. A. Peskovatskii, V. P. Shakhparyan, and V. Shulga, *Fiz. Tverdogo Tela* **17**, 150 (1975).

Korolkov, V. S., and A. K. Potapovich, *Optika i Sepkt.* **16**, 461 (251) (1964).

Korst, N. N., A. N. Kuznetsov, A. V. Lazarev, and E. P. Gordeev, *Teor. Eksp. Khim.* **8**, 51 (1972).

Korst, N. N., V. A. Savel'ev, and N. D. Sokolov, *Soviet Phys. Dokl.* **7**, 1037 (1963).

Korst, N. N., and L. I. Antsiferova, *Uspekh.* **126**, 67 (1978).

Korst, N. N., and T. N. Khazanovich, *Sov. Phys. JETP* **18**, 1049 (1964).

Koryakov, V. I., A. A. Maksimov, and V. A. Banov, *Tr. Inst. Khim. Ural. Nauchn. Tsentr., Akad. Nauk SSSR* **34**, 106 (1975).

Kothe, G., E. Ohmes, F. A. Neugebauer, and H. Zimmerman, *Ber. Bunsenges. Phys. Chem.* **78**, 206 (1974).

Kothe, G., *Mol. Phys.* **33**, 147 (1977).

Kothe, G., A. Naujok, and E. Ohmes, *Mol. Phys.* **32**, 1215 (1976).

Kothe, G., and E. Ohmes, *J. Mol. Struct.* **46**, 481 (1978).

Kotomin, E. A., O. V. Plotnikov, E. Rajavee, and J. Jansons, *Zh. Prikl. Specktrosk.* **22**, 363 (1975).

Kottis, P., and R. Lefebvre, *J. Chem. Phys.* **39**, 383 (1963); **41**, 379, 3660 (1964).

Kottis, P., *J. Chem. Phys.* **47**, 836 (1967).

Kowert, B., and D. Kivelson, *J. Chem. Phys.* **64**, 5106 (1976).

Kozhukhar, A. Yu., A. D. Prokhorov, and G. A. Tsintsadze, *Fiz. Tverd. Tela* **20**, 550 (1978).

Kozhukhar, A. Yu. and G. A. Tsintsadze, *Fiz. Tverd. Tela* **17**, 3414 (1975).

Krebs, P., and E. Sackmann, *J. Magn. Reson.* **22**, 359 (1976).

Krishnamurthy, M. V., *J. Phys. Chem. Solids* **33**, 1645 (1972).

Krupski, M., and J. Stankowski, *Acta Phys. Pol.* **A50**, 685 (1976).

Krusic, P. J., P. Meakin, and J. P. Jesson, *J. Phys. Chem.* **75**, 3438 (1971).

Kubo, R., and K. Tomita, *J. Phys. Soc. Japan* **9**, 888 (1954).

Kubler, L., G. Gewinner, J. J. Koulmann, and A. Jaegle, *Phys. Status Solidi* **B78**, 149 (1976).

Kumano, M., and Y. Ikegami, *Chem. Phys. Lett.* **54**, 109 (1978).

Kumeda, M., Y. Jinno, and T. Shimizu, *Phys. Status Solidi B* **81**, K71 (1977).

Kummer, H., *Helv. Phys. Acta* **35**, 901 (1963).

Kunii, S., and T. Kasuya, *J. Phys. Soc. Japan* **46**, 13 (1979).

Kushida, T., J. C. Murphy, and M. Hanabusa, *Solid State Commun.* **15**, 1217 (1974).

Kuznetsov, A. N., A. Y. Volkov, V. A. Livshits, and A. T. Nirzoian, *Chem. Phys. Lett.* **26**, 369 (1974).

Lagendijk, A., and D. Schoemaker, *Phys. Rev. B* **16**, 47 (1977).

Lagendijk, A., *Physica B & C* **83B & C**, 283 (1976).

Lahiry, S., J. Sharma, G. D. Sootha, and H. O. Gupta, *Phys. Status Solidi* **46**, 153 (1978).

Lamotte, B., and P. Gloux, *J. Chem. Phys.* **59**, 3365 (1973).

Lamotte, B., and J. Gaillard, *J. Chem. Phys.* **57**, 3319 (1972).

Lang, R., C. Calvo, and W. R. Datars, *Can. J. Phys.* **55**, 1613 (1977).

Larin, G. M., *Teor. Eksp. Khim.* **4**, 244 (1968).

Lardon, M., and H. H. Günthard, *J. Chem. Phys.* **44**, 2010 (1966).

Larys, L., J. Stankowski, and M. Krupski, *Acta Phys. Pol. A* **A50**, 351 (1975).

Lazjzerowicz-Bonneteau, J., *Spin Labeling*, 239 (1976).

Lebedev, Ya. S., *Zh. Strukl. Khim.* **4**, 19 (1962).

Lebedev, Ya. S., D. M. Chernikova, N. N. Tikhomirova, and V. V. Voevodskii, *Atlas of Electron Spin Resonance Spectra*, Consultants Bureau, New York, 1963.

Leblond, J., and J. Uebersfeld, *Phys. Rev. A* **4**, 4 (1971).

Leblond, J., J. Uebersfeld, and K. Korringa, *Phys. Rev. A* **4**, 1532 (1971).

LeCraw, R. C., H. Von Philsbrom, and M. D. Storge, *J. Appl. Phys.* **38**, 965 (1967).

Ledgerwood, L. W., Jr., *IEEE Trans. Educ.* **E-19**, 100 (1976).

Lee, S., *J. Magn. Reson.* **31**, 351 (1978).

Lee, S., and C. M. Brodbeck, *Phys. Rev. B* **16**, 4743 (1977).

Lee, J. Y., and M. T. Rogers, *J. Chem. Phys.* **65**, 580 (1976).

Lefebvre, R., *J. Chem. Phys.* **33**, 1826 (1960); **34**, 2035 (1961); **35**, 762 (1961).

Lefebvre, R., and J. Maruani, *J. Chem. Phys.* **42**, 1480, 1496 (1965); see also J. Maruani, *Proc. 12th Colloque Ampere*, Bordeaux, 1963, p. 303.

Leigh, J. S., Jr., *J. Chem. Phys.* **52**, 2608 (1970).

Lemaistre, J. P., and P. Kottis, *J. Chem. Phys.* **68**, 2730 (1978).

Lemaistre, J. P., and A. H. Zewail, *J. Chem. Phys.* **72**, 1055 (1980).

Lepine, D. J., *Phys. Rev. B* **2**, 249 (1970).

Levanon, H., S. Charbinsky, and Z. Luz, *J. Chem. Phys.* **53**, 3056 (1970); **66**, 955 (1977).

Levanon, H., G. Stein, and Z. Luz, *J. Chem. Phys.* **53**, 876 (1970).

Leyko, W., G. Bartosz, and K. Gwozdzinski, *Cancer Ther. Hyperthermia Radiat. Proc. Int. Symp.*, 163 (1978).

Libertini, L. J., A. S. Waggoner, P. C. Jost, and O. H. Griffith, *Proc. Natl. Acad. Sci.* **64**, 13 (1969).

Libertini, L. J., C. A. Burke, P. C. Jost, and O. H. Griffith, *J. Magn. Reson.* **15**, 460 (1974).

Lin, W. L., and J. H. Freed, *J. Phys. Chem.* **83**, 379 (1979).

Lin, D. P., and L. Kevan, *J. Phys. Chem.* **81**, 1498 (1977).

Ling, A. C., *J. Chem. Ed.* **51**, 174 (1974).

Lippe, R., W. Windsch, G. Volkel, and W. Schulga, *Solid State Commun.* **19**, 587 (1976).

Lipiarz, Z., *Chem. Abstr.* **82**, 25330 (1975).

Livingston, R., and H. Zeldes, *J. Chem. Phys.* **24**, 170 (1956).

Livshits, V. A., *J. Magn. Reson.* **24**, 307 (1976).

Livshits, V. A., *Zh. Fiz. Khim.* **50**, 808 (1976).

Lord, N. W., and S. M. Blinder, *J. Chem. Phys.* **34**, 1694 (1961).

Loudon, R., *Phys. Rev.* **119**, 919 (1960).

Loveland, D. B., and T. N. Tozer, *J. Phys. E* **5**, 535 (1972).

Luckhurst, G. R., *Spin Labeling*, 133 (1976).

Luckhurst, G. R., and M. Setaka, *J. Chem. Soc.* **72**, 1340 (1976).

Luckhurst, G. R., and R. N. Yeates, *E.S.R. in Nematic Phases*, 996 (1976).

Luckhurst, G. R., and C. Zannoni, *J. Magn. Reson.* **23**, 265 (1976).

Lundin, A. A., and V. E. Sobov, *J. Magn. Reson.* **26**, 229 (1977).

Lünd, A., and T. Vänngård, *J. Chem. Phys.* **42**, 2979 (1965).

McGavin, D. G., and W. C. Tennant, *Mol. Phys.* **32**, 1477 (1976).

McGregor, K. T., R. P. Scaringe, and W. E. Hatfield, *Mol. Phys.* **30**, 1925 (1975).

McGregor, K. T., and Z. G. Soos, *J. Chem. Phys.* **64**, 2506 (1976).

McGregor, K. T., and Z. G. Soos, *Inorg. Chem.* **15**, 2159 (1976).

McConnell, H. M. and B. G. McFarland, *J. Biophys.* **3**, 91 (1970).

McClintock, I. S., and J. C. Orr, *Phys. State Sol.* **29**, K157 (1968).

McDowell, C. A., H. Nakajima, and P. Raghumathan, *Can. J. Chem.* **48**, 905 (1970).

McDowell, C. A., P. Raghunathan, and J. C. Tait, *J. Chem. Phys.* **59**, 5858 (1973).

McDowell, C. A., and K. Shimokoshi, *J. Chem. Phys.* **60**, 1619 (1974).

McHargue, C. J., and J. L. Scott, *Metall. Trans. A. Symp. Materials Requirements Unconventional Energy Syst.* **9A**, No. 2, Sept. 1976 (Niagara Falls).

McMillan, M., and W. Opechowski, *Can. J. Phys.* **39**, 1168 (1960); **39**, 1369 (1961).

McMillian, J. A., and S. A. Marshall, *J. Chem. Phys.* **48**, 467 (1968).

Mackey, J. H., M. Kopp, E. C. Tynan, and Y. T. Teh Fu Yen, *Proc. Symp. Electron Spin Resonances of Metal Complexes*, Cleveland, Ohio, 1969, p. 33.

Mackowiak, M. M., *Nuovo Cimento B* **29B**, 207 (1975); *Radiospektrosk. Ciala Stalego*, 305 (1975).

Maekawa, S., *J. Phys. Soc. Japan* **33**, 573 (1972).

Mailer, C., H. Thomann, B. H. Robinson, and L. E. Dalton, *RSI* **51**, 1714 (1980).

Malley, M. M., *J. Mol. Spectrosc.* **17**, 210 (1965).

Maniv, S., A. Reuveni, and Z. Luz, *J. Chem. Phys.* **66**, 2285 (1977).

Mao, C. R., and R. W. Kreilick, *Mol. Phys.* **31**, 1447 (1976).

Marguardt, D. W., R. G. Bennett, and E. J. Burrell, *J. Mol. Spectrosc.* **7**, 269 (1961).

Marrone, M. J., F. W. Patten, and M. N. Kabler, *Phys. Rev. Lett.* **31**, 467 (1973).

Marshall, S. A., T. Marshall, and R. A. Serway, *Phys. Status Solidi* **48**, 165 (1978).

Martinelli, M., L. Pardi, C. Pinzino, and S. Santucci, *Phys. Rev.* **16**, 164 (1977).

Marušič, M., and C. F. Schwerdtfeger, *Mol. Cryst. Liq. Cryst.* **38**, 131 (1973).

Maruani, J., J. A. Coope, and C. A. McDowell, *Mol. Phys.* **18**, 165 (1970).

Maruani, J., C. A. McDowell, H. Nakajima, and P. Raghunathan, *Mol. Phys.* **14**, 349 (1968).

Mason, R. P., C. F. Polnaszek, and J. H. Freed, *J. Phys. Chem.* **78**, 1324 (1974).

Mavroyannis, S., *Phys. Rev. B* **17**, 2871 (1978).

Mayer, J. E., *J. Chem. Phys.* **1**, 270, 327 (1933).

Menard, M., and M. B. Walker, *Can. J. Phys.* **52**, 61 (1974).

Mengeot, M., R. H. Bartram, and O. R. Gilliam, *Phys. Rev. B* **11**, 4110 (1975).

Menzel, E. R., and J. R. Wasson, *J. Phys. Chem.* **79**, 366 (1975).

Meservey, R., D. Paraskevopoulos, and P. M. Tedrow, *Physica B and C*, **91B & C**, 91 (1977).

Meservey, R., and P. M. Tedrow, *Phys. Rev. Lett.* **41**, 805 (1978).

Meservey, R., P. M. Tedrow, V. R. Kalvey, and D. Paraskevopoulos, *J. Appl. Phys.* **50**, Pt. 2, 1935 (1979).

Meservey, R., P. M. Tedrow, D. Paraskevopoulos, M. J. G. Lee, J. M. Perz, and E. Fawcett, *Transition Metals, 1977, Toronto, Canada*, 262 (1978).

Meservey, R., and T. Wolfram, *Proc. Int. Conf. Symp. on Electron Tunneling, Columbia, Mo.*, 230 (1978).

Meyer, H. C., J. S. Bennett, P. L. Donoho, and A. C. Daniel, *Bull. Am. Phys. Soc.* **11**, 202 (1966).

Mialhe, P., A. Briguet, and B. Triballet, *J. Phys. Chem. Solids* **32**, 2639 (1971).

Mialhe, P., and A. Erbeia, *Phys. Rev. B* **7**, 4061 (1973).

Mialhe, P., and P. Quedec, *Phys. Rev. B* **14**, 2757 (1976).

Michalik, J., and L. Kevan, *J. Magn. Reson.* **31**, 259 (1978).

Miller, J. R., *J. Chem. Phys.* **56**, 5173 (1972).

Misra, S. K., *J. Magn. Reson.* **23**, 403 (1976).

Misra, B. N., P. Giquere, and G. R. Sharp, *J. Chem. Phys.* **66**, 1758 (1973).

Misra, B. N., S. D. Sharma, and S. K. Gupta, *Acta Phys. Pol. A* **49**, 41 (1976).

Misra, S. K., and G. R. Sharp, *J. Magn. Reson.* **23**, 191 (1976).

Misra, B. N., and S. D. Sharma, *J. Chem. Phys.* **63**, 5322 (1975).

Misra, B. N., S. D. Sharma, and S. K. Gupta, *Acta Phys. Pol. A* **49**, 14 (1976).

Misra, B. N., S. D. Sharma, and S. K. Gupta, *Acta Phys. Pol. A* (*Poland*) **49**, 35 (1976).

Miyadai, T., S. Miyahara, and Teruo Teranishi, *J. Phys. Soc. Japan* **31**, 1591 (1971).

Mizokawa, V., and S. Nakamura, *Japan J. Appl. Phys.* **14**, 779 (1975).

Mohos, B., *Magn. Reson. Chem. Biol. Lect. Ampere Int. Summer Sch.*, 187 (1975).

Mohrmann, L. E., B. B. Garrett, and W. B. Lewis, *J. Chem. Phys.* **52**, 535 (1970).

Möhwald, H., and E. Sackmann, *Solid State Commun.* **15**, 445 (1974).

Monod, P., *J. Phys. Colloque C6, Supplement 8* **39**, C6-1472 (1978).

Mooij, J. J., and E. De Boer, *Mol. Phys.* **32**, 113 (1976).

Mooy, J. H. M., and J. Bolhuis, *Solid State Commun.* **19**, 1005 (1976).

Morantz, D. J., and R. Thompson, *J. Phys. C Solid State Phys.* **3**, 1335 (1970).

Mori, H., *Prog. Theor. Phys.* **30**, 478 (1963).

Mori, H., and K. Kawasaki, *Prog. Theor. Phys.* **38**, 971 (1962).

Morigaki, K., S. Toyotomi, and Y. Toyotomi, *J. Phys. Soc. Japan* **31**, 511 (1971).

Morimoto, Y., and M. Date, *Prog. Theor. Phys.* **28**, 971 (1970).

Moriuchi, S., and J. Sohma, *Mem. Fac. Eng., Hokkaido Univ.* **2**, 147 (1972); 335 (1974).

Moriuchi, S., and J. Sohma, *Mol. Phys.* **21**, 369 (1971).

Mosina, L. V., and Yu. V. Yablokov, *Phys. Stat. Solidi* **72**, K51 (1974).

Mottley, C., L. D. Kispert, and S. Clough, *J. Chem. Phys.* **63**, 4405 (1975).

Mottley, C., L. Kispert, P. S. Pu Sen Wang, E. R. Andrew, and C. A. Bates, *Proc. 18th Ampere Congr. Magn. Reson. Related Phenom.* **II**, Nottingham, England, 385 (1975).

Movaghar, B., and L. Schweitzer, *Phys. Status Solidi B* **80**, 491 (1977).

Muha, G. M., and J. Sakamoto, *J. Chem. Phys.* **68**, 1432 (1978).

Mukai, K., and J. Sakamoto, *J. Chem. Phys.* **68**, 1432 (1978).

Muller, K. A., and W. Berlinger, *Phys. Rev. Lett.* **29**, 715 (1972).

Muller, K. A., W. Berlinger, C. H. West, and P. Heller, *Phys. Rev. Lett.* **32**, 160 (1974).

Muller, K. A., and Th. Von Waldkirch, *Local Properties at Phase Transitions*, IBM Zurich Research Labatory, Ruschlikon, 1975.

Murakami, K., and A. Kawamori, *Solid State Commun.* **22**, 47 (1977).

Murakami, K., S. Namba, N. Kishimoto, K. Masuda, and K. Gamo, *Appl. Phys. Lett.* **30**, 300 (1977); *J. Appl. Phys.* **49**, 2401 (1978).

Muramoto, T., *J. Phys. Soc. Japan* **35**, 921 (1973).

Murthy, S. V., *RSI* **34**, 106 (1963).

Nagamura, T., and A. E. Woodward, *J. Polym. Sci.* **14**, 275 (1976).

Nagasawa, H., *J. Phys. Soc. Japan* **20**, 1808 (1965).

Nagata, K., and M. Date, *J. Phys. Soc. Japan* **19**, 1823 (1964).

Nagata, K., *J. Phys. Soc. Japan* **40**, 1209 (1976).

Nagata, K., T. Nishino, and T. Hirosawa, *J. Phys. Soc. Japan* **44**, 813 (1978).

Nagata, K., I. Yamada, T. Komatsubara, and T. Nishizaki, *J. Phys. Soc. Japan* **43**, 707 (1977).

Nagai, S., S. Ohnishi, and I. Nitti, *Chem. Phys. Lett.* **13**, 379 (1972).

Narayana, P. A., and K. V. L. N. Sastry, *J. Chem. Phys.* **57**, 1805 (1972).

Nazarian, A., Y. H. Shing, D. Walsh, and G. Donnay, *Solid State Commun.* **10**, 1005 (1972).

Neiman, R., and D. Kivelson, *J. Chem. Phys.* **35**, 156 (1961).

Neimark, E. I., and B. E. Vugmeister, *Fiz. Tverd. Tela* **17**, 2193 (1975).

Nelsen, S. F., R. T. Landis II, and J. C. Calabrese, *J. Org. Chem.* **42**, 4192 (1977).

Nguyen, T. T., L. B. Chiu, P. R. Elliston, A. M. Stewart, and K. N. R. Taylor, *Physica* **86-88B**, 181 (1977).

Niarchos, D., A. Kostikos, A. Simopoulos, D. Coucouvanis, D. Piltingsrud, and R. E. Coffman, *J. Chem. Phys.* **68**, 4411 (1978).

Nicklin, R. C., H. A. Farach, and C. P. Poole, Jr., *J. Chem. Phys.* **65**, 2998 (1976).

Nishikawa, T., and K. Someno, *Anal. Chem.* **47**, 1290 (1975).

Nordio, P. L., *Chem. Phys. Lett.* **7**, 250 (1970).

Nordio, P. L., *Spin Labeling*, 5 (1975).

Nordio, P. L., and U. Segre, *Chem. Phys.* **11**, 57 (1976).

Nordio, P. L., G. Rigatti, and U. Segre, *J. Chem. Phys.* **56**, 2117 (1972).

Nothe, D., M. Moroni, and H. J. Keller, *Solid State Commun.* **26**, 713 (1978).

Ochiai, Y., and E. Matsuura, *Phys. Status Solidi A* **38**, 243 (1976).

Ochiai, Y., and E. Matsuura, *Phys. Status Solidi A* **45**, K101 (1978).

Odgaard, E., T. B. Melo, and T. Henriksen, *J. Magn. Reson.* **18**, 436 (1975).

Ohkura, H., Y. Mori, M. Matsuoka, T. Watanabe, and A. Satoh, *Prog. Theor. Phys. Suppl.* (*Japan*) **57**, 68 (1975).

Okuda, K., H. Hata, and M. Date, *J. Phys. Soc. Japan* **33**, 1574 (1972).

Olson, D. W., and T. G. Castner, *Phys. Rev. B* **17**, 3318 (1978).

Onda, M., and K. Morigaki, *J. Phys. Soc. Japan* **34**, 1107 (1973).

Oluwole, A. F., H. B. Olantyt, S. G. Schmelling, and J. B. Aladekomo, *J. Phys. B* **8**, 2615 (1973).

O'Reilly, D. E., *J. Chem. Phys.* **29**, 1188 (1958).

O'Reilly, D. E., and C. P. Poole, Jr., *J. Phys. Chem.* **67**, 1762 (1963).

O'Reilly, D. E., and T. Tsang, *Phys. Rev.* **128**, 2639 (1962); **131**, 2522 (1963).

Oshina, K., K. Okuda, and M. Date, *J. Phys. Soc. Japan* **41**, 475 (1976).

Ovchinnikov, I. V., and V. N. Konstantinov, *Fiz. Tverd. Tela* **18**, 1478 (1976).

Ovchinnikov, I. V., and V. N. Konstantinov, *J. Magn. Reson.* **32**, 179 (1978).

Owens, F. J., *J. Phys. Chem. Solids* **39**, 291 (1978).

Owens, F. J., *Chem. Phys. Lett.* **46**, 380 (1977); *Phys. Stat. Solidi* **B79**, 623 (1977).

Owens, F. J., *J. Chem. Phys.* **57**, 2349 (1972).

Owens, F. J., in *Magnetic Resonance of Phase Transitions*, F. T. Owens, C. P. Poole, Jr., and H. A. Farach, Academic Press, New York, 1979, Chap. 6.

Owens, F. J., C. P. Poole, Jr., and H. A. Farach, *Magnetic Resonance of Phase Transitions*, Academic Press, New York, 1979.

Owen, G. S., and G. Vincow, *J. Chem. Phys.* **34**, 368 (1971).

Pake, G. E., *J. Chem. Phys.* **15**, 327 (1948).

Pake, G. E., and E. M. Purcell, *Phys. Rev.* **74**, 1184 (1948); **75**, 534 (1949).

Panepucci, H., and H. A. Farach, *Med. Phys.* **4**, 46 (1977).

Parmon, V. N., A. I. Kukurin, and G. M. Zhidomirov, *Zh. Strukt. Khim.* **18**, 132 (1977).

Parmon, V. N., A. I. Kukurin, G. M. Zhidomirov, and K. I. Zamaraev, *Mol. Phys.* **30**, 695 (1975).

Parmon, V. N., and G. M. Zhidomirov, *Mol. Phys.* **32**, 613 (1976).

Pedersen, E., *J. Phys. E* **5**, 492 (1972).

Pekcan, O., and E. Keyman, *Chem. Phys. Lett.* **69**, 78 (1980).

Peter, M., D. Shaltiel, J. H. Wernick, H. J. Williams, J. B. Mock, and R. C. Sherwood, *Phys. Rev.* **126**, 1395 (1962).

Petrov, M. P., and S. A. Kizhaev, *Sov. Phys. Solid State* **11**, 1968 (1970).

Petrova, A. G., A. V. Rakov, V. P. Yarigin, E. N. Ivanov, and Yu. I. Pashintsev, *Fiz. Tekh. Poloprov* **5**, 1140 (1971).

Philbrick, S. E., and G. Portinari, *Conference on Distribution Cables and Jointing Techniques for Systems up to 11 Kv.*, London, England, **1**, 67 (1976).

Pifer, J. H., and R. Magno, *Phys. Rev. B* **3**, 661 (1971).

Pijpers, F. W., M. R. Arick, B. M. P. Hendriks, and E. DeBoer, *Mol. Phys.* **22**, 781 (1971).

Pilbrow, J. R., and M. E. Winfield, *Mol. Phys.* **25**, 1073 (1973).

Pilo-Veloso, D., R. Ramasseul, A. Rassat, and P. Rey, *Tetrahedron Lett.*, 3599 (1976).

Plato, M., and K. Möbius, *Nesstechnik* **80**, 224 (1972).

Poggi, G., and C. S. Johnson, Jr., *J. Mag. Res.* **3**, 436 (1970).

Polak, M., *J. Chem. Phys.* **65**, 2785 (1976).

Polnaszek, C. F., and J. H. Freed, *J. Phys. Chem.* **79**, 2283 (1975).

Pomerantz, M., F. H. Dacol, and A. Segmuller, *Phys. Rev. Lett.* **40**, 246 (1978).

Pomponiu, C., and A. T. Balaban, *Computing Program for Interpreting ESR Spectra of Organic Free Radicals*, 1972.

Poole, C. P., Jr., and H. A. Farach, *Relaxation in Magnetic Resonance*, Academic Press, New York, 1971.

Poole, C. P., Jr., and H. A. Farach, *Theory of Magnetic Resonance*, John Wiley, New York, 1972.

Poole, C. P., Jr., and H. A. Farach, *Bull. Magn. Reson.* **1**, 162 (1980).

Poole, C. P., Jr., and H. A. Farach, *J. Magn. Reson.* **4**, 312 (1971); **5**, 305 (1971).

Poole, C. P., Jr., and H. A. Farach, in Electron Spin Resonance, in Vol. 2, *Handbook of Spectroscopy*, J. W. Robinson, Ed., CRC Press, 1974, Chap. 4.

Poole, C. P., Jr., H. A. Farach, and R. C. Nicklin, in Vol. 3, *Handbook of Spectroscopy*, J. W. Robinson, Ed., CRC Press, 1981.

Poole, C. P., Jr., and R. S. Anderson, *J. Chem. Phys.* **31**, 346 (1959).

Poole, C. P., Jr., W. C. Kehl, and D. S. MacIver, *J. Catal.* **1**, 407 (1962).

Poole, C. P., Jr., and J. E. Itzel, *J. Chem. Phys.* **41**, 287 (1964).

Poole, C. P., Jr., and D. S. MacIver, *Adv. Catal.* **17**, 223 (1967).

Popp, C. A., and J. S. Hyde, *J. Magn. Reson.* **43**, in press (1981).

Porumb, T., and E. F. Slade, *J. Magn. Reson.* **22**, 219 (1976).

Posener, D. W., *Aust. J. Phys.* **12**, 184 (1959).

Posener, D. W., *J. Comput. Phys.* **11**, 326 (1973).

Poupko, R., *J. Magn. Res.* **12**, 119 (1973).

Pryce, M. H. L., and K. W. H. Stevens, *Proc. Phys. Soc. London A* **63**, 36 (1950).

Püsnik, F. and M. Schara, *Chem. Phys. Lett.* **37**, 106 (1976).

Rao, K. V. S., C. F. Polnaszek, and H. H. Freed, *J. Phys. Chem.* **61**, 449 (1977).

Raoux, D., *J. Phys. Chem. Solids* **36**, 359 (1975).

Rea, D. K., *Geophys. Res. Lett.* **5**, 561 (1978).

Reddoch, A. H., and S. Konishi, *J. Chem. Phys.* **10**, 2121 (1979).

Reed, G. H., J. S. Leigh, Jr., and J. E. Pearson, *J. Chem. Phys.* **55**, 3311 (1971).

Reibisch, K., H. Kothe, and J. Brickmann, *Chem. Phys. Lett.* **17**, 86 (1972).

Reichert, J. F., N. Jarosik, R. Herrick, and J. Andersen, *Phys. Rev. Lett.* **42**, 1359 (1979).

Reidel, E. F., and R. D. Willet, *Solid State Commun.* **16**, 413 (1975).

Reineker, P., *Z. Naturforsch.* **29a**, 282 (1974).

Reineker, P., *Z. Phys. B* **21**, 409 (1975).

Reineker, P., *Phys. Status Solidi B* **70**, 189 (1975); **74**, 121 (1976).

Reineker, P., *Chem. Phys.* **16**, 425 (1976).

Reineker, P., *Solid State Commun.* **14**, 153 (1974); **25**, 859 (1978).

Reiter, G. F., *Phys. Rev. B* **9**, 3780 (1974).

Rengan, S. K., M. P. Khakhar, B. S. Prabhananda, and B. Venkataraman, *Proc. Nucl. Phys. and Solid State Phys. Symp. BARC, Bombay* **3**, 231 (1972).

Rettori, C., D. Davidov, I. Amity, L. J. Tao, E. Bucher, P. Allen, E. R. Andrew, and C. A. Bates, *Proc. 18th Ampere Congr. Magn. Reson. Relat. Phenom.* **2**, Nottingham, England, Amsterdam, 1974.

Rettori, C., D. Davidov, H. M. Kim, and E. P. Chock, *AIP Conf. Proc. 19th Annual Conference on Magnetism and Magnetic Materials*, **18**, Part II, Boston, Mass., 1973.

Rettori, C., D. Davidov, J. Suassuna, G. E. Barberis, B. Elschner, and B. Born, *Solid State Commun.* **25**, 543 (1978).

Reuter, G. E. H., and E. H. Sondheimer, *Proc. R. Soc. London A* **195**, 336 (1948).

Reuveni, A., and Z. Luz, *J. Magn. Reson.* **23**, 271 (1976).

Richards, P. M., *Phys. Rev.* **13**, 458 (1976).

Richards, P. M., *Phys. Rev. B* **10**, 805 (1974).

Richards, P. M., *Solid State Commun.* **13**, 253 (1973).

Richards, P. M., and M. B. Salamor, *Phys. Rev. B* **9**, 32 (1974).

Richards, P. M., K. A. Muller, H. R. Boesch, and F. Waldner, *Phys. Rev. B* **10**, 4531 (1974).

Richards, P. M., R. K. Quinn, and B. Morosin, *J. Chem. Phys.* **59**, 4474 (1973).

Ritter, J. C., and M. E. de Morton, *J. Aust. Inst. Met.* **22**, 51 (1977).

Rius, G., A. Herve, R. Picard, and C. Santier, *J. Phys.* **37**, 129 (1976).

Robinson, B. H., L. R. Dalton, and L. A. Dalton, *Chem. Phys. Lett.* **29**, 56 (1974).

Robinson, B. H., L. R. Dalton, L. A. Dalton, and A. L. Kwiram, *Chem. Phys. Lett.* **29**, 56 (1974).

Robinson, B. H., L. A. Dalton, A. H. Beth, and L. R. Dalton, *Chem. Phys.* **18**, 321 (1976).

Roch, J., R. Sardos, K. Haye, and R. Chastanet, *Rev. Phys. Appl.* **6**, 243 (1971).

Rogers, R. N., M. E. Anderson, and G. E. Pake, *Bull. Am. Phys. Soc.* **4**, 261 (1959).

Romanelli, M., and L. Burlamacchi, *Mol. Phys.* **31**, 115 (1976).

Ross, R. T., *J. Chem. Phys.* **42**, 2919 (1965).

Rotaru, M., A. Valeriu, and M. Weiner, *Rev. Roumaine Phys.* **9**, 269 (1964).

Royaud, M. M. J., and B. Treville, *C.R. Acad. Sci.* **279**, 215 (1974).

Rukhadze, A. A., and V. P. Silin, *Zh. Tekh. Fiz.* **32**, 432 (307) (1962).

Sagstuen, E., *J. Chem. Educ.* **54**, 153 (1977).

Sames, D., *Z. Phys.* **198**, 71 (1967).

Samokhvalov, A. A., V. S. Babushkin, V. G. Bamburov, and N. I. Labochevskaya, *Sov. Phys. Solid State* **13**, 2530 (1972).

Samokhvalov, A. A., T. I. Arbuzova, A. Ya. Afanasev, V. S. Babushkin, N. N. Loshkareva, Yu. N. Morozov, M. I. Simonova, V. G. Bamburov, and N. I. Lobachevskaya, *Fiz. Tverdogo Tela* **17**, 48 (1975).

Samokhvalov, A. A., T. I. Arbuzova, V. S. Babushkin, B. Z. Gizhevskii, N. N. Efremova, M. I. Simonova, and N. M. Chebotaev, *Fiz. Tverdogo Tela* **18**, 2930 (1976).

Sands, R. H., *Phys. Rev.* **99**, 1222 (1955).

Saraswat, R. S., and G. C. Upreti, *Phys.* **92B**, 253 (1977).

Sardos, R., *Rev. Phys. Appl.* **4**, 29 (1969).

Sargent, F. P., and E. M. Grady, *Chem. Phys. Lett.* **38**, 130 (1976).

Sarles, L. R., and R. M. Cotts, *Phys. Rev.* **111**, 853 (1958).

Sastry, B. A., and G. S. Sastry, *Physica* **54**, 20 (1971); *Indian J. Pure Appl. Phys.* **12**, 632 (1974).

Schaafsma, T. J., *Magn. Reson. Relat. Phenom. Proc. Congr. Ampere, 18th* **2**, 497 (1975).

Schara, M., *Proc. Int. Sch. Phys. "Enrico Fermi"* **59**, 765 (1975).

Scheffler, K., and H. B. Stegmann, *Ber. Bunsengesell. Phys. Chem.* **67**, 864 (1964).

Scheithe, W., J. Kötzler, and P. Radhakrishna, *Phys. Lett.* **66A**, 419 (1978).

Schlenker, C., R. Buder, M. Schlenker, J. F. Houlihan, and L. N. Mulay, *Phys. Status Solidi B* **54**, 247 (1972).

Schlindler, H. and J. Seelig, *Ber. Bunsenges Phys. Chem.* **78**, 941 (1974).

Schmidberger, R., and H. C. Wolf, *17th Congress Ampere Nuclear Magnetic Resonance and Related Phenomena. Turku, Finland* (1972).

Schmidberger, R., and H. C. Wolf, *Chem. Phys. Lett.* **16**, 402 (1972).

Schmidt, L., H. Port, and S. Schmid, *Chem. Phys. Lett.* **51**, 413 (1977).

Schoemaker, D., *Phys. Rev. B* **7**, 786 (1973).

Schoemaker, D., and J. L. Kolopus, *Phys. Rev. B* **2**, 1148 (1970).

Schoffa, G., and G. Burk, *Phys. Status Solidi* **8**, 557 (1965).

Schonland, D. S., *Proc. Phys. Soc.* **73**, 788 (1959).

Schwoerer, M., R. A. Huber, and W. Hartl, *Chem. Phys.* **55**, 97 (1981).

Schouler, M. C., M. Decorps, and F. Genoud, *Mol. Phys.* **32**, 1671 (1976).

Schwarz, V. D., and V. R. Wessel, *ETP* **24**, 531 (1976).

Scott, P. L., H. J. Stapleton, and C. Wainstein, *Phys. Rev.* **137**, A71 (1965).

Searl, J. W., R. C. Smith, and S. J. Wyard, *Proc. Phys. Soc.* **74**, 491 (1959); **78**, 1174 (1961); *Arch. Sci. Fasc. Spec.* **13**, 236 (1960) (9th Colloque Ampere).

Seehra, M. S., *J. Appl. Phys.* **42**, 1290 (1971).

Seehra, M. S., *Phys. Rev. B* **6**, 3186 (1972).

Seehra, M. S., and T. G. Castner, Jr., *Solid State Commun.* **8**, 787 (1970).

Seehra, M. S., and D. L. Huber, *AIP Conf. Proc.* **24**, 261 (1975).

Seehra, M. S., and W. S. Sheers, *Physica* **85B**, 142 (1977).

Seehra, M. S., and D. W. Sturm, *J. Phys. Chem. Solids* **36**, 1161 (1975).

Seehra, M. S., and R. P. Gupta, *Phys. Rev. B* **9**, 197 (1974).

Seelig, J., *Spin Labeling*, 373 (1976).

Seipler, D., and B. Elschner, *Phys. Lett. A* **55**, 115 (1976).

Servant, Y., J.-C. Bissey, and S. Gharbage, *J. Magn. Reson.* **24**, 335 (1976).

Servant, R., *C. R. Acad. Sci.* **280**, 447 (1975).

Shaffer, J. S., H. A. Farach, and C. P. Poole, *Phys. Rev. B* **13**, 1869 (1976).

Shain, A. L., *J. Chem. Phys.* **56**, 6201 (1972).

Shaltiel, D., and W. Low, *Phys. Rev.* **124**, 1052 (1961).

Sharnof, M., and E. B. Aiteb, *Izv. Akad. Nauk SSSR Ser. Fiz.* **37**, 522 (1973).

Sharp-Dent, G., M. Hardiman, J. R. Sambles, and J. E. Cousins, *Phys. Stat. Sol. B* **75**, 155 (1976).

Shcherbakova, M. Ya., and V. E. Istamin, *Phys. Stat. Sol. B* **67**, 461 (1975).

Sherman, L. R., J. S. Mattson, H. B. Mark, Jr., and H. C. MacDonald, Jr., *Computer Assisted Instruction in Chemistry*, Pt. 8, 173 (1974).

Shia, L., and G. Kokoszka, *J. Chem. Phys.* **60**, 1101 (1974).

Shiga, T., A. Lund, and P. O. Kinell, *Int. J. Radiat. Phys. Chem.* **3**, 131 (1971).

Shimada, K., and M. Szwarc, *J. Am. Chem. Soc.* **97**, 3321 (1975).

Shimada, K., and M. Szwarc, *Chem. Phys. Lett.* **34**, 503 (1975).

Shimozato, Y., K. Shimada, and M. Szwarc, *J. Am. Chem. Soc.* **97**, 5831 (1975).

Shiotani, M., and J. Sohma, *Progr. Polym. Phys. Japan* **17**, 505 (1974).

Shiraishi, H., H. Kadoi, Y. Katsumura, Y. Tabata, and K. Oshima, *J. Phys. Chem.* **80**, 2400 (1976).

Shmueli, U., M. Sheinblatt and M. Polak, *Acta Cryst.* **A32**, 192 (1976).

Shrivastava, K. N., *Phys. Lett. A* **55**, 295 (1975).

Shteinshneider, N. Ya., and G. M. Zhidomirov, *Teor. Eksp. Khim.* **7**, 651 (1970).

Shteinshneider, N. Ya., G. M. Zhidomirov, and K. I. Zamaraev, *Zh. Struct. Khim.* **13**, 795 (1972).

Shul'man, L. A., and G. A. Podzyarei, *Sov. Phys. Solid State* **14**, 1521 (1972).

Shul'man, L. A., G. A. Podzyarei, and T. A. Nachal'naya, *Ukr. Fiz. Zhur.* **16**, 443 (1971).

Shumilkina, E. V., and A. I. Obraztsov, *Fiz. Tverd. Tela* **17**, 1234 (1975).

Sillescu, H., *J. Chem. Phys.* **54**, 2110 (1971).

Sillescu, H., and D. Kivelson, *J. Chem. Phys.* **48**, 3493 (1968).

Silver, B. L., *C.R. Collog. Int. Isot. Oxygene*, 21 (1975).

Singer, L. S., *J. Chem. Phys.* **23**, 379 (1955).

Singer, L. S., W. J. Vpry, and W. H. Smith, *Proceedings of the 1957 Conference on Carbon*, Pergamon, New York, 1957, p. 121.

Singstad, I., *Fra Fys. Verden* **37**, 88 (1975).

Sjoblom, R., *J. Magn. Reson.* **22**, 411 (1976); **34**, 393 (1979).

Smigel, M. D., L. A. Dalton, L. R. Dalton, and A. L. Kwiram, *Chem. Phys.* **6**, 183 (1974).

Smith, M. R., H. A. Buckmaster, and D. J. I. Fry, *J. Magn. Reson.* **23**, 103 (1976).

Smith, D. R., and J. J. Pieroni, *Can. J. Chem.* **42**, 2209 (1964).

Smith, I. C. P., in *Biological Applications of Electron Spin Resonance*, J. M. Swartz, J. R. Bolton, and D. C. Borg, eds. Wiley-Interscience, 1972, p. 484.

Smith, I. C. P., S. Schreier-Muccillo and D. Marsh, *Free Radicals Biol.* **1**, 149 (1976).

Soos, Z. G., T. Z. Hyang, J. S. Valentine, and R. C. Hughes, *Phys. Rev. B* **8**, 993 (1973).

Sperlich, G., and H. Muller, *Phys. Status Solidi B* **71**, 305 (1975).

Sperlich, G., P. H. Zimmermann, and G. Keller, *Z. Phys.* **270**, 267 (1974).

Stankowski, J., A. Dezor, B. Sczaniecki, and J. M. Janik, *Phys. Status Solidi* **16**, K167 (1973).

Stasz, J., *Phys. Status Solidi A* **47**, K23 (1978).

Stasz, J., *Acta Phys. Polon.* **A52**, 831 (1977).

Stasz, J., M. Jelonek, and P. Takely, *Acta Phys. Polonica* **A49**, 737 (1976).

Steiner, R. C., *1976 Underground Transmission and Distribution Conference Supplement*, 97 (1977).

Sternlicht, H., *J. Chem. Phys.* **33**, 1138 (1960); **42**, 2250 (1965).

Stesmans, S., and J. Van Meijel, *J. Phys. E* **10**, 339 (1977).

Stevenson, G. R., *Magn. Reson. Rev.* **6**, 209 (1980).

Stillman, A. E., and R. N. Schwartz, *J. Magn. Reson.* **22**, 269 (1976).

Stoklosa, H. S., and J. R. Wasson, *J. Inorg. Nucl. Chem.* **38**, 677 (1976).

Stoneham, A. M., *J. Phys. D Appl. Phys.* **5**, 670 (1972).

Stoneham, A. M., *Phys. Lett.* **14**, 297 (1965).

Stoneham, A. M., K. A. Muller, and W. Berlinger, *Solid State Commun.* **10**, 1005 (1972).

Stoppels, D., R. Valkenburg, and G. A. Sawatzky, *Phys. B and C* **86-88**, 135 (1976).

Storek, W., *Talanta* **23**, 649 (1976).

Strandberg, M. W. P., *Ann. Phys.* **77**, 174 (1973).

Strother, E. F., H. A. Farach, and C. P. Poole, Jr., *Phys. Rev.* **4**, 2079 (1971).

Stryukov, V. B., and E. G. Rozantsev, *Teor. Eksp. Khim.* **7**, 209 (1971).

Stryukov, V. B., L. A. Shul'man, A. V. Zvarykina, and D. N. Fedutin, *Sov. Phys. Solid State* **16**, 187 (1974).

Suassuna, J. F., G. E. Barberis, C. Rettori, and C. A. Pela, *Solid State Commun.* **22**, 347 (1977).

Subramanian, J., and P. T. Narasimhan, *J. Chem. Phys.* **56**, 2572 (1972).

Suchard, S. N., and K. W. Bowers, *J. Chem. Phys.* **56**, 5540 (1972).

Sugano, T., and H. Kuroda, *Chem. Phys. Lett.* **47**, 92 (1977).

Sugihara, K., *J. Phys. Soc. Japan* **38**, 1061 (1975).

Sullivan, P. D., *Magn. Reson. Rev.* **2**, 35, 315 (1974); **3**, 250 (1974); **4**, 197 (1978).

Svare, I., *Phys. Rev. A* **138**, 1718 (1965).

Svare, I., and G. Seidel, *Phys. Rev. A* **134**, 172 (1964).

Swalen, J. D., and T. R. Lusebrink, *Magn. Reson. Rev.* **2**, 165 (1973).

Swartz, H. M., J. R. Bolton and D. C. Borg, Biological Applications of Electron Spin Resonance, Wiley, New York, 1972.

Swarup, P., *Can. J. Phys.* **37**, 848 (1959).

Sweeney, W. V., D. Coucouvanis, and R. E. Coffman, *J. Chem. Phys.* **59**, 369 (1973).

Sýkora, S., and J. Vogt, *Bruker Report* **2**, 15 (1979).

Takagi, S., S. Izumida, and K. Dawabe, *Phys. Lett. A* **57A**(2), 189 (1976).

Takagi, S., and K. Kawabe, *Phys. Lett. A* (*Netherlands*) **57A**(5), 470 (1976).

Takagi, S., and T. Kojima, *Japan J. Appl. Phys.* **13**, 1195 (1974).

Tanaka, M., *Rep. Res. Lab. Surf. Sci., Okayama Univ.* **3**(3), 98, 137, 145 (1969); **3**(4), 177(1970); **3**(4), 251 (1971); **4**(3), 139 (1975).

Tanaka, M., and T. Kawabe, *Rep. Res. Lab. Surf. Sci. Okayama Univ.* **3**(5), 251 (1971).

Tanaka, M., and Y. Kondo, *J. Phys. Soc.* **34**, 934 (1973); **38**, 1539 (1975); **40**, 35 (1976).

Tanaka, M., and S. Watabe, *Rep. Res. Lab. Surf. Sci. Okayama Univ.* **3**, 155 (1969).

Tanemoto, K., and T. Nakamura, *Japan J. Appl. Phys.* **17**, 1561 (1978).

Taylor, P. C., and P. J. Bray, *Am. Ceram. Soc. Bull.* **51**, 234 (1972).

Taylor, P. C., J. F. Baugher, and H. M. Kriz, *Chem. Rev.* **75**, 203 (1974).

Taylor, P. C., and P. J. Bray, *J. Magn. Reson.* **2**, 305 (1970).

Taylor, R. H., and B. R. Coles, *J. Phys. F* **5**, 121 (1972).

Taylor, P. C., U. Strom, and S. G. Bishop, *Phys. Rev. B* **18**, 511 (1978).

Tazuke, Y., and K. Nagata, *J. Phys. Soc. Japan* **30**, 285 (1971); **38**, 1003 (1975).

Tchao, Y. H., *C. R. Acad. Sci.* **251**, 668 (1960).

Teodorescu, M., *Rev. Roum. Phys.* **22**, 73 (1977).

Teslenko, V. V., *Chem. Phys. Lett.* **32**, 332 (1975).

Teslenko, V. V., and B. D. Shanina, *Fiz. Tverd. Tela* **17**, 1978 (1975).

Thomann, H., B. H. Robinson, L. A. Dalton, A. H. Beth, R. C. Perkins, and J. H. Park, *Chem. Phys. Lett.* **73**, 131 (1980).

Thomas, D. D., *Biophys. J.* **24**, 439 (1978).

Thomas, D. D., and H. M. McConnell, *Chem. Phys. Lett.* **25**, 470 (1974).

Thomas, P. A., and D. Kaplan, *AIP Conf. Proc. Int. Conf. Struct. Excit. of Amorph. Solids* **31**, 85 (1976).

Thyer, J. R. W., *8th Aust. Spectrosc. Conf.* **121**, K9 (1971).

Tigyi, J., and K. Hideg, *Acta. Biochim. Biophys. Acad. Sci. Hung.* **11**, 147 (1976).

Tikhomirova, N. N., and V. V. Voevodskii, *Optica Spectrosk*. **7**, 829 (486) (1959).

Tkach, Yu. G., S. N. Dobryakov, and Ya. S. Lebedev, *PTE* **18**, 84 (1975).

Tolkachev, V. A., and A. J. Mikhailov, *PTE* **6**, 95 (1242) (1964).

Tomashpolskii, Yu. Ya., *Fiz. Tverd*. **16**, 3191 (1974).

Tormala, P., and J. J. Lindberg, *Struct. Stud. Macromol. Spectrosc. Methods* (*Meet*), 255 (1976).

Tornero, J. D., F. J. Lopez, and J. M. Cabrera, *Solid State Commun*. **16**, 53 (1975).

Toy, A. D., and T. D. Smith, *J. Chem. Soc. A*, 2600 (1970).

Toy, A. D., T. D. Smith, and J. R. Pilbrow, *J. Chem. Soc. A*, 451 (1970).

Toyata, E., and K. Hirakawa, *J. Phys. Sci. Japan* **30**, 692 (1971).

Trifunac, A. D., and E. C. Avery, *Chem. Phys. Lett*. **27**, 141 (1974).

Truesdale, R. D., C. P. Poole, and H. A. Farach, *Phys. Rev*. **B22**, 365 (1980); in press (1982).

Tsay, F. O., H. B. Gray, and J. Danon, *J. Chem. Phys*. **54**, 3760 (1971).

Tsukerblat, B. S., and M. P. Chobanu, *Sov. Phys. Solid State* **18**, 1232 (1976).

Tsukerblat, B. S., M. P. Chobanu, and A. V. Ablov, *Dokl. Akad. Nauk SSSR* **230**, 1158 (1976).

Tumanov, V. S., *Fiz. Tverd. Tela* **4**, 2419 (1773) (1962).

Turoff, R. D., R. Coulter, J. Irish, M. Sundquist, and E. Buchner, *Phys. Rev*. **164**, 406 (1967); see esp. footnote 16.

Tykarski, L., *J. Phys. D* **7**, 786 (1974).

Tynan, E. C., and T. F. Yen, *J. Magn. Reson*. **3**, 327 (1970).

Uehara, H., M. Tanimoto, and Y. Morino, *Mol. Phys*. **22**, 799 (1971).

Upreti, G. C., *J. Phys. Chem. Solids* **35**, 461 (1974).

Urban, P., D. Davidov, B. Elschner, T. Plefka, and G. Sperlich, *Phys. Rev*. *B* **12**, 72 (1976).

Urban, P., K. Jansen, G. Sperlich, and D. Davidov, *J. Phys. F* **8**, 977 (1975).

Ursu, I., and E. Burzo, *J. Magn. Reson*. **8**, 274 (1972).

Usher, J. L., and J. R. Powell, *1976 IEEE Int. Conf. Plasma Sci., Austin, Texas*, 176 (1976).

Van, S. P., G. B. Birrell, and O. H. Griffith, *J. Magn. Reson*. **16**, 444 (1974).

Van Camp, H. L., and A. H. Heiss, *Magn. Reson. Rev*. **7**, 1 (1981).

Van Gerven, L., and J. Accou, *Magn. Reson. Relat. Phenom. Proc. Congr. Ampere*, 18th, **2**, 567 (1975).

Van Gerven, L., and A. Van Itterbeek, *Arch. Sci., Spec. No*. **14**, 117 (1961) (10th Colloque Ampere).

Vänngård, T., and R. Aasa, *Paramagn. Reson*. **2**, 509 (1963).

Van Vleck, J. H., *Phys. Rev*. **74**, 1168 (1948).

Velichko, V. V., and A. Yu. Gusev, *Izv. Vyssh. Uchebn. Zaved. Fiz*. **19**, 118 (1976).

Verdier, P. H., E. B. Whipple, and V. Schomaker, *J. Chem. Phys*. **34**, 118 (1961).

Vieth, H. M., and K. H. Hausser, *17th Congress Ampere Nuclear Magnetic Resonance and Related Phenomena, Turku, Finland*, 62 (1972).

Vieth, H. M., and K. H. Hausser, *Ber. Bunsenges. Phys. Chem*. **78**, 185 (1974).

Vigouroux, B., J. Rolland, and J. C. Gourdon, *J. Phys. Lett*. **39**, L-261 (1978).

Vigouroux, B., J. C. Gourdon, and J. Pescia, *J. Phys. F Metal Phys*. **6**, 1575 (1976).

Vigouroux, B., J. C. Gourdon, P. Lopez, and J. Pescia, *J. Phys. E* **6**, 557 (1973).

Vignais, P. M., and P. F. Devaux, *Enzymes Biol. Membr*. **1**, 91 (1976).

Villa, J. F., and W. E. Hatfield, *Inorg. Nuclear. Chem. Lett*. **6**, 511 (1970).

Vincow, G., and P. M. Johnson, *J. Chem. Phys*. **39**, 1143 (1963).

Vishnevskaya, G. P., F. M. Gumerov, and B. M. Kosyrev, *Theor. Exper. Chem*. **11**, 162 (1975).

Vitko, J., Jr., and R. E. Huddleston, *Inis Atomindex* **7**(23), No. 274145 (1976).

Vittoria, C., P. Lubitz, and V. Ritz, *J. Appl. Phys.* **49**, 4908 (1978).

Vivien, D., A. Kahn, A. M. Lejus, and J. Livage, *Phys. Status Solidi B* **73**, 593 (1976).

Voget-Grote, U., J. Stuke, and H. Wagner, *AIP Conf. Proc. Int. Conf. Struc.* **31**, 91 (1976).

Volino, F., and A. Rousseau, *Mol. Cryst. Liquid Cryst.* **16**, 247 (1972).

Vollmer, R. T., and W. J. Caspary, *J. Magn. Reson.* **27**, 181 (1977).

Volkoff, G. N., *Can. J. Phys.* **31**, 820 (1953).

Von Waldkirch, Th., K. A. Miller, and W. Berlinger, *Phys. Rev. B* **7**, 1052 (1973).

Von Waldkirch, Th., K. A. Miller, and W. Berlinger, *Phys. Rev. Lett.* **28**, 503 (1972).

Vugman, N. V., M. F. Elia, and R. P. A. Muniz, *Mol. Phys.* **30**, 1813 (1975).

Wagoner, G., *Phys. Rev.* **118**, 647 (1960).

Wakim, F. G., and A. Nolle, *J. Chem. Phys.* **37**, 3000 (1962).

Waller, W. G., and M. T. Rogers, *J. Magn. Reson.* **18**, 39 (1975); **13**, 53 (1974); **9**, 92 (1973).

Walter, R. L., R. S. Codrington, A. F. D'Adamo, Jr., and H. C. Torrey, *J. Chem. Phys.* **25**, 319 (1956).

Warner, F. W., and A. J. McPhate, *J. Magn. Reson.* **24**, 125 (1975).

Wasiela, A., J. Duran, and Y. M. D'Aubigne, *C.R. Acad. Sci. B* **278**, 1099 (1974).

Watanabe, I., Y. Inagaki, and T. Shimizu, *Japan J. Appl. Phys.* **15**, 1993 (1976).

Waugh, J. S., F. B. Humphrey, and D. M. Yost, *J. Phys. Chem.* **57**, 486 (1953).

Webb, R. H., *Phys. Rev.* **158**, 225 (1967).

Weidner, R. T., and C. A. Whitmer, *Phys. Rev.* **91**, 1279 (1953).

Weil, J. A., and J. H. Anderson, *J. Chem. Phys.* **28**, 864 (1958).

Weil, J. A., and H. G. Hecht, *J. Chem. Phys.* **38**, 281 (1963).

Weill, G., J. J. Andre, and A. Bieber, *Mol. Cryst. Liq. Cryst.* **44**, 237 (1977).

Welter, M., and W. Windsch, *Abstracts of the Conference on Magnetic Resonance and Related Phenomena, Bucharest, Romania*, 1971.

Wessel, V. R., and D. Schwarz, *ETP* **23**, 641 (1975); **26** 195 (1978).

Wieckowski, A., *Radiospektrosk. Ciala Stalego*, 69 (1975); *Piz. Dielektr. Radiospektrosk.* **7**, 101 (1975).

Willet, R. D., and M. Extine, *Chem. Phys. Lett.* **23**, 281 (1973).

Wilson, R. C., and D. Kivelson, *J. Chem. Phys.* **44**, 154, 4440 (1966).

Wilson, R. C., and R. J. Myers, *J. Chem. Phys.* **64**, 2208 (1976).

Windsch, W., *Ferroelectrics* **12**, 63 (1976).

Witters, J., *Neo. Tiudschr. Natuurk.* **38**, 177 (1972).

Wolfe, J. P., and C. D. Jeffries, *Phys. Rev. B* **4**, 731 (1971).

Wuu-Jyi, Lin, and J. H. Freed, *J. Phys.* **83**, 379 (1979).

Wyard, S. J., *Proc. Phys. Soc.* **86**, 587 (1965).

Yamada, T., Y. Ajiro, S. Matsukawa, and T. Haseda, *Phys. Lett. A* **51A**, 330 (1975).

Yamada, I., and T. Nishizaki, *J. Phys. Soc. Japan* **43**, 2093 (1977).

Yamada, I., and M. Ikebe, *J. Phys. Soc. Japan* **33**, 979 (1972).

Yo, C. H., R. Poupko, and R. M. Hornreich, *Chem. Phys. Lett.* **54**, 142 (1978).

Yokota, I., *Prog. Theoret. Phys.* **8**, 380 (1952).

Yokozawa, Y., *J. Phys. Soc. Japan* **31**, 1590 (1971).

Young, W. A., *J. Appl. Phys.* **35**, 460 (1964).

Zamaraev, K. I., R. F. Khairutdinov, and J. R. Miller, *Chem. Phys. Lett.* **57**, 311 (1978).

Zaspel, C. E., and J. E. Drumheller, *Phys. Rev. B* **16**, 1771 (1977).

Zegarski, B. R., T. J. Cook, and T. A. Miller, *J. Chem. Phys.* **62**, 2952 (1975).

Zevin, V., D. Davidov, R. Levin, and D. Shaltiel, *Physica B and C* **86–88**, 1158 (1976).

Zevin, U. Ya., *Sov. Phys. Solid State* **3**, 662 (1961).

Zevin, V., D. Davidov, R. Levin, D. Shaltiel, and K. Baberschke, *J. Phys. F* **7**, 2193 (1975).

Zhidomirov, G. M., Y. S. Lebedev, and S. N. Dobryakov, *Interpretation of Complex EPR Spectra*, Nauka, USSR, Moscow, 1975.

Zieger, J., and H. C. Wolf, *Chem. Phys.* **29**, 209 (1978).

Ziegler, E., and E. G. Hoffman, *Frezenius Z. Anal. Chem.* **240**, 145 (1968).

Zimmerman, N. J., J. A. Van Santen, and J. Van Den Handel, *Physica* **57**, 334 (1972).

Zupan, J., M. Penca, D. Hadzl, and J. Marsel, *Anal. Chem.* **49**, 2141 (1977).

CHAPTER 13

Relaxation Times

A. General Description of Relaxation Times

Consider a paramagnetic sample located in a magnetic field. The sample contains two types of electronic spins S_1 and S_2 and two types of nuclear spins I_3 and I_4. The electronic and nuclear spins constitute four spin systems each of which is in thermodynamic equilibrium within itself and possesses a distribution of spin energy corresponding to the characteristic temperatures θ_1, θ_2, θ_3, and θ_4, respectively.* Let these four spin systems be completely isolated from each other, and from the lattice. The spins in each system are distributed among their Zeeman energy levels in accordance with the Boltzmann distribution (see Sec. 11-E). Let us assume that the four spin systems are initially at the same temperature θ_0:

$$\theta_1 = \theta_2 = \theta_3 = \theta_4 = \theta_0. \tag{1}$$

Now if the sample is irradiated at a microwave frequency that corresponds to the resonance condition for the S_1 spins, then these spins will gradually undergo transitions from the lower to the upper energy levels until eventually the populations of all of the S_1 Zeeman levels will be equal.† The S_1 spin system now is said to be saturated, and can absorb no more microwave energy. Quantum mechanically, we say that the Einstein coefficient for induced emission is equal for the transitions up and down between each pair of levels, and thermodynamically we say that the spin system S_1 is at an infinite temperature ($\theta_1 = \infty$). The Einstein coefficient for spontaneous emission may be neglected at microwave frequencies.

If the inhomogeneities in the magnetic field exceed the linewidth, then only some of the S_1 spins will be at resonance, and as a result the remainder cannot absorb energy. The individual spins within each spin system may transfer energy among themselves at the rate $1/T_2$ by means of the dipolar and exchange interactions. This process of energy transfer causes all of the S_1 spins to maintain a constant temperature θ_1 as long as the microwave power is sufficiently low so that the rate at which the spins absorb microwave photons does not exceed $1/T_2$. Physically, we say that the resonant S_1 spins slowly

---

*In this section only, the symbol θ is used for temperature to avoid confusion with the use of T_1 and T_2 for relaxation times. In the remainder of the book, the symbol T is used for temperature.
†Technically speaking, an infinite time is needed to equalize the populations completely.

absorb microwave energy, and rapidly pass it on to the nonresonant S_1 spins. If the spins do absorb energy at a greater rate than $1/T_2$, then the S_1 spin system will not be in internal thermal equilibrium during the irradiation process until saturation sets in. As saturation is approached, the rate of energy absorption will slow down and eventually cease, thereby restoring internal thermal equilibrium at an infinite temperature. Since the other three spin systems do not interact with S_1 spins, and they are not in resonance at the same microwave frequency, it follows that the spin systems are now described by the following temperatures:

$$\theta_1 = \infty \tag{2}$$

$$\theta_2 = \theta_3 = \theta_4 = \theta_0 \tag{3}$$

If we now allow the spin system S_1 to interact with S_2 at the rate $1/T_x^{12}$, then "spin energy" will pass from S_1 to S_2 until both spin systems reach equilibrium with each other. The sample is now characterized by the temperatures

$$\theta_1 = \theta_2 \tag{4}$$

$$\theta_3 = \theta_4 = \theta_0 \tag{5}$$

The characteristic time T_x^{12} is called the cross relaxation time between the spin systems S_1 and S_2 (see Poole and Farach, 1971), and it may arise from dipolar or exchange interactions. In general, the cross relaxation time between the ith and jth spin systems is denoted by T_x^{ij}. Five additional cross relaxation times, T_x^{13}, T_x^{14}, T_x^{23}, T_x^{24}, and T_x^{34} may also be defined for this system, where we assume $T_x^{ij} = T_x^{ji}$.

In a given sample that contains both nuclear and electronic spins, the electronic spin–spin relaxation time T_2^e is usually much less than the nuclear spin–spin relaxation time T_2^n. If these two spin systems interact with the cross relaxation time T_x^{en}, which is also much less than T_2^n, then the measurement of T_2 in a nuclear magnetic resonance experiment will produce a value much shorter than the actual T_2^n. This is because the NMR experiment measures the rate of interchange of energy between the nuclear spins, and the slow process of energy transfer

$$\text{Nucleus} \xrightarrow{T_2^n} \text{nucleus} \tag{6}$$

is now short-circuited by much more rapid processes of the type

$$\text{Nucleus} \xrightarrow{T_x^{en}} \text{electron} \xrightarrow{T_2^e} \text{electron} \xrightarrow{T_x^{en}} \text{nucleus} \tag{7}$$

since

$$T_2^e, T_X^{en} \ll T_2^n \tag{8}$$

If, on the other hand, the electronic spins are saturated so that

$$\theta^e \sim \infty \tag{9}$$

then in some cases the mechanism for energy exchange between the electrons and nuclei permits the nuclei to be polarized by a double resonance experiment. Such experiments are discussed in Chap. 14.

If the only mechanisms for the transferal of spin energy were of the spin–spin and cross-relaxation type, then resonance experiments would serve to heat up spin systems, and they would stay "hot" forever. Actually each spin system is imbedded in a fluid or solid "lattice" characterized by a temperature θ_L, which ordinarily arises from vibrations in a solid and rotations and translations in a fluid. The spin system transfers energy to the lattice at the rate $1/T_1$, where T_1 is the spin-lattice relaxation time. If more than one spin system is present, then one may use the notation T_1^i to distinguish each one. In solids one usually has

$$T_1^i \gg T_2^i \tag{10}$$

for each spin system.

In a liquid, the rate at which molecules migrate is characterized by the correlation time τ_c, which is a function of the viscosity η through the Debye-like relation (vide Eq. 12-0-4)

$$\tau_c = \frac{4\pi\eta a^3}{3k\theta_L} \tag{11}$$

where a is the molecular radius. In liquids, the spin-lattice relaxation time T_1 is given in terms of the magnetic resonance angular frequency ω_0 by

$$\frac{1}{T_1} = C\left(\frac{\tau_c}{1 + (\omega_0\tau_c)^2} + \frac{2\tau_c}{1 + (2\omega_0\tau_c)^2} \right) \tag{12}$$

where C is a constant. In the proton NMR of water, one has explicitly

$$C = \frac{2}{5}\frac{g^4\beta_N^4}{\hbar^2}\frac{I(I + 1)}{b^6} \tag{13}$$

where b is the proton–proton distance in the water molecule and $I = \frac{1}{2}$. The

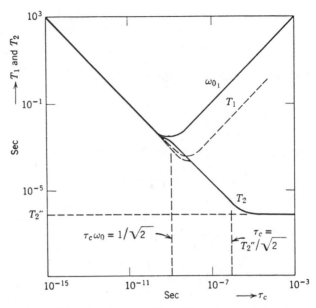

Fig. 13-1.   Dependence of the relaxation times T_1 and T_2 on the correlation time τ_c (Bloembergen, Purcell, and Pound, 1948).

minimum value of T_1 occurs where

$$\tau_2\omega_0 = \left(\tfrac{1}{2}\right)^{1/2} \tag{14}$$

as Fig. 13-1 indicates.

When $\tau_c\omega_0 \ll \left(\tfrac{1}{2}\right)^{1/2}$, then T_1 equals T_2 and is inversely proportional to τ_c. When, on the other hand, $\tau_c\omega_0 \gg \left(\tfrac{1}{2}\right)^{1/2}$, then T_1 becomes directly proportional to τ_c and in addition $T_1 \gg T_2$. In this region the material is a solid, and so other theories of saturation are more appropriately applied (e.g., Clough and Scott, 1968; Bendiashvili et al., 1970; Shul'man et al., 1971).

The presence of paramagnetic gases such as oxygen can have a very pronounced effect on the relaxation rates of spin systems. The usual result of the presence of O_2 is a shortening of T_1 since this paramagnetic molecule provides an efficient relaxation path. Several systems in which this oxygen effect has been observed include free radicals and radical ions in solution, carbons, irradiated solids, and petroleum oils; references to this work are given in the first edition of this monograph. Techniques for removing oxygen from solution are discussed in Sec. 8-G.

Two general methods of measuring relaxation times are emphasized in the present chapter, namely, the continuous or CW saturation method described in Sec. 13-C and the pulse saturation or pulse recovery method described in Sec. 13-D. Other measurement techniques are also available. For long relaxation times, greater than 1 sec, an inversion recovery method may be used (Castle et al., 1960). This employs fast magnetic field sweeps and the measurement of the entire ESR line as it recovers from an adiabatic fast passage. In the slow-motion region ($\tau_c < 10^{-7}$ sec) adiabatic rapid-passage ESR spectra of spin labels have a shape that is sensitive to the rate of the molecular motion (Dalton et al., 1975; Hyde and Dalton, 1972; Robinson et al., 1978, 1979). Hervé (1963), with Pescia (1962, 1963), employed a source modulation method as explained at the end of Sec. 13-C. Various techniques such as ENDOR, ELDOR, DNP, DEP, and acoustic ESR, which are described in the next chapter, also provide information on relaxation times. Rengan et al. (1972) summarized some of the above methods.

It is beyond the scope of this volume to survey experimental relaxation measurements, and we shall be content to cite some representative articles on solids (Alekseev et al., 1972, Baumberg 1973; Brom et al. 1972; Candela, 1970; Cheng and Kemp, 1971; Chau, 1972; Cristea et al., 1971; Deigen et al., 1978; Gerritsma et al., 1978; Gooijer and Blekemolen, 1973; Hyde and Rao 1978; Liuphart et al., 1972; Korolev et al., 1975, 1977; Kurkin et al., 1978; Maniv et al., 1976; Meservey and Tendrow, 1978; Ohkura et al., 1977; Passini et al., 1978; Poupko, 1973; Roest et al., 1973, 1973, 1973; Sanders and Rowan, 1971; Scott et al., 1972; Van Heugten et al., 1978), solutions (Barbarin and Germain, 1975; Dietz et al., 1978; Brunton et al., 1976; Burlamacchi et al., 1971, 1971, 1978; Catterall, 1970; Glausinger and Sienko, 1975; Hudson and Lewis, 1970; Jolicoeur and Bernier, 1975; Leniart et al., 1975, Michalik and Kevan, 1979; Muus and Atkins, 1972; Vishnevskaya et al., 1975; Wilson and Myers, 1976), spin labels (Davis et al., 1975; Freed, 1976; Hudson and Treweek, 1976), triplet states (Aleksandrov and Zapasskii, 1978; Andreev and Sugakov, 1975; Avarmaa and Suisalu, 1979; Avdeenko et al., 1978; Baldacchini et al., 1979; Chachaty and Maruani 1971; Lutz et al., 1978; Eliav and Lebanon, 1975; Kilmer and Kalantar, 1978; Konzelmann et al., 1975; Levanon and Vega, 1974; Minaev and Serebrennikov, 1978; Renk et al., 1978; Tanaka and Kazusaka, 1975; Verbeek et al., 1978, 1979; Vollmann, 1979; Winscom and Dinse, 1978), other types of exited state energy levels (Broer et al., 1978; Misu, 1978, 1979), cross relaxation (Abragam et al., 1976; Ajiro et al., 1975; Atsarkin, 1975; Brik et al., 1974; Buishvili and Giorgadze 1972; Clayman 1979; Dalton et al., 1972; Gokhman and Shanina, 1975; Hoogstraate et al., 1973; Kispert et al., 1975; Moore and Satten, 1973; Shulman and Podzyarei, 1974; Sikri and Narchal, 1979; Srivastava and Dale, 1978; Vasson and Vasson, 1978), exchange (Glinchuk et al., 1971; Gourier et al., 1978; Misra et al., 1975; Murakami and Kawamori, 1977; Tanaka and Kazusaka, 1975; Zaspel et al., 1977), spin temperature (Atsarkin, 1973), Jahn Teller (Lee and Walsh, 1971; Bugai et al., 1978), bottlenecks (Abragam et al., 1976; Antipin et al., 1978; Brya, 1971; Buishvili and Giorgadze, 1972; Gerritsma et al., 1978, Hirst et al., 1973), hindered rotations, (Emid et al., 1978; Michalik and Kevan, 1978, 1978; van Duyneveldt et al., 1979), nuclear polarization (Antipin et al., 1978; Bales and Gough, 1979; Leblond et al., 1971, 1971), phase relaxation (Milov et al., 1972, 1973; Raitsimring et al., 1975), weakly coupled van Vleck paramagnets (Davidov et al., 1977; Antipin et al., 1979), and forbidden transitions (Glarum and Marshall, 1975). Schlick and Kevan (1976) found that main lines and satellites saturate similarly in Cu^{2+}-doped $CaCd(OAc)_4 \cdot 6H_2O$, while the main lines saturate much more readily than the satellites for PO_3^{2-} ions in γ-irradiated $Na_2HPO_3 \cdot 5H_2O$. Kasthurirengan and Kar (1973) used saturation recovery to study relaxation times in anion radicals. Kispert et al. (1975) applied electron–electron double resonance to the study of relaxation in irradiated organic solids. The lineshape for longitudinally detected ESR has been studied as a function of the relaxation rate (Chiarini et al., 1975; Giordano et al., 1978; Martinelli et al., 1977; Vigouroux and Gourdon, 1978). The field has been reviewed by Poole and Farach (1971), Standley and Vaughan (1969), Vaughan (1975), and Geschwind (1972).

Some samples exhibit an orientation dependence of the relaxation times (Bratus et al., 1978; Konchits et al., 1978; Vikhnin et al., 1978). Flokstra et al. (1978) discussed correcting measured relaxation times for the finite thermal conductivity of liquid helium. The measurement of muon spin resonance relaxation has become fashionable in recent years (Castro et al., 1979; Hayano et al., 1978; Denison is preparing a review).

Some recent theoretical studies are the application of stochastic methods to the relaxation of spin labels (Dalton et al., 1975) and liquid crystals (Freed, 1976), the investigation of the role of lattice anharmonism on spin-lattice relaxation (Zevin et al., 1971), and the analysis of the contribution of hyperfine interactions to anisotropic broadening and saturation (Efremov and Kozhushner, 1972). Vugmeister (1978) obtained the shape of the spectral lines arising from fast-relaxing paramagnetic centers.

A number of articles have discussed the theory and presented calculations of pulse recovery dynamics. Some of the earlier work was discussed in the first edition and in books by Poole and Farach (1971) and Standley and Vaughan (1969). Shteinshneider et al. (1978) calculated the discrete saturation spectra and recovery dynamics for transient ESR line profiles in polycrystalline samples for various spin-lattice relaxation mechanisms and system parameters of the $S = \frac{1}{2}, I = \frac{1}{2}$ spin case. Sokolov et al. (1969) published experimental and theoretical signal recovery curves for different saturation factors. Smigel et al. (1974) described the pulsed saturation of very slowly tumbling spin labels. Hanssum et al. (1978) discussed the elimination of systematic errors in fast inversion recovery T_1 measurements. Other typical articles are by Antipin et al. (1979), Dalton et al. (1972), Hyde and Rao (1978), Marshall and Nistor (1973), Sarna and Hyde (1978), and Velter-Stefanescu and Grosescu (1978).

Freed (1974) extended his previous theories (Freed, 1965, 1967; Freed et al., 1967, 1973; Eastman et al., 1969), on steady-state saturation and double resonance of free radicals in solution to the cases of time-dependent experiments such as pulse saturation.

B. Homogeneous and Inhomogeneous Broadening

Homogeneous broadening occurs when the ESR signal results from a transition between two spin levels that are not sharply defined, but instead are somewhat broadened. Several sources of homogeneous broadening are (1) dipolar interaction between like spins; (2) spin-lattice interaction; (3) interaction with the radiation field; (4) motion of carriers in the microwave field; (5) diffusion of excitation throughout the sample; (6) motionally narrowed fluctuations in the local field.

An inhomogeneously broadened resonant line is one that consists of a spectral distribution of individual resonant lines merged into one overall line or envelope, as shown in Fig. 13-2. For example, if the applied magnetic field inhomogeneities over the volume of the sample exceed the natural linewidth $1/\gamma T_2$, then the spins in various parts of the sample find themselves in different field strengths, and the resonance is artificially broadened in an inhomogeneous manner. Other sources of inhomogeneous broadening include unresolved fine structure, hyperfine structure, and the dipolar interaction between spins with different Larmor frequencies.

Portis (1955) has presented a mathematical treatment of inhomogeneously broadened lines under the following assumptions:

$$H_1 < \Delta H_{\text{env}} \tag{1}$$

$$H_m < \Delta H_{\text{env}} \tag{2}$$

$$\gamma H_1 T_2 > 1 \qquad \text{for saturation} \tag{3}$$

$$\frac{\omega_m H_m}{H_1} < \gamma H_1 \qquad \text{adiabatic condition} \tag{4}$$

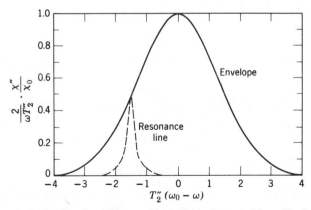

Fig. 13-2.   Absorption envelope and one of its individual resonant lines (Portis, 1953).

$$\frac{1}{H_1}\frac{dH_0}{dt} < \gamma H_1 \qquad \text{adiabatic condition} \tag{5}$$

$$\frac{dH_0}{dt} < \omega_m \Delta H_{env} \tag{6}$$

where H_1 is the applied rf field, H_m and ω_m, respectively, are the modulation amplitude and angular frequency, ΔH_{env} is the overall linewidth of the envelope, γ is the gyromagnetic ratio, T_2 is the spin–spin relaxation time, H_0 is the applied "constant" magnetic field, and t is the time. The amplitude of the dispersion signal χ' for various conditions (cases) is given in Table 13-1, and the actual lineshapes are sketched by Portis (1955).

When individual spin packets are narrow compared to the inhomogeneous broadening, then saturation reduces the intensity of the absorption χ'' in agreement with the experimental results on F centers in alkali halides as shown in Fig. 13-3 (Portis, 1953, 1955). Under the conditions of power modulation, the microwave power P_c incident on the crystal has the form

$$P_c = F\chi_0\omega_0 T_2 P_w \frac{\chi''(\omega, H_1)}{\chi''(\omega, 0)}(1 + \sin \omega_m t) \tag{7}$$

where P_w is the microwave power incident on the cavity, ω is the microwave resonant frequency, and F is a function of the Q of the cavity, the filling factor, and other quantities. The saturation factor $\chi''(\omega, H_1)/\chi''(\omega, 0)$ is the ratio of the imaginary part of the susceptibility at the rf field amplitude H_1 to its value at $H_1 = 0$. The detected signal P_c may be expanded in a Fourier series:

$$P_c = F\chi_0\omega_0 T_2 P_w \sum (a_n \cos n\omega_m t + b_n \sin n\omega_m t) \tag{8}$$

and the lock-in detector is tuned to detect only the $b_1 \sin \omega_m t$ term. The results for the following four cases are:

Case 1

Homogeneous broadening and $\omega_m T_1 < 1$. The system follows the variations in microwave power, and

$$b_1 = \left(\frac{2}{s'^2}\right)\left[\frac{(1 + s')}{(1 + 2s')^{1/2}}\right] \tag{9}$$

Case 2

Homogeneous broadening and $\omega_m T_1 > 1$. The system does not follow the variations in the microwave power, but saturates at the average power level,

TABLE 13-1

The Amplitude of the rf Susceptibility χ' when Recorded under Various Conditions (Portis, 1955)

Case	General Conditions[a]			Amplitude of χ'	Description	Phase
	H_1 and $H_m < \Delta H$; dH_0/dt and $\omega_m H_m < \gamma H_1^2$	$\gamma H_1 T_2 > 1$; $dH_0/dt < \omega_m \Delta H$	$T_2 \Delta H_0/dt < H_1$			
I	$\omega_m T_2 < H_1/\Delta H < 1$	$H_m < H_1/\omega_m T_2$		$\chi_0(\omega/\Delta\omega)(H_m/\Delta H)\cos\omega_m t$	Dispersion derivative	0
IIA	$H_1/\Delta H < \omega_m T_2 < 1$	$H_m < H_1/\omega_m T_2$	$T_2\, dH_0/dt < H_1$	$\chi_0(\omega/\Delta\omega)\omega_m T_2(H_m/H_1)\sin\omega_m t$	Absorption	$\pi/2$
IIB	$H_1/H_m < \omega_m T_2 < 1$	$H_m > H_1/\omega_m T_2$	$dH_0/dt < \omega_m H_m$	$\chi_0(\omega/\Delta\omega)\ln(2H_m\omega_m T_2/H_1)\sin\omega_m t$	Absorption	$\pi/2$
IIIA	$\omega_m T_2 > 1$	$H_m < H_1$	$T_2\, dH_0/dt < H_1$	$-\chi_0(\omega/\Delta\omega)(H_m/H_1)\cos\omega_m t$	Absorption	π
IIIB	$\omega_m T_2 > 1$	$H_m > H_1$	$T_2\, dH_0/dt < H_m$	$-\chi_0(\omega/\Delta\omega)\ln(2H_m/H_1)\cos\omega_m t$	Absorption	π
IV	$\omega_m T_2 > 1$	$T_2\, dH_0/dt > H_1$ and H_m		$\mp\chi_0(\omega/\Delta\omega)(H_m/\Delta H)\ln(2\,\Delta H/H_1)\cos\omega_m t$	Absorption derivative (sign reverses with direction of travel)	

[a] ΔH denotes ΔH_{env}.

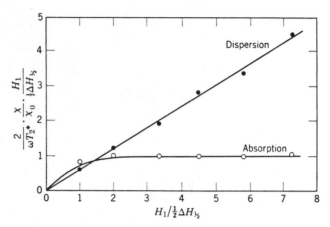

Fig. 13-3. Saturation behavior of the absorption and dispersion modes of F centers in KCl (Portis, 1953).

and

$$b_1 = \frac{1}{(1 + s')} \tag{10}$$

Case 3

Inhomogeneous broadening and $\omega_m T_1 < 1$. The spin packets will individually follow the periodic variations in power level and will saturate individually rather than transferring power at once to the entire spin system. The quantity b_1 will have the form

$$b_1 \cong \frac{(1 + 0.60s')}{(1 + s')^{3/2}} \tag{11}$$

Case 4

Inhomogeneous broadening and $\omega_m T_1 > 1$. This gives

$$b_1 = \frac{1}{(1 + s')^{1/2}} \tag{12}$$

where

$$s' = \gamma^2 H_1^2 T_1 T_2 = -1 + \frac{1}{s} \tag{13}$$

and s is the saturation factor used in the next section. These functions are plotted in Fig. 13-4.

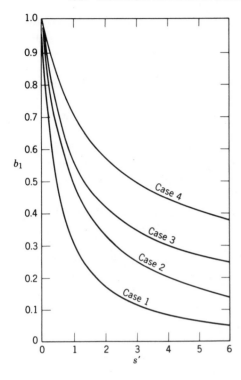

Fig. 13-4. The dependence of the observed signal amplitude b_1 on the parameter s' for the four cases given by Eqs. (9) to (12) (Portis, 1953).

Castner (1959) generalized Portis's theory by omitting the assumption that the individual spin packet width is very much less than the envelope width. He assumed that each spin packet had a Lorentzian shape, and showed that only those spins within $1/T_2$ or γH_1, whichever is larger, of the Larmor frequency will be saturated at sufficient microwave power levels. Spin diffusion (Anderson, 1959) is too slow to modify this. Many workers (e.g., Van Haelst et al., 1978; Moran et al., 1961) obtained experimental evidence in support of Castner's more general theory.

Cullis (1976) treated inhomogeneously broadened lines in terms of a spin temperature in the rotating coordinate frame associated with the spin packets. He obtained expressions that describe the system for both slow- and fast-passage conditions and that reduced to Castner's results (1959) in several cases. Atsarkin and Rodak (1972) had previously used a rotary-frame spin temperature. Strandberg (1972) discussed correlation time narrowing and narrowing by spin stirring experiments with inhomogeneously broadened lines. The relaxation and adiabatic fast-passage behavior was also elucidated.

Hyde and Hyde (1981) showed that the wings of an inhomogeneously broadened line are identical in shape to those that would result from shifting all of the spin packets to the envelope center. They obtained this result after establishing that the far wings of a pair of equally intense lines have the same shape as a single line of twice their amplitude located midway between them.

The relaxation parameter of spin packets can be measured by the continuous-saturation and spin-echo methods. Semenov and Fogel'son (1975) showed that an investigation of the response signal dependence on the intensity and duration of a saturating microwave pulse will provide the saturation factor and width $(1/T_2)$ of spin packets making up inhomogeneous lines. They studied inhomogeneous lines by pulse saturation.

Baumberg et al. (1972) found coherence effects in the inhomogeneous lines of nitrogen in SiC when irradiated with a microwave pulse whose duration was less than T_2. This pulse causes a hole to be burned at the point of application, with the appearance of smaller equally spaced holes located symmetrically about the main one. These arise from the rotation of spin packet magnetization vectors by the coherent microwave pulses.

Epifanov and Manenkov (1971) examined relaxation processes in inhomogeneously broadened lines by a quantum-statistical method. Bugai (1963) studied passage effects and Zhidkov et al. (1967) used continuous saturation to determine the width of the distribution function. Wollan (1976) discussed the theoretical and experimental aspects of the dynamic nuclear polarization of inhomogeneous lines.

It has been found that at liquid-helium temperature the $-\frac{1}{2} \rightarrow \frac{1}{2}$ absorption line of Cr^{3+} in dilute ruby is inhomogeneously broadened when the external magnetic field H_0 is directed along the c axis and homogeneously broadened when the external field is perpendicular to this axis (Boscaino et al., 1973, 1974, 1976). The broadening arises from superhyperfine structure of ^{27}Al nuclei that become magnetically equivalent in groups when H_0 is along c, and that generate spectral spin diffusion in the electron Cr^{3+} spin system, and hence increase the width.

Vugmeister (1978) and Vugmeister et al. (1973) asserted that inhomogeneous broadening is characterized by interactions that lack dynamic terms, that is, terms which cause energy exchange between different spins. The dynamic portion of the dipolar interaction produces homogeneous broadening when the energy exchange between separate spins occurs more rapidly than spin-lattice relaxation. For the case $g_{\parallel} > g_{\perp}$ with H_0 along the symmetry axis the dynamic part of the interaction may be negligible, producing inhomogeneous broadening. At sufficiently low temperature the homogeneity will be angularly dependent.

Dahlberg (1977) separated the homogeneous and inhomogeneous contributions to the line broadening in the ESR of alloys by carrying out measurements at widely separated frequencies.

Mailer et al. (1977) analyzed the theoretical and experimental aspects of rf field nonuniformity, modulation field nonuniformity, and sample length on the CW saturation characteristics of inhomogeneously broadened lines. They presented curves of signal amplitude dependence on various parameters, and illustrated their results with experimental data on biological samples. Coffman (1975) analyzed inhomogeneously broadened high-spin Fe^{3+} spectra from biological specimens by making use of multiple contour integrals.

Bowman et al. (1976) analyzed the saturation behavior of an inhomogeneously broadened line with second harmonic detection under slow-passage conditions. They presented curves that may be employed to extract the width and relaxation times T_1 and T_2 from saturation data on inhomogeneous lines.

Inhomogeneously broadened lines are readily studied by electron–electron double resonance (ELDOR), in which one part of the line can be saturated, and another part can be observed. This technique has been reviewed by Kevan and Kispert (1976), and their book may be consulted for an explanation of the theory and experimental details (compare Secs. 14-D and 14-E).

Mozurkewich et al. (1979) discussed cases in which the inhomogeneity of a line arises mainly from nonuniformities of the sample's demagnetizing field.

References to some of the earlier literature on inhomogeneous broadening may be found in the first edition. The subject has been covered by Talpe (1971) and in a recent review by Poole and Farach (1980).

C. Saturation Methods for Determining Relaxation Times

If one assumes that the Bloch equations are valid, and the broadening is homogeneous, then the magnetic resonance absorption Y and its first derivative Y' have a Lorentzian lineshape, and in the notation of Chap. 12 we may write

$$Y = \frac{y_m^0}{1 + \left[(H - H_0)/\frac{1}{2}\Delta H_{1/2}\right]^2} \tag{1}$$

$$Y' = \frac{\left(\frac{4}{3}\right)^2 (H - H_0) y_m^{0\prime}}{\frac{1}{2}\Delta H_{pp}\left\{1 + \frac{1}{3}\left[(H - H_0)/\frac{1}{2}\Delta H_{pp}\right]^2\right\}^2} \tag{2}$$

for normalized lineshapes. If one writes these relations as functions of the spin-lattice and spin–spin relaxation times T_1 and T_2 and includes the effect of the microwave amplitude H_1, they assume the form

$$Y = \frac{H_1 y_m^0}{1 + (H - H_0)^2 \gamma^2 T_2^2 + H_1^2 \gamma^2 T_1 T_2} \tag{3}$$

and

$$Y' = \frac{16}{3^{3/2}} \frac{(H - H_0)\gamma T_2 H_1 y_m^{0\prime}}{\left[1 + (H - H_0)^2 \gamma^2 T_2^2 + H_1^2 \gamma^2 T_1 T_2\right]^2} \tag{4}$$

where the quantities y_m^0 and $y_m^{0\prime}$ are the maximum amplitudes below saturation

$(H_1^2\gamma^2T_1T_2 \ll 1)$, and γ is the gyromagnetic ratio given by

$$\gamma = \frac{g\beta}{\hbar} = (0.87934 \times 10^7)g \, \text{Hz/G} \tag{5}$$

If the saturation factor s is defined by

$$s = \frac{1}{1 + H_1^2\gamma^2T_1T_2} \tag{6}$$

then Y and Y' have the forms

$$Y = \frac{sH_1 y_m^0}{1 + s(H - H_0)^2\gamma^2T_2^2} \tag{7}$$

and

$$Y' = \frac{16(H - H_0)\gamma T_2 s^2 H_1 y_m^{0'}}{3^{3/2}\left[1 + s(H - H_0)^2\gamma^2T_2^2\right]^2} \tag{8}$$

Note that below saturation $s \sim 1$. The half-power linewidth $\Delta H_{1/2}$ of Y and the peak-to-peak linewidth ΔH_{pp} of Y' are given in terms of their unsaturated $(s = 1)$ values $\Delta H_{1/2}^0$ and ΔH_{pp}^0, respectively, by

$$\Delta H_{1/2} = \left[\Delta H_{1/2}^0\right]s^{-1/2} = \left[\frac{2}{(\gamma T_2)}\right]s^{-1/2} \tag{9}$$

$$\Delta H_{pp} = \left[\Delta H_{pp}^0\right]s^{-1/2} = \left[\frac{2}{(3^{1/2}\gamma T_2)}\right]s^{-1/2} \tag{10}$$

The amplitude y_m of the center of the absorption line is obtained by letting $H = H_0$ in Eq. (7), and has the form

$$\left(\frac{y_m}{H_1}\right) = [y_m^0]s \tag{11}$$

while the peak-to-peak amplitude of the first derivative curve $Y' = y_m'$ at $H - H_0 = \pm\frac{1}{2}\Delta H_{pp}$ [i.e., at $H - H_0 = \pm(3\gamma^2T_2^2s)^{-1/2}$] has the form

$$\left(\frac{y_m'}{H_1}\right) = [y_m^{0'}]s^{3/2} \tag{12}$$

where

$$y_m^0 = \lim_{H_1 \to 0} \left(\frac{y_m}{H_1} \right)$$

(13)

$$y_m^{0\prime} = \lim_{H_1 \to 0} \left(\frac{y_m'}{H_1} \right)$$

(14)

Since below saturation $s = 1$, it follows that under this condition both y_m/H_1 and y_m'/H_1 are independent of the power level. Equations (9)–(12) are all of the general form $A = Bs^n$ and so

$$\frac{1}{s} = \left[\frac{B}{A} \right]^{1/n}$$

(15)

where B is a constant. Below saturation $s = 1$, so that

$$\frac{1}{s} = \left[\frac{\lim\limits_{H_1 \to 0} (A)}{A} \right]^{1/n} = 1 + \gamma^2 H_1^2 T_1 T_2$$

(16)

and one has explicitly

$$\frac{1}{s} = \begin{cases} \left[\dfrac{\Delta H_{1/2}}{\lim\limits_{H_1 \to 0} \Delta H_{1/2}} \right]^2 & \text{using Eq. (9)} \\[3em] \left[\dfrac{\Delta H_{pp}}{\lim\limits_{H_1 \to 0} \Delta H_{pp}} \right]^2 & \text{using Eq. (10)} \\[3em] \left[\dfrac{\lim\limits_{H_1 \to 0} (y_m/H_1)}{y_m/H_1} \right] & \text{using Eq. (11)} \\[3em] \left[\dfrac{\lim\limits_{H_1 \to 0} (y_m'/H_1)}{y_m'/H_1} \right]^{2/3} & \text{using Eq. (12)} \end{cases}$$

(17)

The short-hand notation y_m^0 corresponds to $\lim_{H_1 \to 0} y_m$, and similarly for $y_m^{0\prime}$, $\Delta H_{1/2}^0$, and ΔH_{pp}^0. Note that the terms in parentheses have different exponents (2, 2, 1, and 2/3, respectively). It is best to use the amplitudes y_m/H_1 or y_m'/H_1 for computing $1/s$ in Eq. (16) since they can be more accurately measured than the linewidths.

The peak-to-peak amplitude y_m' is proportional to the microwave magnetic field H_1 below saturation. Experimentally, one observes a linear dependence on the square root of the power, and a power-independent linewidth in this region, as shown in Figs. 13-5a and 13-5b. When the resonant line is strongly saturated, the amplitude y_m' becomes proportional to H_1^{-2} or P^{-1}, and therefore it decreases with increasing power in the manner shown in Fig. 13-5a. At these high power levels the linewidth becomes proportional to H_1, as shown in Fig. 13-5b. The width ΔH_{pp} will follow the dotted straight line for powers greater than those shown.

From Eq. (16) we see that a plot of $1/s$ against H_1^2 will give a straight line with the slope $\gamma^2 T_1 T_2$ and the intercept 1. Figures 13-5c and 13-5d show plots of $1/s$ versus H_1^2 using the derivative amplitude and the linewidth expressions, respectively. [The denominators of the ordinate functions $(1/3)^{2/3}$ and 7 are $(H_1/y_m^{0\prime})^{2/3}$ and ΔH_{pp}^0, respectively.] This is of course equivalent to plotting $1/s$ against the microwave power P because P is proportional to H_1^2. The slope of the line in either Fig. 13-5c or 13-5d may be used to calculate the product $T_1 T_2$.

In order to determine the relaxation times T_1 and T_2, a series of ESR spectra is recorded with the power varying from a condition of negligible saturation ($H_1^2 \gamma T_1 T_2 \ll 1$) to one of pronounced saturation ($H_1^2 \gamma^2 T_1 T_2 \gg 1$). If necessary, one may set the leakage at 4 μA or less for these measurements. The spin–spin relaxation time T_2 is calculated from the linewidth below saturation (i.e., $\Delta H_{1/2}^0$ or ΔH_{pp}^0) by means of the expression

$$
T_2 = \frac{2}{\gamma \Delta H_{1/2}^0} = \frac{2}{3^{1/2} \gamma \Delta H_{pp}^0} = \frac{2.2744 \times 10^{-7}}{g \Delta H_{1/2}^0}
$$

$$
= \frac{1.3131 \times 10^{-7}}{g \Delta H_{pp}^0} \tag{18}
$$

and Eq. (6) allows us to write for T_1

$$
T_1 = \left(\frac{(3)^{1/2} \Delta H_{pp}^0}{2\gamma} \right) \left(\frac{1/s - 1}{H_1^2} \right) \tag{19}
$$

$$
= \frac{0.9848 \times 10^{-7} \Delta H_{pp}^0}{g} \left(\frac{1/s - 1}{H_1^2} \right) \tag{20}
$$

where both ΔH_{pp}^0 and H_1 are in gauss. A similar expression may be written by substituting $\Delta H_{1/2}^0$ for $3^{1/2} \Delta H_{pp}^0$. Equation (20) may be used to calculate T_1 from the data. Note that below saturation the term $(1/s - 1)/H_1^2$ becomes very difficult to measure because $(1/s - 1)$ becomes much less than one, while far above saturation, the amplitude y_m' (or y_m) is very weak and the width ΔH_{pp} (or $\Delta H_{1/2}$) is very large so that the ratio $(1/s - 1)\Delta H_{pp}/H_1^2$ cannot be

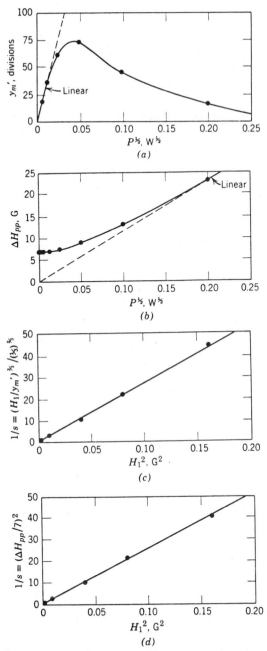

Fig. 13-5. (a) Peak-to-peak amplitude y'_m plotted as a function of the square root of the microwave power P. The dotted line is an extrapolation of the linear dependence at low powers. (b) Peak-to-peak linewidth ΔH_{pp} plotted as a function of the square root of the microwave power P. The dotted line gives the linear asymptotic behavior at very high powers. (c) The normalized quantity $(y'_m/H_1)^{-2/3}$ depends linearly on the square of the microwave field strength H_1^2, as shown. The slope of the line is $\gamma^2 T_1 T_2$ and $1/s = 1$ when $H_1 = 0$. (d) The linear dependence of $(\Delta H_{pp})^2$ normalized relative to the value 7^2 below saturation depends linearly on H_1^2 as shown. The slope of the line is $\gamma^2 T_1 T_2$ and $1/s = 1$ when $H_1 = 0$. The data points shown in Figs. 13-5 are obtained from Table 13-2.

593

accurately evaluated. The best results are obtained when Eq. (20) is used for power levels that are only moderately above saturation. One of the principal experimental advantages of this saturation method of determining relaxation times is that it may be carried out with routine spectrometers using very low leakages (e.g., ~ 4 μA), while the pulse methods to be discussed in the next section require the use of specialized microwave and electronic components.

A sample calculation of T_1 and T_2 by the saturation method is presented in Table 13-2. The raw data consist of the microwave power P_w measured on a

Table 13-2

Typical Saturation Curve Data and Illustrative Calculation of T_1 and T_2

P, W	H_1, G	ΔH_{pp}, G	y'_m Divisions	y'_m/H_1^b	$(1/s)^{3/2}$	$(1/s - 1)/H_1^2$	s
3.9×10^{-5}	0.0063	7.0	19	3.0	0.98	—	0.99
1.56×10^{-4}	0.0125	7.1	36	2.9	1.02	—	0.96
6.25×10^{-4}	0.025	7.4	61	2.4	1.23	—	0.86
2.5×10^{-3}	0.05	8.9	73	1.5	1.97	—	0.59
1.0×10^{-2}	0.1	13.1	46	0.46	6.4	244	0.23
4.0×10^{-2}	0.2	23.	17	0.085	35	242	0.083
8.0×10^{-2}	0.28	33.	8	0.029	103	263	0.044
1.6×10^{-1}	0.4	45.	4	0.01	300	—	0.023
6.4×10^{-1}	0.8	90.	1	1.3×10^{-3}	2000	—	0.006

$$y_m^{0'} = \lim_{H_1 \to 0} \left(\frac{y'_m}{H_1} \right) = \frac{1}{2} \left(\frac{19}{0.063} + \frac{36}{0.0125} \right) \sim 3 \times 10^3 \text{ divisions/G}$$

$$(1/s)^{3/2} = \left(\frac{\lim_{H_1 \to 0} (y'_m/H_1)}{y'_m/H_1} \right) \text{ using amplitude data in the fourth column}^a$$

$$\frac{1/s - 1}{H_1^2} = \frac{1}{3}(244 + 242 + 263) \approx 250$$

$$\Delta H_{pp}^0 = \lim_{H_1 \to 0} \Delta H_{pp} = 7 \text{ G}$$

$$g = 2.00$$

$$T_2 = \frac{1.3 \times 10^{-7}}{g \Delta H_{pp}^0} = \frac{1.3 \times 10^{-7}}{2 \times 7} \sim 10^{-8} \text{ sec}$$

$$T_1 = \frac{0.98 \times 10^{-7} \Delta H_{pp}^0}{g} \left(\frac{1/s - 1}{H_1^2} \right) = 0.98 \times 10^{-7} \times \tfrac{1}{2} \times 7 \times 250 \sim 0.88 \times 10^{-4} \text{ sec}$$

$$\gamma = \frac{g\beta}{\hbar} = 0.88 \times 10^7 \text{ g Hz/G}$$

$$s = \frac{1}{1 + \gamma^2 H_1^2 T_1 T_2} = \frac{1}{1 + 0.78 \times 10^{14} \times 10^{-8} \times 3.5 \times 10^{-4} H_1^2} = \frac{1}{1 + 275 H_1^2}$$

^{a}A less accurate calculation may be made using $(1/s)^{1/2} = (\Delta H_{pp}/\Delta H_{pp}^0)$ in the sixth column.
^{b}Divisions per milligauss.

power meter, the linewidths ΔH_{pp}, and the peak-to-peak amplitudes of the derivative spectrum y'_m in arbitrary units. These data are plotted in Figs. 13-5a and 13-5b. The microwave magnetic field H_1 at the sample may be computed from the microwave power P_w incident on the resonant cavity by the methods described in Sec. 5-H, and will have the functional dependence

$$H_1^2 = KP_w \tag{21}$$

where H is in gauss, P_w is in watts, and in a typical case K is of the order of unity. Equation (21) with $K = 1 \text{ G}^2/\text{W}$ is used to compute the second column from the first column. Then the fifth column is calculated by dividing the values of the fourth column by those in the second column. The two top entries in the third column and also the two top entries in the fifth column are close to each other, which indicates that they correspond to the condition of negligible saturation. Therefore, they are used to determine ΔH_{pp}^0 and $\lim_{H_1 \to 0}(y'_m/H_1)$. The sixth column is calculated by dividing $\lim_{H_1 \to 0}(y'_m/H_1)$ by the entries in the fifth column since

$$(1/s)^{3/2} = \frac{\lim_{H_1 \to 0}(y'_m/H_1)}{y'_m/H_1} = \frac{3.0}{y'_m/H_1} \tag{22}$$

Only three entries are given for $(1/s - 1)/H_1^2$ in the seventh column since at low powers $1/s$ is too close to unity, and at the highest powers the amplitude y'_m is too small for accuracy. The saturation factor s computed from the calculated T_1 and T_2 values is listed in the eighth column for completeness. Figure 13-5c shows values of $1/s$ computed from the sixth column, and data from the third column are plotted in Fig. 13-5d.

A simpler way to calculate T_1 is to determine either y_m or y'_m as a function of power, and ascertain where a maximum is reached. For example, we might deduce from Fig. 13-5a or from the data in Table 13-2 that the maximum value of y'_m is obtained for H_1 between 0.025 and 0.05 G, and somewhat closer to the higher value. A reasonable estimate is 0.04 G. The maxima in Eqs. (11) and (12), respectively, come where

$$\frac{dy_m}{dH_1} = 0 \quad \text{and} \quad \frac{dy'_m}{dH_1} = 0 \tag{23}$$

One may easily show that these maxima occur when

$$s = \tfrac{1}{2} \quad \text{max in } y_m \tag{24}$$

$$s = \tfrac{2}{3} \quad \text{max in } y'_m \tag{25}$$

Using Eqs. (20), (24), and (25), one may show that at the maxima

$$T_1 = \frac{0.57 \times 10^{-7} \Delta H_{1/2}^0}{\left(gH_1^2\right)} \quad \text{max in } y_m \tag{26}$$

$$T_1 = \frac{0.49 \times 10^{-7} \Delta H_{pp}^0}{\left(gH_1^2\right)} \quad \text{max in } y_m' \tag{27}$$

In the example under consideration, $\Delta H_{pp}^0 = 7$ G and we estimated $H_1 = 0.04$ G, so from Eq. (27)

$$T_1 \sim \frac{\left(0.49 \times 10^{-7} \times \frac{1}{2} \times 7\right)}{(0.04)^2} \sim 1 \times 10^{-4} \text{ sec} \tag{28}$$

in reasonable agreement with the estimate obtained from Table 13-2. One may of course determine the maximum in y_m' more precisely by a graphical method with data obtained at intermediate power settings.

Some authors such as Singer and Kommandeur (1961) display saturation data by plotting the amplitude y_m' relative to a (ruby) standard y_{ms}' against the logarithm of the power, as shown in Fig. 13-6 for iodine complexes. This has the advantage of fitting on one graph power data that span several orders of

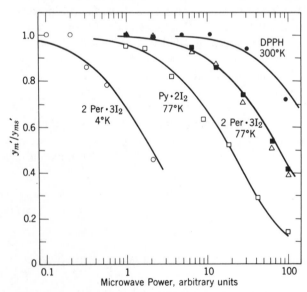

Fig. 13-6. Saturation curves for pyrene and perylene complexes at low temperatures. The ordinate is the radical ion amplitude y_m' divided by the amplitude of the ruby standard y_{ms}' normalized to 1 at low power (Singer and Kommandeur, 1961).

magnitude. Singer and Kommandeur determined the product T_1T_2 of an unknown sample relative to that of solid DPPH from the relation

$$T_1T_2 = \left(\frac{P_{\text{DPPH}}}{P} \right) T_{1(\text{DPPH})}^2$$

where for DPPH Lloyd and Pake (1953) and Bloembergen and Wang (1954) found (see also Berthet, 1955)

$$T_1 = T_2 = 6 \times 10^{-9} \text{ sec}$$

The powers P_{DPPH} and P of DPPH and the unknown are determined at the same amount of saturation, for example, where y'_m/y'_{ms} in Fig. 13-6 reaches 0.5. The value of T_2 for the unknown is deduced from Eq. (18).

Two difficulties involved in the saturation method of determining relaxation times are the conversion of power P_w incident on the cavity to H_1^2 values, and the measurement of P_w. The former is difficult to do accurately, and Sec. 5-I may be consulted for the theory behind it. A calibration of H_1^2 may be carried out experimentally from saturation data on a sample such as DPPH whose relaxation times are known.

The value of P may be measured by means of a power meter and bolometer which monitors the microwave power through a directional coupler. One may also use a nonsaturating standard sample, in the same resonant cavity, and the Singer method (Singer, 1959; Singer et al., 1961), described in Sec. 11-K is useful. This has the advantage of monitoring the actual H_1 field and automatically correcting for the impedance mismatch that is inevitable at low power levels. In other words, when the slide screw tuner is inserted very far to achieve the desired leakage level at low powers, the cavity pip on the mode is raised close to the top, and a much smaller percentage of the incident microwave power actually enters the cavity. This causes the power meter to give a false indication of the energy density $(\mu H_1^2/2)$ in the cavity. A standard sample must be used to check H_1^2 accurately. The use of a very low leakage (e.g., 4 μA) at these low powers will minimize the mismatch difficulty. A dual sample cavity such as the one shown in Fig. 5-38 is convenient for use with the standard sample.

McConnell (1958) generalized the Bloch equations to include a system in which the observed spins are reversibly exchanging between two sites such as might occur in a simple chemical exchange system. Sohma (1962) discusses the effect of free-radical lifetimes on ESR relaxation, and shows how to estimate either T_1 or the radical lifetime when the other is known.

The saturation method of determining relaxation times assumes the validity of the Bloch equations and a Lorentzian lineshape. It should be emphasized that not all systems obey the Bloch equations, and in particular, the dispersion mode χ' often saturates at much higher power levels than the absorption mode χ'' as shown in Fig. 13-3 (see, e.g., Portis, 1953; Redfield, 1955, 1957). When a resonant line is inhomogeneously broadened, then equations such as (9), (10), and (18) are only valid for the individual component spin-packet linewidth, not for the overall (envelope) width.

Hasegawa and Yazaki (1978) showed that an inhomogeneous line with an individual spin packet width $\Delta H_{1/20}$ and an envelope width ΔH^* given by

$$\Delta H_{1/20} = \frac{2}{\gamma T_2}$$

$$\Delta H^*_{1/2} = \frac{2}{\gamma T_2^*} \tag{29}$$

where

$$\Delta H^*_{1/2} \gg \Delta H_{1/20}$$

$$T_2^* \ll T_2 \tag{30}$$

has the following peak-to-peak amplitude:

$$\left(\frac{y'_m}{H_1} \right) = \frac{y_m^{0'} s^{1/2}}{\left[1 + T_2^* / \left(T_2 s^{1/2} \right) \right]^2} \tag{31}$$

where the notation is that of Eq. (13-C-12). This means that for the condition

$$s \gg \left(\frac{T_2^*}{T_2} \right)^2 \tag{32}$$

we have, using Eq. (14)

$$\frac{1}{s} = \left[\frac{\lim_{H_1 \to 0} \left(y_m^1 / H_1 \right)}{y_m^1 / H_1} \right]^2 \tag{33}$$

in analogy with Eqs. (17). As a result experimental saturation data obtained in the range $1 > s \gg (T_2^*/T_2)^2$ can provide the spin packet relaxation time values T_1 and T_2 by the methods described above. At much higher powers where

$$s \ll \left(\frac{T_2^*}{T_2} \right)^2 \tag{34}$$

we obtain

$$\frac{1}{s} = \left(\frac{T_2}{T_2^*} \right)^{4/3} \left[\frac{\lim_{H_1 \to 0} \left(y'_m / H_1 \right)}{y_m^1 / H_1} \right]^{2/3} \tag{35}$$

When a saturation study is carried out with a large sample such as one along the axis of a cavity, then the microwave field will vary throughout the sample. Some spins will be saturated, while others are not. Many samples such as the one shown in Fig. 5-5 are oriented so that one expects a sinusoidal variation of the microwave field strength H_1 of the type $\sin \pi x/a$ given by Eq. (5-H-8). This, however, is not the case since the sample in the waveguide and the presence of Dewar structures perturb the H_1 field, and a typical variation of H_1 across the cavity is given in Fig. 5-28 (Schreurs et al., 1960; Hyde and Brown, 1962; Kooser et al., 1969; Rataiczak and Jones, 1972). Mailer et al. (1977) obtained 35-GHz experimental data on the relative effects of sample length on the signal intensity for nonsaturating and partially saturating conditions. Variations in the cavity Q during the passage through resonance can influence H_1, especially for strongly absorbing samples (Goldberg and Crowe, 1977; Vigouroux et al., 1973). Nonuniformities in the modulation field can alter the CW saturation behavior of inhomogeneously broadened lines (Mailer et al., 1977).

The filling factor is smallest where the saturation is least at the top and bottom of the tube. One should use only the center $\frac{1}{3}$ or $\frac{1}{4}$ of the tube for best results.

The modulation frequency will influence the detected ESR signal, as discussed in Sec. 6-I. When audio modulation frequencies are used that satisfy the conditions $f_{\text{mod}} \ll 1/T_1$, and a fortiori, $f_{\text{mod}} \ll 1/T_2$, then the saturation factor (spin system) will follow the modulation cycle. The data tabulated in Table 13-2 and displayed in Fig. 13-5 were recorded under these conditions. When the modulation frequency becomes comparable to or exceeds the reciprocal of the relaxation time, then the magnetic field changes too rapidly for the spin system to follow it, and the spins "see" an average applied magnetic field. This produces a change in the amplitude of the detected signal. The relaxation time T_1 may be determined by modulating the microwave field H_1 and observing the detected amplitude as a function of f_{mod} in the neighborhood of $f_{\text{mod}} \sim 1/T_1$ (Pescia and Hervé, 1961; Bassompierre and Pescia 1962; Hervé and Pescia, 1962, 1963; Pescia, 1965; Zueco and Pescia, 1965). Hervé (1963) used modulation frequencies up to 30 Mc. Halbach (1954) also described a modulation method for determining relaxation times. Carruthers and Rumin (1965) evaluated the spin-lattice relaxation time from a measurement of the phase and amplitude of the detected ESR signal. This method does not require a knowledge of the power level or linewidth.

Brill et al. (1975) found that the high field resonance arising from ferrimyoglobin frozen in solution at 4.2 K saturates at a much higher power level than the low field resonance. The SO_3^- radical ion produced in X-irradiated $KHSO_4$ exhibited a linewidth that is isotropic and an amplitude that is very anisotropic (Sunandana, 1978). Maruani and Roncin (1973) carried out a continuous saturation study of the main and satellite lines of an oriented free radical. Krishnamoorthy and Prabhananda (1981) studied slow-motional averaging by this method. Kulikov and Likhtenshtein (1974) used saturation curves of spin labels for evaluating distances in biological specimens. Smigel et al. (1974) treated the saturation recovery of spin labels. Schlick and Kevan (1976) used differential saturation to distinguish radial and angular relaxation mechanisms. Brotikovskii et al. (1971) reported modulation effects on the saturation of nonuniformly broadened lines. De Neef and Verbruggen (1974) employed a static torque technique for the measurement of very slow relaxation. Freed et al. (1973) treated the saturation of ENDOR and ELDOR transitions. Other recent typical articles on inhomogeneously broadened spectra are by Bowman et al. (1976), Cullis (1976), Holstein et al. (1976), and Zevin and Brik (1972).

D. Pulse Spectrometers

One of the main experimental methods for measuring relaxation times consists in saturating the spin system with a high power pulse of microwave energy and then measuring the amplitude of the ESR signal at various times after the pulse to monitor the rate at which the magnetization returns to its equilibrium value. This is referred to as pulse saturation, or saturation recovery. In the first edition we described and presented block diagrams of several pulse spectrometers (Bloembergen and Wang, 1954; Davis et al., 1958; Faughnan and Strandberg, 1961; Fletcher et al., 1960; Kaplan 1962; Leifson and Jeffries, 1961; Llewellyn et al., 1962; Pashinin and Prokhovov, 1961; Ruby et al., 1962; Scott and Jeffries, 1962; Zverev, 1961). During the intervening years a number of advances in technology have improved the performance of pulse spectrometers, and so now we describe several more recent designs. The first edition may be consulted for references to the earlier literature (e. g. Pescia, 1966).

Wolfe and Jeffries (1971) designed the 1-cm superheterodyne pulse-recovery spectrometer illustrated in Fig. 13-7. It is similar to earlier versions used in the same laboratory (Scott and Jeffries, 1962; Larson and Jeffries, 1966; Harris and Yngvesson, 1968; see also Fig. 18-11 of the 1st edition). The microwaves from a $\frac{1}{2}$W klystron were repetitively switched from a saturating high power level to a monitoring low power level by a microwave diode switch with a 2-dB insertion loss and 30-dB on–off power ratio.

The switch was driven by a pulse generator, and at high power levels this generator was adjusted for long intervals between pulses and sometimes short pulses to prevent the heating of the cavity walls and the formation of tiny helium bubbles that shift the cavity frequency. An automatic frequency-control circuit using 60 kHz frequency modulation and lock-in detection locked the klystron to the cavity frequency. A 10-dB directional coupler was employed in place of a circulator to couple incident power to the cavity and reflected power to the detector. This arrangement maximized the power in the cavity by sacrificing sensitivity through the 10-dB loss of signal. The noise figure of the receiver was 12 dB.

Wolfe and Jeffries employed the high-Q TE_{011} mode cylindrical resonant cavity illustrated in Fig. 5-58. The crystal was rotatable about a horizontal axis and the magnetic field could be rotated in a horizontal plane to permit the

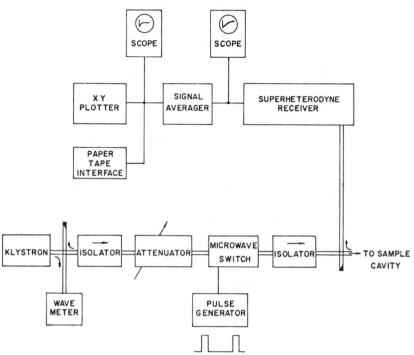

Fig. 13-7.   Bloch diagram of a spectrometer for measuring relaxation times by the pulse recovery method (Wolfe and Jeffries, 1971).

precise alignment of the crystal c axis. A port was available for optical pumping. The entire cavity was immersed in a liquid-helium Dewar located between the polepieces of an electromagnet that produced magnetic fields up to 19 kG.

The pulse spectrometer designed by Huisjen and Hyde (1974) and improved by Percival and Hyde (1975) has the block diagram presented in Fig. 13-8, the data acquisition system illustrated in Fig. 13-9 and the bimodal resonant cavity sketched in Figure 13-10 with the pump and observing modes excited by coherent microwaves from a single klystion. This arrangement offers several advantages: (1) the relative phase between the pump and the reference microwaves at the detector crystal can be adjusted to provide free induction absorption or dispersion; (2) the mode-to-mode isolation suppresses cavity ringing and permits the observation of earlier recovery times, that is, times closer to the end of the pump pulse; (3) the signal can be observed during the pump cycle; and (4) saturation recovery experiments can be carried out without field modulation.

The bimodal resonant cavity shown in Fig. 13-10 consists of two TE_{103} rectangular modes that have a full wavelength section in common. After insertion of the Dewar the frequencies are brought together by adjusting

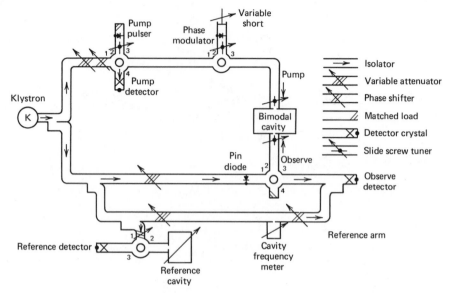

Fig. 13-8. Block diagram of the microwave bridge of a pulse spectrometer. The data acquisition system is shown in Fig. 13-9 (Percival and Hyde, 1975).

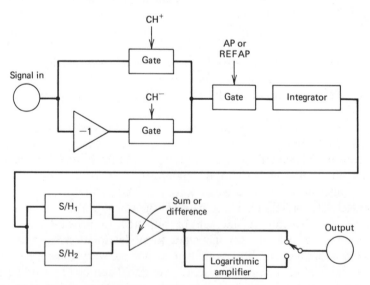

Fig. 13-9. Block diagram of the data acquisition system of a pulse spectrometer showing, from left to right, the sample input, the operational amplifier with −1 gain, the gating circuits controlled by CH + , CH − , AP or REFAP, the boxcar integrater, the sample and hold (S/H) circuits, the differential amplifier, the optional logarithmic amplifier, and the output. The aperture gate is opened by the logic pulse AP for transient signal sampling, and after each AP pulse REFAP initiates reference sampling (Percival and Hyde, 1975).

602

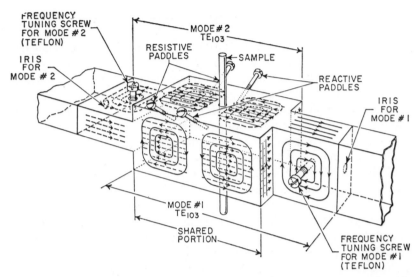

Fig. 13-10. Bimodal resonant cavity in which two rectangular TE_{013} modes are crossed and have two half-wavelength sections in common. The magnetic field lines of mode #1 are indicated by solid lines and those of mode #2 by dashed lines. Fine-frequency tuning is accomplished with the Teflon screws, and mode isolation is achieved by adjusting the resistive and reactive paddles (Hyde et al., 1969; Huisjen and Hyde, 1974).

2-mm-diameter quartz rods (not shown), and fine frequency tuning is accomplished by the Teflon tuning screws shown in the figure. The mode isolation can be adjusted to 65 dB with the aid of the resistive and reactive paddles under operating conditions with the Dewar and sample in place. The unloaded Q of 6000 produces a ringing time of $Q/2\pi\nu \sim 10^{-7}$ sec. The AFC can lock on either cavity mode.

The boxcar integrator shown in Fig. 13-9 increases the signal-to-noise ratio of the repetitive transient waveform by gating a small portion of the waveform into the integrator many times and slowly moving the gate along the waveform. The high pulse repetition rate of 6.25 to 50 kHz helps to overcome $1/f$ crystal noise. Phase-sensitive detection is accomplished at 50 Hz by alternately sampling the integrator output with two sample and hold (S/H) circuits at the end of each half-cycle. The signal next passes through a differential amplifier and RC filter, and then goes to the recorder directly or after passage through a logarithmic amplifier.

The $\frac{1}{2}$-W microwave power output is split into pumping and monitoring channels, and the latter is further split into a portion incident on the cavity and a reference signal at the detector. The bridge of Fig. 13-8 can be operated in the following modes by the suitable adjustment of the microwave phases and amplitudes and the choice of signals driving pin diodes and field modulation coils.

(1) In the CW reflection bridge mode the pump power is suppressed; the klystron is locked to the observing cavity, which acts as a reflection cavity; the PIN diode in the observing arm transmits; and the ESR signal is detected by the observe detector. (2) In the ESR induction mode the monitor power is turned off, the klystron is locked to the pump cavity mode, and the instrument operates as a CW induction spectrometer (compare Sec. 5-L). CW power is incident on the pump cavity, and the ESR signal that emerges from the observe cavity mode is detected by the observe detector. Either induced absorption or dispersion signals can be recorded by the proper adjustment of the relative phase of the microwave power at the crystal in the observing arm. (3) Circularly and elliptically polarized radio-frequency fields may be obtained by adjusting the powers and phases of the microwaves incident on the observing and pumping cavity modes. (4) Free induction spectrometer operation results when the AFC locks to the observing mode with minimal rf power, the pin diode in the pumping arm is pulsed at various duty cycles and repetition rates, and a 50-Hz square wave magnetic field modulation is applied to the sample. (5) The saturation recovery method operates in the manner described above. In this configuration the signal observed following the pumping pulse is a free induction decay at very low observing powers (0.5 μW) and a superposition of saturation recovery and a free induction decay at higher observing powers (e.g., 0.5 mW). The free induction decay can be subtracted from the higher power signal to provide the saturation recovery curve. Another way to suppress the free induction signal is by setting the phase of the pump to detect dispersion and setting the magnetic field at an absorption maximum, since the free induction decay signal is dependent upon the phase while saturation recovery signals are not (Huisjen and Hyde, 1974, 1974; Percival and Hyde, 1975). Both methods can be employed simultaneously. The saturation recovery curve monitors the longitudinal magnetization, which is proportional to the population difference between the two ESR spin levels. The free induction decay, on the other hand, measures the time evolution of the M_x and M_x components of the magnetization. Thus the present spectrometer can measure the time evolution of all three magnetization components.

Another method employed by Percival and Hyde (1975) for separating the free induction decay and saturation recovery signals—called the pump phase modulation technique—is illustrated in Fig. 13-11. Curves 13-11a and 13-11b show the combination of free induction decay and saturation recovery signals obtained with the pump phases set at 0° and 180°, respectively. If the signals integrated in consecutive half-cycles are subtracted the free induction decay (13-11c) is obtained, and if they are added the result is the saturation recovery signal (13-11e). The former may be compared to the free induction decay (13-11d) obtained using field modulation at zero observe power.

Lingam et al. (1972) assembled the pulse spectrometer illustrated in Fig. 13-12. It employs an X-band superheterodyne detection system with broad band isolators inserted to prevent the high power pulse from affecting the AFC level oscillator. A microwave switch produces pulsed microwave power with

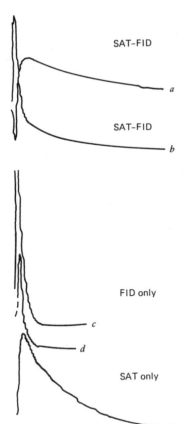

SAT–FID
— a

SAT–FID
— b

FID only
— c
— d

SAT only
— e

Fig. 13-11. Separation of the free induction decay (FID) and saturation (SAT) recovery signals using the pump-phase-modulation technique. The first two curves show the combination of free induction decay and saturation recovery with the phase set at (a) 0° and (b) 180°. The difference of those curves (c) gives the free induction decay signal, and their sum (e) provides the saturation recovery signal. For comparison, curve (d) shows the free induction decay signal obtained using field modulation. The sample was the TCNE radical anion at 31°C with the relaxation times $T_1 = 7.8$ μsec and $T_2 = 0.67$ μsec (Percival and Hyde, 1975).

variable pulse widths and repetition rates. The main power klystron rated for 500 mW is locked to an external cavity. The response time of the i.f. amplifier and the high Q of the cavity limit the spectrometer to the measurement of relaxation times greater than 3 μsec.

A pulse spectrometer operating in the submillimeter region was designed by Date et al. (1975) and modified by Kuroda et al. (1978); it is illustrated in Fig. 13-13. The high magnetic fields are produced by a mutilayer coil (Date, 1975) capable of generating fields up to 1 MG using a 6 kV capacitor energy source charged to 18 kJ. The magnetic field has a pulse width of 0.2 msec and is confined to a space 2 mm in diameter. The H_1 field is produced by a pulsed laser with a pulse width from 0.3 to 0.4 msec that is greater than the magnet pulse width. Electromagnetic waves of 119 and 337 μm are produced by H_2O and HCN lasers, respectively, using the circuit illustrated in Fig. 13-14. A plane mirror and a concave mirror with a radius of curvature of 4 m formed the optical cavity. The maximum output power was about 100 mW. A digital transient recorder was used to record the ESR signal and the magnetic field

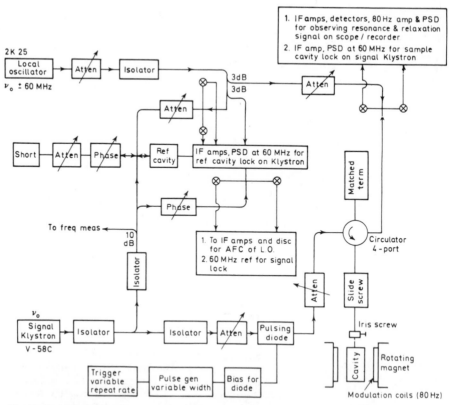

Fig. 13-12. Block diagram of an X-band superheterodyne pulse spectrometer capable of measuring relaxation times longer than 3 μsec (Rengan et al. 1972; Lingam et al., 1972).

strength. Recorder or oscilloscope presentation were provided for. Measurements carried out with NH_4Co Tutton's salt exhibited an appreciable H^3S term in the Hamiltonian, in addition to the usual Zeeman $g\beta HS$ term.

Sorokin et al. (1960) studied cross relaxation in diamond using a double resonance technique with an ELDOR cavity (compare Sec. 14-E). Their experiment consisted in using a high-power microwave pulse to saturate one of the three hyperfine lines of a nitrogen center and a low-power probing signal to monitor the effect of the pulse on one of the other two hyperfine lines. Their article presents a block diagram of the spectrometer and describes the design of the double microwave resonant cavity.

Brown and Sloop (1970) reported the design of a pulsed spectrometer, and Moore and Yalcin (1973) discussed a baseline restoring technique for the measurement of exponential time constants in the presence of noise. Some recent pulsed experimental measurements of relaxation times were reported by Fessenden et al. (1981), Michalik and Kevan (1979), Mims et al. (1978), Rengan et al. (1974, 1974), and Sharnof and Aiteb (1973). Freed (1974) provided a theoretical treatment of saturation recovery.

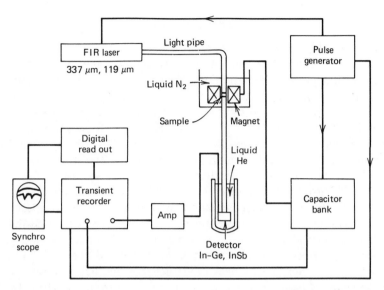

Fig. 13-13. Block diagram of a pulsed high-magnetic-field submillimeter ESR spectrometer (Kuroda et al. 1978).

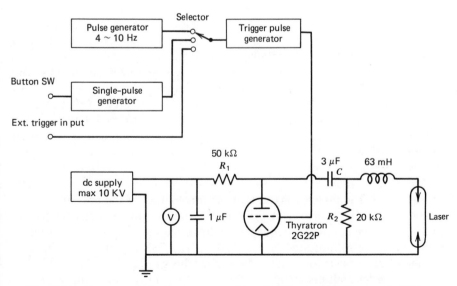

Fig. 13-14. Power supply for laser source used in submillimeter ESR (Date et al., 1975).

E. Time-Resolved ESR

Many free radical species are very short lived, and special techniques must be used to detect them. The usual way to accomplish this is to produce the radicals with a pulse of radiation and then to record or monitor the radical spectrum at various times after the cessation of the pulse. This technique for radical formation is referred to as pulse radiolysis or pulse photolysis, and the radical detection method is called time-resolved ESR spectroscopy. It has various applications such as the measurement of radical lifetimes, the determination of chemical reaction mechanisms, and the study of chemically induced dynamic electron polarization as discussed in Sec. 14-F.

In a series of papers Fessenden (1964, 1973) and co-worker (1976) Verma (1973, 1976) discussed instrumentation for time-resolved ESR spectroscopy, and they described a spectrometer for this purpose. They eschewed the use of high-frequency field modulation to avoid the line broadening and possible rapid-passage effects that can result from it. Instead they used direct ac-coupled amplification from a back diode detector that has an optimum signal to noise characteristic for a frequency in the range from 1 to 10 kHz. Many transient experiments involve radical lifetimes near 100 μsec, which is in this frequency region.

The block diagram of the time-resolved spectrometer is illustrated in Fig. 13-15. The microwave source or klystron and its automatic frequency control circuits are on the left, and the signal processing channel is on the right. The automatic phase control circuit produced pairs of pulses ϕ that adjusted the mixer phases to compensate for frequency changes in the cavity arising from electron beam heating. These continual phase adjustments are required for achieving an optimum signal-to-noise ratio.

The signal-processing channel contains a preamplifier followed by a clipper that removes the signals arising from the phase-control pulses and from the electron beam. The gated amplifier is added after the clipper to permit filtering with the time constant R_1C_1 without generating a tail on the interference pulse R produced by the electron beam. The filter is followed by a baseline-restoring circuit that resets the baseline before the radiolysis pulse in order to remove noise caused by microphonics or 60-Hz pickup. This circuit also eliminates the signal from more stable radicals that live longer than the time interval between the radiation pulses and accumulate in a steady-state population. There are separate channels for recording time profiles and fixed time spectra, and each will be described in turn.

The signal channel processes the detector output in the manner summarized in Fig. 13-16. The three strongest signals in the output of the preamplifier are, in the order of their arrival at the detector, the two phase control signals ϕ_1 and ϕ_2 from the automatic frequency control circuit and the signal R from the radiolysis pulse as indicated in Fig. 13-16a. The ESR signal that comes after the radiolysis pulse is too weak to be seen at this gain, and in Fig. 13-16b it is illustrated after multiplication by a factor of 100, which causes the signals ϕ_1,

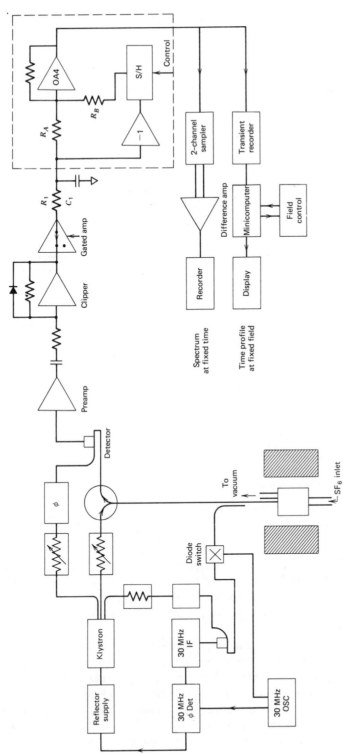

Fig. 13-15. Block diagram of time-resolved ESR spectrometer. The AFC circuits are on the left, and the signal channels are on the right. The latter consists of a preamplifier, clipper, gated amplifier, baseline-restoring circuit (in dashed-line box), and separate channels for the fixed-time and fixed-field modes. The former employs a two-channel boxcar amplifier sampler and difference amplifier followed by a recorder. The fixed-field channel employs a transient recorder and minicomputer to signal-average the output of the baseline restorer before displaying it (Fessenden, 1973; Verma and Fessenden, 1976).

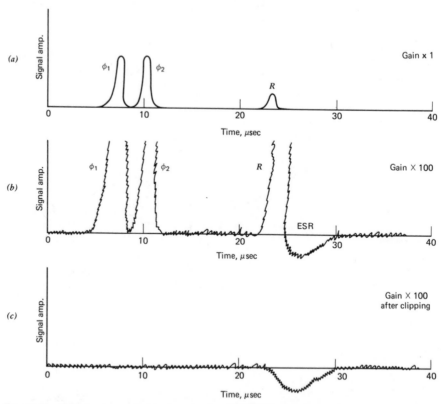

Fig. 13-16.   Signals processed after detection by the time-resolved ESR spectrometer sketched in Fig. 13-15. The detected voltage (top) contains strong signals from the two-phase control pulses ϕ_1 and ϕ_2 and from the beam pulse R, plus a weak ESR signal that appears in (b) after increasing the gain by a factor of 100. The three strong signals ϕ_1, ϕ_2, and R are removed by the clipper to provide the time profile itself as illustrated in (c).

ϕ_2, and R to be considerably off scale. The clipper removes these three strong signals so that the much weaker ESR signal can be seen without interference, as indicated in Fig. 13-16c, which is taken with the spectrometer set at the middle of a free-radical ESR spectral line. This time profile trace that emerges from the clipper shows the growth in time of the ESR signal amplitude as the radicals form immediately after the radiation pulse, and their subsequent slower decay with time.

The signal-to-noise ratio of the trace presented in Fig. 13-16c is ordinarily too weak for use, and the "time profile at fixed field" lower signal channel of Fig. 13-15 is employed to enhance it. A large number (typically 2000 to 10,000) of these traces are obtained in succession at 10-msec intervals corresponding to 20 to 100 sec of recording for each time profile. These profiles are added point by point in a transient recorder controlled by a PDP-3 minicomputer. The shortness of the 10-msec interval between irradiation pulses only permits 100

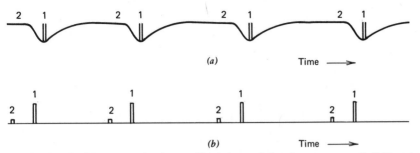

Fig. 13-17. Sketch of four successive time profiles (a) recorded at the same magnetic field setting which produce (b) the relatively large voltage pulses, 1, obtained at a fixed time delay and the very small voltage pulses, 2, obtained from sampling the noise prior to the radiation pulse. These pulses are accumulated in the two-channel sampler and subtracted by the differential amplifier.

points on the line to be read and signal-averaged in the memory of the minicomputer, which has a capacity for handling 2000 points. A recorder displays the time profile, which has the appearance of Fig. 13-16c with a much lower signal-to-noise ratio. The field control circuit shown at the lower right of Fig. 13-15 holds the spectrometer on resonance while carrying out the successive recordings.

To obtain the ESR spectrum at a particular time delay after the radiolysis pulse the upper channel, labeled "spectrum at fixed time," of Fig. 13-15 is employed. The amplitude at the corresponding point on the time profile is recorded as a pulse as indicated in Fig. 13-17, and successive pulses are added in one channel of a two-channel boxcar integrator sampler (compare Blume, 1961). The other channel sums the baseline amplitude pulses obtained at a fixed time (~ 10 msec) before the profile as indicated in Fig. 13-17. The differential amplifier subtracts the background pulses, labeled 2 on the figure, from the signal pulses, labeled 1. This process is repeated at the same time delay with a large number of magnetic field settings to provide an ESR absorption spectrum. The amplitudes of the absorption spectra recorded at various time delays trace out the time profile, as indicated in Fig. 13-18.

The electron beam provided 2.8-MeV electrons in pulses of length 0.5 to 1 μsec with 10-msec intervals between them. The time-averaged electron current in the cavity was 5 to 10 μA, corresponding to an average of 14 to 28 W. About 0.3 μA entered the sample, and it is estimated that typically $10^{-5}M$ radicals/pulse were produced.

The distortion of the cavity mode by the electron beam produces the large pulse R shown on the first trace of Fig. 13-16. To minimize this interference the cavity is evacuated, a low pressure (< 1 torr) of SF_6 is passed through it to scavenger electrons, and some NH_3 is introduced to thermalize the electrons. The end of the radiation pulse was taken as the time zero.

Trifunac et al. (1979) designed a nanosecond time-resolved electron spin-echo spectrometer suitable for detecting the transient radicals produced by a 3-MeV

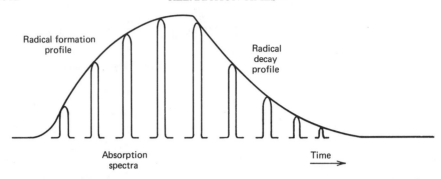

Fig. 13-18. Time profile traced out by the peaks of absorption spectra obtained at various times after the beam radiolysis pulse. At first the curve rises due to radical formation and then it decreases due to radical decay.

van de Graaff electron accelerator. The block diagram of the microwave system is presented in Fig. 13-19 and the timing, microwave switching, and data acquisition components are arranged in the manner illustrated in Fig. 13-20. A direct detection X-band spectrometer (Trifunac et al. (1975) was modified for spin-echo operation by the incorporation of microwave switches and traveling-wave tube amplifiers as will be explained in the next section (see especially Fig. 13-28). The van de Graaff accelerator generates short electron pulses of widths 5, 12, 25, 55, and 100 nsec with peak currents of up to 2 A and repetition rates of up to several kilohertz. The system was used to study the CIDEP from an acetate radical formed during pulse radiolysis, and it measured initial radical concentrations of 10^{-4} to $10^{-5}M$ per electron pulse.

The spin-echo method has the following advantages over other time-resolved ESR techniques: (1) improved time resolution (0.30–50 nsec), (2) less interference from the electron beam pulse at short times, and (3) lack of dependence of transient signals on microwave power or magnet inhomogeneity at short times.

The pulse radiolysis technique described above for electron beam irradiation can also be carried out with other types of exciting radiation. For example, Hamilton et al. (1970) used a rotating sector with two opposed 90° annular sectors cut out to provide equal light and dark periods. The sector was rotated by a motor with a variable speed controller. A system of lenses focuses light into the sector aperture and then into the cavity as illustrated in Fig. 13-21. This technique was employed to block the reaction-initiating light from reaching the ESR detector in a study of the recombination rates of photochemically produced organic free radicals.

The rotating sector just described had equal on–off times. Very often it is more appropriate to use sectors that produce pulses of light separated by relatively long dark intervals. For example, Livingston and Zeldes (1973) employed a 61-cm-diameter sector that produced 3.7-msec pulses at a rate of 60 per second to study the kinetics of radicals exhibiting ESR emission signals during continuous radiolysis.

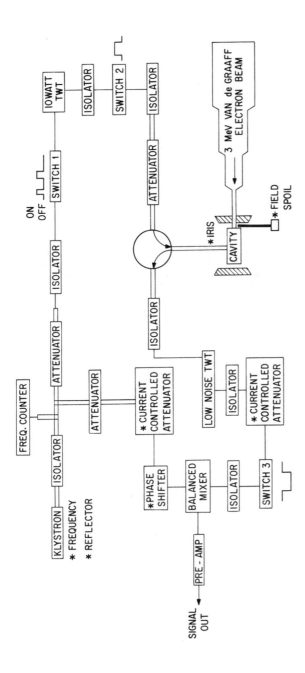

Fig. 13-19. Block diagram of the microwave components of a nanosecond time-resolved electron spin-echo spectrometer for use in pulse radiolysis studies. The output of the preamplifier on the left enters the data acquisition section of Fig. 13-20 (Trifunac et al., 1979).

613

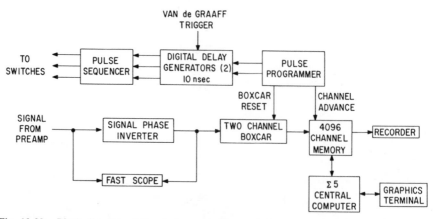

Fig. 13-20. Block diagram of the timing, switching and the data acquisition sections of the time-resolved spin-echo spectrometer illustrated in Fig. 13-19. The signal output from the left side of Fig. 13-19 enters the data acquisition section at the left side of this figure (Trifunac et al., 1979).

Atkins et al. (1970) designed a flash-correlated ESR spectrometer for flash photolysis studies. It is employed to observe transient free radicals formed photolytically by a pulsed ultraviolet laser. Measurements can be made within 1 μsec of the photolysis flash. In one mode of operation the spectrometer yields a series of decay curves that plot the ESR amplitude for a particular magnetic field setting as a function of the time after the flash. Another mode of operation employs a sawtooth, a ramp generator, and a data storage unit,

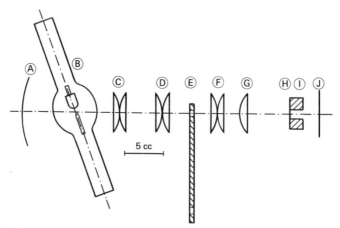

Fig. 13-21. Schematic side view of a rotating sector light modulator showing: A—Spherical mirror of 10-cm focal length; B—2.5-kW lamp; C, D, F—compound lenses of about 5-cm focal length, each made up of two nearly touching quartz lenses of 10-cm focal length; E—rotating sector; G—quartz lens of 7.5-cm focal length; H—1.6-mm-thick quartz plate; I—copper front plate of microwave cavity; J—sample center line. A water-cooled filter solution holder is positioned between C and D but is not shown (Hamilton et al., 1970).

and many successive pulses provide a spectrum corresponding to one time after the pulse.

Livingston et al. (1979) recently described ESR equipment for observing labile free radicals present at high temperatures. It makes use of a fluid pressurized up to 140 kg/cm² (140 atm) and heated up to 560°C that flows through the microwave cavity. The high density quenches the angular momentum of the overall radical rotation, and this reduces the ESR linewidth contribution from the spin–rotational mechanism, which varies inversely as the pressure and directly as $T^{3/2}$ according to the results of Schaafsma and Kivelson (1968). Peak-to-peak linewidths of ~ 370 mG corresponding to $T_2 = 0.18$ μsec were obtained in typical cases.

The flow system employed in these experiments is illustrated in Fig. 13-22, and the heating system is sketched in Fig. 13-23. The sample was contained in a glass reservoir of 100 ml capacity and was circulated by a high-pressure liquid chromatograph pump. The desired pressure was set by the BPR throttling valve. The part of the heating system that was in the cavity resonator was made of high-purity spectrosil and had an outside diameter of 1.1 cm. Air

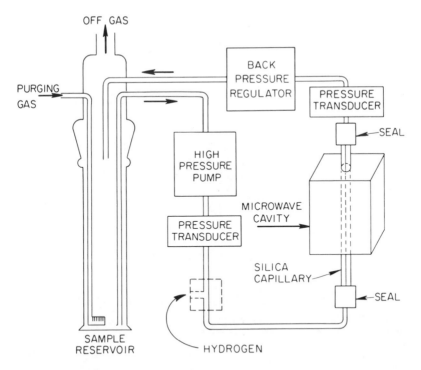

FLOW SYSTEM FOR PYROLYTIC STUDIES

Fig. 13-22. Flow system operating at high temperatures (560°C) and high pressures (140 kg/cm²) for carrying out ESR studies of pyrolytically generated labile free radicals. Figure 13-23 shows the details of the heating system in the cavity resonater (Livingston et al., 1979, 1981).

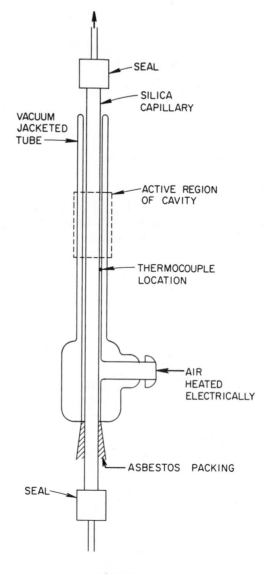

SEAL

SILICA
CAPILLARY

VACUUM
JACKETED
TUBE

ACTIVE REGION
OF CAVITY

THERMOCOUPLE
LOCATION

AIR
HEATED
ELECTRICALLY

ASBESTOS PACKING

SEAL

HEATING SYSTEM
FOR PYROLYTIC STUDIES

Fig. 13-23. Heating system for ESR pyrolytic studies carried out with the flow system illustrated in Fig. 13-22. (Livingston et al., 1979, 1981).

flowing at 30 liters/min was heated by an Iconel wire coil, and the temperature was monitored by a thermocouple.

Norris and Warden (1980) reported how to modify a Varian 100-kHz phase detector which has a minimum system time constant of 300 μsec to provide it with an overall system response of about 20 μsec so that it can be employed for studying transient phenomena. To accomplish this the low-pass network at the output is eliminated, and the tuned circuit assembly of the receiver is replaced by a 0.1- to 1-μF capacitor and a 0- to 2-kΩ ten-turn helipot. These changes produce a phase shift of approximately 90° in the detector. The modified system is tuned by setting the trimpot to match the gain of the unmodified receiver and then adjusting the trimpot setting and the 100-kHz modulation-phase vernier to optimize the signal from a strong pitch sample.

Equipment of the type discussed in this section is useful for the study of radical kinetics and chemical reactions. A number of representative studies of this type have been carried out by Ayscough et al. (1975), Alegria et al. (1975), Bennett et al. (1967, 1969, 1974), Bil'kis et al. (1976), Brunton and Ingold (1976), Bullock (1976), Carlier and Sochet (1975), Chawla and Fessenden (1975), Cookson et al. (1976), Davies et al. (1976, 1976), Elson et al. (1975), Fessenden et al. (1975), Gilbert et al. (1975), Hack et al. (1978), Hellebrand et al. (1976, 1976), Hewgill et al. (1976), Ingold (1975), Iordanov et al. (1976), Kim et al. (1976), Kochi (1975), Levanon and Vega (1974), Maruyama et al. (1976), Meghea and Panait (1976), O'Neill et al. (1975), Prokof'ev et al. (1976), Rakovskii et al. (1976), Rudenko et al. (1976), Russell (1975), Razuvaev et al. (1975), Shiotani et al. (1976), Silver (1975), Smaller et al. (1968), Stevensen et al. (1976, 1976), Stradins and Gavars (1976), Verbeek et al. (1976), and Wan and Wong (1976). Gillies et al. (1979) used pulsed laser excitation and nanosecond resolved phosphorescence detection together with 1-kW microwave pulses to survey triplet-state spin-lattice relaxation in benzil crystals. Time-resolved ESR can also be detected optically (Ponte Goncalves and Gillies, 1980; Smith and Trifunac, 1981; Trifunac and Smith, 1980).

This time-resolved technique has been employed to study chemically induced dynamic electron polarization (CIDEP) effects (Trifunac, 1976; Trifunac et al., 1975, 1977, 1978; Warden and Adrianowycz, 1981; see Sec. 14-F). Time-resolved ESR has been detected via chemically induced dynamic nuclear polarization (CIDNP) (Lawler et al., 1979; Trifunac and Evanochko, 1980). Chemically induced magnetic polarization transient techniques have been reviewed by Nelson et al. (1979) and in the volume edited by Lepley and Closs (1973).

Brown and Richardson (1973) and Matenaar (1973) designed resonant cavities for gas-phase studies. The latter cavity may be employed for making gas kinetic measurements up to 1000 K. Doetschman and Hutchison (1972) employed ESR and ENDOR for investigating rates and mechanisms of chemical reactions. Makino and Hatano (1979) separated and characterized short-lived radicals by a high-speed liquid chromatograph associated with an ESR spectrometer. Pasimini (1978) discussed the effect of field modulation on time-resolved spectra. Wardman (1978) reviewed the application of pulse radiolysis methods to the investigation of biomolecules. Recent articles reporting on short-lived radical species and their kinetics include those of Chong and Itoh (1973), Halliday et al. (1979), Lunsford (1975), McKenzie et al. (1973), Morino (1973), Symons (1976) et al. (1976, 1979), Taplick and Raetzsch (1979), Warden and Bolton (1976), and Warman et al. (1972).

F. Spin Echoes

Spin-echo techniques have been much more widely applied in NMR (compare reviews by Ellett (1971) and Poole and Farach (1971)] than in ESR [compare reviews by Mims (1972, 1976), Mims et al. (1980), and Norris et al. (1980)] and the underlying principle is the same in both cases. The experiment begins with the net electron spin magnetization M aligned along the magnetic field direction, which is selected as the z axis, as indicated in Fig. 13-24. This alignment is brought about in the absence of applied microwave power by

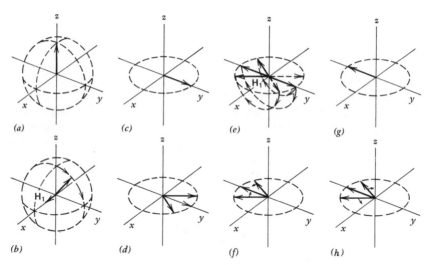

Fig. 13-24. The formation of an echo. Initially the net magnetic moment vector is in its equilibrium position (a) parallel to the direction of the strong external field. The rf field H_1 is then applied. As viewed from the rotating frame of reference the net magnetic moment appears (b) to rotate quickly about H_1. At the end of a 90° pulse the net magnetic moment is in the equatorial plane (c). During the relatively long period of time following the removal of H_1, the incremental moment vectors begin to fan out slowly (d). This is caused by the variations in H_0 over the sample. At time $t = \tau$ the rf field H_1 is again applied. Again the moments (e) are rotated quickly about the direction of H_1. This time H_1 is applied just long enough to satisfy the 180° pulse condition. This implies that at the end of the pulse all the incremental moment vectors begin to recluster slowly (f). Because of the inverted relative positions following the 180° pulse and because each incremental vector continues to precess with its former frequency, the incremental vectors will be perfectly reclustered (g) at $t = 2\tau$. Thus maximum signal is induced in the pickup coil at $t = 2\tau$. This maximum signal, or echo, then begins to decay as the incremental vectors again fan out (h) (Carr and Purcell, 1954).

placing the sample in the applied magnetic field for a time that is long compared to the spin-lattice relaxation time T_1. Since typical values of T_1 are milliseconds or microseconds, the alignment is virtually complete before the first pulse can be applied.

Electron spin-echo experiments are carried out using pulses of microwaves from ~ 3 to ~ 100 nsec long with powers ranging from 10 to 10^4 W. The pulse is applied at the Larmor precession frequency ω_0 with the amplitude H_1 pointing along the x direction. In the rotating frame of reference the spins find themselves in the presence of an applied magnetic field of magnitude H_1 oriented along the x axis, and they begin to precess around it at the frequency ω:

$$\omega_0 = \gamma H_1 \tag{1}$$

in the manner indicated in Fig. 13-24b. If the magnetic field is kept turned on for a time t_w then the spins will precess about H_1 through the following angle θ expressed in radians:

$$\theta = \omega t_w = \gamma H_1 t_w \tag{2}$$

In this spin-echo experiment the product of the field strength H_1 and the duration t_w of the pulse are selected so that the angle θ is $90°$ or $\pi/2$:

$$\gamma H_1 t_w = \frac{\pi}{2} \tag{3}$$

and as a result the magnetization bends through an angle of $90°$ and is aligned in the y direction at the termination of the pulse, as indicated in Fig. 13-24c. After the cessation of the pulse the individual spins are influenced by slightly different magnetic field strengths in the z direction arising from the inhomogeneity of the applied field, and so they spread out with time in the xy plane, as indicated in Fig. 13-24d. After a certain time τ has passed a $180°$ pulse of length t'_w is applied to the sample where

$$\gamma H_1 t'_w = \pi \tag{4}$$

and it causes the individual spins to flip in the manner illustrated in Fig. 13-24e. After this the individual spins begin to rotate toward each other in the xy plane, until they come together fully aligned as indicated in Fig. 13-24g. At this point the spin alignment produces a microwave field H_1 in the cavity corresponding to an emission signal that is referred to as an echo. This echo is picked up by the detector and recorded or displayed on an oscilloscope. Figure 13-25a shows the two pulses separated by the time τ and the echo that appears at the time 2τ after the start of the first pulse.

As the spins precess in the $x y$ plane they also undergo random relaxation processes that disturb their movement and prevent them from coming together

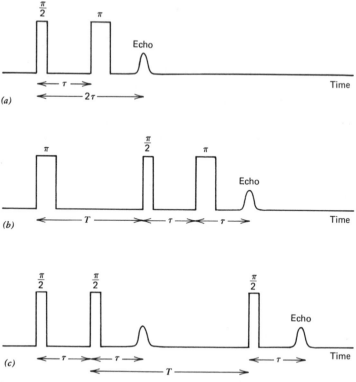

Fig. 13-25. (a) Two-pulse $\pi/2$, π sequence; (b) three-pulse π, $\pi/2$, π sequence; and (c) three-pulse $\pi/2$, $\pi/2$, $\pi/2$ sequence showing the characteristic times τ and T and the echoes.

fully realigned. The longer the time τ between the pulses the more the spins lose coherence and consequently the weaker the echo. Figure 13-26 illustrates how the amplitude of the echo decreases with increasing τ. The gradual increase in randomness can be monitored by applying a 90° pulse followed by a succession of 180° pulses at regular intervals. Each 180° pulse produces an echo, and the echo amplitudes decrease with time in the manner illustrated in Fig. 13-27. The decay curve can be analyzed to provide the relaxation time (Mims, 1972, 1976; Poole and Farach, 1971; Dalton et al., 1972, 1972).

Sometimes three pulse echo sequences are employed—Fig. 13-25 shows a π, $\pi/2$, π sequence and also a $\pi/2$, $\pi/2$, $\pi/2$ sequence. All three sequences provide relaxation rate information. The two-pulse sequence may be employed to measure the longitudinal magnetization M_z, the phase memory time T_m of the transverse magnetization M_{xy}, and the hyperfine spacings through a phenomenon called electron spin-echo envelope modulation, which will be discussed below.

The three pulse sequence π, $\pi/2$, π is used to determine the spin-lattice relaxation time T_1, and the $\pi/2$, $\pi/2$, $\pi/2$ sequence can provide both T_1 and hyperfine information (Norris et al., 1980).

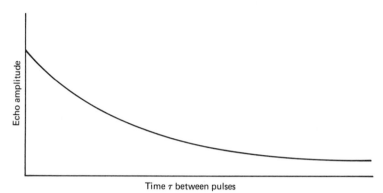

Fig. 13-26. Decay of echo amplitude as a function of the time τ between the pulses in a $\pi/2$, π two-pulse sequence.

The adaptation of an ESR spectrometer for carrying out spin-echo measurements requires the use of pulsed magnetons, klystrons, or traveling-wave tube sources. We see from Eq. (2) that the angle θ is proportional to γ, and hence much shorter times t_w are required in ESR to obtain a 90° or 180° pulse than in NMR since γ is 10^3 larger. In a typical ESR spin-echo experiment pulse lengths from 10 to 100 nsec are employed.

Figure 13-28 presents a block diagram of an electron spin-echo spectrometer system designed by Norris et al. (1980). The microwave power is provided by a mechanically tuned Gunn diode oscillator without any automatic frequency control, since an AFC is not needed, and its frequency is monitored by the counter via directional coupler DC1. The directional coupler DC3 splits the microwave power into two channels. The main power passing through isolator IS 5 is gated with switch S1, set at the desired power level by the attenuator AT1, amplified by the traveling-wave tube amplifier, and proceeds to the

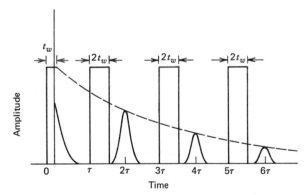

Fig. 13-27. Application of a 90° pulse followed by the successive application of 180° pulses each of which produces an echo with exponentially decaying amplitudes (dotted curve) (Poole and Farach, 1971).

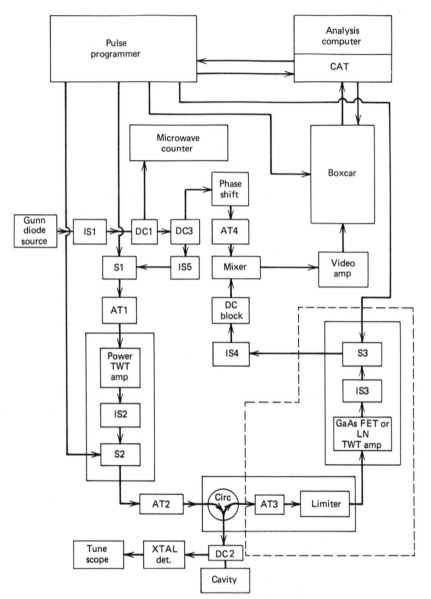

Fig. 13-28. Block diagram of spin-echo spectrometer with an unstabilized GUNN diode micro-wave source and traveling-wave tube (TWT) power amplifier. The GaAs FET low-noise amplifier may be replaced by a TWT, in which case the limiter is not needed. The following notation is used: AT—attenuator, DC—directional coupler, IS—isolator, S—microwave switch (Norris et al., 1980).

resonant cavity through the circulator. Switch S2 turns off the TWT amplifier while the echo is detected. The pulse programer sets the times and widths of the pulses and the on and off times of the TWT.

The resonant cavity is a commercial TE_{110} type with an enlarged iris screw to permit the Q to be decreased by overcoupling. The directional coupler DC2 permits the reflected signal to be displayed on the oscilloscope so that the microwave source can be tuned to the cavity frequency. The ESR signal of the reflected microwaves is amplified by a GaAs FET or low-noise TWT amplifier with a gain of 35 dB. The limiter protects the amplifier from the high-power pulses, and switch S3 turns it off during these pulses and on during the echo period. The signal is processed in the double-balanced mixer and amplified further with the video amplifier before entering the boxcar integrator, which is triggered by the pulse programmer. One boxcar channel integrates the baseline, the other integrates the echo, and the difference between the two constitutes the echo signal. For a typical pulse sequence a number of echoes are averaged, typically 16, and the accumulated value is transferred to the computer memory. The process is repeated for various values of the pulse sequences. At the completion of the experiment the data stored in the data acquisition computer are available for analysis.

The envelope of electron spin echoes (Blumberg et al., 1973; Mims, 1972; Kevan, 1979; Salikhov et al., 1976) is the curve that is obtained when the spin-echo amplitude is plotted against the time interval between pulses τ as shown in Fig. 13-26. This envelope often exhibits a periodic modulation of the type illustrated in Fig. 13-29, which arises from superhyperfine structure that can be obscured by inhomogeneous broadening in ordinary ESR experiments. The literature can be consulted for an explanation of this envelope modulation mechanism (Rowan et al., 1965; Zhidomirov and Salikhov, 1968; Mims, 1972, 1981).

A number of spin-echo modulation studies have been made on particular materials by Bowman et al. (1973), Brown and Kreilick (1975), Dikanov et al.

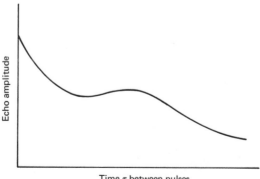

Fig. 13-29.   Envelope of two-pulse echo showing modulation arising from superhyperfine structure.

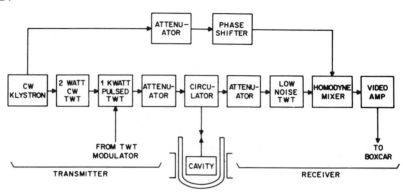

Fig. 13-30. Block diagram of the microwave system of a spin-echo envelope spectrometer. The output signal at the right enters the boxcar integrator of Fig. 13-31 (Blumberg et al., 1973).

(1977), Hess and Rowan (1979), Ichikawa et al. (1979), Kevan et al. (1975), Lee (1978), Liao and Hartmann (1972), Mims et al. (1976, 1977, 1978), and Narayama et al. (1975, 1976, 1978).

The echo decay envelope may be Fourier-transformed to provide a superhyperfine spectrum that is analogous to an ENDOR spectrum (Shimizu et al., 1979; Mims, 1981). This method is referred to as electron spin-echo envelope spectroscopy, or more succinctly by the title echo envelope spectroscopy. Several recent publications utilizing this technique are by Burger et al. (1981), Fee et al (1981), Merks and de Beer (1979, 1980), Mims et al. (1980), and Sloop et al. (1981).

Blumberg et al. (1973) described a spin-echo envelope spectrometer for recording the envelope of electron spin echoes. It consists of the microwave system illustrated in Fig. 13-30 and the pulse timing and data accumulating system presented in the block diagram of Fig. 13-31. The pulsed high microwave power is obtained from a klystron followed by two microwave traveling-wave tube amplifiers in series, namely a 2-W CW type and a 1-kW pulsed 7- to 11-GHz model. Pulses as narrow as 10 nsec can be delivered in the band from 7 to 11 GHz. The detection system consists of a low-noise TWT, a homodyne mixer, and a broad-band video pulse amplifier (0–150 MHz). Homodyne rather than heterodyne detection is employed because of its relative simplicity, its ability to give a linear response for weak pulses and a negative response for negative pulses, and finally the elimination of amplitude fluctuations due to the random time relationship between the echo waveform and the i.f. sampling noise. The observed dead time of ~ 150 nsec following the transmitter pulse was mainly due to ringing down of the resonant cavity from the 1-kW pulse to the thermal noise level.

Davis and Mims (1978) designed the microwave cavity illustrated in Fig. 13-32 for performing ENDOR measurements in conjunction with spin-echo experiments. Its resonant element is a half-wavelength-long stripline that is coupled to the microwave electric field in tapered waveguide sections. It

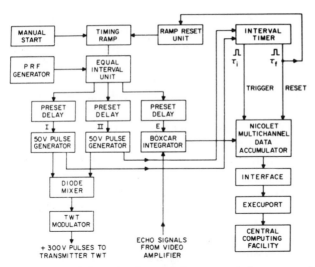

Fig. 13-31. Block diagram of the pulse time and data accumulation system of the spin-echo envelope spectrometer shown on Fig. 13-30 (Blumberg et al., 1973).

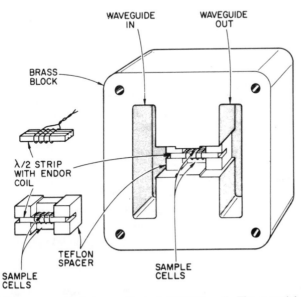

Fig. 13-32. Microwave transmission cavity and ENDOR coil. The essential element in the microwave circuit is a half-wave strip line resonator. This strip is coupled to the microwave electric field in tapered waveguide sections. Narrow grooves are cut in the sides of the strip to provide a return path for the rf magnetic field generated by the ENDOR coil (Davis and Mims, 1978).

operates at 9.4 GHz and has a Q of about 80. The magnetic field at the sample is about 11 G per ampere of coil current, which corresponds to a field of about 8 G/A in the rotating frame. Brown and Sloop (1970) designed the spin-echo spectrometer described in Sec. 14-B, which can also be used for ENDOR.

A typical spin-echo spectrometer has about the same sensitivity as a CW instrument (Mims, 1965; Salikhov et al., 1976). Some paramagnetic centers cannot be studied by spin echo, such as those with phase memory times less than 5 nsec, but these are usually readily studied by CW techniques.

The presence of inhomogeneous broadening of the resonant lines can be partly compensated for by the use of electron spin echoes. Ordinary ESR absorption measurements record the envelope of an inhomogeneously broadened spectral line. In a spin-echo measurement the phase memory time associated with the echo arises from the narrower individual spin packet linewidth rather than the much broader envelope width. A spin packet 10 kHz or 3 mG wide produces an echo that decays in several dozen microseconds. The resolution is much improved by measuring spin packets with spin-echo techniques rather than measuring overall envelopes with conventional CW techniques. Most electron spin-echo experiments are carried out at low temperatures where the spin-lattice relaxation time T_1 is long enough to minimize homogeneous broadening.

In a three-pulse sequence with times $t = 0$, $t = \tau$, and $t = \tau + T$ the echo may be studied as a function of both τ and T. This procedure consists in "burning holes" of simple form and controllable width in an inhomogeneously broadened line and observing the rate at which these holes are filled in. Gorlov et al. (1976, 1978) show some ESR spectra of Gd^{3+} with holes burned in them, and Marko et al. (1970), Shanina et al. (1972) and Zaritskii et al. (1972) give similar hole-burned spectra for P and N donors in SiC. When Baumberg et al. (1972) saturated nitrogen atoms in silicon with strong pulses of duration 10^{-4} to 10^{-5} sec they found two symmetrically located satellite holes burned in the line in addition to the main hole. They explained the satellite holes in terms of the saturation of forbidden transitions of donor electrons.

Electric field effects on ESR spectral lines can be determined by means of a spin-echo method (Mims, 1974). The X-band superheterodyne spectrometer of Zwarts and van Ormondt (1966) was modified for spin-echo operation in the following manner and associated with the pulsed electric field power supply described in Sec. 14-G (deBeer et al., 1976). A PIN diode switch driven by a switch driver produced microwave pulses with an on–off attenuation ratio of 80 dB and rise and decay times of about 100 nsec. The cavity Q of 1800 permitted the 400 mW of microwave power to generate 90° pulses with a pulse time of 150 nsec. This high Q, which is generally excessive for spin-echo experiments, was tolerable for the study of Mn^{2+} in $La_2Mg_3(NO_3)_{12} \cdot 24H_2O$, where the phase memory time was ~ 12 μsec at liquid-helium temperature since the value of H_1 did not exceed the ESR linewidth.

The reviews by Mims (1972, 1976) and Norris et al. (1980) list a number of references to the ESR spin-echo literature. Franke and Windsch (1976) and

Raitsimring et al. (1974) discuss the diffusion of spins, Hu and Walker (1978) calculated spectral diffusions decays, Liao and Hartmann (1972) and Narayana and Kevan (1977) treat the effects of the quadrupole interaction on spin echoes, and Stillman and Schwartz (1976) proposed a theory of electron spin echoes in nonviscous liquids. Some systems on which experimental spin-echo measurements have been made include glassy materials (Bowman et al., 1973; Kevan et al., 1975; Mims et al., 1977), triplet states (Botter et al., 1976; Cheng et al., 1976; Grechishkin, 1976), radicals absorbed in zeolites (Dikanov et al., 1977), ferroelectrics (Brunner et al., 1978; Völkel et al., 1974), and relaxation in molecular crystals (Dalton et al., 1972).

The optical detection of electron spin echoes has been discussed (Breiland et al., 1973; Harris et al., 1973; Kopvillem et al., 1973; Schmidt, 1972; Van't Hof et al., 1973, 1975). Trifunac and Norris (1978) studied kinetic time profiles and CIDEP by a spin-echo method. Other articles that treat electron spin echoes and their measurement include those by Atkins et al. (1974), Bozanic et al. (1970), Breen (1972), Chiba and Hirai (1972), Hurrell and Davies (1971), Kaplan (1975), Shimizu et al. (1979), Watson (1971), Raitsimring and Tsvetkov (1969), and Botter et al. (1976).

G. Saturation Transfer

Conventional electron spin resonance experiments are carried out under continuous wave (CW) conditions at low microwave power levels, where the system responds linearly in the sense that the signal amplitude is proportional to the square root of the microwave power, and the disturbance of the Boltzmann populations of the spin levels is negligible. In Chap. 12 we described how lineshape analyses of such spectra can provide a considerable amount of information on the interactions of the spin system. These lineshapes are strongly influenced by the spin-lattice and spin–spin relaxation times, and for a Lorentzian-shaped homogeneous singlet the latter time T_2 may be obtained directly from the linewidth through the usual expression:

$$T_2 = \frac{2}{\gamma \Delta H_{1/2}}$$

A considerable amount of additional information can be obtained by carrying out nonlinear response experiments at higher power levels, where the spin populations are considerably disturbed from their equilibrium Boltzmann values and saturation effects dominate the spectral response.

When paramagnetic molecules absorb microwave energy at high power levels they can recover by transferring the saturation between different resonant lines before passing the energy on to the Brownian motion or lattice vibrations. The former process of transferring saturation between different parts of the resonant spectrum is called spectral diffusion, and the latter process is spin-lattice relaxation. The study of these two competing mechanisms for processing the flow of energy from the incident microwave power to the ultimate heating of the sample is referred to as saturation transfer spectroscopy. Dalton et al. (1976) and Hyde and Thomas (1980) explained the various factors that are involved in saturation transfer spectroscopy.

Dalton et al. (1976) pointed out that there are three types of ESR experiments that can be performed to probe the details of the transfer of saturation in a spin system. A stationary state or frequency domain experiment measures the steady-state diffusion of the saturation throughout the spectrum. Examples of stationary state experiments are conventional or CW ESR and electron nuclear double resonance. A transient or time-domain experiment consists in pulse-saturating part of the resonant spectrum and monitoring the diffusion of saturation thoughout the spectrum following the pulse. This technique permits the separate determination of the rates of spectral diffusion and spin-lattice relaxation. A modulation or repetitive stimulation experiment permits the separation of the spectral diffusion and spin-lattice relaxation effects by the variation of the phase of the lock-in detector. Typical repetitive stimulation studies are carried out by adjusting the phase of the lock-in detector to the 90° out-of-phase (quadrature) setting so that no signal is seen with an ordinary in phase type sample. Saturation transfer spectra are then recorded using either (1) second-derivative detection at $2f_{mod}$ with the cavity tuned to absorption or (2) first-derivative detection at f_{mod} with the cavity tuned to dispersion. Ordinarily the procedure is repeated at various modulation frequencies from several hundred hertz to 1 MHz, and the results can be compared to computer-simulated spectra. If the sample molecule contains more than one type of nucleus such as nitrogen and protons each nuclear type may respond differently under saturation transfer detection conditions.

Saturation transfer experiments are generally carried out with solutions in the slow-tumbling regime where correlation times are in the range 10^{-7} sec $< \tau_c < 10^{-3}$ sec. Here the lineshape is dependent upon the product $\nu_{mod}\tau_c$ of the modulation frequency ν_{mod} and the rotational correlation time τ_c. In this range of correlation time the spin system cannot follow the high-frequency (100 kHz) field modulation, and strong signals are obtained when the phase of the lock-in detector is set in quadrature (90° out of phase). Figures 13-33 and 13-34 show how the in-phase and out-of-phase absorption spectra respectively, of meleimide-labeled hemoglobin, vary with the correlation time and with the modulation frequency. The correlation time was varied by changing the glycerine-to-water ratio of the solvent. A comparison of Figs. 13-33 and 13-34 shows that the out-of-phase spectra are more sensitive to changes in ν_{mod} and τ_c. These spectra were recorded by detecting at the fundamental frequency (first harmonic) ν_{mod}. Similar spectra were obtained by Hyde and Dalton (1972) and by Perkins et al. (1976). The shape of the resonant line is optimally sensitive to saturation transfer phenomena when one observes out-of-phase dispersion at the fundamental frequency ν_{mod} or out-of-phase absorption at the second harmonic $2\nu_{mod}$ (Hyde and Dalton, 1972; Hyde and Thomas, 1973).

To treat saturation transfer spectra quantitatively Thomas et al. (1976) adopted three parameters L''/L, C'/C, and H''/H, which have the definitions

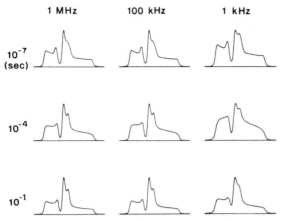

Fig. 13-33.   Simulated first harmonic in-phase absorption spectra at X-band employing source modulation at 1 MHz, 100 kHz, and 1 kHz (left to right) for the rotational correlation times 10^{-7}, 10^{-4}, and 10^{-1} sec (top to bottom). Magnetic and molecular parameters appropriate for modeling maleimide spin-labeled hemoglobin in glycerol/water mixtures were employed: $g_{xx} = g_{yy} = 2.00741$; $g_{zz} = 2.00241$; $A_{xx} = A_{yy} = 7.0$ G; $A_{zz} = 35.0$ G; microwave field intensity, $H_1 = 0.08$ G; modulation amplitude, $H_m = 0.04$ G; $T_1 = 6.6 \times 10^{-6}$ sec; $T_2 = 2.0 \times 10^{-7}$ sec. Display width is 100 G (Balasurbramanian et al., 1978).

given in Fig. 13-35. These parameters exhibit a regular monatomic dependence on the rotational correlation time τ_c as is indicated by Kusumi et al.'s (1980) data, which are plotted in Fig. 13-35. We see from these plots that the three parameters have their greatest sensitivities to changes in τ_c in different regions of correlation time. The data from the two systems hemoglobin and bovine serum albumin (BSA) with the same spin label meleimide (MSL) agree quite well, as indicated by the figure.

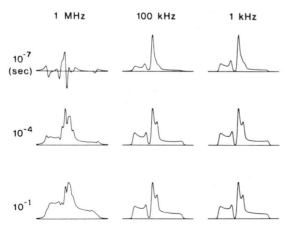

Fig. 13-34.   Simulated first harmonic out of phase (quadrature) absorption spectra. Other input parameters are as in Fig. 13-31. (Balasurbramanian et al., 1978).

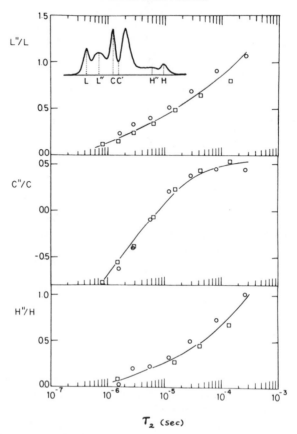

Fig. 13-35. The parameters L, L'', C, C'', H, and H'' are defined in the insert X-band absorption spectrum at the upper left and the dependences of the ratios L''/L, C''/C, and H''/H on the rotational correlation time τ_c are presented in the three graphs. The data are for hemoglobin (○) and bovine serum albumin (□) labeled with maleimide spin labels and dissolved in glycerol/water mixtures (Kusumi et al., 1980).

Mailer et al. (1980) designed the $TM_{110} \times TM_{110}$ bimodal cavity illustrated in Fig. 13-36 for the measurement of despersion and saturation transfer. The use of a bimodal cavity eliminates the source frequency modulation noise that is particularly troublesome when the dispersion mode is utilized.

Some typical experimental studies were carried out by Beth et al. (1979, 1980), Clarkson and Kooser (1978), Hyde and Thomas (1973), Mailer and Miller (1978), Robinson et al. (1980), Swift et al. (1980), and Wilkerson et al. (1978). Hyde and Dalton (1972) used an adiabatic rapid-passage method to study very slowly tumbling spin labels. Gamble et al. (1975) employed spin-echo and saturation recovery techniques modified to ensure quenching of spectral diffusion by applying pulses for times long compared to the spectral diffusion time. Balasurbramanian et al. (1978) analyzed the sensitivity of saturation

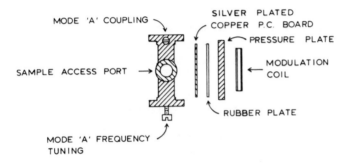

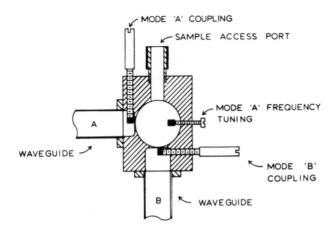

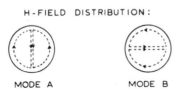

Fig. 13-36. Sketch of a cylindrical bimodal $TM_{110} \times TM_{110}$ resonant cavity showing the coupling screws for the two modes and the frequency-tuning screw for one mode. The rf magnetic field distributions for the two modes are indicated at the bottom of the figure (Mailer et al., 1980).

transfer ESR using source modulation. Johnson and Hyde (1981) obtained saturation transfer spectra at Q band (35 GHz) and noted that the increased resolution is beneficial for motions that preferentially affect specific nitroxide axes. Hyde and Thomas (1980) reviewed various applications of saturation transfer spectroscopy. Rengan et al. (1979) measured saturation transfer by a pulsed method described in Sect. 14-E.

Techniques for the computer simulation of saturation transfer spectra were developed by various workers, and typical simulations by Balasurbramanian and Dalton (1979) are presented in Figs. 13-33 and 13-34. These calculations were made using the modified Bloch equations (Clarkson and Kooser, 1978); Thomas et al., 1974, 1976) or the density matrix formalism (Smigel et al., 1974; Dalton et al., 1976; Galloway and Dalton, 1978, 1978, 1979; Mailer and Miller, 1978; Robinson et al., 1978, 1979, 1980).

H. Muon Spin Rotation

The negative muon μ^- is a spin-$\frac{1}{2}$ lepton particle that acts in all respects like an electron except that it has a mass 206.77 times larger than that of an electron. Its electric charge is identical with that of an electron. There are also positive muons μ^+ that have the same mass and opposite charge of a negative muon and that behave like positive electrons called positrons. A muon has a magnetic moment of 3.183 nuclear magnetrons. Its gyromagnetic ratio

$$\frac{\gamma\mu}{2\pi} = 13.55 \text{ MHz/kG} \tag{1}$$

is about three times that of a proton (4.2577 MHz/kG) and much smaller than that of an electron (2.8026 GHz/kG).

One dramatic difference between a muon and an electron is the fact that a muon is unstable with a lifetime of 2.199 μsec while an electron is a stable particle. A positive muon decays into a positron (e^+) and two neutrinos:

$$\mu^+ \rightarrow e^+ + \nu_e + \bar{\nu}_0 \tag{2}$$

with the positron emitted in a preferential direction relative to the muon magnetic moment. If P is the initial polarization of the muon beam then the number of positrons emitted at an angle θ relative to this direction is given by

$$N_{(\theta)} = \frac{N_0}{4}[3 + P\cos\theta] \tag{3}$$

The muon precession can be monitored with the aid of positron detectors that measure the relative number of muons oriented in various directions.

A muon may be employed as a probe of the magnetic environment of a solid on an atomic scale. Sometimes it remains as a free particle as in metallic conductors, and sometimes it combines with an electron to form muonium as in various insulators and semiconductors. Muonium is an exotic "hydrogen atom" containing an electron in orbit around a positive muon that plays the role of the nucleus.

In a solid the muons precess about the local magnetic field, and the precession is monitored by measuring the angular correlation pattern given by Eq. (3). This measurement of muon spin rotation (μSR) gives important information on the distribution of local magnetic fields in the solid, and it has become a useful tool in solid-state physics.

The μSR technique is limited to a few meson factories that provide large enough fluxes of muons. The main sites for μSR work are the Los Alamos Meson Physics Facility (LAMPF) in New Mexico, the Tri University Meson Facility (TRIUMF) in Vancouver, British Columbia, the Schweizerisches Institut für Nuklearforschung (SIN) in Villigen, Switzerland, and the Russian Institute for Nuclear Studies at Dubna. This subject has been reviewed by Brewer and Crowe (1978), Denison (1980, 1981), and Denison et al. (1981).

References

Abragam, A., V. Bouffard, and Y. Roinel, *J. Magn. Reson.* **22**, 53 (1976).

Ahn, M., *J. Magn. Reson.* **22**, 289 (1976).

Ajiro, Y., S. A. Friedberg, and N. S. Van der Ven, *Phys. Rev. B* **12**, 39 (1975).

Alegria, A. E., R. Concepcion, and G. R. Stevenson, *J. Phys. Chem.* **79**, 361 (1975).

Aleksandrov, E. B., and V. S. Zapasskii, *Fiz. Tverd. Tela* **20**, 1180 (1978).

Alekseev, A. V., U. Kh. Kopvillem, and V. R. Nagibarov, *Phys. Status Solidi* **54**, K91 (1972).

Al'tshuler, S. A., and B. M. Kozyrev, *Electron Paramagnetic Resonance*, translated by Scripta Technica, C. P. Poole, Jr., Ed., Academic Press, New York, 1964.

Anderson, P. W., *Phys. Rev.* **109**, 1492 (1958); **114**, 1002 (1959).

Andreev, V. A., and V. I. Sugakov, *Fiz. Tverd. Tela* **17**, 1963 (1975).

Antipin, A. A., B. N. Kazakov, S. L. Korableva, R. M. Rakhmatullin, Yu. K. Chirkin, and A. A. Fedii, *Izv. Vuz Fiz.* **9**, 93 (1978).

Antipin, A. A., I. S. Konov, S. L. Korableva, R. M. Rakhmatullin, M. S. Tagirov, M. A. Teplov, and A. A. Fedii, *Fiz. Tverd. Tela* **21**, 111 (1979).

Antipin, A. A., L. D. Livanova, and A. A. Fedii, *Fiz. Tverd. Tela* **20**, 1783 (1978).

Atkins, P. W., A. J. Dobbs, and K. A. McLauchlan, *Chem. Phys. Lett.* **25**, 105 (1974).

Atkins, P. W., K. A. McLauchlan, and A. F. Simpson, *J. Phys. E Sci. Instrum.* **3**, 547 (1970).

Atsarkin, V. A., *Zh. Eksp. Teor. Fiz.* **64**, 1087 (1973).

Atsarkin, V. A., *Fiz. Tverd. Tela* **17**, 2398 (1975).

Atsarkin, V. A., and M. I. Rodak, *Sov. Phys. Uspekh.* **15**, 251 (1972).

Ayscough, P. B., T. E. English, D. A. Tong, G. E. Adams, E. M. Fieldem, and B. D. Michael, *Proceedings of the 5th L. H. Gray Conference on Fast Processes in Radiation Chemistry and Biology, Brighton, Sussex, England*, 10 (1975).

Avarmaa, R. A., and A. P. Suisalu, *Fiz. Tverd. Tela* **21**, 177 (1979).

Avdeenko, A. A., A. I. Kononenko, and S. N. Pakulov, *Izv. Acad. Nauk SSR Ser. Fiz. Proc. XXIV Conference on Luminescence (Molecular Luminescence)* **42**, 505 (1978).

Balasurbramanian, K., and L. R. Dalton, *J. Magn. Reson.* **33**, 235 (1979); **33**, 245 (1979).

Balasurbramanian, K., L. R. Dalton, K. D. Schmalbein, and A. H. Heiss, *Chem. Phys.* **29**, 163 (1978).

Baldacchini, G., U. M. Grassano, and A. Tanga, *Phys. Rev. B* **19**, 1283 (1979).

Bales, B. L., M. K. Bowman, L. Kevan, and R. N. Schwartz, *J. Chem. Phys.* **63**, 3008 (1975).

Bales, F. E., and T. E. Gough, *J. Magn. Reson.* **33**, 171 (1979).

Barbarin, F., and J. P. Germain, *J. Phys.* **36**, 475 (1975).

Bassompierre, A., and J. Pescia, *C. R. Acad. Sci.* **254**, 4439 (1962).

Baumberg, I. B., L. L. Buishvili, D. M. Daraselia, T. I. Sanadze, and M. D. Zviadadze, *Physica* **68**, 540 (1973).

Baumberg, I. B., D. M. Daraseliva, and T. I. Sanadze, *Sov. Phys. JETP* **35**, 553 (1972).

Bendiashvili, N. S., L. L. Vuishvili, and G. R. Khutsishvili, *Sov. Phys. JETP* **30**, 671 (1970).

Bennett, J. E., J. A. Eyre, C. P. Rimmer, and R. Summers, *Chem. Phys. Lett.* **26**, 69 (1974).

Bennett, T. J., R. C. Smith, and T. H. Wilmshurst, *J. Phys. E. Sci. Instr.* **2**, 393 (1969); *Chem. Comm.* 513 (1967).

Beth, A. H., R. C. Perkins, Jr., S. D. Venkataramu, D. E. Pearson, C. R. Park, J. H. Park, and L. R. Dalton, *Chem. Phys. Lett.* **69** (1980).

Beth, A. H., R. Wilder, L. S. Wilkerson, R. C. Perkins, B. P. Meriwether, L. R. Dalton, C. R. Park, and J. H. Park, *J. Chem. Phys.* **71**, 2074 (1979).

Berthet, G., *C. R. Acad. Sci.* **241**, 1730 (1955).

Bil'kis, I. I., N. P. Makshanov, M. V. Kansankow, and S. M. Shein, *Izv. Siv. Otd. Akad. Nauk SSR, Ser. Khim. Nauk.*, 1001 (1976).

Bloembergen, N., and S. Wang, *Phys. Rev.* **93**, 72 (1954); *BPP*, ibid **73**, 679 (1948).

Blumberg, W. E., and W. B. Mims, and D. Zuckerman, *RSI* **44**, 545 (1973).

Blume, R. J., *RSI* **32**, 1016 (1961).

Boscaino, R., M. Brai, and I. Ciccarello, *18th Ampere Congr. Notthingham*, 139 (1974).

Boscaino, R., M. Brai, I. Ciccarello, and G. Contrino, *Phys. Lett. A* **46**, 190 (1973).

Boscaino, R., M. Brai, and I. Ciccarello, *Phys. Rev. B* **13**, 2798 (1976).

Botter, B. J., C. J. Nonhof, J. Schmidt, and J. H. Van der Waals, *Chem. Phys. Lett.* **43**, 210 (1976).

Bowman, M. K., L. Kevan, and I. M. Brown, *Chem. Phys. Lett.* **22**, 16 (1973).

Bowman, M. K., H. Hase, and L. Kevan, *J. Magn. Reson.* **22**, 23 (1976).

Bowman, M. K., and J. R. Norris, private communication, 1980.

Bozanic, D. A., D. Mergerian, and R. W. Minarik, *J. Appl. Phys.* **41**, 5041 (1970).

Bratus, V. Ya., I. M. Zaritskii, G. S. Pekar, and B. D. Shanina, *Fiz. Tverd. Tela* **20**, 3140 (1978).

Breen, D. P., *IEEE Trans. Mang. 1972 Intermag. Conf.* **Mag-8**, 704 (1972).

Breiland, W. G., C. B. Harris, and A. Pines, *Phys. Rev. Lett.* **30**, 158 (1973).

Brewer, J. H., and K. M. Crowe, *Ann. Rev. Nucl. Part. Sci.* **28**, 239 (1978).

Brik, A. B., N. P. Baran, S. S. Ishchenko, and L. A. Shulman, *Zh. Eksp. Teor. Fiz.* **67**, 186 (1974).

Brill, A. S., Chin-I Shyr, and T. C. Walker, *Mol. Phys.* **29**, 437 (1975).

Broer, M. M., J. Hegarty, G. F. Imbusch, and W. M. Yen, *Opt. Lett.* **3**, 175 (1978).

Brom, H. B., J. Soeteman, and A. J. Duyneveldt, *17th Congress Ampere Nuclear Magnetic Res. and Related Phenomena, Turku, Finland*, 21 (1972).

Brotikovskii, O. I., G. M. Zhidomirov, V. B. Kazanskii, and B. N. Shelimov, *Teor. Eksp. Khim.* **7**, 245 (1971).

Brown, I. I., and R. W. Kreilick, *J. Chem. Phys.* **62**, 1190 (1975).

Brown, I. M., and R. J. Richardson, *RSI* **44**, 77 (1973).

Brown, I. M., and D. J. Sloop, *RSI* **41**, 1774 (1970).

Brunner, W., G. Volkel, W. Windsch, I. N. Kurkin, and V. I. Shlenkin, *Solid State Commun.* **26**, 853 (1978).

Brunton, G., D. Griller, L. R. C. Barclay, and K. U. Ingold, *J. Am. Chem. Soc.* **98**, 6803 (1976).

Brunton, G., and K. U. Ingold, *J. Chem. Soc. Perkin Trans.* **2**, 1659 (1976).

Brya, W. J., *Phys. Rev. B* **3**, 633 (1971).

Bugai, A. A., *Sov. Phys. Solid State* **4**, 2218 (1963).

Bugai, A. A., V. S. Vikhin, V. E. Kustov, V. M. Maksimenko, and B. K. Krulinkovskii, *Zh. Eksp. Teor. Fiz.* **74**, 2250 (1978).

Buishvili, L. L., and N. P. Giorgadze, *Teor. Mat. Fiz.* **12**, 420 (1972).

Bullock, A. T., *Annual Rep. Prog. Chem. Sect. B* **72**, 88 (1976).

Burger, R. M., A. D. Adler, S. B. Horwitz, W. B. Mims, and J. Peisach, *Biochem.* **20**, 1701 (1981).

Burlamacchi, L., G. Martini, M. F. Ottaviani, and M. Romanelli, *Adv. Mol. Relaxation and Interaction Processes* **12**, 145 (1978).

Burlamacchi, L., and M. Romanelli, *Chem. Phys. Lett.* **10**, 59 (1971).

Burlamacchi, I., and M. Romanelli, *J. Chem. Phys.* **55**, 979 (1971).

Candela, G. A., *J. Chem. Phys.* **52**, 3754 (1970).

Carlier, M., and L. R. Sochet, *Combust. Flame* **25**, 309 (1975).

Carr, H. Y. and E. M. Purcell, *Phys. Rev.* **94**, 630 (1954).

Carruthers, J. A., and N. C. Rumin, *Can. J. Phys.* **43**, 576 (1965).

Castle, J. G., Jr., P. F. Chester, and P. E. Wagner, *Phys. Rev.* **119**, 953 (1960).

Castner, T. G., Jr., *Phys. Rev.* **115**, 1506 (1959).

Castro, M., J. Keller, and H. Schilling, *Hyperfine Interactions, Proceedings of the First International Topical Meeting on Muon Spin Rotation*, **6**, 43 (1979).

Catterall, R., *Phil. Mag.* **22**, 779 (1970).

Chachaty, M. M. C., and J. Maruani, *C. R. Acad. Sci.* **273**, 1119 (1971).

Chawla, O. P., and R. W. Fessenden, *J. Phys. Chem.* **79**, 2693 (1975).

Cheng, J. C., and J. C. Kemp, *Phys. Rev. B* **4**, 2841 (1971).

Cheng, C., and D. J. Lin, *Chem. Phys. Lett.* **44**, 576 (1976).

Chau, C. K., *Phys. Rev. B* **6**, 287 (1972).

Chiarini, F., M. Martinelli, L. Pardi, and S. Santucci, *Phys. Rev. B* **12**, 847 (1975).

Chiba, M., and A. Hirai, *J. Phys. Soc. Japan* **33**, 730 (1972).

Clarkson, R. B., and R. G. Kooser, *Surf. Sci.* **74**, 325 (1978).

Chong, T., and N. Itoh, *J. Phys. Soc. Japan* **35**, 518 (1973).

Clayman, B. P., *Can. J. Phys.* **57**, 1209 (1979).

Clough, S., and C. A. Scott, *Proc. Phys. Soc. (J. Phys. C.)* **1**, 919 (1968).

Coffman, R. E., *J. Phys. Chem.* **79**, 1129 (1975).

Cookson, P. G., A. G. Davies, N. A. Fazal, and B. P. Roberts, *J. Chem. Soc.* **98**, 616 (1976).

Cristea, G., T. L. Bohan, and H. J. Stapleton, *Phys. Rev. B* **4**, 2081 (1971).

Cullis, P. R., *J. Magn. Res.* **21**, 397 (1976).

Dahlberg, E. D., *Phys. Rev. B* **16**, 170 (1977).

Dalton, L. R., P. Coffey, L. A. Dalton, B. H. Robinson, and A. D. Keith, *Phys. Rev. A* **11**, 488 (1975).

Dalton, L. R., A. H. Kwiram, and J. A. Cowen, *Chem. Phys. Lett.* **14**, 4951 (1972).

Dalton, L. R., A. L. Kwiram, and J. A. Cowen, *Chem. Phys. Lett.* **14**, 77 (1972).

Dalton, L. R., B. H. Robinson, L. A. Dalton, and P. Coffey, *Adv. Magn. Reson.* **8**, 149 (1976).

Date, M., *J. Phys. Soc. Japan* **29**, 892 (1975).

Date, M., M. Motokawa, A. Seki, and H. Mollymoto, *J. Phys. Soc. Japan* **29**, 898 (1975).

Davidov, D., V. Zevin, R. Levin, and D. Shaltiel, *Phys. Rev.* **15**, 2711 (1977).

Davies, A. G., and B. Muggleton, *J. Chem. Soc. Perkin Trans.* **2**, 502 (1976).

Davies, A. G., M. J. Parrott, B. P. Roberts, and A. Skowronska, *J. Chem. Soc. Perkin Trans.* **2**, 1066 (1976).

Davis, C. F., Jr., M. W. P. Strandberg, and R. L. Kyhl, *Phys. Rev.* **111**, 1268 (1958).

Davis, M. S., C. Mao, and R. W. Kreilick, *Mol. Phys.* **29**, 665 (1975).

Davis, J. L., and W. B. Mims, *RSI* **49**, 1095 (1978).

de Beer, R., R. Chatterjee, and R. P. J. Merks, *J. Phys. C Solid State Phys.* **9**, 1539 (1976).

Deigen, M. F., S. S. Ishchenko, V. I. Konovalov, and S. M. Okulov, *Fiz. Tverd Tela* **20**, 476 (1978).

De Neef, T., and J. A. M. Verbruggen, *Phys. Lett. A* **46**, 439 (1974).

Denison, A. B., in *Solid State Chemistry, A Contemporary Overview*, S. C. Hott, J. B. Milstein and M. Robbins, Eds., American Chemical Society, 1980.

Denison, A. B., *Magn. Reson. Rev.*, in preparation.

Denison, A. B., H. Graf, W. Kundig, and P. F. Meier, *Physik-Institt der Universitat Zurich*, 1981.

Dietz, F., U. Konzelmann, H. Port, and M. Schwoerer, *Chem. Phys. Lett.* **58**, 565 (1978).

Dikanov, S. A., V. F. Yudanov, R. I. Samoilova, and Yu. D. Tsvetkov, *Chem. Phys. Lett.* **52**, 520 (1977).

Dikanov, S. A., V. F. Yudanov, and Yu. D. Tsvetkov, *Zh. Strukt. Khim.* **18**, 460 (1977).

Doetschman, D. C., and C. A. Hutchison, Jr., *J. Chem. Phys.* **56**, 3964 (1972).

Eastman, M. P., R. G. Kooser, M. R. Das, and J. H. Freed, *J. Chem. Phys.* **51**, 2690 (1969).

Efremov, N. A., and M. A. Kozhushner, *Uches. Eksp. Khim.* **8**, 53 (1972).

Eliav, V., and H. Levanon, *Chem. Phys. Lett.* **36**, 377 (1975).

Ellett, J. D., Jr., M. G. Gibby, U. Haeberlen, L. M. Huber, M. Mehring, A. Pines, and J. S. Waugh, *Adv. Magn. Res.* **5** (1971).

Elson, I. H., M. J. Parrott, and B. P. Roberts, *J. Chem. Soc. Chem. Commun.*, 586 (1975).

Emid, S., R. J. Baarda, J. Smidt, and R. A. Wind, *Physica B and C* **93**, 327 (1978).

Epifanov, A. S., and A. A. Manenkov, *Sov. Phys. JETP* **33**, 976 (1971).

Faughnan, B. W., and M. W. P. Strandberg, *J. Phys. Chem. Solids* **19**, 155 (1961).

Fee, J. A., J. Peisach, and W. B. Mims, *J. Biol. Chem.* **256**, 1910 (1981).

Fessenden, R. W., *J. Chem. Phys.* **58**, 2489 (1973).

Fessenden, R. W., *J. Phys. Chem.* **68**, 1508 (1964).

Fessenden, R. W., *Fast Processes Radiat. Chem. Biol. Proc. 5th L. H. Gray Conf.*, 60 (1975).

Fessenden, R. W., G. E. Adams, E. M. Fieldem, and B. D. Michael, *Proceedings of the 5th L. H. Gray Conference on Fast Processes in Radiation Chemistry and Biology*, 60 (1975).

Fessenden, R. W., and N. C. Verma, *J. Am. Chem. Soc.* **98**, 243 (1976).

Fessenden, R. W., J. P. Hornak and B. Venkataraman, *J. Chem. Phys.* **74**, 3694 (1981).

Fletcher, R. C., R. C. LeCraw, and E. G. Spencer, *Phys. Rev.* **117**, 955 (1960).

Flokstra, J., G. J. Gerritsma, and L. C. Van Der Marel, *Physica B and C* **94**, 53 (1978).

Franke, M., and W. Windsch, *Ann. Phys.* **33**, 393 (1976).

Freed, J. H., *J. Chem. Phys.* **43**, 2312 (1965); **66**, 4183 (1976).

Freed, J. H., *J. Phys. Chem.* **71**, 38 (1967); **78**, 1155 (1974).

Freed, J. H., in *Spin Labeling, Theory and Application*, Academic Press, New York, 1976, Chap. 3.

Freed, J. H., D. S. Leniart, and H. D. Connor, *J. Chem. Phys.* **58**, 3089 (1973).

Freed, J. H., D. S. Leniart, and J. S. Hyde, *J. Chem. Phys.* **47**, 2762 (1967).

Galloway, N. B., and L. R. Dalton, *Chem. Phys.* **32**, 189 (1978); **30**, 445 (1978); **41**, 61 (1979).

Gamble, W. L., L. A. Dalton, L. R. Dalton, and A. L. Kwiram, *Chem. Phys. Lett.* **34**, 565 (1975).

Gerritsma, G. J., J. Flokstra, G. A. Hartemink, J. J. M. Scholten, A. J. W. A. Vermeulen, and L. C. Van der Marel, *Physica B & C* **95**, 173 (1978).

Geschwind, S., Ed., *Electron Paramagnetic Resonance*, Plenum, New York, 1972.

Gilbert, B. C., R. G. G. Holmes, H. A. H. Laue, and R. O. C. Norman, *J. Chem. Soc. Perkin Trans.* **2**, 892 (1975).

Gilbert, B. C., R. O. C. Norman, G. Placucci, and R. C. Sealy, *J. Chem. Soc. Perkin Trans.* **2**, 885 (1975).

Gillies, R., W. U. Spendel, and A. M. Ponte Goncalves, *Chem. Phys. Lett.* **66**, 121 (1979).

Giordano, M., M. Martinelli, L. Pardi, and S. Santucci, *J. Phys. C* **11**, 1893 (1978).

Glarum, S. H., and J. H. Marshall, *J. Chem. Phys.* **62**, 946 (1975).

Glausinger, W. S., and M. J. Sienko, *J. Chem. Phys.* **62**, 1873 (1975).

Glinchuk, M. D., M. F. Deigen, and L. A. Suslin, *Phys. Status Solidi* **46**, 501 (1971).

Gokhman, V. L., and B. D. Shanina, *Fiz. Tverd. Tela* **17**, 1408 (1975).

Goldberg, I. B., and H. R. Crowe, *Anal. Chem.* **49**, 1353 (1977).

Gooijer, C., and T. G. J. J. Blekemolen, *Chem. Phys. Lett.* **18**, 280 (1973).

Gorlov, A. D., Yu. A. Shertkov, and A. P. Potapov, *Fiz. Tverd. Tela* **20**, 2090 (1978).

Gorlov, A. D., Yu. A. Sheretkov, and V. A. Ribakov, *Fiz. Tverd. Tela* **18**, 1848 (1976).

Gourier, D., D. Vivien, J. Thery, R. Collongues, and J. Livage, *Phys. Status Solidi A* **45**, 599 (1978).

Grechishkin, V. S., *Fiz. Tverd. Tela* **18**, 2116 (1976).

Hack, W., A. W. Preuss, and H. G. Wagner, *Ber. Bunsenges. Phys. Chem.* **82**, 1167 (1978).

Halbach, K., *Helv. Phys. Acta* **27**, 249 (1954).

Halliday, J. W., J. M. Caspersen, C. L. Nickerson, C. W. Rees, and I. A. Taub, *IEEE Trans. Nucl. Sci.* **NS-26**, Pt. 2., 1771 (1979).

Hamilton, E. J., Jr., D. E. Wood, and G. S. Hammond, *RSI* **41**, 452 (1970).

Hanssum, H., W. Maurer, and H. Ruterjans, *J. Magn. Reson.* **31**, 231 (1978).

Harris, C. B., R. L. Schlupp, and H. Schuch, *Phys. Rev. Lett.* **30**, 1019 (1973).

Harris, E. A. and K. S. Yngvesson, *J. Phys. C* **1**, 990, 1011 (1968).

Hasegawa, S., and S. Yazaki, *Thin Solid Films* **55**, 15 (1978).

Hayano, R. S., Y. J. Uemura, J. Imazato, N. Nishida, T. Yamazaki, H. Yasuoka, and Y. Ishikawa, *Phys. Rev. Lett.* **41**, 1743 (1978).

Hellebrand, J., H. Kasperski, and P. Wuensche, *Plaste Kautsch.* **23**, 249 (1976).

Hellebrand, J., T. Noebel, and P. Wuensche, *Plaste Kautsch.* **23**, 100 (1976).

Hervé, J., *Paramagn. Res.* **2**, 689 (1963).

Hervé, J., and J. Pescia, *C. R. Acad. Sci.* **255**, 2926 (1962); **256**, 5079 (1963).

Hess, D. W., and R. G. Rowan, *Phys. Rev. B* **19**, 1 (1979).

Hewgill, F. R., and G. M. Proudfoot, *Aust. J. Chem.* **29**, 637 (1976).

Hirst, L. L., W. Schaefer, D. Seipler, and B. Elschner, *Phys. Rev.* **8**, 64 (1973).

Holstein, T., S. K. Lyo, and R. Orbach, *Phys. Rev. Lett.* **36**, 891 (1976).

Hoogstraate, H., J. van Houten, L. A. Schreurs, A. Wenchke, and N. J. Poulis, *Physica* **65**, 347 (1973).

Hu, P., and L. R. Walker, *Phys. Rev. B* **19**, 1300 (1978).

Huisjen, M., and J. S. Hyde, *RSI* **45**, 669 (1974).

Huisjen, M., and J. S. Hyde, *J. Chem. Phys.* **60**, 1892 (1974).

Hudson, A., and J. W. E. Lewis, *Trans. Faraday Soc.* **66**, 1297 (1970).

Hudson, A., and R. F. Treweek, *J. Chem. Soc. Faraday Trans.* **2**, 855 (1976).

Hurrell, J. P., and E. R. Davies, *Solid State Commun.* **9**, 461 (1971).

Hyde, J. S., and H. W. Brown, *J. Chem. Phys.* **37**, 368 (1962).

Hyde, J. S., and L. Dalton, *Chem. Phys. Lett.* **16**, 568 (1972).

Hyde, J. S., J. W. Chien and J. H. Freed, *J. Chem. Phys.* **48**, 4211 (1968).

Hyde, J. S., and D. A. Hyde, *J. Magn. Res.* **43**, 137 (1981).

Hyde, J. S., and D. D. Thomas, *Ann. N. Y. Acad. Sci.* **222**, 680 (1973).

Hyde, J. S., and D. D. Thomas, *Ann. Rev. Phys. Chem.* **31**, 293 (1980).

Hyde, J. S., and K. V. S. Rao, *J. Magn. Res.* **29**, 509 (1978).

Ichikawa, T., L. Kevan. M. K. Bowman, S. A. Dikanov, and Yu. D. Tsvetkov, *J. Chem. Phys.* **71**, 1167 (1979).

Ingold, K. U., *Mag. Reson. Chem. Biol. Lect. Ampere Int. Summer Sch.*, 217 (1975).

Iordanov, N. D., V. Iliev, and D. Shopov, *Chem. Phys. Lett.* **33**, 162 (1976).

Johnson, M. E., and J. S. Hyde, in publication.

Jolicoeur, C., and P. Bernier, *Chem. Phys. Aqueous Gas Solutions*, 135 (1975).

Kaplan, D. E., *J. Phys. Radium Suppl.* **23**, 21A (1962).

Kaplan, D. E., *Nav. Res. Rev.* **28**, 1 (1975).

Kasthurirengan, S., and S. K. Kar, *Proc. Nucl. Phys. and Solid State Phys. Symposium.*, 358 (1973).

Kevan, L., in *Time Domain Electron Spin Resonance*, L. Kevan and R. N. Schwartz, Eds., Wiley-Interscience, New York, 1979.

Kevan, L., M. K. Bowman, P. A. Narayana, R. K. Boeckman, V. F. Yudanov, and Yu. D. Tsvetkov, *J. Chem. Phys.* **63**, 409 (1975).

Kevan, L., and L. D. Kispert, *Electron Spin Double Resonance Spectroscopy*, Wiley-Interscience, New York, 1976.

Kilmer, N. G., and A. H. Kalantar, *Chem. Phys.* **27**, 355 (1978).

Kim, S. J., and T. Takizawa, *Makromol. Chem.* **176**, 1217 (1976).

Kispert, L. D., M. C. Kichoon Chang, P. S. Allen, E. P. Andrew, and C. A. Bates, *Proc. of the 18th Ampere Cong. on Mag. Res. and Related Phenomena*, 279 (1975).

Kochi, J. K., *Adv. Free-Radical Chem.* **5**, 189 (1975).

Konchits, A. A., V. S. Vikhnin, I. M. Zaritskii, and B. K. Krulikovskii, *Fiz. Tverd. Tela* **20**, 2338 (1978).

Konzelmann, U., D. Kilpper, and M. Schwoerer, *Z. Naturforsch. A* **30**, 754 (1975).

Kooser, R. C., W. V. Volland, and J. H. Freed, *J. Chem. Phys.* **50**, 5243 (1969).

Kopvillem, U. Kh., B. P. Smolyakov, and R. Z. Sharipov, *Izv. Akad. Nauk SSR Ser. Fiz.* **37**, 2240 (1973).

Korolev, V. D., S. A. Peskovatskii, V. P. Shakhparyan, and V. M. Shulga, *Fiz. Tverd. Tela* **17**, 150 (1975).

Korolev, V. D., S. A. Peskovatskii, and V. H. Shulga, *Phys. Stat. Sol. B* **84**, 443 (1977).

Krishnamoorthy, G., and B. S. Prabhananda, *J. Chem. Phys.*, in press.

Kulikov, A. V., and G. I. Likhtenshtein, *Biofiz.* **19**, 420 (1974).

Kurkin, I. N., Yu. K. Chirkin, and V. I. Shlenkin, *Izv. Vuz Fiz.* **8**, 141 (1978).

Kuroda, S., M. Motokawa, and M. Date, *J. Phys. Soc. Japan* **44**, 1797 (1978).

Kusumi, A., T. Sakaki, T. Yoshizawa, and S. Ohnidshi, *J. Biochem.* **88**, 1103 (1980).

Larson, G. H., and C. D. Jeffries, *Phys. Rev.* **141**, 461 (1966); **145**, 311 (1966).

Lawler, R. G., D. J. Nelson, and A. D. Trifunac, *J. Phys. Chem.* **83** (1979).

Leblond, J., P. Papon, and J. Korringa, *Phys. Rev.* **4A**, 1539 (1971).

Leblond, J., J. Vebersfeld, and J. Korringa, *J. Phys. Rev.* **4A**, 1532 (1971).

Lee, K. P., and D. Walsh, *Can. J. Phys.* **49**, 1620 (1971).

Lee, S., *J. Magn. Reson.* **31**, 351 (1978).

Leifson, O. S., and C. D. Jeffries, *Phys. Rev.* **122**, 1781 (1961).

Leniart, D. S., H. D. Connor, and J. H. Freed, *J. Chem. Phys.* **63**, 165 (1975).

Lepley, A. R., and G. L. Closs, *Chemically Induced Magnetic Polarization*, Wiley, New York, 1973.

Levanon, H., and S. Vega, *J. Chem. Phys.* **61**, 2265 (1974).

Liao, P. F., and S. R. Hartmann, *Solid State Commun.* **10**, 1089 (1972).

Lingam, K. V., P. G. Nair, and B. Venkataraman, *Proc. Indian Acad. Sci.* **76**, 207 (1972).

Liao, P. F., and S. R. Hartmann, *Phys. Rev. B* **8**, 69 (1973).

Liuphart, E. E., A. C. de Vroomen, and N. J. Poulis, *17th Congr. Ampere Nucl. Mag. Reson. Related Phenom.*, 56 (1972).

Livingston, R., and H. Zeldes, *J. Chem. Phys.* **59**, 4891 (1973); *RSI* **52**, 1352 (1981).

Livingston, R., H. Zeldes, and M. S. Conradi, *J. Am. Chem. Soc.* **101**, 4312 (1979).

Llewellyn, P. M., P. R. Whittlestone, and J. M. Williams, *JSI* **39**, 586 (1962).

Lloyd, J. P., and G. E. Pake, *Phys. Rev.* **92**, 1576 (1953).

Lunsford, J. H., *J. Solid State Chem.* **12**, 288 (1975).

Lutz, D. R., K. A. Nelson, R. W. Olson, and M. D. Fayer, *J. Chem. Phys.* **69**, 4319 (1978).

McConnell, H. M., *J. Chem. Phys.* **28**, 430 (1958).

McKenzie, A., M. F. R. Steven, and J. R. Steven, *J. Chem. Phys.* **59**, 3244 (1973).

Mailer, C., and D. M. Miller, *J. Magn. Reson.* **32**, 289 (1978).

Mailer, C., T. Sarna, H. M. Swartz, and J. S. Hyde, *J. Magn. Reson.* **25**, 205 (1977).

Mailer, C., H. Thomann, B. H. Robinson, and L. R. Dalton, *RSI* **51**, 1714 (1980).

Makino, K., and H. Hatano, *Chem. Lett.*, 199 (1979).

Maniv, S., A. Revvent, and Z. Luz, *J. Chem. Phys.* **66**, 2285 (1976).

Marchand, R. L., *Phys. Rev. B* **9**, 4613 (1974).

Marko, J. R., and A. Honig, *Phys. Rev.* **B1**, 718 (1970).

Marshall, S. A., and S. V. Nistor, *Defects in Insulators*, *Session C*, V. Hovi, Ed., XVII Congress Ampere, North-Holland Publishing Co., 1973.

Martinelli, M., L. Pardi, C. Pinzino, and S. Santucci, *Phys. Rev.* **16**, 164 (1977).

Maruani, J., and J. Roncin, *Chem. Phys. Lett.* **23**, 449 (1973).

Maruyama, K., T. Otsuki, and Y. Naruta, *Bull. Chem. Soc. Japan* **49**, 791 (1976).

Matenaar, H., and R. N. Schindler, *Messtechnik* **81**, 67 (104) (1973).

Meghea, A., and C. Panait, *Rev. Chim.* **27**, 102 (1976).

Merks, R. P. J., and R. de Beer, *J. Phys. Chem.* **83**, 3319 (1979).

Merks, R. P. J., and R. de Beer, *J. Magn. Res.* **37**, 305 (1980).

Meservey, R., and P. M. Tedrow, *J. Phys. Colloq*: *XVth Int. Conf. Low Temperature Phys.* **39**, 683 (1978).

Michalik, J., and L. Kevan, *J. Chem. Phys.* **68**, 5325 (1978); **70**, 2438 (1979).

Michalik, J., and L. Kevan, *J. Magn. Reson.* **31**, 259 (1978).

Milov, A. D., K. M. Salikhov, and Yu. D. Tsvetkov, *Fiz. Tverd. Tela* **15**, 1187 (1973).

Milov, A. D., K. Hm. Salikhov, and Yu. D. Tsvetkov, *Zh. Eksp. Teor. Fiz.* **63**, 2329 (1972).

Mims, W. B., in *Electron Paramagnetic Resonance*, S. Geschwind, Ed., Plenum, New York, 1972, Chap. 4.

Mims, W. B., *Phys. Rev. B* **5**, 2409 (1972).

Mims, W. B., to appear in *Proc. of ACS Symp. on Transform Methods in Chemistry*, A. Marshall, Ed., Plenum, New York, 1981.

Mims, W. B., *Proc. Royal Soc.* **283**, 452 (1965).

Mims, W. B., *RSI* **36**, 1472 (1965); **45**, 1583 (1974).

Mims, W. B., and J. Peisach, Vol. 3, *Biological Magnetic Resonance*, L. J. Berlinger and J. Reuben, Ed., Plenum, New York, 1980.

Mims, W. B., *The Linear Electric Field Effect in Paramagnetic Resonance*, Clarendon Press, Oxford, 1976.

Mims, W. B., and J. L. Davis, *J. Chem. Phys.* **64**, 4836 (1976).

Mims, W. B., and J. Peisach, *J. Chem. Phys.* **69**, 4921 (1978).

Mims, W. B., J. Peisach, and J. L. Davis, *J. Chem. Phys.* **66**, 5536 (1977).

Mims, W. B., J. Peisach, R. W. Shaw, and H. Beinert, *J. Biological Chemistry* **255**, 6843 (1980).

Mims, W. B., G. E. Peterson, and C. R. Kurkjian, *Phys. Chem. Glasses* **19**, 14 (1978).

Minaev, B. F., and Yu. A. Serebrennikov, *Izv. Vuz Fiz.* **5**, 27 (1978).

Misra, B. N., S. D. Sharma, and S. K. Gupta, *Acta Phys. Pol. A* **A49**, 25 (1975).

Misu, A., *J. Magn. and Magn. Mater: Proc. Int. Conf. Solids and Plasmas in High Magnetic Fields* **11**, 161 (1979).

Misu, A., *J. Phys. Soc. Japan* **44**, 1161 (1978).

Moore, C. A., and R. A. Satten, *Phys. Rev. B* **7**, 1753 (1973).

Moore, C. A., and T. Yalcin, *J. Magn. Reson.* **11**, 50 (1973).

Moran, P. R., S. H. Christensen, and R. H. Silsbee, *Phys. Rev.* **124**, 442 (1961).

Morino, Y., *J. Mol. Struct.* **19**, 1 (1973).

Mozurkewich, G., H. I. Ringermacher, and D. I. Bolef, *Phys. Rev. B* **20**, 33 (1979).

Murakami, K., and A. Kawamori, *Solid State Commun.* **22**, 47 (1977).

Muus, L. D., and P. W. Atkins, *Adv. Electron Spin Relaxation in Liquids*, Plenum, New York, 1972.

Narayana, P. A., D. Becker, and L. Kevan, *J. Chem. Phys.* **58**, 652 (1978).

Narayana, P. A., and L. Kevan, *J. Magn. Res.* **23**, 385 (1976); **26**; 437 (1977).

Narayana, P. A., M. K. Bowman, L. Devan, V. F. Yudanov, and Yu. D. Tsvetkov, *J. Chem. Phys.* **63**, 3365 (1975).

Nelson, D. J., A. D. Trifunac, M. C. Thurnauer, and J. R. Norris, *Chemically Induced Magnetic Polarization Studies in Radiation Photochemistry*, Verlag Chemie International, Inc., 1979.

Norris, J. R., M. C. Thurnauer, and M. K. Bowman, *Adv. Biol. Med. Phys.*, in press.

Norris, J. R., and J. T. Warden, *EPR. Lett.* (1980).

Ohkura, H., T. Watanabe, D. Nakamur, and Y. Mori, *J. Phys. Soc. Japan* **41**, 707 (1977).

O'Neill, P., S. Steenken, and D. Shculte-Frohlinde, *J. Phys. Chem.* **79**, 2273 (1975).

Pashinin, P. P., and A. M. Prokhorov, *Zhur. Eksp. Teor. Fiz.* **40**, 49 (33) (1961).

Pasimeni, L., *J. Magn. Reson.* **30**, 65 (1978).

Passini Yuhas, M., D. I. Bolef, and J. G. Miller, *Phys. Rev. B* **17**, 4228 (1978).

Perkins, R. C., Jr., T. Lionel, B. H. Robinson, L. A. Dalton, and L. R. Dalton, *Chem. Phys.* **16**, 393 (1976).

Pescia, J., *Ann. Phys.* **10**, 389 (1965).

Pescia, J., *J. Phys.* **27**, 782 (1966).

Pescia, J., and J. Hervé, *Arch. Sci. (Spec. No.)* **14**, 123 (1961).

Percival, P. W., and J. S. Hyde, *RSI* **46**, 1522 (1975).

Ponte Goncalves, A. M., and R. Gillies, *Chem. Phys. Lett.* **69**, 164 (1980).

Poole, C. P., and H. A. Farach, *Relaxation in Magnetic Resonance*, Academic Press, New York, 1971.

Poole, C. P., Jr., and H. A. Farach, *Bull Magn. Reson.* **1**, 162 (1980).

Poole, C. P., Jr., E. N. DiCarlo, C. S. Noble, J. F. Itzel, Jr., and H. H. Tobin, *J. Catalysis* **4**, 418 (1965).

Portis, A. M., *Phys. Rev.* **91**, 1071 (1953).

Portis, A. M., *Magnetic Resonance in Systems with Spectral Distributions*, Technical Note No. 1, Sarah Mellon Scaife Radiation Laboratory, University of Pittsburgh, 1955.

Poupko, R., *J. Magn. Reson.* **12**, 119 (1973).

Prokof'ev, A. I., N. A. Malysheva, N. N. Bubnov, S. P. Soldovnikov, I. S. Belostotskaya, N. L. Komissarova, V. V. Ershov, and M. I. Kabachnik, *Dokl. Akad. Nauk SSSR* **229**, 13 (1976).

Raitsimring, A. M., K. M. Salikhov, S. A. Dikanov, and Yu. D. Tsvetkov, *Fiz. Tverd. Tela* **17**, 3174 (1975).

Raitsimring, A. M., K. M. Salikhov, S. F. Bychkov, and Yu. D. Tsvetkov, *Fiz. Tverd. Tela* **17**, 484 (1975); **16**, 756 (1974).

Raitsimring, A. M., Yu. D. Tsvetkov, *Fiz. Tverd. Tela* **11**, 1282 (1969).

Rakovskii, S. K., S. D. Razumovskii, and G. E. Zaikov, *Izv. Akad. Nauk SSSR, Ser. Khim.*, 701 (1976).

Rataiczak, R. D., and M. T. Jones, *J. Chem. Phys.* **56**, 3898 (1972).

Razuvaev, G. A., and G. A. Abakumov, and V. K. Cherkasov, *Dokl. Akad. Nauk* **220**, 116 (1975).

Redfield, A. G., *Phys. Rev.* **98**, 1787 (1955); IBM *J. Res. Dev.* **1**, 19 (1957).

Rengan, S. K., V. R. Bhagat, V. S. Suryanarayana Sastry, and B. Venkataraman, *J. Magn. Reson.* **33**, 227 (1979).

Rengan, S. K., M. P. Khakhar, B. S. Prabhananda, and B. Venkataraman, *Pure Appl. Chem.* **32**, 287 (1972); *Panamana* **3**, 95 (1974); *J. Magn. Reson.* **16**, 35 (1974).

Renk, K. F., H. Sixl, and H. Wolfrum, *Semicond. and Insul.*, *Proceedings of the 3rd Specialized Colloque Ampere on Optical Techniques in Magnetic Resonance Spectroscopy* **4**, 265 (1978).

Robinson, B. H., A. H. Beth, P. S. Crooke, and L. R. Dalton, *Chem. Phys.*, 1 (1978).

Robinson, B. H., and L. R. Dalton, *J. Chem. Phys.* **72**, 1312 (1980).

Robinson, B. H., and L. R. Dalton, *Chem. Phys.* **36**, 207 (1979).

Robinson, B. H., L. R. Dalton, and L. A. Dalton, *Chem. Phys. Lett.* **29**, 56 (1978).

Robinson, B. H., G. Forgacs, L. R. Dalton, and H. L. Frisch, **73**, 4688 (1980).

Roest, J. A., A. J. Van Duyneveldt, H. M. C. Eijkelhof, and C. J. Gorter, *Physica* **64**, 335 (1973).

Roest, J. A., A. J. Van Duyneveldt, A. Van Der Bilt, C. J. Gorter, *Physica* **64**, 306 (1973); **64**, 324 (1973).

Rowan, L. A., E. L. Hahn, and W. B. Mims, *Phys. Rev. A* **137**, A61 (1965).

Ruby, R. H., H. Benoit, and C. D. Jeffries, *Phys. Rev.* **127**, 52 (1962).

Rudenko, N. K., L. T. Oxolin, V. I. Ermakov, and P. I. Glabov, *Fiz. I Khim. Protsessy Gorn. Protz-Va*, 127 (1976).

Russell, G. A., K. D. Schmitt, and J. Mattox, *J. Am. Chem. Soc.* **97**, 1882 (1975).

Salikhov, K. M., A. G. Semenov, and Yu. D. Tsvetkov, *Electron Spin Echo and Its Application*, Nauka, Novosivirsk, 1976 (in Russian).

Sanders, R. L., and L. G. Rowan, *Phys. Rev. B* **4**, 2099 (1971).

Sarna, T., and J. S. Hyde, *J. Chem. Phys.* **69**, 1945 (1978).

Schaafsma, T. J., and D. Kivelson, *J. Chem. Phys.* **49**, 5235 (1968).

Schlick, S., and L. Kevan, *J. Magn. Reson.* **22**, 171 (1976).

Schmidt, J., *Chem. Phys. Lett.* **14**, 411 (1972).

Scott, J. F., T. C. Damen, and P. A. Fleury, *Phys. Rev.* **6**, 3856 (1972).

Scott, P. L., and C. D. Jeffries, *Phys. Rev.* **127**, 32 (1962).

Schreurs, J. W. H., G. E. Blomgren, and G. K. Fraenkel, *J. Chem. Phys.* **32**, 1861 (1960).

Semenov, I. T., and M. S. Fogel'son, *Fiz. Tekh. Polufrovodn.* **9**, 1925 (1975).

Shanina, B. D., I. M. Zaritskii, and A. A. Konchits, *Sov. Phys. Solid State* **13**, 2503 (1972).

Sharnof, M., and E. B. Aiteb, *Izv. Akad. Nauk SSSR Ser. Fiz.* **37**, 522 (1973).

Shimizu, T., W. B. Mims, J. Peisach, and J. L. Davis, *J. Chem. Phys.* **70**, 2249 (1979).

Shiotani, M., S. Murabayashi, and J. Sohma, *Int. J. Radiat. Phys. Chem.* **8**, 483 (1976).

Shteinshneider, N. Ya., G. M. Zhidomirov, and V. I. Muromtsev, *Fiz. Tverd. Tela* **20**, 343 (1978).

Shul'man, L. A., A. B. Brik, T. A. Nachal'naya, and G. A. Podzyarei, *Sov. Phys. Solid State* **12**, 2303 (1971).

Shulman, L. A., and G. A. Podzyarei, *Fiz. Tverd. Tela* **16**, 2112 (1974).

Singer, L. S., *J. Appl Phys.* **30**, 1463 (1959).

Sikri, A. K., and M. L. Narchal, *J. Phys. Chem.* **40**, 539 (1979).

Silver, B. L., *C. R. Colloq. Int. Isot. Oxygene*, 88 (1975).

Singer, L. S., and J. Kommandeur, *J. Chem. Phys.* **34**, 133 (1961).

Singer, L. S., W. H. Smith, and G. Wagoner, *RSI* **32**, 213 (1961); *J. Appl. Phys.* **30**, 1463 (1959).

Sloop, D. J., H. L. Yu, T. S. Lin, and S. I. Weisman, *J. Chem. Phys*, in press.

Smaller, B., J. R. Remko, and E. C. Avery, *J. Chem. Phys.* **48**, 5174 (1968).

Smigel, M. D., L. A. Dalton, L. R. Dalton, and A. L. Kwiram, *Chem. Phys.* **6**, 183 (1974).

Smith, J. P., and A. D. Trifunac, *J. Phys. Chem.* **85**, 1645 (1981).

Smith, J. P., and A. D. Trifunac, *Chem. Phys. Lett.*, in press.

Sohma, J., *J. Chem. Phys.* **37**, 2151 (1962).

Srivastava, J. K., and B. W. Dale, *Phys. Status Solidi B* **90**, 391 (1978).

Sokolov, V. N., V. I. Muromtsev, A. N. Grumberg, and B. Ya. Medvedev, *Teor. Exper. Khim.* **5**, 850 (1969).

Sorokin, P. P., G. J. Lasher, and I. L. Gelles, *Phys. Rev.* **118**, 939 (1960).

Standley, K. J., and R. A. Vaughan, *Electron Spin Relaxation Phenomena in Solids*, Plenum, New York, 1969.

Stevenson, G. R., and A. E. Alegria, *J. Phys. Chem.* **80**, 69 (1976).

Stevenson, G. R., R. Concepcion, and I. Oxasio, *J. Phys. Chem.* **80**, 861 (1976).

Stillman, A. E., and R. N. Schwartz, *Mol. Phys.* **32**, 1045 (1976).

Stradins, J., and R. Gavars, *Tezisy Dokl. Vses. Soveshch. Elektrokhim. 5th* **1**, 309 (1976).

Strandberg, M. W. P., *Phys. Rev. B* **6**, 747 (1972): *Phys. Rev.* **110**, 65 (1958).

Sunandana, C. S., *Phys. Status Solidi A* **48**, K19 (1978).

Swift, L. L., J. B. Atkinson, R. C. Perkins, Jr., L. R. Dalton, and V. S. Lequire, *J. Membrane Biol.* **32**, 103 (1980).

Symons, M. C. R., *J. Chem. Soc. Dalton Trans.*, 1568 (1976).

Symons, M. C. R., and Richard L. Petersen, *J. Chem. Soc. Faraday Trans.* **2**, 210 (1979).

Symons, M. C. R., D. X. West, and J. G. Wilkinson, *J. Chem. Soc. Dalton Trans.*, 1565 (1976).

Talpe, J., *Theory of Experiments in Paramagnetic Resonance*, Pergamon, New York, 1971.

Tanaka, K., and A. Kazusaka, *Chem. Phys. Lett.* **39**, 536 (1975).

Taplick, T., and M. Raetzsch, *Acta Polym.* **30**, 24 (1979).

Thomas, D. D., and H. M. McConnell, *Chem. Phys. Lett.* **25**, 470 (1974).

Thomas, D. D., L. R. Dalton, and J. S. Hyde, *J. Chem. Phys.* **65**, 3006 (1976).

Trifunac, A. D., *J. Amer. Chem. Soc.* **98**, 5202 (1976).

Trifunac, A. D., and W. T. Evanochko, *J. Am. Chem. Soc.* **102**, 4598 (1980).

Trifunac, A. D., and D. J. Nelson, *Chem. Phys. Lett.* **46**, 346 (1977).

Trifunac, A. D., D. J. Nelson, and C. Mottley, *J. Magn. Reson.* **30**, 263 (1978).

Trifunac, A. D., and J. R. Norris, *Chem. Phys. Lett.* **59**, 140 (1978).

Trifunac, A. D., J. R. Norris, and G. R. Lawlaer, *J. Chem. Phys.* **71**, 4380 (1979).

Trifunac, A. D., and J. P. Smith, *Chem. Phys. Lett.* **73**, 94 (1980).

Trifunac, A. D., and M. C. Thurnauer, *J. Chem. Phys.* **62**, 4889 (1975).

Trifunac, A. D., M. S. Thurnauer, and J. R. Norris, *Chem. Phys. Lett.* **53**, 471 (1978).

Van Duyneveldt, A. J., J. A. Van Santen, H. A. Groenendijk, and R. L. Carlin, *Physica B and C* **97**, 41 (1979).

Van Haelst, M., J. Broeckx, P. Matthys, and E. Boesman, *Phys. Status Solidi B* **89**, K25 (1978).

Van Heugten, W. F. W. M., F. V. D. Berg, L. M. Caspers, and A. Van Veen, *Delft Prog. Rep.* **3**, 97 (1978).

Van't Hof, C. A., and J. Schmidt, *Chem. Phys. Lett.* **36**, 457 (1975).

Van't Hof, C. A., J. Schmidt, P. J. F. Verbeek, and J. H. Van Der Waals, *Chem. Phys. Lett.* **21**, 437 (1973).

Vasson, A. M., and A. Vasson, *J. Phys. Colloq.*, The XVth Int. Conf. on Low Temperature Physics **39**, Pt. 2, 1005 (1978).

Vaughan, K. A., *Magn. Resonance Rev.* **4**, 25 (1975).

Velter-Stefanescu, M., and R. Grosescu, *Rev. Roum. Phys.* **23**, 369 (1978).

Verbeek, J., W. Berends, and H. C. A. VenBeek, *Recl. Trav. Chim. Pays Bas.* **95**, 285 (1976).

Verbeek, P. J. F., H. J. Den Blanken, and J. Schmidt, *Chem. Phys. Lett.* **60**, 358 (1979).

Verbeek, P. J. F., A. I. M. Dicker, and J. Schmidt, *Chem. Phys. Lett.* **56**, 585 (1978).

Verma, N. C., and R. W. Fessenden, *J. Chem. Phys.* **58**, 2489, 2501 (1973); **65**, 2139 (1976).

Vigouroux, B., and J. C. Gourdon, *Solid State Commun.* **28**, 707 (1978).

Vigouroux, B., J. C. Gourdon, P. Lopez, and J. Pescia, *J. Phys. E Sci. Instrum.* **6**, 557 (1973).

Vikhnin, V. S., *Fiz. Tverd. Tela* **20**, 1340 (1978).

Vikhnin, V. S., Yu. S. Gromovdi, I. M. Zaritskii, and G. Corradi, *J. Phys. Chem. Solids* **39**, 1113 (1978).

Vishnevskaya, G. P., G. M. Gumerov, and B. H. Kozyrev, *Teor. Eksp. Khim.* **11**, 205 (1975).

Völkel, G., U. Bartuch, W. Brunner, and W. Windsch, *Phys. Status Solidi A* **25**, 591 (1974).

Vollmann, W. *Chem. Phys.* **37**, 239 (1979).

Vugmeister, B. E., *Ukr. Fiz. Zh.* **23**, 1724 (1978).

Vugmeister, B. E., *Phys. Status Solidi B* **90**, 711 (1978).

Vugmeister, B. E., V. L. Gokhman, and V. Ya. Zevin, *Sov. Phys. Solid State* **14**, 2280 (1973).

Wan, J. K. S., and S. K. Wong, *Rev. React. Species Chem. React.* **1**, 227 (1976).

Warden, J. T., *Isr. J. Chem.*, in press.

Warden, J. T., and J. R. Bolton, *RSI* **47**, 201 (1976).

Warden, J. T., and O. L. Adrianowycz, *Proc. Fifth Intl. Congr. Photosynth.*, in press.

Wardman, P., *Rep. Prog. Phys.* **41**, 259 (1978).

Warman, J. M., R. W. Fessenden, and G. Bakale, *J. Chem. Phys.* **57**, 2702 (1972).

Watson, J. H., *Nature* **229**, 28 (1971).

Wilkerson, L. S., R. C. Perkins, Jr., R. Roelofs, L. Swift, and L. R. Dalton, *Proc. Natl. Acad. Sci. (USA)* **75**, 838 (1978).

Wilson, R. C., and R. J. Myers, *J. Chem. Phys.* **64**, 2208 (1976).

Winscom, C. J., and K. P. Dinse, *Semicond. and Insul: Proc. of the Third Specialized Colloque Ampere on Optical Techniques in Magnetic Resonance Spectroscopy* **4**, 211 (1978).

Wolfe, J. P., and C. D. Jeffries, *Phys. Rev. B* **4**, 731 (1971).

Wollan, D. S., *Phys. Rev.* **13**, 3671, 3686 (1976).

Zaritskii, I. M., A. A. Konchits, and L. A. Shul'man, *Sov. Phys. Solid State* **13**, 1588 (1972).

Zaspel, C. E., and J. E. Drumheller, *Phys. Rev. B* **16**, 1771 (1977).

Zevin, V. Ya., and A. B. Brik, *Ukr. Fiz. Zh.* **17**, 1688 (1972).

Zevin, V. Ya., V. I. Konovalov, and B. D. Shanina, *Sov. Phys. JETP* **32**, 306 (1971).

Zhidomirov, G. M., and K. M. Salikhov, *Theor. Exptl. Chem.* **4**, 332 (1968).

Zhidkov, O. P., V. I. Mumotsev, I. G. Akhvlediani, S. N. Safronov, and V. V. Kopylov, *Sov. Phys. Solid State* **9**, 1095 (1967).

Zueco, E., and J. Pescia, *C. R. Acad. Sci.* **260**, 3605 (1965).

Zverev, G. M., *PTE* **5**, 105 (930) (1961).

Zwarts, C. M. G., and D. van Ormondt, *JSI* **43**, 317 (1966).

CHAPTER 14

Double Resonance

Since this book is primarily concerned with electron spin resonance, we are mainly interested in those types of double-resonance experiments that include an ESR transition. Chapter 19 of the first edition discussed the characteristics of a number of different polarization schemes such as the overhauser effect, dynamic nuclear polarization, method of parallel fields, the solid effect, and thermal mixing. In the present chapter we concentrate upon the experimental aspects of electron nuclear double resonance (ENDOR), electron electron double resonance (ELDOR), dynamic electron polarization (DEP), electric field effects, optically detected magnetic resonance (ODMR), and acoustic or ultrasonic paramagnetic resonance (UPR).

A. Double Resonance Schemes

The basic principles underlying the ENDOR and ELDOR double-resonance schemes can be understood by an examination of the $S = \frac{1}{2}$, $I = \frac{1}{2}$ energy level system illustrated in Fig. 14-1. These levels are obtained from the spin Hamiltonian (Poole and Farach, 1972).

$$\mathcal{H} = g\beta \mathbf{H} \cdot \mathbf{S} - g_N \beta_N \mathbf{H} \cdot \mathbf{I} - \hbar A \mathbf{S} \cdot \mathbf{I} \tag{1}$$

where a negative sign is placed in front of the hyperfine term $\hbar A \mathbf{S} \cdot \mathbf{I}$ corresponding to the case of a negative hyperfine constant A.

If we adopt the notation

$$\hbar \omega_e = g\beta H$$

$$\hbar \omega_N = g_N \beta_N H \tag{2}$$

and select the symbols g, g_N, H, and A as positive, then we obtain the two ESR transitions

$$\omega_{\text{ESR}} = \omega_e + \tfrac{1}{2}A \tag{3a}$$

$$\omega_{\text{ESR}} = \omega_e - \tfrac{1}{2}A \tag{3b}$$

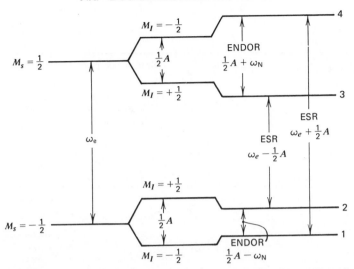

Fig. 14-1. Energy-level diagram for the $S = \frac{1}{2}$, $I = \frac{1}{2}$ case showing the hyperfine splitting alone (center) and the hyperfine and nuclear Zeeman splitting (right). The ESR and ENDOR transition energies are indicated.

which obey the selection rules

$$\Delta M_S = \pm 1$$

$$\Delta M_I = 0 \qquad (4)$$

and the two ENDOR transitions

$$\omega_{\text{ENDOR}} = \tfrac{1}{2}A + \omega_N \qquad (5a)$$

$$\omega_{\text{ENDOR}} = |\tfrac{1}{2}A - \omega_N| \qquad (5b)$$

which satisfy the selection rules

$$\Delta M_S = 0$$

$$\Delta M_I = \pm 1 \qquad (6)$$

as indicated in the figure. The absolute value is used in Eq. (5) to take into account the two cases $\frac{1}{2}A > \omega_N$ and $\frac{1}{2}A < \omega_N$.

In an ordinary ESR experiment the magnetic field is swept through the region of resonance $H = \hbar\omega_e/g\beta$ and a doublet is observed, one line for each of the ESR transitions indicated in Fig. 14-1.

In an ENDOR experiment the magnetic field is set at the value corresponding to resonance for one of the ESR transitions of Eq. (3), and a radio

DOUBLE RESONANCE

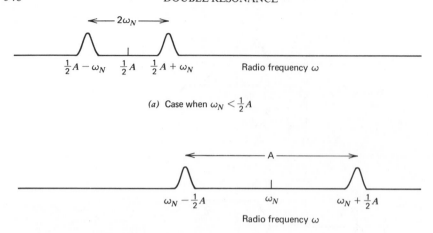

(a) Case when $\omega_N < \frac{1}{2}A$

(b) Case when $\frac{1}{2}A < \omega_N$

Fig. 14-2.   ENDOR spectra shown (a) when $\frac{1}{2}A > \omega_N$ and (b) when $\omega_N > \frac{1}{2}A$. For some systems case (a) occurs at X-band (9 GHz) and case (b) at Q band (35 GHz).

frequency ω is swept through the region corresponding to the ENDOR frequencies of Eq. (5). When ω passes through the values $\frac{1}{2}A + \omega_N$ and $|\frac{1}{2}A - \omega_N|$, the populations of the corresponding pairs of levels are altered, and a change occurs in the magnitude of the detected ESR absorption signal that is plotted by a chart recorder, as indicated in Fig. 14-2. When measurements are made at different microwave frequency bands ω_N is proportional to ω_e, and the frequencies of the ENDOR transitions vary with ω_N in the manner illustrated in Fig. 14-3. Spectra for the two cases of ω_N less than and greater than $\frac{1}{2}A$ are illustrated in Fig. 14-2.

For an ENDOR signal to be observable the ESR and NMR transitions must both be saturable at available powers. Partial saturation is sufficient for the ESR transition, and if this cannot be readily achieved at room temperature then it can generally be accomplished at low temperatures such as 4 K. More complete saturation is required for the NMR transition. These are also other relaxation-time requirements. For example, if ENDOR occurs via the common T_{ie} mechanism, then either the cross relaxation time or the nuclear relaxation time must be comparable to the electron spin saturation time so that the nuclear resonance absorption can produce a measurable ESR signal change. Feher and Isaacson (1972) claim that saturability of the ESR transition is not a rigid requirement for the observation of ENDOR. The ENDOR spectral lines are comparable in width to ESR lines, typically 0.1 mHz for free radicals in liquids. The number of lines in the ENDOR spectrum is considerably less, however, so the effective resolution is much greater, and the spectrum is easier to interpret. This is illustrated in Fig. 14-4 for the case of the diphenylanthracene (DPA) negative ion (Biehl, 1980).

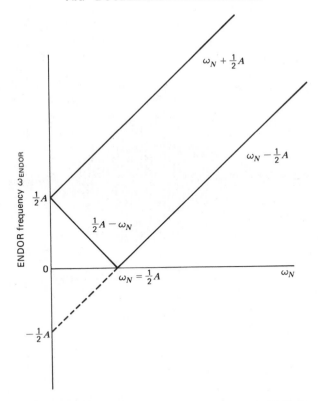

Fig. 14-3.   Dependence of the ENDOR frequency ω_{ENDOR} on the NMR frequency ω_N.

In an ELDOR experiment one ESR transition is saturated, and the other is observed. The ELDOR response is the change in the observed ESR intensity that results from the saturation of the other level. One way to carry out this experiment is to set the magnetic field H at the value H_0 that satisfies the resonance condition

$$\omega_0 = \frac{g\beta H_0}{\hbar} - \frac{1}{2}A \qquad (7)$$

for the observed frequency ω_0. In Fig. 14-5 we show the magnetic field scanned at time t_1 through the hyperfine line at the position $\omega_0 - \frac{1}{4}A$ and then stopped at time t_2 at the peak of the hyperfine line at $\omega_0 + \frac{1}{4}A$ corresponding to the $2 \rightarrow 3$ transition of Fig. 14-1. The magnet setting is maintained at this value where Eq. (7) is satisfied. The pump klystron is then turned on, and its frequency ω_p is scanned at the time t_3, when ω_p passes through the value that satisfies the following resonant condition:

$$\omega_p = \frac{g\beta H_0}{\hbar} + \frac{1}{2}A \qquad (8)$$

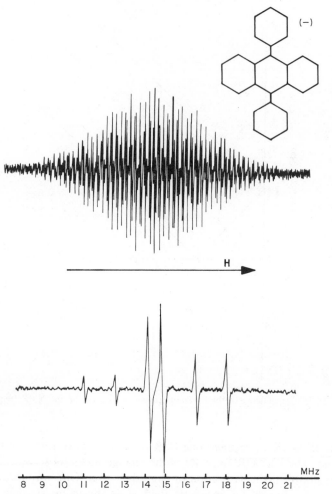

Fig. 14-4. Comparison of (*a*) ESR and (*b*) ENDOR spectra of the diphenylanthracene (DPA) negative ion (Biehl, 1980).

the $1 \rightarrow 4$ hyperfine transition of Fig. 14-1 becomes saturated. The saturation of this transition produces nuclear spin flips that disturb the populations of the levels 2 and 3 and thereby change the intensity of the $2 \rightarrow 3$ transition being monitored by the observe klystron. The result is a decrease in detected signal as the pump frequency ω_p passes through its resonance value at the time t_3. If the term ($g\beta H/h$) is eliminated from Eqs. (7) and (8) by subtraction we obtain the relation

$$\omega_p - \omega_0 = A \tag{9}$$

which provides the hyperfine coupling constant A from the frequency dif-

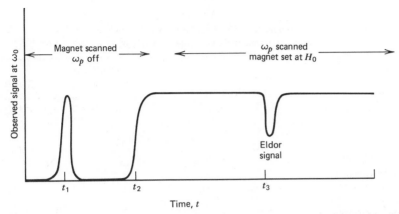

Fig. 14-5. ENDOR experiment showing the amplitude of the signal detected at the observe klystron frequency ω_0. At first the magnet is scanned with the pump power turned off, as shown on the left. At the time t_1 the magnetic field passes through the value $H = \hbar(\omega_0 - \frac{1}{2}A)/g\beta$ corresponding to the $1 \rightarrow 4$ transition of Fig. 14-1, and the scan is stopped at the time t_2 when it reaches the value $H_0 = \hbar(\omega_0 + \frac{1}{2}A)/g\beta$ at the center of the $2 \rightarrow 3$ transition. At this point the pump klystron is turned on and ω_p is scanned for times $t > t_2$, while H is maintained at the value H_0. At time t_3 the pump frequency ω_p passes through the value that satisfies the resonance condition $\omega_p = (g\beta H_0/\hbar) + \frac{1}{2}A$ of the $1 \rightarrow 4$ transition and there is a decrease in the signal detected at the frequency ω_0. This decrease in detected microwave power at ω_0 constitutes the ELDOR signal.

ference $\omega_p - \omega_0$ when both the pump and observing klystrons are on resonance.

Thus in both ENDOR and ELDOR a change is observed in one of the ESR transition intensities when another transition is saturated. The two techniques differ in the type of transition that is saturated.

Double resonance has been reviewed by Atherton (1976), Dalton (1972, 1973), Dorio (1977), Hadley (1980), Kevan and Kispert (1976), Kwiram (1971), and Mieher (1972). Freed et al. (1973) applied the theory of saturation to double resonance. Table 14-1 compares double resonance techniques with NMR and ESR for studying paramagnetic molecules (Dalal, 1982).

B. ENDOR Instrumentation

In the previous section we mentioned that in an ENDOR experiment a nuclear transition ($\Delta M_S = 0$, $\Delta M_I = \pm 1$) is saturated, and its effect on an ESR absorption signal ($\Delta M_S = \pm 1$, $\Delta M_I = 0$) is monitored. As was explained in Sec. 13-C, the degree of saturation is provided by the saturation factor s:

$$s = \frac{1}{1 + \gamma^2 H_1^2 T_1 T_2} \tag{1}$$

where H_1^2 is proportional to the rf power level. The amount of power required

TABLE 14-1

Comparison of Various Techniques of Studying
Paramagnetic Molecules (Dalal, 1982)

Feature	ESR	NMR	ENDOR[a] and Triple	ELDOR
Sensitivity	$\gtrsim 10^{-5}M$ ($\sim 10^{12}$ spins)	$\gtrsim 10^{-2}M$ $> 10^{16}$	$\gtrsim 10^{-4}M$ $\sim 10^{13}-10^{14}$	$\gtrsim 10^{-4}M$ $10^{13}-10^{14}$
Resolution	Poor 1–10 MHz	High ($\sim$ kHz)	High ($\sim$ kHz)	Poor $\sim$ 10 MHz
Identification of nuclei directly	No	Yes	Yes	No
Overlapping from spectra of several species present simultaneously	Yes, severe	Yes	No	No
Relative sign determination	Seldom	Yes	Yes	Not usually
Absolute sign determination	No	Yes	Seldom	No
Intensity proportional to number of nuclei	Yes, if resolved	Yes, generally	Seldom done but possible	Seldom investigated
Study of relaxation phenomena	Not very convenient	Indirect	Indirect	Direct
Time scale	$\sim 10^{10}$ Hz	10^8 Hz	10^8-10^{10} Hz	10^9 Hz
Range of relaxation times	$> 10^{-9}$ sec	$\gtrsim 10^{-11}$ sec	$\gtrsim 10^{-7}$ sec	$> 10^{-9}$ sec
Measurement of small (1 Gauss) couplings	Not possible for complex species	Yes, via FT–NMR	Usually difficult	No
Spectral complexity for N coupling constants	2^N	N lines	$2N$ lines	$\gtrsim 2N$ lines

[a]Some of these requirements are less stringent for spin-echo ENDOR, but such experiments have not yet been reported for solutions.

to produce a given level of saturation, that is, a given value of s, is inversely proportional to the product T_1T_2. The name high-power ENDOR is used to designate measurements made with power levels from 10 to 1000 W on liquid and solid samples with very short relaxation times (< 0.1 msec) and low-power ENDOR experiments employ 0.1 to 1 W for relaxation times from 0.1 to 10 sec. Intermediate relaxation times use intermediate power levels. In a typical case transition-ion-doped solids have long relaxation times below 4 K and require 0.1 W of rf power. The same samples at room temperature relax rapidly and require many watts of power. Free radicals in solution near room temperature can have linewidths less than 60 mG, which necessitates the use of many hundreds of watts of rf power.

To carry out an ENDOR experiment it is necessary to add to an ESR spectrometer a strong variable-frequency rf power source, a resonant cavity with radio-frequency coils, and a signal processing system for the detected microwave power as indicated in the block diagram of Fig. 14-6. The high-power rf oscillator produces an output that is amplified and impressed on the ENDOR coil located around the sample in the resonant cavity. The rf oscillator signal is sent to a frequency counter and a digital-to-analog converter to provide the x drive of the $x-y$ recorder. The oscillator is frequency-modulated, and this modulator also furnishes the reference signal for the right-hand lock-in detector in the figure. A field-modulation source provides the magentic field modulation and also the reference signal for the left-hand lock-in detector. For straight ESR operation the crystal detector output is fed directly through the field-modulation lock-in detector to the $x-y$ recorder that receives its x-axis drive from the magnet scan. For ENDOR operation the microwave detector output goes first through the FM-modulation lock-in detector and then through the field-modulation lock-in to the $x-y$ recorder, which now receives its x-axis drive from the rf oscillator via the digital-to-analog converter.

A typical rf oscillator for this application can be scanned from 5 to 50 MHz at a power level up to 1 kW. To reduce sample heating this oscillator may be pulsed with a duty cycle between 0.1 and 0.5 so that average power levels range from about 100 to 500 W. A pulser (not shown) can be employed to trigger a gate that turns off the signal input to the first amplifier during the oscillator off periods so that noise is not amplified. Field-modulation and phase-sensitive detection can be carried out at a low audio frequency.

For narrow ESR lines the standard automic frequency-control circuits are not adequate, and thus a field frequency lock unit is employed to lock the magnetic field to the resonance of a DPPH sample. Maki et al. (1968) describe an ENDOR spectrometer in which they employ a 70-kHz feedback loop, and they couple part of the microwave power to a microwave helix containing DPPH to produce this frequency lock.

The ENDOR spectrometer just described was a prototype, not an actual one. The first edition of this work contains descriptions of several older rf double-resonance spectrometers (Baker and Williams, 1962; Carver and Slichter, 1956; Hashi, 1961; Müller-Warmuth, 1960; Müller-Warmuth et al., 1958). See also Doyle (1962), Hausser and Reinhold (1962), Holton and Blum (1962), Lambe et al. (1961), Mims (1964), Parker et al. (1960), Rädler (1961), Terhune et al. (1961), Van Steenwinkel and Hausser (1965), and Watkins and Corbett (1964). In the remainder of the section we present the salient features of several more recent ENDOR instrumentation systems.

A high-power ENDOR spectrometer designed by Dalal et al. (1973) is illustrated in Fig. 14-7. It was developed around a Varian V-4205 X-band EPR spectrometer, and is similar to the ones described by Maki et al. (1968) and Allendoerfer et al. (1971). Solid-state amplifiers provide the 200-W rf power. The average maximum intensity of the rf field in the rotating frame varies from

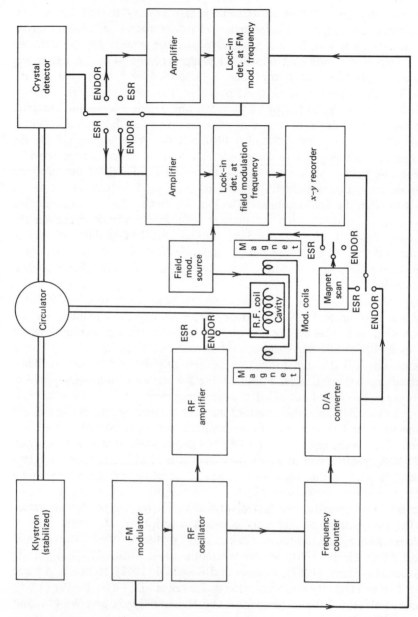

Fig. 14-6. Block diagram of an ENDOR spectrometer. The five switches convert from direct ESR to ENDOR operation.

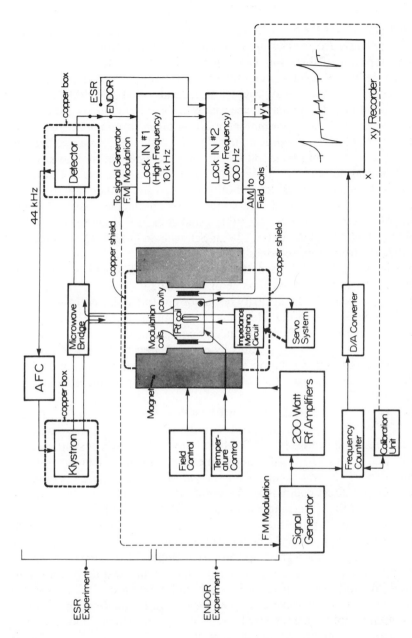

Fig. 14-7. Block diagram of X-band high rf power ENDOR spectrometer (Dalal et al., 1973).

653

$\sim$ 12 to $\sim$ 15 G over the 8- to 25-MHz ranges of the oscillator. Both Zeeman and rf modulation frequencies are continuously variable from 1.5 Hz to 100 kHz, and typically 10 kHz was used for the rf and 100 Hz for the Zeeman field. First-derivative spectra presented on an xy recorder are calibrated with pip markers from the frequency counter at intervals of 10, 100, 1000, ... kHz.

It was found that the rf interference with electronic components is quite high at these power levels, and precautions must be taken to minimize it (Möbius and Dinse, 1972). To accomplish this Dalal et al. (1973) surrounded the magnet, rf amplifier, and tank coil system with a copper sheath and enclosed the klystron and detector unit in separate copper boxes. The use of filters and the proper placement of rf cables and other components further diminished the noise. Some high-power ENDOR systems use continuous wave rf power sources (Allendoerfer and Eustace, 1971; Dalal et al., 1973; Dinse et al., 1973; Maki et al., 1968; Miyagawa et al., 1973; Schmalbein et al., 1972; Yamamoto et al., 1972) instead of pulsed ones (Hyde, 1965).

Dinse et al. (1973) described the cw high-power ENDOR spectrometer illustrated in Fig. 14-8. This is capable of producing cw NMR fields up to 30 G. The TE_{011} mode cylindrical cavity resonator illustrated in Fig. 14-9 contains a double loop in the usual Helmholtz configuration (Seidel, 1961). The resonator was constructed from 10 concentric rings and 2 thin end plates, and it had a Q of about 3500 when critically coupled. In a later work Plato et al. (1976) carried out high-resolution studies with this spectrometer by omitting the Zeeman modulation, and they obtained variable temperature data accurate to within 1 K.

Schmalbein et al. (1972) describe an ENDOR system built around a Bruker 418S ESR spectrometer and equipped with the slow wave or microwave helix resonator presented in Fig. 14-10. The helix provided almost constant sensitivity in the range from 8 to 11 GHz due to the special shield system illustrated in the figure. The rf coil was tunable from 8 to 40 MHz and operated between 40 and 80 MHz by direct coupling to the power amplifier. The article presents a block diagram of the rf tracking circuits used in the spectrometer.

Gruber et al. (1974) published the design of an X-band ENDOR spectrometer that features high rf fields in the range from 3 to 39 MHz and a new microwave cavity design. Their spectrometer employed a minicomputer for controlling the experiment, the data acquisition, the processing for sensitivity enhancement, and the display. Schweiger and Günthard (1978) used the spectrometer with a self-locking magnetometer and a liquid-helium cryostat. Single crystal samples were mounted by means of a two-circle goniometer, and a small laser was employed to measure the sample rotation angle in the magnetic field.

Gazzinelli and Mieher (1968) described an X-band superheterodyne ESR spectrometer and showed how the radio-frequency field is applied to the sample to induce the ENDOR transitions.

Brown and Sloop (1970) described a pulsed spectrometer that made use of a transient-echo method for carrying out electron nuclear double resonance.

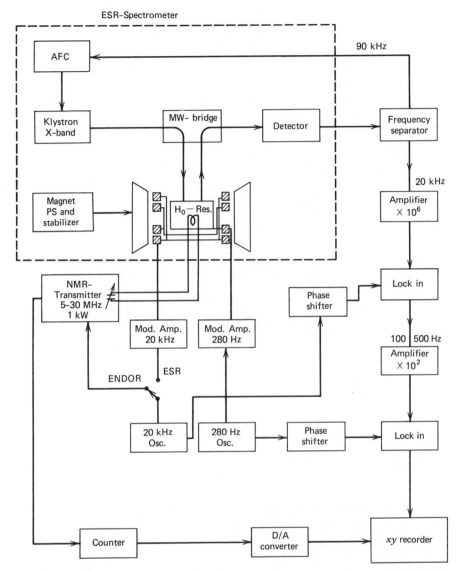

Fig. 14-8.   Block diagram of an ENDOR spectrometer using the cavity illustrated in Fig. 14-9 (Dinse et al., 1973).

Eighteen turns of closely wound No. 18 copper wire mounted on quartz formed the outer boundary of the cavity resonator illustrated in Fig. 14-11, and this same wire also carried the rf currents for the NMR fields. A similar design with modulation coils due to Hyde (1965) is presented in Fig. 14-12. With the aid of the spectrometer Brown and Sloop detected ENDOR effects using both two-pulse and three-pulse echoes generated by rf pulses with rise times of 200

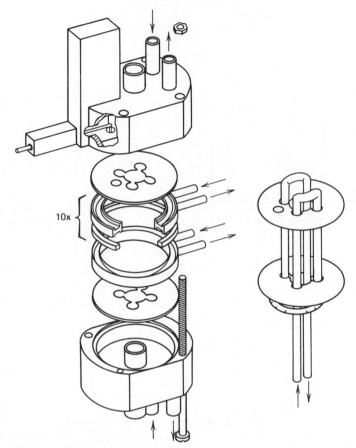

Fig. 14-9. Sketch of ENDOR cavity resonater (left) with the double loop NMR coils (right) (Dinse et al., 1973).

nsec. In a related work Franke and Windsch (1976) detect the ENDOR transition by a modified electron spin-echo experiment (see Sec. 13-F for a discussion of the spin-echo method; see also Liao and Hartmann, 1973).

Miyagawa et al. (1973) described a 10-turn ENDOR coil made of plastic-coated No. 38 copper wire wound on a Teflon core supported by a quartz Dewar tube. A 40-turn coil of No. 40 copper wire was employed for lower rf frequencies.

Radio-frequency fields of the order of several milligauss are required for the study of solids in the liquid-helium temperature range. Baker and Williams (1962), and Box et al. (1967) provided descriptions of instrumentation for use with low-power ENDOR. When slotted or wire-wound cavities are employed it is feasible to place the rf coil outside the cavity.

The ENDOR signal-to-noise ratio is considerably less than the ESR one, and typically it is only 1% of the ESR value. As a result samples should be as

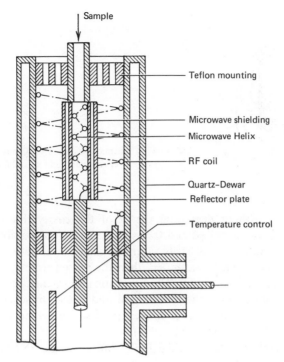

Fig. 14-10. Sketch of slow-wave resonance system for use in an ENDOR spectrometer (Schmalbein et al., 1972).

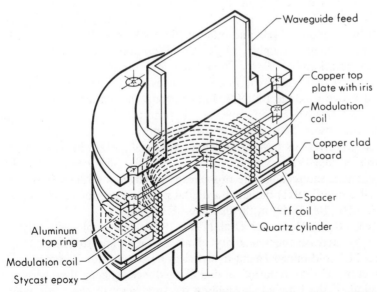

Fig. 14-11. Cavity resonater used for pulsed ENDOR (Brown and Sloop, 1970).

657

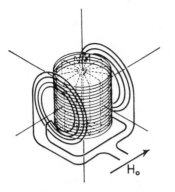

Fig. 14-12. Schematic diagram of wire-wound cylindrical TE_{011} resonant cavity and field modulation coils (Hyde, 1965).

large as possible without degrading the Q of the cavity through dielectric losses. At high rf power pulsed and CW schemes have comparable sensitivity. When the nuclear resonance is easily saturated then CW is better by the factor $\sqrt{10}$ and when it is not easily saturated, then the pulsed technique is better by the factor $\sqrt{10}$ (Hyde et al., 1970). The lineshape and intensity of the ENDOR spectra are dependent upon the modulation and detection schemes (Hyde et al., 1970), and various ones are in use (e.g., Allendoerfer and Eustace, 1971; Miyagawa et al., 1973; Dinse et al., 1973).

There are several limitations on the applicabilies of the ENDOR spectrometers described above. The range from 5 to 50 G at high power is adequate for most proton ENDOR at X band since $\omega_{NMR} = 14$ MHz, and the hyperfine splitting A is generally less than 30 MHz (recall $\omega_{ENDOR} = |\omega_{NMR} \pm \frac{1}{2}A|$). There are some nuclei with small hyperfine splittings and small nuclear Zeeman frequencies at X band such as deuterium (McCalley and Kwiram, 1970), ^{35}Cl (Blumberg and Feher, 1960), and ^{39}K (Feher, 1959) that produce ENDOR lines below 6 MHz and other nuclei such as V (Borcherts and Lohr, 1969; Lawrence and Lambe, 1963), Cr (Woonton and Dyer, 1967), Mn (Lawrence and Lambe, 1963), Fe (Locher and Geschwind, 1965), and Gd (Hurrell, 1965) that have hyperfine frequencies considerably larger than protons, and hence produce ENDOR lines beyond the Varian high-power ENDOR limit of 42 MHz and in some cases beyond the Varian 90-MHz extended range at low power. Several spectrometers reported in the literature operate from 1 to 100 MHz or beyond (Allendoerfer et al., 1971; Maki et al., 1968; Miyagawa et al., 1973; Schmalbein et al., 1972). The commercial Bruker instrument uses interchangeable probeheads to cover the range from 1 to 80 MHz. Witte et al. (1971) designed a digital programmer for producing broad sweeps (5 kHz–100 MHz) in the 0.3–500-MHz range.

In recent articles Hoentzsch et al. (1974, 1978) have employed a microwave amplifier to increase the sensitivity of an ENDOR spectrometer. They obtained an 11.5-fold improvement in signal-to-noise ratio for H centers in KCl. Andrist et al. (1976) reported a digital feedback system that permits the stabilization of the ratio of the microwave frequency to the static magnetic field to within one part in 10^6 while scanning the ENDOR oscillator. Alexander and Pugh (1975) reported a digitally controlled recorder driver for ENDOR.

When the NMR transition is being scanned the ESR absorption signal generally increases in magnitude, an effect that is referred to as positive ENDOR. Some factors such as the use of field modulation, the application of a very intense rf field, and others can result in a decrease in the ESR signal, a phenomenon called negative ENDOR (Atherton et al., 1971; Freed et al., 1967; Helms et al., 1973; McCalley and Kwiram, 1970; Miyagawa et al., 1973, 1978; Shanina, 1979; Suzuki, 1974; Terhune et al., 1960).

Resonant cavities for ENDOR applications may be constructed so that the NMR coil surrounds the cavity, forms the cavity walls, or is inside the cavity (Motchane et al., 1958). The latter two methods are more frequently employed. Section 19-C of the first edition discussed a number of ENDOR cavity designs (Baker et al., 1962; Eisinger et al., 1958; Feher, 1959; Feher et al., 1956, 1959; Grützediek et al., 1965; Hall and Schumacher, 1962; Hardeman, 1960; Holton and Blum, 1962; Jacubowicz et al., 1961; Jeffries, 1961; Kessenikh, 1961; Kramer et al., 1965; Lambe et al., 1961; Leifson and Jeffries, 1961; Motchane, 1962; Parker et al., 1960; Pipkin et al., 1957, 1958, 1958; Reichert and Townsend, 1965; Schacher, 1964; Terhune et al., 1961; Watkins and Corbett, 1964). Hyde (1965) and Yagi et al. (1970) designed wired-wall cavities for ENDOR. Davis and Mims (1978) designed an ENDOR cavity for spin-echo studies. Van Camp et al. (1976) reported a plastic ENDOR cavity for studying biological compounds at 2 K. Kiefte and Harvey (1970) employed a quartz-filled TE_{011} cylindrical cavity similar to that of Baker and Williams (1962), which is sketched on page 758 of the first edition. Other ENDOR cavities were mentioned above. Figure 14-13 presents the ENDOR cavity of Merks and de Beer (1978). Helices have been employed for ENDOR resonators (Hausser,

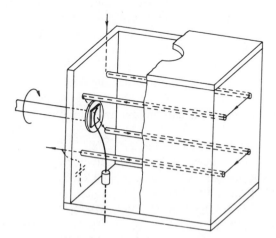

Fig. 14-13. Rectangular cavity resonator with electrode and ENDOR coil configuration (Merks and de Beer, 1978).

and Reinhold, 1962; Kenworthy and Richards, 1965; Richards and White, 1962, 1964; Van Steenwinkel and Hausser, 1965; Werner, 1964; see also Baldin and Stepanov, 1976).

ENDOR has been carried out with circularly polarized radio-frequency fields (Schweiger and Günthard, 1981), and Forrer et al. (1981) described a spectrometer for this purpose. This technique may drastically reduce the number of lines in a complicated ENDOR spectrum.

Some ENDOR work has been carried out at high pressure. Wolbarst (1976) reported a high-pressure cavity that operates at low temperatures. Fainstein and Oseroff (1971) designed a unaxial stress system for ENDOR at 35 GHz. Kasatochkin and Yakovlev (1977) investigated the S-state splitting of Gd in fluorite by ESR and ENDOR at high pressure. Grachev et al. (1972) treated the theory of ENDOR in the presence of external stresses. Dalal and Manoogian (1977) identified multiple quantum transitions and described their use in enhancing the resolution of ENDOR spectra.

Some relaxation work has been carried out with ENDOR (e.g., Abragam and Bleaney, 1970; Brik and Ishchenki, 1976; Gokhman and Shanina, 1975; Hoostraate et al., 1975; McIrvine et al., 1964; Mollenauer and Pan, 1972; Zevin and Brik, 1972). It is important to understand the various relaxation paths since the appearance of the spectrum is dependant on these mechanisms. Tunneling studies have been made involving ENDOR linewidths (Derouane and Vedrine, 1974; Helbert et al., 1972; Kevan et al., 1979) and tunneling methyl groups and their sidebands have been observed (Clough et al., 1974, 1974, 1975). Triplet states have been investigated (e.g., Clarke and Hutchison, 1971; Hutchison et al., 1970, 1971).

We have been discussing conventional ENDOR in which the microwave frequency and the magnetic field are held constant at an ESR resonance point and the radio-frequency field is swept to produce the ENDOR spectrum. In an alternate technique referred to as ENDOR-induced ESR, the microwaves and the radio frequency are set at values corresponding to a particular ENDOR transition, and the static magnetic field is swept through the region of resonance. Schweiger et al. (1981) describe a spectrometer for recording ENDOR-induced ESR, and a number of workers have employed this technique (e.g., Allendoerfer and Papez, 1970; Biehl et al., 1977; Cook, 1966; Hyde, 1965; Niklas and Spaeth, 1981; Robinson et al., 1976).

ENDOR studies have been carried out in solutions (Allendoerfer, 1973, 1975; Allendoerfer et al., 1970, 1971; Dalal et al., 1973, 1974; Hyde and Eyring, 1974; Kirillov et al., 1975; Leniart et al., 1975; Vollmann, 1976), in glasses (Bales et al., 1974; Hase et al., 1975; Helbert et al., 1972; Iwasaki et al., 1972; Volino and Rousseau, 1972), in inorganic solids (Abraham et al., 1974, 1974, 1975, 1975, 1976, 1976; Balzar, 1976; Boldu et al., 1977; Chen et al., 1975; Chu et al., 1969; Dalal et al., 1972, 1972; DuVarney et al., 1969; Gaillard and Gloux, 1976; Gazzinelli et al., 1973; Hale and Mieher, 1969; Hutchison and Kim 1979; Krebs and Jeck, 1972; Maffeo and Hervé, 1976; Marzke and Mieher, 1969; Mollenauer and Pan, 1972; Rubio et al., 1976;

Unruh et al., 1973, 1973, 1977; Van Ormondt et al., 1976; van Veen, 1977; Vedrine and Imelik, 1974; Watkins, 1975; Yang and Kim, 1979), in organic solids (Allendoerfer and Eustace, 1971; Atherton et al. 1975, 1976; Bernhard et al., 1976; Box and Budzinski, 1977; Brenner et al., 1974; Budzinski and Box, 1978; Chacko et al., 1979; Ching et al., 1977; Close et al., 1977, 1979; Doetschman and Hutchison, 1972; Finch et al., 1979; Gloux and Lamotte, 1973; Herak et al., 1976; Madden and Bernhard, 1979; Nelson, 1977; Nelson et al., 1978, 1980; Park and McDowell, 1976; Wells, 1970), in organometallics (Apaydin, 1971; Colligiani et al., 1978; Dalal et al., 1975; Kispert et al., 1979; Rist and Hyde, 1970), and in biologically important compounds (Borg et al., 1976; Box et al., 1975; Kennedy et al., 1974), such as amino acids (Box et al., 1971, 1972, 1974, 1974, 1976, 1979; Castleman and Moulton, 1972; Close et al., 1977, 1979; Hampton et al., 1973, 1975; Muto et al., 1973, 1977; Quoc-Hai et al., 1974; Rustgi, 1977; Kou et al., 1976), peptides (Lee and Box, 1974; Rustgi and Box, 1973), and nucleotides (Adams and Box, 1975; Box et al., 1977; Budzinski and Box, 1975; Herak and McDowell, 1974; Hwang et al., 1979; Pugh and Alexander, 1977), and vitamins (Joerin, 1976). Kwiram (1968) studied randomly oriented free radicals. Dalton and Kwiram (1972) provided a computer analysis of polycrystalline ENDOR spectra and Schweiger et al. (1976) showed how to diagonalize angularly dependent hamiltonians.

C. Triple Resonance

ESR and ENDOR generally only provide the magnitudes of hyperfine coupling constants. To determine the relative signs of the couplings in a hyperfine multiplet a double-ENDOR or triple-resonance method may be employed. This will be illustrated for the three-spin case $S = I_1 = I_2 = \frac{1}{2}$.

The energy-level diagram for these three spins when one nucleus has a positive hyperfine coupling constant A_1 is presented in Fig. 14-12 for the two cases in which the other coupling constant A_2 is positive and is negative. In constructing this diagram it was assumed that both nuclei have the same gyromagnetic ratio γ_N. The four nuclear transitions ω_i that are scanned in ENDOR have the energies

$$\omega_1 = \omega_N - \tfrac{1}{2}A_1$$

$$\omega_2 = \omega_N - \tfrac{1}{2}|A_2|$$

$$\omega_3 = \omega_N + \tfrac{1}{2}|A_2|$$

$$\omega_4 = \omega_N + \tfrac{1}{2}A_1$$

$$(1)$$

where Fig. 14-14 was drawn for the condition $2\omega_N > A_1 > |A_2|$. We notice from the figure that the frequencies ω_2 and ω_3 shift between the upper and lower sets of energy levels when the sign of A_2 changes.

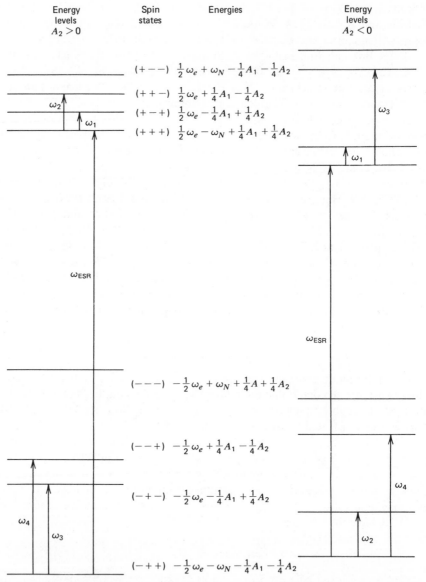

$$
\begin{array}{cccc}
\text{Energy} & \text{Spin} & \text{Energies} & \text{Energy} \\
\text{levels} & \text{states} & & \text{levels} \\
A_2 > 0 & & & A_2 < 0
\end{array}
$$

$(+--)\quad \frac{1}{2}\omega_e + \omega_N - \frac{1}{4}A_1 - \frac{1}{4}A_2$

$(++-)\quad \frac{1}{2}\omega_e + \frac{1}{4}A_1 - \frac{1}{4}A_2$

$(+-+)\quad \frac{1}{2}\omega_e - \frac{1}{4}A_1 + \frac{1}{4}A_2$

$(+++)\quad \frac{1}{2}\omega_e - \omega_N + \frac{1}{4}A_1 + \frac{1}{4}A_2$

ω_{ESR}

$(---)\quad -\frac{1}{2}\omega_e + \omega_N + \frac{1}{4}A + \frac{1}{4}A_2$

$(--+)\quad -\frac{1}{2}\omega_e + \frac{1}{4}A_1 - \frac{1}{4}A_2$

$(-+-)\quad -\frac{1}{2}\omega_e - \frac{1}{4}A_1 + \frac{1}{4}A_2$

$(-++)\quad -\frac{1}{2}\omega_e - \omega_N - \frac{1}{4}A_1 - \frac{1}{4}A_2$

Fig. 14-14. Relative energy-level diagrams for a three-spin system with $S = I_1 = I_2 = \frac{1}{2}$ and a positive hyperfine coupling constant A_1 drawn on the left when the coupling constant A_2 is positive and on the right when A_2 is negative. The spin states (M_s, M_{n1}, M_{n2}) and level energies are given and two ESR transitions ω_{ESR} and and the four ENDOR transitions ω_1, ω_2, ω_3, and ω_4 are indicated. Note that the ω_2 and ω_3 transitions defined by Eqs. (14-C-1) shift between the upper and lower sets of energy levels when the sign of A_2 changes. Figure 14-15 gives the relative populations of the energy levels for various conditions.

662

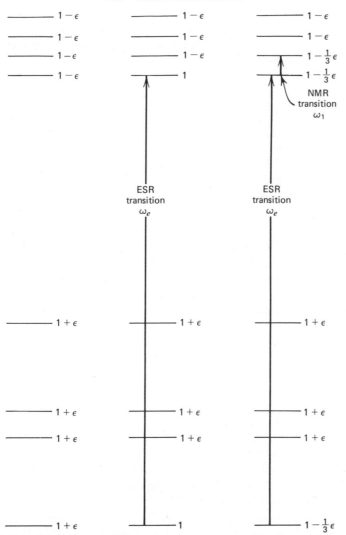

Fig. 14-15. Relative energy-level populations shown in the absence of saturation (left), when the ESR transition is saturated (center) and when the ESR and the ω_1 ENDOR transitions are saturated (right), where $\epsilon = g\beta H/2kT$ and $S = I_1 = I_2 = \frac{1}{2}$.

To carry out a double-ENDOR experiment, one ESR and one NMR transition are saturated, and a separate rf oscillator is employed to scan through other NMR transitions while monitoring the ESR absorption. Figure 14-15 shows the relative populations of the eight levels in the $S = I_1 = I_2 = \frac{1}{2}$ system using the notation

$$\epsilon = \frac{g\beta H}{2kT} \ll 1 \tag{2}$$

TABLE 14-2

Relative Intensities of the Transitions ω_1 to ω_4 Determined by Ordinary and Double
ENDOR for the Cases in Which the Hyperfine Coupling Constants A_1 and A_2 Have the
Same or the Opposite Signs, Where $\epsilon = g\beta H/2kT$

Saturating Transition	Signs of A_1 and A_2	Relative Intensities				ENDOR Type
		ω_1	ω_2	ω_3	ω_4	
ω_{ESR}	Same	ϵ	ϵ	ϵ	ϵ	Ordinary
ω_{ESR}	Opposite	ϵ	ϵ	ϵ	ϵ	Ordinary
$\omega_{ESR} + \omega_1$	Same	—	$\frac{2}{3}\epsilon$	$\frac{4}{3}\epsilon$	$\frac{4}{3}\epsilon$	Double
$\omega_{ESR} + \omega_1$	Opposite	—	$\frac{4}{3}\epsilon$	$\frac{2}{3}\epsilon$	$\frac{4}{3}\epsilon$	Double

for the three situations: (1) static thermal equilibrium, (2) dynamic thermal equilibrium in the presence of saturating microwave power at the ESR frequency ω_{ESR}, and (3) dynamic thermal equilibrium in the presence of saturating power from the ESR frequency ω_{ESR} and one of the ENDOR transition frequencies ω_I. This figure is drawn for the case of positive A_2, but the level populations also apply to the negative A_2 case. Here we are assuming the simple model of transient ENDOR in which relaxation mechanisms do not complicate the picture. If a scan is made through another ENDOR line then the saturated ESR signal will be affected to an extent that is proportional to the population difference between the corresponding ENDOR levels. We see from Table 14-2 that the population difference is the same value ϵ for all four ordinary ENDOR transitions, but that in the case of double-ENDOR the ω_2 and ω_3 population differences differ for the two coupling-constant cases. Thus the experimental determination of the relative intensities of the ω_2 and ω_3 lines by double ENDOR provides the relative signs of the hyperfine coupling constants.

The double-ENDOR technique just described has been used by a number of workers (Baker and Blake, 1970, 1973; Cook and Whiffen, 1964, 1965; Coope et al., 1972; Dalal and McDowell, 1970; Hampton and Moulton, 1975; Ursu, 1972), and Forrer et al. (1977) designed a spectrometer for this application. Double ENDOR is carried out with nuclei having the same gyromagnetic ratios. Biehl et al. (1975) [see also Dinse et al. (1974)] performed a more general electron–nuclear–nuclear triple-resonance experiment by irradiating two nonequivalent nuclei at their respective NMR frequencies. This also provides the relative signs of hyperfine coupling constants. Kevan and Kispert (1976) describe other ways in which ENDOR can assist in determining the relative signs of coupling constants. Biehl et al. (1975, 1977) used double ENDOR to determine deuteron quadrupole coupling constants.

D. ELDOR Reduction Factor

In Sec. 14-A we explained that ELDOR is a double-resonance experiment involving the saturating of one ESR transition and the low-power detection of the signal from another such transition. The saturation factor s defined previously in Sec. 13-C and 14-B is a measure of the level of saturation. One high-power klystron called the pump klystron is employed to produce the saturation, and a second klystron referred to as the signal klystron monitors the ESR absorption about 350 MHz away from the pump frequency. When pumping power saturates the transition $1 \to 4$ in Fig. 14-1 the populations of levels 1 and 4 become more equal, and various relaxation mechanisms that are operative induce transitions that make the populations of the pair of levels 2 and 3 more equal. The result is a decrease in the amplitude of the observed $2 \to 3$ transition that is expressed quantitatively in terms of the ELDOR reduction factor R:

$$R = \frac{I_0 - I}{I_0} \tag{1}$$

where I_0 is the observed ESR intensity with the pump power off, and I is the intensity with the pump power on.

For the $S = \frac{1}{2}$, $I = \frac{1}{2}$ spin system the reduction factor R may be expressed in terms of the six relaxation rates W between the four energy levels that are indicated in Fig. 14-16. These relaxation rates are equal to the reciprocals of the corresponding relaxation times and are given by

$$
\begin{array}{lll}
\text{ESR allowed transition} & \left.\begin{array}{l} 1 \to 4 \\ 2 \to 3 \end{array}\right\} & W_e = \dfrac{1}{T_{ie}} \\[2em]
\text{ESR forbidden transition} & 1 \to 3 \quad & W_{F1} = \dfrac{1}{T_{F1}} \\[1.5em]
\text{ESR forbidden transition} & 2 \to 4 \quad & W_{F2} = \dfrac{1}{T_{F2}} \\[1.5em]
\text{NMR transition} & \left.\begin{array}{l} 1 \to 2 \\ 3 \to 4 \end{array}\right\} & W_N = \dfrac{1}{T_{1N}}
\end{array} \tag{2}
$$

Hyde et al. (1968) showed that the ELDOR reduction factor is given by (see also Freed, 1972; Kevan and Kispert, 1976)

$$R = \frac{W_N^2 - W_{F1}W_{F2}}{W_e(2W_N + W_{F1} + W_{F2}) + (W_N + W_{F1})(W_N + W_{F2})} \tag{3}$$

Table 14-3 lists the reduction factors for ELDOR experiments carried out with various combinations of relaxation conditions and pumped and observed transitions.

The discussion until now treated what is called steady-state ELDOR, which measures an ESR signal amplitude in dynamic equilibrium with the pumping power. If the reduction factor R becomes so small that no signal is detected then one must resort to a transient ELDOR experiment in which the reduction is measured immediately after the application of the pump power but before

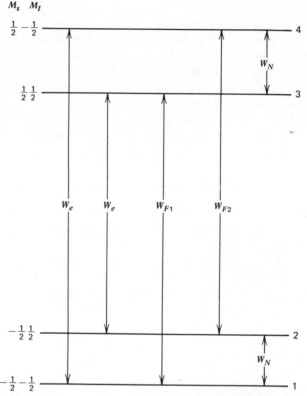

Fig. 14-16. Relaxation rates for the allowed ESR (W_e), the forbidden ESR (W_{F1} and W_{F2}), and the ENDOR W_N transitions.

TABLE 14-2

ELDOR Reduction Factor Expressions as a Function of the Type of Observed and Pumped Transitions and the Relaxation Conditions for an $S = \frac{1}{2}$ and $I = \frac{1}{2}$ Four-Energy-Level System (Kevan and Kispert, 1976, p. 48) This table uses the notation ω_{x1} and ω_{x2} instead of ω_{F1} and ω_{F2}.

Relaxation Condition	Transition Pumped	Transition Observed	Reduction Factor R
$W_{x1} = W_{x2} = 0;\ W_n \neq W_e \neq 0$	1–4 (allowed)	2–3 (allowed)	$W_n/(2W_e + W_n)$
$W_n = W_{x2} = 0;\ W_{x1} \neq W_e \neq 0$ or	1–4 (allowed)	2–3 (allowed)	0
$W_{n1}W_{n2} = W_{x1}W_{x2}$ $W_n = 0;\ W_{x1} \approx W_{x2} \neq W_e$	1–4 (allowed)	2–3 (allowed)	$-W_{x1}W_{x2}/[(W_e(W_{x1} + W_{x2}) + W_{x1}W_{x2})]$
$W_n \approx W_{x1} \approx W_e$	2–4 (forbidden)	2–3 (allowed)	$W_n + W_{x1}/(W_n + W_e + 2W_{x1})$
$W_n \approx W_{x1} \approx W_e$	1–3 (forbidden)	2–3 (allowed)	$(W_n + W_{x2})/(W_n + W_e + 2W_{x2})$
$W_n \neq W_e \neq W_{x1} \neq W_{x2}$	2–4 (forbidden)	1–3 (forbidden)	$(W_n - W_e)/(W_n + W_e + 2W_{x1})$
$W_n \neq W_e \neq W_{x2} \neq W_{x1}$	2–3 (allowed)	1–3 (forbidden)	$(W_n + W_e)/(W_n + 2W_e + W_{x2})$
$W_n \neq W_e \neq W_{x2} \neq W_{x1}$	2–3 (allowed)	2–4 (forbidden)	$\dfrac{(W_n + W_e)(W_n + W_{x1})}{(W_n + 2W_e + W_{x2})(W_n + W_{x2})}$
$W_n \neq W_{ss} \neq 0$	1–4 (allowed)	2–3 (allowed)	$(W_n/W_e) + W_{ss}/2W_e$

the observed signal is reduced to zero intensity. In typical cases this must be done within times of the order of milliseconds after the pump is applied. If the relaxation rates for the ESR forbidden transitions are negligible

$$W_{F1}, W_{F2} \ll W_N \tag{4}$$

then the ELDOR reduction factor becomes

$$R \sim \frac{W_N}{2W_e + W_N} \tag{5}$$

and transient ELDOR methods must be used when $W_e \gg W_N$ since the factor $R \sim W_N/2W_e$ then becomes quite small.

There are two common ways in use for recording ELDOR spectra. In one of them called field-swept ELDOR (Hyde et al., 1968) the pumping and observing microwave frequencies are held fixed at predetermined values, and the magnetic field is swept through the region of resonance. In the other mode referred to as frequency-swept ELDOR (Hyde et al., 1968) the magnetic field and observing microwave frequencies are set at the resonance point of one of the hyperfine transitions, and the pumping microwave frequency is swept. The recorder plots the absorption signal amplitude versus the frequency difference between the pumping and observing klystrons. Table 2-1 of the monograph by Kevan and Kispert (1976) lists the characteristics of these and three other modes of spectral presentation.

ELDOR has been employed to study a number of systems such as inorganic compounds (Iwasaki et al., 1974; Kispert et al., 1973; Mottley et al., 1975; Toyoda and Morigaki, 1977), organic compounds (Huttermann et al., 1976; Kispert et al., 1972, 1973, 1978; Lund et al., 1975), biologically important compounds (Huttermann et al., 1976; Hwang et al., 1979; Kispert et al., 1973, 1976), and glasses (Lin et al., 1976; Yoshida et al., 1973). Molecular motions have been investigated by Geoffroy et al. (1979), Hyde et al. (1975), Mottley et al. (1975, 1975), Vieth and Hausser (1974), and Stetter et al. (1976). Bruno and Freed (1974) discussed ELDOR lineshapes in the slow-motional region.

E. ELDOR Instrumentation

A number of ELDOR spectrometer systems have been reported in the literature (Benderskii et al., 1968; Chiarini et al., 1972; Eastman et al., 1970; Giordmaine et al., 1958; Hyde et al., 1968, 1968, 1969; Sokolov and Benderskii, 1969; Sorokin et al., 1960; Tanaka et al., 1967; Unruh and Culvahouse, 1963; Vieth et al., 1971, 1972), and Moran's (1964) spectrometer is described in the first edition. (Sect. 19-C) of this work. Most ELDOR studies make use of two X-band frequencies. In contrast to this, Bowers and Mims (1959) carried out a microwave double resonance experiment with a pump frequency of 3.9 GHz and a monitor frequency of 4.1 GHz, Cox et al. (1965) describe a double-resonance microwave spectrometer with an X-band pump frequency and a K-band observing frequency, and Tanaka et al. (1967) employed a 22-GHz pump and a 9-GHz observing frequency.

The block diagram of Hyde et al.'s (1968) X-band ELDOR spectrometer is illustrated in Fig. 14-17. The resonant cavity, oscilloscope, and klystron beam and filament supplies are not shown on the figure. The observing klystron was stablized, and the pump klystron was free running, but a pump AFC was added to a later model. The resonant cavity used in the spectrometer is illustrated in Fig. 14-18. It employs a pumping microwave field in a TE_{102} mode and an observing field in a TE_{012} mode. The operation of this cavity is similar to that of the bimodel one illustrated in Fig. 13-10 and described in Sec. 13-D.

In other experiments Hyde et al. (1968, 1969) swept the pumping microwave source using the equipment presented in Fig. 14-19. The difference between the two klystron frequencies is determined by a frequency counter and converted to an analog voltage that is compared to the output of a ramp generator. The resulting difference voltage is fed to a servomotor that scans the frequency by varying the penetration of the dielectric into the unshared portion of the cavity.

Vieth et al. (1971, 1974) and Benderskii et al. (1968) employed the microwave helix illustrated in Fig. 14-20 as the pumping resonator in their traveling-wave tube equipped ELDOR spectrometer. The broad-band feature of the TWT permitted frequency sweeps up to 1 GHz, although in practice sweeps were generally 150 MHz or less. The helix consisted of thin silver tape glued to the inner wall of a quartz tube located perpendicular to the axis of the cylindrical cavity as illustrated in Fig. 14-20. This arrangement provided an isolation of 50 dB between the helical and cavity modes when the difference frequency was 10 MHz. At 8.6 GHz the optimum design parameters were a diameter of 37 mm and a pitch of 3 mm. Larger diameters have the advantage of providing a more homogeneous helical field, with the disadvantage of greater leakage of this field outside of the helix.

Pumping powers between 5 and 220 W were used with duty cycles between 50% and 4% to limit the average power to values below 10 W, and thereby prevent excessive heating of the sample. Cooled gas was blown through the quartz tube to provide temperatures between 173 and 300 K. The sensitivity of about $10^{12}R^{-1}$ spins/G of linewidth with regular spectrometer operation was improved to about $10^{11}R^{-1}$ spins/G with signal averaging and digital filtering, where R is the ELDOR reduction factor.

Stetter et al. (1976) converted their earlier ELDOR spectrometer (Vieth and Hausser, 1974) to a computer-controlled type by linking it to a small computer. The microwave section of the spectrometer is sketched in the lower part of Fig. 14-21, and the interfacing is indicated in the upper part of the figure. The interfaces have D-flip-flop arrays (latches) for temporary storage, address decoders, and status registers, which initiate interface procedures and trigger a program interrupt when data conversion is complete.

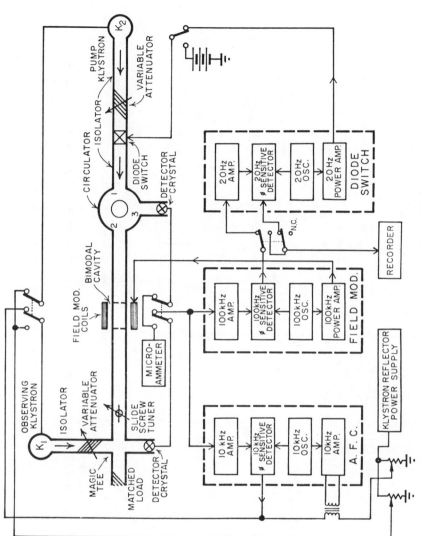

Fig. 14-17. Block diagram of an ELDOR spectrometer (Hyde et al., 1968).

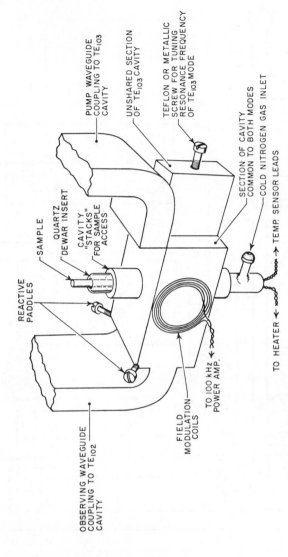

Fig. 14-18. Bimodal resonant cavity used in the ELDOR spectrometer illustrated in Fig. 14-17. It has a TE_{103} rectangular pump mode and a TE_{102} rectangular observing mode (Hyde et al., 1968).

REACTIVE PADDLES

SAMPLE

QUARTZ DEWAR INSERT

CAVITY "STACKS" FOR SAMPLE ACCESS

PUMP WAVEGUIDE COUPLING TO TE_{103} CAVITY

UNSHARED SECTION OF TE_{103} CAVITY

TEFLON OR METALLIC SCREW FOR TUNING RESONANCE FREQUENCY OF TE_{103} MODE

SECTION OF CAVITY COMMON TO BOTH MODES

COLD NITROGEN GAS INLET

TO HEATER ←

→ TEMP. SENSOR LEADS

OBSERVING WAVEGUIDE COUPLING TO TE_{102} CAVITY

FIELD MODULATION COILS

TO 100 kHz POWER AMP.

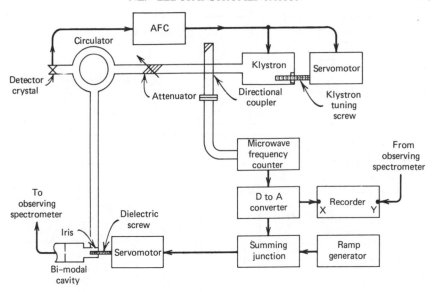

Fig. 14-19. Block diagram of the sweep unit for the pumping microwave frequency of an ELDOR spectrometer (Hyde et al., 1969).

The software controls the spectrometer synchronized by a real-time clock, which also runs the display and the processing of the stored spectra. The various jobs are associated with particular priority levels. Every millisecond the clock pulse initiates the highest level, the pulses are counted, and the computer determines the on–off positions of the PIN modulator (on-time $t_{on} = 1$–15 msec, off-time $t_{off} = 31$ msec $- t_{on}$). At the second highest level of priority the measuring program for simultaneous frequency and field-swept ELDOR spectroscopy operates, and data are accumulated into a register. Two lower-priority

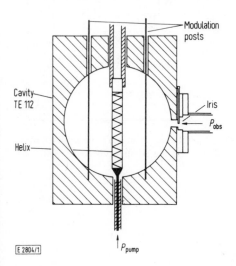

Fig. 14-20. ELDOR cavity that uses a microwave helix for the pump frequency (Vieth et al., 1971, 1974).

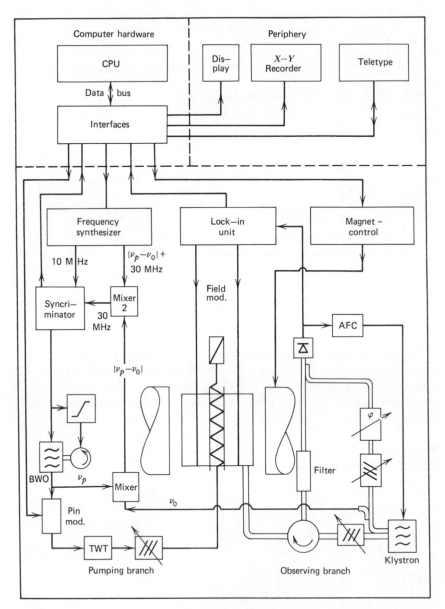

Fig. 14-21. Block diagram of a computer-controlled ELDOR spectrometer. The horizontal dotted line separates the ELDOR spectrometer below from the computer components above (Stetter et al., 1976).

672

levels contain subroutines for averaging the accumulated data and for storing them in the core memory. External messages such as initiate commands and spectrometer settings are introduced at the lowest level. During the run the recorded spectra are displayed on a monitor. Stetter et al. (1976) employed this spectrometer for the study of spin labels (compare Vieth and Hausser, 1972).

Martinelli et al. (1975, 1977) used longitudinal detection of the signal from a spin system irradiated at two X-band microwave frequencies (see also Chiarini et al., 1975). Both frequencies were present simultaneously in a TE_{102} single-mode rectangular resonant cavity with a Q of 4000. To accomplish this it was necessary to operate with a difference frequency between the two klystrons that was not higher than a few hundred kilohertz so that $\Delta\omega/\omega$ for this difference was much less than $1/Q$ for the cavity.

Pristupa (1976) reported on an ESR spectrometer with a crossed resonator operating in the K–X range. Chiarini et al. (1972) describe an ELDOR spectrometer that utilizes a cylindrical bimodal resonator, and they present spectra from the oxypyrrol radical. In addition they comment on the difficulties of applying double-irradiation techniques in ELDOR spectroscopy.

Rengan et al. (1979) devised a method of performing ELDOR spectroscopy by utilizing a pulsed ESR spectrometer operating in the saturating recovery mode. This procedure is to saturate one hyperfine line of a multiplet with a high-power microwave pulse, and then within a time shorter than T_1 to shift the magnetic field to the resonance point of another hyperfine component and monitor the amplitude of the ESR signal as a function of time in the manner of a standard saturation recovery experiment. The amplitude of the ESR signal of the second hyperfine component provides a measure of the saturation transfer (compare Sec. 13-G) that has taken place. Measurements were carried out on the radical ion 2,5-di-t-butyl-p-benzoquinone in ethanol. Data were obtained on the dependence of the measured saturation transfer on the width and height of the saturating microwave pulse. The spectrometer had a linear magnetic field scan rate of 1 G/μsec over a range of 6 G. The response time of 3 μsec made it possible to study transient species with lifetime of 10 μsec or longer.

Nechtschein and Hyde (1970) made use of pulsed ELDOR to study irradiated malonic acid. They observed damped Torrey oscillations about the pumping and observing fields, and they measured directly the spin-lattice and cross relaxation times. Very narrow ESR lines were obtained.

The sensitivity of an ELDOR spectrometer operating in a standard ESR mode is about a factor of 2 less than that of an ordinary spectrometer. This decrease in sensitivity arises from $\sim 60\%$ decrease in the unloaded Q of the bimodal cavity compared to an ordinary one and the use of an isolation filter with ~ 3 dB insertion loss and 40- to 50-dB isolation. In addition the ELDOR reduction factor R typically reduces the observed ELDOR signal to between 5% and 50% of the corresponding signal detected with a standard rectangular cavity resonator.

The intensity of an ELDOR transition can be strongly temperature-dependent. The introduction of a variable temperature Dewar device into the cavity causes the pump frequency scanning range to decrease by a factor of about $\frac{2}{3}$.

Kevan and Kispert (1976) give many references to the ELDOR literature. Annabi and Theobald (1977) discuss ELDOR in terms of the beat of transition probabilities. Hyde et al. (1975) note that the quantitative comparison of calculated and measured spectra requires the measurement of the pumping and observing microwave fields and also the modulation field intensity. Bruno and Freed (1974) and Kispert et al. (1974) discussed saturation in ELDOR, and Clough and Hobson (1975), Gamble et al. (1979), and Mottley et al. (1975) discussed molecular motion and tunneling effects.

F. Dynamic Nuclear and Electron Polarization

In Sec. 14-A we discussed electron nuclear double resonance, in which the ESR transition is monitored while a sweep is made through one of the NMR transitions. In another type of double-resonance experiment called dynamic nuclear polarization (DNP) the NMR signal amplitude is monitored while the ESR signal is saturated. The latter saturation changes the selective populations of the NMR levels, and this alters the intensity Y_{NMR} of the NMR transition. The change in the relative population of the NMR levels is referred to as a polarization, and the extent of the resulting nuclear polarization is expressed in terms of the enhancement factor G:

$$G = \frac{Y_{NMR} - Y_0}{Y_0} \tag{1}$$

where Y_0 is the NMR amplitude in the absence of the microwave power. A negative enhancement corresponds to an emission NMR signal, and it results from the upper of the two NMR levels being populated more than the lower one.

Bendt (1970) designed the cylindrical resonant cavity shown in Fig. 14-22 for carrying out DNP experiments at a microwave frequency of 53 GHz. Each linear dimension of the cavity is about four wavelengths, and so a number of high-order modes are present. The alum crystal itself has dimensions of one to two wavelengths, and as a result the crystal is irradiated uniformly with microwave power, except for particular nodal points in the standing-wave pattern. A Kel-F coil form supported the four-turn NMR coil, which was connected to the oscillator through a coaxial cable. The cavity Q was about 2000.

For more details on the technique of dynamic nuclear polarization one may consult the book of Jeffries (1963), and the reviews which were written by Buishvili et al. (1976), and Hausser and Stehlik (1968). Some representative recent articles on DNP are by Bates (1976, 1977), Brodbeck et al. (1972), Leblond et al. (1971, 1971), Nelson et al. (1979), and Pomortsev (1977).

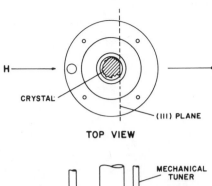

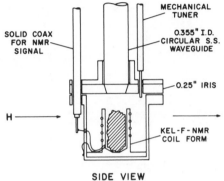

Fig. 14-22. Cylindrical microwave cavity with four-turn NMR coil used for dynamic nuclear polarization measurements (Bendt, 1970).

Sometimes radicals are produced with nonequilibrium populations of the spin levels, a process referred to as chemically induced dynamic nuclear polarization (CIDNP). One mechanism for producing CIDNP is via the decomposition of a precurser S into a radical pair R_a^* and R_b^*:

$$S \rightarrow R_a^* + R_b^* \tag{2}$$

where the asterisks indicate that the radicals are produced in particular nonequilibrium nuclear spin states (Freed and Pedersen, 1976). The observed DNP enhancement depends upon the NMR transition that is monitored, and it also varies with the time, typically reaching a maximum in a time of the order of T_1. The subject of CIDNP has been reviewed by Closs (1975), Lepley and Closs (1973), and Pine (1972). Spencer and O'Donnell (1973) describe a CIDNP experiment using a permanent magnet. Lawler et al. (1979) and Trifunac et al. (1974, 1980) used CIDNP to detect free radicals in pulse radiolysis (see Sec. 13-E).

The electron spin analogue of CIDNP is called chemically induced dynamic electron polarization (CIDEP). In this technique the production of radicals with nonequilibrium spin-state populations causes the intensity ratios of hyperfine components to differ from a binomial type distribution. It occurs, for example, when an excited triplet state is generated with a preferential popula-

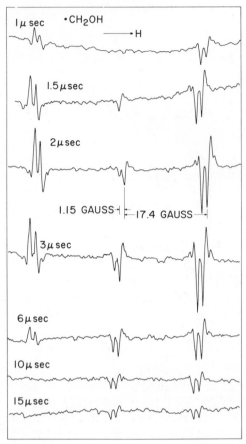

Fig. 14-23.   Polarized ESR spectrum from the CH_2OH radical produced by pulse radiolysis in a 10% aqueous solution of methanol (pH = 2.6) at various times after the cessation of the electron beam pulse. The low field lines are polarized in emission, the high field lines are polarized in absorption, and the central line is not polarized. The polarization decays with time to a final symmetrical ESR absorption spectrum (not shown). (Trifunac and Thurnauer, 1975).

tion of particular M_s levels (Ayscough et al., 1975, 1976; Buchachenko, 1974; Freed and Pederson, 1976; Lepley and Closs, 1973; Paul, 1976; Pederson and Freed, 1975; Wong et al., 1973). The spin-state populations often deviate so much from a Boltzmann distribution that some lines appear in emission, as shown in Fig. 14-23. Sometimes one radical such as R_a^* of Eq. (14-F-2) is produced with inverted spin populations and produces emission lines, in which case the other radical will have spin populations that deviate in the opposite sense from equilibrium and produce enhanced absorption spectra, as illustrated in Fig. 14-23.

The polarization of the electron–nuclear spin levels can arise during radical formation, as described above (Atkins et al., 1973, 1974, 1974; Hutchinson et al., 1972; Kaptein et al., 1977; Wong et al., 1972, 1973, 1973). A mechanism for this is the mixing of the singlet and triplet states of a radical pair as the radicals diffuse away from each other (Atkins, 1976; Dobbs, 1975; Trifunac et al., 1974, 1975; see also Fischer, 1970).

The polarization has also been described in terms of radical–radical reactions in solution (Adrian, 1970, 1971, 1971, 1972; Atkins, 1976; Fessenden, 1973). Freed and Pedersen (1976) present a general stochastic–Liouville approach to the theory that includes the above mechanisms as particular cases (see Leblond et al., 1971, 1971).

Dobbs (1975) lists three experimental arrangements for detecting CIDEP. Two of the methods involve time resolution of the ESR spectrum by a method such as pulse radiolysis with electrons (Eiben and Fessenden, 1971; Fessenden and Schuler, 1963; Smaller et al., 1970, 1971; Trifunac, 1976; Trifunac et al., 1975, 1977, 1978, 1980), or laser photolysis with visible light (Atkins et al., 1970, 1971, 1973; Hack et al., 1978; Trifunac et al., 1978; Warden, 1981; Warden and Adrianowycz, 1981), and the third technique is a continuous-generation method. The best data for the quantitative study of CIDEP are provided by time-resolved methods (compare Sec. 13-E) with a fast response time τ less than the spin-lattice relaxation time T_1, which is typically in the range from 1 to 100 μsec. Usually a pulse of radiation between 20 nsec and 2 μsec in length generates free radicals in a liquid sample situated in the resonant cavity, and the ESR spectrometer monitors the changes in the microwave power as the radicals are created and destroyed.

Figure 14-24 presents the time evolution of the ESR signal amplitude in a CIDEP experiment that exhibits transient absorption and another that corresponds to transient emission. In both cases the signal $Y(t)$ passes through a polarization extremum amplitude Y_p and then decays to an ordinary absorption signal of amplitude Y_0. In each case we can define the polarization factor γ:

$$\gamma = \frac{|Y_p| - |Y_0|}{|Y_0|} \qquad \text{enhanced absorption} \qquad (3)$$

$$\gamma = \frac{|Y_p| + |Y_0|}{|Y_0|} \qquad \text{emission} \qquad (4)$$

The decay from the polarization peak to a normal absorption state takes place at a rate determined by the spin-lattice relaxation time T_1, and the radicals themselves decay at a much slower rate via chemical reactions in the manner illustrated in Fig. 14-25.

The signal-to-noise ratio of a time-resolution ESR experiment can be improved by decreasing the receiver bandwidth through an increase in the response time to the range $\tau \sim 100$ to 200 μsec, which is generally longer than T_1. This use of time-resolved ESR with a slow response obviates the necessity

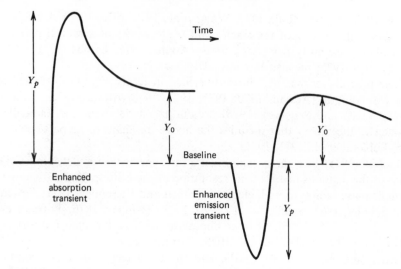

Fig. 14-24. Typical plots of the time evolution of the CIDEP signals for the enhanced-absorption (left) and enhanced-emission (right) cases. Figure 14-25 presents the enhanced-emission case on a broader time scale, which shows the subsequent radical decay process.

of rapidly cutting off the electron beam and hence permits the use of rotating sectors to chop the beam. It has the disadvantage of producing a time-dependent ESR signal $Y_{(t)}$, which is a convolution of the true signal and hence is difficult to interpret quantitatively.

The third technique for observing CIDEP is the continuous-generation method in which the ESR spectra are recorded during the radiolysis and hence

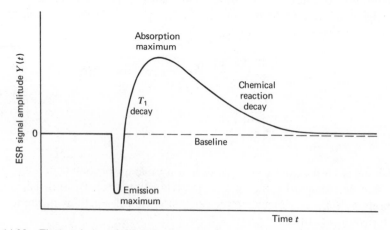

Fig. 14-25. Time evolution of the ESR absorption $Y(t)$ for the CIDEP enhanced-emission case showing the rapid rise of the number of emitting radicals, their rapid decay to the normal absorption with the short time constant T_1, and the subsequent slower decay through radical recombination or chemical reactions. The emission profile is sketched on the right side of Fig. 14-24 on a narrower time scale.

are measured in a dynamic steady state. The interpretation of the spectrum is simplified in the typical case wherein each hyperfine component decays at the same rate and has the same value of T_1. When these conditions are satisfied the polarization of each transition may be estimated from the relative intensities of the hyperfine components.

Another experimental approach to unraveling the dynamics of short lived intermediates is one in which the initiation of the transients is sinusoidally modulated and the Fourier component of the subsequent signal is phase-sensitive-detected at the modulation frequency. The lifetime of the inter-mediate species is evaluated from the frequency dependence of the amplitude and phase of this Fourier component. The relationships between the lifetimes of the intermediates and the frequency dependence of the observed Fourier amplitudes and phases have been worked out for first-order kinetics (Forster et al., 1973; Günthard, 1974; Hexter, 1962; Labhart, 1964; Loth et al., 1974; Wotherspoon et al., 1972) and second-order kinetics (Paul, 1976). In one experimental approach (Paul, 1976) radicals were generated harmonically by modulating the source of uv light in the range 200 Hz $< \omega_L <$ 13 kHz by a rotating sector. In this experiment the standard 100-kHz field modulation served as the carrier frequency. Phase-sensitive detection was carried out at 100 kHz and then at ω_L using a reference signal from the rotating sector.

Wollan (1976) and Matta et al. (1976) discussed various aspects of DNP of an inhomogeneously broadened ESR line. Lampel (1968) reported DNP by optical electronic saturation.

G. Electric Field Effects

In 1961 Ludwig and Woodbury first observed the splitting of an ESR spectral line produced by an applied electric field. Fifteen years later Mims (1976) presented a comprehensive review of electric field effects in ESR. The present section mentions some of the characteristics of these effects and discusses some of the experimental techniques involved in their measurement.

Most of the experimental work that has been carried out has involved linear electric field shifts, although quadratic effects have also been observed (Weger and Feher, 1963; Mims, 1976; Mims and Peisach, 1975; Roitsin, 1968). Linear shifts only occur when the paramagnetic center is located at a site that lacks inversion symmetry, and such sites are quite common in frequently studied host crystals. Sometimes they occur in pairs called inversion image pairs, which produce superimposed spectra in the absence of an electric field and which shift in opposite directions when the field is applied, as indicated in Fig. 14-26 (Royce and Bloembergen, 1963).

Typical experimentally attainable electric field strengths in ESR studies are 10^6 V/cm (Bugai and Roitsin, 1967), and this is two orders of magnitude less than the electric field at a distance of about 5 Å from a single electronic charge. All materials cannot withstand such fields without breaking down electrically. A simple arrangement for applying a strong electric field across a

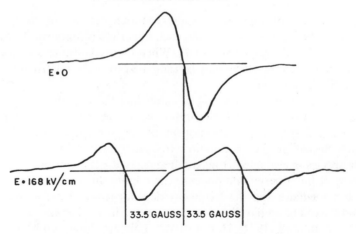

Fig. 14-26. Splitting of the $M_s = \frac{3}{2} \rightarrow \frac{1}{2}$ transition of Cr^{3+} in ruby by the application of an electric field parallel to both the external magnetic field and the c axis (Royce and Bloembergen, 1963).

sample is presented in Fig. 5-51f. In the direct method of measurement, spectra are recorded at successively increasing values of electric field strength. Shifts in the positions of the spectral lines are generally linear up to 10^5 V/cm (Reddy, 1971; Royce and Bloembergen, 1963; de Beer et al., 1976; Merks and de Beer, 1978). Mims (1976) has summarized experimental techniques used to attain high field strengths.

If the shifts are small then they can be measured using electric field modulation and lock-in detection (Marti et al., 1968, 1970; Parrot and Roger, 1968; Weger and Feher, 1963, Wysling and Mueller, 1976). Another experimental method that is particularly useful with lines that are inhomogeneously broadened is the employment of electron spin echos (Mims, 1964, 1972, 1974). The spin-echo spectrometer of de Beer et al. (1976) described in Sec. 13-F was provided with a high voltage power supply continuously variable from 0 to 14 kV. This produced pulse lengths up to ~ 4 μsec with a rise time of ~ 0.4 μsec at 14 kV and a repetition frequency variable from 0 to 1 kHz. The high voltage was connected to the electrodes through a 50-Ω coaxial line.

Electric field shifts are strongly angularly dependent (Roitsin, 1967; Meil'man, 1971; Geifman et al., 1978), and in single crystal work the electric and magnetic fields are generally oriented along particular crystallographic directions. It is more difficult to work with amorphous samples, although some useful information can be obtained from a comparison of spectra with H parallel and perpendicular to E. Neimark and Vugmeister (1975) studied the variation of the linewidth with the electric field strength.

Usmani and Reichert (1969) studied the influence of an applied electric field on the magnetic hyperfine interaction of an F center with its nearest-neighbor nuclei using ENDOR spectroscopy. The high source voltage variable from 0 to 30 kV was brought to the cavity through high voltage vacuum coaxial cables as illustrated in the block diagram of Fig. 14-27. Two high-voltage leads, one

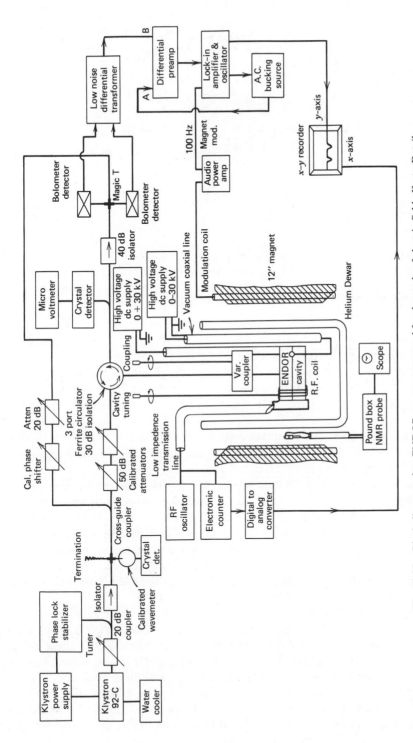

Fig. 14-27. Block diagram of an ENDOR spectrometer used for the study of electric field effects. Details of the resonant cavity are shown on Fig. 14-28 (Usmani and Reichert, 1969).

681

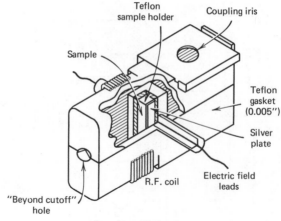

Fig. 14-28.   Details of an ENDOR cavity provided with high-voltage electrical leads for carrying out the electric field effect studies. The cavity is positioned in the magnetic field in the manner illustrated on Fig. 14-27 (Usmani and Reichert, 1969).

positive and one negative, were employed to alleviate electrical breakdown problems through reducing by a factor of 2 the electric field between the leads and the sample. This permits the cavity to be grounded and thereby made safer. The high-voltage leads enter the sides of the cavity illustrated in Fig. 14-28 through two 3.5-mm holes located in its midplane where they do not appreciably disturb the Q. The sample is potted inside the Teflon holder with Dow Corning RTV silicon rubber compound subject to a 24-hr cure. Electric fields up to 10^5 V/cm were applied to the sample, the potential difference being measured by an electrostatic voltmeter.

Truesdale (1979) and Truesdale et al. (1980, 1981) studied hysteresis loops and domain switching in ferroelectrics between room temperature and 77 K using the Dewar system presented in Fig. 14-29. The samples were flat specimens $\sim 2 \times 2 \times 1$ mm, and the electrodes were painted on their flat surfaces with graphite paint. The high-voltage and thermocouple leads come through the top of the Dewar. A home-built power supply (Truesdale, 1982) provided continuously variable voltages between 0 and ± 2 kV and electric field strengths up to 13 kV/cm.

Usmanov (1977, 1978) devised a method for applying strong electric fields (700 kV/cm) at the temperatures 77 and 4.2 K. Edgar (1977) modified a Varian E-266 35-GHz cavity by converting it to the coaxial cylindrical hybrid cavity sketched on Fig. 14-30. The 0.5-mm-thick sample is sandwiched between two 2-mm-diameter brass rods that form the coaxial inner conductor and act as electrodes. These rods are electrically insulated by placing them within a 3-mm OD quartz tube. In this same article Edgar described a method for determining the dielectric relaxation of ionic defect pairs by incorporating a

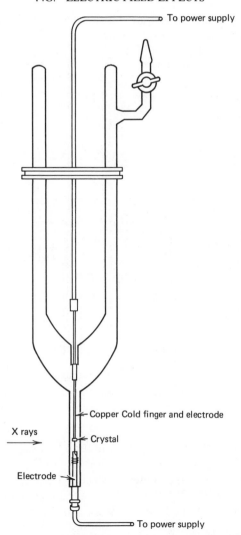

To power supply

Copper Cold finger and electrode

X rays

Crystal

Electrode

To power supply

Fig. 14-29. Sketch of a Dewar for carrying out electric field studies down to 77 K. The leads to the high voltage power supply are shown entering at the top and bottom, and those for the thermocouple enter at the top of the Dewar. The sample crystal is located between the plates of a capacitor at the bottom of the cold finger, and in some cases the plates consist of graphite layers painted on the specimen itself (Truesdale et al., 1979).

paramagnetic ion in the defect pair and recording its ESR spectrum while it is simultaneously modulated by electric and magnetic fields.

Krebs and Sackmann (1976) discussed the application of liquid crystals in the ESR spectroscopy of photoexcited triplet states. They prepared anisotropic glasses from electric-field-oriented nematic mixtures of the liquid crystals cholesteryl chloride and cholesteryl laurate. The article describes the experi-

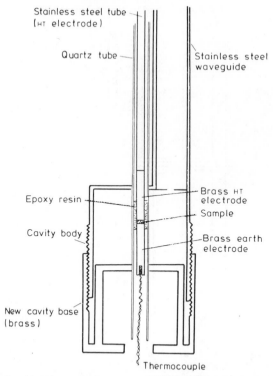

Fig. 14-30. A Varian 35-GHz cylindrical cavity modified for electric field experiments (Edgar, 1977).

mental assembly that brings about the orientation by using electric fields up to 30,000 V/cm.

Some additional articles that describe electric field effects in ESR are by Andreev and Sugakov (1975), Bates (1978), Geifman et al. (1978), Merks et al. (1979), Morigaki et al. (1971), Roitsin (1972), and Teodorescu (1977).

H. Optical Double Magnetic Resonance

A magnetic resonance experiment involves the absorption of a microwave photon that changes the populations of magnetic substates. If one monitors the light absorbed and emitted by an atomic or molecular system during the passage through ESR resonance absorption and there is observed a variation in the intensity of the light, this indicates that the optical transition involves particular magnetic sublevels. This change in light intensity can be employed as a means to detect and study ESR phenomena, and the technique is referred to as optically detected magnetic resonance, or ODMR (Chong and Itoh, 1973; Gillies et al., 1979; Hessler and Hutchison, 1973; Hutchison, 1978; Morigaki and Murayama, 1977; Plato and Möbius, 1972; Shain and Sharnoff, 1973;

Solarz and Levy, 1973; see the review by Denison, 1975). Cova and Longoni (1979) treated optical measuring techniques and signal processing. In a typical ODMR experiment one monitors an optical emission line of a sample placed in a microwave cavity, and when the magnet scans through the region where ESR absorption occurs a change is measured in the optical emission signals. This experiment provides the spin Hamiltonian parameters of the excited optical state. In addition the optical pumping power that populates the excited optical level can bring along a spin memory of the state and thereby preferentially populate particular excited magnetic sublevels. A number of ODMR studies have been carried out with molecular triplet states since they are paramagnetic with spin $S = 1$ (Eliav and Levanon, 1975; Judeikis and Siegel, 1969; Levanon and Weissman, 1971; Lin et al., 1972; Merle-Daubigne, 1975; Schmidt, 1978). Spin memory is also present when triplet states undergo nonradiative transitions to other states. Often the excited state is phosphorescent and the experiment is referred to as phosphorescence microwave double resonance (PMDR) (Breiland et al., 1973, 1975, 1979; Brenner, 1977; Clarke et al., 1973, 1974; El-Sayed, 1973; El-Sayed et al., 1971, 1973; Harris et al., 1972, 1973; Ponte Goncalves and Gillies, 1980; Schmidt et al., 1971; Sixl and Schwoerer, 1971; van Dorp et al., 1973; van Noort et al., 1980).

Coherence effects in optically excited triplet states such as spin echoes, spin locking, echo trains, free induction decay, and adiabatic demagnetization can be observed by detecting the phosphorescence (Breiland et al., 1973, 1975; Brenner, 1977; Brenner et al., 1974; Fayer and Harris, 1974; Harris et al., 1973; Levinsky and Brenner, 1981; Schuch et al., 1975). Two additional ways to measure spin coherence effects in ODMR are "triplet absorption detection of magnetic resonance" (TADMR), in which one observes microwave induced changes in the triplet–triplet absorption intensity, and "fluorescence microwave double resonance" (FMDR), in which microwave-induced changes in fluorescence are detected (Brenner, 1977; Clarke et al., 1973, 1974). A book by Clarke (1982) reviews triplet state ODMR spectroscopy.

Optical double magnetic resonance experiments sometimes make use of polarized light, obtained from lasers (Aleksanarov and Zapasskii, 1977, Decomps et al., 1976; Evenson et al., 1977; Huber et al., 1976; Skrotskaya and Skrotskii, 1973). Some researchers have made use of circularly polarized light and measured circular dichroism (Fukuda et al., 1974, 1977; Imbusch et al., 1967; Izen et al., 1972, 1973; Kiel and Crane, 1968; Mollenauer and Pan, 1972; Moore and Satten, 1973; Takagi et al., 1975). Magnetic circular dichroism (MCD) is the difference in the absorption of light that is left and right circularly polarized when it passes through a crystal located in a magnetic field. The MCD absorption arises from the Zeeman-level splitting due to (1) the population differences between magnetic sublevels, (2) the mixing of m states, or (3) the presence of different energy denominators in the Kramers–Heisenberg formulas for absorption and dispersion (Geschwind, 1972; Shen, 1964). The MCD measurement permits the resolution of small energy splittings, identifies symmetries, and reveals the presence of magnetic moments.

Associated with these polarization phenomena is the Faraday rotation of the plane of linear polarization of the light during passage through the sample in the presence of the magnetic field.

A microwave-optical-magnetic-resonance-induced-by-electrons (MOMRIE) experiment involves the use of an electron beam to raise atoms to excited optical levels and the monitoring of the intensity or polarization of the light emitted by the atoms to detect magnetic resonance absorption induced by microwaves (Freund and Miller, 1972; Miller, 1973). Curl and Oka (1973) reported a microwave infrared double- resonance experiment.

A technique related to ODMR is the measurement of the change in photoinduced conductivity produced by simultaneously scanning through an electron spin resonance absorption line. This is referred to as spin-dependent recombination (SDR), and it has been studied by various authors in silicone single crystals (Honig, 1966; Lepine et al., 1970, 1970; Maxwell and Honig, 1966; Mendz and Haneman, 1978; Morigaki et al., 1971, 1974, 1975; Schmidt and Solomon, 1966; Solomon, 1972; Wosinski et al., 1976; Yamanouchi et al., 1967; Zaitev, 1970). This experimental technique is 100 times more sensitive than ordinary ESR. The SDR results sometimes correlate with photoinduced ESR signals that have also been referred to as photoconductive resonance (PCR) (Caplan et al., 1976; Kurylev and Karyagin, 1974; Lepine, 1972; Ruzyllo et al., 1976; Shiota et al., 1974). Other photoeffects have also been reported (e.g., Baxter and Ascarelli, 1973; Berkovits et al., 1978; Bringer et al., 1979; Cheng and Kemp, 1971; Jaccard et al., 1975; Tani, 1975).

Optical spectroscopy is intrinsically much more sensitive than microwave spectroscopy, and hence the use of optical detection can bring about a large increase in sensitivity. For example, Harris et al. (1973) and Breiland et al. (1973) report that monitoring the microwave-induced modulation of phosphorescent emission is capable of detecting as few as 10^4 excited state spins.

Some ODMR lines are inhomogeneously broadened, and this can be demonstrated by hole burning experiments (Leung and El Sayed, 1972; Zieger and Wolf, 1978). In one such experiment von Schutz and Dietrich (1977) applied fixed microwave power continuously at one frequency within the ODMR line and scanned through the line with a second microwave frequency while monitoring the optical phosphorescence. The results were compared with a scan made in the absence of the fixed microwave frequency. When no hole was present the scans with and without the application of the fixed frequency were identical. When a hole was present the scan made with the fixed microwave power produced a decrease in the detected signal over part of the scan range. Muramoto et al. (1977, 1978) employed a Stark field sweeping technique to observe hole burning of Cr^{3+} ions in ruby sites that lack inversion symmetry.

Dietrich et al. (1973, 1974, 1976) demonstrated the utility of obtaining data from various complementary experimental techniques in their study of the triplet state in $NaNO_2$ single crystals. In this work they carried out measurements using Zeeman spectroscopy in absorption and emission, level crossing spectroscopy, electron spin resonance at 9.5 and 35 GHz, and optically detected magnetic resonance.

In some recent ODMR work Ruedin et al. (1973) investigated F-center pairs in alkali halides, Furrer and Seiff (1978) optically detected the transient ESR of the excited state of ruby, and Takagi et al. (1975) reported the ESR free decay signals in ruby produced by ultrashort light pulses. Bontemps-Moreau et al. (1978) carried out selective optical pumping to study the ESR of states subjected to a definite internal stress. They claim that the ESR measurement of selected centers can be made before relaxation takes place. Hack et al. (1978) used ODMR to determine the rate constant of the reaction

$$OH + HO_2 \rightarrow H_2O + O_2$$

Botter et al. (1976), Breiland and Saylor (1980), Wasiela et al. (1974), and Zieger and Wolf (1978) carried out studies of triplet-state excitons; Levanon and Vega (1974) optically detected the triplet in porphyrins; and Schuch et al. (1974) determined the sign of the zero-field splitting parameter D by high-resolution ODMR. Gillies et al. (1979, 1981), Smith and Trifunac (1981), and Trifunac and Smith (1980, 1981) used optical detection for pulse radiolysis (see Sec. 13-E).

Many researchers have examined samples by both ESR and optical spectroscopy without formally carrying out double-resonance experiments. This is a useful experimental technique since the two methods provide complementary information (e.g., Bates et al., 1977; Bjkharaev and Yafaev, 1976; David et al., 1978; Ensign and Byer, 1973; Graslund et al., 1976; Hochstrasser et al., 1973; Ikeya and Itoh, 1970; Jain et al., 1970, 1971; Kappers et al., 1974; Kato and Shida, 1979; Kawakubo, 1978; Kosky et al., 1972; Stradins and Gavars, 1974; Von der Weid, 1976; Warden and Bolton, 1976; Poole et al., 1967).

I. Optical Double-Resonance Instrumentation

In this section we describe several experimental arrangements that have been employed to carry out combined optical-ESR experiments. Many of these systems adopt commercial optical and ESR components for the combined system.

Figure 14-31 presents the block diagram of a system designed by Imbusch et al. (1967) for the optical detection of ESR in excited electronic states. The sample located in the microwave resonant cavity as shown in Fig. 14-32 is excited by broad-band light from a mercury arc lamp that enters the cavity through a hole in the bottom. The fluorescent light emitted by the sample emerges from the cavity through slots in the side and is focused by a lens into a high-resolution optical spectrometer or a circular polarization analyzer. The output of the spectrometer or analyzer is detected by a photomultiplier and inscribed on a recorder chart after lock-in detection. Microwave power from the source is chopped at an audio frequency by a square wave from a signal generator, which also triggers the multichannel analyzer and lock-in detector.

To carry out an ODMR experiment the applied magnetic field strength at the position of the sample in the cavity is swept slowly through the region of

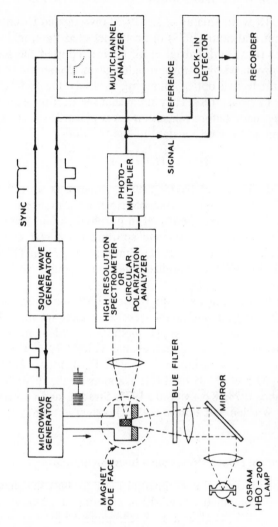

Fig. 14-31. Block diagram of a system for the optical detection of ESR absorption in excited electronic states. Details of the resonant cavity are shown on Fig. 14-32. The optically detected spectra are processed by the lock-in detector and displayed on the recorder. Relaxation studies are carried out with the aid of the multichannel analyzer (Imbusch et al., 1967, 1967).

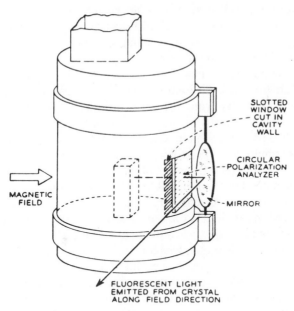

SLOTTED
WINDOW
CUT IN
CAVITY
WALL

CIRCULAR
POLARIZATION
ANALYZER

MAGNETIC
FIELD

MIRROR

FLUORESCENT LIGHT
EMITTED FROM CRYSTAL
ALONG FIELD DIRECTION

Fig. 14-32. Resonant cavity for detecting circularly polarized fluorescence from photolysis of single crystals at low temperature. A mirror system deflects the light to the circular polarization analyzer. The cavity is positioned in the magnet in the manner shown on Fig. 14-31 (Imbusch et al., 1967, 1967).

resonance while simultaneously monitoring the fluorescent light output. At resonance there is an oscillating change in the emitted light intensity that occurs at the modulation frequency, and this change is detected at the lock-in detector and displayed on the recorder. The resonant cavity used in this spectrometer, shown in Fig. 14-32, is designed to operate down to liquid-helium temperature.

For more sensitive studies the high-resolution optical spectrometer is set at the wavelength of the particular fluorescent line to be studied, and a measurement is made of the intensity of this line during the passage through resonance. For more ambitious investigations a circular polarization analyzer consisting of a piece of polaroid and a quarter-wave plate made from mica may be placed in front of the slotted exit window of the cavity as shown in Fig. 14-32. A mirror of the type shown in the figure or a light pipe (Chase, 1970) is used to direct the emitted light out of the cavity. The change in the polarization of the light during the passage through resonance is detected in the circular polarization analyzer, passed through the lock-in detector, and displayed on the recorder.

Konzelmann et al. (1975) published the block diagram of the optical double-resonance spectrometer and flash-photolyzing apparatus that they employed to study the spin-lattice relaxation of triplet states in organic molecules. Moore and Satten (1973) described a double-resonance spectrometer in which the optical absorption of circularly polarized light traversing the crystal was monitored to provide the time-resolved response to a microwave pump pulse. This instrument was used to measure spin population changes at two sites of a mixed crystal and to provide cross relaxation rates.

The optical double-resonance spectrometer designed by Murphy et al. (1975) is presented in the block diagram of Fig. 14-33. This spectrometer employs integrated circuit components that can measure ESR and ODMR either separately or simultaneously. In each mode of operation the microwave source frequency is locked to the sample cavity. The blocks on the figure labeled PIN MOD A, photomultiplier/spectrometer, PAR #2, and microwave optical logic are employed for the ODMR mode alone. The remaining units constitute a double-modulation, homodyne ESR spectrometer with 100-kHz microwave amplitude modulation and audio-frequency magnetic field modulation. The second balanced mixer detects the out-of-phase or quadrature component for use in the automatic frequency-control circuit, and the first balanced mixer selects the in-phase ESR signal, which is further amplified and detected at the field-modulation frequency in the second lock-in amplifier. A linear operational amplifier provides most of the system gain, and stable output signals from the four-quadrant multiplier could be obtained over a 75-dB range.

For ODMR detection the incident microwaves are amplitude-modulated by the PIN modulator A, and the resultant modulation of the optical signal induced by the ESR absorption is detected by the photomultiplier, lock-in amplifier (PAR #2) chain illustrated in Fig. 14-33.

Izen and Modine (1972) constructed the combined magnetic circular dichroism MCD and ESR spectrometer shown pictorially in Fig. 14-34. The ESR

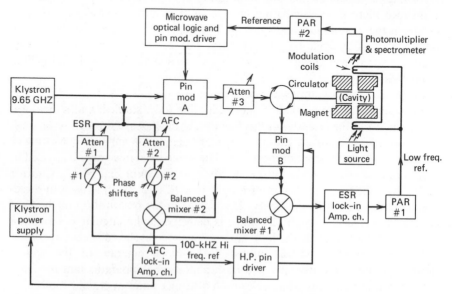

Fig. 14-33. Block diagram of an optical double resonance spectrometer (Murphy et al., 1975).

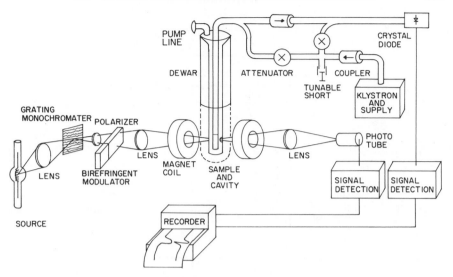

Fig. 14-34. Pictoral layout of a magnetic circular dichroism–ESR spectrometer. The resonant cavity used with this spectrometer is sketched in Fig. 14-35 (Izen and Modine, 1972).

system is a standard reflection arrangement that operates at X band, K band, or K_a band. The rectangular TE_{011} mode cavity resonator sketched in Fig. 14-35 has two optical ports near the sample which as affixed to the wall by low-temperature grease, and in addition it has slits to permit entrance of the modulation field.

The MCD system utilizes a high-pressure xenon arc lamp followed by a monochrometer for the light sources. An optical modulater that becomes birefringent when stressed acts as a quarter-wave plate to convert linear polarization to circular. An oscillating stress applied to the modulater provides polarization-modulated light. When this light beam passes through a dichroic sample the two circular components absorb to a different extent, and the emerging beam detected by the phototube has an amplitude modulation proportional to the dichrosim. A dynode voltage control normalizes the signal and rejects low-frequency noise, and an automatic modulator control corrects for wavelength-dependent factors. A two-pen recorder simultaneously plots the optical and ESR detected signals. The spectrometer can make MCD-ESR double-resonance measurements, or it can be operated as a separate ESR or MCD instrument.

In another experimental arrangement Mollenauer and Pan (1972) used a high-intensity narrow-band circularly polarized pump signal set at one dichroism peak of the absorption band and a low-intensity narrow-band beam polarization modulated at 17 kHz to monitor the other dichroism peak. They had an rf field $H_1 \sim 1$ G present in the cavity.

Fukuda et al. (1974, 1977) and Takagi et al. (1975) employed circularly polarized light pulses from a mode-locked ruby laser to study optically induced

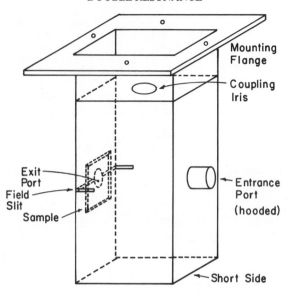

Fig. 14-35 Optical microwave cavity for use in the MCD-ESR spectrometer that is sketched in Fig. 14-34 (Izen and Modine, 1972).

precessing magnetization and free decay signals in ruby. For this technique to be effective the width of the laser pulse must be less than the period of the precessing spins in the ground optical state. The Zeeman splittings can be measured by studying the emitted light intensity at the point of resonance as a function of the microwave frequency. One can use oscilloscope presentation, or enlist the aid of a multichannel analyzer when the signal-to-noise ratio is low (Geschwind et al., 1959).

Both MCD absorption and the Faraday effect can be employed to monitor the transient magnetization in the ground state. Panepucci and Mollenauer (1969) constructed the apparatus illustrated in Fig. 14-36 for the measurement of the spin-lattice relaxation of F centers in KBr and KI using MCD. The ground-state F-center polarization is displaced from equilibrium by a pulse of light from a mercury arc lamp. The recovery of the system is monitored by light from a monochromter (or interference filter) that is switched at 17 kHz between the right and left circular polarizations by means of a stress-modulated quarter-wave plate. Panepucci and Mollenauer used their apparatus together with a superconducting magnet to determine the magnetic field dependence of the spin-lattice relaxation time T_1. Microwave power was not employed in this experiment. Muramoto (1973) employed the experimental arrangement illustrated in Fig. 14-37 to detect ESR in the optical excited state $\bar{E}(^2E)$ of Cr^{3+} in Al_2O_3 by monitoring the fluorescent light emission. UHF was employed since the very low value $g_\perp = 0.0515$ for the optically excited state caused the ESR resonant condition to be satisfied at a frequency of ~ 500 MHz in a field of about 6 kG. A waveform eductor was employed to

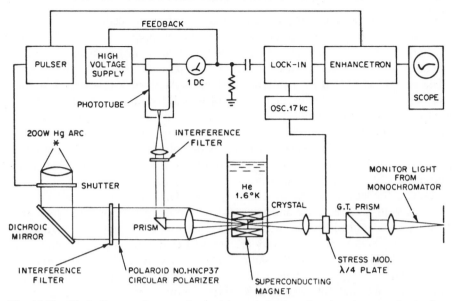

Fig. 14-36. Block diagram of system for studying spin-lattice relaxation via magnetic circular dichroic absorption. Microwave power is not employed (Panepucci and Mollenauer, 1969).

increase the signal-to-noise ratio. Optical techniques can provide relaxation time information (Calviello et al., 1966; Culver et al., 1963; Denison and Fischer, 1969; Fischer and Denison, 1969; Imbusch, 1967; Kiel, 1961, 1963; Robinson and Frosch, 1963; Sims et al., 1970; see review by Geschwind, 1972).

The ODMR experiments described until now involved the measurement of fluorescence in which the emission of light occurs almost instantaneously after the photons are absorbed. Some systems are phosphorescent, i.e. there is an appreciable delay before the light is emitted. It is also of interest to carry out

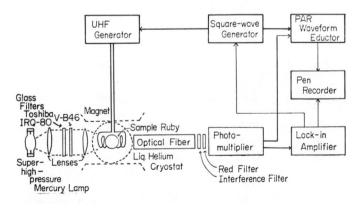

Fig. 14-37. Block diagram of an optical double-resonance system that employs UHF to induce the ESR transitions (Muramoto, 1973).

optical double resonance experiments on such phosphorescent systems, and we will describe an instrumentation system for this purpose.

Vergragt et al. (1977) designed the experimental arrangement illustrated on Fig. 14-38 for the detection of microwave-induced delayed phosphorescence. The sample is placed in a microwave helix located in a cryostat, and it is successively irradiated with microwaves of two different frequencies generated in sweep oscillators labeled 1 and 2 in the figure. The microwaves are frequency-modulated at 100 Hz with a modulation width of 8 MHz by means of a low-frequency sine-wave generator. A dc voltage input at AM operates a PIN diode to turn the microwaves on and off. During the off cycle a dc voltage input at FM changes the microwave frequency to a value off resonance to minimize the leakage of microwaves to the sample. These voltages are timed by a clock with a 12-sec cycle and triggered by a pulse generator. The clock also operates the control for the two shutters. One shutter (1) opens to permit the output of a high-pressure mercury arc lamp filtered by an aqueous solution of $CoSO_4$ and $NiSO_4$ to excite the sample. The other shutter (2) opens to pass the phosphorescent light through a monochrometer to a photomultiplier (PM), where it is detected, followed by a Keithley electrometer, and a signal averager

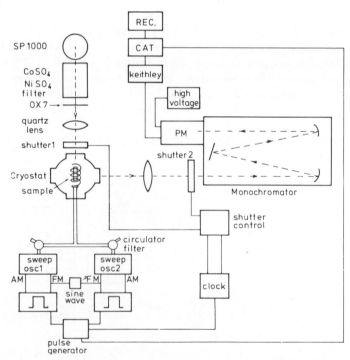

Fig. 14-38. Experimental arrangement for a microwave-induced delayed phosphorescence experiment. The microwaves travel in coaxial lines denoted by double lines, and the light path is indicated by dotted lines (Vergragt et al., 1977).

(CAT) triggered by the pulse generator. Finally, the averaged phosphorescent signal is displayed on a recorder (Rec).

Vertragt et al. (1977) employed this apparatus to study the zero-field levels of the lowest triplet state of p-xylene, which has an overall splitting of 5898 MHz and an E term splitting that provides two additional microwave absorptions at 2577 and 3321 MHz. Saturation of these frequencies caused the phosphorescent intensity to increase, and this increase was found to be independent of the time. Clarke and Hayes (1973) constructed a spectrometer for triplet absorption detection of magnetic resonance (TADMR), and Harris and Hoover (1972) reported a system for the optical detection of adiabatic inversion in phosphorescent triplet states.

Gillies et al. (1979) constructed the nanosecond-resolved optically detected magnetic resonance spectrometer that is presented in Fig. 14-39. Optical excitation is carried out with 10-nsec pulses from a 400-kW nitrogen laser operated at 20 Hz. A PIN diode switch pulses the microwave power from the microwave sweep oscillator with a rise time of 3 nsec and a minimum width of 40 nsec. A driver amplifier followed by a pulsed high-power amplifier provide 1 kW of microwave power in the range from 3.8 to 6.5 GHz. The microwaves are incident on the sample located in a four-turn helix. The phosphorescent output passes through a monochrometer and is detected by a 1P28 photomultiplier (PMT). The signal then passes through a transfer switch to a digitizer with an input bandwidth of 25 MHz and a sampling rate up to 10 nsec per address. Finally, a signal averager is employed to improve the signal-to-noise ratio. The various experimental events are controlled by a homemade timing box, and the event sequences of a typical experiment are indicated in the figure. More recently their spectrometer was improved by the use of a varactor tuned microwave source capable of frequency sweeps as fast as 25 MHz/nsec to provide an ultimate resolution of 10 nsec.

The ODMR-pulsed spectrometer of Breiland et al. (1975), which is sketched in Fig. 14-40, is capable of performing optical spin-echo experiments. It is

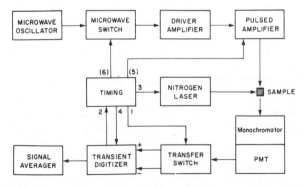

Fig. 14-39. Block diagram of a pulsed ODMR spectrometer capable of nanosecond resolution (Gillies et al., 1979).

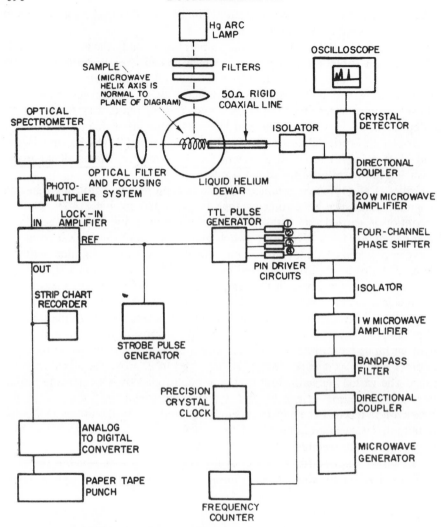

Fig. 14-40.   Block diagram of a pulsed ODMR spectrometer designed for spin-echo operation and spin locking experiments (Breiland et al., 1975).

basically a CW type of spectrometer (Buckley and Harris, 1972) converted to a pulse system by the addition of a pulse generator, a fast microwave switching device and a high-power amplifier. A strobe pulse generator initiates the pulse sequence. The repetition rate depends upon the experiment, and Fig. 14-41 shows how the relevant coherence lifetimes differ for various experiments carried out on 1,2,4,5-tetrachlorobenzene in durene. The final pulse establishes the phosphorescence, which remains in the triplet ensemble, and after the termination of the pulse sequence this intensity decays in the absence of the microwaves. The resulting waveform is time-averaged by a lock-on detector

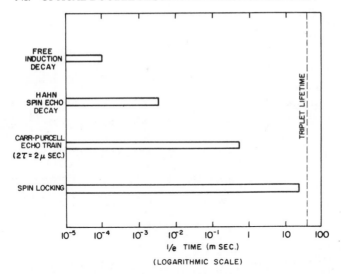

Fig. 14-41. Coherence lifetimes associated with several optically detected magnetic resonance experiments carried out on the triplet state of 1, 2, 4, 5-tetrachlorobenzene in durene (Breiland et al., 1975).

that derives its reference from the strobe pulse generater. The TTL pulse generater contains a digital plus variable delay device capable of producing two pulses with a highly reproducible separation. The strobe pulse opens a gate that admits a train of pulses 1 μsec apart, which originate in the 1-MHz precision crystal clock. The first clock pulse in the train produces a zero time output, and a countercircuit counts the clock pulses until a preset number is reached, when a second output pulse is produced and the gate is shut. The four-channel phase shifter permits either output pulse to be shifted by 0°, 90°, 180°, or 270°. This phase-changing capability is required for various echo-sequence and spin-locking experiments.

Suchard and Bowers (1972) employed the spectrometer system illustrated in Fig. 14-42 to carry out a type of inverse ODMR experiment in which they irradiated a sample of perylene dissolved in sulfuric or fluorosulfuric acid with modulated light and monitored the ESR signal at the modulation frequency. Measurements carried out at different modulation frequencies provided the relaxation rate of the inner solvation shell. Variable-temperature studies yielded information on the outer shells and interaction energies between solvation shells. By measuring these spectra as a function of the wavelength of the irradiating source, ion–electron recombination times in these systems were obtained.

Warden and Bolton (1976) designed the combined optical and ESR kinetic spectrometer illustrated in Fig. 14-43. This spectrometer provides for the simultaneous monitoring of transient optical absorbancy and ESR spectral changes. The optical spectrometer is characterized by a collimated,

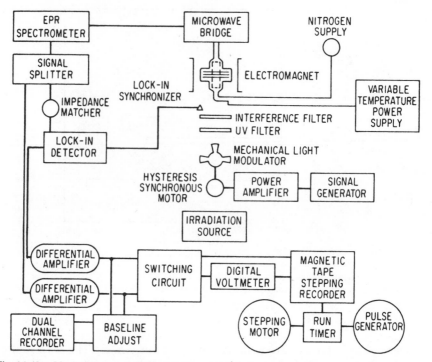

Fig. 14-42. Block diagram of ESR spectrometer designed to irradiate the sample with intensity-modulated light and detect the microwave signal at this modulation frequency (Suchard and Bowers, 1972).

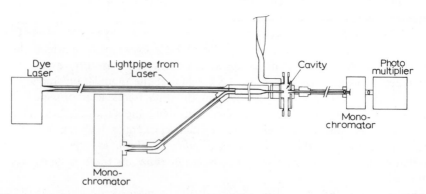

Fig. 14-43. Sketch of the optical arrangement employed in the simultaneous optical ESR spectrometer illustrated in Fig. 14-45. The resonant cavity details are presented in Fig. 14-44 (Warden and Bolton, 1976).

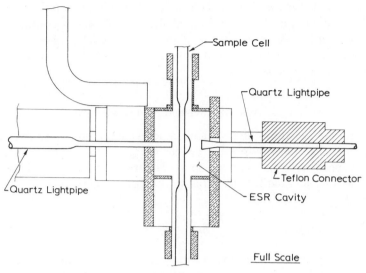

Fig. 14-44.   Details of resonant cavity associated with optical components shown in Fig. 14-43 (Warden and Bolton, 1976).

monochromatic monitoring beam with minimized interference from stray light, an optical range from the ultraviolet to the near infrared, and a photomultiplier protected from the photolyzing or actinic source. A Varian E-232 optical transmission cavity illustrated in Fig. 14-44 constituted the interface of the optical and ESR modes. Light enters and leaves the cavity by means of light pipes, and losses in the optical train are about 50%. The block diagram presented in Fig. 14-45 shows the light paths of the photolyzing and monitor-

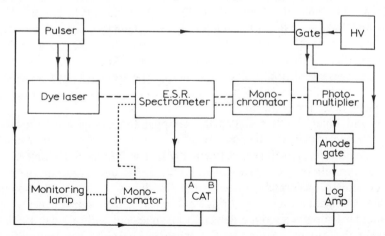

Fig. 14-45.   Block diagram of simultaneous optical ESR spectrometer showing the light paths of the photolyzing (− − −) and monitoring (---) beams as well as the electrical connections (—) between the components (Warden and Bolton, 1976).

ing beams and also the electrical connections between the various components. Warden and Bolton (1972, 1973) used their combined spectrometer for the study of transient intermediates in plant and bacterial photosynthesis. They found a direct correlation between the 865-nm optical absorption and the ESR biphasic kinetic responses.

J. Acoustic Electron Spin Resonance

In an ordinary electron spin resonance experiment, the spins absorb microwave energy and then relax by passing on this energy to the lattice vibrations that are "on speaking terms" with the spins. The inverse process is also possible if one irradiates the sample with ultrasonic energy at the resonant frequency ω_0, where

$$h\omega_0 = g\beta H \tag{1}$$

The effect of ultrasonic energy on magnetic resonance absorption was first observed indirectly by its disturbing influence on the resonance absorption in NMR (Proctor and Tanttila, 1955, 1956). The first direct acoustic excitation of nuclear spins was made by means of an ultrasonic continuous wave (CW) resonance technique in which a marginal rf oscillator was locked in to a high-Q mechanical resonator (Menes and Bolef, 1958, 1961).

The older literature on ultrasonic ESR is reviewed in Sec. 19-F of the first edition of this work. In this section we described the acoustic ESR CW spectrometer designed by Bolef et al. (1962). This instrument operated from 10 to 1000 MHz at temperatures between 1.5 and 300 K. In addition, that section mentioned a number of works that dealt with experimental aspects of acoustic ESR. These and some more recent references are (Baranskii, 1957; Bömmel and Dransfeld, 1960; de Klerk, 1964; Denison et al., 1964; Dobbs, 1975; Dobrov and Browne, 1962, 1963; Dorland, 1963; Ganapol'skii and Chernets, 1964; Jacobsen et al., 1959; Joffrin, 1971; Lewiner, 1972; Lewis, 1965; Pomerantz, 1965; Shiren and Tucker, 1961; Tucker, 1961; Wetsel and Donoho, 1965). The present section will discuss newer spectrometers and associated apparatus and techniques. DuVarney et al. (1981) reported photothermal and optoacoustic detection of ESR.

Miller and Bolef (1969) designed the sampled continuous wave ultrasonic ESR spectrometer that has the block diagram sketched in Fig. 14-46. (see also Bolef and Miller, 1971; Miller and Bolef, 1970). The frequency-locking section shown on the left of the figure ensures that the frequency of the gated rf oscillator remains accurately tuned to the standing-wave acoustic resonance throughout the gating cycle. The bridge and receiver section in the center is a fairly standard superheterodyne detection arrangement with a local oscillator to provide the mixing frequency. For strong signals the spectrum can be recorded directly after passage through a differential voltmeter. Weak signals

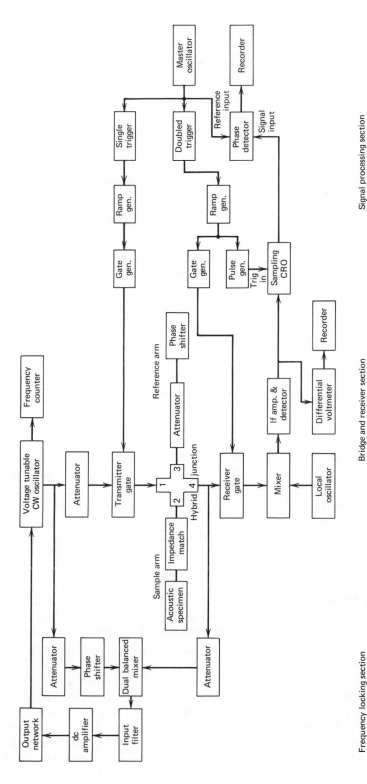

Fig. 14-46. Ultrasonic ESR spectrometer showing the frequency-locking components on the left, the bridge and receiver in the center, and the signal processing and gating circuits on the right (Miller and Bolef, 1969).

701

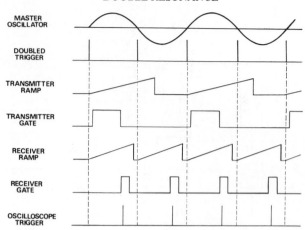

Fig. 14-47. Timing sequence of the signal processing section of the ultrasonic ESR spectrometer presented on Fig. 14-46. Certain time intervals (e.g., "receiver gate") are exaggerated for the sake of clarity (Miller and Bolef, 1969).

are passed through the signal processing section on the right side of the figure where the gating signals, lock-in detector reference and sampling oscilloscope trigger originate in the master oscillator. The oscillator, trigger, ramps, and gate waveforms of the timing sequence are sketched in Fig. 14-47. Miller and Bolef (1969) presented a scheme for frequency locking the CW spectrometer to the standing-wave acoustic resonance of the specimen.

A CW-type ultrasonic spectrometer is inherently monochromatic and can be employed for examining arbitrarily thin specimens (Miller and Bolef, 1968). A disadvantage of a CW spectrometer is the susceptibility to cross-talk, since the transmitter and receiver operate simultaneously. Spin-echo techniques eliminate this cross-talk, but have other disadvantages (Miller and Bolef, 1969). Their lack of monochromaticity causes inhomogeneous broadening of narrow resonant lines, and when thin samples are employed the pulse width cannot be made sufficiently narrow without rendering the carrier frequency undefined. Some workers have modified the ultrasonic pulse echo technique by incorporating into it degrees of monochromaticity and phase coherence (Alers, 1966; Blume, 1963; McSkimin, 1964, 1965; Smith, 1967). This produces a basically pulse spectrometer with some added CW features. The review by Bolef and Miller (1971) compares the sensitivities of the CW, the sampled CW, and the pulse ultrasonic techniques, and it also discusses frequency-stabilization methods for ultrasonic ESR.

Ruby (1969) designed an X-band acoustic transmission probe for use in CW acoustic spectrometers. Figure 14-48 shows the microwave ultrasonic resonator arrangement, and Fig. 14-49 presents a larger view of the low-temperature probe assembly. The specimen under study is placed between two identical transmitting and receiving reentrant cavity resonators (see Fig. 5-22), and the ultrasonic vibrations in the sample are coupled to the microwave fields of the

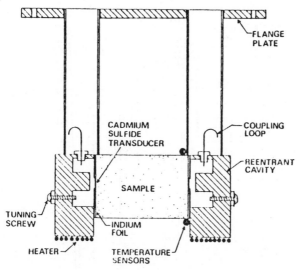

Fig. 14-48. Microwave X-band ultrasonic resonator arrangement showing the sample mounted between two identical reentrant cavity resonators. Figure 14-49 shows the resonators in place on the spectrometer (Ruby, 1969; see also Bolef and Miller, 1971).

cavities through CdS piezoelectric transducers. Adjustable Teflon blocks above the cavities provide the impedance matching.

In strain-modulated electron spin resonance (SMESR), which has also been referred to as ultrasonically modulated electron resonance (UMER), the internal crystal field at the site of the paramagnetic center is modulated by submitting the sample to an alternating mechanical stress (Collins et al., 1971, 1971, 1972; Devine and Robinson, 1977; Henning et al., 1973, 1974, 1974, 1975, 1975, 1976). In 1974 they made the suggestion that the use of an extensional strain mode modulates mainly the D zero-field term, while a flexural mode modulates the off-diagonal elements of the g tensor.

Robinson et al. (1974) strain-modulated spectrometer is shown in Fig. 14-50. This system consists of an ESR spectrometer that can be switched to detect signals with either magnetic field or strain modulation. For the latter application it contains an ultrasonic strain drive unit that applies a sinusoidal strain modulation to the sample. The spectrometer contains switches to turn off the magnetic field modulation, to change the phase-sensitive detector reference from the field modulation to the strain gauge modulation source, and to select between a 100-kHz filter for magnetic field modulation and a 40-kHz filter for strain modulation. The ultrasonic modulation is generated by a piezoelectric composite oscillator excited by a strain drive. In this arrangement the sample crystal forms as integral part of the oscillator.

The strain-modulated ESR spectrometer designed by den Boef and Henning (1974) (Henning et al., 1974) generates flexural vibrations by the trilaminar trausducer sketched in Fig. 14-51. It contains two rectangular piezoxide bars

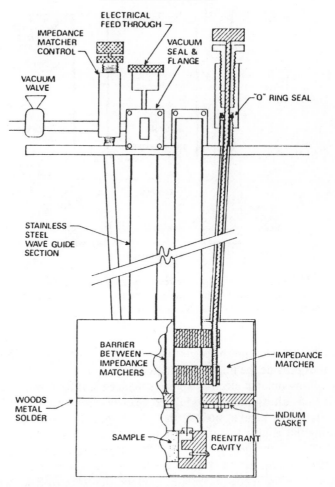

Fig. 14-49. Microwave ultrasonic low temperature probe that incorporates the resonator arrangement presented in Fig. 14-48 (Ruby, 1969; see also Bolef and Miller, 1971).

B_1 and B_2 with opposite bias polarizations cemented on a passive intermediate layer A. A flexural mode is produced by applying alternating voltages of opposite phase to make the bar B_1 elongate when B_2 contracts, and vice versa. The stress field is transferred to a rectangular plate C, which forms part of the microwave cavity wall as shown in Fig. 14-52. The sample is cemented to the center of this plate, and an indium ring is employed to prevent microwaves from leaking out of the cavity. To operate the spectrometer the sample is strain-modulated, and a lock-in detector is used to detect the signal at the first harmonic of the modulation frequency. Second harmonic operation has also been employed.

The review by Bolef and Miller (1971) describes a frequency-modulated transmission spectrometer (Melcher et al., 1968) for measuring very small

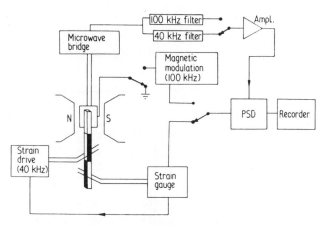

Fig. 14-50. Block diagram of an ultrasonic or strain-modulated ESR spectrometer. The system can be switched to detect signals using either strain or magnetic field modulation (Robinson et al., 1974).

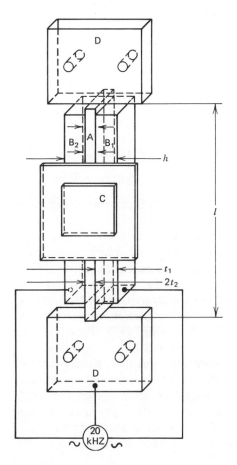

Fig. 14-51. The trilaminar transducer. B_1 and B_2 are piezoxide bars with opposite remanent polarizations (indicated by arrows). Intermediate strip A, end blocks D and tablet C are machined from a single block of molybdenum or tantalum. The transducer is mounted on the resonant cavity in the manner illustrated on Fig. 14-52 (den Boef and Henning, 1974).

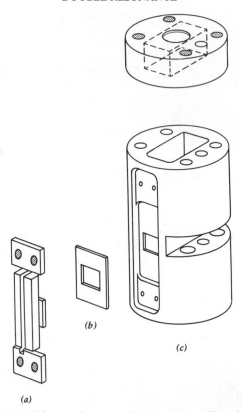

Fig. 14-52.   Exploded view of the transducer + cavity assembly. (*a*) Transducer; (*b*) indium ring; (*c*) cavity. Figure 14-51 shows details of the transducer (den Boef and Henning, 1974).

changes in velocity, a cw instrument for automatically measuring the acoustic velocity (Leisure et al., 1968, 1969), a sensitive transmission spectrometer (Melcher et al., 1968) for measuring absorption and dispersion, a bridge-type microwave transmission spectrometer (Ruby, 1969), and marginal oscillator ultrasonic spectrometers (MOUS) (Miller et al., 1971; Smith et al., 1969, 1971).

Experimental techniques have been developed for carrying out double acoustic magnetic resonance (Golenishchev-Kutuzov et al., 1969, 1970; Saburova, 1971, Shamukov et al., 1970). Ultrasonic measurements can provide relaxation-time information (Bates et al., 1973; Mitin, 1975; Passini et al., 1978).

A number of studies have been made using ultrasonic or strain modulation (e.g., Al'tshuler, 1952, 1955; Al'tshuler et al., 1957, 1960, 1964; Antokol'skii et al., 1975; Averbuch and Proctor, 1963; Bersuker et al., 1975; Brown and Bolef, 1979; Fletcher and Stevens, 1969; Guermeur et al., 1965; Henning and den Boef, 1975, 1976; Jacobsen and Stevens, 1963; Kopvillem and Mineeva, 1963; Lange, 1975; Lewiner et al., 1970; Meyer et al., 1966; Robinson et al., 1975;

1977; Shiren, 1963, 1970; Yuhas et al., 1976, 1978). Yuhas et al. (1975) carried out a nonresonant ($\Delta M_s = 0$) acoustic investigation of Fe^{2+} in MgO. Evora et al. (1980) and Melcher (1982) employed photoacoustic techniques for the thermal detection of the resonant absorption.

References

Abragam, A., and B. Bleaney, *EPR of Transition Ions*, Oxford Univ. Press, London, 1970.

Abraham, M. M., Y. Chen, and J. O. Rubio, *Phys. Rev. B* **14**, 2603 (1976).

Abraham, M. M., Y. Chen, and W. P. Unruh, *Phys. Rev. B* **9**, 1842 (1974); **12**, B4766 (1975).

Abraham, M. M., Y. Chen, L. A. Boatner, and R. W. Reynolds, *Solid State Comm.* **16**, 1209 (1975); *Phys. Rev. Lett.* **37**, 849 (1976).

Abraham, M. M., W. P. Unruh, and Y. Chen, *Phys. Rev. B* **10**, 3540 (1974).

Adams, S. M., and H. C. Box, *J. Chem. Phys.* **63**, 1185 (1975).

Adrian, F. J., *Chem. Phys. Lett.* **10**, 70 (1971).

Adrian, F. J., *J. Chem. Phys.* **53**, 3374 (1970); **54**, 3918 (1971); **57**, 5107 (1972).

Aleksandrov, E. B., and V. S. Zapasskii, *Sov. Phys. Solid State* **19**, 1802 (1977).

Alers, G. A., in *Physical Acoustics*, W. P. Mason, Ed., Academic Press, New York, Vol. 4A, 1966.

Alexander, C., Jr., and H. L. Pugh, Jr., *J. Magn. Reson.* **18**, 185 (1975).

Allendoerfer, R. D. *J. Magn. Reson.* **9**, 226 (1973).

Allendoerfer, R. D., *Int. Rev. Sci. Phys. Chem. Ser.* **2**, 29 (1975).

Allendoerfer, R. D., and R. Chang, *J. Magn. Reson.* **5**, 273 (1971).

Allendoerfer, R. D., and D. J. Eustace, *J. Phys. Chem.* **75**, 2765 (1971).

Allendoerfer, R. D., and R. J. Papez, *J. Am. Chem. Soc.* **92**, 6971 (1970).

Al'tshuler, S. A., *Dokl. Akad. Nauk. SSSR* **85**, 1235 (1952); *Zh. Eksper. Teor. Fiz.* **28**, 49 (1955).

Al'tshuler, S. A., and Sh. Sh. Bashkirov, *Conf. Paramagn. Res. Kazan*, 78 (1960).

Al'tshuler, S. A., and B. M. Koxyrev, *Electron Paramagnetic Resonance*, Trans. by Scripta Technica, C. P. Poole, Jr., Ed., Academic Press, New York, 1964.

Al'tshuler, S. A., M. M. Zaripov, and L. Y. Shekun, *Izv. Akad. Nauk. Ser. Fiz.* **21**, 844 (1957).

Andreev, V. A., and V. I. Sugakov, *Ukr. Fiz. Zh.* **29**, 1998 (1975).

Andreev, V. A., and V. I. Sugakov, *Sov. Phys. Solid State* **17**, 1285 (1975).

Andrist, M., K. Loth, and Hs. H. Gunthard, *Lab. for Phys. Chem.* (*Switz.*) (1976).

Annabi, M., and J. G. Theobald, *C. R. Hebd. Seances Sci. B*2, **31** (1972).

Antokol'skii, G. O., V. S. Baranov, and E. Tolins, *Pis'ma Zh. Eksp. Teor. Fiz.* **23**, 685 (1975).

Apaydin, F., *Rev. Fac. Sci. Univ. Istanbul C* **36**(1-4), 19 (1971).

Atherton, N. M., *Electron Spin Reson.* **3**, 23 (1976).

Atherton, N. M., A. J. Blackhurst, and I. P. Cook, *Chem. Phys. Lett.* **8**, 187 (1971).

Atherton, N. M., and B. Day, *Mol. Phys.* **29**, 325 (1976).

Atherton, N. M., and P. A. Henshaw, *J. Chem. Soc.* (*Perkins Trans.*) **2**, 258 (1975).

Atkins, P. W., *Electron Spin Res.* **3**, 35 (1976).

Atkins, P. W., L. C. Buchanan, R. C. Gord, K. A. McLauchlan, and A. F. Simpson, *Chem. Commun.*, 513 (1970).

Atkins, P. W., and G. T. Evans, *Chem. Phys. Lett.* **25**, 108 (1974).

Atkins, P. W., and G. T. Evans, *Mol. Phys.* **27**, 1633 (1974).

Atkins, P. W., R. C. Gord, K. A. McLauchlan, and A. F. Simpson, *Chem. Phys. Lett.* **8**, 55 (1971).

Atkins, P. W., K. A. McLauchlan, and P. W. Percival, *Mol. Phys.* **26**, 281 (1973).

Atkins, P. W., and K. A. McLauchlan, in *Chemically Induced Magnetic Polarization*, A. R. Lepley and G. L. Closs, Eds., Wiley-Interscience, New York, 1973.

Averbuch, P., and W. G. Proctor, *Phys. Lett.* **4**, 221 (1962).

Ayscough, P. B., T. H. English, G. Lambert, and A. J. Elliot, *Chem. Phys. Lett.* **34**, 557 (1975)

Ayscough, P. B., G. Lambert, and A. J. Elliot, *J. Chem. Soc. Faraday Trans.* I **72**, 1770 (1976).

Baker, J. M., and W. B. J. Blake, *J. Phys. C Solid Stat Phys.* **6**, 3501 (1973).

Baker, J. M., and W. B. J. Blake, *Phys. Lett. A* **31**, 61 (1970).

Baker, J. M., and F. I. B. Williams, *Proc. R. Soc. London A* **267**, 283 (1962).

Baldin, V. I., and A. P. Stepanov, *Prib. Tekh. Eksp.* **19**(5), 168 (1976).

Bales, B. L., J. Helbert, and L. Kevan, *J. Phys. Chem.* **78**, 221 (1974).

Balzer, R., and F. R. Engler, *Phys. Status Solidi B* **74**, K135 (1976).

Baranskii, K. A., *Dokl. Akad. Nauk. SSSR* **114**, 517 (1957); translation in *Sov. Phys. Dokl.* **2**, 237 (1958).

Barnes, D. J., *Nature* **200**, 253 (1963).

Bates, C. W., Jr., A. Salau, and D. Leniart, *Phys. Rev. B* **15**, 5963 (1977).

Bates, C. A., *J. Phys. C, Solid State Phys.* **11**, 3447 (1978).

Bates, C. A., S. C. Jones, V. W. Rampton, and P. Steggles, *XVII Colloque Ampère*, V. Hovi Ed., North Holland, 1973.

Bates, R. D., Jr., *J. Chem. Phys.* **66**, 1759 (1977).

Bates, R. D., B. E. Wagner, and E. H. Poindexter, *J. Phys. Chem.* **80**, 320 (1976).

Baxter, J. E., and G. Ascarelli, *Phys. Rev. B* **7**, 2630 (1973).

Benderskii, V. A., L. A. Bluenfeld, P. A. Stunzas, and E. A. Sokolov, *Nature* **220**, 365 (1968).

Bendt, P. J., *Phys. Rev. B* **2**, 4366, 4375 (1970).

Berkovits, V. L., C. Hermann, G. Lampel, A. Nakamura, and V. I. Safarov, *Phys. Rev. B*, 1767 (1978).

Bernhard, W. A., J. Huttermann, A. Muller, D. M. Close, and G. W. Fouse, *Radiat. Res.* **68**, 390 (1976).

Bersuker, I. B., B. J. Vekhter, and I. Ya. Ogurtsov, *Usp. Fiz. Nauk* **116**, 605 (1975).

Biehl, R. M., private comm.

Biehl, R., M. Plato, and H. Möbius, *J. Chem. Phys.* **63**, 3515 (1975).

Biehl, R., W. Lubitz, K. Möbius, and M. Plato, *J. Chem. Phys.* **66**, 2079 (1977).

Biehl, R., M. Plato, and K. Möbius, *J. Chem. Phys.* **63**, 3515 (1975).

Bjkharaev, A. A., and N. R. Yafaev, *Zh. Prikl. Spektrosk.* **24**, 727 (1976).

Blumberg, W. E., *Phys. Rev.* **110**, 1842 (1960).

Blumberg, W. E., and G. Feher, *Bull. Am. Phys. Soc.* **5**, 183 (1960).

Blume, R. J., *RSI* **34**, 1400 (1963).

Boldu O., J. L., H. J. Stapleton, Y. Chen, and M. M. Abraham, *Phys. Rev. B* **16**, 3875 (1977).

Bolef, D. I., J. De Klerk, and R. B. Gosser, *RSI* **33**, 631 (1962).

Bolef, D. I., and J. G. Miller, *Phys. Acoust.* **8** (1971).

Bömmel, H. E., and K. Dransfeld, *Phys. Rev.* **117**, 1245 (1960).

Bontemps-Moreau, N., A. C. Boccara, and P. Thibault, *Semiconductors and Isulators* **3**, 165 (1978).

Borcherts, R. H., and L. L. Lohr, *J. Chem. Phys.* **50**, 5262 (1969).

Borg, D. C., A. Forman, and J. Fajer, *J. Am. Chem. Soc.* **98**, 6889 (1976).

Botter, B. J., C. J. Nonhof, J. Schmidt, and J. H. van der Waals, *Chem. Phys. Lett.* **43**, 210 (1976).

Bowers, K. D., and W. B. Mims, *Phys. Rev.* **115**, 285 (1959).

Box, H. C., and E. E. Budzinski, *J. Chem. Phys.* **55**, 2446 (1971); **62**, 197 (1975); **64**, 1593 (1976); **67**, 4726 (1977); **70**, 1572 (1979).

Box, H. C., E. E. Budzinski, and H. C. Freund, *J. Chem. Phys.* **61**, 2222 (1974).

Box, H. C., E. E. Budzinski, and K. T. Lilga, *J. Chem. Phys.* **57**, 4295 (1972); **64**, 4495 (1976).

Box, H. C., E. E. Budzinski, and W. R. Potter, *J. Chem. Phys.* **61**, 1136 (1974).

Box, H. C., H. G. Freund, and K. T. Lilga, *J. Chem. Phys.* **46**, 2130 (1967).

Box, H. C., G. Potienko, and E. E. Budzinski, *J. Chem. Phys.* **66**, 342 (1977).

Box, H. C., W. R. Potter and E. E. Budzinski, *J. Chem. Phys.* **62**, 3476 (1975).

Breiland, W. G., T. E. Altman, and J. S. Voris, *Chem. Phys. Lett.* **67**, 30 (1979).

Breiland, W. G., H. C. Brenner, and C. B. Harris, *J. Chem. Phys.* **62**, 3458 (1975).

Breiland, W. G., C. B. Harris, and A. Pines, *Phys. Rev. Lett.* **30**, 158 (1973).

Breiland, W. G., and M. C. Saylor, *J. Chem. Phys.* **72**, 6485 (1980).

Brenner, H. C., *J. Chem. Phys.* **67**, 4719 (1977).

Brenner, H. C., J. C. Brock, and C. B. Harris, *J. Chem. Phys.* **60**, 4448 (1974).

Brenner, H. C., C. A. Hutchinson, Jr., and M. D. Kemple, *J. Chem. Phys.* **60**, 2180 (1974).

Brik, A. B., and S. S. Ishchenki, *Fiz. Tverd. Tela* **18**, 2442 (1976).

Bringer, A., M. Campagna, R. Feder, W. Gudat, E. Kisker, and E. Kullman, *Phys. Rev. Lett.* **42**, 1705 (1979).

Brodbeck, C. N., H. H. Niebuhr, and S. Lee, *Phys. Rev. B* **5**, 19 (1972).

Brown, J., and D. I. Bolef, *Phys. Rev. B* **19**, 5849 (1979).

Brown, I. M., and D. J. Sloop, *RSI* **41**, 1774 (1970).

Bruno, G. V., and J. H. Freed, *Chem. Phys. Lett.* **25**, 328 (1974).

Bubachenko, A. L., *Khimicheskaya Polirarizataiya Electronov*, Yader, Moscow, 1974.

Buckley, M. J., and C. B. Harris, *J. Chem. Phys.* **56**, 137 (1972).

Budzinski, E. E., and H. C. Box, *J. Chem. Phys.* **62**, 2006 (1975); **63**, 4927 (1975); **68**, 5296 (1978).

Bugai, A. A., and A. B. Roitsin, *Sov. Phys. JETP Lett.* **5**, 67 (1967).

Buishvili, L. L., T. G. Vardosanidze, and A. I. Ugulava, *Fiz. Tverdogo Tela* **18**, 558 (1976).

Calviello, J. A., E. W. Fisher, and Z. H. Heller, *J. Appl. Phys.* **37**, 3156 (1966).

Caplan, P. J., J. N. Helbert, B. E. Wagner, and E. H. Poindexter, *Surf. Sci.* **54**, 33 (1976).

Carver, R., and C. P. Slichter, *Phys. Rev.* **92**, 212 (1953); **102**, 975 (1956).

Castleman, B. W., and G. C. Moulton, *J. Chem. Phys.* **57**, 2762 (1972).

Chacko, V. P., C. A. McDowell, and B. C. Singh, *Mol. Phys.* **38**, 321 (1979).

Chase, L. L., *Phys. Rev. B* **2**, 2308 (1970).

Chen, Y., M. M. Abraham, L. C. Templeton, and W. P. Unruh, *Phys. Rev.* **B11**, 881 (1975).

Cheng, J. C., and J. C. Kemp, *Phys. Rev. B* **4**, 2841 (1971).

Chiarini, F., M. Martinelli, L. Pardi, and S. Santucci, *Phys. Rev. B* **12**, 847 (1975).

Chiarini, F., M. Maritnelli, P. A. Rolla, and S. Santucci, *Lett. Nuovo Cimento* **5**, Ser. 2, 197 (1972).

Ching, L. K., and H. C. Box, *J. Chem. Phys.* **67**, 2811 (1977).

Chong, T., and M. Itoh, *J. Phys. Soc. Japan* **35**, 518 (1973).

Chu, Y. H., and R. L. Mieher, *Phys. Rev.* **188**, 1311 (1969).

Clarke, R. H., and J. M. Hayes, *J. Chem. Phys.* **59**, 3113 (1973).

Clarke, R. H., Ed., *Triplet State ODMR Spectroscopy*, Wiley-Interscience, New York, 1982.

Clarke, R. H., and R. H. Hofeldt, *J. Chem. Phys.* **61**, 4582 (1974).

Clarke, R. H., C. A. Hutchison, *J. Chem. Phys.* **54**, 2962 (1971).

Close, D. M., and W. A. Bernhard, *J. Chem. Phys.* **70**, 210 (1979).

Close, D. M., G. W. Fouse, and W. A. Bernhard, *J. Chem. Phys.* **66**, 1534, 4689 (1977).

Close, D. M., G. W. Fouse, W. A. Bernhard, and R. S. Anderson, *J. Chem. Phys.* **70**, 2131 (1979).

Closs, G. L., *Adv. Magn. Reson.* **7**, (1975).

Clough, S., J. R. Hill, and M. Punkkinen, *J. Phys. C* **7**, 3413, 3779 (1974).

Clough, S., and J. Hobson, *J. Phys. C* **8**, 1745 (1975).

Colligiani, A., C. Pinzino, M. Brustolon, and C. Corvaja, *J. Magn. Reson.* **32**, 419 (1978).

Collins, M. A., S. Devine, R. A. Hoffman, and W. H. Robinson, *Abstracts of the 4th International Symposium on Magnetic Resonance*, Rehovot, Israel, 1971.

Collins, M. A., S. D. Devine, R. A. Hoffman, and W. H. Robinson, *J. Phys. C* **4**, 416 (1971); *J. Magn. Res.* **6**, 376 (1972).

Cook, R. J., *JSI* **43**, 548 (1966).

Cook, R. J., and D. H. Whiffen, *Proc. Phys. Soc.* **84**, 845 (1964); *J. Chem. Phys.* **43**, 2908 (1965).

Coope, J. A. R., N. S. Dalal, C. A. McDowell, and R. Spinivasan, *Mol. Phys.* **24**, 403 (1972).

Cova, S. and A. Longoni, *Analytical Laser Spectroscopy*, Dr. Nicolo Omenotto, Ed., Wiley, New York, 1979.

Cox, P., G. W. Flynn, and E. B. Wilson, Jr., *J. Chem. Phys.* **42**, 3094 (1965).

Culver, W. H., R. A. Satten, and C. R. Viswanathan, *J. Chem. Phys.* **38**, 775 (1963).

Curl, R. F., Jr., and T. Oka, *J. Chem. Phys.* **38**, 4908 (1973).

Dalal, N. S. in *Fourier Hadamard and Hilbert Transforms in Chemistry*, A. G. Marshall, Ed., Plenum Press, New York, 1982, p. 327, Chap. XI.

Dalal, N. S., D. E. Kennedy, and C. A. McDowell, *Ber. Bunsenges, Phys. Chem.* **78**, 4 (1974); *J. Chem. Phys.* **59**, 3403 (1973); **61**, 1698 (1974).

Dalal, N. S., and A. Manoogian, *Phys. Rev. Lett.* **39**, 1573 (1977).

Dalal, N. S., and C. A. McDowell, *Chem. Phys. Lett.* **6**, 617 (1970); *Phys. Rev. B* **5**, 1074 (1972).

Dalal, N. S., C. A. McDowell, and J. M. Park, *J. Chem. Phys.* **63**, 1856 (1975).

Dalal, N. S., C. A. McDowell, and R. Srinivasan, *Mol. Phys.* **24**, 417 (1972).

Dalton, L. R., *Magn. Reson. Rev.* **1**, 301 (1972).

Dalton, L. R., and L. A. Dalton, *Magn. Reson. Rev.* **2**, 361 (1973).

Dalton, L. R., and A. L. Kwiran, *J. Am. Chem. Soc.* **94**, 6903 (1972).

Dalton, L. R., and A. L. Kwiran, *J. Chem. Phys.* **57**, 1132 (1972).

David, C., A. Demiddeleer Proumen, and G. Geuskens, *Radiat. Phys. Chem.* **11**, 63 (1978).

Davis, J. L., and W. B. Mims, *RSI*, **49**, 1095 (1978).

de Beer, R., R. Catterjee, and R. P. J. Merks, *J. Phys. C Solid State Phys.* **9**, 1539 (1976).

Decomps, B., M. Dumont, M. Ducloy, and H. Walther, *Laser Spectroscopy of Atoms and Molecules*, Berlin, Germany, 1976, p. 283.

De Klerk, J., *Ultrasonics* **2**, 137 (1964).

den Boef, J. H., and J. C. M. Henning, *RSI* **45**, 1199 (1974).

den Boef, J. H., J. C. M., Henning, P. S. Allen, and E. P. Andrew, *Proc 18th Ampere Congr. Magn. Reson. Relat. Phenom.* **1**, 197 (1974).

Denison, A. B., *Magn. Reson. Rev.* **2**, 1 (1975).

Denison, A. B., L. W. James, J. D. Currin, W. H. Tanttila, and R. J. Mahler, *Phys. Rev. Lett.* **12**, 244 (1964).

Denison, A. B. and P. H. H. Fischer, *XV-th Colloque Ampere*, North Holland, Amsterdam, 1969.

Derouane, E. G., and J. C. Vedrine, *Chem. Phys. lett.* **29**, 222 (1974).

Devine, S. D., and W. H. Robinson, *J. Appl. Phys.* **48**, 1437 (1977).

Dietrich, W., F. Dressler, D. Schmid, and H. C. Wolf, *Z. Naturforsch.* **28a**, 213 (1973).

Dietrich, W., and D. Schmidt, *Phys. Status. Solidi* **74**, 609 (1976).

Dietrich, W., L. Schmidt, and D. Schmid, *Chem. Phys. Lett.* **28**, 249 (1974).

Dinse, K. P., R. Biehl, and K. Mobius, *J. Chem. Phys.* **61**, 4335 (1974).

Dinse, K. P., I. Mobius, and R. Biehl, *Z. Naturforsch.* **28a**, 1069 (1973).

Dobbs, A. J., *Mol. Phys.* **30**, 1073 (1975).

Dobrov, I., and M. E. Browne, *Proc. 11th Colloq. Ampere*, Eindhoven, 1962, p. 129; *Paramagn. Reson.* **2**, 447 (1963).

Doetschman, D. C., and C. A. Hutchison, Jr., *J. Chem. Phys.* **56**, 3964 (1972).

Dorio, M. M., *Magn. Reson. Rev.* **4**, 105 (1977).

Dorland, M., *J. Phys. Suppl. No. 10* **24**, 191A (1963).

Doyle, T., *RSI* **33**, 118 (1962); *Phys. Rev.* **13**, 555 (1963).

DuVarney, R. C., A. K. Garrison, and G. Busse, *Appl. Phys. Lett.* **38**, 675 (1981).

DuVarney, R. C., A. K. Garrison, R. H. Thorland, *Phys. Rev.* **188**, 657 (1969).

Eastman, M. P., G. V. Bruno, and J. H. Freed, *J. Chem. Phys.* **52**, 321 (1970).

Edgar, A., *J. Phys. E Sci. Instrum.* **10**, 1261 (1977).

Eiben, K., and R. W. Fessenden, *J. Phys. Chem.* **75**, 1186 (1971).

Eisinger, J., and G. Feher, *Phys. Rev.* **109**, 1172 (1958).

Eliav, U., and H. Levanon, *Chem. Phys. Lett.* **36**, 377 (1975).

El-Sayed, M. A., *Izv Akad. Nauk SSSR Ser. Fiz.* **37**, 248 (1973).

El-Sayed, M. A., W. R. Moomaw, and J. B. Chodak, *Chem. Phys. Lett.* **20**, 213, 825, 4321 (1973).

El-Sayed, M. A., and J. Olusted, *Chem. Phys. Lett.* **11**, 568 (1971).

Ensign, T. C., and N. E. Byer, *Phys. Rev. B* **7**, 907 (1973).

Evenson, K. M., D. A. Jennings, P. R. Peterson, J. A. Mucha, J. J. Jimenfz, R. M. Charlton, and C. J. Howard, *IEEE J. Quantum Electron.* **QE-13**, 442 (1977).

Evora, C., R. Landers, and H. Vargas, *Appl. Phys. Lett.* **36**, 864 (1980).

Fainstein, and S. B. Oseroff, *RSI* **42**, 547 (1971).

Farach, H. A., and H. Teitelbaum, *Can. J. Phys.* **45**, 2913 (1967).

Fayer, M. D., and C. B. Harris, *Chem. Phys. Lett.* **25**, 149 (1974).

Fedders, P. A., *Phys. Rev. B* **5**, 181 (1972).

Feher, G., *Phys. Rev.* **103**, 500, 834 (1956); **114**, 1219 (1959); *Phys. Rev. Lett.* **3**, 135 (1959).

Feher, G., and R. A. Isaacson, *J. Magn. Reson.* **7**, 111 (1972).

Feher, G., and E. A. Gere, *Phys. Rev.* **103**, 501 (1956); **114**, 1245 (1959).

Fessenden, F. W., *J. Chem. Phys.* **58**, 2489 (1973).

Fessenden, R. W., and R. H. Schuler, *J. Chem. Phys.* **39**, 2197 (1963).

Finch, L. L., J. E. Johnson, G. C. Moulton, *J. Chem. Phys.* **70**, 3662 (1979).

Fletcher, J. R., and K. Stevens, *J. Phys. C* **2**, 444 (1969).

Fischer, H., *Chem. Phys. Lett.* **4**, 611 (1970).

Fischer, P. H. H., and A. B. Denison, *Mol. Phys.* **17**, 297 (1969).

Forrer, J., A. Schweiger, N. Berchten, and H. H. Günthard, *J. Phys. E Sci. Instrum.* **14**, 565 (1981).

Forrer, J., A. Schweiger, and H. H. Günthard, *J. Phys. E Sci. Instru.* **10**, 473 (1977).

Forster, M., U. P. Fringeli, and Hs. Hs. Gunthard, *Helv. Chem. Acta* **56**, 389 (1973).

Franke, M., and W. Windsch, *Ann. Phys.* **33**, 393 (1976).

Freed, J. H., *Electron Spin Relaxation in Liquids*, L. T. Muus and P. W. Atkins, Eds., Plenum, NY, 1972, pp. 503-530.

Freed, J. H., D. S. Leniart, and J. S. Hyde, *J. Chem. Phys.* **47**, 4762 (1967).

Freed, J. H., D. S. Leniart, and H. D. Connor, *J. Chem. Phys.* **58**, 3089 (1973).

Freed, J. H., and J. B. Pederson, *Adv. Magn. Reson.* **8**, 1 (1976).

Freund, R. S., and T. A. Miller, *J. Chem. Phys.* **56**, 2211 (1972).

Fukuda, Y., Y. Takagi, and T. Hashi, *Phys. Lett.* **48A**, 183 (1974).

Fukuda, Y., Y. Takagi, K. Yamada, and T. Hashi, *J. Phys. Soc. Japan Lett.* **42**, 1061 (1977).

Furrer, R., and F. Seiff, *Z. Phys. B* **29**, 189 (1978).

Gaillard, J., and P. Gloux, *J. Phys.* **37**, 407 (1976).

Gamble, W. L., L. A. Dalton, L. R. Dalton, and A. L. Kwiram, *Chem. Phys. Lett.* **34**, 565 (1979).

Ganapol'skii, and A. N. Chernets, *Zh. Eksper. Teor. Fiz.* **47**, 1677 (1964).

Gazzinelli, R., G. M. Ribeiro, and M. L. DeSiqueira, *Solid State Commun.* **13**, 1131 (1973).

Gazzinelli, R., and R. L. Mieher, *Phys. Rev.* **175**, 395 (1968).

Geifman, I. N., M. D. Glinchyk, and B. K. Krylikovskii, *Zhur. Eksp. Teor. Fiz.* **75**, 1468 (1978).

Geoffroy, M., L. D. Kispert, and J. S. Hwang, *J. Chem. Phys.* **70**, 4238 (1979).

Geschwind, S., Chap. 8 in *Electron Paramagnetic Resonance*, S. Geschwind, Ed., Plenum, New York, 1972.

Geschwind, S., R. J. Collins, and A. L. Schawlow, *Phys. Rev. Lett.* **3**, 544 (1959).

Gillies, R., W. U. Spendel, and A. M. Ponte Goncalves, *Chem. Phys. Lett.* **66**, 121 (1979).

Gillies, R., and A. M. Ponte Goncalves, *Chem. Phys. Lett.*, in press.

Giordmaine, J. A., L. E. Alsop, R. R. Nash, and C. H. Townes, *Phys. Rev.* **109**, 302 (1958).

Gloux, P., and B. Lamotte, *Mol. Phys.* **25**, 161 (1973).

Gokhman, V. L., and B. D. Shanina, *Fiz. Tverdogo Tela* **17**, 1408 (1975).

Golenishchev-Kutuzov, V. A., U. Kh. Kopvillem, and V. A. Shamudov, *Zh. Eksper. Teor. Fiz. Pis'ma* **10**, 240 (1969).

Golenishchev-Kutuzov, V. A., R. V. Saburova, and N. A. Shamukov, *Fiz. Tver. Tela* **12**, 3100 (1970).

Grachev, V. G., M. F. Deigen, V. Ya. Zevin, and B. A. Novominskii, *Zh. Eksp. Teor. Fiz.* **62**, 1472 (1972).

Graslund, A., H. Steen, and A. Rupprecht, *Int. J. Radiat. Biol.* **30**, 263 (1976).

Gruber, K., J. Forrer, A. Schweiger, and H. H. Gunthard, *J. Phys. E Sci. Instrum.* **7**, 569 (1974).

Grützediek, H., K. D. Kramer, and W. Müller-Warmuth, *RSI* **36**, 1418 (1965).

Guermeur, R., J. Joffrin, A. Levelut, and J. Penne, *Phys. Lett.* **13**, 107 (1964); **15**, 203 (1965).

Günthard, Hs. H., *Ber. Bunsenges. Phys. Chem.* **78**, 1110 (1974).

Hack, W., A. W. Preuss, and H. Gg. Wagner, *Ber. Bunsenges. Phys. Chem.* **82**, 1167 (1978).

Hadley, J. H., *Magn. Reson. Rev.* **6**, 29 (1980).

Hale, E. B., and R. L. Mieher, *Phys. Rev.* **184**, 739 (1969).

Hall, J. L., and R. T. Schumacher, *Phys. Rev.* **127**, 1992 (1962); erratum, **131**, 2839 (1963).

Hampton, D. A., and C. Alexander, Jr., *J. Chem. Phys.* **58**, 4891 (1973).

Hampton, D. A., and G. C. Moulton, *J. Chem. Phys.* **63**, 1078 (1975).

Hardeman, E. G., *Phillips Res. Rept.* **15**, 587 (1960).

Harris, C. B., and R. J. Hoover, *J. Chem. Phys.* **56**, 2189 (1972).

Harris, C. B., R. L. Schlupp, and H. Schuch, *Phys. Rev. Lett.* **30**, 1019 (1973).

Hase, H., F. Q. H. Ngo, and L. Kevan, *J. Chem. Phys.* **62**, 985 (1975).

Hashi, T., *J. Phys. Soc., Japan* **16**, 1243 (1961).

Hausser, K. H., and F. Reinhold, *Z. Naturforsch.* **16a**, 1114 (1961); *Phys. Lett.* **2**, 53 (1962).

Hausser, K. H., and D. Stehlik, *Adv. Magn. Reson.* **3**, 79 (1968).

Helbert, J., L. Kevan, and B. L. Bales, *J. Chem. Phys.* **57**, 723 (1972).

Helms, H. A., Jr., I. Suzuki, and I. Miyagawa, *J. Chem. Phys.* **59**, 5055 (1973).

Henning, J. C. M., and J. H. den Boef, *Phys. Lett. A* **46**, 183 (1973).

Henning, J. C. M., and J. H. den Boef, *Phys. Rev.* **14**, 26 (1976).

Henning, J. C. M., and J. H. den Boef, *Solid State Commun.* **14**, 993 (1974).

Henning, J. C. M., and J. H. den Boef, *Phys. Status Solidi B* **72**, 369 (1975).

Henning, J. C. M., J. H. den Boef, E. P. Andrew, P. S. Allen, and C. A. Bates, *Proc. 18th Ampere Congr. Magn. Reson. Related Phenom.* **1**, 117 (1975).

Henning, J. C. M., J. H. den Boef, R. P. Van Stapele, and D. Polder, *Solid State Commun.* **15**, 1535 (1974).

Herak, J. N., and C. A. McDowell, *J. Chem. Phys.* **61**, 1129 (1974).

Herak, J. N., D. Krilov, and C. A. McDowell, *J. Magn. Reson.* **23**, 1 (1976).

Hessler, J. P., and C. A. Hutchison, Jr., *Phys. Rev. B* **8**, 1822 (1973).

Hexter, R. M., *J. Opt. Soc. Am.* **53**, 703 (1962).

Hochstrasser, R. M., G. W. Scott, and A. H. Zewail, *28th Symposium on Molecular Structure and Spectroscopy*, 1973.

Hoentzsch, C., J. R. Niklas, and J. M. Spaeth, *RSI* **49**, 1100 (1978).

Hoentzsch, C., M. H. Wagner, ad J. M. Spaeth, *1974 International Conference on Colour Centres in Ionic Crystals*, 1974.

Holton, C., and H. Blum, *Phys. Rev.* **125**, 89 (1962).

Honig, A., *Phys. Rev. Lett.* **17**, 188 (1966).

Hoostraate, H., J. Poot, W. Th. Wenckebach, and N. J. Poulis, *Phys. B and C* **79B and C**, 499 (1975).

Huber, G., R. Klapisch, C. Thibault, T. H. Duong, P. Juncar, S. Liberman, J. Pinard, J. L. Vialle, and P. Jacquinot, *3rd International Conference on Nuclei far from Stability*, 188 (1976).

Hurrell, J. P., *Brit. J. Appl. Phys.* **16**, 755 (1965).

Hutchinson, D. A., S. K. Wong, J. P. Colpa, and J. K. S. Wan, *J. Chem. Phys.* **57**, 3308 (1972).

Hutchison, C. A., Jr., *Proc. 3rd Spec. Colloq. Ampere Opt. Tech. Magn. Reson. Spectrosc.* **3**, 61 (1978).

Hutchison, C. A., and M. D. Cohen, *2nd International Symposium on Organic Solid State Chemistry*, 14 (1971).

Hutchison, C. A., Jr., S. S. Kim, *Phys. Rev. B* **19**, 4454 (1979).

Hutchison, C. A., Jr., J. V. Nicholas, and G. W. Scott, *J. Chem. Phys.* **53**, 1906 (1970).

Huttermann, J., W. A. Bernhard, E. Haindl, and G. Schmidt, *Mol. Phys.* **32**, 1111 (1976).

Hwang, J. S., A. C. Dickinson, and L. D. Kispert, *J. Phys. Chem.* **83**, 3381 (1979).

Hyde, J. S., *J. Chem. Phys.* **43**, 1806 (1965).

Hyde, J. S., T. Astlind, L. E. G. Eriksson, and A. Ehrenberg, *RSI* **41**, 1598 (1970).

Hyde, J. S., J. C. W. Chien, and J. H. Freed, *J. Chem. Phys.* **48**, 4211 (1968).

Hyde, J. S., and H. Eyring, Paramagnetic Relaxation, *Annu. Rev. Phys. Chem.* **25**, 407 (1974).

Hyde, J. S., L. D. Kispert, R. C. Sneed, and J. C. Chein, *J. Chem. Phys.* **48**, 3824 (1968).

Hyde, J. S., M. D. Smigel, L. R. Dalton, and L. A. Dalton, *J. Chem. Phys.* **62**, 1655 (1975).

Hyde, J. S., R. C. Sneed, Jr., and G. H. Rist, *J. Chem. Phys.* **51**, 1404 (1969).

Ikeya, M., and N. Itoh, *J. Phys. Soc. Japan* **29**, 1295 (1970).

Imbusch, G. F., *Phys. Rev.* **153**, 326 (1967).

Imbusch, G. F., S. R. Chinn, and S. Geschwind, *Phys. Rev.* **161**, 295 (1967).

Izen, E. H., and F. A. Modine, *RSI* **43**, 1563 (1972).

Izen, E. H., R. M. Mazo, and J. C. Kemp, *J. Phys. Chem. Solids* **34**, 1431 (1973).

Iwasaki, M., H. Muto, B. Eda, and K. Nunome, *J. Chem. Phys.* **56**, 3166 (1972).

Iwasaki, M., K. Toriyma, and K. Nunome, *J. Chem. Phys.* **61**, 106 (1974).

Jaccard, C., P. A. Schnegg, M. Aegerter, P. S. Allen, E. R. Andrew, C. A. Bates, *Proc. of the 18th Ampere Congress on Magnetic Resonance and Related Phenomena* **1**, 219 (1975).

Jacobsen, E. H., N. S. Shiren, and E. B. Tucker, *Phys. Rev. Lett.* **3**, 81 (1959).

Jacobsen, E. H., and K. W. H. Stevens, *Phys. Rev.* **128**, 2036 (1963).

Jacubowicz, J., J. L. Motchane, and J. Uebersfeld, *Arch. Sci. Spec. No. 14*, 476 (1961).

Jain, S. C., S. K. Agarwal, and G. D. Sootha, *J. Phys. Chem. Solids* **32**, 897 (1971).

Jain, S. C., S. K. Agarwal, G. D. Sootha, and R. Chander, *J. Phys. C* **3**, 1343 (1970).

Jeffries, C. D., *Dynamic Nuclear Orientation*, Interscience, New York, 1963.

Jeffries, C. D., *Phys. Rev.* **106**, 164 (1957); **117**, 1056 (1960); *Prog. Cryog.* **3**, 129 (1961).

Joerin, E., F. Graf, A. Schweiger, and H. H. Guenthard, *Chem. Phys. Lett.* **42**, 376 (1976).

Joffrin, J. P., *Proc. 7th Int. Congr. Acoust.* **1**, 127 (1971).

Judeikis, H. S., and S. Siegel, *Magnetophotoselection: Effect of Depopulation and Triplet–Triplet Absorption*, 1969.

Kappers, L. A., F. Dravnieks, and J. E. Wertz, *J. Phys. C* **7**, 1387 (1974).

Kaptein, R., L. T. Muus, P. W. Atkins, K. A. McLauchlan, and J. B. Pederson, *Introduction to Chemically Induced Magnetic Polarization*, 1 (1977).

Kasatochkin, S. V., and E. V. Yakovlev, *Phys. Technol.* **8**, 615 (1977).

Kato, T., and T. Shida *J. Amer. Chem. Soc.* **101**, 6869 (1979).

Kawakubo, T., *Mol. Cryst. Liq. Cryst.* **46**, 11 (1978).

Kennedy, D. E., N. S. Dalal, and C. A. McDowell, *Chem. Phys. Lett.* **29**, 521 (1974).

Kenworthy, J. G., and R. E. Richards, *JSI* **42**, 675 (1965).

Kessenikh, A. V., *Zh. Eksper. Teor. Fiz.* **40**, 32 (1961); *PTE* **3**, 107, 521 (1961).

Kevan, L., and L. D. Kispert, *Electron Spin Double Resonance Spectroscopy*, Wiley, New York, 1976.

Kevan, L., P. A. Narayana, K. Toriyama, and M. Iwasaki, *J. Chem. Phys.* **70**, 5006 (1979).

Kiefte, H., and J. S. M. Harvey, *Canad. J. Phys.* **48**, 562 (1970).

Kiel, A., *Advances in Quantum Electronics*, Columbia University Press, New York, 1961; *Symposium Paramagnet Resonance*, Vol. 2, Academic Press, New York, 1963, p. 525.

Kiel, A., and G. R. Crane, *J. Chem. Phys.* **48**, 3751 (1968).

Kirillov, S. T., M. A. Kozhushner, and V. B. Stryakov, *Zh. Eksp. Teor. Fiz.* **68**, 2249 (1975); *Chem. Phys.* **17**, 243 (1976).

Kispert, L., and K. Chang, *J. Magn. Reson.* **10**, 162 (1973).

Kispert, L. D., K. Chang, and C. Bogan, *Chem. Phys. Lett.* **17**, 592 (1972); *J. Chem. Phys.* **58**, 2164 (1973); *J. Phys. Chem.* **77**, 629 (1973).

Kispert, L. D., K. Chang, and T. C. S. Chen, *Chem. Phys. Lett.* **44**, 269 (1976).

Kispert, L. D., T. C. S. Chen, J. R. Hill, and S. Clough, *J. Chem. Phys.* **69**, 1876 (1978).

Kispert, L. D., J. R. Hill, and C. Mottley, *J. Magn. Reson.* **33**, 379 (1979).

Kispert, L. D., K. Chang, C. Mottley, P. S. Allen, E. P. Andrew, and C. A. Bates, *Proc. 18th Ampere Congr. Magn. Reson. Related Phenom.* **1**, 279 (1974) publ. 1975.

Konzelmann, U., D. Kilpper, and M. Schwoerer, *Z. Naturforsch.* **30a**, 754 (1975).

Kopvillem, U. Kh., and R. M. Mineeva, *Izv. Akad. Nauk. SSSR Ser. Fiz.* **27**, 93 (98), 95, (101) (1963).

Kosky, C. A., B. R. McGarvey, and S. L. Hott, *J. Chem. Phys.* **56**, 5904 (1972).

Kou, W. W. H., and H. C. Box, *J. Chem. Phys.* **64**, 3060 (1976).

Kramer, K. D., W. Müller-Warmuth, and J. Schindler, *J. Chem. Phys.* **43**, 31 (1965); *Z. Naturforsch.* **19a**, 375 (1964).

Krebs, J. H., and R. K. Jeck, *Phys. Rev. B* **5**, 3499 (1972).

Krebs, P., and E. Sackmann, *J. Magn. Reson.* **22**, 359 (1976).

Kurylev, V., and S. Karyagin, *Phys. Status Solidi A* **21**, K127 (1974).

Kwiram, A. L., *Annu. Rev. Phys. Chem.* **22**, 133 (1971).

Kwiram, A. L., *J. Chem. Phys.* **49**, 2860 (1968).

Labhart, H., *Helv. Chim. Acta* **47**, 2279 (1964).

Lambe, J., N. Laurance, E. C. McLevine, and R. W. Terhune, *Phys. Rev.* **122**, 1161 (1961).

Lampel, G., *Phys. Rev. Lett.* **20**, 491 (1968).

Lange, J., *Phys. Rev. B* **12**, 226 (1975).

Lawler, R. G., D. J. Nelson, and A. D. Trifunac, *J. Phys. Chem.* **83**, 3444 (1979).

Lawrence, N., and J. N. Lambe, *Phys. Rev.* **132**, 1029 (1963).

Leblond, J., and P. Papon, *Phys. Rev. A* **4**, 1539 (1971).

Leblond, J., and J. Uebersfeld, *Phys. Rev. A* **4**, 1532 (1971).

Lee, J. Y., and H. C. Box, *J. Chem. Phys.* **61**, 428 (1974).

Leifson, O. S., and C. D. Jeffries, *Phys. Rev.* **122**, 1781 (1961).

Leisure, R. G., and D. I. Bolef, *RSI* **39**, 199 (1968).

Leisure, R. G., and R. W. Moss, *RSI* **40**, 946 (1969); *Phys. Rev.* **188**, 840 (1969).

Leniart, D. S., H. D. Connor, and J. H. Freed, *J. Chem. Phys.* **63**, 165 (1975).

Lepine, D., *Phys. Rev. B* **6**, 436 (1972).

Lepine, D., V. A. Grazhulis, and D. Kaplan, *Proc. 13th Int. Conf. Phys. Semicond., Rome*, 1081 (1970).

Lepine, D., and J. Prejean, *Proc. Int. Conf. Phys. Semicond., Boston*, 1970.

Lepley, A. R., and G. L. Closs, Eds., *Chemically Induced Magnetic Polarization*, Wiley, New York, 1973.

Leung, M., and M. A. El Sayed, *Chem. Phys. Lett.* **16**, 454 (1972).

Levanon, H., and S. Vega, *J. Chem. Phys.* **61**, 2265 (1974).

Levanon, H., and S. I. Wiessman, *J. Am. Chem. Soc.* **93**, 4309 (1971).

Levinsky, H. B., and H. C. Brenner, *Chem. Phys. Lett.* **78**, 177 (1981).

Lewiner, J., *Cesk. Cas. Fis. A* **22**, 470 (1972).

Lewiner, J., P. H. E. Meijer, J. F. Lally, and R. Meister, *J. Appl. Phys.* **41**, 4070 (1970).

Lewis, M. F., *Phys. Lett.* **17**, 183 (1965).

Liao, P. F., and S. R. Hartmann, *Phys. Rev. B* **8**, 69 (1973).

Lin, D. P., D. F. Feng, F. G. H. Ngo, and L. Kevan, *J. Chem. Phys.* **65**, 3994 (1976).

Lin, C. T., A. A. Gwaiz, and M. A. El-Sayed, *J. Am. Chem. Soc.* **94**, 8234 (1972).

Locher, P. R., and S. Geschwind, *Phys. Rev. A* **139**, 991 (1965).

Loth, K., M. Andrist, F. Graf, and Hs. H. Gunthard, *Chem. Phys. Lett.* **29**, 163 (1974).

Ludwig, G. W., and H. H. Woodbury, *Phys. Rev. Lett.* **7**, 240 (1961).

Lund, A., T. Gillbro, D. F. Feng, and L. Kevan, *Chem. Phys.* **7**, 414 (1975).

McCalley, R. C., and A. L. Kwiram, *Phys. Rev. Lett.* **24**, 1279 (1970).

McIrvine, E. C., J. Lambe, and N. Lawrence, *Phys. Rev. A* **136**, 467 (1964).

McSkimin, H. J., *J. Acoust Soc. Am.* **37**, 864 (1965).

McSkimin, H. J., in *Physical Acoustics*, W. P. Mason, Ed., Academic Press, New York, 1964, Vol. 1A.

Madden, K. P., and W. A. Bernhard, *J. Chem. Phys.* **70**, 2431 (1979).

Maffeo, B., and A. Hervé, *Phys. Rev. B* **13**, 1940 (1976).

Maki, A. H., R. D. Allendoerfer, J. C. Danner, and T. R. Keys, *J. Am. Chem. Soc.* **90**, 4225 (1968).

Marti, C., R. Parrot, G. Roger, and J. Hervé, *C. R. Acad. Sci.* **267B**, 931 (1968).

Marti, C., R. Parrot, and G. Roger, *J. Phys. Chem. Solids* **31**, 275 (1970).

Martinelli, M., L. Pardi, C. Pinzino, and S. Santucci, *Solid State Commun.* **17**, 211 (1975).

Martinelli, M., L. Pardi, C. Pinzino, and S. Santucci, *Phys. Rev. B* **16**, 164 (1977).

Marzke, R. F., and R. L. Mieher, *Phys. Rev.*, **453** (1969).

Matta, M. L., B. D. Sukheeja, and M. L. Narchal, *Indian J. Pure Appl. Phys.* **14**, 748 (1976).

Maxwell, R., and A. Honig, *Phys. Rev. Lett.* **17**, 188 (1966).

Meil'man, M. L., *Sov. Phys. Cryst.* **15**, 705 (1971).

Melcher, R. L., *Photoacoustic Detection of Electron Paramagnetic Resonance*, in preparation.

Melcher, R. L., D. I. Bolef, and J. B. Merry, *RSI* **39**, 1613 (1968).

Mendz, G., and D. Haneman, *J. Phys. C* **11**, 1197 (1978).

Menes, M., and D. I. Bolef, *Phys. Rev.* **109**, 218 (1958); *J. Phys. Chem. Solids* **19**, 79 (1961).

Merks, R. P. J., and R. deBeer, *J. Phys. C Solid State Phys.* **11**, 1673 (1978).

Merks, R. P. J., P. van der Heide, and R. de Beer, *J. Phys. C* **12**, 3097 (1979).

Merle-Daubigne, Y., P. S. Allen, E. R. Andrew, and C. A. Bates, *Proc. 18th Am. Cong. Magn. Res. and Related Phenomena*, 39 (1974) publ. 1975.

Meyer, C., J. S. Bennett, P. L. Donoho, and A. C. Daniel, *Bull. Am. Phys. Soc.* **11**, 202 (1966).

Mieher, R. L., *Magn. Reson. Rev.* **1**, 225 (1972).

Miller, T. A., *J. Chem. Phys.* **58**, 2358 (1973).

Miller, J. G., and D. I. Bolef, *RSI* **40**, 361; 915 (1969).

Miller, J. G., and D. I. Bolef, *J. Appl. Phys.* **41**, 2282 (1970).

Miller, J. G., and D. I. Bolef, *J. Appl. Phys.* **39**, 5815 (1968).

Miller, J. G., W. D. Smith, D. I. Bolef, and R. K. Sandfors, *Phys. Rev. B* **3**, 1547 (1971).

Mims, B., *Proc. R. Soc. London A* **283**, 452 (1964); *Phys. Rev.* **141**, 499 (1964).

Mims, W. B., in *Electron Paramagnetic Resonance*, S. Geschwind, Ed., Plenum, New York, 1972, Chap. 4.

Mims, W. B., *RSI* **36**, 1472 (1965); **45**, 1583 (1974).

Mims, W. B., *The Linear Electric Field Effect in Paramagnetic Resonance*, Clarendon Press, Oxford, 1976.

Mims, W. B., *Phys. Rev. A* **133**, 835 (1964).

Mims, W. B., and J. Peisach, *Magn. Reson. Relat. Phenom. Proc. Congr. Ampere*, *18th* **1**, 275 (1975).

Mitin, A. V., *Akust. Zh.* **21**, 86 (1975).

Miyagawa, I., and Y. Kotake, *Mol. Cryst. Liq. Cryst. Proc. 5th Int. Symp. Org. Solid State Chem.* **50**, 13 (1978).

Miyagawa, I., R. B. Davidson, H. A. Helms, Jr., and B. A. Wilkinson, Jr., *J. Magn. Reson.* **10**, 156 (1973).

Möbius, K., and K. P. Dinse, *Chemia* **26**, 461 (1972).

Mollenauer, L. F., and S. Pan, *Phys. Rev. B* **6**, 772 (1972).

Moore, C. A., and R. A. Satten, *Phys. Rev. B* **7**, 1753 (1973).

Moran, R., *Phys. Rev. A* **135**, 247 (1964).

Morigaki, K., N. Kishimoto, and D. Lepine, *Solid State Comm.* **17**, 1017 (1975).

Morigaki, K., and K. Murayama, *Solid State Phys.* (*Japan*), **12**, 120 (1977).

Morigaki, K., and M. Onda, *J. Phys. Soc. Japan* **36**, 1049 (1974).

Morigaki, K., and S. Toyotomi, *J. Phys. Soc. Japan* **30**, 1207 (1971).

Morigaki, K., S. Toyotomi, and Y. Toyotomi, *J. Phys. Soc. Japan* **31**, 511 (1971).

Motchane, L., *Ann. Phys.* **7**, 139 (1962).

Motchane, L., E. Erb, and J. Uebersfeld, *C. R. Acad. Sci.* **246**, 1833 (1958).

Mottley, C., K. Chang, and L. Kispert, *J. Magn. Reson.* **19**, 130 (1975).

Mottley, C., L. D. Kispert, and S. Clough, *J. Chem. Phys.* **63**, 4405 (1975).

Müller-Warmuth, W., *Z. Angew. Phys.* **10**, 497 (1958); *Z. Naturforsch.* **15a**, 927 (1960); **19a**, 1309 (1964).

Müller-Warmuth, W., and P. Servoz-Gavin, *Z. Naturforsch.* **13a**, 194 (1958).

Muramoto, T., *J. Phys. Soc. Japan* **35**, 679, 92 1 (1973).

Muramoto, T., S. Nakanishi, and T. Hashi, *Opt. Commun.* **21**, 139 (1977).

Muramoto, T., S. Nakanishi, and T. Hashi, *Opt. Commun.* **24**, 316 (1978).

Murphy, J. C., P. R. Zarriello, H. A. Kues, and L. C. Aamodt, *RSI* **46**, 1555 (1975).

Muto, H., and M. Iwasaki, *J. Chem. Phys.* **59**, 4821 (1973).

Muto, H., M. Iwasaki, and Y. Takahashi, *J. Chem. Phys.* **66**, 1943 (1977).

Nechtschein, M., and J. S. Hyde, *Phys. Rev. Lett.* **24**, 672 (1970).

Neimark, E. I., and B. E. Vugmeister, *Fiz. Tverd. Tela* **17**, 2193 (1975).

Nelson, W. H., *Mol. Phys.* **33**, 833 (1977).

Nelson, W. H., and C. D. Gill, *Mol. Phys.* **36**, 1779 (1978).

Nelson, W. H., and D. R. Taylor, *J. Chem. Phys.* **72**, 524 (1980).

Nelson, D. J., A. D. Trifunac, M. C. Thurnauer, and J. R. Norris, *Chemically Induced Magnetic Polarization Studies in Radiation and Photochemistry*, Verlag Chemie International, 1979.

Niklas, J. R., and J. M. Spaeth, *Phys. Stat. Sol. B* **101**, 221 (1980).

Owens, F. J., C. P. Poole, Jr., and H. A. Farach, *Magnetic Resonance of Phase Transitions*, Academic Press, New York, 1980.

Panepucci, H., and L. F. Mollenauer, *Phys. Rev.* **178**, 589 (1969).

Park, J. M., and C. A. McDowell, *Mol. Phys.* **32**, 1511 (1976).

Parker, J., G. A. McLaren, and J. J. Conradi, *J. Chem. Phys.* **33**, 639 (1960).

Parrot, R., and G. Roger, *C. R. Acad. Sci.* **266B**, 1628 (1968).

Passini Yuhas, M., D. I. Bolef, and J. G. Miller, *Phys. Rev. B* **17**, 4228 (1978).

Paul, H., *Chem. Phys.* **15**, 155 (1976).

Pedersen, J., and J. H. Freed, *J. Chem. Phys.* **62**, 1705 (1975).

Pine, S. H., *J. Chem. Educ.* **49**, 664 (1972).

Pipkin, M., *Phys. Rev.* **112**, 935 (1958).

Pipkin, M., and J. W. Culvahouse, *Phys. Rev.* **106**, 1102 (1957); **109**, 1423 (1958).

Plato, M., R. Biehl, K. Möbius, and K. P. Dinse, *Z. Naturforsch.* **31a**, 169 (1976).

Plato, M., and K. Möbius, *Messtechnik* **80**, 224 (1972).

Pomerantz, M., *Phys. Rev. A* **139**, 501 (1965).

Pomortsev, V. V., *Sov. Phys. JETP* **45**, 95 (1977).

Ponte Goncalves, A. M., and R. Gillies, *Chem. Phys. Lett.* **69**, 164 (1980).

Poole, C. P., Jr. and H. A. Farach, *Theory of Magnetic Resonance*, Wiley Interscience, N.Y., 1972.

Poole, C. P., Jr., and J. F. Itzel, Jr., *J. Chem. Phys.* **41**, 287 (1964).

Poole, C. P., Jr., W. L. Kehl, and D. S. MacIver, *J. Catalysis* **1**, 407 (1962).

Poole, C. P., Jr., and D. S. MacIver, *Adv. Catalysis* **17**, 223 (1967).

Pristupa, A. I., *Zavod. Lab.* **42**, 1344 (1976).

Proctor, G., and W. H. Tanttila, *Phys. Rev.* **98**, 1854 (1955); **101**, 1757 (1956).

Pugh, H. L., Jr., and C. Alexander, Jr., *J. Chem. Phys.* **66**, 726 (1977).

Quoc-Hai Ngo, F., E. E. Budzinski, and H. C. Box, *J. Chem. Phys.* **60**, 3373 (1974).

Rädler, H., *Ann. Phys.* **7**, 45 (1961).

Reddy, T. Rs., *Phys. Lett.* **36A**, 11 (1971).

Reichert, F., and J. Townsend, *Phys. Rev. A* **127**, 476 (1965).

Rengan, S. K., V. R. Bhagat, V. S. Suryanarayana, and B. Venkataraman, *J. Magn. Reson.* **33**, 227 (1979).

Richards, E., and J. W. White, *Proc. R. Soc. London A* **269**, 287 (1962); **A279**, 474, 481 (1964).

Rist, G. H., and J. S. Hyde, *J. Chem. Phys.* **52**, 4633 (1970).

Robinson, B. H., L. R. Dalton, L. A. Dalton, and A. L. Kwiram, *Chem. Phys. Lett.* **29**, 56 (1974).

Robinson, B. H., L. R. Dalton, and L. A. Dalton, *Chem. Phys. Lett.* **29**, 56 (1974).

Robinson, B. H., L. A. Dalton, A. H. Beth, and L. R. Dalton, *Chem. Phys.* **18**, 321 (1976).

Robinson, W. H., M. A. Collins, and S. D. Devine, *J. Phys. E* **8**, 139 (1975).

Robinson, G. W., and R. P. Frosch, *J. Chem. Phys.* **38**, 1187 (1963).

Robinson, W. H., and S. D. Devine, *J. Phys. C* **10**, 1357 (1977).

Roitsin, A. B., *Sov. Phys. Usp.* **14**, 766 (1972).

Roitsin, A. B., *Sov. Phys. Solid State* **10**, 751 (1968).

Roitsin, A. B., *Radiofrequency Spectroscopic Methods in Structural Investigations* (in Russian), Nauka, Moscow, 1967, p. 89.

Royce, E. B., and N. Bloembergen, *Phys. Rev.* **131**, 1912 (1963).

Rubio, J., H. T. Tohver, Y. Chen, and M. M. Abraham, *Phys. Rev. B* **14**, 5466 (1976).

Ruby, D. R., Ph. D. thesis, Washington University, St. Louis Missouri (unpublished) (1969).

Ruedin, Y., P. A. Schnegg, C. Jaccard, and M. A. Aegerter, *Phys. Stat. Sol. B* **55**, 215 (1973).

Rustgi, S. N., *J. Chem. Phys.* **66**, 2094, 2234 (1977).

Rustgi, S. N., and H. C. Box, *J. Chem. Phys.* **59**, 4763 (1973); **60**, 3343 (1974).

Ruzyllo, J., I. Shiota, N. Miyamoto, and J. Nishizawa, *J. Electrochem.* **123**, 26 (1976).

Saburova, R. V., V. A. Golenishchev-Kutuzov, and N. A. Shamukov, *Soviet Phys. JETP* **32**, 797 (1971).

Schacher, E., *Phys. Rev. A* **135**, 185 (1964).

Schmalbein, D., A. Witte, R. Roder, and G. Laukien, *RSI* **43**, 1164 (1972).

Schmidt, J., *Semicond. and Insul.* **4**, 193 (1978).

Schmidt, J., and I. Solomon, *C. R. Acad. Sci.* **B263**, 169 (1966).

Schmidt, J., W. G. van Dorp, and J. H. van der Waals, *Chem. Phys. Lett.* **8**, 345 (1971).

Schuch, H., C. B. Harris, P. S. Allen, E. P. Andrew, and C. A. Bates, *Proc. 18th Ampere Cong. On Magn. Res. Related Phenom.*, 223 (1975).

Schuch, H., F. Seiff, R. Furrer, K. Möbius, and K. P. Dinse, *Z. Naturforsch. A* **29a**, 1543 (1974).

Schweiger, A., F. Graf, G. Rist, and Hs. H. Günthard, *Chem. Phys.* **17**, 155 (1976).

Schweiger, A., and H. H. Günthard, *Chem. Phys.* **32**, 35 (1978).

Schweiger, A., and H. H. Günthard, *Mol. Phys.* **42**, 283 (1981).

Schweiger, A., M. Rudin, and J. Forrer, *Chem. Phys. Lett.*, in press.

Seidel, H., *Z. Phys.* **165**, 239 (1961).

Shain, A. L., and M. Sharnoff, *Chem. Phys. Lett.* **22**, 56 (1973).

Shamukov, N. A., and V. A. Golenishchev-Kutuzov, *PTE*, 186 (850) (1970).

Shanina, B. D., *Fiz. Tverd. Tela* **21**, 283 (1979).

Shen, Y. R., *Phys. Rev. A* **133**, 511 (1964).

Shiota, I., N. Miyamoto, and J. Nishizawa, *Surface Sci.* **36**, 414 (1973); *Japan J. Appl. Phys. (Suppl.)* **2**, 417 (1974).

Shiren, S., *Paramagn. Reson.* **2**, 482 (1963).

Shiren, S., and E. B. Tucker, *Phys. Rev. Lett.* **6**, 105 (1961).

Shiren, N. S., *Phys. Rev. B* **2**, 2471 (1970).

Sims, L. J., A. V. Guzzo and A. B. Denison, *Chem. Phys. Lett.* **5**, 370 (1970).

Sixl, H., and M. Schwoerer, *Abstr. Conf. Magn. Reson. Relat. Phenom.*, 1971.

Skrotskaya, E. G., and G. V. Skrotskii, *Izv. Vuz Fiz.* **4**, 35 (1973).

Smaller, B., E. C. Avery, and J. R. Remko, *J. Chem. Phys.* **55**, 2919 (1971).

Smaller, B., J. R. Remko, and E. C. Avery, *Proceedings of the 4th International Congress on Radiation Reports Evian, France, June 29–July 4 1970*, Gordon and Breach, New York, 1970.

Smith, R. T., *Proc. 21st Annu. Freq. Control Symp., Atlanta City, N.J.*, 24 (1967).

Smith, J. P., and A. D. Trifunac, *Chem. Phys. Lett.*, in press.

Smith, W. D., J. G. Miller, D. I. Bolef, and R. K. Sandfors, *J. Appl. Phys.* **40**, 4967 (1969); **42**, 2579 (1971).

Sokolov, E. A., and V. A. Benderskii, *PTE* **2**, 232 (1969).

Solarz, R., and D. H. Levy, *J. Chem. Phys.* **58**, 4026 (1973).

Solomon, I., *Proc. 12th Int. Conf. Phys. Semicond.*, Warsaw, 1972.

Sorokin, P. P., G. J. Lasher, and I. L. Gelles, *Phys. Rev.* **118**, 939 (1960).

Spencer, T. S., and C. M. O'Donnell, *J. Chem. Educ.* **50**, 152 (1973).

Stetter, E., H. M. Vieth, and K. H. Hausser, *J. Magn. Reson.* **23**, 493 (1976).

Stradins, J., and R. Gavars, *Tezisy Dokl. Vses Soveshch Electrokhim 5th* **1**, 309 (1974).

Suchard, S. N., and K. W. Bowers, *J. Chem. Phys.* **56**, 5540 (1972).

Suzki, I., *Solid State Phys.* **9**, 451 (1974).

Takagi, Y., Y. Fukuda, M. Chiba, K. Yamada, and T. Hashi, *J. Phys. Soc. Japan* **38**, 1214 (1975).

Tanaka, S., A. Koma, and M. Kobayashi, *J. Phys. Soc. Japan* **22**, 127 (1967).

Tani, T., *Photogr. Sci. Eng.* **19**, 356 (1975).

Teodorescu, M., *Rev. Roum. Phys.* **22**, 73 (1977).

Terhune, A. L., J. Lambe, C. Kikuchi, and J. Baker, *Phys. Rev.* **123**, 1265 (1961).

Terhune, R. W., J. Lambe, G. Makhov, and L. G. Cross, *Phys. Rev. Lett.* **4**, 234 (1960).

Thomas, P. A., and D. Kaplan, *AIP Conf. Proc. No. 31*, 85 (1976).

Toyoda, Y., and K. Morigaki, *J. Phys. Soc. Japan* **43**, 118 (1977).

Trifunac, A. D., *J. Am. Chem. Soc.* **98**, 5202 (1976).

Trifunac, A. D., and E. C. Avery, *Chem. Phys. Lett.* **27**, 141 (1974).

Trifunac, A. D., and E. C. Avery, *Chem. Phys. Lett.* **28**, 297 (1974).

Trifunac, A. D., and W. T. Evanochko, *J. Am. Chem. Soc.* **102**, 4598 (1980).

Trifunac, A. D., and D. J. Nelson, *Chem. Phys. Lett.* **46**, 346 (1977).

Trifunac, A. D., D. J. Nelson, and C. Mottley, *J. Magn. Reson.* **30**, 263 (1978).

Trifunac, A. D., and J. P. Smith, *Chem. Phys. Lett.* **73**, 94 (1980); *J. Phys. Chem.*, in press.

Trifunac, A. D., and M. C. Thurnauer, *J. Chem. Phys.* **62**, 4889 (1975).

Trifunac, A. D., M. C. Thurnauer, and J. R. Norris, *Chem. Phys. Lett.* **57**, 471 (1978).

Truesdale, R. D., PhD Thesis, Univ. South Carolina, 1981; M.S. Thesis, ibid., 1979.

Truesdale, R. D., C. P. Poole, Jr., and H. A. Farach, *Phys. Rev. B* **25**, 474 (1982).

Truesdale, R. D., H. A. Farach, and C. P. Poole, Jr., *Phys. Rev. B* **22**, 365 (1980).

Tucker, E. B., *Phys. Rev. Lett.* **6**, 183 (1961).

Underhauser, V. M., and A. Abragam, *Phys. Rev.* **98**, 1729 (1955).

Underhauser, V. M., L. H. Bennett, and H. C. Torrey, *Phys. Rev.* **108**, 449 (1957).

Underhauser, V. M., and J. Seiden, *C. R. Acad. Sci.* **245**, 1528 (1957).

Unruh, W. P., and J. W. Culvahouse, *Phys. Rev.* **129**, 2441 (1963).

Unruh, W. P., Y. Chen, and M. M. Abraham, *J. Chem. Phys.* **59**, 3284 (1973); *Phys. Rev. Lett.* **30**, 446 (1973); *Phys. Rev. B* **15**, 4149 (1977).

Ursu, I., *Rev. Roum. Phys.* **17**, 955 (1972).

Usmani, Z., and J. F. Reichert, *Phys. Rev.* **180**, 482 (1969).

Usmanov, R. Sh., *PTE* **20**, 258, 1204 (1977).

Usmanov, R. Sh., *Cryogenics* **18**, 542 (1978).

Van Camp, H. L., C. P. Scholes, and R. A. Isaacson, *RSI* **47**, 516 (1976).

Van Dorp, W. G., T. J. Schaafsma, M. Soma, and J. H. Van der Waals, *Chem. Phys. Lett.* **21**, 221 (1973).

Van Noort, H. M., P. J. Vergragt, J. Herbich, and J. H. Van Der Waals, *Chem. Phys. Lett.* **71**, 5 (1980).

Van Ormondt, D., R. De Beer, C. M. De Jong, M. G. Van der Oord, M. H. Homs, and H. W. Den Hartog, *Physica B & C* **84B & C**, 110 (1976).

Van Steenwinkel, R., and K. H. Hausser, *Phys. Lett.* **14**, 24 (1965).

Van Veen, G., and R. de Deer, *J. Phys. Chem. Solids* **38**, 217 (1977).

Vedrine, J. C., and B. Imelik, *J. Magn. Reson.* **16**, 95 (1974).

Vergragt, Ph. J., J. A. Kooter and J. H. van der Waals, *Mol. Phys.* **33**, 1523 (1977).

Vieth, H. M., H. Brunner, and K. H. Hausser, *Z. Naturforsch.* **26a**, 167 (1971).

Vieth, H. M., and K. H. Hausser, *17th Congr. Ampere Nucl. Magn. Res. Relat. Phenom.*, 1972.

Vieth, M., and K. H. Hausser, *Ber. Bunsenges. Phys. Chem.* **78**, 185 (1974).

Volino, F., and A. Rousseau, *Mol. Cryst. Liq. Cryst.* **16**, 247 (1972).

Vollmann, W., *Mol. Phys.* **32**, 395 (1976).

von der Weid, J. P., *Rev. Bras. Fis.* **6**, 1 (1976).

von Schutz, J. U., and W. Dietrich, *Chem. Phys. Lett.* **51**, 418 (1977).

Warden, J., *Isr. J. Chem.* in press.

Warden, J. T., and O. L. Adrianowycz, *Proc. Fifth Intl. Congr. Photosynth.*, in press.

Warden, J. T., and J. R. Bolton, *J. Amer. Chem. Soc.* **94**, 4351 (1972); **95**, 6435 (1973).

Warden, J. T., and J. R. Bolton, *RSI* **47**, 201 (1976); *J. Magn. Reson.* **24**, 125 (1975).

Wasiela, A., J. Duran, and Y. Merle Daubigne, *C. R. Hebd. Seances Acad. Sci. B* **278**, 1099 (1974).

Wasiela, A., J. Duran, P. S. Allen, E. P. Andrew, and C. A. Bates, *Proc. 18th Congr. Magn. Res. Relat. Phenom.* **1**, 217 (1974).

Watkins, G. D., *Phys. Rev. B* **12**, 5824 (1975).

Watkins, G. D., and J. W. Corbett, *Phys. Rev. A* **134**, 1359 (1964).

Weger, M., and G. E. Feher, *Paramagnetic Resonance II*, W. Lav., Ed., Academic Press, New York, 1963, p. 628.

Wells, J. W., *J. Chem. Phys.* **52**, 4062 (1970).

Werner, K., *Hochfrengztech. Elekt. Akust.* **73**, 115 (1964).

Wetsel, G. C., Jr., and P. L. Donoho, *Phys. Rev. A* **139**, 334 (1965).

Witte, A., D. Schmalbein and G. Laukien, *RSI* **42**, 1534 (1971).

Wolbarst, A. B., *RSI* **47**, 255 (1976).

Wollan, D. S., *Phys. Rev.* **13**, 3671, 3686 (1976).

Wong, S. K., D. A. Hutchinson, and J. K. S. Wan, *J. Chem. Phys.* **58**, 985 (1973).

Wong, S. K., D. A. Hutchinson, and J. K. S. Wan, *J. Am. Chem. Soc.* **95**, 622 (1973).

Wong, S. K., and J. K. S. Wan, *J. Am. Chem. Soc.* **94**, 7197 (1972).

Woonton, G. A., and G. L. Dyer, *Can. J. Phys.* **45**, 2265 (1967).

Wosinski, T., T. Figielski, and A. Makosa, *Phys. Stat. Solid A* **37**, K57 (1976).

Wotherspoon, N., G. K. Oster, and G. Oster, in *Methods of Chem.*, 1, Part IIIb, A. Weissberger, Ed., John Wiley, New York, 1972, p. 482.

Wysling, P., and K. A. Mueller, *J. Phys. C* **9**, 635 (1976).

Yagi, H., M. Ingue, T. Tatsukawa, and T. Yamamoto, *Japan J. Appl. Phys.* **9**, 1386 (1970).

Yamamoto, T., M. Kono, K. Sato, T. Miyamae, K. Mukai, and K. Ishizu, *Jeol News* **10a**, 61 (1972).

Yamanouchi, C., K. Mizuguchi, and W. Sasaks, *J. Phys. Soc. Japan* **22**, 859 (1967).

Yang, C. C., and Y. W. Kim, *Phys. Rev. B* **20**, 1291 (1979).

Yoshida, H., D. F. Feng, and L. Kevan, *J. Chem. Phys.* **58**, 4924 (1973).

Yuhas, M. P., D. I. Bolef, and J. G. Miller, *Ultrason. Symp. IEEE* (1975); *Phys. Rev. B* **17**, 4228 (1978).

Yuhas, M. P., Doctoral dissertation, Washington University, St. Louis, Missouri, 1976.

Yuhas, M. P., P. A. Fedders, J. G. Miller, and D. I. Bolef, *Phys. Rev. Lett.* **35**, 1028 (1975).

Zaitev, A., *Sov. Phys. Semicond.* **3**, 1245 (1970).

Zevin, V. Ya., and A. B. Brik, *Ukr. Fiz. Zh.* **17**, 1688 (1972).

Zieger, J., and H. C. Wolf, *Chem. Phys.* **29**, 209 (1978).

Author Index

Page numbers indicate where the author's work is referred to although the name may not be mentioned in the text. Numbers in *italics* indicate the pages on which the full references are listed.

723

Subject Index

Absorption, 187, 381, 388, 389, 420, 491, 586
  coefficient, 372
  enhanced, 677
  first derivative, 466
  line, 459
  second derivative, 477
  transient, 677
Acoustic ESR, 644, 700
Acoustic excitation, 700
Activation energy, 368
Adiabatic demagnetization, 328, 685
Adiabatic fast passage, 581, 630
Adiabatic inversion, 695
Admittance, 268
Adsorption, 299
Alcohol, 535
Alkali halide, 19, 291
Alkali metal, 18, 214
Alkaline earth, 18
Alum, 483
Alumina, 62, 444, 520
Amino acid, 359, 661
Ammonia, 1, 18, 214
Amorphous distribution, 524
Amplification, narrow band, 414
Amplifier, parametric, 424
Anisotropic g-factor, 505
Anisotropy, 6, 533
Antenna, 97, 118
Antibonding level, 360
Antiferromagnetism, 17, 20, 514
Antireciprocal character, 98
Aqueous sample, 200, 441
Area:
  Gaussian, 463, 464, 475, 477
  Lorentzian, 465, 476, 477
Arrhenius, 368
As, 19
Atom, free, 291
Attentuation, 95
Attenuation constant, 38, 44, 97
Attenuator, 95
Autocorrelation function, 546
Automatic frequency-control, 87

Avalanche diode, 81

Backward diode, 267
Backward wave oscillator, *see* Traveling wave tube
Bakelite, 62
Balanced resonator, 205
Bandwidth, 264
Barite, 357
BARITT, 80
Barretter, 260
Beam voltage, 73
  power supply, 86
Beattie-Bridgeman equation, 293
Benzene, 361
Benzophenone, 435
Bessel function, 35, 52, 162, 249
Bessel roots, 135
Binomial, 13, 539
Biological materials, 18, 359
Biphasic kinetic responses, 700
Biradical, 18, 413
Birefringent modulator, 691
Blank absorption, 423
Bleaching, 369
Bloch equations, 401, 597
Bloch theory, 551
Bohr magneton, 3, 4
Bolometer, 5, 260, 268, 383, 396, 402
Boltzmann distribution, 577
Bond energy, 362, 363
Bottleneck, 581
Broadening function, 245
Brownian motion, 529, 627
Bucking, 267, 397
Burning holes, 626
Burnout, 266
Butane, 325

Cage effect, 364
Calcium chloride, 326
Capacitance, 34
Carbazyl, 511
Carbon, 291
Carbon resistance, 330